BOTANICHESKII INSTITUT AKADEMII NAUK SSSR
Botanical Institute of the Academy of Sciences of the U.S.S.R.

FLORA OF THE U.S.S.R.

(Flora SSSR)

Volume V

Chief Editor and Editor of this Volume
Academician V. L. Komarov

Compiled by
E. G. Bobrov, B. A. Fedchenko, Yu. S. Grigor'ev, N. A. Ivanova,
V. L. Komarov, A. N. Krishtofovich, O. I. Kuzeneva, A. S. Lozina-Lozinskaya,
V. P. Maleev, M. I. Nazarov, V. L. Nekrasova, N. V. Pavlov,
E. V. Vul'f, and A. V. Yarmolenko

Izdatel'stvo Akademii Nauk SSSR
Moskva-Leningrad
1936

Translated from Russian

Israel Program for Scientific Translations
Jerusalem 1970

TT 69-55052

Published Pursuant to an Agreement with
THE SMITHSONIAN INSTITUTION
and
THE NATIONAL SCIENCE FOUNDATION, WASHINGTON, D. C.

IPST Cat. No. 5401

Translated by Dr. N. Landau

Printed in Jerusalem by Keter Press
Binding: Wiener Bindery Ltd., Jerusalem

Available from the
U. S. DEPARTMENT OF COMMERCE
Clearinghouse for Federal Scientific and Technical Information
Springfield, Va. 22151

TABLE OF CONTENTS

* [Russian page numbers appear in the left-hand margin of the text.]

SUBJECTS AND CONTRIBUTORS

Indexes	Editorial Staff
Characteristics of the Dicotyledoneae, key for the comparison of orders and key for the comparison of families of Archichlamydeae, characteristics of the orders; the genera Populus and Alnus, the family Rafflesiaceae, key for determination of the family Polygonaceae and the genus Polygonum, with the exception of the section Aconogonon Meisn.	Arranged by V. L. Komarov
Families Saururaceae and Chloranthaceae, genus Betula	Arranged by O. I. Kuzeneva
Genera Chosenia and Salix	Arranged by M. I. Nazarov
Family Juglandaceae	Arranged by V. L. Nekrasova
Tribe Coryleae; characteristics of the family Fagaceae, key for determination of genera, the genus Castanea; the family Santalaceae	Arranged by E. G. Bobrov
Genus Quercus	Arranged by V. P. Maleev
Genus Fagus	Arranged by E. V. Vul'f
Order Urticales	Arranged by A. V. Yarmolenko
Family Loranthaceae	Arranged by B. A. Fedchenko
Family Aristolochiaceae	Arranged by N. A. Ivanova
Genera Oxyria, Rumex, Rheum, Fagopyrum, and Koenigia	Arranged by A. S. Lozina-Lozinskaya
Genera Atraphaxis and Calligonum	Arranged by N. V. Pavlov
Section Aconogonon Meisn. of the genus Polygonum	Arranged by Yu. S. Grigor'ev
Indication of the locations of fossilized plant remains	Arranged by A. N. Krishtofovich

Addenda – Descriptiones plantarum novarum in tomo V Florae URSS commemoratum.

The plates were drawn by the following artists: A. S. Lozina-Lozinskaya – I, II, X, XXVI – XXVIII; L. M. Kopteva – III, IX, XXV; A. F. Domracheva – IV; A. P. Voronova – V, VI, VIII, XXIII; Z. V. Kobyletskaya – VII, XXXVIII – XLIX; M. S. Mollot – XI; S. A. Moiseeva – XII – XVI, XXIV; T. L. Yuzepchuk – XVII, XVIII, XX – XXII; M. M. Parfenenko – XIX, XXXIXa; M. A. Dobrov – XXX – XXXVII.

SYSTEMATIC INDEX OF SPECIES IN VOL. V OF THE "FLORA OF THE USSR"*

Class II. DICOTYLEDONEAE DC.

Subclass I. Archichlamydeae Engl.

Order 10. Piperales Lindl.

* [This Index has been reproduced photographically from the Russian original.]

** [Russian page numbers appear in the left-hand margin of the text.]

Section 3. **Polares Nas.**

Section 4. **Arcticae Rydb.**

Section 5. **Myrsinites Borr.**

Subgenus 2. **Caprisalix Dumort.**

Section 6. **Glaucae Fries.**

Section 7. **Chrysanthae W. D. Koch.**

Section 8. **Villosae Anderss.**

Section 9. **Phylicifoliae Dumort.**

Section 10. Arbuscloideae Floder.

Section 11. Nigricantes Borr.

Section 12. Capreae Bluff.

Section 13. Lividae Nym.

Section 14. Myrtilloides Borr.

Section 15. Hastatae Fries.

Genus 363. *Corylus* L.

Tribe II. BETULEAE DÖLL.

Genus 364. *Betula* L.

Subgenus 1. Eubetula Rgl.

Section 1. Costatae Rgl.

Section 2. Asperae Nakai.

Section 3. Nanae Rgl.

Section 4. Fruticosae Rgl.

Section 5. Dahuricae Rgl.

Section 6. Albae Rgl.

Order 18. Polygonales Lindl.

Family **LII. Polygonaceae** Lindl.

Genus 389. *Oxyria* Hill.

Genus 390. *Rumex* L.

Section 1. Acetosella Meisn.

Section 2. Acetosa Meisn.

Section 3. Hololapathum A. Los.

Section 4. Odontolapathum A. Los.

Section 2. **Pterygobasis** Borszcz.

Section 3. **Eucalligonum** Borszcz.

Section 4. **Caliphysa** Endl.

Genus **394.** ***Polygonum*** **L.**

Section 2. **Cephalophilon Meisn.**

Section 3. **Persicaria Meisn.**

Section 4. **Aconogonon Meisn.**

PREFACE

The first four volumes of the "Flora of the USSR" comprise a complete section of the flora, namely archegoniate and monocotyledonous plants. This section is concluded with a special issue containing a complete synonymy of these plants.

With the fifth volume, we take up the review of the largest group, namely Choripetalae. We anticipate that the description of this group will require ten volumes; of these three will be devoted to the very important family, Leguminosae.

Volume V contains descriptions of a great many of the USSR's arboreous plants: poplars, willows, birches, alders, oaks, hornbeams, hazelnuts, etc. These plants are of particular economic value and their description broadens the scope of possible exploitation. The treatments of many of these genera, as published here, introduce some extensive modifications and provide more precise information concerning their systematics and geography. However, in some cases the treatments are not yet final. This is due to difficulties involved in the systematic study of woody plants and to scarcity of material, especially from countries adjoining the USSR.

This volume also includes the rather large family of Polygonaceae, the shrubby representatives of which are so characteristic of the deserts of Central Asia. The systematics of the genera concerned has so far been very obscure and our treatment brings them all together for the first time.

The Editors

1 Class II. DICOTYLEDONEAE DC.

Embryo with 2, rarely several cotyledons, or with only a single cotyledon by abortion; stem with open vascular bundles, i. e., having a cambial ring, rarely the cambium undeveloped; owing to the fascicular and interfascicular cambium, the stem is capable of secondary growth, annual rings forming in the case of perennial stems; leaves typically with pinnate venation; flowers of 5-merous cycles, rarely 2-merous, or the number of cycles and their components different; seeds with well-developed endosperm (albuminous seeds) or endosperm suppressed and the materials required for the development of the embryo stored in strongly developed cotyledons (exalbuminous seeds).

Subclass I. ARCHICHLAMYDEAE ENGL. seu Choripetalae EICHLER, Apetalae JUSS.

Flowers without perianth, consisting merely of androecium and gynoecium, or with simple perianth; if perianth consisting of calyx and corolla, then petals quite distinct, or in some closely related plants petals may be united or absent.

Key to Orders of Archichlamydeae*

1. Flowers without perianth or, if perianth present, then its segments bractlike . . . . 2.
\+ Flowers with typically developed perianth . . . . 7.
2 2. Leaves pinnately divided, exstipulate; trees or large shrubs; staminate flowers with 3 or 4 stamens, in long aments; pistillate flowers with 2 carpels united into 1-locular ovary terminating in 2-parted, fringed stigma; fruit a drupe or nut . . . . Order 13. **Juglandales** Engl.
\+ Leaves entire or somewhat lobed, not pinnate . . . . 3.
3. Flowers in small, erect spikes, in USSR flora bisexual; herbs or shrubs, in USSR flora only herbs; stamens 1—10; carpels 1—4 . . . . Order 10. **Piperales** Lindl.

* When using this key, it should be kept in mind that the orders in Engler's system are not based on the principle of resemblance of component families but on phylogenetic relationship. Thus, families and genera belonging to the same order often differ in some distinctive character. What unifies them in such a case is the overall combination of other characters. A good example is provided by Alchemilla which is completely destitute of petals and thus, by perusal of this key, might easily be referred to Centrospermae, even though its characters as a whole place it beyond any doubt in the order Rosales. Similar difficulties may arise in connection with many other genera. The key is nevertheless indispensable for general orientation within the system of Archichlamydeae.

\+ Flowers, especially the staminate, solitary, or in long aments, fascicles, or glomerules, not in erect spikes . . . . 4.

4\. Fruit a 2-valved capsule; seeds comose; trees or shrubs; leaves stipulate; flowers dioecious, in aments, without perianth, but with scalelike, annular or columnar nectariferous glands; staminate flowers with 2, 3, 5, rarely more, stamens; pistillate flowers with 1 ovary terminating in 2, rarely 4, stigmas . . . . Order 11. **Salicales** Lindl.

\+ Fruit a nut, drupe, or samara, 1-seeded; seeds not comose . . . . . 5.

5\. Flowers mostly actinomorphic, with 4 (rarely 5) cruciformly arranged perianth segments; herbs, rarely shrubs or trees; leaves entire, lobed or (in a single case) pinnatisect, stipulate; flowers in false umbels, aments, or some other arrangement; fruit a drupe or nut, wingless or winged . . . . Order 15. **Urticales** Lindl.

\+ Flowers without perianth or with rudimentary annular perianth, bracteate . . . . 6.

6\. Shrubs with aeromatic glands; flowers in aments, without perianth; in USSR flora almost always dioecious; stamens most often 4, rarely 2—6; stigmas 2; fruit a drupe . . . . . . Order 12. **Myricales** Engl.

\+ Trees or large shrubs; staminate flowers in catkins pistillate or solitary; in USSR flora monoecious; fruit a nut or acorn . . . . Order 14. **Fagales** Engl.

7\. Flowers with simple perianth, perianth segments green or colored; more often perianth clearly differentiated into calyx and corolla; often herbs, rarely subshrubs or small trees; leaves opposite or
3 alternate, stipulate or exstipulate; stamens opposite the perianth segments, and numbering the same as them (rarely 1 to many); ovary superior, rarely inferior, many-locular or more often 1-locular; seeds with endosperm and curved embryo . . . . Order 19. **Centrospermae** Engl.

\+ Combination of characters different . . . . 8.

8\. Perianth simple . . . . 9.

\+ Perianth clearly differentiated into calyx and corolla, the latter mostly colored . . . . 11.

9\. Flowers solitary, medium or large; ovary inferior, 3—6-locular; exstipulate; fruit many-seeded; herbs or twining shrubs . . . . Order 17. **Aristolochiales** Lindl.

\+ Flowers small, in inflorescence; fruit 1-seeded . . . . 10.

10\. Herbs or shrubs, parasitic on other plants; leaves exstipulate; flowers bisexual or unisexual, 5-merous or very rarely 4-merous; ovary inferior, or 2 or 3 (rarely 1) carpels, each carpel containing 1 ovule . . . . Order 16. **Santalales** Lindl.

\+ Herbs or shrubs; leaves stipulate, the stipules sheathing the stem in tubular or ringlike ocrea; flowers small; perianth of 3—6 segments; ovary superior with 1 mostly erect ovule; fruit a nutlet [achene] with mealy endosperm . . . . . Order 18. **Polygonales** Lindl.

11\. Carpels not united into ovary, but distinct, each developing into follicle, berry, or some other fruit, thus a group of fruits developing from a single flower; herbs, shrubs, rarely trees; leaves with stipules or exstipulate; flowers with simple, colored perianth or with calyx and corolla, actinomorphic or zygomorphic; stamens as a rule numerous; carpels 1 to many . . . . Order 20. **Ranales** Lindl.

\+ Carpels united into 1 to several ovaries or distinct and borne on more or less convex receptacles .... 12.

12\. Ovary superior or seemingly inferior due to fusion with receptacle and calyx (in Rosaceae) .... 13.

\+ Ovary inferior .... 20.

13\. Flower parts typically in multiples of 2; herbs, rarely shrubs; flowers often in racemes or solitary and large, actinomorphic or zygomorphic; carpels 2 or many, fully united .... Order 21. **Rhoeadales** Engl.

\+ Flower parts typically in multiples of 5 .... 14.

14\. Leaves entire, simple, insectivorous; flowers actinomorphic, in a raceme; carpels 3 — 5; seeds many, small .... Order 22. **Sarraceniales** Engl.

\+ Leaves not insectivorous .... 15.

4 15. Ovary superior or half-inferior when developing on bottom of cup-shaped calyx and then stamens and petals often arising at its margin; herbs, trees or shrubs; leaves often stipulate; flowers cyclic, with inner cycles becoming a spiral, actinomorphic or zygomorphic; sepals and petals clearly contrasted although the petals may be obsolete .... Order 23. **Rosales** Lindl.

\+ Leaves exstipulate or stipulate; stamens and petals attached to receptacle and not to calyx .... 16.

16\. Flowers of 5 or 4 cycles, with cyclic or spiral arrangement; stamens often numerous; carpels with parietal placentation, the placenta sometimes projecting to center of ovary or very rarely the seed attached to ovary base .... Order 28. **Parietales** Lindl.

\+ Flowers of 4 or 5 cycles; placentation central .... 17.

17\. Herbs, shrubs, or trees; flowers cyclic, with calyx and corolla, these sometimes undeveloped but mostly 5-merous; carpels 3 — 5 (rarely more), united into superior ovary, sometimes separating again in maturity; ovules anatropous, with ventral raphe and micropyle up or, when ovules many, with dorsal raphe and micropyle down .... Order 24. **Geraniales** Lindl.

\+ Combination of characters different .... 18.

18\. Shrubs or trees, rarely herbs; ovules reversed or pendulous, with dorsal raphe and micropyle up or erect with ventral raphe and micropyle down .... Order 25. **Sapindales** Lindl.

\+ Stamens 5 or many, opposite petals .... 19.

19\. Only shrubs or small trees; leaves exstipulate or subtended by small stipules; flowers cyclic, actinomorphic, heterochlamydeous though corolla sometimes abortive; carpels 5 or down to 2, normally united into superior ovary, though ovary sometimes becoming inferior; ovules 1 or 2 .... Order 26. **Rhamnales** Lindl.

\+ Herbs, shrubs, or trees; leaves mostly alternate; flowers commonly 5-merous, heterochlamydeous, rarely apetalous, mostly actinomorphic; stamens numerous, carpels 2 or many, united into superior ovary with 1 to many ovules .... Order 27. **Malvales** Lindl.

20\. Herbs, shrubs, or trees; flowers heterochlamydeous or rarely with abortive petals, mostly actinomorphic, with concave calyx; carpels 2 or rarely 1 to many, united into inferior ovary, rarely 1 free .... Order 29. **Myrtiflorae** Endlicher.

5 + Herbs, rarely shrubs or trees; flowers mostly in umbels, heterochlamydeous, though petals or sepals sometimes abortive, or the peripheral flowers of the umbel irregular, frequently 5-merous, rarely 4-merous, or the number of members indefinite; carpels 2, rarely 5—1, very rarely indefinite, united into inferior ovary; ovule 1, rarely 2, anatropous, with copious endosperm . . . . Order 30. **Umbelliflorae** Bartl.

Key to Families of Archichlamydeae Engl.

(or Eleutheropetalae A. Br.)

1. Plants parasitic on other plants and therefore destitute of roots . . . 2.
+ Plants green, rarely differently colored, with roots in soil or in water . . . . 4.
2. Plants parasitic on branches of trees, with green leaves and berry-like fruits; perianth segments and stamens 4—6 . . . . Family XLIX. **Loranthaceae** D. Don.
+ Plants parasitic on roots of other plants, destitute of chlorophyll, often yellow or red, with scalelike leaves . . . . 3.
3. Flowers in terminal racemes; perianth 4- or 5-merous; stamens indefinite, attached below stigma; ovary inferior . . . . Family LI. **Rafflesiaceae** Dumort.
+ Flowers in reddish-brown, fleshy terminal spadix; perianth of 1—5 linear-spatulate segments; stamen solitary . . . . Family CXVII. **Cynomoriaceae** Engl.
4. Perianth none; flowers consisting merely of stamens and pistils, sometimes also containing nectaries, subtended by a bract (scale) or rarely ebracteate . . . . 5.
+ Perianth present, at least in reduced form . . . . 8.
5. Herbaceous plants; flowers perfect, in an upright cluster; fruit a capsule or a green drupe; seeds never hairy-tufted . . . . 6.
+ Trees or shrubs, dioecious, very rarely monoecious; staminate flowers in rather long pendulous aments; fruit a many-seeded capsule containing numerous seeds with tufts of long white hairs, or a wax-coated drupe . . . . 7.
6. Stamens 3—6; fruit a capsule with 3 spreading styles; leaves cordate at base . . . . Family XXXVIII. **Saururaceae** Lindl.
+ Stamens 1—3, the connective longer than the anthers; fruit a 1-styled drupe; leaves cuneate at base . . . . Family XXXIX. **Chloranthaceae** Blume.

6 7. Scales of pistillate aments obtuse and hairy at apex; capsule many-seeded; seeds with tufts of long hairs . . Family XL. **Salicaceae** Lindl.
+ Scales of pistillate aments naked at apex, point-tipped; fruit a coriaceous drupe, coated with waxy bloom . . . . Family XLI. **Myricaceae** Lindl.
8. Perianth rudimentary, or more or less developed, but simple, i. e., consisting of either calyx or corolla . . . . 9.
+ Perianth typically fully developed, mostly consisting of calyx and corolla (for exceptions see individual families) . . . . 35.

9. Perianth rudimentary, mostly green, inconspicuous . . . . 10.
\+ Perianth conspicuous, mostly colored (in the genus Alchemilla green), sometimes corollaceous, but always simple. . . . . 21.
10. Trees or shrubs . . . . 11.
\+ Herbaceous plants . . . . 20.
11. Leaves imparipinnate; staminate flowers with 30—40 stamens, in long pendulous catkins; pistillate flowers paired or solitary; ovary with large fringed stigma; perianth of small toothlike segments; fruit a nut or drupe . . . . Family XLII. **Juglandaceae** Lindl.
\+ Leaves entire or sinuate-toothed . . . . 12.
12. Trees or shrubs, nearly always monoecious; flowers unisexual; staminate inflorescence a catkin; perianth inconspicuous; fruit nutlike, 1-seeded; seeds without endosperm . . . . 13.
\+ Flowers mostly bisexual; perianth well developed . . . . 15.
13. Each pistillate flower surrounded by numerous small bracts which become indurated in fruit, hence the fruit (acorn) enclosed in a cupule; staminate catkins interrupted, greenish . . . . Family XLIV. **Fagaceae** A. Br.
\+ No cupule or, if present, of not more than 3 bracts; staminate catkins compact . . . . 14.
14. Cupule of 3 bracts, these longer than enclosed nut; pistillate catkins small, budlike, with exserted red stigmas; staminate catkins relatively large, pendulous, brownish . . . . Family XLIII. **Betulaceae**, genus **Corylus** L.
\+ No cupule; fruit winged or wingless nut; both staminate and pistillate catkins elongate, cylindric or oblong . . . . Family XLIII. **Betulaceae** C. A. Agardh.
15. Fruit indehiscent . . . . 16.
\+ Fruit dehiscent, bivalvular; flowers in capitate inflorescences; leaves broad, in USSR flora resembling hazelnut leaves . . . . Family LXXV. **Hamamelidaceae** R. Br.
7 16. Perianth of 4 segments becoming fleshy in maturity, individual fruits tightly packed on common receptacle, forming an apparently multiple fruit, or else the receptacle pyriform or globose, hollow within, the walls studded with very small flowers or fruits . . . . Family XLVI. **Moraceae** Lindl.
\+ Perianth 5-merous, drying up and not accrescent in maturity; receptacle inconspicuous . . . . 17.
17. Leaves 2-ranked; perianth 4- or 5-merous; stigmas 2 . . . . 18.
\+ Leaves alternate; perianth 5-merous; stigmas 2, 3, or 4 . . . . . . 19.
18. Flowers mostly bisexual, in clusters; leaves often asymmetric at base; fruit a broad-winged nut [samara] . . . . Family XLV. **Ulmaceae** Mirbel, subfamily **Ulmoideae** A. Br.
\+ Flowers clustered or solitary, unisexual, borne in the leaf axils; fruit a subspherical drupe . . . . Family XLV. **Ulmaceae**, subfamily **Celtoideae** A. Br.
19. Leaves exstipulate; fruit a nut enclosed in the alately accrescent perianth . . . . Family LIII. **Chenopodiaceae** Less.

\+ Leaves stipulate; fruit a nut with winged, crested, setaceous, or otherwise shaped outgrowths, or smooth .... Family LII. **Polygonaceae** Lindl.

20. Flowers dioecious; staminate flowers in loose spreading racemes or terminal panicle; pistillate flowers in axils of leaves or in axils of bracts, crowded at end of stem or forming short cones (fruiting spikes) .... Family XLVI. **Moraceae** Lindl., subfamily **Cannaboideae** Engl.

\+ Flowers rarely perfect, more often unisexual, in some cases dioecious; perianth segments and stamens 4 or 5 (rarely 2—3); flowers in short or long catkinlike axillary cymes; perianth of 2—5 segments; stamens 2—5; pistil solitary; ovary 1-locular; fruit a small nut or drupe .... Family XLVII. **Urticaceae** Engl.

21. Perianth conspicuously developed by union of 3 segments; stamens 6, distinct or united with style below stigma; fruit a many-seeded capsule; leaves alternate, petiolate, cordate or reniform .... Family L. **Aristolochiaceae** Blume.

\+ Flowers small, with obsolescent or 4 or 5 (8)-merous perianth ... 22.

22. Terrestrial plants; if aquatic, with floating leaves, then the flowers pink or white in a dense, submersed spike .... 23.

\+ Aquatic plants, rarely terrestrial and then forming small depressed tufts with opposite leaves .... 31.

8 23. Leaves linear; inflorescence a loose simple raceme; flowers small, yellow or white, in axils of leaflike bracts; ovary inferior; fruit a green subglobose nut, crowned by 5-parted perianth .... Family XLVIII. **Santalaceae** R. Br.

\+ Leaves varying in shape; inflorescence more crowded or flowers solitary in axils of normal leaves; fruit a nut or drupe, not crowned by the perianth .... 24.

24. Leaves stipulate .... 25.

\+ Leaves exstipulate .... 26.

25. Membranous stipules at leaf base united into ringlike or tubular ocrea subtending leaf petiole; fruit a nut enclosed in perianth .... Family LII. **Polygonaceae** Lindl.

\+ Stipulate adnate to leaf petiole; leaves pinnate or rounded-reniform with lobed margin and palmate venation .... Family LXXVII. **Rosaceae** B. Juss.

26. Fruit a black or dark red berry; flowers in large dense racemes .... Family LVII. **Phytolaccaceae** Lindl.

\+ Fruit a many-seeded capsule, drupe, or nut .... 27.

27. Flowers unisexual, the pistillate flowers in axils of leaf petioles; fruit a drupe .... Family LVI. **Thelygonaceae** Eichl.

\+ Flowers bisexual or some unisexual by abortion; fruit a many-seeded capsule or a nut .... 28.

28. Fruit a many-seeded capsule; flowers often with staminodes .... Family LVIII. **Aizoaceae** A. Br.

\+ Fruit a 1-seeded nut .... 29.

29. Perianth corollaceous, strongly accrescent in maturity and transformed into an involucre enclosing the fruit .... Family LV. **Nyctaginaceae** Lindl.

\+ Perianth herbaceous or scarious . . . . 30.

30. Perianth herbaceous or fleshy; styles 2 — 5; mostly herbs, rarely shrubs or even trees (H a l o x y l o n), with abnormal secondary growth; leaves entire, rarely with irregular dentation, mostly fleshy; flowers small, inconspicuous, in glomerules, cymes, or racemes . . . . Family LIII. **Chenopodiaceae** Less.

\+ Perianth dry, scarious or rarely herbaceous; style 1; flowers in small cymes gathered into rather large panicles or racemes . . . . Family LIV. **Amarantaceae** Juss.

31. Paludal or aquatic plants, with verticillate leaves . . . . 32.

\+ As above, but leaves not verticillate . . . . 34.

9 32. Flowers unisexual; staminate flowers with 12 — 24 stamens, filaments thickened; fruit an ovaloid nut, style prolonged into a spine; leaves rigid, with linear toothed divisions . . . . Family LXII. **Ceratophyllaceae** Asa Gray.

\+ Stamens less than 10; leaves soft . . . . 33.

33. Leaves entire, in whorls of 4 — 12, linear, lanceolate, or obovate; flowers solitary in leaf axils, generally also in whorls; stamen 1; fruit an oblong-ellipsoid drupe . . . Family CXVI. **Hippuridaceae** DC.

\+ Leaves pectinately pinnate, in whorls of 5 (rarely 4 or 6); flowers axillary, solitary or in small groups, forming a more or less interrupted terminal spike; fruit 4-angled, separating into 4 drupes . . . . Family CXV. **Halorrhagidaceae** Lindl.

34. Leaves obovate, spoon-shaped, or linear, opposite or partly in rosettes at branch ends; staminate flowers with 1 stamen; fruits axillary, composite, separating into 4 rounded carpels . . . . Family LXXXVII. **Callitrichaceae** Lindl.

\+ Leaves oblong or oblong-ellipsoid; flowers axillary, sessile, perfect; stamens 3 — 6; fruit a many-seeded capsule; seeds straight or curved, ribbed, several in each locule . . . . Family CIII. **Elatinaceae** Lindl.

35 (8). Trees or shrubs, erect or twining . . . . 36.

\+ Herbaceous plants . . . . 76.

36. Twining or scandent plants, climbing up trees to considerable height . . . . 37.

\+ Herbaceous plants, with more or less upright stems and branches, or trailing, tall or dwarf . . . . 42.

37. Plants with prehensile tendrils opposite leaves and inflorescence; flowers actinomorphic, perfect or unisexual; calyx a 3 — 7-toothed flange; petals 3 — 7, at anthesis expanding or united at the tips as a hood; stamens 3 — 7; hypogynous disk with lobes projecting between the stamens; ovary superior, 2 — 6-locular, with 2 — 6 ovules in each locule; fruit a berry . . . . Family XCVIII. **Vitaceae** Lindl.

\+ Plants without tendrils . . . . 38.

38. Leaves much dissected, pinnate or bipinnate, with prehensile petioles . . . .
Genera **Atragene** and **Clematis** of Family LXIII. **Ranunculaceae** Juss.

\+ Leaves simple . . . . 39.

39. Leaves triangular-cordate or rounded-cordate; flowers saccate
10 at base, tubular above; fruits dry, gray, pyriform or oblong, 6-angled; seeds flat . . . . Family L. **Aristolochiaceae** Blume.

\+ Leaves not cordate; perianth parted; fruits not dry . . . . 40.

40. Fruit a coriaceous capsule; seeds with red aril; in one of the species, runners penetrate bark of trees by means of stipules transformed into spines . . . . Genus **Celastrus** of Family XCII. **Celastraceae** Lindl.

\+ Fruit a berry . . . . 41.

41. Flowers unisexual; stamens 5; berries globose, borne on slender, elongate torus, forming a long simple raceme . . . . .Genus **Schizandra** of Family LXVI. **Magnoliaceae** J. St. Hil.

\+ Flowers in short axillary dichasia, perfect or unisexual; calyx and corolla 5-parted; stamens very numerous; stigma stellate; fruit a green oblong many-seeded berry . . . . Family CI. **Actinidiaceae** Van Tiegh.

42. Stamens more than 14 (many) . . . . 43.

\+ Stamens 1—14 (not more) . . . . 49.

43. Ovary apocarpous, i. e., the carpels borne on convex receptacle, distinct or slightly united, so that in maturity each apparently forms a separate fruit; in general, the fruit an aggregate . . . . 44.

\+ Ovary syncarpous, i. e., each flower producing one fruit . . . . 45.

44.* Perianth petaloid, with little differentiation into calyx and corolla; carpels many, spirally arranged on prolonged receptacle, in maturity forming a conelike aggregate fruit . . . . Family LXVI. **Magnoliaceae** J. St. Hil.

\+ Calyx of 4 or 5 lobes; corolla 5-parted; fruit not conelike (an aggregate drupe or achene), each individual fruit with short lateral style . . . . Family LXXVII. **Rosaceae** Juss.

45. Leaves stipulate, alternate; receptacle strongly developed; stamens attached to edge of receptacle at base of sepals; fruit a drupe or pome (fleshy fruit developing from receptacle, with a cartilaginous capsule formed by the gynoecium) crowned by dry remants of the calyx . . . . Family LXXVII. **Rosaceae** Juss.

\+ Leaves mostly opposite; stipules absent . . . . 46.

46. Ovary inferior . . . . 47.

\+ Ovary superior . . . . 48.

11 47. Calyx campanulate, coriaceous, 5—7-lobed; petals 5—7; stamens many; leaves leathery; branches often spiny; fruit large, spherical, with leathery pericarp . . . . Family CXII. **Punicaceae** Horan.

\+ Calyx cup-shaped, 4- or 5-parted, foliaceous; petals 5; fruit dryish or indurated, obconical . . . . Family* **Myrtaceae** Pers.

48. Fruit a many-seeded capsule; small shrubs or undershrubs; flowers in racemes or fascicles of 2—6 or solitary . . . . Family CVI. **Cistaceae** Lindl.

\+ Fruit a 1- or 2-seeded nut; flowers in a cyme attached to a dry whitish winglike bract; sepals 5, caducous; trees with deciduous, mostly cordate leaves . . . . Family XCIX. **Tiliaceae** Juss.

* Repeated for the benefit of those who chose the wrong alternative at Stage 8 [It is not entirely clear whether this footnote belongs to paragraph 44.]

49 (42). Ovary inferior or half-inferior; calyx enclosing ovary and united with it except for tips of sepals . . . . 50.

\+ Ovary superior and not united with calyx . . . . 52.

50. Leaves opposite; stamens 4; fruit a drupe; flowers in umbels or corymbose panicles . . . . Family CXX. **Cornaceae** Link.

\+ Leaves alternate; stamens 5; fruit a berry . . . . 51.

51. Flowers in umbels; ovary 5-locular; leaves palmately divided or trifoliate, simple only in ivy . . . . Family CXVIII. **Araliaceae** Vent.

\+ Flowers solitary or in more or less elongated panicles; ovary 2-locular; leaves simple, slightly lobed . . . . Family LXXIV. **Saxifragaceae**, genus **Ribes** I.

52. Fruit a berry or drupe with juicy pulp . . . . 53.

\+ Fruit dry, a capsule, silique, pod, nut, etc. . . . . 60.

53. Sepals, petals, and stamens 6; flowers yellow, in pendulous racemes; anthers opening by valves or lids, not by slits; branches often with 3-parted spines subtending very short branchlets terminating in a cluster of short-petioled leaves . . . . Family LXIV. **Berberidaceae** Torr. et A. Gray.

\+ Anthers opening by a longitudinal slit; plants without spines or with simple (not 3-parted) ones . . . . 54.

54. Calyx, corolla, and androecium 5-merous; petals much smaller than sepals; stamens attached to petals; flowers solitary or in fascicles of several, rarely in a terminal raceme; fruit a berrylike drupe containing 1—4 nutlets; leaves entire . . . . Family XCVII. **Rhamnaceae** Lindl.

\+ Stamens not attached to petals . . . . 55.

12 55. Perianth petaloid, simple; calyx apparently absent; flower parts in multiples of 2 (mostly 4—8) . . . . 56.

\+ Calyx and corolla conspicuous; flower parts in multiples of 2, 3, 5 or 7 . . . . 57.

56. Perianth 2—4-parted; fruit 1-seeded; leaves opposite, entire, densely covered with silvery scurfy hairs . . . . Family CX. **Eleagnaceae** Lindl.

\+ Perianth tubular-infundibular, the limb 4-lobed; fruit 1-seeded; leaves alternate, rarely opposite, glabrous or hairy, but never with scurfy hairs . . . . Family CIX. **Thymeleaceae** Rchb.

57. Leaves compound, imparipinnate, deciduous; flowers greenish, in flat racemes; berries black, 2-seeded, resinous-scented . . . . Family LXXXIII. **Rutaceae** Juss.

\+ Leaves simple, entire, coriaceous, evergreen . . . . 58.

58. Leaves oblong-linear, revolute; flowers solitary in axils of upper leaves, unisexual; stamens 3, with long slender filaments; fruit a black succulent drupe . . . . . Family LXXXIX. **Empetraceae** Lindl.

\+ Leaves more or less flat at margin; flowers in an inflorescence . 59.

59. Leaves of young shoots spiny-toothed on margin; calyx 3—6-toothed; petals and stamens 4 or 5; flowers small, white, in cymes; fruit a berry . . . . Family XCI. **Aquifoliaceae** DC.

+ Leaf margin slightly undulate, not toothed; perianth 4-toothed; stamens 8—14; flowers in axillary umbels, these at first enclosed in a spherical involucre; fruit a drupe . . . . Family LXVII. **Lauraceae** Lindl.

60. Stamens numerous; some of them indefinite . . . . 61.

+ Stamens 4—10 . . . . 63.

61. Gynoecium of 3—5 carpels, these distinct or united at base, each bearing a pistil; inflorescence an umbel or panicle; flowers numerous, small . . . . Family LXXVII. **Rosaceae** B. Juss., subfamily **Spiraeoideae** Focke.

+ Gynoecium of 1 or several united carpels, these bearing 1 style or a group of styles . . . . 62.

62. Sepals and petals 5—3; leaves opposite, rather densely covered with stellate and glandular hairs; flowers in fascicles or solitary, often large . . . . Family CVI. **Cistaceae** Lindl.

+ Sepals and petals 4 or 5, glabrous or covered with simple or partly stellate hairs; flowers in racemes . . . . Family LXXIV. **Saxifragaceae** DC.

13 63. Fruit a pod, formed by a single carpel with apparently fused margins, dehiscent by a ventral and a dorsal suture or breaking up into 1-seeded segments; leaves compound, trifoliate from a single leaflet articulated at juncture with petiole . . . . Family LXXVIII. **Leguminosae** Juss.

+ Fruit a capsule, samara, or nut, not a pod . . . . 64.

64. Fruit 1- or 2-seeded, winged . . . . 65.

+ Fruit mostly many-seeded, wingless, or nutlike, 1-seeded 66.

65. Fruit separating into 3—5 oblong samaras; flowers in paniculate inflorescence; leaves imparipinnate, strongly odoriferous . . . . Family LXXXIV. **Simarubaceae** Lindl.

+ Fruit 2-winged; flowers in spike, raceme, or panicle; leaves opposite, simple, lobed, or trifoliate . . . Family XCIV. **Aceraceae** Lindl.

66. Fruit a capsule, mostly many-seeded, dehiscent by valves or slits, fruit rarely aggregate, the ovary separating into 4 nutlets . . . . . . 67.

+ Fruit indehiscent, a drupe or nut, always 1-seeded . . . . 75.

67. Leaves composed of 2, 3, 5, or more leaflets, geminate, palmate, or pinnate . . . . 68.

+ Leaves simple . . . . 70.

68. Leaves geminate, with fleshy leaflets; stamens 8—10; fruit a 4- or 5-angled capsule . . . . Family LXXXII. **Zygophyllaceae** Lindl.

+ Leaves palmate or pinnate . . . . 69.

69. Flowers slightly irregular, staminate, pistillate, or hermaphrodite; inflorescence an erect pyramidal raceme; calyx campanulate; petals 4; stamens 7 or 8; ovary coriaceous, often spiny; capsule 3-locular, with 2 ovules in each locule; leaves of 7 leaflets opposite, palmate . . . . Family XCV. **Hippocastanaceae** Torr. et Gray.

+ Flowers regular; capsule membranous, inflated, thin-walled, 2- or 3-lobed; leaves imparipinnate; flowers in drooping racemes . . . . Family XCIII. **Staphyleaceae** (DC.) Lindl.

70. Inflorescence capitate . . . . 71.

+ Inflorescence not capitate or flowers solitary . . . . 72.

71. Calyx 5—7-parted; petals none; stamens 5—7; capsule woody, 2-valvular; leaves ovate, obscurely lobed, often oblique at base . . . . Family LXXV. **Hamamelidaceae** R. Br.

\+ Calyx of 3—6 obtuse segments; petals cuneate or ovate; leaves
14 deeply 5—7-lobed; flower heads 2—4 on a common peduncle; fruit multiple, separating into 4 oblong hairy-tufted nutlets, the hairs rather long but shorter than the body . . . . Family LXXVI. **Platanaceae** Lindl.

72. Leaves very small, imbricated, densely covering young branchlets; flowers small, in rather long racemes gathered into panicles; seeds with a tuft of long hairs . . . . Family CV. **Tamaricaceae** Baillon.

\+ Leaves normal; seeds not hairy-tufted . . . . 73.

73. Capsule 3—5-locular; seeds surrounded by a red aril; inflorescence 2- or many-flowered, a corymb or false umbel . . . . Family XCII. **Celastraceae** Lindl.

\+ Capsule 2- or 3-locular; seeds not arillate; flowers solitary, in fascicles or heads . . . . 74.

74. Capsule ovoid, 3-horned at top; leaves opposite, leathery evergreen . . . . Family LXXXVIII. **Buxaceae** Dumort.

\+ Capsule globose or dorsiventrally flattened, without horns or protruberances; leaves relatively thin, deciduous . . . . Family LXXXVI. **Euphorbiaceae** J. St. Hil.

75. Calyx 3- or 5-parted; petals 3—5 or none; fruit a coriaceous, wingless capsule; leaves simple or pinnate . . . . Family XC. **Anacardiaceae** Lindl.

\+ Calyx 4- or 5-parted; petals 3—5; ovary 2- or 3-locular, forming an indurated dry fruit, surrounded by an orbicular wing; stipules modified to spines . . . . Family XCVII. **Rhamnaceae** Lindl., genus **Paliurus** Mill.

76 (32). Ovary inferior or half-inferior . . . . 77.

\+ Ovary superior . . . . 82.

77. Aquatic plants with floating rosulate leaves; fruit nutlike, with 2—4 spiny horns . . . . Family CXIII. **Hydrocaryaceae** Raimann.

\+ Terrestrial, rarely paludal plants, not floating; fruit without spiny processes . . . . 78.

78. Fruit of 2 carpels [mericarps], flattened commissural face, dorsally convex, with 5 prominent ribs and canals in intervals; styles 2; stamens 5; flowers in simple or compound umbels . . . . Family CXIX. **Umbelliferae** Moris.

\+ Fruit a berry or capsule . . . . 79.

79. Fruit a 2- or 5-locular berry, with 1 seed in each locule . . . . . . 80.

15 + Fruit a capsule . . . . 81.

80. Leaves 5-lobed, palmate, often no more than 2 or 3; petals and stamens 5; umbel 5—16-flowered; involucre insignificant . . . . Family CXVIII. **Araliaceae** Vent.

\+ Leaves simple, entire, opposite; petals and stamens 4; umbel subtended by involucre of 4 large, white bracts . . . . Family CXX. **Cornaceae** Link.

81. Styles 2; receptacle mostly rather flat or patelliform; fruit a 2-locular, 2-beaked capsule, partly embedded in the receptacle . . . . Family LXXIV. **Saxifragaceae** DC.

\+ Style 1; receptacle elongate, tubular; fruit a 4—6-locular capsule or a 1- or 2-seeded nut; stigma simple of 2 or more parts . . . . Family CXIV. **Onagraceae** Lindl.

82. Ovary 1-locular, open; styles 3 or 4; fruit a 3-toothed capsule, open at top from the beginning . . . . Family LXXI. **Resedaceae** DC.

\+ Ovary and fruit closed at top . . . . 83.

83 (76). Stamens indefinite . . . . 84.

\+ Stamens 1—15 . . . . 91.

84. Aquatic plants; sepals 3—12, most often 4; petals numerous; carpels 3; fruit commonly fleshy, indehiscent, many-locular; leaves mostly floating, cordate at base . . . Family LXI. **Nymphaeaceae** DC.

\+ Terrestrial plants, rarely paludal or aquatic, in the latter leaves different from above . . . . 85.

85. All stamens distinct . . . . 86.

\+ Stamens united at base of filaments, all together or in several clusters . . . . 89.

86. Ovary of numerous carpels, each with its own style; fruit compound, of 1- or many-seeded drupes or follicles, rarely 2 or more drupelets contained in a cup-shaped ovary and then fruit abortive . . . . 87.

\+ Ovary of 1—5 carpels; fruit a capsule or berry . . . . 88.

87. Flowers spirally cyclic; calyx green or petaloid; petals with nectaries and additional nectaries often present; number of perianth segments often variable; fruit a dehiscent or indehiscent follicle, in one case (*Actaea*) transformed into a berry . . . . Family LXIII. **Ranunculaceae** Juss.

\+ Flowers cyclic, mostly 5-merous, or more rarely 5—8-merous; calyx green, an outer calyx often present; styles terminal or lateral; sepals, petals and stamens often inserted on edge of cup-shaped receptacle; when petals are plucked, stamens often detached together with them . . . . Family LXXVII. **Rosaceae** B. Juss.

16 88. Styles 5; perianth simple, rather deeply 5-parted; capsule becoming indurated and breaking up at maturity; leaves more or less fleshy; staminodes often present . . . . Family LVIII. **Aizoaceae** A. Br.

\+ Styles 5; fruit a silique-like capsule or a berry; perianth double; stamens 4—6 or many; petals 4 . . . . Family LXIX. **Capparidaceae** Lindl.

89. Stamens in 2 series, lower part of filaments of inner series united into column surrounding ovary; styles many; capsule 4- or 5-valvular or the fruit separating into carpels arranged around base of style . . . . Family C. **Malvaceae** Juss.

\+ Stamens distinct or united at base into 3—5 clusters or fused into a tube . . . . 90.

90. Perianth rather large, 4- or 5-merous; ovary 3—5-locular; fruit a capsule or berry; seeds many, small, beardless; leaves opposite, entire, frequently sessile, flat . . . . Family CII. **Guttiferae** Juss.

\+ Perianth small; sepals and petals 5; ovary subglobose, 5-angled; leaves ovate to linear-subulate, alternate; seeds bearded . . . . Family CV. **Tamaricaceae** Baill.

91. (from 79). Flowers zygomorphic . . . . 92.
+ Flowers actinomorphic . . . . 96.
92. Flowers spurred . . . . 93.
+ Flowers not spurred . . . . 95.
93. Sepals 2, caducous; petals 4, the upper spurred, rarely with 2 symmetrically arranged spurs; stamens 2, 3-parted; flowers in racemes . . . . . . Family LXVIII. **Papaveraceae** B. Juss., subfamily **Fumarioideae** A. Br.
+ Sepals 4 or 5, persistent; petals 5; fruit a capsule . . . . 94.
94. Spur formed by lower petaloid sepal; stamens coherent by the ends of filaments; capsule linear-oblong, dehiscent by 2 valves; leaves alternate, exstipulate . . . . Family XCVI. **Balsaminaceae** Lindl.
+ Spur formed by lower petals; stamens distinct, the 2 anterior stamens penetrating spur; capsule globose or ovaloid, rarely oblong, dehiscent by 3 valves; leaves stipulate . . . Family CVII. **Violaceae** DC.
95. Stamens 10, all the filaments or only 9 of them united into a tube; fruit a pod or its modification; leaves compound, stipulate . . . . Family LXXVIII. **Leguminosae** Juss., subfamily **Papilionatae** Taub.
17 + Stamens 8, adnate to petals, lower petal with fringed crest on back; capsule flat; leaves entire, simple, exstipulate . . . . Family LXXXV. **Polygalaceae** Lindl.
96. Stamens 1—3—5 or more; fruit a capsule separating into 3 small carpels or 3-locular and 3-seeded; stigmas 3, 2-lobed; leaves alternate, rarely whorled or opposite, often containing a milky sap . . . . Family LXXXVI. **Euphorbiaceae** J. St. Hil.
+ Stamens in multiples of 2, 3 or 5; fruit not 3-parted and not 3-celled . . . . 97.
97. Stamens and pistils 3 or many; fruit succulent, baccate, with a single large strongly curved seed; twining plants with alternate, entire or palmate leaves . . . . . . Family LXV. **Menispermaceae** DC.
+ Plants with different characters, not twining . . . . 98.
98. Stamens in multiples of 2, rarely 3 . . . . 99.
+ Stamens in multiples of 5. . . . . 107.
99. Stamens in most genera 6, rarely 2, 8, or 9 . . . . 100.
+. Stamens mostly 4, or 8—12 . . . . 101.
100. Stamens 6, of these 2 short and 4 long (rarely also 2); calyx 4-parted; corolla of 4 petals, rarely none; fruit a 2-parted silique or silicle; flowers commonly in racemes; bracts none; leaves nearly always alternate . . . . Family LXX. **Cruciferae** B. Juss.
+ Calyx and corolla 6-merous or calyx of 4 and corolla of 8 segments; fruit a 1-locular capsule or dry and berrylike; stamens of equal length . . . . Family LXIV **Berbiridaceae** Torr. et A. Gray.
101. Perianth simple, corollaceous, tubular with 4-lobed limb; ovary superior, 1-locular, with 1 ovule; stigma clavate or capitate; fruit a nut; leaves narrow, entire . . . . Family CIX. **Thymeleaceae** Rchb.
+ Perianth clearly differentiated into calyx and corolla, not tubular . . . . 102.
102. Sepals 2, caducous: stamens 4; styles 2; fruit a fleshy silique-like capsule; annuals with bipinnate leaves . . . . Family LXVIII. **Papaveraceae** B. Juss., subfamily **Hypecoideae** Prantl et Küding.
+ Sepals 4 or more, persistent in fruit; stamens 4 or more; capsule globose, ovaloid, etc. . . . . 103.

103. Leaves decussate, subulate-linear or linear-lanceolate; flowers small, greenish or white; styles 2—5; capsule dehiscent by 4—6 teeth or by 4 valves . . . . Family LX. **Caryophyllaceae** Rchb., subfamily **Alsinoideae** Vierh.

\+ Style 1 . . . . 104.

18 104. Calyx tubular or campanulate, toothed at top . . . . 105.

\+ Calyx stellate or cup-shaped, sepals distinct from base . . . . 106.

105. Calyx commonly with accessory teeth between lobes; ovary 2—6-locular or rarely 1-locular; leaves entire, sessile, sometimes auriculate at base but always exstipulate . . . . Family CXI. **Lythraceae** Lindl.

\+ Calyx without accessory teeth; ovary 1-locular; capsule 3—5-valvular; leaves small; plants mealy-pubescent or hairy all over . . . . Family CIV. **Frankeniaceae** DC.

106. Stamens 4; ovary 4-locular; capsule 4-angled; seeds 8; flowers ca. 1 mm in diameter, in cincinni . . . . Family XCII. **Zygophyllaceae** Lindl., genus **Tetradiclys** Stev.

\+ Stamens 8; ovary 6-locular; capsule 4-valvular; flowers in paniculate inflorescence . . . . Family LXXXIII. **Rutaceae** Juss.

107. Leaves opposite . . . . 108.

\+ Leaves alternate . . . . 109.

108. Calyx of 2 sepals; stamens 3—15; style 1, 3—6-parted; capsule 1-locular; leaves more or less fleshy; bracts scarious . . . . Family LIX. **Portulacaceae** Juss.

\+ Calyx of 5 sepals, rarely of 4; stamens as many as or twice the number of sepals; styles 3 or 5, rarely 2; capsule 1-locular or at base 2—5-locular; leaves nearly always exstipulate; in one case fruit a dryish berry . . . . Family LX. **Caryophyllaceae** Rchb.

109. Calyx, corolla and androecium 5-merous; all leaves in a basal rosette, entire, covered with long red glandular hairs; inflorescence a simple raceme; capsule many-seeded; or submersed, aquatic rootless plants, verticillate spoon-shaped leaves folding up when irritated, flowers solitary . . . . Family LXXII. **Droseraceae** DC.

\+ Leaves at least partly (e. g., one leaf) cauline . . . . 110.

110. Pistils several, distinct or in lower part united . . . . 111.

\+ Pistil 1 . . . . 112.

111. Pistils 2, united at base; fruit a 2-locular, many-seeded, 2-horned capsule . . . . Family LXXIV. **Saxifragaceae** DC.

\+ Pistils 4—10, distinct or united at base; fruit a many-seeded follicle; leaves fleshy, succulent; flowers in cymes . . . . .Family LXXIII. **Crassulaceae** DC.

19 112. Flowers mostly dioecious; sepals 3—9; petals 8 or more; stamens 4—25; ovary inferior, of 3—8 carpels; capsule many-seeded; leaves simple or more often pinnate; flowers small, in racemes . . . . Family CVIII. **Datiscaceae** Lindl.

\+ Flowers monoecious; ovary superior . . . . 113.

113. Stigma sessile or nearly so; 5 palmately-parted, glanduliferous staminodes present in addition to 5 stamens . . . . Family LXXIV. **Saxifragaceae** DC., genus **Parnassia** L.

\+ Stigma raised on a distinct style . . . . 114.

114\. Leaves long-petioled, trifoliate; capsule dehiscent by dorsal sutures, widely scattering the seeds; styles 5 . . . . Family LXXX. **Oxalidaceae** Lindl.

\+ Leaves simple, entire, lobed or dissected . . . . 115.

115\. Leaves narrow, entire; stamens 8—10, united at base, often of unequal length; styles 4 or 5; capsule subglobose, seemingly 10-locular, septicidal . . . . Family LXXXI. **Linaceae** Dumort.

\+ Number of ovary locules not more than 5 . . . . 116.

116\. Leaves pellucid-dotted by glands emitting an essential oil; stamens inserted at base of a disk; fruit in USSR species a 4- or 5-valvular capsule, in some cultivated plants a large berry or drupe; leaves spiral or opposite, entire or dissected, exstipulate . . . . Family LXXXIII. **Rustaceae** Juss.

\+ Leaves not pellucid-dotted . . . . 117.

117\. Petals 5, white, pink, mauve, blue, etc., but not yellow; stamens 10—15, often only 5 of them fertile; style 5-parted at end; fruit separating into 5 one-seeded portions, their upturned appendages united with style into a beaklike column; leaves lobed or divided into narrowed, toothed segments . . . Family LXXIX. **Geraniaceae** J. St. Hil.

\+ Petals 5, yellow or reddish; stamens 10—8, rarely 15; fruit a 4- or 5-angled capsule or separating into 3 prickly or tuberculate nutlets; leaves simple, binate, or pinnate . . . . Family LXXXII. **Zygophyllaceae** Lindl.

Order 10. **Piperales** LINDL.

Flowers achlamydeous or monochlamydeous, hermaphroditic or unisexual; stamens 1—10; carpels 1—4, distinct or united; microspores 2-nuclear; flowers very small, in spikes; leaves undivided, stipulate or exstipulate.

20 1. Stamens 6 or less (in USSR flora 3); carpels and stigmas 3 or 4 (in USSR flora 3), capsule many-seeded; leaves alternate, cordate at base . . . . Family XXXVIII. **Saururaceae** Lindl.

\+ Stamens 1—3, seemingly united with one another and with ovary; carpel 1; fruit a drupe; leaves opposite, cuneate at base . . . . Family XXXIX. **Chloranthaceae** Blume.

Family XXXVIII. **SAURURACEAE** LINDL.*

Flowers achlamydeous, perfect, borne on elongate inflorescence axis; stamens 6 or less; anthers dehiscent longitudinally; carpels 3 or 4, distinct or united. Herbaceous plants with alternate petiolate leaves and flowers in a spikelike inflorescence.

* Treatment by O.I. Kuzeneva.

Genus 353. **HOUTTUYNIA** * THUNB.
Thunb. Fl. jap. (1784) 12.

Stamens 3; anthers ovaloid, double yellow; carpels 3 or 4 united; pistils as many, distinct, erect; capsule subglobose, 1-locular, many-seeded; seeds numerous, globose.

1. **H. cordata** Thunb. Fl. jap. (1784) 234; Grossg., Fl. Kavk. II (1930) 1. — Ic.: Thunb. ib. tab. 26.

Perennial; rhizome more or less vertical, with numerous fibrous roots at knots; stem 1 to several, 20—30 cm high, erect or slightly flexuous, sulcate, glabrous; leaves alternate, petiolate, entire, broadly ovate to ovate-lanceolate, point-tipped, deeply cordate at base; petioles shorter than blades, sulcate; stipules paired, adnate at base to petiole (rarely those of lower leaves distinct), entire, oblong, obtuse, slightly ciliate on margin, scarious-tipped; inflorescence spikelike, 10—30 mm long, subtended by 4 large ovate-oblong whitish bracts. May—June.

Wet coastal wasteland, especially in the zone of tea plantations. — Caucasus: W. Transc. **Gen. distr.**: Jap.-Ch. Described from Japan. Type in Uppsala.

Note. Adventive in Transcaucasia from China.

Family XXXIX. **CHLORANTHACEAE** BLUME.**

Flowers perfect or unisexual, achlamydeous; stamens 1—3, united with ovary; filaments connate; anthers oblong, 1-locular, the connective projecting above anthers; pollen spherical; ovary 1-locular, with flat sessile stigma; ovule 1; drupe succulent, 1-seeded.

21 Genus 354. **CHLORANTHUS** SWARTZ.
Swartz. Philos. Transact. LXXVII (1787) 359.

Flowers sessile in a spikelike inflorescence; ovary conical, with flattened sessile stigma; stamens 3; filaments thick, adnate to broad ovary; drupe fleshy, globose.

1. **Ch. japonicus** Sieb. in Nova Acta cur. XIV, 2 (1829) 681; Kom., Fl. Manchzh. II (1904) 7. — Ch. manshuricus Rupr. in Maak, Putesh. na Amur (1859) V. — Ic.: Rupr. ib.; Kom. and Alis., Opred. rast. Dal'nevost. kr. I (1931), Plate 127.

Perennial; rhizome branched, horizontal, creeping; stem herbaceous, glabrous, 15—40 cm high; leaves in 2 or 3 (4) pairs, scalelike, broadly ovate, 5—8 mm long, and 4 decussate, larger leaves on upper part of stem; terminal leaves with petioles 3—10 mm long, obovate or elliptic, at flowering 2—6 cm long and 1—3 cm broad, at fruiting 10—12 (16) cm long and 7 or 8 (10) cm broad, acuminate, sharply sinuate-serrate on margin, glabrous; inflorescence terminal, spikelike, simple, rather loose, 4—6 cm long;

* Named for the Dutch botanist Houttuyn (1774—1883).

** Treatment by O. I. Kuzeneva.

stamens pure white, 5—7 mm long, falling after flowering; fruit a greenish drupe. May to beginning of June.

In groups in shade of rocks or on stony slopes in oak, rarely in mixed woods. — Far East: Uss. **Gen. distr.**: Jap. - Ch. Described from Japan. Type in Leningrad.

Order 11. **Salicales** LINDL.

Flowers achlamydeous, dioecious; nectariferous disk cup-shaped or reduced to toothlike scales; stamens 2 to many; carpels 1 or 2, 1-locular; fruit a capsule; seeds exalbuminous, with basal tuft of white hairs. Trees or shrubs; leaves entire, rarely pedate, stipulate; inflorescence spikelike (a catkin).

Family XL. **SALICACEAE** LINDL.*

Dioecious trees, shrubs, or undershrubs; flowers primitive, unisexual, in lateral or apical, almost simple staminate or pistillate spikes (catkins), these pendulous or upright, deciduous; individual flowers consisting essentially of membranous scale subtending stamens or pistil; perianth vestigial, in genus *Populus* cup-shaped, enclosing stamens or pistil, in genus *Salix* consisting of 1 or 2 nectariferous glands, these homologous to disk or representing the reduced perianth; bracts deciduous or
22 persistent, glabrous or hairy, monochromatic or particolored, in *Salix* entire, in *Populus* dissected; staminate flowers with mostly 2 stamens (majority of *Salix* species), in *Chosenia* 5, in some species of *Salix* 5 (8—12), and in *Populus* up to 30—40—65; stamens distinct or basally or wholly connate, glabrous or hairy, attached at the center of the disk or toward the middle of the bract; anthers 2-locular, in species with connate filaments 4-locular; pistillate flowers with 2 (rarely 1 or 3) coherent carpels; ovary superior, 2-merous, distinct, sessile or stipitate; style 1 or 2, united or distinct; stigmas 2, entire or 2-parted; fruit a 1-locular, many-seeded, dehiscent, 2-valved capsule with parietal placentation; seeds numerous, small, exalbuminous, with orthotropous embryo and basal tuft of hairs; seedling with 2 simple entire cotyledons. It flowers with opening of the leaf or simultaneously with the latter. Plants entomophilous (*Salix*) or anemophilous (*Populus*); bud covered by ribbed scale formed by fusion of 2 scales; leaves alternate, rarely opposite, simple, entire, petiolate, stipulate or exstipulate, deciduous.

Note. Chiefly distributed in the Temperate and Frigid zones and in mountains up to the upper limit of vegetation. The family contains between 350 and 400 species, of these about 200 in the Soviet Union.

Economic importance. Family of great economic importance as a source of timber, carpentry wood, fuel, fodder, bast, tanning agents, and medicinal products. Many of the arboraceous species make very rapid growth and are easily propagated by cuttings. Not fastidious as regards soil but mostly dependent on good supply of moisture.

* Treatment of genera *Chosenia* and *Salix* by M.I.Nazarov (Moscow), who also contributed the description of the family and the key to genera.

The family Salicaceae is one of the most ancient among Dicotyledones to be found in a fossilized state. The genera Populus and Salix occur even in the lower formation of the Cretaceous period, while in Upper Cretaceous layers they are of frequent occurrence.

Key to Genera

1. Styles united or stigmas sessile; flowers containing 1 or 2 nectariferous glands or a cup-shaped disk (torus) in addition to androecium or gynoecium . . . . 2.

\+ Styles 2, distinct, with 2-parted stigmas; flowers consisting merely of a scale and ovary; glands and torus lacking; bracts imbricated, short, membranous, those of staminate flowers 3—5 (6)-nerved; stamens 5, shorter than the bracts and attached at their middle or
23 at 1/3 the length from base; catkins pendulous . . . . 355. **Chosenia** Nak.

2. Buds terminal, with numerous scales; leaves long-petioled; catkins pendulous; catkin-scales dissected or fringed; stamens numerous (30—40—65), with short filaments; capsule 2—4-valvular; base of ovary or androecium surrounded by a cup-shaped torus [disk] . . . . 357. **Populus** L.

\+ Buds lateral, commonly with 1 scale; leaves frequently short-petioled; catkin-scales entire; stamens mostly 2, rarely 3—5 (12), with long filaments; capsule 2-valvular; 1 or 2 nectiferous glands at the base of ovary or androecium, not surrounding the flower . . . . . . 356. **Salix** L.

Genus 355. **CHOSENIA** * NAK.

Nak. in Tok. Bot. Magaz., XXXIV (1920) 68.

Buds with a single 3-nerved scale; branchlets developing from the buds enveloped by the scales; staminate catkins pendulous, pistillate erect or ascending, anemophilous; bracts of pistillate catkins (bracteae) leaflike; scales of catkins (squamae) inflated, imbricated, membranous, 3—5-nerved; ciliate-margined, those of pistillate flowers caducous; perianth none; stamens 5, attached to base of scale; styles 2, distinct, jointed at the middle; stigma 2-parted, feebly jointed to pistil.

1. **Ch. macrolepis** (Turcz.) Kom., the third genus of the family Salicaceae in Yubil. sborn. P. Borodina (1927) 275—281; Kom. and Alis., Opred. rast. Dal'nevost. kr. 1, 418. — Salix macrolepis Turcz. Fl. baic.-dah. II (1854) 98; Anderss., Mon. Salic. (1867) 52 et tab. III, f. 33; Ej. in DC. Prodr. XVI, 2, 213. — S. bracteosa Turcz. in sched. — S. acutifolia Nak. Fl. kor. II (1911). — S. splendida Nak. in Tok. Bot. Mag., XXXII (1918) 215. — Chosenia splendida Nak., ibid. XXXIV (1920) 68. — Ch. eucalyptoides Nak. in Journ. Arn. Arbor. V (1924) 72. — Ch. bracteosa Nak. Fl. sylv. kor., XVIII (1930) 59—63 et tab. III—V.

* From "Chosen," the Japanese name for Korea.

A fast-growing tree, with an upright trunk to 37 m high and usually to 30 cm in diameter at breast height (Anadyr; in Korea to 1.5 m high); bark of young trees light-colored, that of old trunks brownish-cinereous, with deep longitudinal fissures; branches arising nearly at base and appressed to the trunks, erect or spreading, slender, glabrous, reddish, pruinose when young; buds oblong-ovoid, 2—5 mm long, flattened; petioles 5—7 mm long, glaucous, channeled, eglandular; exstipulate; leaves 4—8 mm long and 1—2 cm broad, oblong-lanceolate or oblanceolate, glabrous on both
24 sides, white-dotted by stomata, when young intensely glaucous especially on the lower surface; staminate catkins pendulous, 1—2.5 cm long, pistillate ascending, after flowering 1—2 cm long, borne on a stalk to 1.5 cm long, the stalk and rachis slender; bracts at the base of the catkin resembling the leaves but more often entire; scales imbricated, concave, yellowish-green, broadly obovate, truncate or erose, 3—5-nerved, enclosing the inner flower parts, in pistillate flowers caducous; stamens 5, attached to the base of the bract, glabrous; anthers spherical, yellow; ovary to 2 mm long, borne on a stipe ca. 0.5 mm long, ovoid-oblong, very obtuse, truncate, glaucescent, quite glabrous; style 2-parted, 0.5 mm long; stigmas distinct, as long as or longer than style. May—June.

Pebbly places in valleys, and inundated woods in the mountain zone.—Arctic: An.; E. Siberia: Dau., Lena-Kol.; Far East: Okh., Kamch., Sakh., Ze.-Bu., Uss. Gen. distr.: Japan, Korea, Manch. Described from Transbaikalia. Type in Leningrad.

Economic importance. A tree used for construction purposes, and valuable for its upright trunk. Used in the Far East for construction of bridges, for telegraph poles, and for boats made by gouging out the wood. Also used for fuel and for odd jobs. The bast is suitable for mat weaving, etc. The tree is unfortunately short-lived and often hollow in age. A fast-growing tree.

Genus 356. **SALIX*** L.

L. Gen. plant. (1737) 300.—Chamitea Kerner in Verh. Bot.-Zool. Gesellsch. Wien X (1866) 275.—Toisusu Kimura in Tok. Bot. Magaz. XLII (1928) 287.— Russian names: iva, bredina, vetla, verba, rakita, loza, lozina.

Trees, shrubs, or barely perceptible undershrubs of an almost herbaceous aspect; buds axillary, commonly with a 2-angled scale (formed by fusion of 2 scales); buds and leaves arranged spirally; leaf indentation narrow, crescent-shaped or horseshoe-shaped; leaves simple, alternate, rarely opposite, mostly short-petioled, stipulate, deciduous, with pinnate or reticulate venation; catkins always developing from one of lateral buds on the shoot of the preceding year, upright, spreading, or pendulous, developing before, together with, or after the leaves; 1 or several bracts (bracteae) green or scalelike, persistent or caducous at the base of catkins; flower scales (squamae) simple, monochromatic or particolored; perianth rudimentary, represented by 1 (inner or posterior) nectariferous gland or 2 (inner and outer or anterior and posterior) glands, rarely 3, 4, or 5, more or less united into a cup-shaped or lobed torus [disk]; staminate flowers consisting of scales (glands) and disk; pistil, if present (f. hermaphroditica), usually abortive; styles mostly 2, rarely 3—5

* From Celtic "sal," near, and "lis," water, alluding to the predominant habitat.

25 (to 12); filaments distinct, or connate at base or throughout, glabrous or hairy below; anthers 2-locular, when filaments united seemingly 4-locular; pistillate flowers consisting of bracts, glands, and pistil; ovary mostly ovoid-conical, glabrous or pubescent, sessile or stipitate; style 1, undivided or 2-parted, with entire or 2-parted stigma; capsule 1-locular, invariably 2-valvular; seeds oblong, small, with silvery hairs.

Note. A genus well represented in all republics and regions of the USSR, but particularly widespread in locations with moderately cool and wet climate. Dwarf forms predominate in the polar regions and in the alpine zone of mountains; these are sometimes almost concealed under the cover of mosses. In the forest and steppe regions the genus is mostly represented by rather tall shrubs and trees reaching first-grade size.

Willows occur generally in extensive thickets, forming everywhere an important element of the landscape and participating in a great variety of plant associations, but they show a definite preference for wet places. Being endowed with the ability to form adventitious roots, willows are very easily propagated by cuttings or slips. When cut down, they proliferate readily from the roots and, when the head is removed, thick new growth is produced from the stem; they are therefore particularly amenable to pollarding. Salix species hybridize easily. Because of their large number, the hybrids are not described here separately; more or less detailed descriptions are often included in local floras. The hybrids are rather easily distinguished upon acquaintance with the more important species of the Soviet flora.

Economic importance. Willows are of outstanding importance to man, furnishing timber, fuel, wood for carpentry, material for basket making, tanning agents, and medicaments. They are also of importance as ornamental plants and as a source of fodder for domestic animals.

The presence of willow thickets is biologically associated with certain lignivorous ruminants, such as reindeer, red deer, elk, etc., which feed upon the bark, branches, and leaves. The bark is also eaten by hares, squirrels, beavers, and water voles. Certain species of Lagomyidae collect and dry willow leaves. Birds, such as rock ptarmigan, willow grouse, hazel hen, blackcock, etc., feed in winter and spring upon willow buds and catkins. Willow thickets provide stations for various animals and birds of economic importance that find in them refuge and food. The kinds yielding material for basketry are of outstanding value; they are collected and cultivated both for home use and for export.

Fossil species — S. alba L. in Posttertiary formations (travertines) of Cisc. (Zheleznovodsk, Maikop district), E. Transc. (Khevsha R.), in the Pliocene of E. Transc. (Keramal-Naftalan), and the Pliocene of W. Transc. (Abkhazia). — S. angusta A. Br. in Maiotis layer* of Bl. (Odessa), and Oligocene of W. Transc. (Sochi district); S. Transc. (Argachi, Mount
26 Solyanaya). — S. aurita L. in Postpliocene interglacial formations of U. V. (Likhvin) and postglacial formations of S. Transc. (Georgia, Makart). — S. caprea L. in the Pliocene of V.-Don (Krivobor'e in the Postpliocene interglacial formations of V.-Kama (Galich) and postglacial formations

* [The uppermost Miocene layer of the Black Sea and Caspian Sea basins. "Maiotis" is the ancient Greek name for the Sea of Azov.]

generally, e. g., Lad.-Ilm. (Luga), V.-Don (Dubenshino), Cisc. (Zheleznovodsk), S. Transc. (Makart in Khevsuretiya), L. V. (Balykleya). — S. c i n e r e a L. in the Pliocene of V.-Don (Krivobor'e), in interglacial formations of V.-Kama (Galich), in the Pliocene of E. Transc. (Shiraki), in the Postpliocene of S. Transc. (Makart and Khevsuretiya). — S. ex gr. f r a g i l i s L. in the interglacial formations of V.-Kama (Galich). — S. g l a u c a L. in the Postpliocene of Yen. (Lower Yenisei). — S. h e r b a c e a L. in the Postpliocene of Lad.-Ilm. (Tosno R.), Dv.-Pech. (Vologda), Ob (Demyanskoe), and Yen. (Lower Yenisei). — S. i n t e g r a Goepp. in Sarmatian layers of W. Transc. (Khvteeba). — S. cf. l a p p o n u m Lad.-Ilm. (Volkhov R.). — S. L a v a t e r i Heer in the Tertiary layers of Uss. (near Vladivostok), and Sakh. (Dun, etc.); id. for. m i n o r in the Oligocene of M. Dnp. (Volyanshchina). — S. l o n g a A. Braun in Lower Tertiary formations of Uss. (Rechnoi Peninsula) and Balkh. (Ashutas). — S. m i c a n s Anderss. in the Pliocene of E. Transc. (Shiraki). — S. m a c r o p h y l l a Heer in the Maiotis formations of Bes. (Seimeny). — S. m e d i a Heer in the Pliocene of W. Transc. (Goderskii Pass) and E. Transc. (Khvteeba). — S. m y r s i n i t e s L. in the Postpliocene of Lad.-Ilm. (Volkhov). — S. n i g r i c a n s L. (aut. S. c a p r a e a L.) in the Postpliocene of L. V. (Balykleya). — S. p e n t a d r a L. in the Postpliocene of Georgia, S. Transc. (Makart). — S. p o l a r i s L. in interglacial formations of V.-Kama (Galich), in the Postpliocene of Lad.-Ilm. (Volkhov), and in glacial formations of Ob (Demyanskoe) and Lena-Kol. — S. p u r p u r e a L. in the Pliocene of V.-Don (Krivobor'e) and the Postpliocene of V.-Don (Yamani in northern Voronezh). — S. r e t i c u l a t a L. in postglacial formations of Lad.-Ilm. (Tosno; Volkhov R.). — S. t r i a n d r a L. in the Pliocene of V.-Don (Krivobor'e). — S. v a r i a n s Goepp. in Tertiary formations of Bl. (Khadzhibei near Odessa) and Bes. (Gangury), in the Miocene of Cisc. (Adagum), and in Lower Tertiary formations of Sakh. (Akhzngy, Sertunai). — S. v a r i a n s Goepp. f. B r u c k m a n n i i Heer in Maiotis formations of Bl. (Khadzhibei near Odessa). — S. v i m i n a l i s L. in the Pliocene of V.-Don (Krivobor'e, etc.) and L. V. (near Balykleya), in the Postpliocene of U. V. (Dubenshchina), and Pliocene (?) of Irt. (Chingistai). — S. sp. is reported from many locations for the flora of Tertiary and Quaternary periods from all over the USSR.

Key to Sections

1. Low and dwarf arcto-alpine shrubs, mostly with terminal or subterminal catkins arising side by side from the terminal bud of the shoot of the preceding year; leaves rather small, 2—5 (8)-nerved, the prominent venation often reticulate . . . . 2.
\+ Trees and shrubs of lower latitudes and altitudes, with lateral catkins borne along the shoots or on short lateral branchlets . . . . 6.
2. Leaves large, 2—5 cm long and 1.5—4 cm broad, mostly entire, often (not always) long-petioled; catkins fairly long-stalked, dense, often very large; ovary woolly . . . . 3.
\+ Leaves small, stiff, coriaceous, mostly serrate, spiny-toothed, or entire; petioles usually short; branchlets often subterranean, slender, pale yellow; catkins small, stalked or sessile; ovary glabrous or more or less hairy . . . . 4.

27 3. Leaves orbicular, reniform, or ovate, mostly entire, revolute, long-petioled, exstipulate, rugose, densely and handsomely reticulate-veined beneath, glabrous or silvery-tomentose, with 2—5 pairs of lateral veins; scales of catkins purple, light brown, or pink; style short . . . . (page 28) 1. **Reticulatae** Fries.

\+ Leaves obovate, obovate-oblong, or obovate-cuneate, more rarely broadly lanceolate or ovate-lanceolate, often stipulate, glabrous or more or less pubescent, for the most part entire, with 7 or 8 pairs of riblike arching veins; scales of catkins mostly black at apex, covered with long white hairs; style long . . . (page 35) 4. **Arcticae** Rydb.

4. Dwarf undershrubs, usually with only 2 or 3 leaves above ground, at the ends of branchlets; leaves ovate, obovate, or suborbicular, small, commonly entire, glabrous, pale green on both sides, with 3—5 lateral veins arcuately converging toward apex; catkin-scales rounded and black at apex; ovary commonly densely hoary, rarely with more scattered hairs, brown . . . . (page 34) 3. **Polares** Nas.

\+ Prostrate or erect undershrubs; leaves mostly stiff, brittle, reticulate-veined, varying in size and shape, serrate or incised-serrate, rarely entire, glabrous or pubescent, with 4 or 5 pairs of lateral veins; scales of catkins acute or obtuse, mostly dark, purple or blackish at the tip; ovary glabrous or rarely more or less hairy, reddish-brown or purple . . . . 5.

5. Dwarf undershrubs with procumbent or subterranean branches; leaves mostly orbicular, roundly emarginate, glabrous on both sides, thin, dentate-serrate or crenate-serrate, rarely entire, deciduous, not turning black, mostly without stipules; catkins 4—7-flowered, rarely 10—30-flowered; ovary always glabrous, green or brownish, conical-subulate; style short but distinct . . . . (page 30) 2. **Herbaceae** Borr.

\+ Larger prostrate, depressed or upright shrubs; leaves stiff, coriaceous, mostly coarsely glandular-serrate or incised-spiny, rarely entire, sometimes turning black; in some species dead leaves persistent on the branchlets for 1 or 2 subsequent seasons; stipules present or absent; ovary more or less hairy or glabrous, commonly purple; style brown, elongated; catkins ranging from very small, subsessile and few-flowered to very large, densely many-flowered, and borne on a long stalk, as long as the catkin . . . . (page 40) 5. **Myrsinites** Borr.

28 6. Scales of catkins mostly particolored, light-colored at base, dark at apex; pistillate and staminate flowers with only 1 (posterior) gland; mainly shrubs . . . . 7.

\+ Scales of catkins mostly monochromatic, light-colored; staminate and pistillate flowers with 2 (or several) glands or the staminate with 2 glands (anterior and posterior), the pistillate with 1 (posterior); mainly tall trees . . . . 25.

7. Stamens distinct (more or less connate stamens occurring only as deformity) . . . . 8.

\+ Stamens always connate throughout or up to the middle . . . . 19.

8. Grown leaves more or less densely hairy on both sides or only beneath . . . . 9.

\+ Grown leaves glabrous or nearly so on both sides . . . . 13.

9. Grown leaves commonly rather broad, lanceolate, elliptic, or orbicular, not much longer than broad, the lower surface dull or slightly lustrous, gray or snow-white, woolly, velutinous, tomentose, or rarely silky, the dense short hairs without definite direction, straight or curved . . . . 10.

\+ Grown leaves commonly narrow, linear-lanceolate, lanceolate, or oblong-lanceolate, much longer than broad, the lower surface mostly silvery or lustrous-sericeous, without dense appressed hairs . . . . . 22.

10. Subarctic and subalpine shrubs; branches mostly knotted, dark brown, often stout and woolly-tomentose; ovary commonly short-stipitate . . . . 11.

\+ Shrubs and trees of lower latitudes and altitudes; ovary long- or short-stipitate . . . . 13.

11. Catkins borne on long or elongating leafy stalks; scales almost uniformly light-colored; stamens woolly at base; leaves mostly entire, without prominent veins . . . . (page 47) 6. **Glaucae** Fries.

\+ Catkins sessile or short-stalked; scales black-tipped; stamens glabrous; leaves glandular-serrate or entire, the lower surface with prominent and often reticulate venation . . . . 12.

12. Leaves mostly oblong, spatulate, ovate-oblong, or ovate-elliptic, glabrate; branches relatively slender; ovary tomentose or woolly . . . . (page 53) 8. **Villosae** Anderss.

\+ Leaves mostly ovate-orbicular, elliptic, broadly obovate, ovate-elliptic, or rarely sublanceolate, glabrous or woolly beneath; branches stout, covered with gray or dark tomentum; ovary commonly glabrous . . . . (page 50) 7. **Chrysanthae** W. D. Koch.

29

13. Leaves blackening, smooth, glabrous or more or less pubescent on both sides, green or glaucous beneath, or dull glaucous on both sides; ovary commonly glabrous . . . . 14.

\+ Leaves not blackening, glabrous or densely pubescent beneath; ovary glabrous or tomentose . . . . 15.

14. Large forest shrubs; leaves large, serrate, green above, commonly green beneath or glaucous but green-tipped; style long . . . . (page 68) 11. **Nigricantes** Borr.

\+ Low shrubs of marshes and peat bogs; leaves small, entire, revolute, dull above, glaucescent beneath; style short . . . . (page 88) 14. **Myrtilloides** Borr.

15. Ovary short-stipitate; leaves commonly glabrous beneath, glaucous . . . 16.

\+ Ovary long-stipitate; leaves rugose, dull green above, the lower surface tomentose, more or less pubescent, or glabrate, commonly with a prominent network of veins . . . . (page 70) 12. **Capreae** Bluff.

16. The conical ovary and its stipe glabrous; leaves commonly thin, sharply toothed, glaucous beneath, with a large stipule . . . . (page 90) 15. **Hastatae** Fries.

\+ Ovary and its stipe tomentose or more or less pubescent, rarely glabrate . . . . 17.

17. Branches slender, commonly grayish-brown, yellowish-brown, or reddish, dull, pubescent or glabrous; leaves mostly thin, 5—7-nerved; catkin-scales light-colored or more or less dark-tipped, sparingly hairy . . . . (page 83) 13. **Lividae** Nym.

\+ Branches stouter, commonly dark red or castaneous, mostly glabrous, lustrous; leaves frequently firm, coriaceous, glossy above, 7—12-nerved; scales of catkins commonly black at apex, rarely almost monochromatic . . . . 18.

18\. Leaves firmer, coriaceous, large, often stipulate; scales of catkins mostly particolored, commonly long-hairy . . . . (page 57) 9. **Phylicifoliae** Dumort.

\+ Leaves frequently thin, rather small, commonly exstipulate; scales commonly light-colored, almost monochromatic, less pubescent . . . . (page 63) 10. **Arbusculoideae** Floder.

19\. Stamens mostly connate throughout or at least by their filaments; leaves frequently lanceolate, linear-lanceolate, or oblanceolate, acute, much longer than broad . . . . 20.

\+ Stamens connate throughout or merely at base; leaves commonly short, very broad, often narrowed toward the base from a broad apex, rarely lanceolate to broadly lanceolate . . . . 21.

20\. Ovary long and narrow; style elongated, twice as long as ovary; anthers free, 2-locular; leaves broadly lanceolate, elliptic-lanceolate, or or oblanceolate, 2—3 cm broad . . . (page 101) 18. **Subviminales** O.v.Seem.

30 + Ovary commonly small, ovoid to ovoid-spherical; style absent or more or less evident; anthers mostly connate, seemingly 4-locular; leaves commonly narrower, lance-linear, lanceolate, or oblanceolate . . . . (page 118) 20. **Helix** Dumort.

21\. Leaves obovate-oblong, rounded at apex or short-pointed, cuneately narrowed toward base, to 2.5 cm broad, cinereous, glaucous, or lustrous-sericeous beneath; stamens free or connate at base; style elongate . . . . (page 100) 17. **Sieboldianae** O. v. Seem.

\+ Leaves oblong, elliptic-lanceolate, or oblanceolate, rather short, broad, the lower surface mostly glaucous, glabrous, with numerous slender veins; rarely lanceolate, pilose; stamens completely or partly connate; style mostly short . . . . (page 139) 21. **Caesiae** Kern.

22\. Bark of branches naked, pruinose, lemon-colored on the inside; leaves and ovary glabrous; style as long as or longer than ovary . . . . (page 141) 22. **Daphnoides** Dumort.

\+ Bark of branches not pruinose; ovary and the lower surface of leaves sericeous or silky-pubescent . . . . 23.

23\. Style short; stigma short, oblong; young (and often grown) leaves silvery-lustrous, on both sides or only beneath glabrate, cinerescent, sometimes turning black, from ovate to linear, the petioles short . . . . (page 96) 16. **Incubaceae** Dumort.

\+ Style more or less elongated; grown leaves more or less lustrous-sericeous . . . . 24.

24\. Grown leaves obovate-oblong, rounded or short-pointed at apex, cuneately narrowed toward base, to 2.5 cm broad, dark green above, the lower surface silvery or sericeous or shiny-tomentose, with prominent lateral veins; gland ovate . . . . (page 100) 17. **Sieboldianae** O. v. Seem. (**S. sitchensis** Sans.).

\+ Grown leaves commonly ranging from elongate and linear-lanceolate to broadly lanceolate, point-tipped, the lower surface lustrous-sericeous to glabrate and glaucous, the midrib always prominent, brownish or stramineous, the lateral veins prominent or rather inconspicuous; style, stigma, and gland commonly long, linear . . . . (page 103) 19. **Viminales** Bluff et Fingerh.

25. Arcto-alpine shrubs; branches stout, woolly-tomentose; leaves oblong, entire or subentire, rather long-hairy; staminate flowers with 1 or 2 glands; pistillate with 1 posterior gland (cf. stages 7—11) . . . . (page 47) 6. **Glaucae** Fries.

\+ Trees and tall waterside shrubs of the forest and steppe regions (or corresponding altitude zones); branches glabrous or puberulent; leaves long, lanceolate; staminate and pistillate flowers with 2 or rarely more glands . . . . 26.

31 26. Staminate flowers with 2 glands, pistillate with 1 (inner) gland . . . . . 27.

\+ Staminate and pistillate flowers with 2 (or more) glands . . . . 29.

27. Mostly waterside shrubs; androecium as a rule of 3 stamens; branches mostly glabrous; grown leaves glabrous, thin, rather large, serrate or dentate, the two surfaces concolor or the lower glaucous; catkins sparsely flowered; scales persistent to maturity; ovary long-stipitate, ovoid-conical, green, glabrous . . . . (page 145) 23. **Triandrae** Dumort.

\+ Mostly trees; androecium of 2—3—5—6—8 stamens; young branches often pubescent; grown leaves lance-elongate, glabrous or before aging more or less silvery and silky, serrate or rarely subentire; scales mostly falling before maturity; ovary sessile or short-stipitate, mostly glabrous . . . . 28.

28. Androecium as a rule of 2 stamens; leaves rather thin, sericeous (especially when young) or glabrate, the vesture appearing on both surfaces or confined to one of them; scales of catkins falling before maturity; ovary ovoid-conical, sessile or short-stipitate . . . . (page 147) 24. **Albae** Borr.

\+ Stamens 3 (rarely 2), 4, 6, or 8; leaves stiff, glabrous on both sides, glaucous beneath, entire or serrulate; catkins dense; ovary often thick, short-stipitate . . . . (page 153) 25. **Acmophyllae** Anderss.

29. Androecium of 2 or 3 (4) stamens . . . . 30.

\+ Androecium of 5 (many) stamens . . . . 31.

30. Stamens 2; style and stigma or stigma alone mostly long . . . .(page 155) 26. **Subfragiles** O.v.Seem.

\+ Stamens 2 or 3 (4); style and stigma short . . . .(page 159) 27. **Fragiles** C. Koch.

31. Style lacking or short, simple or 2-parted, persistent; scales of catkins yellowish-green or lemon-colored, shorter than ovary or stamens; leaves green beneath . . . (page 161) 28. **Pentandrae** Dumort.

\+ Style rather long, 2-parted, its lobes falling together with style; scales of catkins light brown or yellowish, pellucid, 5-nerved, imbricated, almost concealing the stamens or ovary; leaves glaucous beneath . . . .(page 162) 29. **Urbanianae** O.v.Seem.

Subgenus 1. **CHAMAETIA** Dumort. Verhandel. in Bijdr. Nat. Wetensch. Amsterd. I (1826) 56. — Low, mostly prostrate shrubs of the Arctic Region and the alpine zone; branches often procumbent, rooting; leaves short, more or less rounded, coriaceous, brittle, mostly green, lustrous, glabrous, reticulate-veined, with 2—5 lateral veins; catkins terminal on annotinous, often gemmiferous and leafy shoots; scales mostly monochromatic,

concave, enclosing the ovary base; ovary commonly subsessile; style mostly brown, long, simple or cleft; stigma more or less divided; glands
32 1 or 2, cleft or simple.

Section 1. **RETICULATAE** Fries, Physiogr. Sällsk. Årsb. (1825) 36. — Leaves suborbicular or ovate, long-petioled, coriaceous, exstipulate, mostly entire, beautifully and densely reticulate-veined beneath; lateral veins 2—5; lower surface glabrous, glaucous or on the veins densely silvery-tomentose; upper surface glaucescent-green, glabrous, rugose or more or less densely silvery-tomentose; catkins terminal, long-stalked, cylindric; scales round, purple, light brown, or pink; glands cup-shaped, lobed; stamens and ovary woolly; style short, 2-parted.

1. Leaves glabrous, reniform, cordate at base, the petiole to 7—8 cm long . . . . 3. **S. kurilensis** Koidz.
\+ Leaves glabrous or pubescent, different in shape, the petiole shorter . . . . 2.
2. Leaves oblong-orbicular, dark green above, densely covered beneath with long silvery-white hairs; an upright shrub of the subalpine zone, to 75 cm — 1 m high . . . . 4. **S. vestita** Pursch.
\+ Leaves different; prostrate shrubs of the Arctic Region and the alpine zone . . . . 3.
3. Leaves orbicular or obovate, glabrous on both sides, revolute-margined . . . . 2. **S. orbicularis** Anderss.
\+ Leaves ovate, elliptic-orbicular or obovate, entire, undulate-margined, or rarely denticulate, mostly glabrous or with scattered hairs . . . . 1. **S. reticulata** L.

1. **S. reticulata** L. Sp. pl. (1753) 1018, No. 1446; Turcz. Fl. baic.-dah. II, 1, 122 (ex p.); Ldb. Fl. Ross. III, 623 (ex p.); Kryl., Fl. Zap. Sib. IV, 773 (ex p.). — Chamitea reticulata Kern. N. Oest. Weid. (1860) 153. — Ic.: Fl. Dan., t. 212; Rchb. Ic. Fl. Germ. XI, t. 557, f. 1184. — Exs.: Fries, Herb. norm. IX, 62; Wimm. et Kr. Herb. Salic. 80, 93; Coll. Salic. 123, 124; Enand. Salic. Scand., 1, 2.

A depressed undershrub; branches procumbent, often buried and rooting, glabrous, stout, reddish-brown (rarely grayish-brown or yellowish-brown), to 50—75 cm long; young branches puberulent; stipules none or sometimes replaced by faintly purplish glands; petioles 0.3—1.5 cm long, pinkish or purple; leaves ovate, elliptic-orbicular, or obovate, rounded or rarely cuneately narrowed at base, rounded or slightly retuse at apex, 1.5—5 cm long and 1—3.5 cm broad, typically entire with wavy and revolute margin or denticulate (f. denticulata Lundstr.), coriaceous, finally glabrous (f. glabra O. Bus.) or, especially beneath, hairy (f. sericea Gaud.), when young mostly covered with silky hairs, dark green and densely reticulate-
33 rugose with impressed veins above, glaucescent-white beneath, with dense reticulation of pinkish or silky veins; lateral veins 3—5 pairs; catkins serotinous, on leafy terminal branchlets, their stalks to 5 cm long; scales suborbicular or obovate, light brown, pinkish, or yellowish, glabrous on the back, short-ciliate on the margin, to 1—1.5 cm long; stamens 2, distinct, short-hairy in lower part; anthers violet or yellow; ovary sessile or

very short-stipitate, oblong-ovate, faintly purplish, covered with crisped white hairs, to 3—6 mm long; stipe, when present, much shorter than the 2-parted gland; style very short or obsolescent; stigma with 4 short oblong lobes. June—July. (Plate II, Figure 3).

Stony, gravelly or lichen Arctic and alpine tundra. — Arctic: Arc. Eur., Nov.Z., Arc. Sib., Chuk., An.; European part: Kar.-Lap., Urals (N. and S.); W. Siberia: Alt.; E. Siberia: Yen., Ang.-Say., Lena-Kol.; Far East: Kamch., Okh. Gen. distr.: Arctic, Scand., mountains of Centr. Eur., N. Mong., Ber., N. Am., Greenland. Described from Sweden. Type in London.

Hybridizing with S. arbuscula, glauca, hastata, herbacea, Krylovii, kurilensis, lapponum, nigricans, nummularia, polaris, rotundifolia, vestita.

Economic importance. Leaves and tips of branches are eaten in winter by deer which dig them up from under the snow. Very ornamental.

2. **S. orbicularis** Anderss. in DC. Prodr. XVI, 2 (1868) 300. — S. reticulata subsp. orbicularis Floder. in Arkiv för Bot. 20 A, No. 6 (1926) 5; Hultén, Fl. of Kamtch. II, 19; Kom., Fl. Kamch. II, 33.

Prostrate or wide spreading shrub; branches long, creeping, dark, glabrous, rooting, only the tips with 2 or 3 leaves rising above ground; buds blackish-brown, glabrous, obtuse, subappressed; petioles slender, somewhat flexuous, to 2—3 cm long; leaf blades suborbicular to obovate-orbicular, to 2—3 cm in diameter, glabrous on both sides, retuse at apex, mostly rounded, narrowed or subcordate at base, entire, revolute-margined, rather thin, the veins impressed above, conspicuously raised beneath, the upper surface dark green, dull, the lower intensely glaucous, quite hairless, densely reticulate with brownish veins; catkins borne on a long glabrous stalk, slender, dense, upright; scales rounded-obovate, brown, glabrous; ovary ovoid-conical, slightly pointed, finely hoary-tomentose; style obsolescent; stigmas brown, 2-parted, divergent. June—July. (Plate I, Figure 1).

Alpine and subalpine zones, and dry stony tundra. — Arctic: Chuk.; Far East: Okh., Kamch. **Gen. distr.**: N. Am. Described from Kamchatka. Type in Leningrad.

34 3. **S. kurilensis** Koidz. in Tokyo Bot. Mag. XXXII (1918) 62. — S. longipetiolata Floder. Arkiv för Botan. 20 A No. 6 (1926) 28—29; Hultén, Fl. of Kamtch. II, 15; Kom., Fl. Kamch. II, 28—19.

A prostrate shrub; branches procumbent, rooting, ascending at the ends, 10—15 cm long, with quite smooth reddish-brown bark; buds obovoid, medium-sized or small, inflated, glabrous; stipules lacking; petioles to 7—8 cm long, often erect, dilated toward base, channeled above, glabrous; leaves reniform or reniform-orbicular, rounded or retuse at apex, cordate at base, in luxuriant specimens resembling the leaves of Oxyria digyna (L.) Hill., 1.5—4 cm long and 1.4—5 cm broad, slightly glandular-serrulate on the margin, glabrous on both sides, dark green with impressed veins above, grayish and densely reticulate beneath with conspicuously raised yellowish-brown veins, the leaves turning somewhat brownish in drying; pistillate catkins lateral, appearing after the leaves, dense, 2—4 cm long, the glabrous stalk to 1 cm long in fruit; scales obovate-orbicular, monochromatic, purple, covered with long white hairs; stamens 2, distinct, glabrous, with purple filaments and round dark purple anthers, gland inner, shorter than the stipe, purple; ovary yellowish-green, glabrous, the stipe to 1 mm long; style ca. 1 mm long; stigma 2-lobed. July. (Plate I, Figure 6).

Far East: Kamch. (maritime tundra and at the S. extremity). **Gen. distr.:** Kurile Islands. Described from Shumshu Island. Type in Tokyo.

Note. Closely related to S. venusta Anderss. described from Sitka Island (Anderss. in DC. Prodr., XVI, 2, 288). A hybrid of S. kurilensis Koidz. and S. Pallasii Anderss. is occasionally encountered.

4. **S. vestita** Pursch, Fl. Amer. sept. II (1814) 610; Anderss. in DC. Prodr. XVI, 2, 300. — S. reticulata β villosa Trautv. in Ldb. Fl. Alt. IV (1833) 291. — S. reticulata α vestita Anderss. Salic. bor.-amer. (1858) 133. — S. erecta Ldb. in sched. — Ic.: Rchb. Ic. Fl. Germ., XI, t. 560, f. 1195.

A shrub 50 — 75 cm high; branches upright or ascending, dense, glabrous, shiny gray or castaneous-brown, rugose; summer shoots very short, with 2 or 3 leaves and catkins; buds large, ovoid, obtuse, light brown, silky-hairy; petioles to 0.4 — 0.6 cm long, channeled and glabrous above, bearded with white hairs beneath; leaf blades oblong-orbicular, elliptic, ovate, or obovate, 4 — 8 cm long and 2 — 6 cm broad, subobtuse, almost round or subcordate at base, entire or slightly crenulate, firm, the upper surface bright green and somewhat rugose from the impressed veins, the lower surface densely covered with long slender milky-white hairs, the hairs not concealing the reticulation and the 5 or 6 pairs of lateral veins prominent; senescent leaves somewhat glabrescent; catkins terminal, borne on a leafy tomentose
35 stalk, rather sparsely flowered almost from base, narrowly cylindric, 1.5 — 2 cm long and 0.3 — 0.4 cm broad, the pistillate to 4 cm long in fruit; scales rufescent or light-colored, obtusely obovate, strongly villous on both sides; stamens 2, distinct, densely covered on the inside with flexuous white hairs; anthers round, yellow; ovary oblong-ovoid, short, obtuse, sessile, hoary-tomentose, rarely with scattered pubescence and purple, the style obsolescent; stigmas 2-parted, divergent, yellowish-brown; gland 1, internal, rectangular, 2-lobed, ca. 1 mm long; capsule to 4 mm long, more or less pubescent. June. (Plate 1, Figure 1).

Forming thickets on slopes and at the foot of barren heights in the subalpine zone. — W. Siberia: Alt.; E. Siberia: Ang.-Say., Dau.; Far East: Okh. **Gen. distr.:** Mong., N. Am. Described from North America. Type in America.

Hybridizes with S. alatavica and S. reticulata.

Economic importance. Very ornamental.

Section 2. **HERBACEAE** Borr. ap. Hook. Brit. Fl. (1830) 432. — Very small undershrubs, often almost herbaceous in aspect, the woody underground branches 7 — 1 mm [?] thick; leaves short-petioled, rather thin, green on both sides, orbicular or oblong-elliptic, coriaceous, brittle, glabrous, small, prominently reticulate-veined; lateral veins 4 or 5; catkins borne on a gemmiferous stalk, subterminal, subtended by 2 or 3 leafy bracts, short, loosely flowered; scales round, sparsely covered with short flexuous hairs;

glands 1 or 2, simple or lobed; ovary conical-subulate, reddish-brown, glabrous, stamens glabrous; style distinct.

1. Compact undershrubs, herbaceous in aspect, the very short brittle branchlets terminating in 2—4 leaves . . . . 2.
\+ Branches serpentine, to 50 cm long, leafy throughout . . . . 4.
2. Catkins 10—30-flowered; leaves 1.5—3.5 cm long, rather long-petioled, mostly cuneately narrowed toward base, minutely denticulate . . . . 7. **S. Turczaninowii** Laksch.
\+ Catkins 4—7-flowered; leaves small, short-petioled, dentate-serrate or crenate-serrate . . . . 3.
3. Leaves 1.2—1.8 cm long, reniform or orbicular, mostly emarginate or somewhat torn at apex, usually broader than long, dentate-serrate; scales of catkins more or less dark-tipped, light-colored below . . . . 5. **S. herbacea** L.
\+ Leaves obcordate, orbicular or elliptic, ca. 0.3 — 0.9 cm long, sometimes almost 3-lobed; scales of catkins monochromatic dark purple; all parts smaller than in the preceding species . . . . 6. **S. liliputa** Nas.
36 4. Branches rooting; leaves mostly orbicular or rounded-cordate, to 1.5 cm in diameter, flat, entire, rarely obscurely denticulate . . . . 8. **S. nummularia** Anderss.
\+ Branches not rooting; leaves smaller, orbicular or elliptic, mostly folded at apex, sometimes glandular-dentate . . . . 9. **S. rotundifolia** Trautv.

5. **S. herbacea** L. Sp. pl. (1753) 1018; Ldb. Fl. Ross. III, 624; Kryl., Fl. Zap. Sib. IV, 776. — Ic.: Rchb. Ic. Fl. Germ. XI, t. 557, f. 1182. — Exs.: Fries, Herb. norm. V, No. 67, VIII, No. 64; Wimm. et Kr. Herb. Sal. No. 40 Enand. Salic. Scand. Nos. 20—21.

A prostrate undershrub, to 30—50 cm in head diameter; branches woody, subterranean, glabrous, rooting, brittle, yellow when young, dark brown or almost black in age; buds narrowly ovoid, glabrous; stipules none; petioles commonly ca. 0.5 cm long, glabrous; leaves crowded, at the ends of branchlets in pairs, small, 0.5—2.5 cm long and 0.4—2 (2.5) cm broad, mostly broader than long, orbicular, reniform, or rarely ovate, rounded at apex or often retuse, rounded or cordate at base, crenulate or sometimes incised, glabrous or at first subsericeous, bright green, dull above, lustrous beneath, densely and prominently reticulate-veined; lateral veins 4—6 pairs; catkins with leaves, at the ends of leafy branchlets, small, 4—18-flowered, borne on a short leafy puberulent stalk; scales to 1.5 mm long, broadly obovate, obtuse or truncate, concave, greenish-yellow or pale violet, dark at apex; stamens 2, glabrous, distinct or at base united (v. s y n a n d r a Bus.), with violet or yellowish-violet anthers; ovary ovoid-conical, ca. 3.3 mm long, glabrous, green or reddish-purple, the stipe as long as the gland; gland 1 or rarely 2, oblong, ca. 1 mm long, obtuse or emarginate or 2- or 3-lobed; style very short; stigma 2-lobed, the lobes divergent, narrow, often recurved; ripe capsule to 5—6 mm long. June—August. (Plate II, Figure 3).

Polar and alpine, stony, gravelly, or lichen tundra. — Arctic: Arc. Eur. (Kola Peninsula, Kanin, Kolguev I., Vaigach I., Komi ASSR, Arctic Urals). **Gen. distr.**: Arctic, Scand., Centr. Eur., Atl. Eur., N. Am., Greenland. Described from Scandinavia. Type in London.

Hybridizing with S. arbuscula, arctica, aurita, caprea, coaetanea, glauca, glandulifera, hastata, lanata, lapponum, myrsinites, nigricans, nummularia, polaris, phylicifolia, reticulata, rosmarinifolia, rotundifolia.

Economic importance. Readily eaten by deer. Stands up to browsing and replaces lichen destroyed by grazing. An aqueous infusion of leaves is used for tanning of hides by peoples inhabiting the polar regions.

37 6. **S. liliputa** Nas. sp. nova in Addenda IV, p. 539. — S. herbacea f. pygmaea Laksch. in sched.

An inconspicuous, practically subterranean undershrub with a herbaceous aspect; branches subterranean, slender (0.5—1 mm thick), yellowish-brown, rooting, very brittle, producing above ground 2 or 3 leaves at the ends of branchlet 0.6—1.2 cm long; leaves exstipulate, with petiole ca. 1 mm long, varying greatly in shape, obcordate, reniform, orbicular, or rarely elliptic, 0.3—0.9 cm long and 0.3—0.7 cm broad, mostly broadest above the middle, obtuse, retuse, or rarely pointed at apex, narrowed or subcordate toward base, crenulate or sometimes almost 3-lobed, coriaceous, brittle, glabrous on both sides, bright green, prominently veined, with 3 pairs of lateral veins; catkins inconspicuous (staminate unknown), appearing with the leaves, at the ends of leafy branchlets, 2—5-flowered, the faintly white-tomentose stalk with 1 or 2 leaves at base; scales dark purple, truncate, sparingly white-hairy on the margin; ovary ovoid-conical, outstretched toward apex, ca. 3 mm purple, glabrous, the stipe ca. 0.5 mm long, subsequently elongating and exceeding the scale; glands 2, the outer linear, reddish, the inner 2- or 3-lobed; style very short; stigmas 2-lobed, divergent; capsule to 4 mm long. July. (Plate II, Figure 5).

E. Siberia: Dau. — The alpine zone of barren heights near the mouth of the Upper Angara R. (Mount Kiren). Endemic. Described from Mount Kiren. Type in Leningrad.

Note. Closely related to S. herbacea L. which does not occur in Asia. Differs from it in the smaller number of leaves on the branchlets and fewer flowers in catkins, the plain dark purple scales, the miniature dimensions of the plant as a whole, of the branches and leaves, as well as leaf shape.

7. **S. Turczaninowii** Laksch. Schedae ad HFR VIII (1914) 50; Kryl., Fl. Zap. Sib. IV, 778. — S. herbacea × myrsinites Wolf in litt. ex Kryl., Fl. Alt. V (1909) 1235. — S. herbacea Turcz. (non L.) Fl. baic.-dah. II (1854) 124. — S. herbacea f. altaica Görz in sched. — exs.: HFR No. 2495.

A dwarf undershrub of barren heights, herbaceous in aspect, 2—5 cm high; subterranean parts strongly branched; branches mostly subterranean, yellow, glabrous, rooting, obliquely ascending, 5—10 or rarely to 30 cm long, green and herbaceous when young, woody and dark brown in age; buds rufescent, glabrous, small; stipules none; petioles slender, glabrous, often

reddish, 0.3 — 1 cm long; leaf blades usually varying in shape on the same plant, ovate, elliptic, or broadly obovate, rounded or pointed at apex, narrowed, sometimes cuneate or subcordate at base, glabrous on both sides,
38 pale green, prominently reticulate-veined, 1.5 — 3.5 cm long and 0.7 — 2 cm broad, with 4 — 6 pairs of lateral veins, the margin dentate throughout; catkins terminal, loosely 20 — 40-flowered, cylindric, 1 — 2 cm long and 0.3 — 0.7 cm broad, elongating in fruit to 5 cm, borne on a stalk to 1 cm long with 2 leafy bracts developing with or after the leaves; scales obovate, rounded at apex and purple-tipped, yellowish below, glabrous outside, with scattered hairs on the inside, ciliate on the margin, 1.2 — 1.8 mm long; stamens 2, distinct, the pale filaments to 3 mm long, hairy at base; anthers oblong, yellow; ovary ovoid-conical, glabrous, brownish-purple, 1.5 — 2 mm long, the stipe ca. 0.7 mm long; style obsolescent; stigmas short, 2-parted, divergent; glands 2, the inner oblong, obtuse or emarginate, 4 — 5 times as long as the stipe, the outer mostly absent; capsule to 4.5 mm long. June — July. (Plate II, Figure 2).

Moss-and-lichen tundra of the alpine zone. — W. Siberia: Alt.; E. Siberia: Lena-Kol. (Stanovoi Range), Ang.-Say., Dau.; Centr. Asia: Dzu.-Tarb. **Gen. distr.**: Mongolian Altai. Described from Altai. Type in Leningrad.

A hybrid with S. *nummularia* is known.

8. **S. nummularia** Anderss. in DC. Prodr. XVI, 2 (1868) 298; Kryl., Fl. Zap. Sib. IV, 777. — S. *retusa* β *rotundifolia* Turcz. Fl. baic.-dah. II (1854) 213.

A prostrate undershrub; branches flagelliform, creeping, rooting, ca. 30 — 50 cm long, yellow or brown; stipules 3 — 4 mm long; leaves 0.5 — 1.5 cm in diameter, orbicular, elliptic, or orbicular-cordate, rarely oblong-obovate and cordate at base, retuse on rounded apex, glabrous on both sides, pale green, prominently reticulate-veined, entirely or very rarely obscurely denticulate; catkins after the leaves, on lateral branches bearing 2 or 3 leaves, small, 2 — 5 — 7-flowered, the short glabrous stalk ebracteate; scales 1 — 1.2 mm long, round or truncate, concave, yellow or light brown, glabrous or the margin sparingly white-ciliate; glands 2, to 1.5 mm long, rather deeply lobed; stamens 2, with distinct glabrous filaments and violet anthers; ovary ovoid-conical, ca. 1 mm long, glabrous, at first green, becoming brown, sessile or stipitate, the stipe shorter than glands; style very short; stigmas lobed, the lobes divergent. June — July. (Plate II, Figure 6).

Stony or gravelly Arctic and alpine tundra. — Arctic: Arc. Eur., Arc. Sib., Chuk., An.; W. Siberia: Alt.; E. Siberia: Lena-Kol., Ang.-Say., Dau.; Far East: Okh. **Gen. distr.**: N., NW, and E. Mong. Described from Transbaikalia.
39 Hybridizes with S. *arctica, herbacea, reticulata, torulosa, Turczaninowii*.

9. **S. rotundifolia** Trautv. Nouv. Mém. Soc. natur. Mosc. II (VIII) (1832) 304; Kryl., Fl. Zap. Sib. IV, 775. — S. *serpyllifolia* Scop. Fl. carn. ed. II, 2 (1772) 555. — S. *polaris* var. *lejocarpa* Cham. in Linnaea VI (1831) 541, p. p. — S. *lejocarpa* Coville, in Proceed. Washingt. Ac. Sc. XIII (1901) 338 et t. XLI, f. 2. — Ic.: Anderss. Sal. Lapp. t. 26; Trautv., l. c., t. II; Lundstr. Weid. Nov. Seml., t. III. — Exs.: HFR No. 2494; Enand. Salic. Scand. No. 191.

Prostrate undershrub to 5 cm high, with a cone-shaped thickened crown above ground; branches mostly densely and extensively spreading, serpentine, to 30 — 75 cm long, knotted, not rooting, yellowish-brown or often reddish-brown, glabrous; summer shoots very slender, the shoots and their buds covered with short recurved hairs, stipules mostly lacking; petioles 1 — 5 mm long; leaves 0.5 — 2 cm long and 0.4 — 1.5 cm broad, coriaceous, orbicular or ovate, truncate or cordate at base, rounded, retuse, or short-pointed at apex (in var. uralensis Korsch. elliptic or ovate, short-pointed, 0.8 — 1.2 cm long and 0.5 — 0.8 cm broad), entire, or from base to the middle with up to 10 recurved glandular teeth (this mostly being an indication of hybrid derivation), often impressed-folded at apex, mostly glabrous and dull green above, paler beneath, glabrous or with sparse long somewhat recurved hairs, both surfaces slightly mottled, the veins prominent, on the lower side slightly raised; catkins terminal, 3 — 4 to 7-flowered, borne on short branchlets with 2 or 3 developed reniform leafy bracts; scales obovate, broad, obtuse, monochromatic yellow or light yellowish-brown, glabrate or, mostly in male flowers, sparsely covered with long hairs; outer gland lacking; inner gland oblong, narrowed; stamens 2, distinct, the filaments glabrous; ovary subsessile, glabrous, green or faintly reddish-brown, the stipe shorter than the gland; style one-third as long as the ovary, thick; stigmas small, thick, commonly cleft and divergent; capsule reddish-brown, glabrous. June — July. (Plate II, Figure 4).

Stony, gravelly, sandy, and lichen tundra of the Arctic Region. — Arctic: Arc. Eur., Nov. Z., Arc. Sib.; Far East: Okh., Kamch. **Gen. distr.**: Eur. and N. Am. Arc., Ber., Korea. Described from North America. Type in Leningrad.

Hybridizing with S. herbacea, hastata, lanata, lapponum, polaris, reptans, reticulata.

Economic importance. Eaten by reindeer, especially when young; dug up in winter from under the snow.

Section 3. **POLARES** Nas., sect. Nova. — Small shrubs, mostly of
40 herbaceous aspect, with slender subterranean branches; leaves mostly small, entire, with 3 — 5 lateral nerves; catkins terminal, 3 — 17-flowered; scales round, glabrous or with long white hairs; ovary mostly densely hoary-tomentose, rarely covered with scattered hairs, purple, short-stipitate; gland 1, oblong-linear; style 0.5 — 0.8 mm long.

10. **S. polaris** Whlb. Fl. Lapp. (1812) 261; Ej. Fl. Suec. 659; Ldb. Fl. Ross. III, 625; Kryl., Fl. Zap. Sib. IV, 777. — S. pseudopolaris Floder, in Arkiv för Bot. 20 A, 6 (1926) 8. — Ic.: Whlb. Fl. Lapp., t. 13, f. 1; Anderss. Salic. Lapp. f. 28; Rchb. Ic. Fl. Germ. XI, t. 557, f. 1183; Kom., Fl. Kamch. II, Plate I. — Exs.: Fries, Herb. norm. V, 68.

A diminutive undershrub of herbaceous aspect; branches subterranean, 3 — 5 cm long, rooting, yellow; producing above ground few small leaves; stipules lanceolate, mostly lacking; petioles to 1 cm long, glabrous; leaf blades orbicular or broadly obovate, sometimes reniform, rarely broadly elliptic-lanceolate, mostly broadest above the middle, rounded or often retuse at apex, rounded or cordate or rarely cuneate at base, entire or occasionally with 1 or 2 shallow denticulations at base (presence of more teeth is a borrowed character), pure green on both sides, dull above, somewhat lustrous beneath, 0.8—2.5 cm long, 0.6—1.3 cm broad; veins on the upper side impressed, only the midrib and lateral veins prominent, the latter arcuately converging toward apex; the lower side of leaves with subsidiary lateral veins also evident though less prominently than in S. herbacea; catkins terminal, on leafy spurs, 3—17-flowered, commonly oblong or ovoid, ca. 1.6×0.9 cm (in fruit up to 2.5 cm and more), borne on a hairy-pubescent stalk; scales ovate or obovate, rounded at apex, concave, dark brown, sometimes crenulate, mostly merely long-ciliate or the margin beset with long white hairs; stamens 2, distinct, glabrous, anthers dark; gland oblong-ovate, narrowed, ca. 1 mm long; ovary conical, to 5 mm long, typically hoary-tomentose, at length glabrescent, rarely sparingly hairy, greenish or purple, the stipe 0.5 mm long; style 0.5—0.8 mm long; stigmas as long, 2-parted, divergent; gland oblong-linear, 3 times as long as stipe, ca. 1.5 mm long. June—July. (Plate II, Figure 1).

Grassy, gravelly, or clayey, Arctic and alpine tundra. — Arctic: Arc. Eur., Nov. Z., Arc, Sib., Chuk., An.; E. Siberia: Ang.-Say. (Tunkinskie
43 Gol'tsy [barren heights]); Far East: Kamch., Komandorskie Islands; **Gen. distr.**: alps of Scandinavia, Arc. Eur., America, Greenland. Described from Lapland. Type in Stockholm.

Hybridizing with S. arbuscula, arctica, glauca, hastata, herbacea, lanata, lapponum, nigricans, phylicifolia, reptans, reticulata, rotundifolia.

Economic importance. The leaves are retained for a long time and are often covered by snow while still green; they provide summer and winter feed for reindeer. In the north of Yakutia the leaves are used by the local population as a tea substitute.

Note. The species S. pseudopolaris described from Kamchatka, does not display any substantial specific differences as compared with S. polaris Whlb.; the latter also has 3—9—15 flowers, and in both cases the ovary may be either white-tomentose or brownish, with scattered hairs. Thus, S. pseudopolaris Floder. has been included among the synonyms of S. polaris Whlb.

Section 4. ARCTICAE Rydb. in Bull. N. York. Bot. Gard. I (1899) 2631 (pro parte). — Prostrate or extensively spreading shrubs; branches stout, mostly dark red or brownish; leaves firm, coriaceous, rather large, the upper surface commonly glossy, glabrous, rarely hairy, dark, the lower paler, with 7 or 8 pairs of prominent veins; buds large; stipules developed, sometimes exceeding the petioles; catkins terminal or lateral; scales mostly dark at apex; ovary mostly densely hairy, short-stipitate; style elongated.

1. Stipules 3—10 mm long, greatly exceeding the petioles, especially those of terminal leaves narrow and densely glandular-toothed; catkins with the leaves, borne on a bractless stalk . . . . . . 14. **S. pulchra Cham.**

PLATE I. 1. Salix vestita Pursh. – 2. S. orbicularis Anderss. – 3. S. reticulata L. – 4. S. ovalifolia Trautv. – 5. S. phlebophylla Anderss. – 6. S. kurilensis Koidz. – 7. S. Chamissonis Anderss.

+ Stipules shorter than petioles, often inconspicuous; catkins usually after the leaves, borne on a leafy-bracted stalk . . . . 2.

2. Leaves ovate-lanceolate to broadly lanceolate, uniformly narrowed toward both ends, sparsely serrulate . . . . 15. **S. divaricata** Pall.

+ Differing from the above . . . . 3.

3. Leaves ovate or obcordate to obovate-oblong and elliptic, thin, entire, rather densely covered with more or less persistent soft hairs; young branches pubescent . . . . 13. **S. Pallasii** Anderss.

+ Leaves firm, coriaceous, pubescent only when young, at length glabrescent; leaves and branches glossy . . . . 4.

4. Leaves long-cuneate, obovate in upper part, large, to 5—6 cm long; catkins usually after the leaves . . . . 12. **S. torulosa** Trautv.

44 + Leaves ovate or obovate, little narrowed or more or less rounded at base, 3—4 cm long; catkins usually slightly before or together with the leaves . . . . 11. **S. arctica** Pall.

11. **S. arctica** Pall. Fl. Ross. 1, 2 (1788) 86; Ldb. Fl. Ross. III, 619; Kryl., Fl. Zap. Sib. IV, 770 (ex prt.). — Ic.: Rchb. Ic. Fl. Germ. XI, t. 566, f. 2007; Coville in Proceed. Wash. Acad. III (1901) tab. XL; Pall., l. c. — Exs.: HFR No. 2493.

A prostrate shrub, with a strong root system spreading out in a plane parallel to the soil surface and an abbreviated cone-shaped trunk, this giving rise to numerous branches; branches short, stout, knotted, grayish-brown, glabrous, sprawling on the ground surface, to 50 cm long; stipules rarely evident, ovate, sometimes developing into a broad upper leaf; petioles 3—11 mm long, stout, strongly dilated toward base; leaf blades mostly ovate or obovate, to 1.5—5 cm long and 1—2.5 cm broad, mostly rounded or rarely pointed at apex, entire, the upper surface lustrous, dark green, somewhat hairy when young, becoming glabrous, the lower pale green, slightly glaucescent, rather faintly reticulate-veined, the midrib mostly covered with slender upward-pointing fugacious hairs; catkins with or just before leaves, terminal, 2—3 cm long, the long stalks commonly bearing 2—4 leaflike bracts, elongating in fruit to 8—12 cm and erect; scales broad, dark reddish-fulvous or brown, obtuse, long-hairy; gland 1, inner, obtuse, in staminate flowers 2-lobed; stamens 2, distinct, glabrous; ovary covered with short fugacious hairs, in maturity commonly dark reddish-brown, the short white-tomentose stipe as long as the gland; style well developed, 2-parted, to 1 mm long; stigmas often cleft down to the middle, reddish; capsule 6—7 mm long. June—July. (Plate III, Figure 6).

Polar tundra. — Arctic: Arc. Eur., Nov. Z., Arc. Sib., An., Chuk.; European part: Urals; Far East: Kamch. **Gen. distr.**: European Arctic. Related species in America. Described from the lower reaches of the Ob.

Note. Hairy leaves with veins impressed above, light-colored scales, separate glands, and short style, are characters foreign to this species in its pure manifestation and are borrowed chiefly from S. glauca L. or from S. reptans with which S. arctica is often confounded. Hybrids are also formed with S. arbuscula, Chamissonis, cuneata, herbacea, kurilensis, nummularia, parallelinervis, polaris, pulchra, reticulata.

Economic importance. Eaten by reindeer. This willow, like some others, is used by the Yakuts as a substitute for tea and is known under the name "chai-talak."

45 12. **S. torulosa** Trautv. in Mém. Soc. Nat. Mosc. II (1832) 309. — S. altaica Lundstr. Krit. Bemerk. über die Weiden Nov. Seml. (1877) 36. — S. arctica Ldb. Fl. Alt. IV, 283, non Pall. — S. arctica f. altaica E. Wolf in Kryl., Fl. Alt. V (1909) 1229. — S. arctica var. altaica E. Wolf in Kryl., Fl. Zap. Sib. IV (1930) 771. — Ic.: Ldb. Ic. Fl. Ross. t. 460.

A prostrate shrub; branches yellowish-brown, glabrous, lustrous, to 50 cm long; buds large, yellow, glabrous, pointed; petioles to 1.5 cm long; glabrous, yellowish-green; leaves typically obovate-cuneate, rounded or, often obliquely, short-pointed at apex, rarely obovate or obovate-elliptic, long-tapering toward base, firm, entire, 5—6 cm long, to 2—5 cm broad, the upper surface dark green, typically glabrous, the lower glaucescent, prominently veined, covered with rather long scattered white hairs, commonly somewhat hairy when young; lateral veins 7 or 8 pairs; stipules rarely present, lanceolate; catkins together with or after the leaves, 2—2.5 cm long and ca. 0.7 cm broad, on leafy lateral branchlets, the short stalks later becoming much elongated; pistillate catkins (excluding stalk) 5—6 cm long and to 1.5 cm broad in fruit, upright, densely flowered, very hairs; scales 3 × 1.5 mm, broadly ovate, pointed or obtuse, brown, darker and sometimes almost black at apex, covered on both sides with long white hairs; gland 1, oblong, in staminate flowers 2-lobed; stamens 2, distinct, glabrous, the rounded-oblong anthers yellow, finally darker; ovary 3—5 mm long, subsessile (the stipe ca. 0.7 mm long), densely white-tomentose, with a long simple or cleft style and 2-lobed stigmas; capsule 6—7 mm long, soft-hairy. May—June. (Plate III, Figure 7).

Alpine zone of mountains. — W. Siberia: Alt.; E. Siberia: Ang.-Say.; Far East: Ze.-Bu. (Stanovoi Range). **Gen. distr.**: Mong. Described from Altai. Type in Leningrad. Hybridizing with S. nummularia.

13. **S. Pallasii** Anderss. in DC. Prodr. XVI, 2 (1868) 285. — S. diplodictya Trautv. in Nouv. Mém. Soc. Natur. Mosc. II (VIII) (1832) 307, tab. 14. — S. crassijulis Trev. ex Trautv., ibid., 308, tab. 15. — S. arctica × glauca Floder. Arkiv för Bot. 20 A, (1926) 22—27; Hultén, Fl. of Kamtch. II, 5; Kom., Fl. Kamch. II, 24. — Ic.: Kom., ibid., Plates II and III.

A prostrate, depressed, or upright shrub, to 1 mm high; branches obliquely rising or ascending, rooting, pubescent when young, glabrous, dark or yellowish, lustrous in age; buds yellowish, sparsely hairy; stipules narrow, entire; petioles ca. 1—1.5 cm long; leaf blades obovate or obcordate to obovate-oblong or elliptic, to 5 cm long and to 2.5 cm broad
46 above the middle, rounded or short-apiculate at apex, obtusish or tapering at base, thin, flat or somewhat revolute, entire, rather densely covered with soft hairs, at length ciliate or glabrate, the lower surface glaucous, distinctly reticulate-veined; catkins lateral or subterminal, stalked, elongated, stout, rather dense, upright, ca. 2.5 —3 cm long, the pistillate elongating to 9—12 cm in fruit, the stalk naked above, with 2 or 3 leaflike bracts below; scales black, obtuse, covered with long white hairs; stamens 2, distinct, glabrous,

the pink anthers finally dark; ovary conical, covered with long gray hairs, the stipe exceeding the gland; style very elongated, brown; stigmas 2-parted, with elongated filiform straight divergent lobes. June—July.

Far East: Kamch. (the forest zone and lichen tundra of the peninsula and of the Komandorskie Islands). Described from Kamchatka. Type in Leningrad.

Hybridizing with S. Chamissonis, cuneata.

14. **S. pulchra** Cham. in Linnaea VI (1831) 543; Hultén, Fl. of Kamtch. II, 19; Kom., Fl. Kamch. II, 16. — S. taimyrensis Trautv. in Middend. Sibir. Reise I (1847) 24. — S. boganidensis var. latifolia Trautv. Fl. terr. Tschuktschor. (1878) 34; Ldb. Fl. Ross. III, 616. — S. arctica ε taimyrensis Anderss. in DC. Prodr. XVI, 2 (1868) 28. — Ic.: Trautv., l. c., tab. 5, 6.

A trailing or semiprostrate shrub; trunk buried; branches virgate, fulvous-brown, glabrous, the young branches and buds commonly puberulent (in var. eriocládos Laksch. in sched., the annotinous and 2-year-old branches, buds, the midrib of leaves beneath, and petiole, densely clothed with grayish-yellow tomentum); stipules greatly exceeding the petiole, those of the large terminal leaf long, narrow, densely glandular-toothed; petioles 3—10 mm long; leaf firm, obovate, elliptic-obovate, or broadly lanceolate, rounded or pointed at apex, narrowed toward base, 2—4 cm long and 0.6—2 cm broad, commonly entire, rarely with 1—3 (10) irregularly spaced sharp teeth or serrate, glabrous even when young, dark green above, light-colored with a prominent reddish midrib beneath; catkins before the leaves, lateral, sessile, leaflike bracts none; scalelike bracts caducous; staminate catkins 2—3 cm long, pistillate 2—8 cm long and to 2 cm broad, interrupted below; scales black or dark fulvous, covered with long white hairs, broad, slightly pointed; gland 1, inner, oblong, narrow; stamens 2, distinct, glabrous, with dark filaments and oblong dark anthers; ovary short-stipitate or sessile, grayish-violet, rather densely silky-pubescent; style to 2—2.5 mm long, slender, fulvous, naked; stigmas 0.6—0.7 mm
47 long, slender, simple or cleft, straight. May—July. (Plate III, Figure 8).

Grassy, mossy, boggy, or gravelly tundra. — Arctic: Arc. Eur., Nov. Z., Arc. Sib., Chuk., An.; Far East: Kamch. **Gen. distr.**: Ber., N. Am. Described from North America.

Hybridizing with S. arctica and S. Seemannii.

The leaves are easily retained on the branches until the following spring.

Economic importance. Flowers and leaves are eaten by reindeer.

15. **S. divaricata** Pall. Fl. Ross. I, 2 (1788) 80; Turcz. Fl. baic.-dah. II, 1, 115; Ldb. Fl. Ross. III, 625. — S. dasycarpa Turcz. in sched.

A shrub up to 20—35 cm in height; branches spreading or procumbent, stout, knotty, dark red, becoming grayish-brown in age, glabrous, lustrous; buds dark brown to almost black, glabrous, lustrous; stipules fugacious, very small, lanceolate, toothed; petioles 0.2—0.3 cm long, glabrous; leaves ovate-lanceolate or broadly lanceolate, narrowed at both ends, firm, dark green above, glaucescent beneath, to 3 cm long and 1.5 cm broad, sparsely serrulate, glabrous or at first with scattered appressed hairs, midrib reddish beneath, lateral veins 8—10 pairs; catkins with leaves, 1.3—1.5 cm long, 0.5 cm broad, in fruit to 6 cm long and 0.7 cm broad, subsessile, the staminate always sessile; scales ovate to orbicular, black, covering the ovary up to the middle, long-hairy; gland obtriangular, obtuse; stamens 2, distinct, glabrous; anthers yellow; ovary tomentose, the stipe very short; style elongated, sometimes 2-parted; stigmas 2-parted, the lobes ovate; gland 1, inner, broad, obtuse; capsule 4—6 mm long, inflated, with scattered pubescence. June. (Plate III, Figure 9).

Alpine zone. — E. Siberia: Dau.; Far East: Ze.-Bu. (Stanovoi Range). **Gen. distr.**: Mongolia. Described from Mount Sokhondo in Transbaikalia. Type in Leningrad.

Section 5. MYRSINITES Borr. in Hook. Brit. Fl. (1830) 431. — Prostrate or rarely upright shrubs; leaves stiff, coriaceous, lustrous, prominently reticulate-veined, with 3—5 pairs of lateral veins, often coarsely glandular-serrate, sometimes spiny-incised or prickly-serrate, often turning brown upon wilting and often persisting on the branches for one or two subsequent seasons; catkins lateral or subterminal, the stalk without buds, glabrous above, stoutish; scales dark, black or purple at apex, commonly long-hairy; gland 1, simple; stamens glabrous; ovary more or less hairy, rarely glabrous, commonly purple; style brown, elongated.

1. Leaves entire, very small or medium-sized . . . . 2.

\+ Leaves dentate, serrate, or spiny-incised, glandular, small or medium . . . . 5.

48 2. Leaves oblong-obovate, acuminate, mostly recurved at the tip; lateral veins parallel, prominent and riblike on both sides but more so beneath, on decaying of the leaf the network of veins persisting in subsequent seasons . . . . 24. **S. phlebophylla** Anderss.

\+ Veins not riblike; network of veins not persisting in subsequent seasons . . . . 3.

3. Leaves mostly broadly elliptic, pointed at both ends, the petiole and adjoining portion of midrib red; stipules small, ovate-lanceolate, narrowed at both ends, dentate . . . . 21. **S. saxatilis** Turcz.

\+ Leaves obovate . . . . 4.

4. Leaves mostly obovate, small or medium-sized, firm, coriaceous, entire, revolute, exstipulate, dark green above, glaucous or glaucescent beneath, very prominently reticulate-veined on both sides, generally resembling the leaves of bog whortleberry . . . . 19. **S. ovalifolia** Trautv.

\+ Leaves very small, dense, very stiff, lustrous above, retuse; stipules obtuse, broad, scarious, glabrous . . . . 23. **S. erythrocarpa** Kom.

5. Leaves obovate, long-tapering toward base, to 5 cm long, the prominent veins arching beneath; catkins upright, up to 4 cm long in fruit . . . . 18. **S. cuneata** Turcz.

\+ Leaves different . . . . 6.

6. Leaves obovate or elliptic, small, stiff, bright green on both sides, spiny- or glandular-serrate to incised-spiny, the recurved teeth glandular-tipped . . . . 22. **S. berberifolia** Pall.

\+ Leaf margins different . . . . 7.

7. Stipules ovate or elliptic or narrowly ovate-lanceolate, narrowed at both ends, obtuse or pointed at apex, dentate-margined; leaves ovate or elliptic, remotely serrate or entire, green above, plumbeous-bluish beneath, blackening . . . . 20. **S. fumosa** Turcz.

+ Stipules developed, but different from above . . . . 8.

8. Leaves of the preceding year not persistent; leaves thin, almost transparent, green, obovate, rounded or short-apiculate at apex, glabrous on both sides, sharply glandular-serrulate, reticulate-veined . . . . 17. **S. Chamissonis** Anderss.

49 + Foliage mostly including last year's leaves, firm, coriaceous, not translucid, elliptic to obovate, glabrous or pubescent, bright or dull green, glandular-serrulate, prominently veined . . . 16. **S. myrsinites** L.

16. **S. myrsinites** L. Sp. pl. (1753) 1018; Ldb. Fl. Alt. IV, 284; Ej. Fl. Ross. III, 620; Turcz. Fl. baic.-dah. II, 1, 117; Kryl., Fl. Zap. Sib. IV, 771. — S. rectijulis Ldb. ex Trautv. in Nouv. Mém. Soc. Nat. Mosc. II (1832) 313. — S. arbutifolia Seringe, Sal. helv. (1815) 44. — Ic.: Fl. Dan. 6, t. 1054; Rchb. Ic. Fl. germ. XI, t. 559, f. 1088; Enand. Salic. Scand. No. 57 b. — Exs.: Enand. l. c., 53, 57b; Fries Herb. Norm. No. V, 66; Wimm. et Kr. Herb. Sal. No. 92.

An upright (up to 50 cm high) or prostrate shrub; branches beset with remnants of wilted leaves; leaves 1.3 — 3.5 cm long and 0.7 — 2 cm broad, elliptic to obovate, rounded or short-pointed at apex, tapering or rounded at base, firm, stiff, mostly bright green, prominently veined, glandular-serrulate (var. serrata Neiler., also characterized by thick catkins and styles, and obtusish ovary) or entire, hairy, grayish-green (var. Jacquiniana (Willd.) Koch) with more slender catkins, pointed glabrate ovary, and more slender style; lateral veins 5 — 8 pairs; petiole 0.3 — 0.5 cm long; stipules rarely present, ovate or linear; catkins densely flowered, terminal, borne on short leafy spurs; scales obovate or ovate, dark purple or blackish, covered with long white hairs; stamens 2, distinct, naked; anthers dark violet; ovary rather densely covered with crisped hairs, usually glabrous at base, short-stipitate; gland about twice the length of the stipe; style one-fourth to one-half as long as ovary, purple, sometimes cleft; stigmas 2-lobed, the lobes divergent. May — June. (Plate III, Figure 5).

Forest-tundra, tundra, and the alpine and subalpine mountain zones. — Arctic: Arc. Eur., Arc. Sib.; European part: Kar.-Lap., Dv.-Pech., Urals; W. Siberia: Ob, Alt.; E. Siberia: Yen., Lena-Kol., Ang.-Say., Dau.; Far East: Okh., Ze.-Bu. **Gen. distr.**: Arctic, Scand., Centr. Eur., Atl. Eur., Med., Mong., N. Am. Described from Europe ("alps of Lapland, Switzerland, and Italy"). Type in London.

Hybridizing with S. arbuscula, glauca, hastata, herbacea, Krylovii, myrtilloides, nigricans, phylicifolia.

Economic importance. Insignificant. The bark contains about 3 — 4% of tannins. Usually not eaten because of the stiff foliage.

17. **S. Chamissonis** Anderss. in DC. Prodr. XVI, 2 (1868) 290; Ldb. Fl. Ross. III, 620 (ex parte); Kom., Fl. Kamch. II, 23; Floder. in Arkiv för Bot. 20 A, No. 6, 29. — Ic.: Coville, Will. of Alaska, f. 23.

50 Mostly a prostrate shrub; branches short, greenish-yellow, blackening in drying, rooting, ascending, glabrous; young branches white-hairy; stipules well developed or small, semiovate, glandular-serrate; petioles 2 — 10 mm long, glandular; leaf blades obovate, rounded at apex or shortly

tapering to a point, sharply glandular-serrulate, thin, almost translucent, green, paler beneath, glabrous on both sides, reticulate-veined; catkins lateral, slightly spreading, oblong-cylindric, pink, stalked, the stalk naked above, with 2—4 leafy bracts below; catkins rather sparsely flowered at base, to 3—4 cm long, the pistillate to 7 cm long in fruit; scales brownish-black, acutish, covered with grayish hairs; stamens 2, distinct; anthers blackish-violet; ovary conical, pointed, violet-brown, covered with gray hairs, the stipe twice the length of the gland; style elongated, filiform, simple, brown; stigmas 2-parted, brown, the lobes divergent. June—July. (Plate I, Figure 7).

Dry, gravelly, stony, and meadow-alpine tundra; also subalpine slopes and valleys. — Arctic: Chuk., An.; E. Siberia: Lena-Kol.; Far East: Okh., Kamch. **Gen. distr.**: Ber., N. Am. Described from North America. Type in Stockholm.

Hybridizing with S. Pallasii.

18. **S. cuneata** Turcz. Fl. baic.-dah. II, 1 (1854) 122; Ldb. Fl. Ross. III, 623; Kom., Fl. Kamch. II, 27. — S. psilostachya Turcz. pl. exsicc. 1835. — S. arctica α nervosa subvar. cuneata Anderss. in DC. Prodr. XVI, 2 (1868) 286.

A prostrate undershrub; branches few, glabrous, 20—30 cm long and ca. 0.5—1 cm thick, arising from a cone-shaped thickened crown; bark of branches grayish-yellow; buds small, yellow, glabrous; stipules lanceolate, dentate, sometimes lacking; petioles to 2 cm long, channeled, glabrous; leaf blades to 5 cm long and to 1.8 cm broad, obovate, elliptic, or cuneately spoon-shaped, short-pointed or obtusely rounded at apex, long-tapering toward base, entire or serrulate, dull green and somewhat rugose above, pale glaucescent-green beneath, sparsely covered with white hairs when young, becoming glabrous, prominently veined on both sides, very much so beneath; lateral veins 5—7 pairs, arching off at an acute angle; midrib pale yellow on both sides; catkins with leaves, on lateral branchlets, mostly upright or slightly spreading, to 3—4 cm long, borne on a soft-hairy leafy stalk, interrupted in lower part; scales spoon-shaped or elliptic, 1.5 × 2.2 mm, rounded at apex, brown, the reddish tip often covered with long white hairs; stamens 2, distinct, naked, the red filaments 5—7 mm long, the small dark purple anthers linear; ovary ovoid-conical, brown, becoming bright purple or violet, glabrous, the stipe ca. 5 mm long; gland 1, internal, 0.8—1 mm long, the simple style 0.6—1.5 mm long, often red; stigmas linear, divergent,
51 red, simple or 2-parted; capsule 3—7 mm long. June—July. (Plate III, Figure 10).

Alpine tundra, in rocky and stony places. — Arctic: Chuk., An.; W. Siberia: Alt.; E. Siberia: Lena-Kol., Ang.-Say., Dau.; Far East: Okh., Ze.-Bu., Uda. **Gen. distr.**: Kurile Islands. Described from the E. shore of Lake Baikal. Type in Leningrad.

Hybridizing with S. arctica, fumosa, Seemannii, pulchra.

19. **S. ovalifolia** Trautv. in Nouv. Mém. Soc. Nat. Mosc. II (VIII) (1832) 306; Coville, The Willows of Alaska, 331; Ldb. Fl. Ross. III, 620. — Ic.: Trautv., l. c., t. 13; Coville, l. c., f. 26.

A prostrate undershrub; branches to 30 cm long, hairy when young; becoming glabrous, brown; buds small, brown, ciliate becoming glabrous; stipules lacking; petioles slender, channeled above, ciliate, to 0.4 mm long; leaves commonly broadest above the middle, obovate or elliptic, mostly obtuse, rarely pointed, coriaceous, glabrous or ciliate-margined, somewhat hairy beneath, in f. pubescens Anderss. with long hairs at leaf base and on the petiole, sparingly gray-pubescent, mostly entire, revolute, rarely sparsely glandular-denticulate, to 1.2 — 2 cm long and 0.6 — 1.8 cm broad, the slender reticulate veins very prominent on both sides; catkins lateral, 1 — 2 cm long and ca. 1 cm broad, elliptic, oblong, or ovate, rather dense, the staminate ca. 1.7 cm long, the pistillate in fruit to 2.2 cm long and 1.5 cm broad; scales brownish-rufous or violet-brown, obovate-orbicular or spoon-shaped, white-hairy or ciliate, tightly enveloping the base of ovary; stamens 2, distinct, naked, 3 — 4 times the length of the scale; filaments yellow; anthers yellow, rounded-oblong; ovary to 3 — 4 mm long, violet-brownish, oblong-lanceolate, 3 — 4 times as long as the scale, glabrous, subsessile; gland barely equaling the stipe; style distinct, to 0.5 — 0.7 mm long; stigmas brown, 2-parted, the short lobes divergent; capsule to 8 mm long, glabrous. June — July. (Plate I, Figure 4).

Lichen, polar, and alpine tundra. — Arctic: Chuk., An.; E. Siberia: Dau., Lena-Kol.; Far East: Kamch., Uda. Gen. distr.: NW America and islands of the Pacific Ocean. Described from America. Type in Leningrad.

A hybrid with S. fumosa is known.

20. **S. fumosa** Turcz. Fl. baic.-dah. II, 1 (1854) 111. — S. arctica β Brownei 3° fumosa Anderss. in DC. Prodr. XVI, 2 (1868) 286. — S. phylicifolia β majalis Ldb. Fl. Ross. III (1851) 612. — S. Arnelli Lundstr. in Bot. Notis. (1888) 26, 31 et in Bot. Centralbl. XXXV (1888) 31.

52 A prostrate shrub; branches ascending, to 0.5 m long, brownish, becoming gray, pubescent when young; stipules not more than 0.5 cm long and 0.3 cm broad, ovate or elliptic or ovate-lanceolate, narrowed at both ends, obtusish or pointed at apex, the margin dentate; petioles to 0.3 cm long, pubescent; leaves obovate or elliptic, obtuse, remotely serrate or entire, slightly revolute, green above, plumbeous-blue beneath, at first sericeous, becoming glabrous, prominently veined, blackening in drying, to 3.5 cm long and 2 cm broad (in var. angustifolia Traut. leaves lanceolate, to 2 cm long and 0.4 — 0.6 cm broad); lateral veins 8 or 9 pairs; catkins after the leaves, lateral, borne on a short leafy stalk, to 2.8 cm long, in fruit to 7 cm long and 1.4 cm broad, interrupted below; scales spatulate-ovate, obtusish, dark purple or blackish, small, with long white cilia, exceeding the stipe; stamens 2, distinct; anthers oval-oblong, yellow, at length darkening; ovary ovoid-conical, purple, more or less hoary-pubescent; stipe hairy, to 1 mm long, about twice as long as the obtriangular obtuse inner gland; style ca. 1 mm long, the style and the 2-parted stigma purple; capsule to 4 mm long. June — July. (Plate III, Figure 3).

Alpine and subalpine mountain zone, whence descending onto the pebbles in wooded valleys of mountain streams, forming underbrush. — Arctic: Chuk., An.; E. Siberia: Yen., Lena-Kol., Ang.-Say., Dau.; Far East: Kamch., Ze.-Bu., Uda. Endemic. Described from SW Baikal area (Tunkinskii Territory). Type in Leningrad.

Hybridizing with S. cuneata, glauca, ovalifolia.

(53)

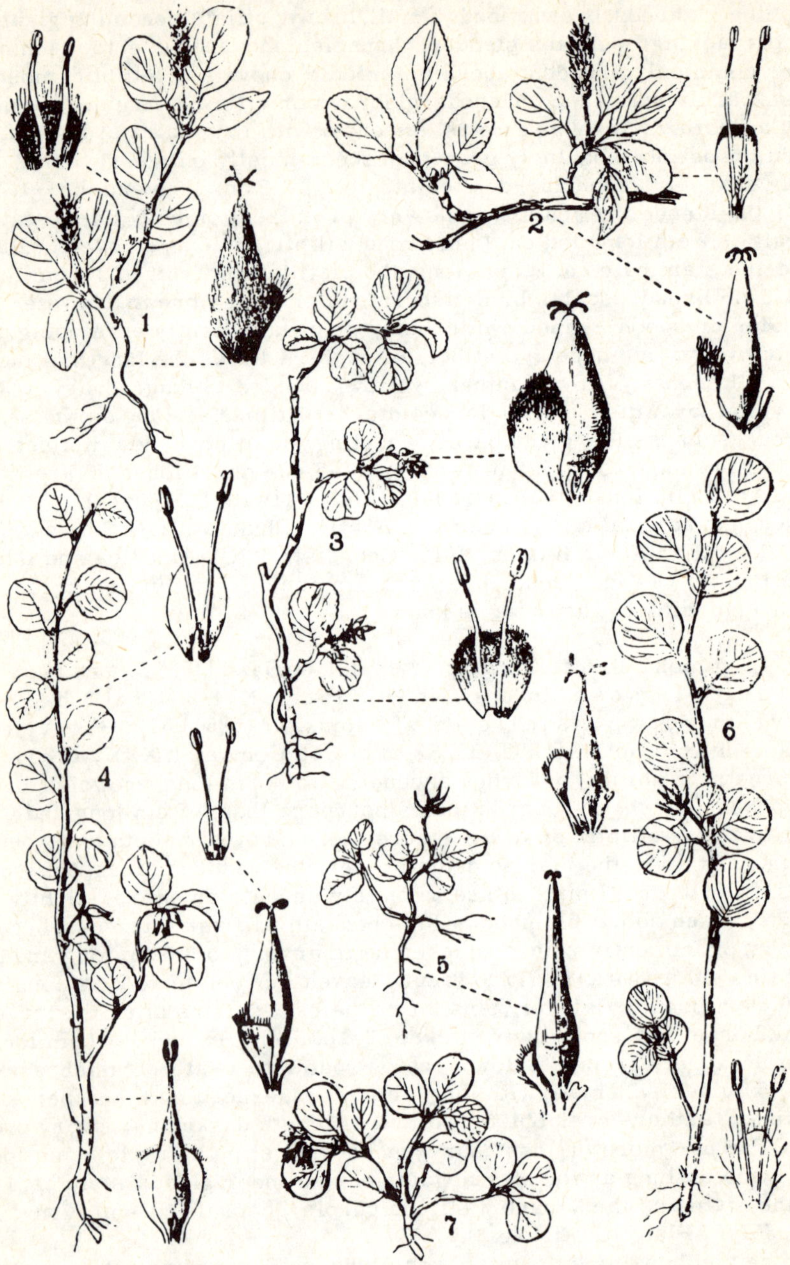

PLATE II. 1. Salix polaris Whlb. – 2. S. Turczaninowii Laksch. – 3. S. herbacea L. – 4. S. rotundifolia Trautv. – 5. S. liliputa Nas. – 6. S. nummularia Anderss. – 7. S. erythrocarpa Kom.

21. **S. saxatilis** Turcz. ex Ldb. Fl. Ross. III (1851) 621; Turcz. Fl. baic.-dah. II, 1 (1854) 118. — S. fumosa var. saxatilis Trautv. in A. H. P. V (1877) 109.

A prostrate undershrub; branches flexuous, reddish, glabrous, to 15—20 cm long; buds small, pointed, glabrous, brown; stipules small, ovate-lanceolate, narrowed at both ends, dentate; petioles 0.2—0.3 cm long, red as also the adjoining portion of the midrib; leaves 0.8—1.2 cm long and 0.5—0.9 cm broad, broadly elliptic, tapering toward both ends, entire, subrevolute, green above, plumbeous-blue beneath, sericeous when young, at length glabrous or the midrib white-hairy beneath; catkins to 3 cm long, borne on a short leafy stalk; scales dark brown, elliptic, acuminate, profusely covered with white hairs; stamens 2, distinct, glabrous, strongly exserted from behind the scales; filaments and anthers
55 purple; ovary ovoid-conical, purple, slightly covered with white hairs, short-stipitate; style ca. 1 mm long, red; stigmas 2-lobed. June—July. (Plate III, Figure 4).

Alpine and subalpine zone and pebbles in the valleys of mountain streams. — E. Siberia: Ang.-Say., Lena-Kol., Dau.; Far East: Ze.-Bu. Endemic. Described from SW Baikal area (Tunkinskii Territory). Type in Leningrad.

Hybridizing with S. berberifolia.

22. **S. berberifolia** Pall. Reise III (1776) 321, Anhang 759, No. 134, t. Kk., f. 7; Ej. Fl. Ross. II, 84 et t. 82; Turcz. Fl. baic.-dah. II, 1, 119; Kom., Fl. Kamch. II, 32. — S. Brayi Ldb. Fl. Alt. IV (1833) 289 nom Nym.; Ej. Fl. Ross. III, 621; Kryl., Fl. Zap. Sib. IV, 772. — S. ilicifolia Willd. herb. No. 18223 (legit Adams in Dahuria). — Ic.: Ldb. Ic. pl. Fl. Ross. t. 449.

A prostrate undershrub; stem partly buried; branches rooting, densely clothed with remnants of withered leaves, to 35 cm long, the annotinous yellowish-brown; petioles 1—4 cm long; stipules rarely present, lanceolate, glandular-dentate, shorter than petioles; leaf blades 1—2 cm long and 0.3—1 cm broad, stiff, bright green on both sides, glabrous, lustrous, prominently veined, varying in size and shape, typically elliptic, serrate or prickly-serrate or incised-spiny with recurved glandular-tipped teeth, or (v. Brayi Ldb. pro spec.) obovate, narrowed toward base, prickly-serrate; catkins with or after the leaves, on lateral or terminal branchlets, leafy-stalked, upright, ca. 1—2 cm long, ovoid, becoming cylindric, dense, the pistillate to 4 cm long in fruit; scales obovate, obtuse, sometimes retuse, dark brown or cherry-red, rather densely covered with white hairs, rarely glabrate, purple; stamens 2, distinct, glabrous; ovary conical-cylindric, obtuse, green, becoming purple, typically glabrous (var. lejocarpa Trautv.), rarely hairy (var. eriocarpa Trautv.), the stipe exceeding the elongated obtuse or emarginate gland; style to 0.5 mm long, often 2-parted down to base, the divergent lobes cleft. June—July. (Plate III, Figures 1 and 2).

Alpine moss-and-lichen or gravelly tundra. — Arctic: Chuk., An.; W. Siberia: Alt.; E. Siberia: Ang.-Say., Dau., Lena-Kol.; Far East: Kamch., Okh., Ze.-Bu. **Gen. distr.**: Mongolian Altai, Korea. Described from Transbaikalia. Type in Leningrad.

23. **S. erythrocarpa** Kom. in Fedde, Repert. sp. nov. XIII (1914) 165; Ej., Fl. Kamch. II, 29; Hultén, Fl. of Kamtch. II, 11; Floder. Arkiv för Botan. 20 A, 12. — S. r u b r i c a p s u l a Toepff. in Oest. Bot. Zeitschr. LXVI (1916)
56 403. — Ic.- Kom., l. c., tab. V.

A prostrate shrub, 5 — 15 — 30 cm in diameter; main stem subterranean; branches creeping, yellowish-brown, strongly flexuous, very densely covered with fresh foliage and remnants of dead leaves; buds small, obovoid, obtuse, glabrous; stipules obtuse, broad, scarious; petioles 2 — 4 (to 8) mm long, dilated toward base, glabrous, slender; leaves approximate, small, 0.8 — 1.4 cm long and 0.5 — 0.7 cm broad, obovate to oblong, more or less retuse, rounded or narrowed at base, glabrous, coriaceous, mostly entire, slightly revoluted, the upper surface dark green and glossy, the lower pale glaucous, prominently veined on both sides, the veins impressed above, raised beneath; catkins terminal, short-ovoid, 4 — 10 — 15-flowered, the short stalk sparsely covered with long hairs, elongating in fruit to 7 mm, commonly recurved; scales small, truncate, purple-brown, glabrous or the margin beset with long white cilia; stamens 2, distinct, glabrous; filaments purple, 5 — 8 times the length of the scale; anthers yellow, oblong; ovary ovoid-conical, obtuse, glabrous, in maturity violet-purple or reddish; stipe to 1 mm, finally to 2 mm long; style ca. 0.5 mm long; stigmas with broad simple lobes; gland internal, linear, truncate. June — July. (Plate II, Figure 7).

Lichen and stony alpine tundra. — Far East: Kamch. Endemic. Described from mountains beyond Lake Nachikinskoe. Type in Leningrad.

24. **S. phlebophylla** Anderss. Salic. boreali-amer. (1858) 132 et in DC. Prodr. XVI, 2, 290; Coville in Proceed. Washingt. Ac. III (1901) 336, fig. 28; Kom., Fl. Kamch. II, 31. — S. p a l a e o n e u r a Rydb. in Bull. N. Y. Bot. Gard. I (1899) 267. — S. p o l a r i s var. l e j o c a r p a Cham. in Linnaea VI (1831) 542 pp.

A dwarf undershrub; branches 7 — 10 cm long, procumbent, yellow, becoming dark, densely leafy; leaves approximate, oblong-obovate, acuminate, often folded and recurved at the tip, rarely obtusish, coriaceous, minutely dotted, plain, entire, densely covered with long white fugaceous hairs beneath, lustrous above, the lateral veins parallel, very prominent and riblike on both sides, more so beneath, upon death and decay of the leaf parenchyme the network of veins persistent up to three years, the leaf skeletons densely covering the lower part of branches; catkins borne on leafy terminal or lateral branchlets, ovoid, ca. 0.4 — 0.7 cm long, densely flowered, upright; scales rounded-obovate, black, bearded with very long white hairs; stamens 2, distinct, glabrous, twice the length of the scale; filaments yellowish-brown or violet, sparsely tomentose or pilose,
57 glabrescent, sessile; gland surpassing the ovary base; style brown, elongated, the brown 2-parted stigmas divergent. June — July. (Plate I, Figure 5).

Moss-and-lichen and stony Arctic and alpine tundra. — Arctic: Chuk., An.; E. Siberia: Lena-Kol.; Far East: Kamch. (Koraginskii Island), islands of the Bering Sea. **Gen. distr.**: polar coast of NW America. Described from North America.

Note. Leaves often dark-spotted with sclerotia of the fungus Rhytisma (Kom., Fl. Kamch.).

Subgenus 2. **CAPRISALIX** Dumort. Verhandel. Bijdr. Nat. Wetensch. Amsterd. I (1826) 55. — Shrubs or trees; bud scales falling with leaf emergence, the eglandular petioles rather long and slender; catkins appearing before or together with the leaves, lateral, commonly with small leaflike bracts on the stalk; scales mostly particolored, darker at apex than at base; gland commonly 1, internal, mostly simple; stamens 2, mostly distinct; style 1, commonly simple. Alpine, subalpine, and forest species.

Section 6. GLAUCAE Fries, Phisiogr. Sällsk. Årsb. (1825) 36, emend. Floder. — Shrubs with stout woolly-tomentose branches; leaves short-petioled, firm, entire, 5—7-nerved, grayish-woolly with long hairs; catkins borne on a long leafy stalk; scales almost plain, light-colored; glands 1 or 2, simple or lobed; stamens woolly; ovary covered with silky hairs, short-stipitate; styles 2.

1. Prostrate shrubs with procumbent branches; scales of catkins almost black, obovate or ovate, long-ciliate . . . . 29. **S. reptans** Rupr.

\+ Mostly upright shrubs to 2—4 cm high; scales yellowish-brown to rufous or fulvous, never black . . . . 2.

2. Branches, buds, and petioles glabrous; leaf blades oblong-ovate, green on both sides, glabrous when full grown, serrulate; ovaries crowded, inflated-ovoid, strongly convex, sessile, finely tomentose . . . . 28. **S. alatavica** Kar. et Kir.

\+ Different from above . . . . 3.

3. Stipules large, 0.4—2.4 cm long and 0.2—0.8 cm broad, lanceolate . . . . 26. **S. stipulifera** Floder.

\+ Stipules insignificant, shorter than petiole, ovate-lanceolate . . . . 4.

4. Shrubs to 1.5 m high; full-grown leaves rather densely covered above gray hairs, glaucescent beneath; catkins loose; scales obovate or liguliform, yellowish-brown . . . . 25. **S. glauca** L.

58 + Shrubs up to 3—4 m high; full-grown leaves bright green, covered with silvery hairs when young, finally glabrate, the lower surface densely covered with persistent white or grayish silky hairs . . . . 27. **S. Seemannii** Rydb.

25. **S. glauca** L. Sp. pl. (1753) 1019; Ldb. Fl. Ross. III, 619; Turcz. Fl. baic.-dah. II, 2, 117; Kryl., Fl. Zap. Sib. IV, 768; Kom., Fl. Kamch. II, 22. — Ic.: Fl. Dan. 1058; Rchb. Icon. Fl. Germ. XI, t. 571, f. 1214. — Exs.: Fries, Herb. norm. II, 52; HFR No. 2491.

An upright spreading shrub to 1.5 m high (in high latitudes prostrate); branches reddish or dark brown, glabrous or hairy; summer shoots gray-villous; buds ovoid, yellowish-brown, at first densely hairy, at length with scattered hairs; petioles 2—7 mm long; leaves obovate-lanceolate or oblong-obovate, narrowed at both ends or more or less rounded at apex with a short often folded point, mostly cuneate at base, entire, ciliate, blackening in drying, 3 — 6 cm long and 1.2 —2.8 cm broad, the upper surface rather densely covered with gray hairs, the lower glaucescent and covered with varying amount of recurved hairs; lateral veins 6 — 14 pairs, prominent on both sides; catkins after the leaves, borne on a leafy stalk, 2 — 4 cm long, the pistillate up to 8 cm long in fruit, relatively loose at base, the staminate dense; scales obovate or liguliform, mostly obtuse, rarely pointed, yellowish-brown, light-colored at base, densely covered on both sides with white hairs; stamens 2, distinct, rarely connate at base, to 8 mm long, hairy; anthers dark violet, becoming brown; glands often 2, the inner broader; ovary oblong-ovoid, obtuse, white-tomentose, short-stipitate; style simple, often 2-parted down to base, ca. 0.5 —1 mm long, reddish; stigmas with 2-lobed divergent lobes; gland oblong, often cleft, as long as or rarely longer than the stipe; capsule to 7 — 10 mm long. June — July. (Plate V, Figure 4).

Arctic tundra, forest-tundra, and the subalpine zone of mountains. — Arctic: Arc., Eur., Nov. Z., Arc. Sib.; European part: Kar.-Lap., Dv.-Pech., N. and S. Urals; W. Siberia: Ob, Alt.; E. Siberia: Yen., Lena-Kol., Ang.-Say., Dau.; Far East: Okh., Ze.-Bu. **Gen. distr.**: Arctic, Scandinavia, N. Eur., Atl. Eur., Mong., N. Am. Described from Europe (alps of Lapland and Pyrenees). Type in London.

Hybridizing with S. arbuscula, arctica, Chamissonis, cuneata, fumosa, hastata, herbacea, kurilensis, lanata, lapponum, myrsinites, myrtilloides, nigricans, phylicifolia, polaris, pulchra, reptans, reticulata, rossica, stipulifera, viminalis s. l.

Economic importance. Eaten by reindeer. Contains 9.6 — 14.45% of tannins.

59 26. **S. stipulifera** Floder. in Bot. Notis. (1930) 328; Ej. in Holmb. Skand. Fl. I, b, Hf. 1, 35. — S. appendiculata Vahl, Fl. Dan. 18 (1792) 7. — S. glauca β appendiculata Whlb. Fl. lapp. (1812) 264. — Ic.: Fl. Dan., t. 1056; Floder. in Bot. Notis. (1930) 329.

A shrub up to 1.5 m high, mostly with a dense round head; annotinous branches pale brown, unevenly gray-tomentose; stipules large, 0.4 — 2.4 cm long and 0.2 — 0.8 cm broad, lanceolate; leaves tough, rather large, oblanceolate to obovate, entire, covered on both sides, though more so beneath, with more or less appressed long, commonly straight, shiny light gray hairs, reticulate-veined, the veins somewhat impressed above, lateral veins 5 — 7 pairs; catkins after the leaves, long-stalked, nodding, tomentose-hairy, subtended by large leaflike bracts, the pistillate 25 — 50-flowered; scales pale brownish-yellow or pale yellowish-green, the margin of staminate scales often faintly purple with long crisped hairs; gland in staminate flowers outer, dentiform, in pistillate flowers absent, the inner gland more or less disk-shaped, the lobes often strongly drooping, 2- or 3-parted; stamens 2, distinct, pale golden or purple, hairs at base; ovary ovoid-conical, obtuse, gray-tomentose, the tomentose-hairy stipe commonly to 0.5 mm long; styles 2, distinct, spreading, to 1 — 1.5 mm long, straight or recurved at the ends, hairy below; stigmas almost cleft, the recurved lobes narrow. June — July.

Tundra and forest-tundra. — Arctic: Arc. Eur., Arc. Sib.; European part: Kar.-Lap., Dv.-Pech.; W. Siberia: Ob; E. Siberia: Yen., Lena-Kol. **Gen. distr.**: Scand. Described from Scandinavia. Type in Stockholm.

Hybridizing with S. glauca, herbacea, myrsinites, nigricans, phylicifolia, polaris.

27. **S. Seemannii** Rydb. in Bull. N. Y. Bot. Gard. II, No. 6 (1901) 164; Floder. in Arkiv för Bot. 25 A, No. 10, 5 (offprint). — Ic.: Coville in Proceed. Washingt. Acad. III (1901) t. XXXIX, sub S. glauca L.

Shrub to 3—4 m high; branches glabrous, light brown when young, dark brown in age; summer shoots pubescent; leaves oblong to oblong-lanceolate, narrowed at both ends, 3—7 cm long, rather tough, firm, entire, covered with silvery hairs when young, at length glabrescent, bright green, the lower surface covered with persistent whitish or grayish pubescence; catkins lateral, at the ends of short spurs, borne on a 3—5-leaved stalk, the staminate 2—3 cm long, the pistillate 4—7 cm long, loose; scales oblong, obtuse, light fulvous, somewhat pubescent, ca. 2 mm long; stamens 2, distinct, slender, ca. 8 mm long; ovary 3—4 mm, finally to 8 mm long, densely white-hairy, subsessile; style 0.5—1 mm long; stigmas slender,
60 ca. 1 mm long, 2-lobed. June—July.

Stony, lichen, and dryad tundra; also thickets of Japanese stone pine. — Arctic: An. **Gen. distr.**: N. Am. Described from America. Type in New York.

Hybridizing with S. cuneata and S. pulchra.

Economic importance. Summer feed for reindeer.

28. **S. alatavica** Kar. et Kir. ex Stschegl. in Bull. Soc. Nat. Mosc. XXVII I (1854) 196. — S. prunifolia Kar. et Kir. in sched. — S. spissa Anderss. in DC. Prodr. XVI, 2 (1868) 283. — S. Karelini Turcz. Fl. baic.-dah. II, 2 (1854) 120 (ex p.); O. v. Seem. in Bot. Jahrb. XXI, Beibl. No. 52 (1895) 10.

A spreading shrub of medium size; branches stout, lustrous, glabrous, reddish-brown when young, grayish-brown or yellow in age; buds dark red, pointed, lustrous; petioles ca. 2—4 mm long, dilated toward base, glabrous; stipules very small, ca. 2 mm long, brown, scarious, often lacking; leaf blade 4—6.5 cm long and 2— 2.8 cm broad, somewhat lustrous, thin, typically oblong-ovate, evenly narrowed at both ends or elliptic, rounded at base and at apex, green on both sides, sericeous when young, finally glabrous, pale glaucous beneath, serrulate, the veins prominent beneath, the midrib brownish, the lateral veins 7 or 8 pairs; catkins to 4—6 cm long and ca. 1.5—1.7 cm broad, on a 2—4-leaved stalk, stoutly cylindric, obtuse, spreading, dense, woolly-tomentose; scales oblong, obtuse, rufous or brown, crisped-hairy at apex; stamens 2, distinct; anthers round, yellow; ovaries crowded, strongly inflated at base, ovoid, obtuse, sessile, finely tomentose, grayish-rufous; style short, yellow, or lacking; gland surpassing the leaf base; stigmas brownish, rather deeply divided, 2-lobed; the lobes cruciformly spreading. May. (Plate IV, Figure 1).

Alpine zone; stony taluses, sandy terraces, often in spruce mountain-forests. — W. Siberia: Alt.; E. Siberia: Ang.-Say.; Centr. Asia: Dzu.-Tarb., T. Sh., Pam.-Al. Described from Dzungarian Ala Tau. Type in Leningrad.

Hybridizing with S. caesia, vestita, xerophila.

29. S. **reptans** Rupr. Fl. samojed. cisural. in Beitr. Pfl. d. Russ. Reich. II (1845) 54; Ldb. Fl. Ross. III, 619; Kryl., Fl. Zap. Sib. IV, 775. — Ic.: Rupr., l. c., t. III. — Exs.: HFR No. 2492 ♀♂.

A prostrate undershrub, 5—15 cm high; with mostly buried trunk, rarely a shrub up to 40 cm high; branches flagelliform, rooting, to 25—50 cm long, grayish-fulvous-green or yellowish-brown, the subterranean parts often coral-purple; young branches always covered with long hairs; the old, like the buds, glabrous or densely hairy; stipules lanceolate to
61 elliptic, entire, 1—3 mm long, or more often lacking; petioles 1—3 mm long, broad, hairy below; leaves ovate, elliptic, obovate, or lanceolate, rounded or pointed at apex with a recurved tip, slightly tapering or obtuse or sometimes cordate at base, 2—3 cm long and 0.5—2.5 cm broad, entire or, in hybrids, with several teeth at base, dull green above, paler grayish-green beneath, densely covered with flexuous hairs on both sides, the lateral veins impressed above and raised beneath; catkins on 2—5-leaved lateral branchlets, cylindric-ovaloid, the staminate to 1.8 cm long, the pistillate to 2—4.5 cm long, 0.5—1.5 cm broad; scales almost black at apex, obovate or ovate, paler below, long-ciliate; stamens 2, distinct, naked; anthers dark; outer gland lacking in staminate flowers; inner glands 3, connate at base and often with deeply cleft lobes; ovary sessile, ovoid-conical, obtuse, tomentose-villose; style very short; stigmas fulvous, elongate, often cleft down to base into long narrow lobes. (Plate V, Figure 2).

Arctic: Arc. Eur., Nov. Z., Arc. Sib., An., Chuk.; Far East: Kamch. **Gen. distr.:** Arc. Eur. Described from Bol'shezemel'skaya Tundra. Type in Leningrad.

Hybridizing with S. arctica, glauca, hastata, lanata, polaris, taimyrensis.

Economic importance. Flowers and leaves eaten by reindeer.

Section 7. **CHRYSANTHAE** W. D. Koch, Comment. (1828) 52. — Shrubs, often fairly high, with stout brown branches; leaves ovate-orbicular or broadly lanceolate, reticulate-veined beneath; catkins sessile, subterminal; scales black at apex; ovary conical, cuneate, commonly glabrous; anthers yellow after flowering.

1. Stipules glandular; grown leaves more or less glabrous . . . . 31. **S. glandulifera** Floder.
+ Stipules not glandular; leaves pubescent or glabrous . . . . 2.
2. Catkins golden-yellow; leaves woolly . . . . 30. **S. lanata** L.
+ Catkins covered with silvery or gray hairs; leaves glabrate . . . . . . 3.
3. Leaves undulate, broadly obovate to suborbicular; stipules broadly semiovate . . . . 32. **S. Hookeriana** Barr.
+ Leaves quite flat; stipules narrowly lanceolate, acute . . . . 33. **S. Richardsonii** Hook.

30. **S. lanata** L. Sp. pl. (1753) 1019; Ldb. Fl. Ross. III, 616; Kryl., Fl. Sib. VI, 764. — S. lanuginosa Pall. Fl. Ross. II (1778) 83. — S. chrysan-
62 thos Vahl, Fl. Dan. fasc. 18 (1792) 6. — Ic.: Fl. Dan. tab. 1057.

A depressed or upright shrub, from 0.1 to 2 — 3 m high; branches stout, knotty, grayish-brown, villous or tomentose; buds large, ovoid, brown, woolly; stipules obliquely ovate or semicordate, shorter than to nearly as long as the petioles (0.2 — 0.7 cm); leaf blades elliptic to rounded-elliptic, rounded at apex with a short often folded point, rounded, cordate, or broadly cuneate at base, entire, firm, densely covered on both sides with slender silky hairs, dull green above, paler beneath with scattered hairs, 2 — 7.5 cm long and 2 — 4 cm broad, the veins beneath prominently raised, lateral veins 6 — 12 pairs; catkins sessile, solitary or in 2's — 4's at the ends of shoots of the preceding year, subtended by scalelike bracts, 3 — 6 cm long, the pistillate in fruit to 1 cm, dense, golden-yellow; scales black, ovate or obovate, acuminate, 2— 3 mm long, densely covered with golden hairs; anthers 2, distinct, rarely connate at base, naked, with yellow ovate anthers; gland 1, oblong, obtuse; ovary conical, compressed laterally, to 7 mm long, light green, glabrous, minutely tuberculate near the top; style to 4— 5 mm long, often obliquely inserted; stigmas shorter than style, oblong, simple or 2-lobed; stipe ca. 0.5 mm long, slightly hairy in upper part, as long as or shorter than the gland. June — July. (Plate V, Figure 3).

Arctic tundra, forest-tundra, and the subalpine zone. — Arctic: Arc. Eur., Nov. Z., Arc. Sib.; European part: Dv.-Pech., Kar.-Lap.; W. Siberia: Alt., Ob.; E. Siberia: Lena-Kol., Ang.-Say., Dau. **Gen. distr.**: Scand., Atl. Eur., N. Am. Described from the alps of Lapland. Type in London.

Note. Hybridizing with S. arbuscula, caprea, coaetanea, fulcrata, glauca, hastata, herbacea, lapponum, polaris, reptans.

Economic importance. The leaves and young branches are eaten by reindeer. The bark contains 6.02—10.1% of tannins.

31. **S. glandulifera** Floder. in Sv. Vet. Ak. Årsb. (1924) 192 (nomen); Ej. in Lindm. Sv. Fanerog. fl., ed. 2 (1926) 212; Ej. in Bot. Notis. (1930) 338; Ej. in O. Holmberg, Skand. Fl. I, Hf. 1 (1931) 127. — S. lanata γ glandulosa Whlb. Fl. Lapp. (1812) 259 pp; Hartm., Handb. ed. 1 (1820) 351. — Ic.: Floder. in Bot. Notis. (1930) 339.

A shrub to 3 m high, or higher; annotinous branches erect, very stout, straight, white-woolly when young, leafy at base, becoming glabrate; buds large, obtuse, more or less hairy; stipules large, narrowly pointed, glandular-serrate near apex, on withering readily persistent together with
63 petiole; leaf blade to 6 cm long and to 2.5 cm broad, oblanceolate or obovate, obtuse, flat, entire, densely beset on the margin with globose glands, the upper surface green, with impressed reticulate venation, paler beneath, glabrate; young leaves rather sparsely covered on both sides with very long straight appressed hairs, in age glabrate, only the petiole and the lower part of the midrib often covered with short spreading recurved white hairs; lateral veins 6 — 8 pairs, arising at an acute angle; scales oblong, blackish-brown at apex, covered with straight long hairs, upon frequent cutting back or browsing of the branches, becoming copper-colored, lustrous; gland inner, long, simple; stamens 2, distinct, naked, long; anthers

large, orange-red; ovary ovoid-subulate, yellowish-green, glabrous, short-stipitate, covered with short cinereous hairs; style to 3 mm long, glabrous; stigmas long, slender. June—July.

Tundra and forest-tundra. — Arctic: Arc. Eur.; European part: Kar.-Lap., Dv.-Pech.; W. Siberia: Ob; E. Siberia: Yen., Lena-Kol.; Far East: Kamch. **Gen. distr.**: Scand. Described from Scandinavia. Type in Stockholm.

Note. Hybridizing with S. caprea, hastata, and herbacea. The distribution area coincides with that of S. herbacea from which S. glandulifera differs chiefly in the glanduliferous stipules and leaves.

32. **S. Hookeriana** Barr. in Hook. Fl. bor.-amer. II (1840) 145, tab. 180; Anderss. Ofwers. Wet. Akad. Förh. (1858) 119 et in DC. Prodr. XVI, 2, 274; Floder. in Arkiv för Botan. 25 A. 10, 9 (offprint).

A rather lower but vigorous shrub with stout dark-tomentose branches; stipules broadly semiovate; petioles 1—3 mm long, gray-tomentose; leaf blades 5—6 cm long and 2—5—3 cm [sic] broad, broadly obovate to suborbicular, rather stiff, wavy-margined or sparingly serrulate, generally resembling the leaves of S. caprea L., dark green above with only the midrib white-tomentose, woolly-tomentose beneath; catkins 2.5—5 cm long, the pistillate longer, narrow, sessile, with few leaflike or scalelike bracts, stoutly cylindric, dense, erect, very hairy; scales brownish-black, rounded at apex, densely villous with long hairs, these silky yellow in staminate, gray in pistillate flowers; ovary conical, long-tapering, glabrous, dark green, rufescent in drying, the stipe twice the length of the gland; stigmas linear, simple, straight. June—July.

Forest-tundra. — Arctic: An.; E. Siberia: Lena-Kol. (Zhigansk). **Gen. distr.**: America. Described from North America. Type in New York.

33. **S. Richardsonii** Hook. Fl. bor.-amer. II (1840) 1, 47, t. 182; Anderss. in DC. Prodr. XVI, 2 (1868) 273; Coville in Proceed. Washingt. Acad. III (1901) 315 et fig. 19; Floder. in Arkiv för Botan. 25 A, No. 10, 8 (offprint). —
64 S. lanata americana Anderss. Ofwers. Wet. Akad. Förh. (1852) 119.

A shrub to 1—2.5 m high (in the USSR usually lower); branches spreading in all directions, stout, flexuous, gray-tomentose; stipules narrowly lanceolate, acute; leaves 2.5—3 cm long and ca. 0.7 cm broad, quite flat, prominently veined, glabrous, somewhat lustrous, green above, paler green or glaucous beneath, ovate-elliptic and subentire (f. latifolia Anderss.) or lanceolate to narrowly lanceolate and mostly minutely glandular on the margin (f. angustifolia Anderss.), parallel-nerved; catkins sessile, terminal, hairy, erect, obtuse, stoutly cylindric, scales black, acutish, covered with rather long silvery hairs; ovary conical-cuneate, glabrous or very slightly silvery-pubescent, sessile; style elongated, brownish; stigmas shortly divided, divergent. June—July.

Birch and alder waterside thickets. — Arctic: An. **Gen. distr.**: N. Am. Described from North America. Type in New York.

Section 8. VILLOSAE anderss. in DC. Prodr. XVI, 2 (1868) 275. — Shrubs of medium size, with slender branches; leaves lanceolate, entire, prominently veined beneath; catkins sessile; scales green at apex: anthers yellow after flowering; ovary conical, tomentose.

1. Petioles densely woolly-tomentose; midrib white-tomentose also above . . . . 2.
\+ Petioles sparingly pubescent; midrib without white tomentum above . . . . 3.
2. Leaves mostly liguliform or oblong-lanceolate, the upper surface commonly dark green, glabrate or sparsely covered with scattered hairs, the lower surface densely clothed beneath with appressed silvery-white tomentum; the young leaves densely ciliate-hairy on the margin; style and stigma short; ovary long-stipitate . . . . 35. **S. Krylovii** E. Wolf.
\+ Leaves oblong-ovate to lanceolate, dull grayish-green above; style long; stigmas oblong; ovary short-stipitate . . . . . 34. **S. lapponum** L.
3. Leaves 7 — 10 cm long, broadest in lower or upper part, very densely clothed beneath with snowy-white tomentum which mostly conceals the veins; scales densely covered with long silver or yellow hairs; style very long, filiform, yellow; stigmas long, linear, mostly simple. . . . . 36. S. speciosa Hook. et Arn.
\+ Leaves 5—6 cm long, broadest in upper part, prominently veined on both sides, sparsely hairy or glabrous and glaucescent beneath with a white-tomentose midrib; scales commonly covered with gray hairs; style short, dark, hairy below; stigmas short, 2-parted . . . . 37. **S. anadyrensis** Floder.

65

34. **S. lapponum** L. Sp. pl. (1753) 1019; Ldb. Fl. Ross. III, 617; Shmal'g., Fl. II, 434; Kryl., Fl. Zap. Sib. IV, 765. — Ic.: Fl. Dan. t. 1058; Rchb. Ic. fl. Germ. t. 572, f. 1215. — Exs.: HFR No. 2338—2340, 2490; Wimm. et Kr. Herb. Sal. 6, 39, 110—112.

Shrub to 1—1.5 m high, with yellowish-brown trunk and dark red glabrous branches; young branches pubescent or tomentose; buds ovoid, obtuse, yellowish-brown, glabrous or sparsely hairy; stipules small, obliquely ovate or falcate, rarely present; petioles 0.4—1 cm long; leaf blades 5 — 8 cm long and 2 — 3 cm broad, oblong, oblong-ovate, or lanceolate, rarely narrowly lanceolate (f. *angustifolia*), tapering to a frequently oblique and folded point, rounded or broadly cuneate at base, entire, revolute, sometimes sparingly glandular-toothed or serrate (f. *serrata* Floder.) or on vigorous shoots undulate-sinuate (f. *marrubifolia* Tausch.), dull green and slightly rugose above, white-tomentose beneath, with 7 — 12 pairs of lateral veins; catkins lateral, dense, 2 — 4 cm long, subsessile; scales obovate or liguliform, pointed or obtuse, almost black above, brown and long-hairy beneath; gland 1, posterior, oblong, obtuse; stamens 2, distinct, naked; anthers violet-yellow, darkening; ovary to 5 mm long, ovoid-conical, white-tomentose, the stipe very short; style to 1.5 mm long; stigmas half as long as the style, oblong, emarginate or 2-lobed; capsule 7 — 8 mm long. May — June.

Wet boggy meadows, peatbogs of the forest zone, and shrub thickets of the forest-tundra. — Arctic: Arc. Eur.; European part: Kar.-Lap., Dv.-Pech., Lad.-Ilm., U. V., V.-Kama, U. and M. Dnp., V.-Don; W. Siberia: Ob, U. Tob., Irt.; Alt.; E. Siberia: Yen., Lena-Kol., Ang.-Say. **Gen. distr.**: Scand., Centr. and Atl. Eur. Described from Sweden. Type in London.

Note. Hybridizing with S. arbuscula, aurita, caprea, cinerea, coaetanea, dasyclados, glauca, herbacea, lanata, livida, myrtilloides, nigricans, phylicifolia, polaris, purpurea, reticulata, rosmarinifolia, rotundifolia, sibirica and viminalis s.l., xerophila.

Economic importance. The foliage is readily eaten in summer by reindeer. An ornamental and nectariferous plant; the bark contains 9—14.27% of tannins. The buds constitute the main food for willow grouse and alpine ptarmigan; hence the vernacular W. Siberian name for S. lapponum, "kuropatnik" [derived from "kuropatka," partridge] (Skalozubov).

35. **S. Krylovii** E. Wolf in A. H. P. XXVIII, 4 (1911) 537; Kryl., Fl. Zap. Sib. IV, 767. — S. baicalensis Turcz. ex Anderss. in DC. Prodr. XVI, 2 (1868) 276, nom.; Trautv. Increm., No. 4707; Radde, Reis. im. Süd. von Ost.-Sibir., 121. — S. speciosa δ Trautvetteriana Anderss. in
66 DC. Prodr. XVI, 2,(1868) 276. — S. arenaria Turcz. Catal. baic.-dah., 387, No. 1042. — S. Gmelini et S. Gmeliniana Turcz., pl. exsicc. — S. lapponum Turcz. Fl. baic.-dah. II, 2, 116, non L. — S. pseudolapponum E. Wolf in Kryl., Fl. Alt. V (1909) 1226, non O. v. Seem. (1901). — S. polia C. K. Schn. in Sarg. pl. Wilson. III, 1 (1916) 174.

A shrub to 1—2.5 m high, with sturdy dark castaneous glabrous branches; summer shoots white-tomentose, the annotinous only in upper part hoary-tomentose; stipules lanceolate; petioles 3—5 mm long; leaf blades oblong-elliptic, on suckers liguliform, obtuse, narrowed toward base, subentire, when young silvery-tomentose beneath and white-ciliate on the margin, in age dark green and sparingly pubescent above, tomentose with appressed white hairs beneath, rarely dull hoary above with fine sparse tomentum (f. velutina Laksch.), 5—7 cm long and ca. 2 cm broad, the base of midrib reddish, lateral veins ca. 12 pairs; catkins appearing about the same time as the leaves, 3—3.5 cm long and ca. 1 cm broad, the pistillate up to 5 cm in fruit, oblong-cylindric, dense, looser at base, borne on a short white-tomentose leafy stalk; scales to 2.8 mm long and ca. 1 mm broad, ovate, acutish, black, brown at base, covered with long white hairs; stamens 2, distinct, naked; anthers yellow, oblong; ovaries to 2.5 cm long, strongly spreading, inflated ovoid-conical, obtusish, the stipe to 1.4 mm long, densely white-tomentose with crisped hairs; style ca. 0.5 mm long, simple; stigmas ca. 0.6 mm long, divergent, 2-parted, reddish; gland ca. 1 mm long, shorter than the stipe, oblong, emarginate. May—June. (Plate IV, Figure 8).

Subalpine zone, moss-and-lichen tundra, and stream banks. — Arctic: Chuk., An.; W. Siberia: Alt.; E. Siberia: Lena-Kol., Ang.-Say., Dau.; Far East: Okh., Ze.-Bu. Described from the alps of central Altai. Type in Leningrad.

Hybridizing with S. caprea, myrsinites, and reticulata.

PLATE III. 1. Salix berberifolia L. – 2. S. berberifolia var. Brayi Ldb. – 3. S. fumosa Turcz. – 4. S. saxatilis Turcz. – 5. S. myrsinites L. – 6. S. arctica Pall. – 7. S. torulosa Trautv. – 8. S. pulchra Cham. – 9. S. divaricata Pall. – 10. S. cuneata Turcz.

Economic importance. Summer food for reindeer. In the alps of Cis- and Transbaikalia the branches are cut and dried and are often incorporated into hay made of pika (Okhotona). An ornamental plant.

36. **S. speciosa** Hook. et Arn. in Beech. Voy. Botan. (1831) 130; Anderss. Salic. bor.-amer. 119, No. 21. — S. speciosa β alaxensis et v. ajanensis Anderss. in DC. Prodr., XVI, 2, 275; Ldb. Fl. Ross. III, 625. — S. alaxensis (Anderss.) Coville in Proceed. Washingt. Acad. II (1900) 280 et III (1901) 311, tab. XXXIV. — S. longistylis Rydb. in Bull.
69 N. Y. Bot. Gard. II (1901) 163. — S. lapponum γ ajanensis Trautv. in sched. — Ic.: Coville, l. c.; O. v. Seem. Bot. of Herald, 40, No. 188, t. 10.

A shrub to 2 — 2.5 m high (in North America a tree up to 3 — 9 m high); branches densely covered with yellowish woolly tomentum; stipules persistent, linear or subulate, scarious, exceeding the woolly-tomentose petiole; leaf blade 7 — 10 cm long and 2.4 — 3.6 cm broad, varying in shape, obovate, short-pointed at base and narrowed toward base, or broader, almost round at base and gradually long-tapering, lanceolate, or oblong-lanceolate; or obovate-spatulate with a short oblique point (v. ajanensis Anderss.); or long-tapering toward both ends (v. alaxensis Anderss.), the upper surface commonly dark green, the lower densely covered with snow-white tomentum or (in v. alaxensis) alveolate, the anastomosing veins woolly, impressed above the midrib, often white-woolly also above, lateral veins 8 — 15 pairs; catkins sublateral, sessile, subtended by obsolescent bracts, stoutly cylindric, much elongated, the pistillate up to 10 cm in fruit, slightly recurved, more silvery-hairy, the staminate shorter, 2.5 — 3 cm long, conspicuously yellow-hairy; scales acutish, black at apex, very densely covered with yellowish hairs the length of ovary; stamens 2, golden-colored; ovary ovoid-conical, sessile, woolly-sericeous, 3 — 4 mm long; style very long, filiform, yellow; stigmas straight, long, linear, mostly simple. June — July. (Plate IV, Figure 7).

River valleys, shores, boggy meadows, and shrub thickets. — Arctic: Chuk., An.; E. Siberia: Yen. (Khatanga), Lena-Kol., Ang.-Say., Dau.; Far East: Okh., Uda, Kamch. (Bering, Koraginskii). **Gen. distr.:** Arctic islands of NW America. Type in London.

Economic importance. Very ornamental, nectariferous. The leaves and young branches are eaten by deer.

37. **S. anadyrensis** Floder. in Arkiv för Bot. 25 A, No. 10 (1933) 9 et icon. No. 3.

An upright shrub, 1.5 — 2 m high; branches stout, ribbed, dark gray-tomentose; bark dark, almost black; buds large, to 1.2 cm long, appressed, conical, densely covered with whitish tomentum; petioles stout, 0.7 — 1.2 cm long, also densely covered with whitish tomentum; stipules as long as or longer than petiole, lanceolate or semiovate, glandular-toothed, white-hairy, often caducous; leaf blade to 5 — 6 cm long and ca. 2 cm broad, coriaceous, broadly lanceolate, elliptic, or obovate, short-pointed at apex, cuneate at base, dark green above, light green beneath, entire when young, in age
70 glandular-toothed, prominently veined on both sides, the midrib white-tomentose, especially beneath; lateral veins 7 — 9 — 14 pairs, the upper surface covered with scattered hairs, sometimes glabrate, the lower

sparingly tomentose or glabrate (except the midrib), glaucescent; catkins (pistillate) stout, to 1.6 cm long, the stalk to 1.5 cm long, with 1—3 bracts; [pistillate catkins ?] to 16 cm long, spreading, dense (containing up to 300 flowers); scales blackish-brown, covered with long gray hairs; ovary to 8 mm long, ovoid-conical, obtusish, borne on a very short thick hairy stipe, at first densely hairy, finally greenish with sparse hairs; style up to half the length of ovary, dark, hairy below; stigmas commonly straight, short, 2-parted, dark red; gland 1, inner, to 0.5 mm long, slender. June—July.

Riverbanks. — Arctic: An. (Anadyr, Penzhina), whence described. Endemic. Type in Leningrad.

Section 9. PHYLICIFOLIAE Dumort. Fl. Belg. Prodr. (1827) 12; Anderss. Mon. Sal. (1867) 125. — Low or tall; branches short, mostly dark red; leaves ovate-lanceolate or elliptic-lanceolate to obovate or subspatulate, mostly glabrous, rarely pubescent, with green, pale or glaucous underside, distantly recurved-serrate, sometimes blackening on wilting; catkins ovoid-cylindric, sessile or short-stalked; scales bicolor; ovary more or less covered with silky hairs, rarely glabrous; style evident; gland always 1, posterior; stamens mostly glabrous.

1. Ovary commonly densely pubescent up to the top, hoary or gray-hairy, silky, or glabrate . . . . 2.
\+ Ovary glabrous or sparingly hairy all over or merely at base . . . . . 7.
2. Leaves entire; petioles to 1.6 cm long; ovary sessile; gland ovate . . . . 41. **S. tontomussirensis** Koidz.
\+ Leaves more or less serrate or dentate; petioles 2—10 mm long; ovary short-stipitate; gland oblong, linear, or ribbon-shaped . . . . . 3.
3. Grown leaves quite glabrous beneath; branches mostly dark fulvous, castaneous, or reddish-brown, lustrous . . . . 4.
\+ Grown leaves covered with scattered hairs or somewhat sericeous; branches yellowish-fulvous, grayish-brown, or dingy brown . . . . 6.
4. Leaf blades mostly obovate-elliptic or oblanceolate-elliptic; catkin-scales ferruginous-brown at apex . . . . 5.
\+ Leaf blades elliptic, obovate, or lanceolate; catkin-scales almost black at apex . . . . 38. **S. phylicifolia** L.
5. Leaf blades oblanceolate, elliptic, or oblong; lateral veins connected beneath by vertical subsidiary veins parallel to midrib . . . . 40. **S. parallelinervis** Floder.
71 + Leaf blades oblong-elliptic or obovate-elliptic, without subsidiary connecting veins beneath . . . . 39. **S. oblongifolia** Trautv. et Mey.
6. Branches tawny; petioles 3—5 mm long; leaf blades narrowly lanceolate, narrowed at both ends, somewhat sericeous beneath, serrulate, often entire about the middle; stipe of ovary as long as or shorter than the gland . . . . 43. **S. kolymensis** O. v. Seem.
\+ Branches dingy brown; petioles 6—15 mm long; leaf blades elliptic or oblong-elliptic, covered beneath with scattered hairs, irregularly erose-serrate, subentire at base and at apex; stipe of ovary half as long again as the gland . . . . 46. **S. abscondita** Laksch.

7. Stipules 2—4 times as long as the petiole, supporting the blade . . . . 46. **S. fulcrata** Anderss.

\+ Stipules small, shorter than the petiole . . . . 8.

8. Leaf blades coriaceous, firm, spatulate-lanceolate or oblong, coarsely crenate-dentate; ovary greenish-yellow, finely silky, or merely pubescent at base, or glabrate; stipe of ovary twice as long as the gland . . . . 44. **S. chlorostachya** Turcz.

\+ Leaf blades thin, liguliform-spatulate, sinuate-crenate; ovary latericious-brown, glabrous, the stipe barely exceeding the gland . . . . 45. **S. podophylla** Anderss.

38. **S. phylicifolia** L. Sp. pl. (1753) 1016; Ldb. Fl. Ross. III, 611; Shmal'g., Fl. II, 435; Kryl., Fl. Zap. Sib. IV, 750. — S. Weigeliana Willd. Sp. pl. ed. IV, 2 (1805) 678. — Diplima Waigeliana Rafin., Alsol. Am. (1838) 13. — Wimen bicolor, Wim. Wulfeniana, Usionis Dicksoniana Rafin., ibid., 14. — Ic.: Anderss. Mon., t. 70; Rchb. Ic. XI, t. 564, f. 2003. — Exs.: HFR No. 2309—2312; Enand. Salic. Scand. No. 116, 117.

A fine shrub, 0.5—3.5 m high, mostly not more than 1 m, forming beautiful dense clumps; wood under the bark not ridged; branches yellowish-brown or reddish, stoutish, glabrous, lustrous; buds oblong, yellowish, sparsely covered with appressed hairs; stipules small, semicordate, on vigorous shoots broadly falcate, glandular-dentate or serrate, promptly caducous; petioles much longer than stipule, one-eighth to one-third the length of the blade; leaf blades elliptic or obovate (f. latifolia Anderss.) to lanceolate or narrowly lanceolate (f. angustifolia Anderss.), 4—9 cm long and to 2—4 cm broad, broadest at or above the middle, acute, narrowed or rounded at base, shallowly and irregularly dentate or sinuate to subentire, rather conspicuously bicolor, lustrous
72 dark green above and glaucous-white or pale green beneath, quite glabrous on both sides or with short hairs only on the midrib beneath or when young, laurine, rather stiff, sometimes blackening in drying, more often somewhat yellowing and assuming a shiny olivaceous hue above, the midrib rufescent beneath, thickened at base, the lateral veins rarely prominent; catkins before or almost together with the leaves, sessile or borne on a short petiole with scalelike bracts at base, the staminate ovoid, 1.5—2.5 cm long and ca. 1 cm broad, the pistillate up to 9 cm long in fruit; scales ovate-oblong, obtuse or acute, sometimes toothed or sinuate, light brown, blackish at apex, ca. 1.5 mm long, covered with long white hairs (in f. irrigua Flod.) denticulate, purple, almost white-hairy; stamens 2, distinct, quite glabrous, 3—4 times as long as the scale; anthers yellow, becoming brown; gland 1, posterior, oblong; ovary always tomentose, ovoid-conical, the stipe as long as or half as long again to twice as long as the gland; style developed, one-third the length of ovary and longer than the simple or bifid stigma. Fl. May; fr. June. (Plate V, Figure 8).

Not rising above the forest zone and not penetrating beyond the forest and forest-tundra belts. Meadows, coppices, forest glades and margins, and swampy river valleys. — Arctic: Arc. Eur.; European part: Kar.-Lap., Dv.-Pech., Lad.-Ilm., U. V., V.-Kama, U. Dnp., V.-Don, L. Don; W. Siberia: Ob, U. Tob., Irt.; E. Siberia: Yen., Lena-Kol.; Caucasus: W. Transc. **Gen. distr.**: Scand., Atl. and Centr. Eur. Described from Sweden. Type in London.

Note. Hybridizing with the following species: S. aurita, caprea, cinerea, dasyclados, glauca, hastata, lanata, lapponum, livida, myrsinites, myrtilloides, nigricans, pentandra, sibirica, viminalis s. l., xerophila.

Economic importance. Eaten by reindeer. Suitable for tanning as it contains 6.0—17.4% tannins. Ornamental.

39. **S. oblongifolia** Trautv. et Mey. in Middend. Sibir. Reis. I, 2, Bot. Abt. 2 (1856) 81; Anderss. Mon. Salic., 150 et in DC. Prodr. XVI, 2, 248.

A shrub to 2 m high; branches short, sturdy, straight or flexuous, knotty, reddish-brown or castaneous, glabrous, lustrous, when young slightly pubescent; buds ovoid, glabrous or at tips pubescent; petioles 4—10 mm long, mostly glabrous; stipules small, ovate-lanceolate to lance-linear, glandular-dentate, caducous; leaves oblong-elliptic, narrowed at both ends or obovate-elliptic, tapering toward base, on upper shoots to 4—5 cm long, on the lower 2.5—4 cm long and 2.5—3.5 cm broad, sharply and sparingly serrate, often subentire, the upper surface sublustrous or dull,
73 glabrous or covered with scattered hairs, the lower light green or glaucescent, glabrous or on the stramineous midrib slightly hairy; lateral veins 7—12 pairs, very prominent, glabrous, at first parallel to midrib, then spreading at an angle of 40—50°; catkins appearing about the same time as the leaves or just before, lateral sessile, the pistillate borne on a short stalk with 2—4 very small oblong bracts, dense, the staminate ovoid-cylindric, to 2.5—3 cm long and 1 cm broad, the pistillate to 4—5 cm long and 0.7 cm broad, becoming longer and thicker in fruit, the rachis clothed with gray hairs; scales of staminate flowers lanceolate or obovate, those of pistillate flowers liguliform, ferruginous-brown, hairy; stamens 2, distinct, glabrous; anthers ovate, yellow; gland oblong or linear, truncate or subcapitate; ovary conical from an ovoid base, silky-hairy, the very short stipe equaling or shorter than the gland; style elongated, 2-parted at the end or entire. Fl. May—June; fr. June.

Forest-tundra and the forest belt; swampy river valleys and larch woods. — E. Siberia: Ang.-Say. (Kachug on Lena), Dau. (Barguzin, Nerchinskii Zavod, and Chita areas), Lena-Kol. (Verkhoyansk area and on the border of Ze.-Bu., in Stanovoi Range); Far East: Kamch., Okh., Uda, Ze.-Bu., Uda, Sakh. Endemic. Described from Uda Territory. Type in Leningrad.

Hybridizing with S. fuscescens.

Note. An East Asian vicariad of the European S. phylicifolia L.; in the Arctic Region it is replaced by S. pulchra Cham. The next species, the Kamchatkan S. parallelinervis Floder., represents no more than an insignificant local modification of S. oblongifolia Trautv. et Mey. The species S. tontomussirensis Koidz. (which the present author has not seen), described from Sakhalin, is extremely closely related to it (if not identical).

Economic importance. Young shoots and leaves are eaten by reindeer and domestic stock.

40. **S. parallelinervis** Floder. in Ark. för Bot., Bd. 20, A (1926) 6, 35, fig. 3; Hultén, Fl. of Kamtch. II, 16; Kom., Fl. Kamch. II, 17.

A much branched shrub to 1.5—2 m high; branches short, upright, dark fulvous or reddish, glabrous; stipules soon caducous, lance-linear, glandular-

serrate; petioles 2—5 mm long; leaf blades oblanceolate, elliptic, or oblong, 5—6 cm long and 1.5—2 cm broad, serrulate or unevenly serrate or subentire, the upper surface lustrous, dark green, glabrous, sometimes with pubescent midrib, the lower pale green or glaucous, the veins prominent on both sides; young leaves pubescent, becoming glabrous, midrib yellow; lateral veins 9—12 pairs; decurrent as main veins of third order forming a conspicuous network and often connecting the
74 veins of second order by riblike vertical ramifications parallel to the midrib like the decurrent veins of second order (hence the name of the species); catkins before the leaves, the pistillate 4—5 cm long and ca. 1 cm broad, subsessile, with 2—4 softly hairy caducous foliaceous bracts at base, dense, with pubescent rachis; scales covered with short hairs, the pistillate liguliform, the staminate lanceolate or obovate, fulvous at apex; gland inner, oblong, truncate; stamens 2, distinct, glabrous; anthers yellow; ovary ovoid-conical, gray-tomentose, the short stipe pubescent; style ca. 1 mm long, glabrous; stigma thickish, with divergent lobes. Fl. June—beginning of July; fr. from end of June.

Wet meadows, shores of rivers and lakes, forming thickets. — Far East: Kamch. Endemic. Described from Petropavlovsk area and from Opal cone-shaped mountain. Type in Stockholm.

Hybrids with S. arctica, S. fuscescens, and S. sachalinensis, have been reported.

Note. Galls produced by small insects are often formed on the branches in the shape of "willow roses." A species closely allied to S. oblongifolia Trautv. et Mey. and together with it replacing in Kamchatka the European S. phylicifolia L. and the Americo-Siberian S. pulchra Cham. Partial fusion of stamens is sometimes observed (f. cladostemma).

Economic importance. Summer food for deer.

41. **S. tentomussirensis** Koidz. in Tok. Botanic. Magaz. XXX (1916) 81. — Ic.: Acta Phytotax. et Geobot. II (1933) 227.

A shrub with flexuous branches; young branchlets brownish-fulvous, lustrous; old branches grayish-brown; stipules mostly lacking; petioles yellowish, glabrous, to 1.6 cm long; leaf blades coriaceous, glabrous above, oblong-obovate or elliptic-obovate, rarely narrowly oblong or broadly elliptic, rounded, obtuse, or pointed at apex, broadly cuneate or very rarely obtuse or almost rounded at base, entire, slightly revolute, to 6 cm long and 4 cm broad; catkins on leafy terminal branchlets, elongate-cylindric, in fruit to 7.5 cm long, dense, the pubescent stalk to 10 mm long; scales rounded-obovate, dark, covered outside below the middle with long silky hairs, sparsely hairy on the inside; style elongate, glabrous; stigmas deeply 2-parted, the divisions 2-lobed; gland 1, ovate; capsule to 7 mm long, slightly hairy, sessile. Fr. July—August.

Far East: Sakh. Endemic. Described from S. Sakhalin. Observations are needed to determine its distribution in N. Sakhalin. Type in Tokyo.

Note. The lack of herbarium specimens makes it impossible to establish identity of this species with S. oblongifolia Trautv. et Mey., but, judging by the description, the resemblance between the two is very great.

75 42. **S. fulcrata** Anderss. Monogr. Salic. (1867) 139 et tab. VII, fig. 73; Ej. in DC. Prodr. XVI, 2 (1868) 244.

A robust shrub to 0.75 — 1 m high; branches rigid, straight, yellowish-brown becoming brown in age, glabrous; stipules large, 2 — 4 times as long as the petiole, linear to linear-lanceolate, glandular-serrate, subtending the leaves, soon caducous; leaves 4 — 7.5 cm long and 1.5 — 2 cm broad, lance-obovate, long-tapering toward base, acuminate at apex, serrulate or rarely subentire, thin, brittle, dark green and lustrous above, glaucous beneath, glabrous on both sides, the midrib brownish beneath, very prominent, the relatively obscure lateral veins 12 — 15 pairs, strongly upcurved toward apex; catkins after the leaves, the pistillate to 7 cm long and to 0.8 cm broad, sessile, upright, cylindric, dense, ebracteate or with few small bracts at base; scales acutish, dark at apex, covered with long gray hairs; ovary long conical-cuneate, finely gray-hairy, the stipe barely exceeding the gland; style strongly elongate; stigma yellow, with filiform undivided divergent lobes. June.

Forest-tundra and river valleys. — Arctic: An.; E. Siberia: Lena-Kol.; Far East: Okh., Kamch. **Gen. distr.**: N. Am. Described from the Okhotsk area. Type in Leningrad. — Hybridizing with S. lanata L.

Economic importance. Eaten by deer. The leaves contain 4.05 total nitrogen, 25.31 crude protein, and 11.28 cellulose (Rabotnov).

43. **S. kolymensis** O. v. Seem. in Fedde, Repert. V (1908) 18. — S. boganidensis Trautv. (ex p.) in Middend. Sib. Reis. 1, 2, Bot. Abth. 1 (1847) 154. — S. boganidensis var. angustifolia Herd. in A. H. P. XI (1891) 434. — Ic.: Middend., l. c., tab. 2, 3.

A fairly tall shrub with slender branches; young branchlets tawny, glabrous or hairy, becoming reddish-brown and glabrous in age; buds appressed, small, acute, reddish-brown, glabrous or slightly hairy; stipules linear, glandular-dentate, promptly caducous, as long as or longer than the petiole; petioles 3 — 5 mm long, pubescent; leaves narrowly lanceolate, narrowed at both ends, to 5 — 6 cm long and 0.8 — 1.5 cm broad, flat, sharply serrulate, at first finely pubescent, becoming glabrous, dark green above, glaucescent or somewhat paler and subsericeous beneath; catkins just before the leaves, dense; staminate upright or spreading, sessile or nearly so, with narrowly lanceolate foliaceous bracts at base, narrowed at both ends, cylindric, to 4 cm long and 1.2 cm broad; pistillate borne on a stipe to 0.4 cm long or sessile, narrowly cylindric, ca. 4 cm long and 0.4 cm
76 broad, hoary; rachis sparingly covered with gray hairs; scales ovate, acute, dark fulvous or yellowish-fulvous, light-colored at base, the pistillate covered with long whitish or grayish hairs, the staminate glabrate; stamens 2, distinct, glabrous, twice as long as the scale; anthers rounded-conical; gland 1, posterior, narrow, ribbon-shaped, one-fourth the length of scale and equaling or slightly exceeding the stipe; ovary ovoid-conical, ca. 2 mm long, sparingly canescent; style one-third to half as long as ovary; stigma with narrowly ovate 2-parted divergent lobes; stipe pubescent, ca. 0.5 mm long; Fl. June; fr. June — July. (Plate VI, Figure 7).

Tundra, forest-tundra, thin larch woods and swampy river valleys. — Arctic: Arc. Sib., Chuk.; E. Siberia: Lena-Kol. Endemic. Described from the Boganida River (Taimyr). Type in Leningrad.

Note. Study of the original S. bogadinensis Trauv. disclosed that this name embraces flowering specimens of S. pulchra Cham. and leaf-bearing specimens of S. kolymensis O. v. Seem. Thus S. boganidensis should be referred to partial synonyms.

Economic importance. The bark of this willow is used in Kolyma for making fishing nets and ropes and as lining for boats (Avgustinovich).

44. **S. chlorostachya** Turcz. Fl. baic.-dah. II, 2 (1854) 100; Anderss. Mon. Salic. 143, tab. VII, f. 78 et in DC. Prodr. XVI, 2, 246. — Exs.: HFR No. 2313, 2314.

A shrub to 2.3 m high; branches glabrous, straight, brownish or latericious, becoming brown, rugose, and profusely ridged; buds small, acute, appressed, glabrous, recurved at the tip; stipules lanceolate, very small, half the length of the petiole, at first pale green, becoming brown, promptly caducous; petioles 0.3 — 0.4 cm long; leaves spatulate-lanceolate or oblong, long-tapering toward base, obtusish or short-pointed at apex, somewhat silky beneath when young, becoming quite glabrous on both sides, glaucous beneath, on vegetative shoots acute, serrate, on flowering shoots entire, typically 3 — 8 cm long and ca. 1.5 cm broad, rarely to 4 cm broad, deeply and coarsely crenate-dentate (f. latifolia Nasar.), the midrib yellowish beneath, lateral veins 10 — 12 pairs at an angle of 60 — 80°, venation on both sides prominent; catkins soon after or almost together with the leaves, narrowly cylindric, dense at first, the staminate ca. 2.5 cm long and 0.6 cm broad, the pistillate to 6 cm long and 1 cm broad, slender, the stalk with 4 or 5 small foliaceous bracts at base, the rachis sparingly hairy; scales oblong-liguliform, obtuse, yellow, dark brown at apex, ciliate or beset with long white hairs; stamens 2, distinct, glabrous; anthers round, small, yellow, becoming brown; ovary ovoid-conical, ca. 1.8 mm long, glabrous or finely silky or merely at base puberulous, the
77 stipe twice the length of the gland; style evident, sometimes elongated, rather deeply cleft; stigma almost 2-parted, with straight lobes; capsule 4 — 6 mm long, glabrous, inflated, greenish-yellow. May. (Plate V, Figure 7).

Subalpine zone, shrub thickets, wood margins, and valleys of mountain streams and rivulets. — E. Siberia: Yen., Lena-Kol., Ang.-Say., Dau.; Far East: Kamch., Okh., Ze.-Bu. **Gen. distr.**: N. Mongolia. Described from the Baikal area (Angara, Tunkinskii Territory). Type in Leningrad.

Hybridizing with the following species: S. abscondita, arbuscula (?), brevijulis (?), dasyclados, viminalis coll.

45. **S. podophylla** Anderss. Mon. Salic. (1867) 142, tab. VII, f. 77; Ej. in DC. Prodr. XVI, 2, 246.

Apparently a fairly tall shrub, with straight pitch-dark glabrous branches; buds small, conical, flattened, acute, glabrous; stipules small, falcate; leaves 2.5 — 7.5 cm long, above the middle ca. 1.5 cm broad, liguliform-spatulate, short-acuminate, stiff, often green above even in drying, lustrous, intensely glaucous beneath, sinuate-crenate, the yellow midrib and the subparallel lateral veins riblike; catkins borne on a stalk with small leaflike bracts, in fruit ca. 2.5 cm long, rather dense; scales small, yellowish, obtuse, sparingly ciliate at apex; ovary fairly large, ovoid-conical, somewhat narrowed at base, latericious-brown, glabrous, the very short stipe barely exceeding the gland; style very short, yellow; stigma short, straight, entire or 2-parted. Fr. June.

Coppices. — E. Siberia: Ang.-Say. Endemic. Described from Irkutsk. Type in Leningrad.

Note. Allied to S. chlorostachya Turcz. from which it differs in the leaves that are enlarged upward and short-pointed, the small ciliate scales, and the glabrate ovary.

46. **S. abscondita** Laksch. in Sched. ad Herb. Fl. Ross. VIII (1914) No. 2471. — S. subphylicifolia Laksch. in sched.

A shrub; branches dusky brown, castaneous, or ferruginous-purple, glabrous, sublustrous, the annotinous mostly hoary; buds ovoid, yellowish, at length glabrous, lustrous; petioles pubescent, 0.6 — 1.2 cm long; leaf blades elliptic to oblong-elliptic, 3.5 — 7.5 cm long and 1.5 — 3.5 cm broad, irregularly erose-serrate, subentire at base and at apex, the upper surface dark green, pubescent at first becoming glabrous, sparingly hairy on the veins, the lower surface glaucous, covered with scattered crisped hairs; veins reticulate, prominent, the lateral 8 — 10 pairs, the midrib stramineous;
78 catkins before the leaves, sessile, with 2 — 3 foliaceous bracts at base, dense, the staminate ovaloid or oblong, 2 — 2.5 cm long and 1.3 cm broad, the pistillate 2 — 3 cm long and ca. 1 cm broad; scales liguliform or obovate, dark brown, covered with light-colored hairs, to 2.2 mm long and 0.6 — 1.2 mm broad; gland oblong; stamens 2, distinct, glabrous, 0.5 — 1 cm long; anthers oval; ovary 2.5 — 3.5 cm long, ovoid-conical, hoary-tomentose; style one-fourth the length of the ovary; stigmas oblong-ovate, straight, divergent; stipe half as long again to twice as long as the gland, up to 1.4 — 1.8 mm; capsule to 7.5 mm long. May.

River valleys, slopes, coppices, and wood margins. — E. Siberia: Yen., Lena-Kol., Ang.-Say., Dau. Endemic. Described from the vicinity of Chita. Type in Leningrad.

Hybrids with S. chlorostachya and S. dasyclados have been recorded.

Section 10. **ARBUSCULOIDEAE** Floder. in O. Holmb. Skandin. Fl. I, H. 1 (1931) 112. — Low or medium-sized shrubs; branches and buds short, more or less reddish-brown, glabrous; leaves relatively small, elliptic, crenate, finally glabrous, glossy, glaucous beneath, commonly exstipulate, 7 — 12-nerved; catkins sessile or borne on a short stalk with scalelike bracts; scales light-colored, almost monochromatic or darker at apex; stamens 2, with free glabrous filaments and yellow anthers; gland 1, posterior in staminate and pistillate flowers; ovary hairy, subsessile or short-stipitate; style evident.

1. Leaf blades 3.5 — 6 cm long, oblanceolate, gradually narrowed toward base, minutely ciliate-denticulate; catkin-scales liguliform, covered with long tangled hairs . . . . 51. **S. mezereoides** E. Wolf.
\+ Leaf blades short, obtusish at base, rounded or short-attenuate at apex; catkin-scales obovate, elliptic, or rounded-spatulate, more or less covered with straight hairs . . . . 2.
2. Ovary glabrous or sparingly hairy . . . . 3.
\+ Ovary covered with silky hairs or tomentose . . . . 4.
3. Ovary glabrous; leaves lanceolate, dilated toward apex, remotely serrate or subentire . . . . 50. **S. macilenta** Anderss.

\+ Ovary sparingly hairy; leaves elliptic or obelliptic, densely glandular-serrate . . . . 52. **S. tianschanica** Rgl.

4\. Ovary sessile or subsessile . . . . 5.

\+ Ovary stipitate, the stipe 4 times as long as the gland and about equaling the ovary . . . . 49. **S. leptoclados** Anderss.

79 5. Leaves obovate-oblong or broadly lanceolate, serrulate to subentire, revolute; pistillate catkins sessile, ca. 2 cm long, slender; scales acuminate; gland trapezoid . . . . 48. **S. brevijulis** Turcz.

\+ Leaves ovate, oblong, or lanceolate, distantly dentate to serrulate or subentire; pistillate catkins borne on stipe 2 — 3 cm long elongating in fruit to 5 — 7 cm; scales obtuse; gland filiform or clavate . . . . 47. **S. arbuscula** L.

47. **S. arbuscula** L. Sp. pl. (1753) 1018; Ldb. Fl. Ross. III, 622; Turcz. Fl. baic.-dah. II, 1, 119; Kryl., Fl. Zap. Sib. IV, 752. — S. *Waldsteiniana* Willd. Sp. pl. ed. 4, IV, 2 (1805) 679. — S. *prunifolia* Sm. Fl. brit. (1800—4) 1055. — S. *formosa* Willd. Sp. pl. ed. 4, IV, 2 (1805) 680. — Ic.: Fl. Dan. tab. 1055; Rchb. Ic. Fl. Germ. f. 1196—2000, 2006; Anderss. Monogr. t. VII, f. 145. — Exs.: Fries, Herb. norm, No. 61; Wimm. et Kr. Herb. Sal., 75 — 76.

A wide-spreading subarborescent upright shrub to 1—1.5 m high (v. *erecta* Anderss.) or else depressed or prostrate, with a short, often half-subterranean stem (v. *humilis* Anderss.); branches glabrous, lustrous, castaneous, yellowish-brown or greenish-yellow; buds glabrous, small, acute, reddish-yellow; wood under the bark not ridged; stipules semiovate or ovate-lanceolate, small, glandular-dentate, soon caducous; petioles short, glabrous; leaf blades 2—5 cm long and to 2 cm broad, ovate to broadly ovate (f. *ovalifolia* Anderss.) or oblong to lanceolate (f. *angustifolia* Anderss.), obtusish at apex and at base, remotely dentate to subentire (f. *subintegra* Toepff.), not blackening, dark green and glossy above, glaucescent or green with a yellow midrib beneath; lateral veins slender and rather inconspicuous; vesture none or young leaves with long hairs, mainly beneath; catkins with or after the leaves, the staminate subsessile, to 2.5 cm long, the pistillate on lateral branchlets, borne on a fairly long leafy-bracted stalk, at first 2—3 cm long, elongating to 5 — 7 cm, upright or spreading, dense; scales elliptic or obovate, obtuse and dark brown at apex, light at base, covered with crisped white hairs, in pistillate flowers covering the ovary up to the middle; stamens 2, distinct, glabrous; anthers golden-reddish; ovary thickish, ovoid-conical, silky, rarely glabrate, brownish or reddish (f. *lejocarpa* Görz), subsessile; style one-third the length of ovary, commonly deeply 2-parted; stigma with linear-lobes; gland 1, posterior, filiform or clavate, about as long as the
80 stipe. June. (Plate VI, Figure 5).

Subarctic region; alpine and subalpine zone; moss-and-lichen tundra and banks of mountain streams. — Arctic: Arc. Eur. (to Urals); European part: Kar.-Lap., Dv.-Pech., N. Urals; W. Siberia: Ob, Alt.; E. Siberia: Yen., Lena-Kol., Ang.-Say., Dau.; Caucasus: Cisc., W. Transc., Dag.; Centr. Asia: Dzu.-Tarb. **Gen. distr.**: Scand., Centr. and Atl. Eur. Described from Sweden. Type in London.

Note. Hybridizing with the following species: S. *arctica*, *argyrophylla*, *brevijulis*, *caprea*, *coaetanea*, *glauca*, *hastata*, *herbacea*, *lanata*, *lapponum*, *myrsinites*, *polaris*, *reticulata*.

It differs from the related S. phylicifolia in the subsessile ovary, light-colored almost monochromatic scales, longer style, and leaves broadest at or below the middle (not above the middle as in S. phylicifolia).

Economic importance. Leaves eaten by deer and domestic stock, and buds by arctic partridge and rock ptarmigan. Nectariferous. The bark contains up to 8 — 9% tannins.

48. **S. brevijulis** Turcz. Fl. baic.-dah. II, 2 (1854) 114.

A small, upright or prostrate shrub; branches tuberculate, dark red becoming gray, to 30 — 45 cm long, pubescent when young; buds small, appressed, sparingly hairy; stipules very small, to 5 mm long, ovate-lanceolate, serrate; petioles to 6 mm long, typically 1.7 — 2.8 cm long and 0.9 — 1.8 cm broad, rarely to 5.8 cm long and to 2.5 cm broad, obovate-oblong or broadly lanceolate, mostly obtuse or acuminate at apex, rounded or narrowed at base, slightly revolute, entire or serrulate, green above, glaucous beneath, at first sparingly hairy, finally glabrous or with short straight white cilia; veins prominent on both sides, the lateral 5 — 8 pairs; catkins (pistillate) after the leaves, lateral, sessile, with 2 silky scalelike bracts at base, ovoid, at length cylindric, in fruit to 3 cm long and 0.8 cm broad; scales obovoid, acuminate, dark brown at apex, lighter and long-hairy beneath; staminate flowers unknown; ovary sessile, tomentose, ovoid or ovoid-oblong; gland 1, inner, trapezoid, with a broad truncate upper margin, reaching the base of ovary; style ca. 1.5 mm long, sometimes
83 2-parted; stigmas 2-lobed. May — June. (Plate VI, Figure 6).

From the subalpine zone to the sandy shores of Lake Baikal. — E. Siberia: Ang.-Say., Dau. Endemic. Described from the sandy shores of Lake Baikal at the Turkinskie Springs. Type in Leningrad.

49. **S. leptoclados** Anderss. Monogr. Salic. (1867) 144, t. VII, f. 79; Ej. in DC. Prodr. XVI, 2, 242.

A low shrub; branches spreading, slender, shapely glabrous, lustrous, tarry-brown; buds obtusish, triangular, appressed, glabrous or covered with ferruginous hairs; stipules mostly lacking; leaves ca. 2.5 cm long, obovate-liguliform or lanceolate, above the middle to 1.5 cm broad, narrowed toward base, pointed at apex, entire or undulate-crenate, somewhat appressed hairy when young, becoming glabrous, sublustrous, stiff, dark green above, dull glaucescent beneath, flat, inconspicuously veined, lateral veins 9 — 12 (14) pairs; catkins sessile, ca. 2.5 cm long, the pistillate barely 4 mm broad, narrowly cylindric, cinereous-hairy, with few appressed bracts; scales obtusish, dark at apex, cinereous-hairy, enveloping the ovary; ovary ca. 2 mm long, ovoid, whitish or grayish from silky tomentum, the stipe nearly 4 times the length of the gland and about equaling the ovary; style the length of ovary, often cleft down to base; stigmas brownish, rather thick, divergent. June.

Riverbanks and coppices.— E. Siberia: Lena-Kol., Ang.-Say., Dau.; Far East: Ze.-Bu. (dubious). Endemic. Described from Transbaikalia. Type in Leningrad.

50. **S. macilenta** Anderss. Mon. Salic. (1867) 141 et tab. VII, f. 75; Ej. in DC. Prodr. XVI, 2, 245. — S. phylicifolia Cham. in Linnaea VI (1831) 542, non L.

PLATE IV. 1. Salix alatavica Kar. et Kir. – 2. S. Barclayi Anderss. – 3. S. hastata L. – 4. S. thamnifolia Pall. – 5. S. viridula Anderss. – 6. S. pyrolifolia Ldb. – 7. S. speciosa Hook. et Arn. – 8. S. krylovii E. Wolf.

A lower shrub; branches slender, glabrous, brownish-castaneous, somewhat tuberculate; buds triangular, acute, flat on the inside, appressed, castaneous, glabrous; petioles short; leaves lanceolate, to 2 — 3 cm long and ca. 0.7 cm broad, slightly dilated toward apex, thin at first, becoming stiff, commonly remotely serrate or subentire, small, apparently revolute, the upper surface light green, flat, distinctly veined, the lower intensely glaucous, glabrous or initially the yellow lustrous midrib subsericeous; catkins lateral, short-stalked, subtended by 2 — 4 narrow bracts (these resembling or smaller than foliar leaves), upright or spreading to drooping, the pistillate slender, cylindric, ca. 2.5 cm long and 2 — 5 mm broad, rather loose; scales liguliform-obovate or rounded-subspatulate, yellowish-brown, glabrous on the back, gray-ciliate on the margin, reaching above the ovary base; ovary ca. 2 mm long, narrowly conical, obtusish, glabrous,
84 brownish, the short stipe about equaling the gland; style short or moderate; stigma small, entire, thickish, pale.

Described by Andersson as from "Redovski land" (Chamisso), most probably from the Okhotsk Coast. Type apparently in Berlin.

Note. Related in any case to S. arbuscula L. and also to S. phylicifolia L., but its position is dubious and needs further study. It may possibly be a hybrid between S. arbuscula or S. phylicifolia and S. hastata or S. myrsinites. The possibility of kinship with another insufficiently clarified species of Andersson's, Salix fulcrata, cannot be excluded.

51. S. mezereoides E. Wolf in A. H. P. XXVIII, 4 (1911) 529 — 531; Kom. and Alis., Opred. rast. Dal'nevost. kr. I, 426.

A sparingly branched shrub; young shoots pubescent, the annotinous reddish-brown, glabrous, lustrous; leaves crowded at the ends of branches; stipules elliptic-lanceolate, more or less glandular-serrate, shorter than the sturdy hairy petioles; leaf blades at anthesis 3.2—6 cm long, to 1.5 cm broad, oblanceolate, acuminate or subobtuse, gradually narrowed toward base, remotely denticulate, the upper surface bright green, lustrous, the lower glaucescent, pubescent at first, becoming glabrous, the midrib pale fulvous beneath, lateral veins 7 — 14 pairs, pale violet, at an angle of 50 — 55°; catkins appearing about the same time as the leaves, pistillate unknown, staminate 1.2 — 2.2 cm long, sessile, with green white-silky bracts at base; scales liguliform, obtuse, brownish, darker at apex, covered with long tangled hairs, these at length becoming sparse; stamens 2, distinct, glabrous or with odd hairs at base, 4 times as long as the scale; anthers whitish-yellow; gland 1, posterior, or 2, anterior and posterior, oblong, 4-angled, half as long as the scale. June.

Mountain slopes. — Far East: Uss. Described from the Khabarovsk area. Type in Leningrad.

Note. This dubious species was described from undeveloped branches with only staminate flowers. Judging by the leaves, also incompletely developed, it may be concluded the species does not belong, in any case, to the section Hastatae where it was placed by E. L. Vol'f, but is more likely affiliated to S. fulcrata Anderss. and S. leptoclados Anderss. Additional investigation is needed on the site (the slopes of the Yana Range, N. of Khabarovsk).

52. **S. tianschanica** Rgl. in A. H. P. VII, 2 (1880) 471. — S. nigricans Rgl. Pl. Semenow, No. 984, non Sm.

A low much-branched shrub; branches reddish-brown, glabrous, lustrous; buds small, appressed, acute, yellowish-brown; leaves elliptic or obelliptic, glabrous on both sides, green above, pale green or glaucous beneath, closely glandular-serrate, prominently veined, lateral veins to 12 pairs; catkins
85 sessile, ebracteate, cylindric, the pistillate to 5 cm long and ca. 1 cm broad in fruit; scales ovate-orbicular, obtuse, brown or dark purple, ca. 2.7 mm long and 1 mm broad, glabrous on the back or hairy throughout; ovary 2.2—2.5 mm long, subsessile, greenish, sparingly hairy; stipe ca. 1 mm long, about equaling or slightly exceeding the gland, this 0.7—1 mm long; style ca. 1 mm long; stigmas ca. 0.5 mm long; capsule to 5—5.5 mm long. Fl. June; fr. July.

Subalpine zone of mountains. — Centr. Asia: T. Sh., Pam.-Al. Endemic. Described from Tien Shan. Type in Leningrad.

Note. The seriously damaged original bears an inscription by Lakschevitz: "S. phylicifolia L. vix differt." Dr. Goerz referred it to S. hastata L, but Regel's species differs from it in the hairy ovary. In any case, a dubious species, described from inadequate and badly preserved material. New material in perfect condition is needed from the natural habitat.

Section 11. **NIGRICANTES** Borr. ap. Hook. Fl. brit. (1830) 426. — A depressed or arboraceous shrub with more or less woolly dark branches; leaves petiolate, mostly oblong-elliptic, crenate-serrate, glaucescent beneath except for the green apex, blackening on wilting, 7—12 (—18)-nerved; catkins appearing with or soon after the leaves, sessile or stalked; scales light-colored, commonly dark at apex, acute, sparsely hairy; stamens 2, distinct, hairy; ovary long-stipitate, in pure species glabrous, with a long style.

1. A low or medium-sized shrub; branches more or less covered with scattered hairs; stipules small; leaves of medium size; the midrib glabrous or sparingly pubescent above, the lower leaf surface mostly glaucescent, glabrous or pubescent, the midrib and lateral veins not covered with white tomentum; pistillate catkins 3—5 cm long, borne on a short sparingly pubescent stalk; scales sparingly hairy; ovary conical, pricklelike, borne on a long glabrous stipe . . 53. **S. nigricans** Sm.

\+ A tall or arborescent shrub of the extreme northern forest belt; branches stouter, the young shoots white-tomentose; leaves and stipules larger; midrib always white-tomentose on both sides and often also the lateral veins beneath; pistillate catkins to 10 mm long, thicker (to 1.5 cm in diameter), recurved, subpendulous, looser at base than the preceding; scales dark at apex; ovary very large, thick, borne on a very long hairy stipe; style very long . . . . . 54. **S. borealis** Fries.

53. **S. nigricans** (Sm.) Enand. Salic. exs. III (1910) No. 103. — S. nigri-
86 cans Sm. in Trans. of Linn. Soc. VI (1802) 120 (♂ tantum); Ldb. Fl. Ross. III, 608 (ex p.); Shmal'g., Fl. II, 435 (ex p.); Kryl., Fl. Zap. Sib. IV, 748 (ex p.). S. nigricans campestris (Fr.) Anderss. Sal. Lapp. (1845) 59 ex p. — S. phylicifolia β L. Sp. pl. (1753) 1016. — Ic.: Fl. Dan. tab. 2553; Anderss. Mon. Salic., t. VII, f. 69; Enand. l. c. — Exs.: HFR, No. 1136, 2301, 2307; Enand. No. 103.

A much branched shrub or tree, 0.5 — 8 m high, in W. Siberia even higher; young branches reddish, gray-tomentose, becoming dark pubescent or glabrous, brownish-green to dark brown; buds long, recurved at tips, densely hairy; exposed wood in the pure species not striate; stipules mostly present only on vigorous shoots, semireniform, semiovate, or ovate-lanceolate, dentate, white-dotted; petioles commonly short, channeled, dilated at base, pubescent; leaf blades varying extremely in size, shape and vesture: elliptic, oblong, lanceolate, sometimes inequilateral, commonly cuneately narrowed toward base, thin, green or glaucous on both sides or dark green above, and green or glaucescent white-dotted green-tipped beneath, blackening, 2.5 — 10 cm long and 1.2 — 4 cm broad, commonly broadest about the middle, serrate, often concave or folded at the tip, heavily pubescent when young, with age glabrous or variously pubescent, the midrib greenish-yellow and more or less hairy but not tomentose; lateral veins 8 — 12 pairs, very prominent beneath, at an angle of 40 — 80°, the veins of third order reticulately anastomosing; catkins with leaves, lateral, with a short stalk and 1 — 4 small toothed bracts, the staminate 2 — 2.5 cm long, the pistillate to 7 cm in fruit; scales more or less dark-colored at apex, yellowish-green or rufescent at base, rarely monochromatic, light, hairy; stamens 2, hairy at base, 2 — 3 times as long as the scale; anthers yellow, becoming brown; gland 1, posterior, oblong; ovary subulate from an ovaloid base, both ovary and stipe glabrous (in hybrids hairy or stipe hairy and ovary glabrous; all forms with a densely hairy ovary should be regarded as hybrids mostly derived from crossing with S. cinerea); style 1 — 1.8 mm long; stigma 2-parted, with vertically divergent lobes; stipe 3 — 4 times as long as the pale posterior gland.

Numerous varieties and forms are described in local floras. End of April — May. (Plate VI, Figure 1).

Woods, coppices, wood margins, meadows, in the subarctic and forest belts. — European part: everywhere except Bl.; W. Siberia: everywhere; E. Siberia: Yen. **Gen. distr.**: Scand., Centr., Atl. and E. Eur. Described from Sweden. Type in London.

Note. S. nigricans var. jenisseensis F. Schmidt in Mém. Acad. Petrop. XVIII (1866) No. 211; N. J. Scheutz in K. Sv. vet. Akadem. Handl. Band 22, No. 10 (1888) 201 — this variety differs from the European
87 S. nigricans Sm. in having long-petioled leaves and longer catkins; style very long; capsule often hairy, partly glabrous. Reported for the banks of the Yenisei from Eniseisk to Dudinka.

Further observations should determine whether this is not a vicarious species.

Appears as often in hybrid as in pure form. Hybridizes with the following species: S. acutifolia, caprea, cinerea, dasyclados, glauca, livida, myrtilloides, lapponum, phylicifolia, purpurea, rosmarinifolia, triandra, viminalis s.l.

Economic importance. Suitable for tanning purposes, containing 6.08 — 13.7% tannins. Young branches and leaves are eaten by stock. The bark is used for coarse netting, the branches for brushwood.

54. **S. borealis** Fries, Mant. in Bot. Notis. (1840) 193; Ej. Summa veget. Scand. 206. — S. nigricana α borealis Anderss. in DC. Prodr. XVI, 2, (1868) 241. — Ic.: Fl. Dan. 1053. — Exs.: Fries, Herb. norm. VII, 63; Enand. Salic. Scand. No. 121 — 124 1/2.

An arboraceous shrub or tree to 6 — 8 m high; trunk to 15 cm in diameter, with rather dense round head and sturdy wide-spreading branches; young shoots white-tomentose; buds obtuse, long-hairy; stipules large; petioles stout, hairy; leaf blades large, 6 — 12.5 cm long and to 3 — 5 cm broad, ovate, elliptic, obovate-lanceolate, or oblong, commonly broadest about or above the middle, serrulate or sinuate-serrulate, often hairy, stiff, tomentose-hairy when young, glabrous or hairy in age on both sides or merely beneath, green on both sides or glaucescent beneath, the midrib always densely white-tomentose on both sides, lateral veins very prominent, 12 — 20 pairs, often as in S. caprea L.; catkins after or together with leaves, borne on a terminal elongated leafy-bracted white-tomentose stalk, to 10 cm long, commonly to 1.5 cm thick, spreading or subpendulous, loose toward base; scales mostly dark at apex, with long white hairs; ovary large, thick, cuneate, with a very long hairy stipe and a very long style, typically glabrous or more or less pubescent to subtomentose; gland oblong, emarginate, one-fourth to one-third the length of the stipe; capsule to 8 mm long. June. (Plate VI, Figure 2).

Northern section of the forest belt, forest-tundra, and the subalpine zone of mountains. — European part: Kar.-Lap., Dv.-Pech.; W. Siberia: Ob; E. Siberia: Yen. Described from Sweden. Type in Stockholm.

Economic importance. As for the preceding species.

88 Section 12. **CAPREA** Bluff. et Fingerh. Consp. Fl. Germ. II (1825) 565. — Trees or shrubs; leaves ovate, elliptic, or obovate-lanceolate, mostly with impressed veins and rugose above, prominently reticulate beneath, initially or also in age more or less hairy, mostly sinuate-serrate, rarely subentire; catkins mostly precocious, sessile, thick; scales particolored; stamens 2, mostly distinct; anthers yellow; gland 1, posterior in both sexes; carpel rather densely hairy, long-stipitate; style absent or very short; stigma short or oblong, with entire or divided lobes.

1. Exposed wood of 2- and 3-year-old branches without ribs or striations . . . . 2.
\+ Exposed wood with rather numerous long or short striations . . . . . . 9.
2. Lower side of grown leaves mostly covered throughout with gray or silky tomentum, or with short silky hairs . . . . 4.
\+ Grown leaves mostly glaucous, glabrous or covered with scattered hairs, the veins hairy or tomentose . . . . 3.
3. (See also stage 14). Leaves lanceolate, glabrous on both sides except for the hairy midrib beneath; lateral veins 15 — 20 pairs, anastomosing . . . . 68. **S. paracaucasica** Görz.
\+ Leaves ovate-oblong or obovate, narrowed toward base, the glaucescent lower surface covered with scattered hairs; lateral veins 10 — 12 pairs, arcuately upcurved at the margin to form wide loops; midrib subtomentose beneath . . . . 58. **S. Olenini** Nas.
4. Catkins appearing before the leaves . . . . 5.
\+ Catkins appearing with or after the leaves . . . . 7.
5. Catkins small, the staminate 2 — 3 cm, the pistillate 3 — 4 cm long, elongating in fruit to 6 cm; scales obovate, broad, obtuse; gland one-tenth the length of the stipe . . . . 56. **S. Hultenii** Floder.

\+ Catkins large, the pistillate to 10 cm long in fruit; scales acutish; gland one-fourth to one-third the length of stipe . . . . 6.

6\. Buds obtusish, glabrous; leaves to 11 — 17 cm long and 5 — 8 cm broad, mostly elliptic-oblong or ovate, gray-pubescent beneath, the margin irregularly sinuate-dentate; lateral veins 6 — 9 pairs, forming wide loops at the margin . . . . 55. **S. caprea** L.

\+ Buds acutish, pubescent; leaves ca. 3.5 — 7 cm long and 3 — 4 cm broad, mostly obovate-orbicular, the margin entire, revolute or sinuately glandular-dentate; lateral veins to 8 pairs, not forming
89 wide loops at the margin . . . . 57. **S. Raddeana** Laksch.

7\. Catkins subcoaetaneous or somewhat precocious, few, very short, the pistillate 2 — 3.5 cm long; expanding leaves flat, in age covered beneath with short cinereous hairs; catkin-scales pale brown or tawny . . . . 59. **S. coaetanea** (Hartm.) Floder.

\+ Catkins serotinous or subcoaetaneous, longer; expanding leaves curled . . . . 8.

8\. Catkins after or almost together with the leaves; emerging leaves involute, silvery-hairy, in age subsericeous-lustrous or silky grayish-green beneath or on both sides; catkin-scales rufous or dark brown; style to 2 mm long, glabrous; stigma 1 — 1.5 mm long . . . . 70. **S. pantosericea** Görz.

\+ Catkins after the leaves, the pistillate 4 — 6.5 cm long, elongating in fruit to 12 — 14 cm, very thick; emerging leaves strongly revolute, densely white-tomentose, at length glaucescent and gray-tomentose beneath; catkin-scales light brown, fuscous at apex; style and stigma ca. 0.5 mm long . . . . 61. **S. Kusnetzowii** Laksch.

9\. Young branches densely velutinous; leaves firm, subcoriaceous, tomentose or more or less hairy beneath . . . . 10.

\+ Young branches sparingly pubescent or sparingly hairy, after the first year glabrous; leaves thin, glabrous on both sides, commonly glaucescent beneath . . . . 14.

10\. Buds very large; catkins large, very close together on the shoots; leaves to 12 cm long and 4 — 5 cm broad, ovate-oblong, oblong-lanceolate, or broadly oblanceolate, the margin undulate or irregularly and coarsely sinuate with remote glandular teeth; catkin-scales rufescent-yellow, white-hairy; ovary densely covered with white silky hairs . . . . 60. **S. aegyptiaca** L.

\+ Buds moderate; catkins rather loosely spaced; leaves 4 — 12 cm long and 1 — 3 cm broad, obovate-lanceolate, lanceolate, obovate, obcordate, etc., the margin serrulate or dentate, undulate or erose; catkin-scales liguliform or lobed, monochromatic or particolored; ovary gray-tomentose . . . . 11.

11\. Tall or medium shrubs; young branches densely cinereous-velutinous, at length glabrous or hairy, commonly grayish-brown; leaves lanceolate or oblong-lanceolate, rather long and narrow, serrulate,
90 not rugose . . . . 12.

\+ Low or medium shrubs; young shoots covered with short hairs, becoming glabrous, reddish; leaves 0.8 — 4 cm long and 0.5 — 3 cm broad, rhombic-suborbicular, rounded-ovate, or oblong-obovate, always broadest in upper part, always rugose, the margin undulate, dentate or sinuate; stipules large . . . . 65. **S. aurita** L.

12. A tall, densely leafy shrub; leaves revolute when expanding, the lower surface softly sericeous, becoming cinereous-hairy . . . . 62. **S. cinerea** L.

\+ Shrubs of moderate height; leaves flat when expanding, more or less silky-tomentose beneath . . . . 13.

13. Grown leaves broadly lanceolate, ca. 8 cm long and to 2.2 cm broad, hairy beneath, the hairs fairly long and straighter than in S. cinerea; lateral veins about 8 pairs . . . . 63. **S. alifera** Görz.

\+ Grown leaves spatulate-lanceolate to lanceolate, ca. 4 — 6 cm long and 1 — 2 cm broad, turning brown, farinaceous, greenish-gray, short-hairy lateral veins 10 — 13 . . . . **S. fuscata** Görz.

14. (See also S. paracaucasica Görz.). Exposed wood profusely striate; leaves at emergence slightly curled, full grown broadly lanceolate; lateral veins 10 — 15 pairs . . . . 67. **S. Palibini** Görz.

\+ Exposed wood with scattered or few striations; leaves at emergence flat, at length obovate-oblong or oblanceolate; lateral veins 13 — 18 pairs . . . . 15.

15. Grown leaves commonly elliptic-oblong, oblanceolate, or obovate, green or glaucous beneath; lateral veins not forking at the margin; catkin-scales long-hairy; ovary glabrous or gray-tomentose . . . . 66. **S. caucasica** Anderss.

\+ Grown leaves commonly oblanceolate or elongate-obovate; catkin-scales covered with few straight hairs; ovary glabrous . . . . 69. **S. daghestanica** Görz.

55. **S. caprea** L. Sp. pl. (1753) 1020; Ldb. Fl. Ross. III, 609; Turcz., Fl. baic.-dah. II, 2, 110; Anderss. Mon. Salic., 75 et tab. V, fig. 45; Ej. in DC. Prodr. XVI, 2, 222; Shmal'g., Fl. II, 436; Kryl., Fl. Zap. Sib. IV, 746. — Neetopix Caprea Rafin. Alsol. Am. (1838) 14. — Caprea vulgaris Opiz, Sezn. (1852) 25. — Ic.: Rchb. Ic. Fl. Germ. XI, t. 577, f. 1224. — Exs.: HFR No. 892, 893, 893a.

91 A fast-growing tree of medium size, 6 — 10 m high and to 0.75 m in diameter, or a fairly low arboraceous shrub; bark smooth, greenish-gray, often splitting near the base of the trunk; exposed wood smooth, without striations, reddening; branches stout, spreading, gray-pubescent when young, at length dark, brown or gray, knotty, brittle; buds very large (especially the flower buds), glabrous, brown, ca. 5 mm long and 3 mm broad; stipules reniform, 4 — 7 mm long, serrate and lobed, soon caducous; petioles to 2 cm long, sturdy, dilated toward base; leaves varying in size and shape: ovate, suborbicular (f. rotundata Anderss.), subcordate (f. subcordata Anderss.), elliptic (f. elliptica Kern.), oblong-lanceolate, obovate-oblong (f. obovato-oblonga Anderss.), or rarely lanceolate (f. lanceolata Anderss.), to 11 — 18 cm long and 5 — 8 cm broad, the margin hairy, irregularly dentate, the upper surface glabrous, rugose, dark green, the lower gray-tomentose, rarely diffusely puberulent or glabrate, with a lurid network of veins (f. atomochlena Nasar.); lateral veins 6 — 9, forming wide round loops at the margin; network of veins prominent, with large alveoles; midrib and lateral veins mostly densely hairy; expanding leaves flat; hairs on the lower surface recurved; young leaves silky-pubescent; catkins precocious, dense,

large, the staminate sessile, subtended by few small bracts, 5—6 cm long and 1.6—2 cm broad; pistillate short-stalked, numerous, littering the soil when shedding, in fruit to 10 cm long, the rachis pubescent; scales lanceolate, blackish or dark brown at apex, covered with long white hairs; stamens 2, glabrous (in f. borealis Enand. hairy), 2—3 times as long as the scale; anthers yellow; ovary ovoid-conical, villous-tomentose, the stipe one-half to two-thirds the length of ovary; style short or very short, yellow as the stigma lobes; gland 1, posterior, one-third as long as the stipe. Fl. April; fr. May. (Plate VI, Figure 3).

Slopes, wood margins, coppices, and mixed woods. Throughout the Soviet Union except the Arctic Region and the alpine zone. — **Gen. distr.**: Eur. and Centr. Asia. Replaced in Korea, Japan, and North America by related species. Described from Europe. Type in London.

Hybridizing with the following species: S. acutifolia, arbuscula, aurita, cinerea, coaetanea, dasyclados, glandulifera, hastata, Krylovii, lanata, lapponum, livida, myrtilloides, nigricans, phylicifolia, purpurea, pyrolifolia, rorida, rosmarinifolia, rossica, schugnanica, silesiaca, viminalis s.l., xerophila.

Economic importance. An outstanding early nectariferous plant. Bark is used as tanning agent (tannin content 5.24—13.1%) and for production of a black dye. Employed in popular medicine as an astringent remedy for scurvy and fever. The foliage is used as feed for sheep and goats (hence the Russian name "iva koz'ya" [from "koza," goat]) and as an adulterant for tea.
92 The twigs are used for charcoal making and for fuel; they are not suitable for wickerwork. The wood is reddish or yellowish-brown, shiny, rather firm and tough; harder, heavier, and more tensile than that of many other European species of willow; it splinters easily. It can be used as firewood, as a source of charcoal, as a component of gunpowder, and as construction timber. The European form of this willow is propagated only by seed. Cuttings are difficult to root. The plant is frost-resistant. It consolidates the soil and is used for planting on slopes. Its hybrids with S. viminalis s.l. are distinguished by strong growth and are used as rootstocks for weeping varieties. In Far East Asia the species gradually loses its typical features. In Kamchatka, the related species S. Hultenii Floder. occurs together with S. caprea.

56. **S. Hultenii** Floder. in Arkiv för Botan. 20 A, No. 6. (1926) 51; Hultén, Fl. of Kamtch. II, 14. — S. caprea ssp. Hultenii Kom., Fl. Kamch. II (1927) 11. — S. caprea β orbicularis Anderss. in DC. Prodr. XVI, 2 (1868) 223.

A tree 5—6 m tall; branches spreading, at first pubescent, becoming glabrous, thicker and more angled and brittle than in the preceding species; buds ovoid, obtuse, more rounded, castaneous, glabrous; stipules none; petioles more slender, ca. 2.5 cm long; leaves obovate, orbicular, ca. 5 cm long and 4—5 cm broad, entire, thin, green, smooth and glabrous above, finely tomentose or covered with appressed gray hairs and scarcely rugose beneath; midrib glabrous; lateral veins 9—13, at first heavily pubescent; catkins borne on a short stalk subtended by bracts, stoutly cylindric, shorter than in the preceding, the staminate 2—3 cm long, ca. 1.8 cm broad, the pistillate 3—4 cm long, elongating in fruit to 6 cm; scales obovate, broader and more obtuse than in the preceding, dark brown, covered with long white hairs; stamens hairy at base; anthers ovoid; ovary smaller, conical-subulate, finely sericeous, becoming glabrate, the very slender stipe 10 times the length of the gland; style commonly longer than the almost 2-parted stigma.

Birch and spruce forests, wood margins, and glades. — Arctic: An.; Far East: Kamch. **Gen. distr.**: Sakh., Japan. Described from the vicinity of Petropavlovsk in Kamchatka. Type in Stockholm.

57. **S. Raddeana** Laksch. in sched. 1913 (emend. Nasar.) in Addenda IV, p. 539. — S. villosa Siuz. (in sched., nom. tantum).

Apparently a tall shrub or tree; young shoots densely gray-pubescent, 2- and 3-year-old glabrous, dark brown; buds large, appressed, acute, ribbed, dark brown, pubescent-pilose; stipules large, 6—7 mm long, glabrous or pubescent, acute; petioles 5—10 mm long, hairy; leaf blades firm, coriaceous, ca. 3.5—7 cm long and 3—4 cm broad, obovate-orbicular, obovate, or obovate-elliptic, obliquely short-acuminate, rounded or
93 narrowed at base, entire and revolute or sinuate and coarsely glandular-toothed, dark green above, softly velutinous-tomentose beneath; young leaves covered with white to silvery tomentum; lateral veins to 8 pairs, prominent on both sides, not forming wide loops at the margin as in S. caprea; catkins precocious, sessile, with sericeous scalelike bracts at base; scales ovate-elliptic, acuminate or subobtuse, dark brown, light-colored at base, ca. 2.5 mm long and 0.6—0.8 mm broad, covered with long gray hairs; ovary narrowly conical, long-tapering at the top, to 3.5—4 mm long, gray sericeous-pilose; style glabrous, ca. 1 mm long; stigmas brown, erect or spreading, 4-parted, ca. 1 mm long; stipe hairy, 2—2.5 mm long; gland 1, posterior, oblong, one-fourth to one-third the length of stipe, ca. 0.7 mm long; capsule to 6.5 mm long. Fl. May; fr. June.

Coppices and slopes. — E. Siberia: Dau.; Far East: Ze.-Bu., Uss. **Gen. distr.**: Manchuria. Described from the Far East. Type in Leningrad.

58. **S. Olenini** Nas. sp. nova in Addenda IV, p. 540.

A tall shrub or tree; young branches densely gray-tomentose, 3- and 4-year-old reddish, glabrous; buds ovoid to semiconical, pointed, ribbed, to 5 mm long and 3 mm broad, gray-tomentose; petioles 8—10 mm long, hairy-tomentose; stipules none; leaves 5—8 cm long and 2.5—4.5 cm broad, stiff, brittle, ovate, tapering at both ends or obovate narrowed toward base, remotely and coarsely serrate, the teeth gland-tipped, the upper surface darkish green, glabrous except for the white-pubescent and therefore conspicuous veins, the lower surface glaucous or glaucescent, covered with light appressed variously directed hairs; midrib subtomentose; lateral veins 10—12 pairs, prominent on both sides, at an angle of 40—50°, at the margin arcuately upcurved toward apex; network of veins forming wide loops; flowers unknown.

River valleys. — E. Siberia: Lena-Kol. Endemic. Described from Zhigansk. Type in Leningrad.

Note. Belonging to the same section as the preceding species. Somewhat resembling S. caprea L. but distinguishable by the heavy tomentum on young shoots, buds, and petioles. Distinguishing characters of the leaves are the glaucous lower surface, the scattered fine hairs, and the coarsely serrate-glandular margin.

59. **S. coaetanea** (Hartm.) Floder. in Botan. Notis. (1930) 331. — S. caprea γ coaetanea. Hartm. Handb. Scand. Fl. ed. 3 (1838) 236. — S. caprea sericea arborescens Anderss. Mon. Salic. (1867) 78. —
94 Ic.: Botan. Notis. (1930) 333.

A handsome tree to 3 — 5 m high, with short shoots; annotinous branches spreading, easily breaking off at their base, knotty, dark brown, sparsely cinereous-hairy; young branches densely coated with short cinereous hairs; flower buds large, broadly obovoid to subspherical, commonly with lateral ribs, later smooth; stipules small, narrow, tomentose-pilose, caducous, often absent; petioles ca. 5 (—8) mm long; leaves 5 — 6 cm long and to 3 cm broad, obovate, mostly narrowed toward base, obtuse at apex, the upper surface smooth, densely hairy, the straight appressed hairs parallel to lateral veins, these 5 — 8 (—10) pairs, cinereous-hairy, the lower surface prominently veined, covered with short cinereous hairs; catkins approximately together with or somewhat before the leaves, few, 1 or 2 per branchlet, the staminate 1.5 — 2.5 cm long, the pistillate 2 — 3.5 cm long and to 1.5 cm broad, loose, the rather short stalk with 2 or 3 small bracts; scales liguliform, obtuse or rounded at apex, ca. 2.5 — 3 mm long, pale brown or tawny, rather sparsely covered with short slender wavy hairs; gland 1, posterior, entire; stamens 2, mostly distinct, 7 — 8 mm long, crisp-hairy at base; anthers yellow; ovary ovoid-conical, slender, obtusish, very densely covered with more or less flexuous whitish hairs, these at length thinning out; style obsolescent or ca. 2 (—4) mm long, often bifid, densely hairy; stigma mostly with spreading broadly parted lobes; stipe ca. 2 (—4) mm long, densely covered with cinereous hairs; capsule 4 — 6 (—8) mm long. May — June.

Woods and wood margins. — European part: Kar.-Lap., Dv.-Pech.; W. Siberia. **Gen. distr.**: Scand. Described from Sweden. Type in Stockholm.

Note. Hybridizing with the following species: S. arbuscula, aurita, caprea, lanata, lapponum, livida, xerophila.

Catkins often abortive; flowers rather frequently hermaphrodite (Floder.).

60. **S. aegyptiaca** L. Cent. pl. I (1755) 33; Floder. in Arkiv för Bot. Bd. 25 A, No. 11 (1933) 1 — 44 cum tab. — S. calaf Hasselqu. ap. L. ex Hasselqu. It. 'palaest. (1757) 115. — ? S. nitida J. F. Gmel. Syst. Nat. II (1791) 74. — S. phlomoides M. B. Fl. taur.-cauc. II (1808) 415; III (1819) 628. — S. Medemii Boiss. Diagn. pl. or. I, VII (1846) 100. — S. cinerea β Medemii Boiss. Fl. Or. IV (1879) 1189. — S. sp. (Korolkovi) Rgl. in
97 A. H. P., VI, 2 (1880) 468. — S. pseudomedemii E. Wolf, ibid. XXVIII, 3 (1909) 307. — S. cinerea var. cuneata et var. erosa E. Wolf in Izv. Lesn. Inst. XIV (1906) 189 — 191. — Ic.: E. Wolf, l. c.; Floder., l. c. — Exs.: Kotschy, Pl. pers. austr. No. 113; Toepff., No. 230, 230a, 432, 473, 474, 584, 585.

(95)

PLATE V. 1. Salix fuscescens Anderss. – 2. reptans Rupr. – 3. S. lanata L. – 4. S. glauca L. – 5. S. Bebbiana Sarg. – 6. S. cinerea L. – 7. S. chlorostachya Turcz. – 8. S. phylicifolia L. – 9. S. aegyptiaca L.

A large shrub; branches stout, vigorous, commonly gray or yellow tomentose-pilose to black-velutinous; exposed wood striate; buds approximate, to 8—9 mm long and to 5—6 mm thick, obtuse, dark castaneous or rufescent, densely gray-hairy or woolly, rarely puberulous; stipules large, stalked, reniform or semicordate, rarely inequilateral-lanceolate, palmately nerved, glandular-denticulate, one-third to half the length of petiole; petioles 0.8—1.2 cm long, velutinous, strongly dilated toward base; leaves to 15 cm long and 4—5 cm broad, elliptic, obovate, oblong-elliptic, oblong-lanceolate, or broadly oblanceolate, nearly always broadest about the middle, acuminate, at base arcuate or cuneately long-tapering, the margin coarsely sinuate-erose and glandular-dentate, the teeth longer toward the leaf base, the upper surface dark green, glabrous, the lower glaucescent or grayish-green, velutinous or canescent or merely hairy on the nerves and then glabrate elsewhere; upper leaves more densely hairy; lateral veins 13—18 (—20) pairs, slender, set at an angle of 60—75° and forming loops at the margin; veins of third order evident; emerging leaves yellowish- to whitish-tomentose, revolute; catkins numerous and closely spaced, coaetaneous, dense, thick; pistillate to 3—4.5 cm long and ca. 1.2 cm broad, borne on a long leafy-bracted stalk, the bracts crenulate-serrulate, silky-tomentose; staminate subsessile; scales liguliform, obtuse, rarely lobed or acute, rufescent-yellow, dark at apex, with long white hairs; stamens 2, distinct, glabrous or at base crisp-hairy; anthers oblong, yellow; ovary ovoid-conical, 6—8 mm long, densely whitish-sericeous; style and stigma short; stigma brown, with 4 spreading lobes; stipe half the length of scale, 2—4 as long as the entire or rarely 2-lobed gland. Fl. March—April. (Plate V, Figure 9).

Cultivated in gardens of the Crimea, Transcaucasia, Soviet Central Asia, [former] Lower Volga Territory, Ukraine, and possibly also growing wild. — European part: L. V., Bl., Crim.; Caucasus: Dag., W., S., and E. Transc., Centr. Asia: Kara K., Irt., Amu D. **Gen. distr.**: Med., Iran.

Hybridizing with S. caprea, caucasica, cinerea, daghestanica.

Note. Needs further field study, as species later described, such as S. phlomoides M. B., S. Korolkowi Rgl., S. pseudomedemii E. Wolf, S. cinerea erosa E. Wolf., S. cinerea cuneata E. Wolf., are on the whole very closely related to S. aegyptiaca L.; sometimes
98 differing slightly, chiefly in such characters as shape, general outline, and vesture of the leaves. They represent apparently various hybrid combinations of the three principal species: S. caprea, S. aegyptiaca, and S. cinerea which either grow wild in the Caucasus and in Soviet Central Asia, or have been cultivated there for a very long time.

Economic importance. In the East the flowers are made into an infusion called "makhalar" which is used in treatment of facial rash; as a tincture they are used against gastric ailments. In Azerbaijan, where the plant is known as "bed-myusk" (musk willow), the flowers, and apparently also the leaves, find medicinal applications (oral communication by Grossgeim).

61. **S. Kusnetzowii** Laksch. in sched. 1909; Görz in Grossg., Fl. Kavk. II (1930) 9 et in Fedde Repert. sp. nov. XXXVI (1934) 231 (295).

A shrub, apparently fairly tall; exposed wood without striations; branches stout, in age ranging from rich brown to dark brown, glabrous, when young tomentose-pilose with white hairs; buds yellowish-brown or fulvous, glabrous or more or less pubescent, to 5 mm long and 4 mm broad, subacute; stipules only at the ends of vigorous shoots, very small, semicordate, densely hairy, sometimes tomentose as the midrib beneath; leaf blades oblong, obovate, or rarely narrowly elliptic, 7—13 cm long and 3—4.2 cm broad, 3—3.5 times as long as broad, the margins entire or irregularly dentate or undulate, somewhat revolute; expanding leaves strongly involute, densely white-tomentose; grown leaves dark green whitish-veined above, more or less glaucescent and grayish-tomentose beneath with recurved hairs; veins very prominent beneath, rather insignificant above; lateral veins 10—12 pairs; catkins after the leaves, borne on a leafy-bracted hairy stalk to 2 cm long; staminate ovoid, to 3 cm long; pistillate cylindric, 4—6.5 cm long and 1.5 cm broad, in fruit to 12—14 cm long and 2.5 cm broad, flexuous, loose, interrupted below; rachis densely hairy; scales to 2—3 mm long, liguliform, light brown, sometimes fulvous at apex, more or less hairy, the staminate more heavily; stamens 2, distinct, hairy at base; anthers oblong, yellow; ovary ca. 5 mm long, white-tomentose, ovoid-conical, the stipe 3—4 mm long; style ca. 0.5 mm long, reddish-brown; stigma with 4 spreading lobes, ca. 0.5 mm long, reddish-brown like the style; gland interior, oblong, ca. 0.6 mm long; capsule valves helically twisted after dehiscence. Fl. mid-June; fr. June—July.

Subalpine coppices, birch and beech woods, at 1800—2400 m. — Caucasus: Cisc., Dag., W. and E. Transc. Endemic. Described from the Caucasus. Type in Leningrad.

Note. Old collections of this species, from the years 1825—1835, were partly designated in the herbarium as S. phlomoides.

99 62. **S. cinerea** L. Sp. pl. (1753) 1021; Ldb. Fl. Ross. III, 607; Shmal'g., Fl. II, 436; Kryl., Fl. Zap. Sib. IV, 744. — Vimen cinerea Rafin., Alsol. Am. (1838) 14. — V. acuminata Rafin., ibid., 13. — Caprea cinerea Opiz, Sezn. (1852) 25. — Ic.: Rchb. Ic. fl. Germ. XI, tab. 576, f. 1222, 1223; tab. 579, f. 1227. — Exs.: HFR No. 940, 2266, 2267, 2268, 2460.

A shrub to 5 m tall, with stout rather brittle leafy branches; young and 2-year old branches densely coated with dark, sometimes almost black velvety tomentum; exposed wood with striations to 1.5 cm long; buds divergent, flattened, obtuse, brown, gray-pubescent, to 4 mm long and 2 mm broad; stipules reniform or semicordate, dentate; petioles short, pubescent; leaves obovate, almost subulately short-acuminate (f. obovata Gaud., f. aquatica Sm.), obovate-lanceolate (f. longifolia Anderss.), lanceolate and long-acuminate (f. spuria Wimm.), or narrowly lanceolate (f. angustifolia Döll.), dingy green with impressed veins above, grayish-green and prominently veined beneath, covered on both sides with a short tomentum, more heavily beneath, rarely glabrate on both sides (f. denudata Tepl., f. deserticola Görz., pro spec.), serrulate or more or less coarsely erose-serrate, 4—12 cm long and 1.3 cm broad, at emergence revolute and softly sericeous; lateral veins 10—16, at an angle or 60—80°; veins of third order prominent, forming narrow alveoles; catkins precocious or subcoaetaneous, dense, subsessile, subtended by small bracts, slender, ca. 2 cm long, staminate ovoid, pistillate cylindric, 3—4 cm long; scales spatulate, brown, at apex blackish, long-hairy; stamens 2, distinct, 2—3 times as long as the scale; anthers golden-yellow, turning brown; gland 1, posterior, oblong, recurved at apex; ovary elongate-conical, gray-tomentose; style short, sometimes cleft; stigma with erect or divergent lobes; stipe long, to two-thirds the length of ovary. Fl. April; fr. May. (Plate V, Figure 6).

Swampy alder groves and grassy bogs, canals, and wet mixed woods. — European part: from the N. forest limit to Bl., L. Don, and L. V.; W. Siberia: almost everywhere, except Arc. Sib.; Caucasus: E. Transc. (N. Azerbaijan); Centr. Asia: Ar.-Casp., Balkh., Kyz. K. **Gen. distr.**: Europe, Bal.-As. Min. Described from Europe.

Hybridizing with S. *acutifolia*, *arbuscula*, *aurita*, *caprea*, *dasyclados*, *lapponum*, *livida*, *myrtilloides*, *nigricans*, *phylicifolia*, *purpurea*, *rosmarinifolia*, *tenujulis*, *triandra*, *viminalis* s.l., *xerophila*.

Economic importance. The principal source of willow bark for tanning, as it yields the thickest bark, convenient in use. Average contents of TH:
100 9.2—11.1, HTH 7.7—12.5; quality grade 46.6—54.2. Difficult to propagate by cuttings. Needs peaty soil. Nectariferous. The fragrant bark is used for tanning of morocco and glove leather and other delicate hides. The twigs are used for charcoal, firewood, coarse wickerwork, and brushwood. The foliage provides feed for sheep and goats.

63. ***S. alifera** Görz. in Grossg., Fl. Kavk. II (1930) 8.

Shrub, apparently of medium size; exposed wood striate; branches cinereous-brown, densely velutinous when young, becoming glabrous or puberulous; buds small, appressed, pubescent; stipules large, semi-reniform or semicordate, dentate, hairy, palmately nerved; petioles 3—4 mm long, hairy; leaves broadly lanceolate, to 8 cm long and 2.2 cm broad, dull green above, hairy beneath, the hairs rather long and straighter than in S. *cinerea* L.; expanding leaves flat, initially subsericeous-tomentose beneath; lateral veins about 8 pairs. Flower unknown.

Caucasus: S. Transc. (near the Turkish frontier between the borders of Soviet Armenia and Bayazit [Agri] Province).

64. ***S. fuscata** Görz in Grossg., Fl. Kavk. II (1930) 7.

Shrub, apparently of medium size; wood striate, turning reddish; branches appressed to trunk or obliquely ascending (as in S. *cinerea* L.), densely leafy, when young grayish-velutinous or pubescent, the annotinous glabrous, the old grayish-brown, rugose, profusely ribbed; buds small, cinereous-brown, pubescent; stipules small, semicordate, glandular-dentate, hairy; petioles 4 — 6 mm long, base broadened, hairy, young leaves whitish-tomentose, flat when expanding, in age grayish-green or dark green, glabrate except the white-hairy midrib or covered with scattered short hairs above, greenish-gray, farinaceous and short-hairy beneath, turning brown in drying, stiff, lanceolate or spatulate-lanceolate, narrowed at both ends; terminal leaves sharply glandular-serrate, the lower revolute, remotely denticulate, 4—6 cm long and 1.2—2 cm broad, the midrib and the 10—13 pairs of lateral veins very prominent; veins of third order forming an alveolate network; flowers unknown.

Caucasus: S. Transc. (border of Turkey and Iran), whence described. Type in Leningrad.

Note. R. Görz (or the author of "The Flora of the Caucasus"?) assumes that this willow "is apparently closely related to S. pseudomedemii E. Wolf." In our own view, judging by the authentic specimens, it is undoubtedly akin to S. cinerea L. or some other species of the same section. The characters of S. cinerea L. predominate in this
101 plant and the general aspect is reminiscent of the European and Central Russian hybrid S. cinerea × nigricans, even though S. nigricans Sm. does not occur in the Caucasus.

65. **S. aurita** L. Sp. pl. (1753) 1019; Ldb. Fl. Ross. III, 610; Shmal'g., Fl. II, 436; Kryl., Fl. Zap. Sib. IV, 755. — S. rugosa Seringe, Essai (1815) 18. — Usionis aurita Raf. Alsol. Am. (1838) 14. — Caprea aurita Opiz, Sezn. (1852) 25. — Ic.: Rchb. Ic. fl. Germ. t. 575, f. 1220. — Exs.: HFR No. 2271, 2461; Fries, Herb. norm. No. VII, 60; Wimm. et Kr. Herb. Salic., 25, 99, 143; Coll. Sal. 33—36.

A shrub to 0.5—2 m high, with short spreading branches; young branches pubescent, the annotinous glabrous, reddish-brown, the old dark; buds small, ovoid, reddish, glabrous or puberulous; exposed wood striate; stipules always present, reniform or cordate, dentate, persistent till fall; petioles short, hairy; leaf blades 0.8—4 cm long and 0.5—3 cm broad, commonly broadest above the middle, rhombic-suborbicular (f. rhomboidalis Wimm.), suborbicular-obovate, oblong-obovate, or rarely oblong-lanceolate (f. spathulata Wimm.), mostly folded at apex and cuneate at base, sometimes cordate at base (f. cordifolia Wimm.), the margin revolute, deeply dentate, denticulate, or sinuate, often wavy, the upper surface rugose, with impressed veins, the lower gray-tomentose or pubescent, closely and prominently reticulate-veined; catkins precocious or subcoaetaneous, dense; staminate sessile, short-ovoid, 1—2 cm long; pistillate borne on a short scaly-bracted stalk, in fruit to 2.5—3 cm long; catkin-scales liguliform, hairy, plain light brown or darker at apex; stamens 2, distinct, hairy, 3—4 times as long as the scale; anthers yellow; gland 1, posterior, oblong; ovary ovoid-subulate, long-stipitate, the stipe about equaling the ovary and 2—3 times the length of the solitary posterior gland, obtusish, sericeous; style obsolescent; stigma capitate, entire or bifid, light-colored. Fl. May; fr. June. (Plate VI, Figure 4).

Grassy marshes; wet coppices; deciduous and mixed woods. — European part: all except Transv., L. V., Crim., and Cisc. **Gen. distr.**: most of W. Eur., except Med. Described from Sweden. Type in London.

Hybridizing with S. caprea, cinerea, coaetanaea, hastata, lapponum, livida, myrtilloides, nigricans, phylicifolia, purpurea, rosmarinifolia, viminalis s. l., xerophila.

Economic importance. Nectariferous; suitable for tanning, the tannin content being 11%.

66. **S. caucasica** Anderss. Monogr. Salic. (1867) 68, tab. IV, fig. 42. —
102 S. silesiaca β caucasica Anderss. in DC. Prodr. XVI, 2 (1868) 219. — Ic.: Rchb. Ic. Fl. Germ. XI, 1219. — Exs.: Wimm. et Kr. Herb. Sal. No. 1, 3, 26, 27; Toepff. No. 321.

Shrub to 1 — 2 m high, with arched-recurved or erect glabrous dark brown or gray branches; young branches pubescent; wood with scattered indistinct striations; buds triangular or ovoid, appressed, obtuse, brown, at first pubescent, becoming glabrous; stipules reniform or semicordate, dentate, glabrous, persistent, rather large, shorter than petiole; leaves obovate or elliptic-oblong to oblanceolate, in f. abchasorum Görz broadly lanceolate, rounded at base, short-acuminate at apex, 5 — 7.5 cm long and 2 — 3 cm broad, thin, irregularly dentate, glabrate on both sides, monochromatic or beneath glaucescent, quite glabrous on the veins, slightly blackening in drying; expanding leaves flat; catkins precocious or coaetaneous, narrowly cylindric, rather dense, on lateral shoots, 4- or 5-bracted; staminate erect, 2 — 3 cm long; pistillate spreading to pendulous, loose, long stalked, in fruit to 5 cm long; rachis hairy; scales obovate, 0.8 — 1.2 mm long, loose, dark at apex, with few straight rather long hairs (not villous); stamens 2, glabrous, distinct (in f. heterandra Dode (pro spec.) stamens 2 — 5, more or less connate); anthers yellow, before dehiscence red or reddish; ovary ca. 2 — 4 mm long, narrowly conical, pointed, glabrous, green, long-stipitate (f. lejocarpa Anderss.) or with stipe not exceeding 1 mm, gray-tomentose (f. hebecarpa Anderss.), the stipe 2 — 4 times the length of the gland; style evident, ca. 0.5 mm long, cleft, rufescent like the long parted stigma, the stigmatic lobes spreading. Fl. May; fr. June.

Wet alpine and subalpine meadows, coppices, and coniferous and mixed woods. — Caucasus: W. Transc. (up to 1,850 m). **Gen. distr.**: Centr. and Atl. Eur., Med. Described from the Caucasus. Type in Leningrad.

Hybridizing with S. aegyptiaca, caprea, and Gmelini.

Economic importance. Nectariferous; suitable for tanning purposes and for woodwork.

67. **S. Palibini** Görz in Fedde, Repert. Beih. Bd. LII (1928) tab. 3 et in Grossg., Fl. Kavk. II (1930) 7.

A shrub, apparently of medium size; young branches profusely hairy, becoming quite glabrous; bark dark brown or olivaceous-brown, dull; wood with numerous striations; stipules rather well developed, ovate or broadly subcordate, 4 — 8 mm long, nearly equaling the petiole, glabrous, glandular-dentate, palmately nerved; petioles 4 — 6 mm long, glabrous; leaves 6 — 9 cm long and 2 — 2.3 cm broad, broadly lanceolate (1 : 3 — 4), more or less rounded at base, short-cuspidate at apex, when expanding revolute, slightly pubescent when young, becoming quite glabrous on both sides, dark green or green above, glaucous or whitish-glaucescent beneath, more or less
103 erose-dentate; lateral veins 10 — 15 pairs, at an angle of 60° or more, prominent beneath, less so above; reticulation rather obscure. Flowers unknown.

Caucasus: Cisc., W. Transc., whence described. Endemic. Type in Leningrad.

Note. Differs from the typical S. caucasica Anderss. only in the numerous pronounced striations on the exposed wood, glabrous branches, and the broadly lanceolate leaves, the expanding leaves but slightly revolute.

68. **S. paracaucasica** Görz in Fedde Repert., Beih. Bd. LII (1928) 28, tab. 2 et in Grossg., Fl. Kavk. II (1930) 7. — S. caucasica Laksch. in sched. (ex p., non Anderss.).

A shrub of medium size; branches slender, the young rather densely hairy, the annotinous glabrous, fulvous, sublustrous; wood without striations; buds at first sparsely hairy, becoming glabrous, fulvous, to 1 — 2.5 — 3 mm long, subacute, slightly convex on the back, flat on the inside; stipules semicordate or semireniform, large, to 4 — 7 mm long, equaling the petiole, irregularly dentate, acute; petioles pubescent, to 5 mm long; expanding leaves revolute, glabrate above, white-tomentose beneath, finally lanceolate, 1:4 — 6, broadest at the middle, more or less rounded at base, subacuminate, dark green and wholly glabrous above except for the midrib, glaucescent or grayish-green beneath, hairy on the veins and especially on the midrib, the hairs rather long, recurved, pointing in various directions, the margin serrulate or irregularly sharply dentate; lateral veins 15 — 20 pairs, at an angle of 90 — 70 — 50°, often forking at the margin, the leaves blackening in drying; pistillate catkins as in S. caucasica Anderss. Fr. June.

Caucasus: W. and S. Transc., whence described. Type in Leningrad.

Note. The species differs little from S. caucasica Anderss. Leaves lanceolate, at emergence revolute; grown leaves hairy on the midrib; lateral veins numerous.

69. **S. daghestanica** Görz in Grossg., Fl. Kavk. II (1930) 6 et in Fedde Repert. sp. nov. XXXVI (1934) 236 (300).

A shrub of medium size, with recurved brownish-gray glabrous branches; stipules large, reniform, sharply dentate, palmately nerved; petioles 7 — 8 mm long, glabrous; leaves on terminal shoots oblanceolate, 7 — 9 cm long and 2 — 3 cm broad, the lower elongate-obovate, rounded at base, irregularly dentate, slightly revolute, glabrous, dark green with a greasy sheen above, glaucescent beneath with stramineous midrib; lateral veins
104 15 — 17 pairs, slender, clearly evident, often bifurcate at the margin; catkins with the leaves, cylindric, borne on very short leafy branchlets; scales covered with few straight hairs, bicolor, the staminate unknown; ovary glabrous, stipitate, the stipe one-third to half the length of ovary; style about one-third as long as ovary; gland 1, posterior. May.

Caucasus: Dag., whence described. Endemic. Type in Leningrad.

Note. Dr. Görz's authentic samples were described at one time by P. A. Lakshevits as S. caucasica Anderss. They do not differ from S. caucasica Anderss. to a greater extent than does f. abchasorum Görz, described for that species.

70. **S. pantosericea** Görz in Fedde Repert. sp. nov. XXXVI (1934) 229 (293). — S. argyrophylla Laksch. (nom. tantum) in sched. Herb. Acad. Scient.; Görz in Grossg., Fl. Kavk. II (1930) 8, non Nutt. (1842).

A handsome tall shrub; young shoots yellowish, sparingly covered with scattered pubescence; old branches dark reddish-brown, glabrous; buds small, 2.5 mm long and 2 — 3 mm broad, ovoid, obtuse, yellowish-brown or dark brown, glabrous, lustrous or hairy and dull; stipules very small, lanceolate or ovate, scalelike, scarious, caducous; petioles 4 — 8 mm long, hairy or glabrate, brown; leaf blades from 4.5× 2.5 to 9× 2.8 cm, narrowly elliptic to elongate-oblanceolate or obovate, glandular-denticulate, shiny velutinous beneath from short antrorse hairs or subsericeous, grayish-green (f. decalvescens Laksch.), dull green above it with scattered pubescence or as densely silvery-hairy as beneath (f. splendens (Andrz.) Laksch.), the margin subrevolute, the midrib yellowish-brown beneath, lateral veins 12 —16 pairs, visible through the pubescence beneath or also above; catkins lateral, the pistillate ca. 5 cm long and 1.4 cm broad, with hairy rachis, the hairy stalk ca. 1 cm long, with 2—4 small sericeous-pilose bracts; scales 2.5×1.2 mm, ovate or liguliform, obtusish, rufous or dark fuscous to almost black, with a tuft of straight rather short lurid hairs; ovary ca. 5 mm long, ferruginous-tomentose; style 1—2 mm long, glabrous; stigma ca. 1 mm long, entire or bifid, linear, spreading, reddish as the style; stipe 0.5—0.9 mm long, hairy; gland 1, posterior, ca. 0.7 mm long; capsule 7—8 mm long. June.

Alpine zone; moraine screes and sources of mountain streams. Caucasus: Cisc. and W. Transc. Endemic. Described from the Caucasus. Type in Leningrad.

Note. Somewhat reminiscent of S. glauca L., but the silvery sericeous pubescence of the leaves, as well as the long style and stigma, tend to refer it to another section.

105

Section 13. **LIVIDAE** Nym. Consp. III (1881) 668. — Much branched, sometimes tall shrubs, 3—8 m in height; shoots slender; buds pointed, glabrous; leaves short-petioled, thin, with 5—7 lateral nerves, commonly pale green, glabrous or more or less pubescent to finely tomentose; catkin-scales light-colored; ovary sericeous, borne on a very long stipe; style short.

1. Grown leaves rather thin, glabrous on both sides or nearly so . . . . 2.
\+ Grown leaves stiff, sericeous, canescent, or velutinous-tomentose . . 4.
2. Exposed wood without ridges; shrubs of medium size; leaves 3—6 cm long and 2—3 cm broad . . . . 3.
\+ Exposed wood ridged; tall shrubs; leaves to 6—8 cm long and 4—5 cm broad . . . . .72. S. Starkeana Willd.
3. Branches erect, slender, glabrous, castaneous; catkins loose; scales mostly monochromatic, yellow; filaments glabrous; stipe 3—5 times the length of the gland . . . . 71. **S. livida** Whlb.
\+ Branches spreading to recurved, knotty, yellowish-gray or grayish-tawny, at first tomentose, at length sparsely hairy or glabrous; catkins dense; scales brown, dark at apex; filaments hairy at base; stipe twice the length of the gland . . . . 75. **S. iliensis** Rgl.
4. Grown leaves dingy green and sparsely appressed-hairy above, hairy or velvety beneath; catkins before the leaves, loose, in fruit to 5—7 cm long; stipe 5 times as long as the gland . . . . 73. S. xerophila Floder.
\+ Grown leaves rather densely covered on both sides or only beneath with white shiny tomentum; catkins with or soon after the leaves, very loose, in fruit to 10 cm long; stipe 7—10 times as long as the gland . . . . .74. **S. Bebbiana** Sarg.

71. **S. livida** Whlb. Fl. lapp. (1812) 272, tab. 16, fig. 7. — S. depressa β livida Fries, Mant. I (1842) 58; Shmal'g., Fl. II, 437. — Ic.: Rchb. Ic. Fl. Germ. XI, t. 567, f. 2009; Anderss. Monogr. Salic., tab. V, f. 51. — Exs.: HFR No. 2277—2280.

A small or moderate shrub, ca. 0.5 m high, with slender glabrous castaneous flexuous spreading branches; exposed wood without striations; stipules reniform or obliquely ovate, dentate; petioles rather short, glabrous; leaf blades reddish, thin, pellucid in spring, becoming coarser
106 in fall, obovate, at base cuneate or broadly elliptic and at both ends uniformly tapering to narrowly elliptic, mostly entire or denticulate, green above, plumbeous-glaucous beneath, rather heavily pubescent when young, becoming glabrous on both sides, 2—6 cm long and 1.5—3 cm broad, sometimes smaller: 1—3 cm long and 0.5—1 cm broad (f. microphylla Korsh. in sched.), in E. Siberia 2—3 times as large as in the type, on flowering shoots entire, on vegetative shoots irregularly and coarsely serrate (v. sibirica Laksch.); catkins precocious or subcoaetaneous, slender, loose, ca. 2.5 cm long, the stalk short in staminate, longer in pistillate catkins, subtended by small bracts; scales monochromatic pale yellow, rarely at apex faintly fulvous or (in v. sibirica) brown or almost black, in all varieties always hairy; stamens 2, 2—3 times as long as the scale, with glabrous filament and yellow anthers; gland 1, posterior, oblong; ovary ovoid-subulate, white-sericeous, long-stipitate, the stipe as long as or longer than the ovary and 3—5 times the length of the gland; style short or very short; stigma short, with spreading lobes. May—June. (Plate VI, Figure 8).

Inundated and dry-valley meadows, coppices, wood margins, slopes, and mixed woods. All over the Soviet Union. — **Gen. distr.**: Scand., Centr. Eur., Mong. Described from Scandinavia. Type in Stockholm.

Note. Hybridizing with the following species: S. aurita, arbuscula, caprea, cinerascens, cinerea, hastata, lapponum, myrtilloides, nigricans, phylicifolia, purpurea, rosmarinifolia, viminalis s.l.

This species is particularly susceptible to attacks by the midge Cecidomyia rosaria which result in the formation of so-called "willow roses" at the ends of branches. These are rosulate galls which promote proleptic phenomena involving the development not merely of shoots initiated for the current year but also sprouting of shoots which would not normally develop until the following year.

Economic importance. Young branches and leaves are eaten by domestic animals.

72. **S. Starkeana** Willd. Sp. pl. ed. 4, IV (1805) 677; Trautv. in Ldb. Fl. Alt. IV, 275. — S. depressa var. Starkeana Rchb. Ic. Fl. Germ. XI (1849) tab. 567. — S. malifolia Bess. Primit. fl. Galic. II (1809) 313. — S. vagans γ livida 3° Starkeana Anderss. in DC. Prodr. XVI, 2 (1868) 228.

A tall shrub, with spreading branches; young shoots pubescent, becoming glabrous, dark brown; exposed wood with striations ca. 5 mm long; buds yellow or brownish, ovoid, pointed, recurved at the tip, glabrous; stipules often large, to 1 cm long, semicordate, dentate, mostly absent; petioles to 1 cm long, glabrous; leaf blades to 8 cm long

and to 4 — 5 cm broad, commonly ca. 6 × 3.5 cm, rounded-elliptic to ovate-lanceolate, acuminate, obliquely tipped, rounded at base,
107 stiff, glabrous on both sides, grayish-green or glaucescent beneath, dark green above, mostly reddish when young, the margin reflexed and often coarsely and sharply serrate; veins stramineous, prominent on both sides, lateral veins 6 or 7 pairs, at an angle of 50 — 70°, forking at the margin; catkins serotinous, very thick; scales dark to almost black at apex; ovary sericeous-pilose, with short style and stigma; stipe 4 — 5 times the length of the gland, finally doubling in length; in other characters resembling S. livida Whlb. Fl. May; fr. June.

W. Siberia: Alt.; E. Siberia: Ang.-Say.; Far East: Uss. **Gen. distr.**: Centr. Eur. Described from Europe. Type in Berlin.

Note. Apparently observed in the S. part of the European part of the Soviet Union.

73. **S. xerophila** Floder. in Bot. Not. (1930) 334. — S. Floderusii Nak. Fl. sylv. Kor. XVIII (1930) 123, tab. XXIII. — S. livida β cinerascens Whlb. Fl. Lapp. (1812) 273, tab. XVI, f. 6. — S. depressa α cinerascens Fries, Mant. I (1832) 57. — S. sphacelata Sm. Fl. Brit. III (1805) 1066. — Ic.: Whlb., l. c. — Exs.: HFR No. 2276.

A tree of third size-grade, to 6 m high and ca. 10 cm in trunk diameter, or a rather tall shrub, with a gray trunk and pubescent annotinous and aestival shoots; young branches green, in winter reddish or brownish, the striations of exposed wood not more than 5 mm long; stipules present only on vigorous shoots, reniform, auriculate, lunate, or broadly ovate, 2 — 7 mm long, entire or serrate, green, more or less hairy; petioles 6 — 8 mm long, hairy or glabrate; leaf blades 1 — 8.5 cm long and 0.5 — 4.5 cm broad, stiff, obovate (f. obovata Anderss.), elliptic (f. elliptica Anderss.), or lanceolate, acuminate, entire or undulate-dentate, the upper surface dull green, sparsely appressed-pilose, with slender impressed veins, the lower surface sparingly covered with subappressed tomentum, rarely grayish-green or dingy with sparser vesture (f. obscura Anderss.), or more densely coated with longer hairs and inconspicuously veined (v. kamtschatica Kom.); catkins precocious, sessile or subsessile, at length with 3 — 5 small bracts, lateral, borne on short branchlets, loose; staminate ca. 1.2 — 3 cm long and 1 cm broad, cylindric, golden-yellow; pistillate ca. 1 — 2 cm, in fruit to 5 — 7 cm long; rachis covered with silky hairs; scales persistent, lanceolate, obtusish, dark brown at apex, yellow at base, sparsely hairy, as long as the stipe; stamens 2, distinct; anthers elliptic, yellow; ovary ovoid-oblong, 3 — 3.5 mm long, sericeous, at length glabrescent; stipe 5 times the length of the gland; style short; stigma with large 4-parted lobes; capsule 5 — 6 mm long. May — June.

Wet coppices, boggy valleys, and mountain slopes in the forest zone of
108 mountains and in the forest belt. — Arctic: An.; European part: Kar.-Lap., Dv.-Pech., Lad.-Ilm., U. Dnp., U. V., V.-Kama, V.-Don, Transv.; W. Siberia: throughout; E. Siberia: throughout; Far East: throughout; Centr. Asia: Balkh., Ar.-Casp. **Gen. distr.**: Scand., Centr. Eur., Mong., Jap.-Ch. Described from Scandinavia. Type in Stockholm.

Note. Hybridizing with the following species: S. aurita, caprea, cinerea, coaetanea, hastata, lapponum, livida, myrtilloides, phylicifolia, pyrolaefolia, rosmarinifolia, sibirica, viminalis s. l.

74. S. Bebbiana Sarg. in Gard. and For. VIII (1895) 463 et in Sylva X (1896) 131, t. 477; Britt. a. Brown. Ill. fl. N. U. St. a. Can. I, 498; Coville in Proc. Washingt. Acad. III (1901) 306 et fig. 17. — S. rostrata Richards. in Frankl. Narr. Journ. Pol. Sea, Bot. App. (1823) 753, nec Tuill. (1799). — S. vegans β rostrata Anderss. Monogr. Salic. (1867) 87 et in DC. Prodr. XVI, 2 (1868) 227. — S. vagans cinerascens occidentalis Anderss. Salic. bor.-amer. (1858) 121. — Ic.: Sarg. in Sylva IX, tab. 477; C. K. Schn. Laubholzk. I, 61, fig. 11-w-w^2, 125.

A shrub from 2 to 5 m or a tree to 8 m high; branches grayish-brown, rarely reddish-brown, at first rather heavily pubescent and often tomentose; buds small, pointed, appressed, yellowish, hairy; stipules semicordate, acute, caducous; petioles 0.4 — 1.2 cm long, more or less hairy; leaf blades rounded or cuneate at base, stiff, ovate, elliptic, oblong-lanceolate, lanceolate (f. lanceolata Anderss.), or obovate-oblong (f. obovata Anderss.), short-acuminate, 2.5 — 7.5 cm long and 2 — 4 cm broad, rarely to 5 cm broad (f. latifolia Anderss.), on both sides densely white-tomentose (f. lanata Anderss.), sometimes glabrous with pale veins above, sparsely hairy and prominently reticulate-veined beneath, or hoary above and slightly shiny-tomentose beneath; catkins subcoaetaneous or serotinous, on short lateral branchlets, with a leafy-bracted stalk, 2 — 4 cm long, elongating in fruit to 10 cm, very loose, the rachis hairy; scales liguliform or narrowly ovate, light-colored, hairy, persistent or caducous; stamens 2, distinct, glabrous; anthers yellow; ovary subulate-conical, 5 — 6 mm long, sericeous; stipe slender, 4 — 5 mm long, 2 — 3 times the length of the scale; style short or obsolescent, the subsessile stigma 2 — 4-parted. May. (Plate V, Figure 5).

Forest-tundra. — Arctic: An.; E. Siberia: Yen.; Far East: Uda. **Gen. distr.**: N. Am. Described from North America.

Economic importance. Ornamental.

111 75. S. iliensis Rgl. in A. H. P., VI, 2 (1880) 464; Vol'f in A. H. P. XXI, 2 (1903) 174. — S. pseudolivida Görz* in Tr. Tadzh. bazy Akad. Nauk II (1935) 170.

A shrub with spreading flexuous knotty branches; young shoots tomentulose or tomentose, the annotinous glabrous or nearly so, grayish-fulvous or yellowish-gray; wood without striations; buds conical, pointed, tawny, at first more or less tomentose, becoming glabrate; stipules semicordate, small, ca. 2 — 4 mm long, dentate; petioles 3 — 5 mm long, more or less hairy; young leaves short-pilose, becoming glabrous or nearly so, with puberulous midrib above, 3 — 6 cm long, elliptic to obovate-elliptic (f. typica Görz) or oblong-obovate (f. angustifolia Görz) or broadly elliptic to broadly rhombic (f. latifolia Görz), narrowed at base, short-acuminate at apex, entire or about the middle irregularly toothed, rather dull dark green above, pale green beneath; lateral veins 6 — 10 pairs, at an angle of 40 — 50°, distinct above, less so beneath; catkins somewhat

* S. pseudolivida is apparently a distinct species. — Ed.

PLATE VI. 1. Salix nigricans Sm. – 2. S. borealis Fries. – 3. S. caprea L. – 4. S. aurita L. – 5. S. arbuscula L. – 6. S. brevijulis Turcz. – 7. S. kolymensis O. v. Seem. – 8. S. livida Whlb.

precocious or subcoaetanecus, lateral, subsessile or borne on a short stalk with 4—6 scalelike bracts, dense, the pistillate to 2 cm long and ca. 0.7 cm broad, in fruit to 4 cm long, the rachis hairy; scales ovate-oblong, obtuse, ca. 1.7 mm long and 0.5—0.8 mm broad, brown, darker at apex, rather sparsely hairy; stamens 2, distinct, hairy at base; filaments to 7.5 mm long; anthers round, ca. 0.6 mm long; gland 1, posterior, subquadrate, emarginate at apex, ca. 0.5 mm long, half the length of the stipe; ovary conical from an ovate base, ca. 3 mm long, covered with appressed silky white hairs; style scarcely evident, ca. 1 mm long, finally elongating; capsule to 5 mm long, rather sparsely hairy. Fl. April; fr. May—June.

Subalpine and alpine mountain zones (1,100—3,500 m); slopes and stream beds. — Centr. Asia: Dzu.-Tarb., Pam.-Al. Endemic. Described from the Dzungarian Ala Tau. Type in Leningrad.

Section 14. MYRTILLOIDES Borr. ap. London, Arboret. Brit. III (1838) 1587. — Diminutive shrubs 0.5—1 m high, with slender branches; leaves delicate, thin, ovate or ovate-lanceolate to lanceolate, or obovate-spatulate, glaucous, quite glabrous, reticulate-veined, pinkish-transparent when young, sometimes blackening in drying; catkins loose, borne on a leafy-bracted stalk; ovary long-stipitate, quite glabrous, commonly reddish brown; style very short.

1. Leaves small, ovate, ovate-oblong, or elliptic, acuminate; catkin-scales yellow, at apex tawny; ovary long-stipitate; branches erect . . . . 76. **S. myrtilloides** L.

112 + Leaves spatulate or obovate, narrowed at base, rounded or short-acuminate at apex, blackening; catkins thicker than in the preceding; scales brownish, ciliate; ovary short-stipitate; branches mostly spreading . . . . 77. **S. fuscescens** Anderss.

76. **S. myrtilloides**, L. Sp. pl. (1753) 1019; Ldb. Fl. Ross. III, 613; Turcz. Fl. baic.-dah. II, 2, 112; Shmal'g., Fl. II, 439. — Ic.: Rchb. Ic. Fl. Germ. XI, t. 593, f. 1244. — Exs.: HFR No. 989, 2464 a, b.; Fries Herb. norm. VIII, No. 63; Wimm. et Kr. Herb. Salic. No. 69; Coll. Sal. No. 110.

An upright shrub to 30—80 cm high, rarely to 2 m and to 3 cm in trunk diameter; bark gray at base; young shoots glabrous or at tips puberulous; annotinous tawny or reddish-brown, glabrous; buds ovoid, obtuse, glabrous; stipules small, lanceolate or ovate, sometimes absent; petioles 2—4 mm long; blades 1—3.5 cm long and 0.7—1.5 cm broad, ovate, elliptic, or oblong-elliptic, rounded or rarely narrowed at both ends, entire, rarely somewhat dentate, commonly quite glabrous, only when young sometimes puberulous with straight hairs, thin, dull glaucescent or violet-flushed above, somewhat resembling the leaves of Vaccinium uliginosum L., the lower surface mostly densely reticulate-veined, the veins prominent on both sides; lateral veins 8—10 (12) pairs; leaves slightly blackening in drying; catkins with the leaves, the staminate cylindric, the pistillate ovoid, stalked, loose, to 1.5—2.5 cm long, scarcely elongating in fruit, the bracts foliaceous or scalelike, the stalks transformed at length into short shoots; scales to 1—1.2 mm long, ovate or spatulate-suborbicular, yellow to rufescent with pink or purple apex, or purple throughout, glabrate or covered with short reddish hairs; stamens 2, distinct, hairy, 3—4 times the length of the scale; anthers at first purple, turning yellow, finally dark violet; gland 1, posterior oblong, dilated at base; ovary 3—4 mm long, ovoid-conical, obtusish, glabrous, violet-green or purple, with a very short style; stigmas entire or parted, short, oblong-ovate, rose-purple; stipe ca. 1.5 mm long, in lower part of the catkin longer; gland 1, posterior, one-fourth to one-third the length of the stipe, oblong or yellowish; capsule reddish, glabrous, to 6 mm long. May—June.

Peat bogs and grassy swamps. — Arctic: Arc. Eur., Arc. Sib., An.; European part: Kar.-Lap., Dv.-Pech., Lad.-Ilm., U. V., V.-Kama, U. Dnp., M. Dnp., V.-Don; W. Siberia: Ob; E. Siberia: Yen., Lena-Kol., Ang.-Say., Dau.; Far East: Uda, Ze.-Bu., Uss. **Gen. distr.**: Scand., Centr. Eur.,
113 Jap.-Ch., N. Am. Described from Scandinavia. Type in London.

Hybridizing with S. aurita, brachypoda, caprea, cinerea, fuscescens, glauca, hastata, lapponum, livida, myrsinites, nigricans, phylicifolia, purpurea, rosmarinifolia, sibirica, xerophila.

Note. Eaten by reindeer and domestic animals. It is known to Yakuts by the name "kyl-sirtalak" and is used by them as feed for animals.

77. S. fuscescens Anderss. Monogr. Salic. (1867) 97 et in DC. Prodr. XVI, 2, 230; Coville in Proc. Washingt. Acad. III (1901) 329, fig. 25; Kom., Fl. Kamch. II, 13. — ? S. arbutifolia Pall. Fl. Ross. I, 2 (1788) 79. — S. rhamnifolia Hook. et Arn. in Beech. Voy. (1831) 117, tab. 26, non Pall.

A depressed shrub to 50 cm high, often prostrate; branches spreading or ascending, virgate, slender, glabrous, with dark lustrous bark; buds small, obtuse, appressed, brownish-castaneous; stipules none; petioles 3—4 mm long; blades 1—2.5 cm long and 1—2 cm broad, stiff, brittle, obovate-spatulate, rounded at apex or very short-acuminate, broadest about the middle, very strongly narrowed toward base, glabrous on both sides, dark green above, glaucescent beneath, becoming somewhat darker in drying, entire or faintly hairy, the veins slightly impressed above, raised beneath, lateral veins 10—12 pairs; catkins solitary on a stalk to 2.5 cm long with 4—6 bracts, oblong-cylindric, thicker than those of S. myrtilloides, loose, 2—5 cm long, rigid, spreading or pendulous; scales obovate-lanceolate, becoming brown, glabrous or at apex and on the margin beset with short cinereous cilia; stamens 1, distinct, glabrous; anthers brown; ovary ovoid-conical, 0.7—1 mm long, glabrous or pubescent (var. dasycarpa Trautv.), reddish-brown or purple, twice the length of the scale and four times as long as the gland; style obsolescent; stigmas thick, entire or almost 2-lobed, the lobes straight; capsule to 1.5 cm long. June. (Plate V, Figure 1).

Moss and grass bogs. — Arctic: Chuk., An.; E. Siberia: Lena-Kol., Dau.; Far East: Kamch., Okh., Ze.-Bu., Uda, Sakh. **Gen. distr.**: Ber., N. Am. (Alaska), N. China (Kansu). Described from Siberia. Type in London.

Note. A hybrid with S. myrtilloides occurs in the Zeya-Bureya region.

Section 15. **HASTATAE** Fries, Physiogr. Sällsk. Arsb. (1825) 34. — Shrubs of the subarctic and forest belts, and of the forest and subalpine zones in mountains; leaves glabrous, sharply serrate, variously shaped, green on both sides or the lower surface pale glaucous, with very prominent veins; stipules large; catkins precocious or coaetaneous, sessile or stalked; scales yellowish or brownish, darker at apex, white-
114 hairy; ovary glabrous, dark green, stipitate; style evident; stamens 2, distinct, with yellow anthers; gland 1 in both staminate and pistillate flowers.

1. Leaves orbicular, obovate-orbicular, obovate-oblong, or broad-elliptic; stipules rather large . . . . 2.
\+ Leaves elliptic, lanceolate, or elliptic-lanceolate; stipules small or absent . . . . 8.
2. Leaves blackening, flocculate-hairy; young shoots and midrib white-tomentose; style long; stigmas divergent, often annularly rounded . . . . 86. **S. Barclayi** Anderss.
\+ Leaves not blackening, glabrous beneath; young shoots and midrib more or less hairy or glabrous; style short or evident but not long; stigmas short . . . . 3.
3. Petioles slender, often like the midrib beneath brown or rose-colored, to 1—3 cm long; stipules very large, to 1—1.5 cm broad, rounded-reniform; leaves thin, orbicular, rounded-ovate, or rounded-elliptic, whitish-glaucous and prominently veined beneath, the lateral veins at an angle of 60—85° . . . . 78. **S. pyrifolia** Ldb.
\+ Petioles thicker, short; stipules smaller; leaves stiffer, different in shape, green or glaucescent but not whitish beneath; midrib stramineous or yellowish-green; lateral veins at a more acute angle . . . . 4.
4. A tree or tall shrub; leaves large, to 7—10 cm long and to 5.5 cm broad, elliptic-lanceolate or obovate-oblong, the margin coarsely serrate or dentate-sinuate; petioles 5—12 mm long; nerves very prominent; catkins to 7.5—9 cm long; scales ciliate, crisp-hairy at base . . . . 83. **S. viridula** Anderss.
\+ Low or procumbent shrubs; leaves smaller, short-petioled; catkins much shorter . . . . 5.
5. Young shoots woolly with white hairs or densely pubescent; leaves stiff, brittle, pure green, strongly reticulate, obovate to oblong-obovate, narrowed at base, covered beneath with scattered hairs; catkin-scales rufous or tarry, turning brown, glabrous, bearded at apex with white hairs . . . . 84. **S. rhamnifolia** Pall.
\+ Young shoots more or less hairy; leaves less stiff, obovate-lanceolate, oblong-lanceolate, or elliptic, mostly glabrous beneath; catkin-scales brown, dark at apex, crisp-hairy . . . . 6.
115 6. Stipules large, reniform or semicordate; lateral veins 7—14 pairs; catkins coaetaneous or precocious; scales oblong-elliptic or ovate-lanceolate . . . . 7.
\+ Stipules smaller, ovate-lanceolate; lateral veins 7 or 8 pairs; catkins serotinous or coaetaneous; scales rounded-obovate . . . . 81. **S. psiloides** (Floder.) Kom.
7. Stipe short; style 0.5—0.7 mm long; stigmatic lobes oblong-ovate, entire or rarely 2-parted . . . . 79. **S. hastata** L.

\+ Stipe very short, the ovary subsessile; style evident, ca. 1—2 mm long; stigmatic lobes oblong-linear, entire .... 80. S. apoda Trautv.

8. Buds subspherical; leaves elliptic or lanceolate, concolor on both sides, glabrous, often lustrous beneath, the margin sparingly glandular-toothed .... 82. **S. Fedtschenkoi** Görz.

\+ Buds semiovoid; leaves elliptic-lanceolate with subparallel margins, dark green above, paler beneath, uniformly covered toward fall with soft appressed hairs, the margin to one-third crenate-dentate .... 85. **S. Litwinowi** Görz.

78. **S. pyrifolia** Ldb. Fl. Alt. IV (1833) 270; Ej. Fl. Ross. III, 613; Turcz. Fl. baic.-dah. II, 1, 112; Kryl., Fl. Zap. Sib. II, 757. — S. *alnoides* Schang. ex Siev. in Pall. N. Nord. Beitr. VII (1796) 347, 349, nom. — S. *corylifolia* Turcz., pl. exs. (1828). — S. *sabulosa* Turcz., pl. exsicc. 1830 et Cat. baic.-dah. No. 1036. — Ic.-Ldb. Ic. Fl. Ross. t. 476; Anderss. Mon., tab. IX, 101. — Exs.: HFR No. 990, 991, 2316—2318, 2475.

A shrub or tree to 10 m high, with yellowish-brown or grayish-castaneous commonly glabrous branches; young branches covered with spreading hairs; buds ovoid, pointed, yellowish-brown; stipules rounded-reniform or rounded-cordate, dentate, often very large; petioles 1—3 mm long, yellow, rarely rose, commonly glabrous (in var. *pubescens* Nas. petioles, buds, midribs, lower side of leaves, and young branches pubescent); leaf blades thin, glabrous, prominently veined, green above, whitish-glaucous beneath, ovate to ovate-elliptic, short-acuminate (f. *ovata* Ldb.), or quite orbicular (f. *orbiculata* Ldb.), or orbicular with subcordate base and pink petiole (f. *cordata* Ldb.), serrate, 2—8 cm long and 1.5—6 cm broad; lateral veins 8—14 pairs, at an angle of 60—85°; catkins precocious, sessile or short-stalked, with scalelike caducous bracts or ebracteate, cylindric, compact, to 3.5 cm long, pistillate in fruit to 7 cm long; scales ca. 2 mm long, oblong, obtuse or acute, pale brown, at apex darker, long-hairy;
116 stamens 2, distinct, glabrous, to 8 mm long; anthers yellow; ovary 5 mm long, narrowly conical, often recurved, glabrous, green, the style as long as the 2—4-lobed stigma; stipe 2—4 times as long as the gland and one-fourth to one-third the length of ovary; capsule to 7 mm long. May— June. (Plate IV, Figure 6).

The forest belt, forest-tundra, and the forest and subalpine zones of mountains. — Arctic: Arc. Sib., Chuk.; European part: Kar.-Lap., Dv.-Pech.; W. Siberia: Ob, Irt., Alt; E. Siberia: Yen., Lena-Kol., Ang.-Say., Dau.; Far East: Ze.-Bu.; Centr. Asia: Dzu.-Tarb. **Gen. distr.**: Mongolia. Described from Altai. Type in Leningrad.

Note. Hybridizing with S. *caprea*, *glandulifera*, *rorida*, *xerophila*.

Economic importance. Suitable for tanning, with tannin content up to 10.87%. Leaves eaten by reindeer and by domestic animals. Yakuts use the base of this willow for wickerwork. An ornamental plant.

79. **S. hastata** L. Sp. pl. (1753) 1017; Ldb. Fl. Ross. III, 612; Turcz. Fl. baic.-dah. II, 1, 111; Kryl., Fl. Zap. Sib. IV, 756; Kom., Fl. Kamch. II, 87. — Ic.: Anderss. Mon. Sal. t. IX, 102; Rchb. Ic. Fl. Germ. XI, t. 570, f. 2013. — Exs.: Fries, Herb. norm. III, 53; Wimm. Herb. Sal. No. 71, 72; Coll. Salic. No. 82, 83.

A shrub to 1.5 — 2 m high, mostly lower, in the alpine zone subprostrate; young branches pubescent; annotinous dawny to dark brown, glabrate; wood without striations; buds ovoid, obtuse, hairy at tips; stipules semicordate, obliquely ovoid, or reniform, rarely hastate-acuminate, serrate; petioles 2 — 7 mm long, glabrous; leaf blades thin, ovate, oblong, elliptic, or lanceolate, 1.5 — 7 cm long and 1 — 3.5 cm broad, commonly serrulate, rarely entire (v. subintegra Floder.), glabrous, dull green above, light green or glaucous beneath; in the forest zone leaves and stipules larger, leaf blade broad, suborbicular to elliptic, often dentate (v. vegeta Anderss.); in the subalpine zone leaves narrower, stipules smaller and more rarely present (v. subalpina Anderss.); in the alpine zone leaves small, lanceolate, subentire, often exstipulate (var. alpestris Anderss.); lateral veins 7 — 14 pairs, at an angle of 40 — 80°; catkins in the forest zone precocious, in the more elevated zones coaetaneous, cylindric, compact, borne on short leafy stalks and branchlets, the stalks of pistillate catkins longer; scales oblong or oblong-spatulate, obtuse or acute, light brown, monochromatic or at apex darker, with white crisped tangled hairs; stamens 2, distinct (exceptionally connate; see also S. apoda Trautv. and S. psiloides Floder.), glabrous, to 7 mm long; anthers yellow; ovary to 4 mm long, conical, obtusish, glabrous, green or brownish-green, the stipe slightly longer than the ovary; style short or moderate, sometimes bifid; capsule to 7 mm long, glabrous. Fl. June; fr.
117 July. (Plate IV, Figure 3).

Forest belt, forest-tundra, subalpine and alpine zones. — Arctic: Arc. Eur., Arc. Sib., Chuk., An.; European part: Kar.-Lap., Dv.-Pech.; Caucasus: Cisc., W. Transc.; W. Siberia: Ob, Alt.; E. Siberia: Yen., Lena-Kol., Ang.-Say., Dau.; Far East: Kamch., Okh., Ze.-Bu., Sakh.; Centr. Asia: Dzu.-Tarb., Pam.-Al. **Gen. distr.**: Scand., Centr. and Atl. Eur., Bal.-As. Min. Him., Mong. Described from Sweden. Type in London. In North America replaced by S. cordata Mühlb.

Note. Hybridizing with S. arbuscula, aurita, caprea, glauca, herbacea, glandulifera, lanata, livida, myrtilloides, phylicifolia, polaris, reptans, reticulata, rorida, rosmarinifolia, rotundifolia, xerophila.

Economic importance. Tamin content about 11%. The leaves are eaten by reindeer.

80. **S. apoda** Trautv. in Ind. sem. Horti Petrop. (1865) 37; Boiss. Fl. Or. IV, 1192; Medved., Der. i kust. Kavk. 296.

A rather low robust shrub, with short stout dark brown glabrous branches; buds large, obtuse, latericious-yellow, glabrous, lustrous; stipules reniform or semicordate, glandular-serrate, large; petioles 2 — 4 cm long, dilated toward base, glabrous; leaf blades 5 — 7 cm long and 2.5 — 4 cm broad, obovate, narrowed toward base, mucronate at the broad obtuse apex, rarely elliptic or oblong-lanceolate, tapering toward apex, glandular-serrate, light green above, glaucescent or pale green beneath, thin, brittle, glabrous; midrib pale yellow beneath, almost whitish above, very prominent; lateral veins 10 — 12 pairs, very marked beneath and often clearly visible above; catkins coaetaneous; pistillate 3 — 5 cm long and 0.8 cm thick, dense, in fruit to 10 cm long, the hairy stalk ca. 0.6 cm long, with small bracts; scales lanceolate to ovate-lanceolate, brown, sometimes darker at apex, densely covered with long wavy white hairs; stamens 2, distinct, exceptionally connate to one-half their length; ovary ovoid-lanceolate, glabrous, subsessile; style evident, ca. 1—2 mm long, the stigmatic lobes oblong-linear, undivided, short; capsule inflated, glabrous. June.

Subalpine zone. — Caucasus: Cisc., W. and E. Transc. Endemic. Described from the Caucasus. Type in Leningrad.

Note. A hybrid with S. arbuscula L. is recorded.

81. **S. psiloides** (Floder.) Kom. Fl. Kamtsch. II (1929) 18. — S. hastata subsp. psiloides Floder. in Ark. för Bot., Bd. 20 A No. 6 (1926) 54; Hultén, Fl. of Kamtch. II, 14.

A shrub to 2—3 m high, with dark red or reddish-brown glabrous bark; young branches mostly sparsely pubescent like the buds, or glabrous;
118 stipules ovate-lanceolate, glandular-toothed, scarious, yellowish-green, shorter than the petiole; petioles short; leaf blades obovate-lanceolate, irregularly serrate, glabrous on both sides, green above, glaucous beneath; midrib pubescent beneath; lateral veins ca. 7 pairs; catkins coaetaneous or serotinous; staminate 2—3 cm long and 1—1.2 cm broad, the stalk to 0.5 cm long with 2 or 3 bracts; pistillate not collected; scales projecting, small, rounded-obovate, dark tawny at apex, covered with long crisped tangled hairs; stamens long, slender, connate in lower part, rusty, glabrous; anthers round, turning brown, small; gland 1, interior, somewhat shorter than the scale. June.

Far East: Kamch., described from the vicinity of Klyuchi village. Endemic. Type in Stockholm.

Note. The partial or almost complete fusion of the filaments, that accounts for the separation of this form, is fairly often encountered also in other species, e. g., in the Caucasian S. apoda Trautv., in the North American species of this section, S. cordata Mühlb., as well as in certain species of other sections which normally have free stamens, e. g., S. rosmarinifolia L., S. herbacea L., S. lanata L., etc. It is of topical importance to determine whether such metamorphosis represents a fortuitous deviation, a deformity inherent only in some individuals and commonly designated as f. cladostema, f. synandra, etc., or a regular phenomenon, predetermined by systematic or geographic factors.

82. **S. Fedtschenkoi** Görz, Salic. Asiat. I (1931) 21 et in Fedde Repert. sp. nov. XXXII, I—X (1933) 121—123.

A rather low spreading shrub with stout brownish lustrous branches; young branches glabrous, in age often with whitish excretions; buds 5—8 mm long, spherical, obtuse, plump, always glabrous; stipules of lower leaves obliquely ovate, the others lanceolate, glandular-toothed, glabrous; petioles short, dilated at base, channeled, glabrous or those of upper leaves pubescent; leaf blades glabrous, green on both sides, often lustrous beneath, somewhat hairy on the veins, elliptic (f. typica Görz.) or lanceolate (f. angustifolia Görz), 1.5—4 times as long as broad, with the very variable breadth greatest at or above the middle, obtuse, sometimes oblique at base and at apex, on sterile shoots rounded at base, sparingly glandular-toothed on the margin; lateral veins 8—12 pairs, scarcely forking; catkins subprecocious or coaetaneous; staminate ovoid-cylindric, dense; pistillate stoutly cylindric, ca. 3 cm long and 1 cm thick, in fruit to 5 cm long, the short stalk with scalelike bracts or rarely profusely leafy-bracted; scales ovate, subobtuse, covered with long white hairs; stamens 2, to 4—5 mm long, 2—3 times the length of the scale; ovary gradually tapering from inflated base, sessile or nearly so, glabrous; glands narrow. June.

119 Subalpine and alpine zones (3,000—3,500 m). — Centr. Asia: Dzu.-Tarb., T. Sh., Pam.-Al. Endemic. Described from the mountains of Soviet Central Asia. Type in Leningrad.

Note. Very closely related to S. hastata L, but differing in the concolor leaf surfaces, with often lustrous lower surface, the leaf dentation, the very short or absent stipe, the noncurly hairs on catkin-scales, and the narrow glands (Görz).

83. **S. viridula*** Anderss. ex Asa Gray in Mem. Amer. Academ. VI (1858—1858) 451. — S. padifolia β viridula Anderss. in DC., Prodr. XVI, 2 (1868) 256.

A tall shrub or a third size-grade tree, with slender straight or flexuous branches; young branches greenish, in age rusty-castaneous, glabrous, sublustrous; buds large, yellowish, glabrous; stipules small, semiovoid, crenate, mostly absent; petioles short, glabrous; leaf blades ovate-lanceolate, elliptic-lanceolate, or obovate-oblong, to 7.5 cm long and ca. 2.5 —3 cm broad at the middle, long-acuminate, narrowed or rounded or rarely subcordate at base, thin at first, becoming stiffish, the margin coarsely serrate or glandular-toothed, both surfaces glabrous, the upper dark green and lustrous, the lower paler or glaucescent; midrib yellowish, prominent; lateral veins evident, anastomosing, light-colored or brownish, forming a conspicuous finely alveolate network; catkins precocious, subsessile, subtended by 1 or 2 small bracts, to 7.5—9 cm long, stoutly cylindric, upright or spreading, showy, rather dense; rachis rather densely covered with gray hairs, rarely glabrate; scales ovate, tapering or oblong-liguliform or oblong-spatulate, stramineous or latericious-brownish, obscurely nerved, persistent, about covering the base of ovary or reaching up to its middle, glabrate or ciliate, sparsely covered below with crisped hairs; stamens 2, distinct; filaments twice as long as the scale, pale; anthers rather large, round; ovary plump, ovoid-conical to subcylindric, ca. 2 mm long, obtuse, glabrous or rarely puberulent, green, becoming brownish, the stipe twice the length of the gland; style evident; stigmas brownish-rusty, short, the erect lobes entire or parted. Fl. June; fr. July. (Plate IV, Figure 5).

Subalpine deciduous and mixed woods. — E. Siberia: Ang.-Say., Lena-Kol., Dau.; Far East: Ze.-Bu. **Gen. distr.**- Jap.-Ch. Described from Japan.

120 84. **S. rhamnifolia** Pall. Fl. Ross. I, 2 (1788) 84; Turcz. Fl. baic.-dah. II, 1, 111; Ldb. Fl. Ross. III, 612; Anderss. Monogr. Salic. 169, tab. IX, f. 106 et in DC. Prodr. XVI, 2, 257; Gmel., Fl. Sib. I, 159, No. 13, t. 35, f. A.

A rather stout spreading or procumbent shrub to 50 — 75 cm high, with stoutish tuberculate flexuous dingy or yellowish branches; young branches

* The editors believe that the Siberian plants are not identical with the Japanese and deserve to be separated as a distinct species.

white-woolly, in age glabrous; stipules rather large, semiovate, serrate; leaves in tufts at the ends of branches; petioles short, dilated, toward base sometimes yellowish-green or lemon-yellow, coriaceous, stiff, brittle, commonly obovate to oblong-obovate, to 3.5 — 4 cm long and 2.5 cm broad, narrowed toward base, short-acuminate; upper surface bright green, glabrous, lustrous; lower surface glaucescent or pale, glabrous or sparingly hairy, more densely on the midrib and in lower part, prominently reticulate-veined, the margin rather shallowly and obliquely glandular-serrate; catkins lateral or subterminal, plump, cylindric, 3 — 5 cm long and 1 — 1.5 cm thick, rather dense, upright, stiffish; rachis stout, very sparsely pubescent; stalk with 2 — 5 bracts, glabrous above; scales enveloping the base of ovary, rusty pitch-colored, obtusish, glabrous, bearded at apex with white hairs; ovary ovoid-conical, nearly 4 mm long, turning brown, quite glabrous, short stipitate, the stipe barely exceeding the gland; style very short; stigmas small, entire. June — July. (Plate IV, Figure 4).

Subalpine zone. — E. Siberia: Lena-Kol., Ang.-Say., Dau. Endemic. Described from E. Siberia. Type in Leningrad.

85. **S. Litwinowii** Görz (mnscr. 1933); Addenda VI, p. 540.

A shrub; young branches pubescent; annotinous glabrous, brown; buds scarcely hairy, semiovoid, brown; stipules none; petioles 2 — 3 mm long, sparingly hairy, becoming glabrate; grown leaves elliptic-lanceolate (ca. 1 : 3), with subparallel margins, rounded at base, short-acuminate at apex, the margin to one-third crenate-dentate with glandular teeth; upper surface glabrous but midrib on both sides hairy, prominent, dark green; lower surface paler, uniformly covered toward fall with soft appressed hairs; lateral veins about 10 pairs, prominent on both sides and together with the veins of third order forming wide alveoles. Flowers absent.

Banks of springs and streams. — Centr. Asia: Mtn. Turkm. (Central Kopet Dagh). Described from Turkmenistan (near Chuli). Type in Leningrad.

121 86. **S. Barclayi** Anderss. in Öfvers. Handl. Wet. Akad. Förh. (1858) 125; Ej. Mon. Salic. (1867) 164, tab. VIII, f. 96; Ej. in DC. Prodr. XVI, 2, 254; Coville in Proceed. Washingt. Ac. III (1901) 316, t. XXXVI; Hultén, Fl. of Kamtch. II, 8; Kom., Fl. Kamch. II, 19. — S. c o n j u n c t a Bebb. Bot. Gaz. XIII (1888) 111. — Ic.: Anderss. Monogr. Salic. VIII f. 96; Coville, l. c.; Britt. a. Br., Ill. Fl. N. U. St. a Can. I, 504.

A shrub to 3 m high, with sturdy glabrous dark brown branches; young branches glabrous or more or less white-tomentose; buds small, brown; stipules ovate-lanceolate, semiovate or lanceolate, acute, glandular-toothed, about equaling the petiole, sometimes absent (mostly in f. a n g u s t i f o l i a); petioles 2 — 8 mm long; leaf blades blackening, variously shaped or broadly ovate, 5 — 7 cm long (f. g r a n d i f o l i a Anderss.), dark green sublustrous above, markedly glaucescent beneath, coarsely serrulate; or ca. 2.5 cm long and 2 cm broad, obovate-orbicular (f. r o t u n d i f o l i a Anderss.), rounded or subcordate at base, short-pointed at apex, covered sparsely above and more densely beneath with rather long appressed or floccose hairs, or glabrate, the white-tomentose midrib paler beneath but not glaucous, the reticulate venation prominent, the margin sparingly denticulate-serrulate, in small leaves entire; or leaves 2—3 cm long and to 0.6 cm broad, lanceolate (f. *angustifolia* Anderss.), thin, glabrous, entire, almost monochromatic, mostly exstipulate; catkins subcoaetaneous, lateral, dense; staminate stoutly cylindric, ca. 2.5 cm long; pistillate cylindric, ca. 3—4 cm long, spreading or recurved, at first white-hairy, finally rufescent, borne on a hairy stalk; bracts 4—8, stiff, narrow, or elliptic, glabrate, entire, often stipulate; scales ovate-lanceolate, acute, brownish or dark, densely bearded with long whitish hairs; stamens 2, distinct, glabrous; anthers golden-yellow; ovary conical-subulate, typically glabrous, sometimes hairy when young, at length glabrescent, green or brunescent, prolonged into a long yellowish style, subsessile or with stipe twice the length of the gland; stigmas more or less parted, the lobes mostly spreading, often curly. June. (Plate IV, Figure 2).

Far East: Kamch. (Koraginskii Island). **Gen. distr.:** Ber., N. Am. Described from Sitka Island. Type in Leningrad.

Note. In Andersson's opinion, this is a kind of a western modification of S. *cordata* Mühlb., but differs in its leaves being shorter, broad, tomentose on the midrib above, often floccose, and subentire.

Section 16. **INCUBACEAE** Dumort. Fl. Belg. Prodr. (1827) 12. — *Argenteae* Koch, Comm. (1828) 46. — *Repentes* Wimm. Fl. Schles.
122 (1841) 335. — Low or medium-sized shrubs, often with ascending trunk; leaves alternate or subopposite, ranging from ovate-oblong to linear-lanceolate, mostly entire, with or without stipules, on both sides or only beneath silvery or dark cinerescent, often blackening; sometimes leaves, especially when young, clothed with golden-yellow tomentum; catkins precocious or coaetaneous, with short bracted stalks or sessile, short-cylindric, often subspherical; scales particolored; gland 1, posterior, in both sexes; stamens 2, distinct, rarely partially connate; anthers yellow or reddish; ovary glabrous or sericeous, more or less stipitate; style short; stigmas oblong.

1. Stipules lanceolate, large, sometimes exceeding the petiole . . . . 2.
\+ Stipules narrowly lanceolate, small, mostly absent . . . . 3.
2. Young shoots, buds, catkins, and young leaves clothed with golden-yellow tomentum, rarely silvery-sericeous; stigmas 0.4 — 0.6 mm long, erect or divergent . . . . 87. **S. brachypoda** (Trautv. et Mey.) Kom.
\+ Young shoots and buds hoary-silky or pubescent; catkins white-hairy; leaves elliptic-lanceolate, 3.5—6 cm long and 1—1.5 cm broad, green glabrous or hoary above, glaucescent or more or less silvery-pilose beneath, strongly blackening on drying; stigmas to 1.2—1.5 mm long, linear, spreading . . . . 89. **S. sibirica** Pall.
3. Branches rather stout; young branches and buds at first densely white-tomentose, becoming glabrate; stipules none; leaves dull and covered with short hairs above, scarcely silky beneath, commonly broadly lanceolate, 2—4.5 cm long and ca. 1—1.5 cm broad; lateral veins 8—12 pairs . . . . 90. **S. schugnanica** Görz.

\+ Branches slender, when young lanate, finally glabrous; buds at first pubescent, at length glabrous; stipules narrowly lanceolate, promptly caducous; leaves linear-lanceolate to oblong-lanceolate, when full grown commonly dark green above, glaucescent or sericeous with short hairs beneath; lateral branches 10—12 pairs . . . . 88. **S. rosmarinifolia** L.

87. **S. brachypoda** (Trautv. et Mey.) Kom. in A. H. P. XXXIX (1923) 49 (nomen tantum), Addenda IV, p.541. — S. repens var. brachypoda Trautv. et Mey. in Middend. Sibir. Reise II, 2 (1856) 79; Maxim., Primit. Fl. Amur. (1859) 245. — S. repens var. flavicans Anderss. in DC. Prodr. XVI, 2 (1864) 238. — S. speciosa Hook. et Arn. Bot. Beech. Voy. (1831—41) 130, non Host (1828). — S. chrysea Anderss., ibid., 276 (ex p.). — Ic.: Nakai, Fl. sylv. Kor., XVIII, tab. XXXVI. — Exs.: HFR No. 2466.

A shrub to 1 m high, with slender straight virgate yellow or rusty branches; annotinous branches densely covered above with white or golden-yellow tomentum, or glabrous; buds semiovate, beaked, at first pubescent,
123 becoming glabrous, rufescent; petioles 1—4 mm long, silky; stipules on vegetative shoots strongly developed, lanceolate, 3—7 mm long and 2—3 mm broad; leaves mostly crowded at the ends of branches, alternate or subopposite, elliptic-oblong, elliptic-lanceolate, or lanceolate, 2.5—5.7 cm long and 0.4—1.5 cm broad, subrevolute; upper surface green, sparsely covered with short appressed down, later glabrate; lower surface golden-yellow (f. flavicans Anderss.) or glaucous with sparse tomentum, or both surfaces (f. argentea Nas.) or merely the lower silvery-silky; lateral veins 5—10 pairs, at an acute angle; catkins precocious, sessile or nearly so, subtended by yellow-hairy bracts, oblong or cylindric, dense, the staminate 2—3 cm long and 0.6—0.9 cm broad, the pistillate 1—1.5 cm long and ca. 0.5 cm broad; scales liguliform or obovate, to 1.5—2 mm long, the upper half black, the hairs golden-yellow; rachis silky; stamens 1, distinct, glabrous, 4—5 mm long; anthers ovate, yellow, turning brown; gland 1, posterior, linear or oblong, ca. 0.6 mm long, shorter than the stipe; ovary ovoid-conical, 2—2.5 mm long, yellowish-tomentose; stipe to 1 mm long, one-fourth to one-third the length of ovary; style not more than 0.2—0.4 mm long; stigmas 2—4-parted, 0.4—0.6 mm long, erect or divergent; capsule to 4—5 mm long. May—June.

Bogs, stands of dwarf arctic birch, swampy meadows, and muddy valleys. — E. Siberia: Lena-Kol., Dau.; Far East: Kamch., Okh., Ze.-Bu., Uda, Uss. **Gen. distr.**: Jap.-Ch. (Korea, Manchuria). Described from the Uda region. Type in Leningrad.

Hybridizing with S. lapponum, myrtilloides, and viminalis s. l.

Economic importance. Suitable for wickerwork; used for feed in areas of NE Siberia. Forms with pronounced golden-yellow vesture and very ornamental.

88. **S. rosmarinifolia** L. Sp. pl. (1753) 1020; Ldb. Fl. Ross. III, 615. — Turcz. Fl. baic.-dah. II, 1, 114; Anderss. Mon. Salic. 115, fig. 64. — S. repens var. rosmarinifolia Koch, Comm. (1828) 48; Shmal'g., Fl. II, 437. — Usionis rosmarinifolia Rafin. Alsol. Am. (1838) 14. — Ic.: Anderss.

Mon. Salic. fig. 64; Rchb. Ic. Fl. Germ. XI, t. 1242. — Exs.: Ehrh. Arbor. No. 119; Fries, Herb. norm. VI, 56; HFR No. 2289 — 2293, 2467, 2468.

A shrub to 0.75 cm — 1 m high, with virgate slender branches; young branches dark, woolly, becoming brown or yellowish-brown, glabrous; buds ovoid, obtuse, reddish-brown, at first pubescent, finally glabrous; stipules narrowly lanceolate, promptly caducous, often absent; petioles short; leaf blades flat, narrowed at both ends, 2 — 8 cm long and 0.3 — 1 cm broad, typically linear-lanceolate, 5 — 10 times as long as broad (f. angustifolia Kern.), or oblong-lanceolate, 2 — 5 times as long as
124 broad (f. latifolia Kern.), when young (and sometimes in age) silky-pubescent on both sides, at length commonly dark green above, glaucescent and glabrous beneath, in drying slightly blackening; lateral veins 10 — 12; catkins precocious or subcoaetaneous, lateral subsessile or borne on a short sparingly leafy-bracted stalk, numerous, dense, small; staminate ovoid, 1.5 — 2 cm long; pistillate at first subspherical, finally short-cylindric; scales obovate, obtuse, at apex dark brown, hairy; stamens 2; filaments distinct, reddish (connate in abnormal forms); anthers yellow or purple, turning brown; gland 1, posterior, oblong; ovary turbinate from an ovoid base, tomentose, long-stipitate, 3 — 4 times the length of the gland; style short; stigma reddish, the lobes entire or bifid. Fl. May; fr. June. (Plate VII, Figure 1).

Meadows, damp thickets, sandy places, and peatbogs. — European part: almost everywhere, except the N. part of Kar.-Lap. and Dv.-Pech. and Crimea; reported for S. Transc. (not seen); W. Siberia: almost everywhere; E. Siberia: Ang.-Say., Dau.; Centr. Asia: Ar.-Casp., Syr D., Balkh., Dzu.-Tarb. **Gen. distr.**: Scand., Centr. and E. Eur. Described from Europe. Type in London.

Hybridizing with S. acutifolia, aurita, caprea, cinerea, caspica, lapponum, livida, myrtilloides, nigricans, phylicifolia, purpurea, sibirica, viminalis s.l., xerophila.

Economic importance. Use for feed and for wickerwork. The bark contains up to 8.92% tannins. Sand fixing, exceptionally resistant to blowing out and drift.

Note. In W. Europe replaced by S. repens L.; in W. Siberia contiguous with the distribution area of S. Sibirica Pall.; in E. Siberia replaced by S. brachypoda (Trautv. et Mey.) Kom., in Korea and Japan by S. dohlia C.K. Schn. Scandinavian and Central European authors separate S. rosmarinifolia L. from S. repens L., and some also from S. arenaria L., even indicating triple hybrids (S. arenaria × repens × rosmarinifolia). S. repens which, among other areas, grow also on coastal sands of the Baltic region, does not occur in the USSR S. rosmarinifolia needs further study since, in its three main habitats (wet meadows, pinetum sands, and peatbogs) it varies greatly in shape and in leaf vesture and apparently forms three ecological types: 1) var. pratensis (Host. pro spec.) Nas. — a broad-leaved meadow form, with leaves dark green above, glaucescent beneath, blackening; corresponds to f. latifolia Kern. and, as regards leaf shape, representing a transition toward S. sibirica Pall; 2) var. turfosa Nas., growing in peatbogs and corresponding to S. rosmarinifolia f. angustifolia Kern.; and 3) var. arenosa Nas., growing chiefly in pinetum sands. The structure of flowers and fruits, however, points against such differentiation, as it

does not display any corresponding differences. A lack of constancy applies also to other characters, such as length, breadth, and vesture of
125 the leaves, density and shape of catkins, etc. The narrow-leaved form becomes readily transformed into the broad-leaved as a result of felling, clearing, or burning.

89. **S. sibirica** Pall. Fl. Ross. I, 2 (1788) 78, tab. LXXXI, f. 3; Kryl., Fl. Zap. Sib. IV (1930) 761. — S. livescens Turcz. pl. exsicc. — Exs.: HFR No. 2463 a, b.

An upright shrub, 0.5 — 2 m high; branches slender, dark brown or castaneous, at first hoary-silky, becoming glabrous; buds pubescent; petioles 4 — 10 mm long; stipules lanceolate, entire, rather large, sometimes exceeding the petiole; leaf blades elliptic-lanceolate, apiculate, at base mostly cuneate, 3.5 — 6 cm long and 1 — 1.5 cm broad (on large apoblasts), entire, slightly revolute, blackening in drying; upper surface green and glabrous (f. glabrescens Kryl.) or hoary with appressed hairs (f. pubescens Kryl.), the lower sparsely hairy glaucescent or silvery-hairy; lateral veins 8 — 10 pairs; catkins coaetaneous; staminate sessile or nearly so, 1 — 2.5 cm long and 1 — 1.3 cm broad, dense, leafy-bracted at base; pistillate 1.5 — 2 cm long and 0.7 — 0.8 cm broad, finally elongating to 3 cm, short-stalked, cylindric-oblong or cylindric; scales ovate to ovate-lanceolate, at base yellowish, at apex black or brown, white-hairy, ca. 2 — 2.5 mm long; stamens 2, to 8 mm long, distinct, glabrous; anthers yellow, turning black; gland 1, posterior, oblong, 0.5 — 0.6 mm long; ovary 2 — 2.5 mm long, ovoid-conical, tomentose-pilose; stipe 0.8 — 1 mm long, barely exceeding to twice as long as the gland; style 0.5 — 0.8 mm long; stigmas linear, divaricate, to 1.2 — 1.5 mm long, yellow or purple; capsule to 7 mm long. Fl. May; fr. June. (Plate VII, Figure 2).

Meadows, coppices, and depressions in steppes. — W. Siberia: all; E. Siberia: Yen., Lena-Kol., Ang.-Say., Dau.; Centr. Asia: Balk., Dzu.-Tarb., Syr D., T. Sh. **Gen. distr.**: Mongolia. Described from W. Siberia. Type in Leningrad.

Hybridizing with S. brachypoda, caprea, caspica, cinerascens, lapponum, livida, myrtilloides, rosmarinifolia, viminalis s. l.

Note. Replaced in the W. by S. rosmarinifolia L. and S. repens L., in E. Siberia by S. brachypoda (Trautv. et Mey.) Kom., in Korea and Japan by S. subopposita Miq., in Pamir-Alai by S. schugnanica Görz, and in Middle Asia by S. dohlia C. K. Schn. It differs from S. rosmarinifolia in the large dimensions of all parts, more particularly the leaves, the short stipe, and especially the long divergent stigmas.

Economic importance. Serves as feed; because of its slenderness and suppleness, useful for wickerwork and fencing. According to Skalozubov, it takes very long to produce regrowth after fire.

126 90. **S. schugnanica** Görz in Tr. Tadzh. bazy Akad. Nauk II (1935), 173. — S. repens var. pamirica O. Fedtsch in sched.

A shrub to 1 m high; branches spreading to suberect, rather stout, tuberculate, when young densely covered with white hairs, annotinous quite glabrous; bark fulvous or brown, rarely grayish-brown or castaneous, sublustrous, 2—3 year old sometimes bright castaneous, very shiny (f. caloclados Görz); exposed wood without true striation but covered with small tubercles; buds oblong-conical, often with a beak 6—7—10 mm long, appressed, at first white-tomentose, at length glabrous fulvous; stipules commonly none; petioles 2—3 mm long, white-tomentose; expanding leaves flat, slightly hairy and green above, silky beneath; grown leaves broadly lanceolate, obtuse at base, slightly acuminate at apex, 2—4.5 (—12) cm long and ca. 1—3 cm broad, entire (on luxuriant shoots denticulate), the dull upper surface clothed with short hairs, the lower densely covered with longer nonsilky or scarcely silky hairs; lateral veins 8—12 (—14) pairs, at an angle of 30—45°, indistinct above, pubescent beneath; catkins precocious, numerous, sessile, dense, subtended by scalelike caducous bracts; staminate ovoid, 1.5 cm long and 1 cm thick; pistillate ovoid-oblong, to 2 cm long and 0.9 cm thick; rachis canescent; scales 1.2—2 cm long, obovate, mostly dark brown, bearded with long almost straight hairs; gland 1, the pistillate oblong, 0.5 mm long, one-fourth the length of stipe; the staminate linear-oblong, attenuate toward apex, to 0.9 mm long; stamens 2, distinct, glabrous, 3—4 times as long as the scale; anthers broadly ovate, yellow (at first apparently reddish); ovary ca. 2 mm long, conical from an ovoid base; style 0.3 mm long; stigmas about the length of the style, the connate lobes spreading, brown; stipe ca. 1.8 mm long; capsule to 5 mm long. Fl. May.

Alpine zone; mountain stream beds, meadows and slopes (3,000—3,500 m). — Centr. Asia: Pam.-Al. (Darvaz, Shugnan). Endemic. Described from Pamir. Type in Leningrad.

Hybridizing with S. caprea L.

Note. Very closely related to S. dohlia C. K. Schn. of the interior of Asia (Inner Mongolia, China). This species differs from S. sibirica Pall. and S. rosmarinifolia L. in its stout brownish branches; dull and scarcely hairy leaves, clothed with shorter hairs; commonly larger, dense catkins; narrower glands, especially the staminate; glabrous stamens; and shorter stipe.

Section 17. **SIEBOLDIANAE** O. v. Seem. Salic. Japon (1903) 21. — Rather low trees and shrubs; leaves ovate or obovate-oblong, entire or crenate, cinereous or lustrous beneath, silky when young (or also grown); catkins precocious or subcoaetaneous, short-stalked, cylindric, dense; gland 1,
127 posterior, more or less elongate-ovate; stamens 2, distinct or connate at base or rarely wholly united; anthers yellow; ovary stipitate or sessile, silky; style and stigma short or moderate.

91. **S. sitchensis** Sanson ex Bong. in Mém. de l'Acad. de St. Pétersb. IV sér. II (1831) 162; Ldb. Fl. Ross. III, 609; Anderss. Salic. bor.-amer. (1858) 126; Coville in Proc. Washingt. Acad. III (1901) 307 et tab. XXXIII. — S. ajanensis Anderss. Monogr. Sal. (1867) 107. — S. sitchensis γ ajanensis Anderss. in DC. Prodr. XVI, 2 (1868) 233.

A shrub of medium size (in North America a tree 3—9 m high); branches subvirgate, when young more or less gray-tomentose, in age glabrous; stipules at ends of shoots large, semicordate, glandular-toothed, woolly.

often absent; petioles 0.4 — 0.6 mm long, pubescent like the small pointed buds; leaf blades 4 — 5 cm long, above the middle 1.5 — 2.5 cm broad, oblong-obovate, mostly cuneate at base, rounded and mucronate at base, flat, blackening, the margin entire or undulate or very sparingly serrulate; upper surface dark green, at first tomentulose, finally glabrate, the impressed veins white-hairy; lower surface uniformly clothed with short shiny silvery tomentum, the veins very prominent; lateral veins 12 — 16 pairs, equidistant, rather straight, at an angle of 35 — 40°; catkins precocious, subsessile, the pistillate short-stalked, obliquely ascending or recurved, 7 — 10 cm long and ca. 1 cm broad, oblong-cylindric, dense, the stalk later elongating, with 1 — 3 small bracts; scales obtusish, yellow or rufescent, brown at apex, glabrous or on the back hairy; stamens 2, distinct, yellow; anthers turning brown; ovary to 2 mm long, thickish at base, covered with short silvery tomentum; style elongated, yellowish at base, the upper part and short thick subentire erect stigmas dark brown; stipe twice as long as the gland. June. (Plate VII, Figure 3).

Forest-tundra. — Far East: Okh. (Ayan). **Gen. distr.**: N. America. Described from Sitka Island.

Note. S. sitchensis γ ajanensis Anderss. (not to be confounded with S. speciosa γ ajanensis Anderss.) is no more than a slight western modification of S. sitchensis Sanson. It is represented in the herbarium by such a poor specimen that it provides no information as to the range of variability of the type or the consistency of deviations.

Section 18. SUBVIMINALES O. v. Seem. Salic. Japon. (1903) 20. — Trees
128 or tall shrubs; branches pubescent only when young, becoming glabrous; leaves large, lanceolate or oblanceolate, acute or long-acuminate, with a cartilaginous serrate-dentate margin, at first sericeous-pilose, becoming glabrous, the upper surface dark green, the lower cinerescent or glaucescent and very prominently reticulate-veined; catkins precocious, sessile, cylindric, dense; scales ovate, rounded or pointed at apex, dark fulvous in upper part, light-colored at base; stamens 2, wholly connate; anthers distinct; ovary sessile or stipitate, densely hairy; style about the length of the ovary; stigmas short, oblong, spreading; gland 1, posterior, oblong, ovate or narrowly linear.

1. Leaves mostly lanceolate or lance-oblong, broadest at or below the middle; stipules small, lanceolate; petioles slightly dilated toward base; catkin-scales rounded or obtuse at apex; gland ovate, narrowed toward apex, one-third as long as ovary . . . . 92. S. Pierotii Miq.
\+ Leaves mostly oblanceolate, broadest above the middle; stipules semicordate, rather large; petioles commonly much dilated toward base; catkin-scales long-acuminate; gland narrowly linear, very long, only by about one-third shorter than ovary . . . 93. **S. gracilistyla** Miq.

92. S. Pierotii Miq. in Ann. Mus. Bot. Lugd. Bat. III (1867) 27; Franch. et Savat. Enum. pl. Japon. I (1875) 461; II (1879) 306; O. v. Seemen, Salic. Japon. (1903) 60, tab. XIII; Kom. and Alis., Opred. rast. Dal'nevost. kr. I, 426.

A tree or a tall shrub, with dark grayish-yellow rugose bark; branches fulvous or grayish-yellow, at first with whitish-gray pubescence, becoming

glabrous; buds small, ovoid, initially pubescent; stipules lanceolate, rather straight to lunate, pointed, glandular-serrate; petioles 2—5 mm long, pubescent; leaves lanceolate or lance-oblong, pointed or obtuse at base, long-acuminate at apex, serrulate, when young pubescent, later glabrous or merely with pubescent veins, glaucous or glaucescent-whitish, to 8—12 cm long and 2—3 cm broad; lateral veins 16—18 pairs, very prominent; catkins precocious or subcoaetaneous, sessile, with small serrate bracts at base, upright, dense; staminate to 2.5 cm long and 0.5 cm broad; pistillate ellipsoid-oblong, in flower 0.8—1.5 cm long and 0.5 cm broad, flexuous, yellowish-gray, with dull hairs; scales short-ovate or ovate-oblong, shorter than the ovary, round-tipped or obtuse, dark brown, coated within and on the back with rather short hairs; stamens 2, distinct, hairy at base, 2—3 times as long as the scale; anthers short,
129 ovate, purple, turning brown; gland 1, posterior, ovate or ovate-oblong, truncate, half the length of the scale and one-third as long as the sessile inflatedly ovate gray-sericeous ovary; style about half the length of ovary or longer; stigmas spreading, flat, broad, short, 2-lobed. April.

Offshore islands. — Far East: Uss. (near Vladivostok). **Gen. distr.**: Japan. Described from Japan.

93. **S. gracilistyla** Miq. in Ann. Mus. Bot. Lugd. Bat. III (1867) 26; Ej. Prolus. Fl. Japon. (1867) 214; Nakai, Fl. Sylv. Kor. XVIII, 104—106, tab. XVI. — S. foliis lanceolatis nudis, raro serrulatis Thunb. Fl. Jap. (1784) 369. — S. Thunbergiana Blume (ex herb. reg. berol.) ap. Anderss. in DC. Prodr. XVI, 2 (1868) 271; Kom., Fl. Manchzh. II, 1, 30; Kom. and Alıs., Opred. rast. Dal'nevost. kr. I, 422, Plate 129; O. v. Seem. Salic. Jap. (1903) 61 et XIV, fig. A—E.

A spreading shrub to 2—3 m high or a tree; young and flowering branches sericeous-pilose, becoming glabrate, dark tawny; buds slightly hairy; stipules semicordate or oblong-elliptic, rather large, acute, denticulate; petioles to 10 mm long, at first pubescent, becoming glabrate, strongly dilated toward base; leaf blades oblanceolate to lance-oblong, 10—12 cm long and to 3.5 cm broad, broadest above the middle, short-acuminate at apex, rounded or narrowed at base; upper surface dark green, subsericeous, at length glabrous; lower surface glaucous, the prominent veins almost silvery white-silky lustrous; margin sharply denticulate; lateral veins 12— 18 pairs; catkins precocious, sessile, approximate, appressed to branches, ebracteate, upright, short, cylindric, staminate 3—3.5 cm long, pistillate 2.5 cm long and ca. 1—1.5 cm thick, dense, elongating in fruit to 8 cm; scales ovate, at apex acuminate, dark brown or blackish, white-hairy, ca. 2—2.5 mm long, the pistillate enveloping the fruit; gland 1, posterior, subulate, to 1.5 mm long, about one-third shorter than the ovary and about half as long as the scale; stamens 2, wholly connate, glabrous, 5—6 mm long; anthers apparently 4-locular, elliptic, yellow, turning brown; ovary small, ovoid-conical or oblong, silky with white or gray hairs, subsessile; style very long, filiform, nearly twice as long as ovary, yellow; stigma very short, 4-parted, with upright lobes, or entire. April—May. (Plate VII, Figure 4).

Banks of mountain streams. — Far East: Uss. **Gen. distr.**: Jap.-Ch. Described from Japan. Type in the Netherlands.

Section 19. VIMINALES Bluff et Fingerh., Comp. Fl. Germ. II (1825) 562; Koch, Salic. europ. comment. (1828) 12 et 27. — Trees or tall shrubs, with long flexible virgate branches; leaves narrow, long, lanceolate to
130 linear, mostly dark above, silky-pilose or lustrous-sericeous, revolute, entire or more or less serrate, with a prominent midrib; catkins before or with the leaves, both pistillate and staminate elongate-cylindric, dense, sessile or short-stipitate; scales particolored; gland 1, posterior, stamens 2; distinct; anthers golden-yellow; ovary silky-pilose, sessile or stipitate; style elongated; stigmas bifid, mostly spreading, or entire.

Note. Species of this section and their hybrids provide almost everywhere available ample material of high quality for wickerwork.

1. Branches mostly slender, virgate, straight, mostly brownish or light-colored; growing on riverbanks . . . . 2.
\+ Branches rather stout, strongly flexuous, often hooked, knotty, dark castaneous; growing on rocks and stony screes of the subalpine zone . . . . 102. **S. sajanensis** Nas.
2. Lower surface of grown leaves lustrous-sericeous or silvery-silky and tomentose . . . . 3.
\+ Lower surface of grown leaves glabrous or slightly pubescent . . . . 15.
3. Grown leaves of terminal shoots commonly elongate-lanceolate, narrowly lanceolate, or linear-lanceolate, the upper side mostly with tuberculate glands near the margin . . . 94. **S. viminalis** L. s. l. and stage 4.
\+ Grown leaves of terminal shoots lanceolate, oblong-lanceolate, or broadly lanceolate, mostly without papillose glands on the upper side near the margin . . . . 11.
4. Catkin-scales large, concealing the ovary, slightly pillulose; style more or less bifid, often cleft to base . . . . . 101. **S. strobilacea** E. Wolf.
\+ Catkin-scales moderate or short, not concealing the ovary, densely clothed with long hairs; style undivided . . . . 5.
5. Catkins precocious . . . . 6.
\+ Catkins subcoaetaneous . . . . 7.
6. Expanding leaves narrowly lanceolate, green above, silky beneath, finally to 1.5 cm broad; catkin-scales light brown, dark at apex . . . . 94a. **S. veriviminalis** Nas.
\+ Expanding and grown leaves linear-lanceolate, 0.2—0.4 cm broad or slightly broader, initially silky on both sides; catkin-scales short, almost wholly blackish . . . . 99. **S. pseudolinearis** Nas.
131 7. Lower surface of grown leaves densely silvery-tomentose or velutinous . . . . 8.
\+ Lower surface of grown leaves with less developed vesture, commonly lustrous-sericeous . . . . 9.
8. Grown leaves oblong or ovate-oblong, to 4 cm broad, broadest in lower part, dull green or grayish-green with scattered hairs above, more heavily clothed with whitish-silvery tomentum beneath; branches latericious or yellow with fugacious whitish tomentum . . . . 100. **S. turanica** Nas.
\+ Grown leaves lanceolate or narrowly lanceolate, to 1.5—2 cm broad, broadest at the middle, similarly silky on both sides or the upper surface almost hoary or dull green and silky, the lower densely silvery, tomentose; branches brown, with less developed vesture . . . . 96. **S. splendens** Turcz.

9. Style short, thick, commonly shorter than stigma; grown leaves lanceolate, grayish-green and puberulous above . . . 95. **S. rossica** Nas.

\+ Style about as long as stigma; grown leaves lanceolate or oblong-lanceolate, more or less silky above . . . . 10.

10. Leaves distinctly serrate already when young, rarely entire, subsericeous above; lateral veins evident, arising at a wide angle and then strongly ascending, at the margin often bifurcate; catkin-scales short, blackish-brown at apex; style of uniform thickness . . . . 97. **S. burjatica** Nas.

\+ Leaves entire, the upper surface dull, cinereous, with less developed vesture than in the preceding species, subsericeous becoming glabrate; lateral veins not evident; catkin-scales at first concealing, later shorter than the ovary, rusty or light brown, scarcely darker at apex . . . . 98. **S. rufescens** Turcz.

11. Branches rather stout; young shoots mostly densely gray-woolly; buds very large, woolly; stipules very large, mostly equaling the petiole, falcate, often with a vertically recurved lobe; petioles thick, dilated toward base; leaves broadly lanceolate or oblong-lanceolate, inequilateral, the lower surface grayish-velutinous or silky, with brown midrib . . . . 109. **S. dasyclados** Wimm.

\+ Different from above . . . . 12.

12. Grown leaves of terminal shoots oblong-lanceolate, broadly lanceolate, or rarely lanceolate, to 2.5 cm broad . . . . 13.

\+ Grown leaves of terminal shoots more or less lanceolate or elongate-lanceolate . . . . 14.

13. Petioles to 1 cm long, dilated at base; leaf blades broadest below the middle, bright green and glabrous above, the lower surface pale,

132 whitish-silky from numerous small tubercles; lateral veins 18—23 pairs; catkins serotinous; stipe of ovary twice the length of the gland . . . . 104. **S. semiviminalis** E. Wolf.

\+ Petioles 5—6 mm long; leaf blades commonly broadest above the middle, the upper surface dark green, glabrous or covered with scattered hairs; lower surface lustrous-sericeous; lateral veins 10—12—18 pairs; catkins precocious; stipe of ovary half the length of the gland . . . . 105. **S. argyracea** E. Wolf.

14. Grown leaves to 8 cm long and 1.8 cm broad, lanceolate, elongate-lanceolate, or linear-lanceolate, remotely denticulate, or slightly undulate-sinuate, often entire, revolute, the upper surface green or grayish-green, glabrous or merely on the veins white-hairy, the lower surface lustrous-sericeous from longitudinal hairs . . . . 103. **S. jacutica** Nas.

\+ Different from above . . . . 15.

15. The surfaces of grown leaves concolor, glabrous or glabrate . . . . 16.

\+ The lower surface of grown leaves glaucescent or glaucous-green, glabrous or covered with short appressed hairs . . . . 17.

16. A tree to 30 m high; leaves subentire or undulate and crenate, glabrous above, with few scattered cilia beneath; lateral veins 20—25 pairs, at an angle of 40—50—70°; ovary densely silvery-hairy; style filiform, very long, equaling the ovary . . . . 107. **S. sachalinensis** Fr. Schmidt.

+ A rather tall shrub; leaves sharply serrate, on both sides green and glabrous; lateral veins 10—12 pairs, at an angle of 30—40°; ovary greenish or brownish, sparingly pubescent, inflated; style ca. 1 mm long . . . . 108. **S. udensis** Trautv. et Mey.

17. Upper surface of grown leaves dark green, dull, slightly white-silky; ovary subsessile or short-stipitate; style 1—1.2 mm long . . . . 110. **S. opaca** Anderss.

+ Upper surface of grown leaves dark green, glabrous, lustrous; stipe of ovary equaling to twice as long as the gland; style 0.2—0.6 mm long . . . . 106. **S. Siuzewi** O. v. Seem.

94. **S. viminalis** L. Sp. pl. (1753) 1021 (sensu amplo); Gmel. Fl. sibir. I (1747) 162, No. 16; Pall. Fl. Ross. I, 2 (1788) 76; Ldb. Fl. Ross. III, 605; Turcz. Fl. baic.-dah. II, 1,105; Shmal'g., Fl. II, 434. — Vernacular Russian names: iva korzinochnaya, belotal, verboloz, loza, korzina loza, kuzovitsa.

A much branched shrub to 5—6 m high or rarely a tree to 8—10 m high with straight long slender appressed branches; young shoots short, grayish-pubescent or glabrate, finally glabrous or covered with very short
133 hairs; exposed wood without striations; buds 3—5 mm long, ovate-oblong, flattened, hooked at apex, yellowish or reddish-brown, mostly sparingly pillulose; stipules narrowly lanceolate, sometimes lobed or falcate, long-acuminate, glandular-dentate, commonly promptly caducous, mostly shorter than petiole, often absent; petioles typically 0.5—1.2 cm long, more or less silky, sometimes not more than 0.1—0.4 mm long (f. a b b r e v i a t a Doell.); grown leaves of terminal shoots narrowly lanceolate or linear-lanceolate, 15—20 cm long and 0.3—2—4 cm long, broadest below the middle, broad or cuneate at base, attenuate toward apex, acute, exrevolute margins entire or slightly undulate-sinuate, rarely distinctly serrate; upper surface from dark green glabrate or sparsely pubescent to hoary-pubescent with tuberculate glands near the margin; lower surface densely covered with silky hairs parallel to the lateral veins, lustrous-sericeous or silvery-lustrous; midrib prominent; lateral veins 25—30 pairs, at an angle of 60—80°, faint near the margin; catkins precocious or coaetaneous, sessile or nearly so, ebracteate or with more or less developed bracts at base; staminate elongate-ovoid, 2—3 cm long and to 1.5 cm broad; pistillate 3—4 cm long, elongating in fruit to 6 cm, commonly upright, dense; scales oblong-ovate or broadly rounded, obtuse or more or less acute, light brown throughout or merely at base, darker to almost black at apex, commonly with long pale hairs on both sides or more sparingly pubescent; stamens 2; filaments slender, distinct or rarely somewhat connate; anthers golden-yellow, finally turning darker; ovary ovoid or ovoid-conical, sessile or borne on a very short stipe, densely silky; style 0.3 to 2—2.5 mm long; stigmas 1—2 mm long, filiform, pale, mostly entire or parted, as long as or longer than the style; gland 1, posterior, linear-lorate, to 0.8—1.5 mm long, reaching one-third of the ovary; capsule silky, inflated, 4—5 mm long. Fl. March—May; fr. April—June.

Riverbanks. Nearly all over the Soviet Union, from forest-tundra to the desert-steppe region, exclusively along riverbanks, where it forms dense and often extensive thickets. Does not ascend above the forest zone. Apparently it is not to be found in the Crimea and does not occur in the desert-steppe region of Central Asia (Kara K., Kyz. K., partly

Ar.-Casp., etc.), where it is replaced on the banks of rivers by S. songarica Anderss., S. Wilhelmsiana M. B., etc. **Gen. distr.**: Scand., Centr. and Atl. Eur.; E. Med., Dzu.-Kash., Mong., Jap.-Ch., India. Cultivated in North America and in Chile. Described from Sweden. Type in London.

Hybridizing with S. acutifolia, alba, aurita, caprea, caucasica, chlorostachya, cinerea, dasyclados, fragilis, glauca, hastata, lapponum, livida, nigricans, phylicifolia,
134 purpurea, rosmarinifolia, sibirica, tenuijulis, triandra.

Economic importance. One of the best willows for consolidation of banks and dikes. Yields excellent material for wickerwork. Easily propagated by cuttings. Yield of shoots comparable with that of S. triandra. Sprouts very profusely from stumps and cuts. It is used for pollarded willow culture, for hoops, rigging, and fine wickerwork (as shaved slivers). The wood is white, light, soft; on boiling in water it turns rose. The bark is used for production of salicin and for tanning; tannin content 6.18—14.5% (on the Kama River up to 20%); tannin content of leaves 2.4—9.59%. In the extreme northern parts of W. Siberia and in the Far East, the bark of this willow is used for making fishing nets; Yakuts feed the foliage to livestock. The young twigs, bark, buds, flowers, and leaves provide the chief food of the water vole (Arvicola amphibius L.) which destroys stands of this willow in the lower reaches of the Volga by gnawing the bark.

In cultivation, this willow produces marketable twigs the year after planting. These are distinguished by an outstandingly high yield of material for wickerwork (up to 12.6 tons per hectare), of this about 30—40% of first and second grade. Highly valued for hoops and for basketry. Among the drawbacks, one has to mention the thick loose heartwood and susceptibility to insect damage. Very ornamental and shedding its leaves late.

Hybrids of S. viminalis with S. dasyclados, S. caprea, and S. cinerea (e. g., S. stipularis Sm.) are characterized by their vigor and very rapid growth. Hybrids with S. triandra (e. g., S. mollissima Ehrh.) yield the best tanning bark and the best twigs for wickerwork.

Note. Extremely variable over the enormous distribution area that stretches from the Atlantic to the Pacific Ocean, and from the Kanin Peninsula, the Lena estuary, and Anadyr to the Himalayas, and undoubtedly represents a whole series of minor species as yet insufficiently known and obscured by numerous hybrid forms. More distinct from the systematic and phytogeographic points of view are the following minor entities of the aggregate species S. viminalis L.:

94a. **S. veriviminalis** Nas. nom. nov. — S. viminalis L. (s. str.); Wimm. Salic. europ. 36. — S. longifolia Lam. Fl. Franc. II (1778) 232. — Ic.: Rchb. XI, tab. 597, fig. 1248; Fl. Dan., No. 2485. — Exs.: Kern., Exsicc. No. 70; Wimm., Herb. Salic., 126, 127; Coll. Salic., 101—103.

Young branchlets and buds always rather densely canescent; branches of the preceding year glabrous; expanding leaves narrowly lanceolate, acuminate, green above, silvery beneath; grown leaves longer and narrower, 10—18 times as long as broad, commonly linear-lanceolate, gradually narrowed from the middle toward apex, tapering to a long tip, cuneate

at base, the upper surface dark green, the lower lustrous-sericeous all over; lateral veins (25 — 30 mm) obscure at the ends, at an angle of 60 — 80°;
135 catkins precocious, sessile or nearly so; pistillate ebracteate or with few scalelike bracts at base; staminate leafy-bracted; scales oblong or oblong-ovate, acute to subobtuse, light brown, darker at apex, to blackish-brown; ovary sessile or borne on a very short stipe, elongate-conical, densely clothed with short silvery hairs; style 1.5 — 2 — 2.5 mm long, slender; stigma with filiform lobes, not longer than the style (1 — 1.7 — 2 mm). March — April.

A West European species, replaced in the European part of the Soviet Union by the next species, but may be found near the western border.

95. **S. rossica** Nas., nom. nov. — ? S. *serotina* Pall. Reise III, Anh. (1776) 759, tab. Nn, non Schur (1866). — S. *viminalis* γ *Gmelini* Anderss. in DC. Prodr. XVI, 2 (1868) 266. — S. *Gmelini* Teplouch. in Tr. S.-Peterb. O. Est., XXXI, 3 (1901) 26 — 30; Idem. auct. pl. Fl. Ross., non Pall.; Petun. et Syreishch., Ill. Fl. Mosk. gub. II, 33 — 34 (cum fig.); Kryl., Fl. Zap. Sib. IV, 740; Kom., Fl. Kamch. II, 18. — Exs.: HFR, No. 2479.

A tall shrub or tree to 8 — 20 m high, 20 — 50 cm in trunk diameter, attaining the age of 30 — 40 (and up to 70) years;* branches virgate; the young gray-pubescent or glabrous; the annotinous yellow or olivaceous-latericious and mostly glabrous; bark on the trunk gray, on the inside white or greenish; buds at first pubescent, becoming glabrous, rufescent, conical, to 5 mm long and 1.5 mm broad; stipules mostly confined to elongated shoots, promptly caducous, small, linear, bristle-shaped, lunate, or falcate, commonly shorter than the petiole; petioles 4 — 14 mm long, pubescent; expanding leaves oblong-obovate, short-acuminate to subobtuse, pale and sericeous on both sides; grown leaves on short shoots 5 — 7 cm long and 0.7 — 1.2 cm broad, lanceolate, mostly broadest above the middle, narrowed toward base, rarely broad at base (f. *typica* Görz), obtusish or acuminate at apex; leaves on long vigorous shoots 8 — 10 (20) cm long and 1.5 — 2 cm broad (f. *vulgaris* Kryl.), rarely 4 — 6 cm broad (f. *latifolia* E. Wolf) or narrower, 0.8 — 1.3 cm broad (f. *angustifolia* Turcz.), commonly 7 — 14 times as long as broad, linear-lanceolate, often with margins parallel for a stretch, equally narrowed toward base and apex, or almost round at base, terminating in a short often oblique point, entire or undulate sinuate, slightly revolute, with tuberculate glands near the margin above, slightly pubescent, rather densely clothed beneath with appressed silky hairs, ca. 0.3 — 0.4 mm long, sometimes densely covered
136 with snow-white silky tomentum (f. *pellita* Görz); veins prominent; midrib stramineous; lateral veins 20 — 28 pairs, at an angle of 60 — 70°; catkins coaetaneous, cylindric, subsessile or short-stalked; pistillate 1.5 — 2.5 cm long and ca. 0.7 cm broad, elongating in fruit to 6 cm, subtended by 2 or 3 oblong-obovate leaflets, staminate 2 — 2.5 cm long and ca. 1 — 1.2 cm broad, commonly with foliaceous or scalelike bracts; scales broad, orbicular, rusty or brown, sometimes pale (f. *pallida* E. Wolf), covered on both sides with pale hairs; style 0.3 — 0.5 or rarely to 0.8 mm long,

* Often becoming hollow in age. In the European part of the Soviet Union often 6 — 8 m high and correspondingly thinner.

very rarely longer (f. longistyla E. Wolf), thickish, the filiform stigmas sometimes (not always) several times the length of style, up to 1—1.5 mm long, mostly entire, rarely 2-parted (f. schizmostigma Görz); gland linear, ca. 1—1.5 mm long. Fl. April—May; fr. May—June. (Plate VII, Figure 6).

Nearly the whole of the European part (not seen in the Crimea); W. and E. Siberia, and Far East.

Note. Beyond the Urals it gradually loses its characteristic features and E. of the Ob (to the Lena) catkins are encountered which appear before the leaves, as in S. veriviminalis, or after the leaves. The same can be observed to some extent in the Far East and in Korea; thus, Nakai, for instance, regards the Far Eastern S. viminalis as identical with the West European. Characters indicating close affinity to the West European form are also to be found in S. Schwerini E. Wolf from the banks of the Amur (1929), of which the type specimen is represented by only one leaf. The species S. yezoensis Kimura (1931), reported for Sakhalin and Japan, differs from the preceding chiefly in the shorter stigma, only one-quarter to one-half the length of the style; from S. rossica it differs in the precocious catkins, which bring it close to the European S. veriviminalis Nas. It is impossible to separate the distribution area of S. yezoensis from that of S. rossica which extends as far as the Pacific coast.

96. **S. splendens** Turcz., Fl. baic.-dah. II (1854) 107. — S. viminalis ξ Turcz., l.c. — S. viminalis β splendens Anderss. in DC. Prodr. XVI, 2 (1868) 265.

A tree or shrub; leaves lanceolate to elongate-lanceolate, entire, about equally silky on both sides, or the upper surface dull, grayish-green or somewhat hoary, the lower surface beautifully lustrous-sericeous or densely silvery-tomentose; lateral veins at an angle of ca. 60°; catkins coaetaneous, both staminate and pistillate with bracts at base, elongated, spreading or slightly recurved; scales almost monochromatic, rufescent, short; ovary slightly beaked; style shorter than the undivided or rarely 2-parted stigma; in other characters resembling the preceding. Fl. May.

E. Siberia: Yen., Ang.-Say., Lena-Kol., Dau. Endemic. Described from the vicinity of Irkutsk and from Transbaikalia. Type in Leningrad.

137 97. **S. burjatica** Nas. nom. nov. — S. nitens Turcz. Fl. baic.-dah. II, 1 (1854) 107, non al. — S. viminalis L. γ et δ ibid.

Shrub; leaves lanceolate or oblong-lanceolate, flat, distinctly serrate already when young, rarely entire, the upper surface subsericeous with rather long hairs; lateral veins arising at a wide angle then upcurved toward apex, often bifurcate near margin; catkins sessile, precocious or coaetaneous; staminate ebracteate; pistillate with scalelike bracts at base; scales short, blackish-brown or brownish-black at apex; style moderate, about as long as or shorter than the entire stigmas. Fl. May.

E. Siberia: Ang.-Say. Described from the vicinity of Irkutsk. Endemic. Type in Leningrad.

98. **S. rufescens** Turcz. Fl. baic.-dah. II, 1 (1854) 107. — S. viminalis L. ε Turcz., l. c. — S. viminalis γ Gmelini 2° rufescens Anderss. in DC. Prodr. XVI, 2 (1868) 266. — S. viminalis β splendens 2° brevifolia Anderss. (ex pte.), ibid., 265.

Shrub; branches commonly glabrous, latericious-green or brown; young leaves obtuse, mucronate, dark green, silky; grown leaves flat, oblong-lanceolate, narrowed toward base, or narrowly lanceolate, entire; upper surface dull, cinereous or subsericeous from scattered rather long hairs, with thinner vesture than that of S. nitens Turcz., at length glabrate; lower surface silvery-sericeous; lateral veins not evident; catkins slender; scales subovate, rufous or subferruginous to light brown, scarcely darker at apex, sparsely pillulose, at first equaling and concealing the ovary, finally somewhat shorter; ovary obtusish; style abruptly attenuate, pointed, about as long as or barely longer than the entire stigma. Fl. May.

W. Siberia: Alt.; E. Siberia: Ang.-Say., Dau. Endemic. Described from the vicinity of Irkutsk.

Note. In shape of flowers and fruit more closely related to the European S. viminalis L. than to any of the other Central Siberian representatives of this series.

99. **S. pseudolinearis** Nass. nom. nov. — S. linearis Turcz., Fl. baic.-dah. II, 1 (1854) 106 (non Forb.). — S. viminalis β angustifolia Turcz., ibid.

A shrub 2 — 5 m high, with branches breaking off at their base; grown leaves linear-lanceolate, typically 2 — 4 mm broad or broader, cuneately long-tapering toward base, borne on a petiole ca. 1 cm long, revolute, entire, glabrous or scarcely hairy above; catkins precocious, sessile; staminate ebracteate; pistillate with few scalelike bracts, rather short at anthesis;
138 scales short, almost wholly dark, acutish, in staminate flowers ovate-lanceolate, long-hairy; ovary at first slightly longer than the scale, ovoid, acutish, finally inflated and obtuse; style longer than or about as long as the stigmas. Fl. May. (Plate VII; Figure 6a).

European part: U. V., V.-Kama, V.-Don, Transv.; W. Siberia: U. Tob.; E. Siberia: Ang.-Say., Dau.; Far East: Ze.-Bu.; Centr. Asia: Ar.-Casp., Balkh. **Gen. distr.**: Mongolia. Described from Verkhneudinsk. Type in Leningrad.

Note. Almost indistinguishable from the narrow-leaved form of the European S. viminalis L. (coll.), it occurs rather rarely in W. Europe, while S. pseudolinealis Nas. plays a conspicuous role in the landscape of the Lake Baikal area and in Mongolia, which strengthens the assumption of of its independent standing.

Economic importance. Branches slender, flexible, brittle at the juncture, suitable for wickerwork. Very ornamental. Apparently a very cold-resistant species.

100. **S. turanica** Nas., sp. nova in Addenda IV, p. 541.— S. viminalis β splendens 1° songarica Anderss. in DC. Prodr. XVI, 2 (1868) 265.

A tall shrub, with latericious or yolk-yellow branches clothed with fugacious whitish tomentum; petioles short, white-silky; stipules mostly none; leaf blades stiff, oblong to ovate-oblong, 11 — 14 cm long and 2 — 4 cm

PLATE VII. 1. Salix rosmarinifolia L.– 2. S.sibirica Pall.– 3. S.sitchensis Sans.– 4. S.gracilistyla Mig.– 5. S.dasyclados Wimm.– 6. S.rossica Nas.– 6a. S.pseudolinearis Nas.– 7. S.sajanensis Nas.– 8. S.Siuzewii O.v.Seem.

broad, broadest in lower part, toward base abruptly cuneate-attenuate, toward apex short- or long-acuminate, the margin slightly wavy, becoming flat, entire or obscurely sinuate; upper surface dull green or grayish-green, clothed with scattered short appressed, variously directed hairs or glabrate; lower surface more heavily clothed with whitish-silvery tomentum, lustrous, with prominent tomentose veins; lateral veins 16—18 (20) pairs, at an angle of 50—90°; catkins wholly sessile. Fl. April.

Riverbanks. — Centr. Asia: Balkh., Dzu.-Tarb., T. Sh. **Gen. distr.:** Dzu.-Kash. Described from Semireche. Type in Leningrad.

Economic importance. Very ornamental. Used for basketmaking. Foliage is eaten by domestic livestock.

141 101. **S. strobilaceae** (E. Wolf) Nas. comb. nova, Addenda IV, p. 542. — *S. viminalis* var. *strobilacea* E. Wolf in Izvest. S.-Peterb. Lesn. Inst. XIII (1905), offprint, p. 3—4, tab. 1.

A tall shrub; young shoots whitish-velutinous; annotinous grayish-velutinous; stipules falcate, acuminate, glandular, half as long as to equaling the petiole; petioles ca. 1 cm long; leaf blades narrowly lanceolate, broadest below the middle, gradually arcuate-attenuate toward apex, from 6—8 to 14 cm long and to 2 cm broad, entire, the revolute margin slightly sinuate; upper surface sparsely glandular, near the margin dull green, sparingly pubescent; lower surface whitish-silky from dense coarse shiny hairs; midrib prominent beneath, pale brown or pale yellow; lateral veins 13—15—22 pairs, subparallel, at an angle of 50—60°, very prominent; catkins coaetaneous, short-stalked, with bracts at base; scales large, concealing the ovary, liguliform, acute or obtuse, light brown, darker at apex, sparsely pillulose; ovary sessile or borne on a very short stipe, ovoid-conical, clothed with shiny white hairs; style more or less bifid, often down to base; stigmas with narrow entire or 2-parted lobes; gland oblong, rather long. Fl. April.

Waterside thickets. — W. Siberia: Irt. Endemic. Described from the Karkaralinskoe forest. Type in Leningrad.

102. **S. sajanensis** Nas. sp. nova in Addenda IV, p. 542. — *S. viminalis* L. *α genuina* Turcz. (ex prt.) in Fl. baic.-dah. II, 2 (1854) 379.

A shrub or tree 2—5 m high; bark gray and smooth near base, reddish-brown dull or shiny on the branches; branches often fancifully twisted and curved, rather stout, strongly gnarled; buds to 7 mm long and 4 mm broad, dark castaneous, obtuse, at first silky, becoming glabrous; petioles 4—5 mm long, silky; stipules small, glandular, silky, early caducous; leaf blades obovate to elongate- or linear-lanceolate, 4—9 cm long and 0.5—1.5 cm broad, on strong shoots to 15 cm long and to 3 cm broad, narrowed at both ends, somewhat recurved, mostly entire, the margins revolute or remotely glandular-dentate; upper surface dark green, dull, covered with scattered short pubescence, becoming glabrous; lower surface pale green and slightly silky to lustrous-sericeous; young leaves silky and lustrous on both sides; lateral veins 12—15 pairs, evident beneath, at an angle of 30—40°; pistillate catkins borne on a very short stalk, with scalelike bracts at base, compact, cylindric, 2.5 — 3 cm long and to 0.5 cm thick, in fruit to 4 — 6 cm long and 0.8 cm thick; scales ovate, acutish, dark brown to almost

black, paler at base, densely clothed with silvery hairs; gland
142 oblong; ovary subsessile or borne on a very short stipe, ovoid, greenish and densely hairy; style elongated; stigmas long, linear, sometimes 2-parted, as long as or longer than the style. May—June. (Plate VII, Figure 7).

Subalpine zone, 1,700—2,100 m, on stony screes and on rocks. — W. Siberia: Alt. (Biisk—Krylov; Ust-Kamenogorsk); E. Siberia: Ang.-Say. (herb. Turczaninowii sub. n. S. viminalis L. var. in rupibus subalpinis pr. Tsagan-Gol, leg. 1834, etc.). **Gen. distr.:** N. Mongolia. Described from E. Sayans (Tunkinskie Gol'tsy). Type in Leningrad.

Note. This willow needs to be traced in Yakutia and the Okhotsk area, since the specimen in the Herbarium of the Academy of Sciences, collected by Middendorf, is of uncertain origin. In addition, Seroshevskii writes that in Yakutia, "in the mountains and by the lakes one comes across a misshapen black willow ("kharatalak") with rugged bark and fancifully twisted branches (referred by Maack to S. viminalis). The twigs, owing to their flexibility and strength, may be used as a substitute for ropes." Might not this be S. sajanensis?

103. **S. jacutica** Nas. sp. nova in Addenda IV, p. 543.

A tree? Young branches stoutish, pubescent, becoming glabrous, dark brown; buds to 5—6 mm long, ovoid-conical, obtusish, the upper pubescent, the lower glabrous; stipules small, rarely up to half the length of petiole, lanceolate or falcate, crenulate, soon caducous; petioles ca. 6—10 mm long, dilated toward base, pubescent; leaf blades to 8 cm long and to 1.8 cm broad, lanceolate or elongate-lanceolate to linear-lanceolate, acuminate, rounded or narrowed at base, the margin distantly denticulate or slightly undulate-sinuate, often subentire, revolute; upper surface green or grayish-green, glabrous or white-hairy merely on the veins or covered all over with scattered hairs visible only with a magnifier; lower surface lustrous-sericeous from dense short longitudinal hairs; midrib yellowish, strongly thickened toward the petiole; lateral veins 12—14 pairs, at an angle of 40—80°, clearly evident on the upper surface; young leaves silky-pilose, revolute; flowers unknown.

Islands of the Lena and its tributaries. — E. Siberia: Lena-Kol. (Yakutsk and Verkhoyansk districts). Described from Zhigansk and Bestyakh). Type in Leningrad.

Note. Represented in the Herbarium of the Academy of Sciences by numerous specimens (all without flowers), some of which are linked by transitional forms with S. kolymensis O.v. Seem. or with S. viminalis s.l. There is a definite possibility of hybrid origin, but this seems to be counterindicated by the distribution in the Lena basin. Very similar to S. sajanensis, but the latter grows on stony screes in the subalpine zone of the Altai and Sayan mountains, and is distinguished by such characters as the often oblanceolate leaves with lateral veins at an acute
143 angle of 30—40°, gnarled branches, etc.

104. **S. semiviminalis** E. Wolf in Izv. Lesn. Inst. XIII (1905), offprint, p. 7 et tab. IV. — S. subviminalis E. Wolf ibid.

A shrub, apparently tall; branches slender, the young pubescent, the annotinous slightly hairy, brownish; petioles dilated at base, to 1 cm long;

stipules shorter than petiole, subfalcate, dentate, on weak shoots absent; leaves lanceolate to broadly lanceolate, uniformly narrowed at both ends, to 9—13 cm long and to 2.5 cm broad; the revolute margins undulate or at the middle with few very distant small dentations; upper surface bright green, sublustrous, glabrous; lower surface pale, whitish-silky with slightly crisped white, indefinitely transversal shiny hairs; lower leaves with sparser vesture, studded with numerous small tubercles visible under magnification; midrib strongly attenuate toward apex; lateral veins on the upper larger leaves 18—23, on the leaves of short shoots ca. 10 pairs, arising at an angle of 40—50—70°, upcurved at the margin and convergent; catkins serotinous; staminate unknown; pistillate cylindric, dense, at base loose, stalked, with 2—4 small bracts, 4—4.5 cm long, elongating in fruit to 8 cm, the rachis hairy; scales liguliform, acute or obtuse, light brown, darker at apex, rather sparsely covered with white hairs; ovary ovoid-conical, white-silky, ca. 2.2 mm long; stipe ca. 1 mm long, twice the length of the strap-shaped gland; style ca. 0.6 mm long, hairy; stigma dark, with oblong entire or 2-parted lobes, ca. 0.4—0.6 mm long, erect or spreading; ripe capsules retaining scale and gland. Fl. beginning of May; fr. June.

W. Siberia: U. Tob., Irt. Endemic. Described from Borovskoe forest in Akmolinsk Region. Type in Leningrad.

Note. Very similar to this willow is the hybrid, described by F. A. Teploukhov under the name S. caprea × Gmelini (in Petunn. Kritich. obz. III, 31; Petunn. and Syreishch., Ill. Fl. Mosk. gub. III, 35, with illustration), but here the annotinous shoots are yellow and glabrous; leaves inequilateral, elongate-lanceolate, glaucescent and sparsely woolly above, without small tubercles beneath; catkins coaetaneous; scales obovate, acuminate, dark brown, densely hairy; ovary woolly-tomentose; style about as long as ovary; stipe as long as or shorter than the ovary.

105. **S. argyracea** E. Wolf in Izv. Lesn. Inst. XIII (1905), offprint 6. — Ic.: ibid, tab. III. — Exs.: Görz, Sal. As. III No. 61.

A shrub, apparently fairly tall, upright; branches slender, glabrous, brownish, when young more or less silky; buds small, ovoid, obtuse,
144 brown, glabrous, 5—6 mm long; stipules to 6 mm long, ovate-lanceolate to lanceolate, glandular-serrulate, soon caducous; petioles 4—10 mm long, glabrous or clothed with short hairs, brownish; leaf blades oblong-lanceolate or broadly lanceolate, 4—8 (—11) cm long and 1.5—2.2 cm broad, on short shoots but 1.5—3 cm long, at both ends arcuately narrowed, mostly broadest above the middle, the slightly revolute margins remotely glandular-dentate; upper leaves commonly glandular-serrulate, those at the base of shoots subentire or entire; upper surface dark green, glabrous or covered with scattered appressed hairs; lower surface lustrous-sericeous from compact short hairs parallel to lateral veins; midrib pale brown, strongly dilated toward base, whitish above; lateral veins 9—12 (in large leaves up to 18) pairs, raised beneath, impressed above, at an angle of 50—60° (to 80°); catkins precocious, dense, upright, the stalk scaly-bracted; scales ovate, obtuse or acute, blackish at apex, covered with long white silky hairs; ovary ovoid-conical, densely silky; stipe half the length of the oblong gland; style half as long as ovary, the style and stigma yellowish or brownish; stigma lobes entire or 2-parted, sometimes recurved. May.

Riverbanks. — Centr. Asia: Dzu.-Tarb. **Gen. distr.:** Dzu.-Kash. Described from Semireche [Dzhety-su]. Type in Leningrad.

Economic importance. Ornamental.

106. **S. Siuzewi** O. v. Seem. in Fedde, Repert, sp. nov. V (1908) 17; E. Vol'f in A. H. P. XXVIII, 4 (1911) 527; Siuzew in Tr. Bot. Muz. Ak. Nauk IX (1912) 90, fig. 1. — S. phylicoides var. attenuata Anderss. (ex prt.) Monogr. Salic. (1867) 141 et in DC. Prodr. XVI, 2 (1868) 245. — Exs.: HFR No. 2488♀, 2489♂.

A shrub to 5 m high; branches virgate, glabrous, yellowish-gray, latericious, or olivaceous; buds oblong, pubescent, becoming glabrous; stipules setiform, linear or lanceolate, sometimes serrate, soon caducous, half as long as the petiole; petioles 2—10 mm long; leaves lanceolate, narrowed at both ends, 4.5—6.5 cm long and 0.6—1.2 cm broad, on strong shoots to 14 cm long and to 2 cm broad, the slightly undulate or almost flat margins crenate-sinuate or subentire; upper surface dark green, lustrous, glabrous; lower surface glaucous or glaucescent, glabrous or clothed with scattered appressed hairs; midrib conspicuously rufescent; lateral veins 15—20 pairs, prominent, at an angle of 40—50°; catkins precocious, sessile, dense; staminate oblong-cylindric, upright, to 3 cm long and 1 cm broad; pistillate cylindric, 2 cm long and 0.6 cm broad, upright or pendulous, sometimes recurved, ebracteate or with small bracts at base; scales lanceolate or liguliform, brownish, blackish at apex, hairy; stamens 2, distinct, glabrous; anthers golden, turning brown;
145 gland 1, interior, oblong-linear, 0.6—1.1 mm long; ovary ovoid-conical, ca. 1.5—1.7 mm long, hoary or glabrate; style ca. 0.2—0.6 m long; stigmas elongated or [?] divaricate, as long as or longer than the style (0.7—1 mm); stipe twice as long as the gland (0.6 — 1.5 mm). May. (Plate VII, Figure 8).

Riverbanks. — E. Siberia: Dau., Lena-Kol.; Far East: Ze.-Bu., Uss., Kamch. **Gen. distr.:** Mong., Jap.-Ch. Described from Ussuri Territory. Type in Leningrad.

Note. Regularly confused in herbaria with S. sachalinensis F. Schmidt which is a tall fast-growing tree to 30 m high and 20—35 cm in trunk diameter, with yellowish-brown lustrous naked bark; leaves dull green and sometimes finally lustrous above, with few scattered cilia beneath; catkins coaetaneous, with 2 or 3 bracts at base; ovary densely gray-hairy; style long; stipe shorter than the gland. In areas of convergence of the two species, the distinguishing characters become obscured, and the similarity of the leaves renders identification difficult in the absence of flowers.

Economic importance. Nectariferous; providing material for wickerwork.

107. **S. sachalinensis** F. Schmidt, Reise in Amurl. in Mém. Acad. S.-Pétersb. sér. VII, vol. XIII, No. 2 (1868) 173, 219; O. v. Seem. Salic. Japon., 53 et tab. X; Floder. in Ark. f. Bot. 20 A, No. 6 (1926) 40; Kom., Fl. Kamch. II, 20; Kom. and Alis., Opred. rast. Dal'nevost. kr. I, 426. — S. phylicoides var. attenuata Anderss. (ex. p.) Monogr. Salic. (1867) 141 et in DC. Prodr. XVI, 2 (1868) 245. — S. kamtschatica ad interim) Anderss. in sched., pro parte. — S. Korsakoviensis Leveil. in Bull. Soc. Botan. de France, LVI, No. 5 (1909) 302.

A tree, reaching the age of 50 years, 30 m high and 20—35 cm in trunk diameter; bark smooth, yellowish-brown; young branches pubescent, grayish; buds appressed, ovoid, acutish or [?] glabrous; stipules small, semicordate, acute; petioles to 7 mm long, at first pubescent, becoming glabrous; leaf blades lanceolate, elliptic-lanceolate, or obovate-lanceolate, narrowed toward base, obtusish or pointed at apex, 5—10 cm long and 0.5—3 cm broad, subentire with revolute or undulate margin, undulate, sinuate-denticulate, or crenate, the two sides almost concolor, the upper dark green, toward fall apparently varnished, the lower with scattered cilia; young leaves velutinous-pubescent; midrib strong, at first stramineous, finally rufescent; lateral veins 20—25 pairs, prominent, at an angle of 40—50—70°; catkins coaetaneous, dense, cylindric; staminate sessile, 2—3 cm long and 0.5 cm broad; pistillate 2—4 cm long, elongating in fruit to 6 cm, borne on a stalk ca. 0.5—1 cm long, with 2—4 small
146 pubescent bracts, loose or interrupted below, to 1 cm thick; rachis silky; scales lanceolate or oblong-ovate, densely clothed with gray hairs, light brown or blackish; stamens 2, distinct, glabrous, 3 times as long as the scale; anthers ovate, yellow; gland 1, interior, linear, enlarged toward apex, three-fourths or two-thirds the length of the scale and longer than the stipe of the subsessile ovoid-conical densely gray-pilose ovary; style as long as ovary, filiform; stigmas yellow, ca. half as long as the style, linear, entire or 2-parted, divergent. Fl. June; fr. July.

River valleys, damp slopes, and wood margins, in pure stands or mixed with other plants. — Arctic: An.; E. Siberia: Lena-Kol.; Far East: Kamch., Okh., Sakh., Uss. **Gen. distr.**: Japan, Kurile Islands. Described from the Amur. Type in Leningrad.

Economic importance. The bark, branches, and thin stems are used in large amounts for erection of fishing enclosures in rivers, for construction of bridges, dikes, fences, and for making cordage, etc. A thin-stemmed species, valued for its rapid growth, the stem reaching 47—67 cm in circumference at the age of 36—46 years (Hultén).

108. **S. udensis** Trautv. et Mey. Flor. ochot. in Middend. Sibir. Reise I, 2 (1856) 81; Anderss. Monogr. Salic., 149 et in DC. Prodr. XVI, 2, 249; Kom. and Alis., Opred. rast. Dal'nevost. kr. I, 426.

A tall shrub with straight declinate branches; young branchlets silky; grown branches glabrous, brown; buds small, ovoid, obtusish, light brown, appressed, glabrous; stipules very small, linear, promptly caducous; petioles 2—4 mm long, slender, at first silky; expanding leaves silky beneath, with scattered cilia above, revolute; grown leaves unknown, the young oblong, long-tapering toward base, mostly short-cuspidate at apex, rarely obtusish, to 2.5—4 cm long and to 1 cm broad, both surfaces green and glabrous, the margin sharp-serrate or acute [?]; lateral veins ca. 10 pairs, pale beneath, faint above, at an angle of 30—40°; catkins subcoaetaneous, subsessile, with 2 caducous small bracts at base, cylindric, to 3 cm long and 0.7—0.8 cm broad, at first rather loose, at length closely packed with inflated carpels; rachis hairy; scales to 2 mm long, ovate-lanceolate, acuminate, dark brown above, hairy; ovary ca. 2 mm long, initially conical from an ovoid base, appressed silky, finally strongly inflated, obtuse at the top, greenish or brownish, sparingly pubescent; stipe 0.2—0.3 mm long, shorter than the strap-shaped gland; style undivided, ca. 1 mm long; stigmas 2-parted, with linear divergent lobes, to 0.7 mm long, both style and stigmas yellow. May—June.

Riverbanks. — Far East: Uda, whence described. Endemic. Type in Leningrad.

147 109. **S. dasyclados** Wimm. in Flora XXXII (1849) 35; Petunn., Krit. obz. Mosk. fl. III, 28; Kryl., Fl. Zap. Sib. IV, 743; Syreishch., Ill. Fl. Mosk. gub. II, 35. — S. stipularis ap. Ldb., Turcz., Trautv., Maxim., pp. (non Smith). — S. longifolia Host, Fl. austr. II (1831) 645)(non Lam., non Muhlb.); Wimm., Salic. europ., 42. — S. Gmelini Pall., Fl. Ross. I, II (1788) 77 (non Teplouch., Petunn. et auct. pl. fl. ross.). — ? S. serotina Pall. Reisse, III (1766), Anh., 759, No. 135 et tab. Nn, et in Fl. Ross., l. c. (non Schur). — Ic.: Rchb. Ic. Fl. Germ., t. 601, f. 1252; Host, Sal. 19, t. 62, 63. — Exs.: Wimm. et Kr. Herb. Sal. 7; Coll. Sal. 99; HFR No. 739, 2486, 2487.

A tree of third size-grade or often a tall shrub to 5—8 m, with brown or buff bark and stout branches; wood without striations; young shoots densely gray- or white-woolly (in f. denudata Nas. vesture of young shoots very sparse), dark brown; annotinous glabrous, greenish or olivaceous; buds large, ovoid, beaked, dark brown, pilose or woolly; stipules very large, falcate or falcate-lanceolate, serrate, often lobed with each lobe quite distinct, horizontally spreading; petioles short, pubescent, dilated toward base; young leaves elliptic, covered with appressed white tomentum; grown leaves broadly lanceolate, oblong-lanceolate to lanceolate, 8—20 cm long or longer, 2—3.5 cm broad (f. latifolia Wimm.) or narrower (f. angustifolia Heid.), mostly inequilateral, short-acuminate, revolute, entire or glandular-dentate; upper surface dark green, glabrous, with a yellowish impressed midrib; lower surface grayish-sericeous or dull-silky with short hairs; lateral veins 10—12; midrib stramineous or brownish; catkins precocious, large, closely disposed on the shoots, subsessile, with caducous bracts at base; staminate 3.5—4 cm long and ca. 1.8 cm broad; pistillate stalked, stoutly cylindric, 4—5 cm long and ca. 1.2 cm broad, strongly elongating in fruit, or much longer and more slender (v. baltica Laksch. in sched.) and then also the branches glabrous, lustrous, brown or olivaceous-green; scales particolored, dark brown or almost black at apex, faintly brown at base, obovate, in staminate flowers acute, in pistillate obtuse, long-hairy; stamens 2, distinct, glabrous; anthers yellow; gland 1, interior, linear to linear-oblong, about twice the length of the stipe; ovary ovoid at base, attenuate above, densely white-hairy, subsessile or borne on a short pubescent stipe, finally inflated; style long to very long, pubescent at base; stigmas divergent, linear, recurved. April—May. (Plate VII, Figure 5).

Riverbanks and lake shores. — Arctic: Arc. Eur., Arc. Sib.; European part: Kar.-Lap., Dv.-Pech., Lad.-Ilm., U. Dnp., U. V., V.-Kama, V.-Don, L. Don, Urals; W. Siberia: throughout; E. Siberia: throughout; Far East:
148 Okh., Ze.-Bu. **Gen. distr.**: Scand., Centr. and Atl. Eur., Mong., Jap.-Ch. Described from the [former] German Province of Silesia. Type in Berlin.

Hybridizing with S. abscondita, acutifolia, chlorostachya, cinerea, nigricans, phylicifolia, purpurea, triandra and viminalis s.l.

Economic importance: Fast growing, as are its hybrids with S. viminalis s.l. and with species of the section Capreae (e.g., S. stipularis Sm.). Twigs are used for hoops, fishing implements, and miscellaneous wickerwork for fishing purposes. The populations of Arctic Siberia and also the inhabitants of the Amur Valley weave fishing nets from the bark. Tannin content 6 — 12%. Nectariferous, ornamental, suitable for consolidation of banks, barrage planting, etc. The foliage is eaten by livestock.

110. S. opaca Anderss. ex. O. v. Seem. Sal. Japon. (1903) 53 et tab. 10. — S. stipularis Maxim. Prim. fl. Amur. (1859) 243, non Sm.; F. Schmidt, Reise in Amurl. in Mém. Acad. Sc. Pétersb. VII (1868) 172; Kom. and Alis., Opred. rast. Dal'nevost. kr. I, 425. — S. amnicola E. Wolf in A. H. P. XXVIII, f. 4 (1911) 531.

A tall shrub or a tree, with naked smooth yellowish-brown bark and a dense head of numerous short fulvous branches; young branches sparsely puberulous at the ends; buds fulvous, short-hairy; stipules lanceolate or narrowly semiovate, acute, dentate, not more than half the length of the petiole, glabrous or pubescent; petioles ca. 0.3 — 0.8 cm long, white-hairy; young leaves lanceolate to elliptic-lanceolate; grown lanceolate, narrowed at both ends, 6 — 15 cm long and 1.5 — 3.5 cm broad, the flat margin subentire or irregularly crenate; upper surface dark green, dull, slightly white-silky; lower surface grayish-green or glaucescent, glabrous or sparsely covered with short appressed hairs; midrib very prominent beneath, pale brown; lateral veins to 18 pairs, more or less evident, at an angle of 45 — 50°; catkins precocious, sessile, with small scalelike bracts at base, dense, cylindric, at first to 1.6 — 2 cm long, finally to 3 — 3.5 cm long and 0.7 — 1 cm broad, appressed or divergent, recurved; rachis gray-hairy; scales short-lanceolate or ovate, obtusish or acute, dark brown to almost black or sometimes pale at apex, with long white hairs; stamens 2, distinct, glabrous, 4 — 6 times as long as the scale; anthers ovate, yellow; gland 1, interior, narrowly strap-shaped to linear, ca. 1 mm long, half to three-quarters the length of ovary and scale; ovary 1.6 — 1.9 mm long, subsessile or short-stipitate, ovoid at base, abruptly narrowed into style, silky-pilose, finally somewhat greenish; style 1 — 1.2 mm long, as the stigma yellowish-brown, glabrous; stigmatic lobes ca. 0.6 — 0.7 mm long, filiform, commonly entire, spreading. May — June.

149 Riverbanks — Far East: Okh., Ze.-Bu., Uss., Sakh., and apparently Uda. **Gen. distr.:** Jap.-Ch. Distinguished by Andersson among material from Amur, but described much later by Seemen from Japan. Type in Leningrad.

Note. We believe that there is a misunderstanding concerning S. opaca Anderss. and the synonymous S. amnicola E. Wolf, where the ovarian stipe is described as being about as long as or but little shorter than the gland. The relationship was apparently established for ripe capsules in which considerable elongation of the stipe occurs.

Section 20. **HELIX** Dumort. Verh. Gesl. Wilgen in Bijdr. Naturk. Wetensch. (1825) 15. — Shrubs and trees with rather slender flexible virgate glabrous branches; bark sometimes lemon-yellow on the inside, rarely pruinose outside; leaves from broad- to linear-lanceolate, 4—12 times as long as broad, mostly acute, stiff, glabrous, glaucous, short-petioled, entire or serrate; rarely elliptic or oblanceolate, shorter and broader or with well developed silky pubescence; catkins precocious or subcoaetaneous, sessile or borne on a very short stalk, narrowly cylindric or cylindric, dense; stamens mostly connate throughout or rarely merely the filaments united; anthers apparently 4-locular, mostly yellow, blackening; gland 1, posterior, linear or oblong; ovary small, sessile or subsessile, silky or glabrous; style and stigma short.

Growing mostly along riverbanks in the desert-steppe, steppe, or rarely the forest belt, not rising to great altitudes.

A section containing a large number of species which provide excellent material for wickerwork.

1. Grown terminal leaves large, 8—15 cm long and 0.7—3.2 cm broad . . 2.
+ Grown terminal leaves much smaller; lower leaves and those of lateral shoots usually smaller still . . . . 11.
2. Leaf blades commonly 2—3.2 cm broad . . . . 3.
+ Leaf blades commonly to 1.5 cm or rarely 2 cm broad. . . . . 5.
3. Buds large or very large, glabrous; stipules large; petioles 0.5—1.5 cm long, stout, dilated toward base; leaf blades 2.5—3.2 cm broad; lateral veins at an angle of 35—50° . . . . 4.
+ Buds small, at first hairy, becoming glabrous; stipules small; petioles 0.2—0.4 cm long, slender; leaf blades 1.5—2 or rarely to 2.5 cm broad; lateral veins at an angle of 25—40° . . . 138. **S. oxycarpa** Anderss.
4. Leaf blades ca. 3.2 cm broad, short-tipped, pure green on both sides . . . . 141. **S. ferganensis** Nas.
+ Leaf blades ca. 2.5 cm broad, tapering to a long point, lustrous above,
150 glaucous beneath . . . . 139. **S. lepidostachys** O. v. Seem.
5. Stipules strongly developed, commonly longer than or at least as long as the petiole, rarely absent . . . . 6.
+ Stipules commonly shorter than the petiole . . . . 8.
6. Stipules of upper leaves to 3—4 cm long, those of other leaves 0.8—1.2 cm long; leaves alternate, lanceolate to linear-lanceolate, not blackening, stiff, sharply dentate or sharply serrate . . . . 140. **S. Korshinskyi** Görz.
+ Stipules often absent or, if present, much smaller, ca. 1—1.5 cm long; leaves subopposite, often blackening, commonly oblanceolate, thin, bitter in taste, entire or toward apex sharply serrate . . . . 111. **S. purpurea** L.
7. Leaves linear-lanceolate, 4—9 cm long and 0.6—1.2 cm broad, mostly broadest above the middle, coarsely and irregularly serrate throughout or near base entire, when full grown gray above, glaucescent beneath; stipules linear or lanceolate . . . . . . 113. **S. dahurica** (Turcz.) Laksch.
+ Leaves linear, to 15 cm long and 0.3—1.2 cm broad, mostly narrowed at both ends or enlarged toward apex, remotely sinuate and glandular-serrate; stipules linear-falcate, prominently reticulate-veined . . . . 114. **S. mongolica** Siuz.

8. Ovary densely clothed with shiny silky hairs; stipules often ovate or ovate-lanceolate, sharply serrate . . . . . 176. S. **margaritifera** E. Wolf.

\+ Ovary rather sparsely pubescent or glabrous; stipules mostly narrow-lanceolate to linear or subulate . . . . 9.

9. Lateral veins 7—12 pairs, long-extended toward apex and convergent; ovary quite glabrous, with a long style and 4 long linear recurved stigma lobes . . . . 128. S. **Komarovii** E. Wolf.

\+ Lateral veins 10—20 pairs; ovary more or less hairy at least at base; style and stigmas short . . . . 10.

10. Young shoots glabrous; leaves coarsely sinuate-glandular throughout; catkin-scales obovate or rounded-obovate, obtuse, quite black at apex; ovary short-pilose, glabrescent . . . . 127. S. **serrulatifolia** E. Wolf.

\+ Young shoots white-silky; leaves mostly entire at base, serrulate elsewhere; catkin-scales liguliform, acute or somewhat obtusish, light brown or at apex somewhat darker; ovary sparsely hairy, often almost glabrous except for a collar of hairs at base . . . .
151 . . . . 129. S. **Przewalskii** E. Wolf.

11 (1). Leaves 8—10 times as long as broad . . . . 12.

\+ Leaves 4—6 (7) times as long as broad . . . . 18.

12. Young shoots rather densely silky, rarely glabrate; branches very slender, flexible, often arising at a right angle to the stem; leaves mostly exstipulate, very short-petioled, linear, entire or glandular-serrulate, at first densely silky and lustrous on both sides, at length diffusely silky or rarely glabrate . . . . 13.

\+ Different from above . . . . 14.

13. Ovary glabrous; catkin-scales mostly erose at apex . . . .
. . . . .125. S. **microstachys** Turcz.

\+ Ovary densely sericeous-pilose; catkin-scales acutish . . . .
. . . . 126. S. **Wilhelmsiana** M. B.

14. Leaves gradually long-attenuate toward base, regularly and sharply serrate, commonly blackening . . . . .112. S. **tenuifolia** Turcz.

\+ Different from above . . . . 15.

15. Leaves entire; ovary densely white-hairy . . . . .131. S. **Olgae** Rgl.

\+ Leaves more or less serrate, rarely subentire; ovary glabrate or pubescent and finally glabrescent . . . . 16.

16. Ovary large, 4—5 mm long (the capsule 6—7 mm), narrowly conical, pointed toward apex, reddish or reddish-brown, sparsely covered with short hairs, at base glabrate . . . . 130. S. **linearifolia** E. Wolf.

\+ Ovary smaller, ovoid-conical . . . . 17.

17. Branches commonly lurid, lustrous; young branches and branchlets brownish; network of veins on the upper leaf surface very prominent; ovary white-silky, at length somewhat glabrescent, subsessile or borne on a very short hairy stipe . . . . 115. S. **caspica** Pall.

\+ Branches reddish-yellow, fulvous when young; branchlets yellow; network of veins on the upper surface of leaves less prominent; ovary glabrous or slightly pubescent, borne on a glabrous stipe . . . .
. . . . 117. S. **Michelsoni** Görz.

18 (11). Pistillate catkins up to 10—12 cm long in fruit; capsule glabrous, greenish or brownish-green, inflated at base and abruptly narrowing above . . . . 132. S. **Lipskyi** (Görz) Nas.

\+ Different from above . . . . 19.

152 19. Shoots mostly covered with bluish bloom; catkin-scales pale yellow, caducous; ovary conical, greenish, glabrous . . . 118. **S. coerulea** E. Wolf.
+ Shoots without bluish bloom; ovary glabrous or pubescent . . . . . . 20.
20. Leaves sharply sinuate-serrate, the teeth tapering to a subulate gland-tipped spine, approximate toward apex, more distant toward base . . . . 121. **S. spinidens** E. Wolf.
+ Leaf margin without subulate spiny teeth . . . . 21.
21. Ovary white-silky . . . . 22.
+ Ovary glabrous or slightly pubescent . . . . 26.
22. Pistillate catkins dense, in fruit to 7 cm long and ca. 1 cm broad; catkin-scales brown; ovary large, to 5 mm long . . . . 23.
+ Different from above . . . . 24.
23. Leaves mostly glandular-serrulate, rarely entire; catkins in fruit very plump, to 1—1.2 cm long, the rachis clothed with long hairs . . . . 134. **S. Albertii** Rgl.
+ Leaves mostly entire or obscurely toothed; catkins in fruit to 0.8 cm thick, the rachis covered with short hairs . . . . 135. **S. macrostachya** E. Wolf.
24. Leaf blade to 1.7 cm broad, commonly silky or densely silky on both sides (as in the silvery form S. a l b a); stipules sometimes large, to 1.1 cm long; leaves at the base of shoots obtuse . . . . 133. **S. pseudoalba** E. Wolf.
+ Leaf blades about half as broad, finally glabrous or slightly pubescent; stipules absent or small; leaves at base of shoots acute . . . . 25.
25. Leaves sharply denticulate; ovary silky, sessile or short-stipitate . . . . 116. **S. tenuijulis** Ldb.
+ Leaves mostly entire or minutely serrulate; ovary sericeous-tomentose or tomentose, the short stipe twice the length of the gland . . . . 119. **S. serawschanica** Rgl.
26. (21). Leaves mostly oblanceolate or broadly lanceolate, silky on both sides; ovary sessile . . . . 27.
+ Leaves mostly lanceolate or narrowly lanceolate, glabrous or pubescent; ovary sessile or stipitate . . . . 28.
27. Leaves to 8 cm long, broadly lanceolate or oblanceolate, obtuse or almost round at base, gradually attenuate toward apex, covered on both sides with rather long hairs; catkins loose, borne on a stalk to 1 cm long . . . . 137. **S. holargyrea** Bornm. et Görz.
153 + Leaves to 5 cm long, oblanceolate, mostly obtuse, narrowed toward base, on both sides glabrous, dark green or more or less silky, more heavily beneath; catkins sessile, dense . . . . . . 124. **S. pycnostachya** Anderss.
28. Leaves green and quite glabrous on both sides, sharply serrulate; rachis of catkin sparingly hairy . . . . 120. **S. issykiensis** Görz.
+ Leaves more or less silky, especially beneath, rarely glabrous when full grown, entire or serrulate; rachis of catkin hairy . . . . 29.
29. Leaves lanceolate; at first lustrous-sericeous beneath, glabrate above, finally glabrous on both sides, dark green above, paler beneath; stalk of catkin to 1.5 cm long; catkin-scales pale; stipe of ovary 0.8 — 1.5 cm long; catkin-scales pale; stipe of ovary 0.8—1.5 mm long . . . . 122. **S. Niedzwieckii** Görz.

+ Leaves narrowly lanceolate, at first commonly silky on both sides, finally more or less pubescent, especially beneath; catkins subsessile or borne on a stalk ca. 0.6 cm long; scales brown, in pistillate flowers darker at apex; ovary subsessile . . . . 123. S. blakii Görz.

111. **S. purpurea** L. Sp. pl. (1753) 1017; M. B. Fl. taur.-cauc. II, 412; Ldb. Fl. Ross. III, 502; Shmal'g., Fl. II, 433. — Ic.: Rchb. Ic. Fl. Germ. XI, fig. 1230, 1232, 1235; Ldb. Ic. Fl. Ross., 453. — Exs.: HFR No. 789, 2342, 2343.

A graceful shrub to 2—4 (—10) m high, mostly ca. 1 m high, reaching up to 30 years of age, with slender branches; bark lemon-yellow on the inside, with scattered bluish bloom on the outside; branches flexible, glabrous, smooth, the white smooth wood not turning yellow on exposure to air; buds 3 — 5 mm long, small, appressed, reddish-brown (the flower buds yellow), often opposite, glabrous; stipules rarely present, linear-lanceolate, to 1.5 — 1.8 cm long, serrate; petioles 3 — 6 mm long; leaves dense, alternate or opposite, 3 — 13 cm long and 0.8—1.5 cm broad, oblanceolate, shorter in v. helix (L.) Trautv. and longer in v. Lambertiana Sm., commonly spine-tipped at apex, thin, coerulescent or glaucescent above, rarely pure green on both sides (v. virescens Anderss.), blackening in drying, bitter in taste, entire or toward apex sharply serrate, at first puberulous, finally glabrous; catkins lateral, precocious or subcoaetaneous, sessile, with 2—5 bracts at base, dense, cylindric, 2.8 cm long, the pistillate sometimes narrowly cylindric (v. Smithiana Trautv.), 2—4 mm thick; ovary very short, erect or recurved and semipendulous, gray at first, becoming dark red;
154 staminate scales obovate, pale, dark at apex, sericeous-pilose; stamens 2, wholly connate; filaments hairy, to 3.5 mm long; anthers 4-locular; purple (hence the name of the species), turning black; gland 1, posterior, oblong-ovate; style obsolescent; stigmas very short, red, 2-parted or later 4-parted, divergent. Fl. March — May; fr. May — June.

Shores, canals, wet meadows, coppices, and sands. — European part: Lad.-Ilm., U. V., V.-Kama, U. and M. Dnp., V.-Don, Transv., Bl., Crim., L. Don, L. V., S. Urals; Caucasus: Cisc., Dag., W. and S. Transc.; W. Siberia: U. Tob., Irt.; Centr. Asia: Ar.-Casp., Balkh. **Gen. distr.**: Scand., Atl. and Centr. Eur., Med., N. Afr., Bal.-As. Min., Arm.-Kurd., Dzu.-Kash., Mong., Jap.-Ch. Introduced into N. America. Described from S. Europe. Type in London.

Hybridizing with S. acutifolia, aurita, caprea, cinerea, dasyclados, livida, lapponum, nigricans, myrtilloides, rosmarinifolia, triandra, viminalis s.l.

Note. The bark of this willow is used in preference to other species for salicin extraction (content 0.6 — 1.5%); less suitable for tanning (tannin content 1.7 — 7.81%). Particularly valued for basketry due to fineness, flexibility, and whiteness of the straight smooth lustrous twigs, called "basket hair," and amenable to wickerwork. More valued than others are var. Lambertiana Sm. and v. uralensis hort., which closely resembles v. helix (L.) Trautv. The latter is considered by horticulturists as best for staking fruit trees and grapevine. It is also believed to stand up to annual removal of shoots better than other varieties. Particularly deserving of mention among hybrids used for wickerwork are S. rubra (purpurea × viminalis), S. americana (purpurea × triandra), and S. dasyclados × purpurea. S. purpurea is very ornamental and often adorns gardens and pleasure grounds, especially f. pendula Rgl. grafted on S. caprea L. The species is suitable for hedges and for consolidation of coastal sands. It is easily established on sandy soils. Frost-hardy and resistant to insect damage (bitter leaves!). A very valuable species, extremely easy to cultivate and recommended for very wide distribution both for home use and for export.

112. **S. tenuifolia** Turcz. ex Laksch. in Schedis ad HFR VIII (1914) No. 2497 and 2498; Kom. and Alis., Opred. rast. Dal'nevost. kr. I, 423. — S. tenuifolia Turcz., Catal. baic.-dah. in Bull. Soc. Nat. Mosc. 1838, 1, No. 1020 (nom. nud.). — S. purpurea β foliis angustis etc. Turcz.
155 Fl. baic.-dah., II, 1, 103. — S. purpurea Ldb. Fl. Ross. III, 602 (ex prt.). —

A shrub; branches slender, flexible, olivaceous-latericious, lustrous, glabrous; buds glabrous; stipules on vegetative shoots bristlelike or narrowly lanceolate to linear, serrate, equaling or exceeding the petiole; petioles 1—3 (—5) mm long; leaf blades linear, 3.5—7 cm long and 4—6 mm broad, on vigorous shoots to 9 cm long and 7—9 mm broad; cuspidate, narrowed toward base, regularly and sharply serrate, quite glabrous, faintly green above, glaucescent beneath, commonly blackening in drying, prominently veined on both sides, lateral veins 15 —25 (30) pairs; catkins coaetaneous, dense, upright, with entire bracts at base; staminate oblong-cylindric, subsessile, ca. 2 cm long and 6 mm broad; pistillate cylindric, slim, 1.5—2 cm long and 2—3 mm broad; rachis hoary; staminate scales oblong-obovate, rounded at apex, pale, turning brown, with ciliate margin; hairy at base; pistillate obovate, truncate at apex, brownish, glabrous, pale and pubescent at base; stamens 2, divergent, hairy at base; anthers yellow, subspherical, 4-locular; gland 1, posterior, linear or linear-oblong, ca. 0.4—0.5 mm long; ovary ca. 1.5 mm long, sessile, ovoid, obtuse, canescent, without style; stigmas 0.2—0.4 mm long; capsule to 3 mm long. Fl. May; fr. June. (Plate VIII, Figure 5).

Waterside thickets.— E. Siberia: Ang.-Say., Dau.; Far East: Ze.-Bu., Uss. **Gen. distr.**: S. Siberia, Far East, Mong., Manch. Described from Dauria. Type in Leningrad.

Economic importance. An excellent source of material for basketry. Suitable for consolidation of banks and for ornamental use.

113. **S. dahurica** Turcz. Fl. baic.-dah. II, 1 (1854) 104, nom. nud., descr. apud Laksch. in Schedis ad HFR VIII (1914) 51; Kom. and Alis., Opred. rast. Dal'nevost. kr. I, 423. — S. rubra Turcz. (non Huds.) Fl. baic.-dah. II, 2 (1856) 108. — S. purpurea Ldb. Fl. Ross. III (1849) 603 (ex. p.). — S. pseudopurpurea Laksch. in sched., 1910. — Exs.: HFR No. 2496.

A tall shrub or a small tree; branches virgate, yellowish, glabrous, lustrous; buds oblong, appressed, acute, glabrous, yellowish or latericious-brown; stipules linear to lanceolate, glandular-serrate, equaling or exceeding the petiole; petioles 3 —9 mm long, strong, glabrous; leaf blades 4—9 cm long and 0.6 —1.2 cm broad, firm, linear-lanceolate, narrowed toward base, short-cuspidate at apex, broadest above the middle, coarsely and irregularly serrate throughout or near base entire, glabrous on both

sides, the upper surface somewhat glaucescent becoming gray or pale green, the lower glaucescent or coerulescent; midrib stramineous; lateral veins 14 — 18 pairs, evident; catkins precocious, sessile; staminate
156 2 — 2.5 cm long and ca. 1 cm thick, with scalelike bracts at base; pistillate ebracteate or with a scalelike bract, 2 — 3.5 cm long and 5 — 7 mm thick, cylindric, dense, upright, white-hairy; scales obovate, obtuse or truncate, black from the middle, covered with pale hairs; stamens wholly connate, pubescent up to the middle; anthers apparently 4-locular, round, yellow; gland posterior, oblong, subcapitate, 0.7 — 0.9 mm long; ovary 1.6 — 2 mm long, ovoid to ovoid-conical, subsessile or borne on a stipe ca. 0.3 mm long, clothed with pale hairs; style 0.3 — 0.5 mm long; stigmas oblong, divergent, ca. 0.4 — 0.6 mm long; capsule inflated, ca. 3 — 3.5 mm long. Fl. May; fr. June. (Plate VIII, Figure 4).

Waterside thickets. — E. Siberia: Ang.-Say., Dau.; Far East: Ze.-Bu., Uss. **Gen. distr.**: Mongolia. Described by Turczaninow from the Irkut Valley near the city of Irkutsk. Type in Leningrad.

Note. Easily confused with this species is S. tenuifolia (Turcz.) Laksch. that grows in the same areas. It differs in having linear sharply serrulate leaves to 0.5 cm broad; bristlelike or linear-lanceolate stipules; almost monochromatic catkin-scales, turning brown upon drying; small sessile styleless ovary, and very short stigmas.

Economic importance. Suitable for wickerwork. Nectariferous and ornamental.

114. **S. mongolica** Siuz. in Tr. Botan. Muz. Akad. Nauk IX (1912) 90; Ej. in Fedde, Repert. XIII (1914) 328; Toeppf. Salicol. Mitt., No. 5 (1912) 248; Kom. and Alis., Opred. rast. Dal'nevost. kr. I, 423. — Ic.: Siuz. ibid. f. 2.

A tall shrub; branches silvery, longitudinally striped, pale yellow, glabrous, lustrous; buds very large, ovoid, ribbed, recurved-tipped, light brown and yellowish, glabrous; stipules narrow, linear-falcate, to 1.2 cm long and 1 — 2 mm broad, on vigorous shoots to 3 cm long and 3 mm broad, exceeding the petiole, glandular-serrate, prominently reticulate-veined; petioles 0.8 — 1.2 cm long, glabrous; leaf blades quite glabrous, very stiff, glaucous, linear to linear-lanceolate, narrowed at both ends or toward apex enlarged, remotely sinuate-serrate, to 15 cm long, 1 — 1.2 cm broad (f. latifolia Nas.) or 3 — 4 mm broad (f. gracilior Siuz.), toward fall slightly revolute; lateral veins to 20 pairs, at an angle of 25 — 30°, not evident; catkins coaetaneous, with small bracts at base, sessile, narrowly cylindric, upright, dense; pistillate to 4 cm long and 0.4 cm broad; scales short, broadly obovate, black-tipped, pale below, clothed with long white hairs; ovary ca. 1.3 mm long, ovoid-conical, white-silky, with a short style and short dark brown stigmas. May — June.

157 Riverbanks. — Far East: Uss. May be found in Dauria. **Gen. distr.**: Mong., Manch. Described from Manchuria. Type in Leningrad.

Economic importance. Yielding wicker; ornamental.

115. **S. caspica** Pall. Fl. Ross. I, II (1788) 74; Ldb. Fl. Ross. III, 604; E. Vol'f in A. H. P. XXVIII, 3, 405. — S. pallida Ldb. Fl. Alt. IV (1833) 261. — S. Ledebouriana Trautv. in Bull. sc. Acad. Pétersb. I (1836) 132 (nomen) et in Mém. Sav. Ac. Sc. Pétersb. III (1837) 631; Ldb. Fl. Ross. III,

603; Turcz. Fl. baic.-dah. II, 2, 104; Kryl., Fl. Zap. Sib. IV, 739. — S. volgensis Anderss. in DC. Prodr. XVI, 2 (1868) 314. — Ic.: Gmel. Fl. sib. I, tab. 34, f. 2; Ldb. Ic. pl. Fl. Ross. t. 454. — Exs.: HFR No. 2499.

A graceful loose-branched shrub to 3—5 m high and 3—4 cm in trunk diameter, with smooth gray bark; branches long, straight, slender, glabrous, lustrous, lurid, rarely purple and heavily pruinose (f. pruinosa Anderss. in DC. Prodr. 309); buds appressed, flattened, acute, to 5—6 mm long; stipules linear, exceeding the petiole, soon caducous; petioles 2—5 mm long; leaf blades linear-lanceolate to linear, narrowed at both ends or toward apex somewhat enlarged, entire or in upper part serrulate, at first slightly hairy, becoming quite glabrous, stiff, dull above, glaucous beneath, typically 5—7.5 cm long or (f. longifolia Anderss.) up to 12 cm long and 0.4—0.6 cm broad; catkins subcoaetaneous, subsessile or borne on a very short stalk, subtended by small bracts, dense, narrowly cylindric, upright, staminate 2.5—3 cm long and ca. 0.8 cm broad, pistillate 1.5—3 cm long and 0.3—0.5 cm broad; scales pale brown, darker at apex, obtuse, sparsely hairy; stamens connate throughout, hairy at base; anthers yellow; gland 1, posterior, ovate or oblong; ovary ovoid at base, conical at the top, small, subsessile or short-stipitate, white-silky, at length glabrescent; style short or very short; stigmas thick, dark purple, short, mostly 2-parted; gland entire (posterior) or lobed and divided (anterior and posterior), about equaling or slightly shorter than the stipe. Fl. May; fr. June. (Plate VIII, Figure 2).

Riverbanks in the desert and steppe zone; loose sands and sand dunes amply supplied with fresh groundwater; in blown-out depressions; in mountain valleys up to 2,000 m. — European part: L. V., Transv., L. Don; Caucasus: Cisc., Dag., S. Transc.; W. Siberia: U. Tob., Irt., Alt; E. Siberia; Ang.-Say., Dau.; Centr. Asia: Ar.-Casp., Balkh., Dzu.-Tarb., Syr D. **Gen. distr.:** Mongolia, Iran. Described from the Ryn Sands. Type in Leningrad.

Hybridizing with S. rosmarinifolia, sibirica, tenuijulis, viminalis s. l.

158 **Economic importance.** An outstanding willow for wicker production and for ornament, yielding straight flexible slender twigs to 2 m long or even longer. Requires plenty of light.

116. **S. tenuijulis** Ldb. Fl. Alt. IV (1833) 262; Kryl., Fl. Zap. Sib. IV, 740; E. Vol'f in A. H. P. XXI, 2 (1903) 146. — S. purpurea Ldb. Fl. Ross. III, 603 (ex prt.). — S. Regelii Anderss. in DC. Prodr. XVI, 2 (1868) 309. — S. Schrenkiana Anderss. in sched. — ? S. Capusii Franch. in Ann. Sc. Nat. VI sér., Bot., XVIII (1884) 251. — Ic.: Ldb. Ic. fl. Ross. t. 453.

A tall, sometimes arborescent shrub; branches straight, divaricate, yellowish-gray or reddish-brown, the branchlets initially often purple; young shoots clothed with short appressed hairs, finally glabrous; buds small, ovoid, acute, appressed, yellowish or brownish, glabrous; stipules small, not exceeding the petiole, narrowly lanceolate, serrulate, soon caducous; petioles 2—5 mm long, pubescent; leaf blades oblanceolate, oblong-spatulate, or linear-lanceolate, 3.5—7 cm long and 0.5—1 cm broad, narrowed at both ends, acuminate, sharply denticulate and toward base entire, or dentate throughout (var. Roopii Görz) and then more silky than in the type and with larger linear stipules, pale green or yellowish-green above, glaucescent or cinerescent beneath, at length glabrous on both sides, initially slightly sericeous-pillulose, rarely grown leaves also silky-puberulous on both sides; veins slightly prominent beneath, evident above; catkins coaetaneous or somewhat precocious, lateral, dense, borne on a short leafy-bracted stalk; pistillate slim, recurved, slenderly cylindric, 2.5 — 3.5 mm long and 0.4 — 0.5 cm thick; scales elliptic or broadly lanceolate, yellowish-brown, darker at apex, initially hairy, finally monochromatic and glabrate; ovary sessile or short-stipitate, silky; style obsolescent; stigma short, thick, blackish-brown, deeply parted; gland entire, truncate, longer than stipe, slightly overtopping the ovary base. Fl. May, fr. June.

Waterside thickets. — Caucasus: (var. Roopii Görz) S. Transc.; W. Siberia: SW Altai; Centr. Asia: Ar.-Casp., Balkh., Dzu-Tarb., T. Sh., Syr D., Pam.-Al. Endemic. Described from Altai. Type in Leningrad.

Note. Hybridizing with S. coerulea, oxycarpa, viminalis s.l.

117. S. Michelsoni Görz sp. nova (in sched., nom. tant.); Addenda IV, p. 543.

(Description not published by the author; material inadequate). A tall shrub; branches slender, at first fulvous, finally reddish-yellow; branchlets yellow, very slender, glabrous, lustrous; buds small, yellowish-brown, appressed, acute, puberulous; stipules narrowly lanceolate, serrate,
159 one-fourth to half as long as the petiole; petioles 3 —5 mm long, dilated toward base, glabrous, sometimes twisted; leaf blades firm, glaucous-waxy on both sides, linear-lanceolate, with subparallel margins or else broadest at or above the middle, 4 — 7 cm long and 0.3 —0.8 cm broad, spiny-tipped, narrowed toward base, entire or serrate, cartilaginous-margined, at first slightly puberulous on both sides, finally glabrate or with scattered short cilia; midrib slender, stramineous; lateral veins 13 — 15 pairs, at an angle of 15 —25°; network of veins finely-alveolate, prominent; catkins (only pistillate with open capsules) in fruit to 4 cm long and ca. 0.8 cm broad, borne on a short leafy-bracted stalk, dense, cylindric; rachis slightly hairy; scales caducous; ovary ovoid-conical, glabrate or sparsely hairy, the glabrous stipe ca. 0.6 mm long; style brown, 0.5 — 0.6 mm long; stigmas 2-lobed, brown, ca. 5 mm long; capsule to 6 mm long, glabrous, white. Fr. July.

Riverbanks. — Centr. Asia: Dzu.-Tarb. Described from Dzharkent. Endemic. Type in Leningrad.

Note. Very closely related to S. capsica Pall.

118. S. coerulea E. Wolf in A. H. P. XXI (1903) 157 (with illustr.). — S. egberti-wolfi Toeppf. in Oest. Bot. Zeitschr. (1916) 402.

A low shrub; shoots brownish, mostly covered with bluish bloom; buds linear-lanceolate, acute, small, reddish-brown, glabrous, appressed; stipules soon caducous, filiform, shorter than the petiole; petioles 2 — 4 mm long; leaf blades commonly 3 — 7 cm long and 0.4 — 0.8 cm broad, at first silky or almost silvery, finally glabrous on both sides, glaucous, narrowly lanceolate to linear-lanceolate, with midrib produced into a short point, long-tapering toward base, entire or serrulate; lateral veins 12 — 17 pairs, slightly prominent, at an angle of 30—35°; catkins coaetaneous, loose, the short stalk with 2 or 3 normal leaves; rachis hairy; staminate catkins ca. 2.2 cm long and 4—5 mm broad, mostly upright or divergent; pistillate often recurved, 1.2—1.8 cm long and ca. 4 mm broad, in fruit to 2.5 cm long and 1 cm broad; scales caducous, pale yellow, the pistillate obovate-lanceolate, the staminate hood-shaped and dorsally convex, all sparingly covered at base (and sometimes also on the margin) with short hairs or glabrous; stamens connate throughout; anther apparently 4-locular, yellow, round, the filaments often hairy below; ovary conical, 2—3 mm long, glabrous, greenish, the pubescent stipe twice the length of the gland; style short, often bifid; stigma with 2 more or less 2-parted broad lobes; capsule glabrous, ovoid and strongly inflated, to 6 mm long. Fl. April—May; fr. May—July. (Plate VIII, Figure 3).

Mountain valleys, meadows, river floodland woods, 1,500—2,500 m.—
160 Centr. Asia: Balkh., Dzu.-Tarb., T. Sh., Pam.-Al., Amu D. **Gen. distr.**: Dzu.-Kash. Described from former Samarkand Region. Type in Leningrad.

Hybridizing with S. Niedzwieckii, oxycarpa, tenuijulis?.

Economic importance. Ornamental.

119. **S. sarawschanica** Rgl. in A. Fedch., Putesh. v. Turkest. III, No. 18 (1881) 80; Ej. in A. H. P. VI, 2 (1888) 466; E. Vol'f in A. H. P. XXI, 2 (1903) 173.

A shrub with long glabrous yellowish branches; leaves exstipulate; petioles 3—4 mm long; leaf blades lanceolate, to 5 cm long and 1 cm broad, commonly broadest below the middle, gradually tapering toward apex, narrowed toward base, entire or minutely serrulate, at first silky, becoming glabrous, bright green above, glaucescent beneath; midrib yellowish-brown; network of veins finely alveolate, costately prominent; catkins precocious, borne on a short leafy-bracted stalk; scales yellowish or brownish-yellow, glabrate on the back, villous at base and on the margin; ovary tomentose or silky-tomentose; style moderate; stigmas entire; stipe short, twice the length of the gland. Fl. May; fr. June.

Riverbanks. — Centr. Asia: Pam.-Al., Dzu.-Tarb.?, T. Sh.? Endemic. Described from the Zeravshan Valley. Type in Leningrad.

Note. A dubious species, described from inadequate material.

120. **S. issykiensis** Görz. sp. nova (in sched., nom. tant.); in Addenda IV, p. 544.

A shrub with slender virgate reddish-brown glabrous branches; buds ovoid, obtuse, small, appressed, glabrous, dark brown; petioles slender, 4—5 mm long; stipules none; leaf blades to 5 cm long and 1 cm broad, lanceolate, broadest above or rarely at or below the middle, narrowed toward base, spiny-tipped at apex, sharply serrulate, on both sides green and glabrous; midrib yellowish-brown; lateral veins 10—12 pairs, at an angle of 25—35°, riblike; network of veins of the third order forming fine alveolation, prominent beneath; catkins (available specimen with open and partly shed capsules; staminate unknown) stalked; rachis sparsely hairy; scales ovate, acutish, scarious, brown, with 3—5 forked nerves, glabrous or at base hairy; capsule glabrous, ca. 5 mm long, brown; style slender, ca. 1 mm long; stipe glabrous, ca. 1 mm long; stigma to 0.5 mm long, 4-parted, the shortly oblong lobes slightly divergent.

Shores. — Centr. Asia: T. Sh. Endemic. Described from Lake Issyk-kul'; collected 13 September, 1915, by Titov. Type in Leningrad.

Note. Author's description not published. Present description based on insufficient material.

161 121. **S. spinidens** E. Wolf in A. H. P. XXVIII, 3 (1909) 403.

A tall shrub, with dark brownish, glabrous branches; young shoots more or less hairy at the ends, yellowish like the small appressed buds; stipules lanceolate, glandular-serrulate, half as long to as long as the petiole, on lateral shoots mostly absent; petioles 0.2 — 0.9 mm long, glabrous; leaf blades lanceolate, narrowed at both ends, point-tipped, sharply sinuate-serrate; the teeth prolonged into a glandular-tipped point, toward the leaf base remote and relatively short-pointed; young leaves silvery-sericeous, finally beset on both sides with long white hairs or sometimes glabrate, densely white-dotted, pale or dark green above, glaucescent beneath, 5.5 — 7 cm long, on short lateral shoots 3 — 4 cm long, 0.7 — 1.3 cm broad; lateral veins to 15 pairs, at an angle of 35 — 55°; leaves at the base of shoots linear-lanceolate, glabrate, less sharply serrate; pistillate catkins (staminate unknown) serotinous, slender, 2 — 2.5 cm long and 0.5 — 0.6 mm broad, loose at base, the stalk with 2 or 3 narrowly lanceolate silky entire or glandular-serrulate bracts; rachis hairy; scales liguliform or obovate, obtuse, retuse, crenate (in lower flowers acute) at apex, paler at base, white-hairy; ovary ovoid-conical, short-stipitate, densely clothed with long very shiny white hairs, mostly enclosing the short entire or bifid style; stigma brown, its 4 lobes united in pairs; gland 1, posterior, entire, very rarely 2-lobed, dilated toward base, twice as long as the stipe. May. (Plate VIII, Figure 8).

Centr. Asia: Dzu.-Tarb. Endemic. Described from Lepsinskoe forest in Kazakhstan. Type in Leningrad.

122. **S. Niedzwieckii** Görz, Salic. asiat. I (1931) 18 et in Fedde, Repert. XXXII, No. 1 — 8, 120. — S. coerulea E. Wolf in A. H. P. XXI, 1 (1903) 157 (ex parte).

A shrub with slender graceful glabrous yellowish or brownish branches; young shoots sparsely covered with very shiny silky hairs; buds narrowly conical, 3 — 5 mm long, castaneous, at first pubescent, finally glabrous; stipules soon caducous, lanceolate to linear, glandular-dentate, shorter than petiole, mostly absent; petioles 2 — 5 mm long, initially puberulous; young leaves lustrous-sericeous beneath, glabrate above, finally on both sides glabrous and glaucous, darker above and paler beneath or subconcolor, lanceolate, to 5 cm long and 0.6 cm broad, entire or serrulate; lateral veins 10 — 12 pairs, at an angle of ca. 40°, not evident; catkins coaetaneous, borne
162 on a leafy-bracted stalk to 1.5 cm long; pistillate ca. 5 cm long; staminate shorter, often twisted, rather loose; rachis slender, hairy; scales oblong-obovate, crenate, pale, glabrous, pubescent only at base; gland small, ca. 0.4 mm long, rectangular or ovate, one-third to half the length of the stipe; stamens connate, ca. 3 mm long, hairy at base; anthers yellow; ovary 2 — 2.5 mm long, inflated at base, conical, glabrous, the glabrous stipe 0.8 — 1.5 mm long; style very short, incised at the end; stigmas ca. 0.4 mm long, parted, capsule to 4 mm long, turning brown. Fl. April — May.

Mountains and foothills. — Centr. Asia: Balkh., Dzu.-Tarb., T. Sh., Pam.-Al. Gen. distr.: Dzu.-Kash. Described from W. Tien Shan. Type in Leningrad.

Note. Hybridizing with S. coerulea and S. tenuijulis.

123. **S. Blakii** Görz in Fedde, Repert. sp. nov. XXXVI (1934) 31. — Exs.: Görz, Sal. Asiat. III, No. 67—69.

A shrub, apparently of medium size; young branches reddish-brown, the old grayish or greenish-olivaceous, glabrous, short, slender, closely (to 1 cm) marked with ridges left by preceding year's leaves; young summer branchlets very slender, yellow, brittle; buds small, 3—4 mm long, ovoid, appressed, glabrous or nearly so; stipules rarely present, soon caducous, lanceolate, small; petioles ca. 1—3 mm long, channeled, more or less silky; leaf blades 2—4 cm long and 0.3—0.8 cm broad, narrowly lanceolate, narrowed at both ends, shortly spine-tipped, flat or subrevolute, subentire or minutely and sharply glandular-serrulate, initially silky, the vesture at length almost disappearing; catkins coaetaneous or serotinous, subsessile or borne on a stalk ca. 0.6 cm long, ebracteate, cylindric, dense; staminate to 2.5—4 cm long and 0.4—0.7 cm broad; pistillate 1.5—2 cm long, in fruit to 2—3 cm long and ca. 1 cm thick; rachis densely villous; scales scarious, ovate or obovate, 3-nerved, brown, in pistillate flowers darker at apex, glabrous on the back, the margin ciliate, the base on the inside densely ciliate; stamens connate throughout, the filaments densely villous up to two-thirds from base; anthers yellow, apparently 4-locular, spherical, turning black; gland linear; ovary subsessile, sparsely to densely pubescent, sometimes glabrate; style obsolescent or very short; stigmas short, red, oblong, 2—4-parted. Fl. May; fr. June.

Riverbanks. — Centr. Asia: Balkh., Dzu.-Tarb., T. Sh., Pam.-Al. Endemic. Up to 1,500 m. Described from Tien Shan. Type in Leningrad.

Hybridizing with S. Wilhelmsiana and S. Niedzwieckii.

Economic importance. Ornamental.

163 124. **S. pycnostachya** Anderss. in Journ. Linn. Soc. IV (1860) 44 et in DC. Prodr. XVI, 2 (1868) 309. — Exs.: Görz, Sal. As. III No. 72, 73.

A shrub to 2 m high or (v. alpina Anderss.) procumbent; branches divaricate, dark violet, glaucescent, or yellowish, lustrous; buds fairly large; leaves up to 5 cm long, above the middle scarcely 1—2 cm long [broad ?], lanceolate, obtusish, narrowed toward base, flat, subentire or remotely glandular-serrulate, at first silky, especially beneath, finally glabrous on both sides, pale green, in var. alpina barely 2.5 cm long and 3—5 mm broad, hairy on both sides; midrib and lateral veins pale or brownish; catkins sessile, with small bracts at base, stoutly cylindric, dense, elongated, obtuse, staminate to 4 cm long, pistillate 4—5 cm long; scales rufescent-brownish, clothed with white appressed hairs, the pistillate brownish-tarry above; stamens golden, connate; ovary to 2.5 mm long, ovoid and inflated at base, acutish at the top, sessile, greenish, glabrous, almost styleless; stigmas latericious, short, erect, 2-parted. Fl. May; fr. June.

Shores of mountain streams (up to 4,000 m). — Centr. Asia: T. Sh., Pam.-Al. Gen. distr.: Ind.-Him. (Kashmir). Described from India. Type in London.

125. **S. microstachya** Turcz. ap. Trautv. in Mém. pres. Acad. Pétersb. par div. sav. III (1837) 628; Turcz. Fl. baic.-dah. II, 1 (1854) 104. — S. stenophylla Sukacz. in Tr. i issled. po lesn. opyt. delu, No. 10 (1931), with illustr. — Ic.: Trautv. l. c., tab. IV. — Exs.: HFR No. 2350.

A shrub 3—4 m high or a tree up to 6 m; branches sparse, obliquely ascending, semipendulous, very slender, very long, flexible, divaricate, cinereous-brown; young shoots sericeous-pilose, later more or less pubescent, finally glabrous; buds ovoid, obtuse, appressed, silky, to 4—5 mm long; stipules none or very small, ovate-lanceolate, dentate or entire, soon caducous; petioles 3—5 mm long, dilated toward base, glabrous or silky; leaf blades sublinear or narrowly lanceolate, narrowed at both ends, often falcate, appressed to branches, subentire or serrulate, slightly revolute, silky on both sides, at length glabrate; midrib prominent beneath; lateral veins to 30 pairs, at an angle of 10—15°, faint; catkins serotinous, upright, cylindric, to 1.5—2 cm long, borne on a short leafy-bracted stalk; rachis densely hairy-tomentose; scales ovate, broadly lanceolate, or obovate, erose-dentate at apex, pale brown or yellowish-green, rather densely hairy at base, the back and margins mostly glabrous or in staminate catkins hairy; stamens connate up to the anthers or throughout, glabrous; anthers
164 yellow; gland oblong or subquadrate, very small; ovary ovoid-conical, somewhat elongate toward apex, ca. 2.5 mm long, sessile, glabrous, brownish-green; style evident or short; stigmas reddish-brown with short entire oblong lobes. Fl. beginning of June; fr. end of June and July.

Waterside thickets and valley pebbles. — E. Siberia: Ang.-Say., Dau. **Gen. distr.**: Mong. Described from the Irkut Valley near Irkutsk. Type in Leningrad.

Note. Differs from related S. Wilhelmsiana M. B. in having completely naked ovary, commonly erose bract scales and veins, arising not at a right angle but at a very acute angle; the boundary separating the ranges of the two species has not been exactly traced.

Economic importance. An outstanding source of material for basketry; annual twigs reach up to 2 m in length, slender (2.5 mm thick), very shrinkable, highly flexible, resistant to stretching. Wood white, with a satiny sheen. Tannin content 8%. Easily propagated by cutting; producing adventitious roots, growing rapidly, standing up well to cold winters. Preferring pebbles and sands with a high water table. Suitable for sand fixing (Sukachev).

126. **S. Wilhelmsiana** M. B. Fl. taur.-cauc. III (1819) 627; Trautv. in Mém. pres. Acad. Pétersb. par div. sav. III (1837) 607 et tab. III. — S. angustifolia Willd. Sp. pl. IV (1805) 699, non al.; M. B., ibid. II, 414; Ldb. Fl. Ross. III, 604; Shmal'g., Fl. II, 433. — S. Trautvetteriana Rgl. in A. Fedch., Put. v. Turk. III, 18 (1881) 79 et in A. H. P. VI, 2 (1888) 465. — S. dracunculifolia Boiss. Diagn. pl. orient. ser. I, 7 (1846) 99. — Vernacular name: kubatal (Kazakh).

A shrub or small tree to 6—7 m high; branches very slender, long, virgate, rather wide apart along the stem and on its lower part, arising at nearly a right angle or suspended; branches and branchlets grayish-brown, pubescent or silky-pilose, rarely glabrate; buds ovoid, obtuse, at the tip pubescent; stipules none; petioles 1—2 mm long, dilated at base,

glabrous or pubescent; leaves closely approximate; blades sublinear or narrowly linear-lanceolate, narrowed at both ends, entire or rarely glandular-serrulate, 2—6 cm long and 0.4—0.8 cm broad, at first densely and later sparsely silky-pilose and lustrous on both sides, finally often glabrous above; midrib prominent beneath; lateral veins less evident, at an angle of 10—15°; catkins closely covering the branches of the preceding year; staminate subsessile; pistillate borne on a short stalk, subtended by 2—5 small silky bracts, in flower 2— 3 cm long, cylindric, slender, upright,
165 dense, in fruit to 0.8 cm thick, subcoaetaneous or serotinous; scales ovate, flat or the staminate navicular, mostly subacute, lurid or greenish-yellow, at base, like the rachis, densely woolly-tomentose, on the back and on the margins glabrate or with scattered hairs; stamens 2, small, wholly connate; filament glabrous; anther round, yellow, apparently 4-locular; ovary fully sessile, ovoid-conical, densely silky-pilose; stigma short, reddish-brown, with 2 entire or 2-parted suberect lobes. Fl. May; fr. June. (Plate VIII, Figure 1).

River floodland espalier woods and waterside scrub in semideserts. — European part: L. V. (Lessing), Transv. (Samarskoe); Caucasus: W. and S. Transc.; Centr. Asia: throughout. **Gen. distr.**: Iran., Dzu.-Kash., Tibet, India (Kashmir). In S. Siberia and Mongolia replaced by the related S. microstachya Turcz. Described from Georgia. Type in Leningrad.

A hybrid with S. triandra L. has been recorded.

Note. Although the epithet "Wildenow" is older, its use is not recommended, due to the existence of a whole series of homonyms.

Economic importance. An outstanding source of material for basketry; suitable for fine and delicate wickerwork.

127. **S. serrulatifolia** E. Wolf in A. H. P. XXI, 2 (1903) 163 (cum fig.).

A shrub, apparently fairly tall, with grayish bark below; shoots long, vigorous, reddish-brown, rarely yellowish-brown, glabrous, lustrous; buds large, to 8 mm long, reddish-brown, glabrous, lustrous, ovoid-elongate, acute, appressed, divergent at the tip; stipules from narrowly lanceolate serrate to subulate, shorter than petiole, soon caducous; petioles ca. 1 cm long, reddish, glabrous; grown leaves to 10 cm long and 1—2 cm broad, lanceolate to broadly linear-lanceolate, narrowed at base, stiff, deeply sinuate and glandular-serrate, green above, glaucescent beneath, glabrous, somewhat dull; young leaves turning brown, slightly silky, more finely serrate; leaf margin cartilaginous; midrib stout, very prominent beneath, yellow; lateral veins 10—20 pairs, at an angle of 20—40°, forking at the margin and confluent; network of veinlets conspicuous on both sides; catkins coaetaneous, stoutly cylindric, dense, borne on a short leafy-bracted stalk, staminate unknown, pistillate to 2—3 cm long and 1 cm thick; scales obovate, obtuse, truncate or subacuminate, brown at base, quite black at apex, densely covered on both sides with long white hairs; ovary ca. 4 mm long, ovoid-conical, brownish, short-hairy, at length glabrescent; style and stigma short, the erect lobes entire or 2-parted; stipe very short, about half the length of the gland; capsule to 6 mm long, glabrate or with scattered hairs. Fl. May; fr. June.

166 Riverside thickets, up to 2,000 m. — Centr. Asia: Dzu.-Tarb., T. Sh., at the border of Kyz. K. and Ar.-Casp. (Syr Darya Delta, Bol'shoi Stanskii Island). **Gen. distr.**: Dzu.-Kash. Described from Centr. Asia. Type in Leningrad.

Economic importance. Yielding wicker.

128. S. Komarovii E. Wolf in A. H. P. XXI, 2 (1903) 195 (cum. fig.) et in Engl. Jahrb. XXXII, 279.

A rather low shrub with pale brownish-gray glabrous (not pruinose) branches; young shoots glabrous, turning black; buds narrowly ovoid, appressed, glabrous, reddish-brown; stipules soon caducous, linear, serrulate, shorter than the slender glabrous brownish petioles, these 0.6—1 cm long; leaf blades lance-linear to linear, similarly narrowed at both ends or the lower ones short-acuminate at apex and long-tapering toward base, often oblique, remotely and obscurely denticulate, near base mostly entire, glaucescent or dull green on both sides, glabrous, when young white-silky, 5—9 cm long and ca. 0.8 cm broad, the cartilaginous margins slightly revolute; midrib light brown beneath, abruptly attenuate toward apex; lateral veins 7—9—12 pairs, very slender, at an angle of 20—50°, prominent on both sides, at the margins long-ascending and confluent; network of veinlets very conspicuous beneath, forming small alveoles; catkins coaetaneous, borne on a long stalk with 1—4 bracts, loose, pistillate to 4—5 cm long and ca. 0.7—0.8 cm broad, staminate unknown; scales oblong-ovate, acute or rarely round-tipped, dark brown at apex, paler at base, rather sparsely covered with white hairs; ovary 4—5 mm long, narrowly ovoid-conical, prolonged into a long style; stigma brown, with 4 long, linear, curved or rounded lobes; gland linear, up to one-fourth the length of ovary (excluding style and stigma). Fl. June; fr. July. (Plate VIII, Figure 7).

Waterside thickets. — Centr. Asia: Pam.-Al., 2,000—3,000 m. Endemic. Described from the Zeravshan Basin. Type in Leningrad.

Note. This species cannot be placed in the section Daphnoides Dum. (Pruinosae Koch) to which it was referred by E. L. Vol'f. From all species of the section Purpureae it differs essentially in the long style and stigma; glabrous ovary occurs also in other species, even though rarely. An insufficiently clarified species. More ample material is needed from the original site.

169 129. S. Prszewalskii E. Wolf in Izv. SPb. Lesn. Inst. XV (1907) 179.

A shrub; young shoots whitish-silky, finally glabrous, dark castaneous or greenish-brown; stipules soon caducous, lanceolate to linear-lanceolate, straight or subfalcate, shorter than petiole; petioles 4—6 mm long, slender, glabrous or silky; leaf blades oblong-lanceolate to linear-lanceolate, 6—9.5—13 cm long and 0.7—1 cm broad, narrowed toward base and toward apex, terminating in a short subulate point; the margins often parallel over a considerable stretch, cartilaginous, serrulate, entire near base; upper surface glaucescent, the lower paler, the veins prominent on both sides; terminal leaves more or less sericeous-pilose, the lower ones mostly glabrous; young leaves silvery-lustrous with silky more or less fugacious hairs; lateral veins 15—20 pairs, at an angle of 30—45°, the upper ones ascending toward apex; network of veinlets not evident; catkins coaetaneous, solitary, few, borne on a short leafy-bracted stalk, slender, loose, ca. 4 cm long including the stalk; rachis covered with long white hairs; scales liguliform, acute or obtusish and retuse, brownish and darker at apex or light brown (f. dilata E. Wolf), hairy outside on the margins or (v. glabra E. Wolf) glabrous on the back and ciliolate on the margin; stamens 2, connate up to the anther or throughout; filaments covered with long white

(167)

PLATE VIII. 1. Salix Wilhelmsiana M.B. – 2. S. caspica Pall. – 3. S. coerulea Pall. – 4. S. dahurica Turcz. – 5. S. tenuifolia Turcz. – 6. S. integra Thunb. – 7. S. Komarovii E. Wolf. – 8. S. spinidens E. Wolf.

hairs from base up to the middle; gland 1, posterior, longer than the stipe, simple or lobed, or sometimes glands 2, anterior and posterior; ovary ovoid-conical, borne on a very short stipe, sparsely hairy, or in var. *glabra* pale green with hairs only in a basal tuft or in 2 longitudinal rows; style evident, ca. 0.5 mm long, slender; stigmas dark, entire or 2-parted, about the length of the style; capsule to 3—5 mm long. Fl. May.

Riverbanks. — Centr. Asia: T. Sh. Endemic. Described from Karakol. Type in Leningrad.

130. **S. linearifolia** E. Wolf in A. H. P. XXI, 2 (1903) 160 (cum fig.).

A shrub; annotinous shoots brown or brownish-yellow, glabrous; summer shoots slender, glabrous; stipules lance-subulate, shorter than petiole, soon caducous; petioles short; leaf blades linear-lanceolate to linear, tapering from the middle toward both ends or with almost parallel margins, glandular-serrulate or entire, dull green above, glaucous beneath, at first white-silky, at length glabrous on both sides, 4—7.5 cm long and 0.6—0.8 cm broad (8.5—10 times as long as broad); lateral veins 10—13 pairs, at an angle of 20—40°; catkins coaetaneous, staminate unknown, pistillate borne on a leafy-bracted stalk, in fruit ca. 3.5 cm long and to
170 1.7 cm broad; rachis hairy; scales soon caducous, oblong, 3-nerved, monochromatic brown, glabrous on the back, white-hairy at base and on the margin; capsule 6—7 mm long, narrowly conical, enlarged at base, tapering toward apex, reddish or reddish-brown, sparingly covered with short whitish hairs, at base glabrate; style ca. 0.6 mm long; stigma 0.3—0.4 mm long, with 4 upright brown lobes; stipe short, pubescent, in fruit about twice the length of the gland. Fl. March; fr. April.

Banks of mountain streams, 700—1,500 m. — Centr. Asia: Balkh., Pam.-Al. Endemic. Endemic. Described from Baba-Tag Mountains and W. Tadzhikistan. Type in Leningrad.

Note. Described by Vol'f from undeveloped specimens collected in April. Grown leaves unknown. Material inadequate.

131. **S. Olgae** Rgl. in A. Fedchenko, Put. v. Turkest. III, 18 (1881) 79 et in A. H. P. VI, 2 (1888) 465; E. Vol'f in A. H. P. XXI, 2 (1903) 152.

A shrub with brown bark; shoots reddish, glabrous, strongly forking; branchlets glabrous or puberulous, densely leafy; stipules none; petioles 2—4 mm long, glabrous; leaf blades linear-lanceolate, short-acuminate, to 3.5 cm long and 4—5 mm broad (7—11 times as long as broad), cuspidate, narrowed at base, glaucescent on both sides, glabrous, closely white-dotted, entire; young leaves silky-pilose; midrib pale beneath, slender, obsolescent toward apex; lateral veins 11 or 12 pairs, scarcely evident; catkins serotinous, borne on a short densely white-hairy leafy-bracted stalk; scales lurid or brownish, unicolor, obtuse, narrowly ovate, covered with short white crisp hairs, at base pubescent; ovary ovoid-conical, 3—4 mm long, densely clothed with white hairs; style ca. 0.4—0.5 mm long, the upper part and the stigmas red; stigmas short, oblong, broad, divergent, 0.2—0.3 mm long; gland 0.3—0.4 mm long; stipe shorter than the gland. Fl. March—April.

Shores. — Centr. Asia: Syr. D. (Salar, in the vicinity of Tashkent) whence described. Endemic. Type in Leningrad.

Note. Additional material is needed.

132. S. Lipskyi (Görz) Nas., sp. nova in Addenda IV, p. 544. — S. macrolepis var. Lipskyi Görz, in sched.

A tree; branches slender, straight or arched, reddish-brown, not pruinose, glabrous when young; buds very small, 2 × 3 mm, flattened, appressed, brown, dull, glabrous; stipules none; petioles ca. 5 mm long, glabrous, like the midrib yellowish-green or reddish-brown; expanding leaves silky-pilose beneath, involute; blades of grown leaves commonly
171 5.5 × 1 cm, the terminal to 8 × 1.5—2 cm, oblanceolate or oblong-oblanceolate, narrowed toward base, rounded at apex and shortly acicular-tipped, glaucous on both sides, paler beneath, brittle, entire or sharply glandular-serrulate; lateral veins 18 or 19 pairs, at an acute angle, barely perceptible above, indistinct beneath; staminate catkins unknown; pistillate very long in fruit, slenderly cylindric, to 10—12 cm long and 0.7 cm broad, strongly divergent but not pendulous, flexuous, the stalk 1—2 cm long, with 1 or 2 bracts resembling the leaves but somewhat smaller; rachis hairy; scales persistent or caducous, scarious, mitriform, 5-nerved, light brown (in dry state), glabrous on the margin and at apex, with scattered white cilia on the back at base and on the inside in lower part; capsule glabrous, green, to 3 mm long, pandurate or cochleiform, abruptly contracted above the inflated base and more or less pointed at the top; stipe 0.6—0.8 mm long, hairy at base, 3—4 times as long as the posterior oblong gland; style 1, short, persistent; stigma short, ca. 0.3 mm long, 2-parted, with short thickened lobes. Fl. May; fr. June.

Mountain gorges (banks of mountain streams ?). — Centr. Asia: T. Sh. Described from Kordonnoe Gorge (near Alma-Ata, 1909, Lipskii). Endemic. Type in Leningrad.

Note. In general aspect and in the shape of catkins, somewhat similar to Chosenia macrolepis (Turcz.) Kom. (the distribution area of which is situated far to the east), but in the latter species the branches are covered with bluish bloom, the catkins are 1—2 cm long and in fruit up to 5 cm long; rachis glabrous; bracts 4 or 5; catkin-scales caducous; ovary ovoid-oblong; styles 2, distinct, caducous; glands and torus absent.

133. S. pseudalba E. Wolf in A. H. P. XXI, 2 (1903) 167 (cum fig.).

A tall shrub or tree; branches pale brown or yellowish-gray, closely marked with scars from fallen leaves; annotinous shoots yellowish-brown, glabrous; young shoots slender, pale yellowish-brown, white-pubescent; buds oblong, brownish-yellow, glabrate; stipules soon caducous, subulate or narrowly lanceolate, on vigorous shoots sometimes to 1.1 cm long, lanceolate, glandular-serrate, densely silky; leaf blades 4—6.5 cm long and 0.8—1.7 cm broad, lanceolate to narrowly lanceolate, arcuately narrowed at both ends, those at base of shoots obtuse, entire, the terminal often remotely glandular-serrate, on both sides dull from rather dense longitudinal hairs or subsericeous and silvery as in S. alba (hence the name), the vesture at length thinning out; lateral veins 11—15 pairs, at an
172 angle of 30—40°, evident on both sides, especially on leaves at the base of shoots; catkins apparently coaetaneous, staminate unknown, pistillate on short leafy shoots, borne on leafy-bracted stalks; scales liguliform acute or obtuse, dentate or sinuate, brownish, pale at base or monochromatic, light-colored or brownish, sparsely covered with short hairs, more densely

in lower part; ovary ca. 3—4 mm long, ovoid-conical, densely silky-villous; style very short; stigma with 2—4 divergent lobes, ca. 0.3—0.4 mm long; stipe very short; gland entire or 2-lobed, longer or shorter than the stipe; capsule to 5—7 mm long. Fl. May—June; fr. June—July.

Shores of bodies of water and banks of mountain streams. — Centr. Asia: Pam.-Al. Endemic. Described from the Zeravshan Valley (according to Komarov's collections). Type in Leningrad.

Note. Further study needed. Material poor.

Economic importance. Ornamental.

134. **S. Alberti** Rgl. in A. H. P. VI, 2 (1880) 462; E. Vol'f, ibid. XXI, 2 (1903) 170.

A tall shrub or possibly a tree; grown branches reddish-brown or buff, glabrous; young shoots at first silky-pilose; buds oblong, appressed, medium size, yellowish-brown or light castaneous, glabrous; stipules narrowly lanceolate, glandular-serrate, soon caducous; young leaves silky-pilose, at length glabrous or with scattered hairs, pale green above, glaucous beneath, lanceolate, acuminate, narrowed at base, rarely broadest above the middle, 3—7 cm long and 0.6—1.2 cm broad, on terminal shoots to 4 cm long and to 1.5 cm broad, entire or glandular-serrulate, the glands stalked; lateral veins 12—15 pairs, at an angle of 30—45°; midrib yellowish; catkins on a stalk ca. 0.5—0.7 cm long, with 2 or 3 small lanceolate silky bracts, curved, pistillate in fruit ca. 7 cm long and ca. 1 cm broad, dense; rachis long-hairy; scales ca. 2 mm long, ovate or elliptic, brown, darker at apex, obtusish, covered with pale hairs or glabrous on the back, hairy on the margin and on the inside; ovary large, ca. 5 mm long, ovoid at base, tapering toward apex; style 0.2—0.5 mm long; stigmatic lobes ca. 0.5 mm long, upright, reddish-brown, oblong, entire or 2-parted; gland narrowly oblong, ca. 0.5—0.7 mm long; capsule to 7 mm long. Fl. April May; fr. May—June.

Banks of mountain streams. — Centr. Asia: Balkh. Endemic. Described from Kazakhstan. Type in Leningrad.

Economic importance. Cultivated in Alma-Ata, apparently for decoration.

135. **S. macrostachya** E. Wolf in A. H. P. XXI, 2 (1903) 165 (cum fig.).

A shrub with glabrous pale brownish-yellow branches; young shoots
173 sparingly pubescent; buds ovoid, glabrous, brownish-yellow; stipules soon caducous, subulate, much shorter than the glabrous petiole, this not more than 4—5 mm long; blade of young leaves linear-lanceolate, at the base of shoots lanceolate, often falcate, tapering at both ends, entire or obscurely toothed, at first silvery silky, becoming glabrate, dull pale green on both sides, 3.5—4.5 cm long and ca. 0.7 cm broad (5—7.5 times as long as broad); midrib strongly attenuate toward apex; lateral veins 9—14 pairs, very slender, at an angle of 25—40°, ascending at the margins toward apex, inconspicuous; catkins subcoaetaneous or serotinous, to 5—7 cm long and 0.7 cm thick, the stalk ca. 0.6 cm long, with scalelike bracts or ebracteate; rachis hairy; staminate catkins dense, pistillate looser; scales obovate-oblong to ovate, at apex rounded, truncate, very short-acuminate, or toothed, brown or yellowish-brown, monochromatic or paler at base, rather sparsely covered with short white hairs; stamens to 3.5 mm

long, wholly connate; filaments hairy at base; anthers yellow; ovary ca. 4.5 mm long, ovoid-conical, pointed at apex, imperceptibly passing into the style, which is short, slender and covered with appressed white silky hairs; stigma brown, 4-lobed; stipe short, equaling or finally exceeding the gland. Fl. June; fr. July.

Shores, ca. 2,860 m. — Centr. Asia: Pam.-Al. (Zeravshan Basin). Described from Zeravshan. Endemic. Type in Leningrad.

Note. Because of the limited and incomplete material, it is impossible to decide for the time being whether this willow is synonymous with S. margaritifera E. Wolf (and S. verticilliflora of the same authority) or a distinct species.

136. **S. margaritifera** E. Wolf in A. H. P. XXI, 2 (1903) 162 (cum fig.) — S. verticilliflora E. Wolf, ibid. XXVIII, 3 (1909) 400. — Exs.: Görz, Sal. As. Ill, No. 71.

A rather low shrub, with yellowish-gray, dark brown, or brownish-olivaceous branches; young shoots pubescent or villous; buds ovoid-oblong, rufescent, sparingly pubescent; stipules from narrowly lanceolate to ovate-lanceolate, glandular-serrate, half as long as petiole; petioles 0.4 — 1.2 cm long, slender, channeled, glabrous; leaf blades on vigorous shoots narrowly oblong-lanceolate to linear-lanceolate, 6 — 8 — 10.5 cm long and 0.8 — 1.4 cm broad (the lateral and the weak terminal ones shorter), narrowed at both ends, remotely glandular-serrulate, at first white-silky, later sparsely hairy on both sides and finally glabrous, closely white-dotted, dull pale green above, somewhat lighter and glaucescent beneath; midrib strongly attenuate toward apex; lateral veins 13 — 20 pairs, prominent on both sides, at an angle of 20 — 60°, steeply ascending, at the margins often spread out; catkins coaetaneous
174 or subserotinous; peduncle hairy like the rachis, ca. 1 cm long, with 2 or 3 entire bracts; pistillate catkins slender, 3 — 5 cm long and 0.4 — 0.5 cm broad, with loosely verticillate flowers, often curved; scales liguliform, at apex acute, obtuse, or truncate and toothed, pale at base, rufous or brownish to dark brown at apex, or monochromatic yellowish-brown, covered at base with short hairs, sparsely hairy or glabrate above; staminate flowers unknown; ovary ovoid-conical, subsessile or borne on a stipe half the length of the gland, white-silky, lustrous; style short or moderate; stigmas short, brown, 2- or 4-lobed, divergent. Fl. May; fr. June.

Waterside thickets in mountain valleys. — Centr. Asia: Pam.-Al., T. Sh., Balkh. Endemic. Described from the Kshtut River (Zeravshan). Type in Leningrad.

Note. S. margaritifera was described by Vol'f from a wild-growing specimen from the Zeravshan Valley with underdeveloped spring leaves, while the description of S. verticilliflora was based on ample material taken from shrubs grown in the Forest Institute from cuttings obtained from Frunze; hence some of the trivial differences observed by Vol'f in repeated description. We consider the two willows to be identical. Very similar to these plants is S. macrostachya E. Wolf, also from the Zeravshan Basin, and likewise described from undeveloped spring samples.

137. **S. holargyrea** Bornm. et Gorz in Trudy Tadzh. bazy Ak. Nauk II (1936) 175.

A shrub to 1 m high; young branches white-hairy; annotinous glabrous, dingy fulvous or brown; buds the same color, hairy, becoming glabrous; stipules broadly lanceolate, the length of petiole or shorter; petioles 3—5 mm long, white-tomentulose; expanding leaves flat; grown leaves broadly lanceolate or oblanceolate, to 8 cm long, obtuse or sometimes almost round at base, the entire margins arching toward apex; both surfaces markedly clothed with rather long hairs, the lower white-silky, the upper greenish-silky; midrib pale; lateral veins 9 or 10 pairs, at an angle of 45°; catkins (fruiting) borne on a leafy-bracted stalk to 1 cm long, cylindric, to 5 cm long, rather loose; rachis white-hairy; scales round-tipped, to 0.8 × 1.5 mm long, largely brown, clothed with short hairs; gland 1, oblong, to 0.5 mm long, exceeding the stipe; capsule (open) subsessile, conical from inflated base, glabrous; style and stigmas to 0.3 mm long; stigmas upright or somewhat divergent; staminate flowers unknown.

Banks of mountain streams in the subalpine zone, at an altitude of 2,500—3,000 m. — Centr. Asia: Pam.-Al. Described from Zeravshan. Epidemic. Type in Leningrad.

Hybridizing with S. oxycarpa and viminalis s.l.

Note. In the silvery pubescence approaching closely S. pseudoalba E. Wolf, but that species is distinguishable by the smaller leaves and the tomentose ovary.

175 138. **S. oxycarpa** Anderss. in Journ. Linn. Soc. IV (1860) 45 (excl. var. serratifolia); Ej. in DC. Prodr. XVI, 2 (1868) 310 (excl. var. serrata); Hook. Fl. Brit. Ind. V, 636. — S. iranica Bornm. in sched. ad coll. 1892, No. 2559; Bornm. et Toeppf. Salic. exsicc. No. 471. — S. caramanica Bornm. in Verh. Zool. Bot. Ges. Wien, LX (1910) 174 (nomen) et in Beih. Bot. Centralbl. XXXIII (1915) II, 202.

A shrub to 2 m high; branches softly tomentose or glabrate, brownish or olivaceous; buds at first hairy, at length glabrous; stipules small, semilanceolate or lanceolate, glandular-dentate; petioles 2—4 mm long, slender, glabrous or puberulous, channeled; leaf blades 5—7.5 cm long (rarely up to 10 cm) and 1.5—2 (2.5) cm broad, or shorter and narrower (v. breviuscula Anderss.), the lower oblanceolate, similarly narrowed at both ends, pointed at apex, flat, entire or sparsely serrulate, the young transparent, thin, glabrous, grayish-green, almost waxy, dull glaucescent beneath, on both sides covered with short appressed shiny hairs, closely dotted; midrib yellowish or brownish, prominent beneath, glabrate above; lateral veins at an angle of 25—40°, evident; catkins precocious, sessile or borne on a very short stalk, very elongated, widely spreading, rather loose, with bracts or small leaves at base; staminate stoutly cylindric, to 5 cm long; pistillate erect, 7.5—10 cm long, acutish; scales brown or rufescent, obtusish, hairy; stamens connate up to the middle; ovary nearly 3 mm long, sessile, conically elongated, acute, sparingly silvery-pubescent; style very short; stigma subclavate. Fl. May; fr. June.

Mountain-stream valleys, at 1,700—3,200 m. — Centr. Asia: Pam.-Al. **Gen. distr.:** Tibet, Ind.-Him. Described from NW India. Type in London.

Hybridizing with S. coerulea.

139. **S. lepidostachys** O. v. Seem. in Engl. Bot. Jahrb. XXI, Beibl. 53 (1896) 51; Ej. Salic. Japon. (1903) 58 et tab. XII; Kom. and Alis., Opred. rast. Dal'nevost. kr. I, 425. — S. Miyabeana O. v. Seem. in Engl. Bot. Jahrb. XXI, Beibl. 53 tab. 12 et ejusd. Salic. Jap. 57.

A tall shrub with yellowish-gray bark; branches fulvous, glabrous, lustrous; young branchlets puberulous; buds very large, reddish-brown, glabrous; stipules narrow, lanceolate, dentate, somewhat shorter than petiole; petioles to 1.5 cm long, glabrous, initially hairy; leaf blades lanceolate to elongate-lanceolate, prolonged into a long point, to 10—12 cm long and to 2.5 cm broad, coarsely dentate, at first pubescent, becoming glabrous; lustrous above, glaucous beneath; lateral veins about 20 pairs, at an angle of 35—45°; catkins precocious, sessile, with 1 or 2 scalelike bracts at base, upright or spreading, cylindric, to 4—5 cm long and 1—1.5 cm
176 broad (pistillate ca. 0.9 cm broad); rachis hairy; scales oblong, round-tipped, dark brown, clothed with long pale silky hairs; stamens 2, connate, scarcely hairy in lower part, barely exceeding the scale; anthers often parted, grayish-fulvous, rounded-ovate; gland 1, posterior, narrowly turbinate, equaling the stipe; style short, glabrous; stigma elongated, upright, forked. Fl. May; fr. June.

Waterside thickets. — Far East: Uss. **Gen. distr.**: Japan. Described from Japan. Type in Berlin.

140. **S. Korshinskyi** Görz in Trudy Tadzh. bazy Ak. Nauk II (1936) 178.

A shrub, apparently tall, or a tree; branches long, virgate, at first covered with scattered hairs, slender, glabrous, buff; buds ovoid, sparingly hairy; stipules strongly developed, ovate-lanceolate, sharply dentate, those of terminal leaves to 3—4 cm, others 0.8—12 cm long; petioles glabrous, to 0.8—1.2 cm long; leaf blades lanceolate, to 11 cm long and to 1.4 cm broad, 8—9 times as long as broad, obtuse at base, regularly attenuate toward apex, at first covered with appressed silky hairs, finally glabrous or covered on both sides with scattered hairs, remotely and sharply toothed, the teeth apparently subulate-tipped, in lower part of the blade, 5—7 mm, toward apex 2—3 mm long; upper surface green, lower scarcely paler; midrib reddish; lateral veins 15—20 pairs, at an angle of 45—60°; flowers unknown.

Cultivated in gardens and in proximity of fields; possibly also growing wild. — Centr. Asia: Pam.-Al. Endemic. Described from Roshan, where it was collected by S. Korzhinskii in the Shadau River system, 1897, No. 4846. Type in Leningrad.

Note. Dr. Görz, influenced by the article on S. aegyptiaca L. by Dr. Floderus, apparently had doubts concerning this species and was inclined to consider it as a naked-leaved form of S. aegyptiaca L., which is certainly erroneous. S. Korshinskyi Görz is a perfectly original and indisputably new species which, except for the large leaves, has nothing in common with the Egyptian willow.

141. **S. ferganensis** Nas. sp. nova in Addenda IV, p. 545.

A tall shrub or a tree with olivaceous-brown naked bark and light brown glabrous lustrous branches; buds plump, large, appressed, glabrous; stipules semicordate or lunate, large, dentate; petioles 5—7—10 mm long, glabrous,

brown, dilated toward base, channeled; leaf blades oblong or oblong-lanceolate, arcuately attenuate toward a short-tipped apex, abruptly narrowed toward base, to 11.5 cm long and 3.2 cm broad, remotely glandular-
177 serrate or slightly sinuate, wavy-margined, pure green on both sides, somewhat darker above, slightly lustrous, essentially glabrous, but under 10-power magnification very scattered short white cilia may be seen on the lower surface; midrib strongly raised, thickened toward base, light brown; lateral veins 11—13 pairs, at an angle of 40—50°; network of veinlets prominent on both sides; flowers unknown.

Banks of mountain streams. — Centr. Asia: T. Sh. Endemic. Described from the Fergana Range, from the banks of the Kug-art River, 4 March 1895, No. 5554, S. Korzhinskii. Type in Leningrad.

Note. The general aspect, as well as the very large, pale green, and almost glabrous leaves, place this species quite apart from other USSR species of this section. In spite of some doubts, it is nevertheless more than likely that the species belongs here.

Section 21. **CAESIAE** Kern. in Verh. Zool.-Bot. Ges. Wien X (1860) 205. — Shrubs and undershrubs, mostly with naked branches; leaves oblong, elliptic, lanceolate or obovate to oblanceolate, mostly short, bicolor (conspicuously glaucous beneath), often turning black on drying, finely reticulate-veined; catkins with the leaves, subsessile, ovoid-cylindric or short-cylindric, densely flowered; gland 1, oblong; filaments of stamens partly or wholly connate; ovary subsessile or short-stipitate; style short; the subsessile stigmas oblong.

1. Leaves relatively small, 2—3 cm long and 0.6—1.2 cm broad, lanceolate, or narrowly elliptic, acuminate, narrowed toward base, glabrous or hairy, with 5—10 lateral veins; catkins small; mostly a depressed shrub, often with prostrate branches . . . . 142. **S. caesia** Vill.

\+ Leaves larger, 3—7 cm long and 2—3 cm broad, oblong, oblong-ovate, or obovate, obtuse or short-pointed at apex, broad at base, quite glabrous, with 12—20 lateral veins; catkins larger; upright shrubs . . .2.

2. Leaves subopposite, about equally broad at both ends, amplexicaul at base; lateral veins up to 20 pairs; ovary ovoid . . . 144. **S. integra** Thunb.

\+ Leaves alternate, broadest above the middle, narrowed toward base, not amplexicaul; lateral veins 12—14—17 pairs; ovary conical. . . . . 143. **S. Kochiana** Trautv.

142. **S. caesia** Vill. Hist. Pl. Dauph. III (1789) 768, tab. 50; Wimm. Salic. europ., 100; Turcz. Fl. baic.-dah. II, 1 (1851) 121. — S. minutifolia et
178 S. minutiflora Turcz. pl. exsicc. (1829); Kryl., Fl. Zap. Sib. IV, 737. — S. caesia ε minutiflora Anderss. in DC. Prodr. XVI, 2 (1868) 317. — S. sibirica α glabra Ldb. Fl. Ross. III (1851) 622. — Ic.: Rchb. Ic. Fl. Germ. XI, tab. 565, fig. 2005. — Exs.: Wimm. Coll. Sal. No. 116.

A depressed, sometimes trailing shrub; young branches reddish-brown, often densely clothed with silky hairs, finally yellowish-gray and glabrous; buds small, flattened, appressed, obtusish, yellowish or reddish, glabrous; stipules mostly none; petioles 2—5 mm long, glabrous or pubescent; leaf blades small, 2—3 cm long and 1.5 cm broad, ovate, elliptic, or lanceolate, mostly acute, the entire margins revolute; grown leaves stiff, grayish-green, glaucous beneath, glabrous (f. glabra Turcz.) or dull with appressed hairs (f. pubescens Turcz.); lateral veins 5—12 pairs, at an angle of 30—70°, paler below than above; catkins slightly serotinous, small, dense, borne on a short leafy-bracted stalk; staminate 0.8—1.2 cm long and ca. 0.5 mm broad, pistillate to 1.5—2 cm long and 0.9 cm broad; rachis covered with short silky hairs; scales liguliform or obovate, light brown, terminating in a fulvous point, hairy or dark purple and glabrate; stamens 2, wholly connate, hairy below; anthers distinct, yellow; gland 1, linear-oblong, obtuse or emarginate or 2- or 3-lobed, rufescent, twice as long as the stipe; ovary ca. 3 mm long, ovoid-conical, densely clothed with silky hairs, at length reddish, glabrescent, the stipe very short; style ca. 0.5 mm long; stigma entire or emarginate; capsule to 5 mm long. May—June.

Banks of mountain streams. — European part: S. Urals; W. Siberia: U. Tob., Alt.; E. Siberia: Ang.-Say., Lena-Kol., Dau.; Centr. Asia: Dzu.-Tarb., T. Sh., Pam.-Al., Syr D. **Gen. distr.**: mountains of Centr. Eur., Mong. Described from France. Type in Paris.

Hybridizing with S. alatavica and S. Wilhelmsiana.

143. **S. Kochiana** Trautv. in Mém. pres. à l'Acad. de Pétersb. par. div. sav. III (1837) 632, tab. 1; Anderss. in DC. Prodr. XVI, 2, 314; Ldb. Fl. Ross. III, 502; Turcz. Fl. baic.-dah. II, 1, 108; Kryl., Fl. Zap. Sib. IV, 738. — S. Pontederana Trautv. in Ldb. Fl. Alt. IV (1833) 263, non Willd. — S. loniceraefolia Turcz., pl. exsicc. — Ic.: Trautv., l. c. — Exs.: HFR No. 2348, 2349, 2500.

A shrub 1—2 m high, rarely higher; branches slender, greenish or pale yellow, glabrous, lustrous; buds small, appressed, acute, glabrous; stipules mostly none; petioles 2—5 mm long, glabrous; leaf blades obovate-elliptic, 2.5—7 cm long and 2—3 cm broad (f. latifolia Laksch. in sched.), elliptic, oblong-elliptic, or attenuate-oblong (f. angustifolia Anderss.), usually broadest above the middle, obtuse or point-tipped at apex, somewhat narrowed toward base, with entire and sometimes slightly revolute margins, at times obscurely serrate, glabrous on both sides, dull green above, glaucous beneath, usually turning black on drying; lateral veins 12—14—17
179 pairs, very prominent above, less so beneath, at an angle of 35—50°; catkins serotinous or coaetaneous, on lateral branchlets, mostly short-stalked (in v. curtiflora Anderss. long-stalked), with small bracts at base, dense, 2.5—3 cm long and ca. 0.8 cm broad, pistillate in fruit to 3—5 cm long; scales oblong or spatulate, mostly dark brown or at base rufescent, rather dark at apex, villous at base; stamens 2, more or less connate, the filaments hairy; anthers 2, 2-locular, or 1 apparently 4-locular, round, yellowish; gland 1, varying in shape from linear to oblong or quadrate, entire or 2-lobed, ca. 0.5 mm long; ovary conical, acutish, clothed with silky or silvery hairs; style ca. 0.5 mm long or nearly twice as long (v. curtiflora Anderss.), dark like the 2-lobed stigma; stipe equaling or slightly exceeding the gland; capsule covered with scattered hairs, to 5—6 mm long. Fl. May; fr. June.

Wet valley meadows, scrub, and banks of rivers and lakes. — W. Siberia: Alt.; E. Siberia: Ang.-Say., Dau., Lena-Kol. (Yakutsk District); Far East: Ze.-Bu. **Gen. distr.**: Mong., Manch. Described from the Baikal area (Tunkinskii Territory, Transbaikalia). Type in Leningrad.

Economic importance. Ornamental. The leaves are eaten by Siberian chipmunk, pika (Ochotona Swatoschi), and the mouse Microtus Michnoi; the last two cut off branchlets, dry them, and keep them in store for the winter.

144. **S. integra** Thunb. Fl. Japon. (1784) 24; Willd. Sp. pl. IV, 2 (1805) 686; Nakai, Fl. Sylv. Kor. XVIII, 113 et tab. XX; Kom. and Alis., Opred. rast. Dal'nevost. kr. I, 425, Plate 129. — S. multinervis Franch. et Sav. Enum. pl. Jap. II (1876) 504; Kom., Fl. Manchzh. II, 1, 25. — S. purpurea ssp. amplexicaulis var. multinervis C. K. Schn. in Sarg. Pl. Wills. III, 1 (1916) 168. — Ic.: O. v. Seem. Salic. Jap. III, t. XI; Nakai, l. c.; Kom. and Alis., l. c.

A shrub 1 — 3 m high, with spreading branches of about equal length; 2-year-old branches yellow or reddish, from the beginning glabrous and lustrous; leaves subopposite; petioles ca. 2 mm long; leaf blades 2 — 7 cm long and 1 — 1.8 cm broad, narrowly oblong, rounded at both ends, mostly terminating in a short point, slightly sinuate at base, subsessile and amplexicaul, often reddish when young, finally dark green above, conspicuously glaucous beneath, with a brownish midrib; lateral veins 18 — 20 pairs, slender, evident, also above (hence the synonym S. multinervis), at an angle of 50 — 60°; margins entire or sharply serrulate; catkins precocious, dense, with 3 or 4 small bracts at base; pistillate to 1.5 cm long, elongating in fruit to 2 cm, recurved, slender; rachis hairy;
180 scales obovate, dark brown to almost black, to 1.5 mm long, slightly hairy, sometimes glabrate; gland 1, posterior, ovate, lageniformly narrowed toward apex; stamens 2, wholly connate, with glabrous filament; style 0.3 mm long, glabrous; stigmas small, flattish, divergent, 2- or 4-parted; capsule to 3 mm long, clothed with appressed silky hairs. Fl. beginning of May. (Plate VIII, Figure 6).

Wet meadows, valley coppices, and riverbanks. — Far East: Uss. **Gen. distr.**: Jap.-Ch. Described from Japan. Type in the Urals.

Economic importance. Ornamental and yielding material for wickerwork and tanning.

Section 22. **DAPHNOIDES** Dumort. Fl. Belg. Prodr. (1827) 12. — Trees or tall shrubs, mostly ochthophilous or ammophilous. Branches elongated; bark mostly covered with easily rubbed-off bluish bloom, lemon-yellow on the inside; leaves long, broad or narrow, lanceolate or oblong-lanceolate, crenate, glabrous, mostly with large stipules or exstipulate; catkins precocious, stoutly cylindric, white-silky or silvery-hairy; scales sometimes glandular-toothed at base, commonly particolored, black-tipped, densely white-hairy; gland 1, posterior in both sexes; stamens 2, distinct; anthers yellow; ovary subsessile or stipitate, mostly glabrous; style long; stigma elongated, entire.

1. Stipules developed; style one-fourth to one-half as long as ovary (excluding stipe) . . . . 2.

\+ Stipules mostly none or little developed; catkin-scales nerveless, entire; style 1.5 times as long as ovary (excluding stipe); possibly occurring in the Soviet Far East . . . . **S. roridiformis** Nak.

2. Stipules obliquely ovate or reniform; scales of female catkins glandular-verrucose on the margin at base . . . . . 147. **S. rorida** Laksch.

\+ Stipules lanceolate or ovate-lanceolate or ovate, more or less acute; catkin-scales entire, without wartlike glands . . . . 3.

3. Branches stout, yellowish-green or dark brown; stipules ovate or ovate-lanceolate, one-third to one-half the length of petiole; leaves oblong, ovate-lanceolate, or oblong-lanceolate, 3—5 times as long as broad; catkins approximate; stipe of ovary shorter than the gland; style one-fourth to one-third the length of ovary . . . . 145. **S. daphnoides** Vill.

\+ Branches slender, reddish-brown; stipules lanceolate, long-acuminate, equaling or exceeding the petiole; leaves lanceolate or narrowly lanceolate, terminating in a point, 6—7 times as long as broad; catkins distant; stipe of ovary as long as the gland; style one-third to one-half the length of ovary . . . . 146. **S. acutifolia** Willd.

145. **S. daphnoides** Vill. Hist. pl. Dauph. III (1789) 765; Anderss. in DC. Prodr. XVI, 2, 261 (ex prt.); Wimm. Salic. europ., 4; Ldb. Fl. Ross. III. 602 181 (ex prt.); Shmal'g., Fl. II, 434. — Ic.: Rchb. Ic. Fl. Germ. XI, t. 602, fig. 1253. — Exs.: HFR No. 894, 895, 939, 2319, 2320.

A tree to 15 m high and to 20 cm in trunk diameter, sometimes with a pyramidal head; branches stoutish, the young white-silky, the annotinous pale green, the older olivaceous-brown or dark castaneous (not red), pruinose; bark bitter, lemon-yellow on the inside; stipules ovate, ovate-lanceolate, or ovate-falcate, glandular-serrate, soon caducous, short, one-third to one-half the length of petiole; petioles 0.4—1.4 cm long, pubescent, stramineous, dilated at base, with or without glands; leaves oblong-lanceolate, or ovate-lanceolate or oblong, 7—10 cm long and 1.5—3 cm broad, narrowed or rounded at base, short-acuminate at apex, mostly broadest at the middle, with a flat or somewhat revolute margin or glandular-serrate, pure green; upper surface lustrous with stramineous midrib, at first slightly hairy, finally glabrous or on the midrib pubescent; lower surface paler, dull glaucous or glaucescent-green, glabrous, with yellowish veins; catkins very precocious, sessile, closely approximate, stout, mostly ebracteate or with scalelike bracts at base, ca. 2—5 cm long and ca. 1—1.2 cm broad, densely hairy due to scales being very densely clothed on both sides with straight silvery-white hairs; scales ovate, acute or obtusish, dark red to almost black at apex; stamens 2, distinct, glabrous; anthers oblong, golden; ovary ovoid-conical, acute, flattened, yellowish-green, glabrous, the stipe half as long as the linear-oblong gland; style one-fourth to one-third the length of ovary; stigmas oblong, with upright or divergent lobes. Fl. March—April; fr. April—May. (Plate IX, Figure 1).

In the USSR probably only cultivated, near houses and churches, and in graveyards. Native in the mountains of Centr. Eur. whence it spread over the plains. — European part: Lad.-Ilm., U. and M. Dnp. **Gen. distr.**: Scand., Atl. and Centr. Eur., Med. Described from France.

Hybridizing with S. acutifolia, caprea, phylicifolia.

Economic importance. A good early honey-plant. The bark is used for tanning (tannin content 5.76—11.55%) and for salicyl extraction. Suitable for wickerwork and for ornament. Characterized by rapid growth; easily propagated by cuttings.

146. **S. acutifolia** Willd. Sp. pl. IV, 2 (1805) 668; Ldb. Fl. Ross. III, 601; Shmal'g., Fl. II, 434; Kryl., Fl. Zap. Sib. IV, 735. — S. *daphnoides* β *acutifolia* Anderss. in DC. Prodr. XVI, 2 (1868) 262. — Ic.: Rchb. Ic. Fl. Germ. XI, t. 603, fig. 1255. — Exs.: HFR No. 133, 2324, 2476.

Vernacular names: krasnaya [red] verba, shelyuga, krasnotal, verboloz.

A tree to 10 — 12 m high or an arborescent shrub; branches long,
182 slender, virgate, flexible, reddish-brown or rarely (v. *tatarica* Nas.) bright red with a bloom or egg-yellow without bloom (v. *vitellina* Nas., in spring often greenish; buds appressed, elongated, glabrous, blackish-brown or reddish, to 6 mm long; stipules lanceolate, acute, serrate, equaling or exceeding the petiole; petioles yellowish-red, 0.5 — 1.5 cm long, eglandular; leaves lanceolate, long-acuminate, cuneate at base, 6 — 15 cm long and 0.7 — 1.2 cm broad, rarely linear or lanceolate, 0.4 — 0.6 cm broad (v. *tatarica* Nas.), glandular-serrate, glabrous, lustrous above, glaucous or greenish beneath; catkins very precocious, remote, sessile or nearly so, mostly ebracteate, 2.5 — 3.5 cm long, staminate ovoid, pistillate cylindric, densely silvery-hairy before anthesis; scales acute, almost black at apex, covered with long white hairs, persistent; stamens 2, 2 — 3 times as long as the scale, with glabrous filaments and yellow anthers; gland 1, posterior, strap-shaped, about equaling or somewhat shorter than the stipe; ovary ovoid-conical, compressed laterally, short-stipitate or subsessile, glabrous; style long; stigma oblong, with divergent lobes. Fl. March — April; fr. April — May. (Plate IX, Figure 2).

Riverside and inland sands. — European part: from Kar.-Lap. and Dv.-Pech. to Bl. inclusive (not seen in the Crimea); Caucasus: Cisc.; W. Siberia — rare: U. Tob., Irt.; E. Siberia: Yen.; Centr. Asia: Ar.-Casp., Balkh., T. Sh., Pam.-Al. **Gen. distr.**: E. Prussia, Finland, Baltic States. Described from Germany. Type in Berlin.

Note. In W. Europe replaced by S. *daphnoides* Vill., in E. Siberia by S. *rorida* Laksch., and in Korea by S. *roridiformis* Nak.

Hybridizing with S. *caprea*, *cinerea*, *daphnoides*, *dasiclados*, *purpurea*, *nigricans*, *rosmarinifolia*, and *viminalis* (collect.).

Economic importance. The earliest honey-plant. The bark is used for extraction of salicyl and for tanning (tannin content from 0.69 to 11.44%). It is eaten by hares and sometimes by livestock. The twigs are used for hoops, cart bodies, baskets, simple furniture, and for finer wickerwork as shaved slivers. The flexible roots, reaching a length of 10 — 15 m, are also used in wickerwork. Grows well on sands and stands up easily to sand blowing; hence used for sand fixing. The leaves are retained longer than in other species. Very frost-hardy and also withstanding successfully the hot summer of the sandy deserts of Central Asia. Resembling weeping willows more than other USSR species; it is therefore used for ornamental planting on pond margins. USSR var. *tatarica* is particularly ornamental and is known by horticulturists under the name S. *caspica* hort. (the use of the name is not recommended, since it is liable to lead to confusion with S. *caspica* Pall. of another section); var. *tatarica* also yields the most flexible, slender, shrinkage-resistant twigs.

147. **S. rorida** Laksch. in Sched. ad HFR VII (191) No. 2329 (non
183 Gandoger, 1890, nom. nud. in litt.); Kryl., Fl. Zap. Sib. II, 736; Kom. and Alis., Opred. rast. Dal'nevost. kr. I, 424, Plate 129; Nakai Fl. Sylv. Kor. XVIII, 92

et tab. XII. — S. coerulescens Turcz., pl. exs. a. 1828 (herb. Academ.), non Döll., Fl. Bad. IV (1859) 517. — S. acutifolia Turcz. (non Willd.) Fl. baic.-dah. I (1854) No. 1021; Ldb. Fl. Ross. III, 601 (ex p.). — S. daphnoides Ldb. l. c. 602 (ex p.). — S. praecox Turcz. pl. exs.— S. Lakschewitsiana Toepff. in Oesterr. Bot. Zeitschr. LXVI (1916) 402. — Ic.: O. v. Seem. Salic. Japon. 49, tab. IX, A — E. — Exs.: HFR No. 2329.

A tree to 8 — 15 m high, to 1 or even 2 m in trunk diameter; bark rather deeply cracking, falling off in patches, yellow on the inside; branches slender, virgate, dark brown, smooth, pruinose or reddish-yellow without bloom; young branches and buds glabrous; stipules obliquely ovoid or reniform, glandular-dentate on the margin, about equaling the petiole, ca. 0.4 — 0.8 cm long; petioles to 0.8 cm long; leaves lanceolate, acuminate, to 10 — 12 cm long and 0.7 —3.2 cm broad, 4 —4.5 times as long as broad, regularly glandular-serrate on the margin, dark green and lustrous above, glaucous beneath with slightly prominent stramineous veins; young leaves glabrous or at first covered with fugacious tomentum; lateral veins 15 — 17 pairs; catkins precocious, cylindric, dense, often recurved, sessile, leafy-bracted at base, staminate 1.5 — 3.5 cm long and 1.8 —2 cm broad, pistillate 3 — 4 cm long and 1 — 1.5 cm broad, elongating in fruit to 5 cm; rachis silky; scales obovate, acuminate, often 2-pointed at apex, entire or glandular-margined, pistillate oblong, acute, glandular-crenate on the margin toward base, both staminate and pistillate bearded on both sides, to 3 — 4 mm long; stamens 2, to 7 — 8.5 mm long, with distinct glabrous filaments and ovate yellow anthers; gland 1, interior, oblong or subquadrate, ca. 0.6 — 0.7 mm long, shorter than the stipe; ovary 2 — 3 mm long, ovoid-conical, glabrous, green, the stipe ca. 1 — 1.5 mm long; capsule to 4 —6 mm long, the stipe twice the length of the gland; style ca. 1 — 1.5 — 2 mm long; stigmas ca. 1 mm long, oblong-linear, upright or divergent, yellow like the style. Fl. May; fr. June. (Plate IX, Figure 3).

Riverbanks. — W. Siberia: Alt., Irt.; E. Siberia: Yen., Ang.-Say., Dau., Lena-Kol.; Far East: Ze.-Bu., Uss., Sakh. **Gen. distr.:** Jap.-Ch. Described from Buryat-Mongol area (former Balagansk County, Irkutsk Province). Type in Leningrad.

Hybridizing with S. caprea, hastata, and pyrolifolia.

Economic importance. Nectariferous; used for tanning and for salicyl extraction. The Goldi and other peoples of the Far East used the trunks to make canoes called "omorochka."

Note. A closely related vicariant S. roridiformis Nak. (growing in Korea) may occur in Ussuri Territory and in Sakhalin. This species has
184 lanceolate leaves, sericeous when young, little-developed stipules, style 1.5 times as long as the ovary (excluding stipe), and entire eglandular scales.

Subgenus 3. **AMERINA** Dumort. Bijdr. tot. naturk. (1825) 56. — Trees or tall shrubs; bud-scales falling after the unfolding of leaves; leaves stipulate, often with glandular petioles; veins 8 — 17 pairs; catkins stalked, with bracts at base; catkin-scales monochromatic, pale, caducous; glands in staminate flowers 2, in pistillate flowers gland sometimes solitary, or glands 2 in both sexes; stamens 2 or 3, rarely more, distinct; filaments mostly hairy at base; ovary mostly glabrous.

Section 23. TRIANDRAE Dumort. Bijdr. tot. naturk. (1825) 58. — Trees or tall shrubs, with elongated glabrous branches; bark of old branches peeling off in thin patches; leaves mostly with stipules, lanceolate, acute, serrate, glabrous, concolor or glaucous beneath, not viscous when young; catkins subcoaetaneous, borne on a leafy stalk; scales pale, monochromatic, persistent; staminate flowers with 2 glands and usually with 3 distinct stamens; pistillate flowers with 1 gland; ovary long-stipitate, glabrous, almost without a style, with short stigmas.

This section is represented in the USSR by only one aggregate species S. triandra L., from which, however, two somewhat different minor species, S. Medwedewi Dode and S. nipponica Franch. et Sav., have been separated.

1. Leaves lanceolate, to 2—3.5 cm broad, glandular-serrulate; catkins commonly long; scales pubescent, especially in lower part . . . . 2.

\+ Leaves linear-lanceolate, ca. 0.5—0.6 cm broad, remotely and rather shallowly glandular-serrate; catkins only ca. 3 cm long; scales more densely hairy . . . . 148. **S. Medwedewi** Dode.

2. Young shoots tomentose; young leaves hairy; catkins densely flowered or at base loose; scales sometimes short, sometimes half the length of the ovary . . . . 150. **S. nipponica** Franch. et Sav.

\+ Young shoots glabrous or, like the young leaves, sparsely covered with promptly fugacious hairs; catkins loosely flowered; scales commonly twice as long . . . . 149. **S. triandra** L.

148. **S. triandra** L. Sp. pl. (1753) 1016. — S. amygdalina L. l. c., 1016; Gmel. Fl. sibir. I, 155, No. 9, tab. 34, fig. 3; Ldb. Fl. Ross. III, 600; Shmal'g., Fl. II, 2, 432; Kryl., Fl. Zap. Sib. IV, 733; Turcz. Fl. baic.-dah. II, 1, 99. Ic.: Anderss. Mon. Salic. 17; Fl. Dan. 2558; Rchb. Ic. Fl. Germ. XI, f. 1256, 1259, 1260. — Exs.: HFR No. 487, 2260, 2263, 2459. — Vernacular names:
185 beloz [white willow], belotal, loza, lozina.

A shrub 5—6 m tall and 7—8 cm in stem diameter, rarely a tree to 8—10 m tall; branches slender, upright, flexible, olivaceous green or buff-green, dull, slightly hairy when young; bark of stems and old branches peeling off in thin patches; bark of young branches not bitter; buds ovoid, acute, ribbed, appressed, glabrous, light brown; stipules reniform, ovate, or ovate-lanceolate, dentate, often rather conspicuous, long persistent; scales 1—1.5 cm long, commonly with 2 glands at the top, at first hairy, finally glabrous, often rufescent or brown; leaves approximate, oblong-lanceolate to lanceolate (v. vulgaris Wimm.), ovate-elliptic (v. Villarsiana Wimm.), narrowly elliptic (v. latifolia Toeppf.), elliptic (v. elliptica Toepff.), or narrowly lanceolate (v. angustifolia Ser.), 4—15 cm long and 0.5—3.5 cm broad, acuminate, rounded or cuneate at base, glandular-serrate or remotely dentate; upper surface dark green; lower surface green (f. viridis Spenn.) or glaucous to whitish-glaucous (f. glaucophylla Ser.); expanding leaves puberulous, nonviscous, becoming quite glabrous; midrib prominent, sometimes slightly colored; lateral veins 8—15 (23), pale, at an angle of 40—80°; catkins serotinous, at the ends of short annotinous shoots, with entire or serrate bracts at base, upright or recurved, staminate 3—10 cm long, pistillate 1.5—4 cm long, slender, loose, often verticillate, aromatic; rachis covered with appressed hairs; scales oblong, concave, 1.5 — 3 mm long, monochromatic lurid, persistent, in pistillate flowers 3-nerved, commonly covered only at base with crisp hairs, ciliate on the margin, glabrous on the back, or else rather densely pubescent; stamens 3 (exceptionally 2, 4, or 5), to 5 mm long, distinct, densely covered at base with crisp hairs; anthers yellow; glands in staminate flowers 2, often 2-parted or forming a 4- or 5-lobed disk, the inner trapeziform or rounded-ovate, the outer small and narrow, in pistillate flowers the 2 glands commonly united into a posterior gland, one-fifth to one-third the length of the ovary; ovary ovoid-conical, glabrous, olivaceous-green, more or less glaucescent, ca. 4 — 5 mm long; style short, stout, bifid; stigma short, thick, with divergent lobes; stipe half as long to as long as the ovary; capsule glabrous, to 5 — 6 mm long and 2 — 2.5 mm broad. Fl. April — May, sometimes flowering again in a warm sunny fall or from July to September; fr. June. (Plate IX, Figure 7).

Waterside thickets in the forest, steppe, and desert-steppe regions; not rising into the mountains and absent from the Arctic Region and Kamchatka; European part: throughout; Caucasus: throughout; W. and E. Siberia: throughout; Far East: Okh., Ze.-Bu., Uss.; Centr. Asia: Ar.-Casp., Balkh., Dzu.-Tarb., T. Sh. **Gen. distr.** Scand., Atl. and Centr. Eur., Med., Bal.-As. Min., Iran., Mong., Jap.-Ch. In North America replaced by S. nigra Marsh. Described from Europe. Type in London.

Hybridizing with S. alba, australior, cinerea, fragilis, pentandra, Wilhelmsiana, viminalis s.l.

186 **Economic importance.** Nectariferous. The bark is used for salicyl extraction; considered one of the best for tanning (tannin content from 4.6 to 18 — 20%, fluctuating with season, age, and locality; the highest percentage of tannin is along the Kama River). The bark and a decoction of young leaves give a yellow dye. The wood is firm, white, with a beautiful shiny iridescence, tenacious, but not durable and not very flexible; inferior as fuel. The twigs are somewhat brittle; cleaved and wetted twigs and shaved strips are used for wickerwork; the wood of the twigs assumes a pink color when boiled. The twigs of this willow are suitable for both coarse and fine wickerwork; widely used for wicker fences, baskets, fishing tackle, etc. Inferior in wicker quality to its hybrids with S. purpurea L. and S. viminalis L. (coll.). Yields of wicker material are particularly high on peaty soil. The plant is somewhat susceptible to frost damage. One of the best willows for consolidation of soils subject to shifting or erosion, for barrage planting, etc. Ornamental, especially f. glaucophylla Ser. The bark, leaves, buds, and young twigs of this and other waterside willows, provide favorite food for the water vole (Arvicola amphibius L.).

149. **S. Medwedewi** Dode in Bull. Soc. Bot. Franc. LV (1908) 652 (cum fig. 9). — S. triandra var. Medwedewi Görz in Grossg., Fl. Kavk. II (1930) 5.

A shrub or a low tree; branches upright, very slender, fulvous, glabrous even when young; buds small, narrow, reddish, glabrous; petioles 0.4 — 0.6 cm long, reddish, glabrous, channeled; leaf blades to 10 cm long,

5 — 6 mm broad, linear-lanceolate, remotely dentate, glabrous from unfolding, glandular at base, whitish-glaucous beneath with yellowish veins; catkins serotinous, at the ends of branchlets, ca. 2 cm long, with 4 — 6 leaflike bracts, rather dense, ca. 3 m long and 3 mm broad, cylindric, obtusish; scales more densely hairy than those of S. *triandra* L., elongated, 0.8 × 2.5 mm, lanceolate, acute, brownish, hairy on the inside right up, on the outside hairy below and glabrate at apex; ovary glabrous, elongated, green at the top, yellowish toward base; style pale; stigmas curved, reddish; stipe one-third to one-half as long as the ovary; gland ca. 0.4 — 0.7 mm long, lurid. Fl. April — May; fr. May — June.

Waterside thickets. — Caucasus: Dag., W. and S. Transc. **Gen. distr.**: Asia Minor. Described from the Caucasus. Type in Paris.

150. **S. nipponica** Franch. et Sav. Enum. pl. Japon. I (1875) 459. — S. *triandra* var. *nipponica* O. v. Seem. Salic. Japon. (1903) 27; C. K. Schn. in Sarg. Pl. Wilson. VII (1916) 106; Kom. and Alis., Opred. rast. Dal'nevost. kr. II, 424. — S. *triandra* var. *discolor* Nakai, Fl. sylv.
187 Kor. XVIII (1930) 87, non Anderss. — S. *Kinashii* Leveil. in Bull. Soc. Bot. Franc. LII (1905) 141.

A tree to 10 m high or an arborescent shrub, with stem diameter 7 — 8 cm; bark brown; branches virgate, greenish-brown; young shoots densely hairy; petioles 0.4 — 1 cm long, rufescent-tawny; leaf blades lanceolate, to 7 — 10 cm long and 2 — 3 mm broad, short-tapering to a point, finely serrate with oblique teeth, at first pubescent and often dotted, finally glabrous, lustrous green, the lower surface pale or glaucous; catkins of both sexes borne on a leafy-bracted stalk to 3 — 4 cm long, the bracts mostly entire, rarely dentate; staminate catkins upright, cylindric, cylindric, little more than 2.5 cm long; rachis pubescent; scales monochromatic yellowish, initially clothed with silky hairs, at length glabrate; staminate yellowish, initially clothed with silky hairs, at length glabrate; staminate obovoid, rounded at apex; pistillate oblong, obtuse, yellowish, hairy chiefly below, sometimes (v. *microlepis* Franch. et Sav.) only about half as long as the ovary; stamens 3, distinct, with yellow anthers; ovary glabrous, fusiform, slightly longer than the scale, the stipe 1 — 2 mm long; style barely 0.5 mm long. Fl. May; fr. June.

Waterside thickets. — Far East: Uss. **Gen. distr.**: Jap.-Ch. Described from Japan. Type in Paris.

Section 24. ALBAE Borr. in Hook. Brit. Fl. (1830) 418. — Large trees; bark thick, splitting, but not falling off in patches; young branches rather flexible, long, glabrous or covered with silky hairs; leaf petioles mostly eglandular; blades narrowly lanceolate, long-tipped, not viscous when young; peduncle with bracts at base; catkin-scales monochromatic, falling before fruit ripening; stamens 2, distinct; anthers yellow; glands in staminate flowers 2, in pistillate 1, posterior, or rarely 2, anterior and posterior; ovary sessile or short stipitate, mostly glabrous.

1. Grown leaves mostly broadly lanceolate or ovate-lanceolate, ca. 1—3.5 cm broad, rather long-acuminate, the margin mostly coarsely and sharply serrate; catkins borne on a pubescent stalk, with 3—6 well developed bracts at base; rachis hairy . . . . 2.

\+ Grown leaves linear-lanceolate, ca. 6 mm broad, subentire or rarely serrate; catkins subsessile or short-stipitate, ebracteate or the bracts soon caducous; rachis yellowish . . . 153. **S. Kirilowiana** Stschgl.

2. Branches flexible, mostly brownish; leaves commonly broadest in lower half, often more or less rounded at base, attenuate toward apex, heavily silvery-sericeous beneath or on both sides, rarely dull whitish-gray above, glaucescent beneath, glabrate; catkins loose, pistillate in fruit rarely more than 0.7 cm thick; stipe of ovary rather short; stamens 4—5 mm long . . . . 3.

188 + Branches more brittle, mostly bright-colored; leaves broadest at the middle, mostly narrowed toward base, long-acuminate toward apex, glabrate on both sides, green or beneath glaucescent, with bristlelike appressed hairs; catkins large, denser, much thicker; ovary subsessile; stamens longer . . . . 4.

3. Leaves relatively thin, mostly lanceolate, short-or long-acuminate, silvery-sericeous or glabrate on both sides, green or rarely glaucescent beneath; serrulate; bracts mostly entire; ovary greenish, the short stipe as long as the gland . . . . 151. **S. alba** L.

\+ Leaves firm, coriaceous, mostly ovate-lanceolate, short-acuminate, uniformly narrowed at both ends, dark green and glabrate above, densely covered beneath with short silvery silky strongly appressed hairs, the margin coarsely and sharply serrate; bracts mostly dentate; scales latericious-brown; ovary brown; stipe as long as or longer than the gland . . . . 152. **S. micans** Anderss.

4. Grown leaves green and glabrous on both sides; catkins in fruit to 4—5 cm long; style short, bifid from the middle; stigmas 2-parted, with rather long divergent lobes . . . . 5.

\+ Grown leaves mostly glabrous pale green and slightly lustrous above, azure-glaucous and covered with bristlelike appressed hairs beneath; catkins in fruit to 8 cm long and 1 cm thick; style obsolescent, the subsessile stigmas reddish-fulvous . . . . 156. **S. oxica** Dode.

5. Bracts obovate, abruptly narrowed toward base, point-tipped; catkin-scales mostly retuse or erose at apex; stigmatic lobes oblong, curved; branches mostly brown or yellowish-brown . . . 155. **S. euapiculata** Nas.

\+ Bracts lanceolate to broadly lanceolate or obovate-lanceolate, not point-tipped; catkin-scales obtuse or acute, entire; stigmatic lobes long, divergent; branches mostly orange-red . . . . 154. **S. australior** Anderss.

151. **S. alba** L. Sp. pl. (1753) 1021; Anderss. Mon. Salic., 47, tab. III et in DC. Prodr. XVI, 2, 211; Ldb. Fl. Ross. III, 598; Shmal'g., Fl. II, 432; Kryl., Fl. Zap. Sib. IV, 730. — Ic.: Rchb. Ic. Fl. Germ. XI, tab. 608, fig. 1263. — Exs.: HFR No. 586, 2251, 2252, 2451, 2452. — Vernacular names: (Serebristaya)
189 [silvery] iva, vetla, beloloz [white willow].

A tree reaching great size, 25—30 m high and to 3 m in trunk diameter, up to 100 years of age, but usually dying earlier due to core-rot; bark dark gray, deeply splitting; young branches silvery-pubescent at the ends, the old glabrous, upright, flexible, brown, in winter reddish, in spring greenish; buds appressed, elongated, acute, closed, ca. 6 mm long and 1.5 mm broad, reddish-yellow, silky; stipules small, narrowly lanceolate, glandular, soon caducous, silvery-pubescent; petioles of larger leaves with glands at the top, 0.2 — 1 cm long; leaf blades 5 — 12 (15) cm long and 1 — 3 cm broad (1 : 3 — 7), typically lanceolate (f. typica Anderss.), linear-lanceolate (f. angustifolia Laksch.), acuminate, sometimes prolonged into a long point (v. acuminata Rgl.), closely serrulate, on both sides silvery-sericeous (v. sericea Gaud. = v. argentea Wimm. = v. splendens (Bray) Anderss.) or glabrous above, puberulous beneath, glaucescent (f. coerulea Koch) or on both sides greenish (f. concolor Rgl.), initially mostly covered with longitudinal silvery hairs; lateral veins 12 — 15 pairs, at an angle of 30 — 40 (60)°; catkins coaetaneous, 3 — 5 cm long, rather loose, stalked, with entire oblong-obovate obtusish bracts, the staminate lemon yellow; rachis densely pubescent; scales yellowish or greenish, ciliolate on the margin, crisp-hairy on the back below, in pistillate flowers soon caducous; stamens 2, distinct, hairy at base, twice as long as the scale; anthers bright yellow, turning reddish; glands 2, anterior and posterior, sometimes 2-parted; ovary ovoid-conical, obtuse, glabrous, subsessile or short-stipitate, the stipe as long as or shorter than the posterior gland (the 2 glands of pistillate flowers point to characters of S. fragilis or S. pentandra); style short or very short, often somewhat bifid; stigma yellow, 2-parted, with oblong lobes; capsule to 5 mm long. Fl. April — May; fr. May — June. (Plate IX, Figure 4).

River "plavni,"* banks of rivers and ponds; often cultivated near human habitations. — European part: almost everywhere, except Kar.-Lap., and Dv.-Pech. (N. part); Caucasus: throughout; W. Siberia: everywhere from 59° to 61° N. lat.; Centr. Asia: throughout (mostly cultivated). **Gen. distr.**: almost all of Europe (except Scand.), Med., Bal.-As. Min., Arm.-Kurd., Iran., Ind.-Him., China. Introduced into N. Am. and Nov. Z. Described from Europe. Type in London.

The actual distribution area of this willow is not easy to determine as it has long been widely grown in many places.

Hybridizing with S. acmophylla, australior, babylonica, fragilis, pentandra, songarica, triandra.

Economic importance. Nectariferous. Young branches and leaves are readily eaten by goats. The bark is used for medicinal purposes, for tanning (tannin content 3.96 — 11.02%) for dying silk, wool and kid leather reddish-brown, for production of a buff varnish for painting, and for rope
190 making. The branches are outstandingly flexible (especially v. vitellina); they are used as vine stakes, basketry material, brushwood, and props. The wood provides excellent material for bows (like hybrids of S. alba with S. fragilis, especially S. excelsior Host), hoops, canoes, troughs, wooden kitchen utensils, sandals, etc.; it is also used for charcoal and provides staves for cement and other industries. The trunks are used in construction of log-cabins (in former Ryazan and Chernigov provinces) which are noted for their warmth and, as regards durability, are not inferior to those made of aspen. The amount of heat given off on burning

* [Russian for long-flooded areas with Phragmites, Typha, and Carex, in river deltas and bottomlands (southern USSR).]

is half that of beech wood. The tree lends itself to pollarding; it is suitable for barrage strips, house enclosures, as well as street and roadside planting. Grows satisfactorily everywhere. Particularly ornamental is v. sericea Gaud. In Central Asia, under special care of local population ("aulie-agach") often attains gigantic size. Hybrids with S. excelsior are reputed to grow most rapidly and to produce a straight trunk that is particularly esteemed for shaft-bows and long poles. S. alba is susceptible to frost and very subject to insect damage. Rosette galls ("willow-roses") occur in this species, while clusters of tangled branchlets, often of considerable size, may be seen, with pathogenically undeveloped leaves; the former are caused by the midge Cecidomyia rosaria, the latter by a mite. The trunk unfortunately often becomes hollow with age.

152. **S. micans** Anderss. Mon. Salic. (1867) 49. — S. alba μ micans Anderss. in DC. Prodr. XVI, 2 (1868) 212. — S. varifolia Freyn et Sint. in Bull. Herb. Boiss. 2 sér. II (1902) 307.

A tall tree with splitting bark; branches and branchlets flexible, brownish-castaneous, lustrous, when young densely tomentose; buds small, castaneous, white-hairy above; stipules lanceolate, soon caducous; petioles 0.5—1 cm long, buff, pubescent; leaves silky at first, at length dark green, glabrate or covered with scattered hairs, coriaceous, 5—7.5 cm long and 2—2.5 cm broad, on flowering shoots to 7—9 cm long and 3—4 cm broad, ovate-lanceolate, uniformly narrowed at both ends, terminating in a straight point; closely, coarsely and sharply serrate, very densely clothed beneath with appressed silvery lustrous tomentum (duller on sterile shoots); lateral veins 14—16 pairs, at an angle of 40—60°; catkins appressed, upright, flexuous, ca. 2—5 cm long; pistillate finally loose, to 5 cm long; peduncle ca. 1 cm long; bracts 4—6, foliaceous, obovate-elliptic, dentate, somewhat smaller than foliage leaves; rachis hairy; scales ca. 2.6—3.2 mm long and 1.5 mm broad, ovate, obtuse, latericious-brown, white-hairy from base to the middle, caducous; stamens 2, distinct, hairy at base, ca. 4 mm long; ovary 2—3 mm long, thickened at base, conical, glabrous, brownish in dry state; stipe ca. 0.5—1.2 mm long, equaling or exceeding the gland, this
191 0.6—1 mm long; style 0.8—1.2 mm long; stigmas ca. 0.5 mm long, distinct, divergent; capsule 4—6 mm long. Fl. April; fr. May.

Shores; cultivated. — Caucasus: W. and S. Transc.; Centr. Asia: Ar.-Casp., Kara K. **Gen. distr.**: Bal.-As. Min., Iran. Described from the Caucasus. Type in Leningrad.

Economic importance. Nectariferous; used for woodworking and for fuel; ornamental.

153. **S. Kirilowiana** Stschegl. in Bull. Soc. Nat. Mosc. XXVIII, 1 (1854) No. 198; E. Vol'f in A. H. P. XXI, 2, 190. — S. viminalis Kar. et Kir. Enum. Song. No. 766, non L. — S. alba-viminalis Rgl. in A. H. P. VI, 2 (1880) 460. — S. alba var. angustifolia Anderss. in sched. (herb. Petrop.).

A shrub or tree? Branches tawny, at first densely pubescent, finally glabrous; stipules small, ovate to lanceolate, much shorter than petiole, commonly absent; leaf blades linear-lanceolate, 3.5—5.2 cm long and ca. 6 mm broad, mostly rounded at base, long-tapering from the middle toward apex, entire or rather sparingly serrate, at first appressed-pubescent on both sides (more densely beneath), lustrous-sericeous, at length glabrescent above, prominently veined; staminate catkins unknown; pistillate subsessile or borne on a short glabrous stalk, with leaflike or scalelike caducous bracts; rachis yellowish; scales monochromatic tawny, hairy, glabrate at apex; ovary ovoid-lanceolate, 2 — 3 mm long, quite glabrous; style very short; stigmas 2-parted; stipe shorter than scale, or ovary subsessile. May.

Banks of mountain streams. — Centr. Asia: Dzu.-Tarb. Described from the Ala Tau Mountains. Endemic. Type in Leningrad.

154. **S. australior** Anderss. Monogr. Salic. (1867) 43. — S. fragilis δ *australis* Anderss. in DC. Prodr. XVI, 2 (1868) 210. — Vernacular: Kara-tal.

A tall much-branched tree; branches orange-red like the buds and midrib of leaves; young branchlets and buds at first appressed-hairy, becoming glabrous; stipules semicordate, mostly absent; petioles 3 — 4 mm long; leaf blades broadly to narrowly lanceolate, 5 — 85 cm long and 1.3 — 2 cm broad (f. *latifolia* Anderss. et f. *angustifolia* Görz), long-acuminate, sometimes ovate at base (f. *cuspidata* Görz), coarsely serrate, initially silky, becoming glabrous on both sides, green; lateral veins 11 or 12 pairs, at an angle of 40 — 60°; catkins coaetaneous, borne on a short leafy-bracted stalk, plump, 1.2 — 4.4 cm long or shorter, stoutly cylindric (f. *brachystachya* Görz), staminate densely flowered, pistillate looser and borne on a stalk with 3 or 4 bracts; scales ovate, obtuse, 3.5 — 4 mm long and to 2.2 mm broad, in pistillate flowers caducous, plain lurid, on the back faintly fulvescent and glabrate or slightly hairy,
192 at base pubescent, rarely narrow-lanceolate and acute (f. *stenoloepis* Görz); stamens 2, often villous at base; filaments 5 — 7 mm long; anthers ovate, yellow; glands 2, in staminate flowers the inner oblong and slightly longer than the outer, in pistillate flowers the inner subquadrate and the outer mostly absent; ovary ovoid-conical, obtuse, ca. 3 mm long, glabrous, subsessile, toward maturity minutely stipitate; style short, stout, bifid from the middle; stigma 2-parted, with divergent lobes; capsule to 5 mm long. Fl. March — April; fr. April — May.

Shores; cultivated. — Caucasus: W., S. and E. Transc.; Centr. Asia: Balkh., T. Sh., Pam.-Al., Mtn. Turkm., Amu D., Kara K., Ar.-Casp. **Gen. distr.:** Bal.-As. Min., Arm.-Kurd., Iran. Described from the Caucasus. Type in Leningrad.

Hybridizing with S. *alba*, *babylonica*, *fragilis*, *micans*, *pentandra* and *triandra*.

Economic importance. Nectariferous; used for woodworking and for fuel; ornamental.

155. **S. euapiculata** Nas. sp. nova in Addenda IV, p. 545. — S. *apiculata* Nas. in schedis, non Anderss. (1850). — S. *australior* var. *apiculata* Laksch. in Schedis ad HFR VIII (1913 — 1922), 29, No. 2454 ♀. — S. *australior* f. *apiculata* Görz in Grossg., Fl. Kavk. II (1930) 5. — Vernacular: Kara-tal (Uzbek).

A tall tree; annotinous branches brown or yellowish, glabrous, slightly lustrous, when young glabrous or sparsely covered with fugacious pubescence; buds at first pubescent, becoming glabrous; petioles 3 — 4 mm long, glabrous; stipules none; leaves initially whitish-silvery, finally glabrate, lanceolate, 5 — 8.5 cm long and 1.3 — 2 cm broad, narrowed toward base, long-tapering toward the apiculate apex, glandular-serrate on the margin, green and glabrous on both sides; midrib stramineous; lateral veins 11 or 12 pairs; leaves on lower shoots and on peduncles obovate, narrowed toward base, almost rounded at apex and short-apiculate, entire, glabrous on both sides or silky beneath; pistillate catkins 1.5 — 3 cm long, ca. 0.5 cm thick, elongating in fruit to 4 cm, cylindric, dense, coaetaneous; staminate 3 — 4.5 cm long and ca. 1 cm thick, borne on a stalk with 3 or 4 leafy bracts; brachis hairy; scales yellowish; staminate ovate, retuse; pistillate large, obovate, obtusish, erose at apex; all scales pubescent at base, glabrous on the back; stamens 2, distinct, hairy at base, ca. 5 — 7 mm long; anthers ovate, yellow; inner gland in pistillate flowers subquadrate, in staminate oblong, 0.6 — 0.7 mm long, the outer to 1 mm long, in pistillate flowers often absent; ovary ovoid-conical, obtusish, ca. 3 mm long, subsessile, glabrous; capsule ca. 5.5 mm long, borne on a very short stalk; style thick, ca. 0.5 — 0.7 mm long, semibifid; stigmas 2-parted,
193 ca. 0.5 — 0.7 mm long, with oblong curved lobes. Fl. March — April; fr. May.

Irrigation ditches and riverbanks. — Caucasus: E. Transc.; Centr. Asia: Balkh., T. Sh., Pam.-Al., Amu D. Endemic. Described from Farab. Type in Leningrad.

156. S. **oxica** Dode in Bull. Soc. Bot. de France, LV (1909) 653, fig. B; Schedae ad HFR, VIII, No. 2453. ♂. — Vernacular: Ak-tal (Uzbek).

A tree reaching great size with upright or subdivaricate branches; young branches glabrous, with only the ends of shoots silky, finally faintly fulvescent-reddish, lustrous; buds 3.6 mm long, acute, pubescent; stipules mostly absent; petioles 4 — 8 mm long, reddish, pubescent; lower leaves entire, obovate, flat, more or less silky beneath, long-acuminate; others at first silvery-silky on both sides, at length glabrous pale green and somewhat lustrous above, azure-glaucous and covered with long appressed bristlelike hairs beneath, ciliate, narrow- to broad-lanceolate (but not obovate-lanceolate), attenuate toward base, always long-acuminate, to upward of 12 cm long, closely and sharply serrate; catkins at the ends of lateral branchlets, ca. 1 cm long, often with 4 or 5 obovate entire strongly ciliate bracts, initially very dense, cylindric, narrowed toward apex and toward base, greenish, finally 4 — 6 cm long and to 1 cm broad, elongating in fruit to 8 cm or more; rachis hairy; staminate scales narrowly lanceolate, acute; pistillate obtuse or obtusish, pale, fulvescent at apex, at base and on the inside hairy, at apex with scattered hairs; stamens 2, distinct, hairy at base, about twice the length of the scale; anthers tawny; glands 2, short, yellow; ovary glabrous, green, ovoid-elongate; stigmas reddish-fulvous, subsessile; gland in pistillate flowers 1, lurid, shorter than the stipe, this ca. 1 — 1.5 mm long. Fl. March — April; fr. April — May.

Shores. — Centr. Asia: Ar.-Casp., Kara K., Amu D., Pam.-Al. Described from a cultivated specimen. Type in Paris.

Section 25. ACMOPHYLLAE Anderss. Mon. Salic. (1867) 7. — Trees; leaves long, lance-elongate or narrowly lanceolate, glandular-toothed or subentire, stiffish, glabrous, glaucescent beneath; catkins rather dense; scales yellowish; stamens (2) 3 — 5 — 8; gland in pistillate flowers 1, inner, sometimes almost enveloping the stipe; glands in staminate flowers 2, mostly separate, lobed or emarginate, sometimes the outer slightly larger than the inner; ovary thick at base, ovoid-conical, short-stipitate, with a short broad stigma.

Note. All species of this section are represented by few and incomplete specimens and they are in need of more study in situ. Apart from Central Asia, they may be encountered in S. Transcaucasia. Andersson admits that
194 S. acmophylla Boiss. approaches closely S. persica Boiss. and S. dealbata Anderss.; the latter was described by him in 1850 (from India), but in later publications (Mon. Salic., 1867 et in DC., Prodr., 1868) it is not recorded.

1. Branches slender, elongated, flexible; grown leaves narrowly lanceolate or linear-lanceolate; catkins short-stalked, 1.5 — 2.5 cm long; stamens 4 — 6 . . . . 2.

\+ Branches stoutish, abbreviated, fragile; grown leaves oblong-lanceolate, tapering at apex to a short point, the two surfaces concolor; catkins subsessile, 2.5 — 3 cm long; scales large; stamens 2 or 3 . . . . 158. **S. Daviesii** Boiss.

2. Branches ferruginous-red, virgate; leaves long-tapering to a subulate point, 5 — 12.5 cm long, glaucous or pale green beneath; scales caducous; stamens 4 or 5; filaments short; anthers lurid . . . . 157. **S. acmophylla** Boiss.

\+ Branches red, almost pendulous; leaves shorter, apiculate, the two sides concolor, green; catkin-scales tardily caducous; stamens 4 — 8, mostly 6; filaments twice the length of the scale; anthers golden-yellow . . . . 159. **S. persica** Boiss.

157. **S. acmophylla** Boiss. Diagn. pl. or. ser. I, VII (1846) 98 et in Fl. Or. IV (1879) 1183; Anderss. Monogr. (1867) 7. — Ic.: Anderss. l. c., tab. 1, fig. 6.

A rather tall tree with slender virgate ferruginous-red glabrous branches; buds small, acute; stipules small, caducous, on grown branches absent; petioles ca. 3 mm long, yellow; leaf blades linear-lanceolate, very long-tapering to a subulate point, glabrous and dark green above, glaucous or pale green beneath, 5 — 12.5 cm long and 0.8 — 1.2 cm broad, subentire or slightly serrate, on lower branches obtusish and acicularly serrate; midrib somewhat prominent beneath; lateral veins evident, almost parallel to the margin, riblike; catkins upright, the stipe ca. 0.8 — 1.2 cm long; bracts 2 — 4, scarcely smaller than foliage leaves; staminate catkins stoutly cylindric, ca. 2.5 cm long, golden-yellow, dense; pistillate 2 — 2.5 cm long, subovoid; rachis hairy; scales caducous, ovate-mitriform or oblong, yellowish, crisp-hairy on the outside or in pistillate flowers glabrate; stamens 4 or 5, short; anthers lurid; ovary ovoid-oblong, thickened at base, finally subspherical, lurid, glabrous, ca. 2 mm long, the stipe 2 — 3 times the length of the gland; style obsolescent, the sessile stigmas thick. Fl. May.

Banks of rivers and of irrigation ditches. — Centr. Asia: Kara K., Amu D. **Gen. distr.**: Bal.-As. Min., Arm.-Kurd., Iran., Ind.-Him. Described from Iran. Type in Geneva.

Hybridizing with S. alba L.

195 158. S. Daviesii Boiss. in Kotschy Pl. Pers. austr. (1845) 335; Boiss. Diagn. pl. or. ser. I, VII (1846) 98; Anderss. Monogr. (1867) 9 et in DC. Prodr. XVI, 2 (1868) 195; Boiss. Fl. Or. IV, 1184.

A rather tall tree; branches upright, glabrous, red, brittle; branchlets straight, rather short, 0.5 — 0.7 cm thick; buds small, subspherical, castaneous, glabrous; petioles short; leaf blades 5 — 7.5 cm long and ca. 1.2 — 2 cm broad, oblong-lanceolate, narrowed toward base, tapering at apex to a point 4 — 6 mm long, the lower obtuse, initially slightly appressed-ciliate on the margin, finally quite glabrous, the two surfaces concolor, bright green, the lower not paler and not glaucous, entire or toward base sparingly serrulate; veins brownish, at first hairy, becoming glabrous; catkins fairly upright, obtuse, stoutly cylindric, densely flowered, ca. 2.5 — 3 cm long and 0.8 mm thick; bracts 3 — 5, obtuse, spatulate, one-third to one-half the length of foliar leaves, cuspidate, covered beneath and on the margin with white hairs like the rachis; scales ovate, acuminate or acutish, rather large, to 4 mm long, mitriform, yellowish, glabrous outside, long-hairy within; stamens 2 or 3, yellowish, 3 times the length of the scale; anthers golden-yellow. May.

Shores, gardens, and homesteads. — Centr. Asia: Amu D. **Gen. distr.**: Iran. Described from the Ku-Daen Mountains and from the vicinity of Shiraz. Type in Geneva.

Note. Closely related to S. acmophylla Boiss., but the leaves green on both sides, shorter, broader, more shortly acuminate; catkins long, subsessile; scales much larger, acutish, glabrous; branches more brittle, thicker, and shorter.

*159. **S. persica** Boiss. in Kotschy, Pl. Pers. austr. (1845) No. 101; Boiss. Diagn. pl. or., ser. I, VII (1846) 99; Anderss. Mon. Salic., 8. — Ic.: Anderss. l. c., tab. I.

A tree; branches very slender, almost pendulous, red, glabrous; stipules none; leaves subsessile, narrowly lanceolate or linear-lanceolate, glabrous and green on both sides, entire; young leaves barely 2.5 cm long and 0.6 cm broad, elliptic-lanceolate, narrowed toward base, obtusish and mucronate at apex, sparsely appressed-ciliate beneath and on the margin, becoming glabrous; catkins numerous, lateral; staminate golden-yellow, straight or flexuous, borne on a stalk with 3 or 4 small leaflike bracts, oblong-cylindric, ca. 2.5 cm long, dense; pistillate upright, cylindric, obtuse, 1.2 — 2 cm long; scales ovate or oblong; pistillate hairy within, enveloping about half the ovary, glabrate on the back in both species, tardily caducous; stamens mostly 6 (rarely 4 — 8); filaments yellow, twice as long as the scale, densely hairy from base to the middle; anthers rather large,
196 golden-yellow; ovary ovoid-conical or oblong, tapering to a short style, greenish-yellow, glabrous, the stipe twice the length of the gland; stigmas thick, undivided. May.

Banks of mountain rivulets. So far unknown in the USSR, but may possibly occur near the Iranian frontier. An insufficiently explored species, very closely akin to S. acmophylla, also resembling in aspect S. babylonica L., but differing from the latter in having 6 stamens, a stipitate ovary, and shorter, less tapering leaves. — **Gen. distr.**: Iran. Described from the mountains between Abutir and Shiraz. Type in Geneva.

Section 26. SUBFRAGILES O. v. Seem. Salic. Japon. (1903) 15. — Trees with fragile often pendulous branches; bark of old branches peeling off in flakes; catkins with or before the flowers; catkin-scales persistent; stamens 2; filaments hairy at base, sometimes connate; glands in both staminate and pistillate flowers 2 (very rarely the outer gland of pistillate flowers obsolescent); ovary sessile or subsessile; style rather short or elongated; stigma 2- or 4-parted.

1. Branches long, slender, hanging down to the ground; leaves narrowly lanceolate, long-tapering toward apex, mostly oblique; ovary glabrous or at base with few hairs; style very short . . . . . *160. **S. babylonica** L.
+ Different from above . . . . 2.
2. Stipules undeveloped; leaves broadest above the middle, deeply and sharply serrate; catkins coaetaneous; scales obtuse, light brown; anthers latericious or red; style with 2 long often twisted lobes . . . . 161. **S. dolichostyla** O. v. Seem.
+ Stipules to 3—6 mm long and 2—3.5 mm broad; leaves broadest below the middle, crenate or glandular-serrate; catkins precocious to subcoaetaneous; scales acute, pale green to almost white; anthers yellow; stigmas with 4 short divergent lobes . . . . 162. **S. koreensis** Anderss.

*160. **S. babylonica** L. Sp. pl. (1753) 1017; Anderss. Mon. Salic., 50, tab. III et in DC. Prodr. XVI, 2, 212; Ldb. Fl. Ross. III, 599; Shmal'g., Fl. II, 432. — S. pendula Moench, Meth. (1794) 336.

A very handsome tree of medium size, 10—12 m and rarely up to 18 m tall; with a thick trunk and a picturesque open head; branches long, slender, flexible, hanging down to the ground, reddish or yellowish-green, glabrous, lustrous, often with enormous pathogenic galls; stipules obliquely lanceolate or ovate, toothed or subulate, sometimes transformed into a spine; petioles
197 ca. 1 cm long, often glandular, always hairy; leaf blades oblong or narrowly lanceolate, tapering at apex to a long oblique point, gradually narrowed toward base, glandular-serrulate on the margin, dark green above, glaucescent beneath, initially slightly hairy, sometimes silvery-pubescent, finally glabrous, ca. 9—16 cm long and 1—25 cm broad; lateral veins 15—30 pairs, thin, at an angle of 45—70°; veins of third order prominent above; catkins precocious to serotinous, borne on abbreviated branchlets, slenderly cylindric, sometimes curved, at first dense, on a short leafy-bracted stalk, staminate to 2.5 cm, pistillate to 5 cm long, always pendulous; scales caducous, ovate-lanceolate, yellowish-green or pale stramineous, monochromatic, slightly hairy below, glabrous on the back above; stamens 2, distinct, hairy at base; anthers reddish-yellow; glands in staminate flowers 2, large, brownish, the anterior leaflike and folded; gland in pistillate flowers 1,

oblong, sometimes 2-lobed, slightly longer than the stipe; ovary ovoid, glabrous or with few hairs at base, pale green or yellowish, subsessile; style very short, slightly bifid at the top; stigma thickened, entire, yellowish or with 2 or 4 broad lobes. Fl. March—April; fr. April—May. (Plate IX, Figure 6).

Cultivated in the Transvolga area (city of Kuibyshev), in Soviet Central Asia, in Crimea, and in the Caucasus. In Transcaucasia possibly growing wild. **Gen. distr.**: Not clear, since it is cultivated throughout the East (China, Japan, India, and Java), S. Europe, and N. America. V. L. Komarov maintains that the plant is probably native in China or Iran, more likely the latter (Fl. Manchzh. II, pt. 1, 20). Described from the East. Type in London.

Hybridizing with S. alba, australior, fragilis.

Economic importance. Outstandingly ornamental. Not standing up to the climate to the north of Moscow. In the warm climate of the southern USSR it is grown in gardens and pleasure grounds, by the side of ponds, and in cemeteries. Nectariferous. An infusion of the leaves, shoots, and catkins is used in popular medicine.

Note. The name of Babylon applied to this willow is apparently due to a misunderstanding. It does not grow wild in Mesopotamia. The tree in the shade of which the Jews bewailed their captivity in Babylon — "We hanged our harps upon the willows" (Psalm 136 [137]) has been identified as Populus euphratica Oliv. that grows widely in Mesopotamia. The branches of Salix babylonica (which may have been grown in the gardens of Babylon) are so slender and flexible that it would have been hardly possible to hang musical instruments upon them. A very striking form of weeping willow with revolute-margined leaves rolled up into a ring (s. babylonica annularis Forb., seu crispa, seu Napoleonis hort.) is often grown in gardens also under the name Napoleon's willow. The tree was reputedly planted on the tomb of Napoleon I on Saint Helena Island; in actual fact, the tree that grows there is a typical form of S. babylonica L. and samples of it, taken at Napoleon's tomb, are preserved in the Herbarium of the Academy of Sciences.

198

*__S. Matsudana__ Koidz. in Tok. Bot. Mag. XXIX (1915) 312; C. K. Schn. in Sarg. Pl. Wilson. III, 1, 107.

A tall shrub or a tree, 3.5—13 m high; branches fragile, upright or pendulous, olivaceous or yellowish, becoming fulvous, pubescent at first, finally glabrous; buds hairy; stipules short, lanceolate, glandular-serrate, often absent; petioles 2—8 mm long, slightly hairy along the groove above; leaf blades narrowly lanceolate, 5—8—10 cm long and 1 — 1.5 cm broad, obtuse or rarely acute at base, long-acuminate at apex, distantly and coarsely glandular-serrate on the margin; upper surface bright green, becoming glabrous or sometimes tomentose on the midrib at base; lower surface glaucescent or whitish, at first with scattered silky hairs, finally glabrous, both sides slightly reticulate with slender veins; catkins precocious, small, 1— 1.5 cm long and ca. 0.6 cm (staminate) or 0.4 cm (pistillate) broad; peduncles 2—5 mm long, hairy like the rachis, with 2 or 3 bracts at base, these linear or lanceolate, ca. 2.5 cm long and 0.3 cm broad, entire, obtuse or subacuminate, silky or glabrous; scales ovate, obtuse, yellowish-green, mostly hairy at base only on the outside; stamens 2, hairy at base; anthers yellow, ovate; glands 2, ovate, obtuse; ovary sessile, oblong, glabrous; style obsolescent or very short; stigmas parted, short, ovate, almost lobed; glands in pistillate flowers 2, the outer ovate-oblong, attenuate toward apex, rather thick, the outer small.

Erroneously reported for Dauria (Chita), Uss. Growing in S. Manchuria, Korea, and in China as far as Kansu. Described from China. Type in Tokyo.

Note. The detailed description leaves no doubt that the willow concerned is very closely related to S. babylonica and its hybrids.

161. **S. dolichostyla** O. v. Seem. in Engl. Bot. Jahrb. XXX (1902) Beibl. 67 (1901) 39; Ej. Salic. Japon. (1903) 26, tab. II. — S. mixta Korsh. forma in A. H. P. XII (1892) 391; Schedae ad HFR No. 2255; Kom. and Alis., Opred. rast. Dal'nevost. kr. I, 424.

A tall tree with a thick trunk or an arborescent shrub; old branches yellowish-gray or tawny, glabrous, very brittle, often nodding, knotty on the shoots; young branches covered with short silky hairs; buds oblong, beaked, glabrous, hairy at the tip; stipules none; petioles 0.5 — 1 cm long; leaf
201 blades 6.5 — 9 cm long and 1 — 1.7 cm broad (1:5 — 7), lanceolate, oblong-lanceolate, or ovate-lanceolate, narrowed at both ends, terminating at apex in a long point, closely and sharply serrate on the margin, green above, glaucous beneath; midrib covered beneath with long hairs; lateral veins 12 — 15 pairs, at an acute angle; catkins subcoaetaneous, lateral, upright or recurved, dense; staminate subsessile, to 2 — 4 cm long and 0.5 cm thick; pistillate to 1.5 — 2 cm long and 0.5 cm thick, short-stalked; bracts 2 or 3; small, lanceolate or oblong, dentate, silky beneath and on the margin; rachis densely hairy; scales almost covering the ovary, up to half the length of the stamens, oblong-ovate or ovate, obtuse, coriaceous, concave, light brown, covered in lower part with long hairs; anthers ovate, yellow or purple; in staminate flowers glands 2, the anterior linear, the posterior short-ovate or oblong; in pistillate flowers the linear anterior gland to one-third the length of ovary or often absent, the posterior short-ovate and just as long; ovary sessile, short-ovoid, hairy; style about two-thirds as long as ovary, thick, glabrous; stigmas long, thick, often incurved. Fl. May; fr. June.

Waterside thickets. — Far East: Ze.-Bu., Uss., Sakh. **Gen. distr.**: Japan. Described from Japan. Type in Berlin.

*162. **S. koreensis** Anderss. in DC. Prodr. XVI, 2 (1868) 271; Kom., Fl. Manchzh. II, 1, 24; Koidz. in Bot. Mag. Tok. XXVII (1913) No. 89; Nakai, Fl. Sylv. Kor. XVIII, 164 et tab. XXXVIII. — S. Feddei, S. pogonandra, S. pseudo-gilgiana, S. pseudo-jessoensis Leveillé in Fedde, Repert. X (1912) 436.

A tree to 10 — 20 m high and to 1 — 1.5 m in trunk diameter; bark thick, irregularly splitting lengthwise; branches grayish-brown or brownish-green, fragile, the annotinous appressed-hairy or glabrous; buds ovoid, 2.5 mm long; stipules on vigorous shoots obliquely ovate, 3 — 6 mm long

(199)

PLATE IX. 1. Salix daphnoides Will. – 2. S. acutifolia Willd. – 3. S. rorida Laksch. – 4. S. alba L. – 5. S. songarica Anderss. – 6. S. babylonica L. – 7. S. triandra L. – 8. S. pentandra L. – 9. S. Maximowiczii Kom.

and 2—3.5 mm broad, glandular-serrate or subentire, green above, glaucous beneath; petioles 0.6—1.3 cm long, channeled above, appressed-hairy, becoming glabrate; leaf blades linear-lanceolate or lance-oblong, 9—13 cm long, 1.6—3.8 cm broad, green and appressed-hairy or glabrate above, pruinose beneath and hairy on the midrib, glandular-serrate, long-acuminate; catkins precocious to subcoaetaneous, slenderly cylindric, 6—8 mm thick, dense, sessile, subtended by 1—5 small bracts, staminate 1—3 cm, pistillate 0.7—1.5 cm long; scales ovate-oblong, acutish, almost white, hairy on both
202 sides; pistillate oblong or elliptic, whitish-green, concave, hairy; stamens 2; filaments almost white, distinct or at base connate, below the middle hairy; anthers latericious or red, 2-locular; glands in both sexes 2, the posterior ca. 0.7—0.8 mm, the anterior ca. 0.5—0.7 mm long (in pistillate flowers sometimes absent); ovary ovoid, sessile, hairy; style shorter than ovary; stigmas 4-parted, red. Fl. May.

River valleys and slopes. — Unknown in USSR. Described from Korea. Type in Leningrad.

Section 27. **FRAGILES** C. Koch, De Salic. europ. comment. (1828) 15 (pro parte). — Trees with glabrous, sometimes very fragile, elongated branches; old bark more or less split; petioles often glandular; leaf blades lanceolate, ovate-lanceolate, or narrowly lanceolate, rarely entire, concolor or the lower surface glaucous; young buds not viscous; catkin-scales monochromatic, yellowish, caducous; stamens 2 or 3 (rarely 4); anthers yellow; ovary stipitate or sessile, glabrous, greenish; style short or fairly long; stigmas ovate, thickish; both sexes normally with 1 anterior and 1 posterior gland.

1. Stamens 2; glands 2 (posterior and anterior) or only 1 (posterior); leaves ovate-lanceolate, broadest below the middle, commonly (5)—10—15 cm long and 1.2—2.5 cm broad, dark green above, glaucous beneath; branches olivaceous-green, lustrous, very fragile at base . . . . 163. **S. fragilis** L.

\+ Stamens 3 (rarely 4); glands 2, posterior (or rarely 1 anterior); leaves narrowly lanceolate, commonly 3—7.5 cm long and ca. 0.6 cm broad, broadest below the middle, the two surfaces concolor, green; branches pale brown . . . . 164. **S. songarica** Anderss.

163. **S. fragilis** L. Sp. pl. (1753) 1017; Anderss. Sal. Mon., 41 et in DC. Prodr. XVI, 2, 209; Ldb. Fl. Ross. III, 598; Shmal'g., Fl. II, 432; Kryl., Fl. Zap. Sib. IV, 729. — S. decipiens Hoffm. Hist. Salic. II, 1 (1791) 9. — Ic.: Anderss. Mon., f. 28; Rchb. Ic. Fl. Germ. XI, t. 609, f. 1264. — Exs.: HFR No. 1187 and 2253.

A tree to 15—20 m high, 0.75—1 m in trunk diameter, and up to 60—75 years of age, with a wide head; bark of old trunks deeply fissured; branches spreading or somewhat drooping, quite glabrous, lustrous, olivaceous-green or grayish-green, sometimes somewhat reddish, with rather loose and easily peeling epidermis (but not flaking as in S. pentandra); 4-year-old branches already with longitudinal fissues, always very fragile at base; rarely bark of branches latericious or reddish-brown (f. decipiens (Hoffm.) Anderss.), even more rarely (in S. Transc.) pale yellow or almost

203 ivory-colored (f. roksensis Görz.); buds appressed, long, curved, dark brown, glabrous, lustrous; stipules ovate, semicordate or reniform, on vigorous shoots long persistent; stipules 0.9 — 1.4 cm long, with isolated white glands above; leaf blades narrowly ovate-lanceolate, broadest below the middle, rounded at base, gradually tapering to a point, 5 — 7.5 cm long and 1.2 cm broad (f. angustifolia Anderss.) or 10 — 15 cm long and 2.5 cm broad (f. latifolia Anderss.), commonly dark green above, pale green or glaucous beneath, at first slightly pubescent and somewhat viscous, becoming quite glabrous, the margin coarsely and irregularly serrate; midrib yellow; catkins coaetaneous, long-stalked, with 3 — 5 entire bracts, arching, 4 — 5 cm long; pistillate in fruit to 6 — 7 cm long, at length pendulous; scales yellowish-green or lurid, monochromatic, with long straight hairs at apex, rarely puberulous, glabrate, in pistillate flowers falling at maturity; stamens 2, distinct, hairy at base, twice as long as the scale; anthers yellow, turning brown; glands 2 in both sexes or sometimes the anterior absent; ovary quite glabrous, ovoid-conical, the stipe not shorter than and sometimes 2 — 3 times as long as the gland; style evident or short, bifid at apex; stigma 2-parted, with short divergent lobes. Fl. May; fr. June.

Wet places, flooded areas, and riverbanks; cultivated near houses. Rather rare in pure state. — European part: almost throughout, except the Arctic Zone; in Crimea in hybrid forms; Caucasus: Cisc., W. and W. Transc.; W. Siberia: U. Tob., Alt. (very rare). Gen. distr.: Scand., Atl. and Centr. Eur., Med., Bal.-As. Min., Arm.-Kurd., Iran. Cultivated in North America. Described from N. Europe. Type in London. The actual distribution area is difficult to determine, since the plant is cultivated in many localities (also in the USSR).

More often than in pure state, it occurs in the form of hybrids with S. alba, australior, babylonica, pentandra, and triandra.

Economic importance. A good honey-plant. The bark is used for salicyl production and for tanning (tannin content 4.6 — 11.86%). Used as astringent in popular medicine. Young branches and leaves provide feed for goats. The twigs are not durable; they are used for coarse wickerwork and as brushwood. The wood is used for arches, shafts, troughs, and log-hives. Superior to other willows as fuel. The tree provides construction timber at the age of 15 — 20 years; hence favored for cultivation in parkland and steppe areas of the Soviet Union.

This willow lends itself to pollarding and is easily propagated by cuttings. It suffers to some extent from cold spring winds, when its branches break off easily and litter roadways and streets. Ornamental; used for planted areas at barrages, ponds, streets, and around beehives. All the hybrids with other species are as useful as the true species or even more so; they are also more hardy.

204

164. **S. songarica** Anderss. Mon. Salic. (1867) 53 et tab. III, fig. 34; Ej. in DC. Prodr. XVI, 2 (1868) 213; E. Vol'f, in A. H. P. XXI, 2, 181. — Vernacular: ishke-tal.

A tree to 8 — 10 m high, reaching ages of up to 75 years, with trunk to 35 — 40 cm in diameter and a dense round top; annotinous branches slender, pale brown, the young sometimes pendulous; buds small, appressed, oblong, tawny; stipules lanceolate, rarely ovate, sometimes with large lobes at base, glandular, soon caducous, mostly absent; petioles glandular, 0.4 — 1 cm long; leaf blades narrowly lanceolate, 3 —7.5 cm long and ca. 0.6 cm broad, broadest at or below the middle, broadly cuneate at base, acuminate at apex, glandular-serrulate or entire on the margin, glabrous, lustrous, on both sides concolor green, slightly silky when young; midrib and lateral veins prominent beneath, riblike; lateral veins 12 — 16 pairs, at an angle of 40 — 45°; catkins subcoaetaneous or somewhat serotinous, to 5 —7 cm long, stalked, with 2 or 3 bracts, commonly terminating in short lateral shoots, slender, long, curved, the pistillate loose; rachis puberulous or glabrate; scales caducous; pistillate spatulate or narrowly ovate; staminate broadly obovate, truncate, monochromatic pale yellow, white-woolly at base, puberulous or glabrate toward apex; stamens 3 (rarely 4), villous at base; filaments short; anthers small, yellow, round; glands 2, posterior (or sometimes 1 anterior), one-third as long as the ovary; ovary ca. 4 mm long, ovoid-conical, obtuse, glabrous, green, becoming brownish; style very short; stigma thickish, entire or 2-parted, with short broad lobes; stipe ca. 1.5 mm long; capsule to 5 — 5.5 mm long. Fl. May; fr. June. (Plate IX, Figure 5).

The most widespread willow in the river floodland woods of Centr. Asia: everywhere except Mtn. Turkm. **Gen. distr.**: Dzu.-Kash. Described from the shores of the Chu River. Type in Leningrad.

Hybridizing with S. alba L.

Economic importance. Ornamental. Used for woodworking and for feed. Fast-growing; adapted to banks of rivers and irrigation ditches.

Section 28. **PENTANDRAE** Dumort. in Bijdr. Naturk. Wet. I (1825) 58. — Trees or tall shrubs; leaves mostly large, lanceolate or ovate-lanceolate, very often terminating in a long point, glandular-serrate on the margin, lustrous, when young somewhat viscous; petioles often glandular; staminate catkins opening almost simultaneously with leaves, densely flowered, 5 — 12 (24) [?]; glands 2; catkin-scales monochromatic; pistillate catkins long-stalked, often much elongated, mostly loosely flowered; ovary stipitate, mostly glabrous; style none or very short; stigmas rather short, thick, divergent or parted; glands 2, separate.

205

165. **S. pentandra** L. Sp. pl. (1753) 1016; Anderss. Mon. Salic. 34, f. 23; Ej. in DC. Prodr. XVI, 2, 206; Ldb. Fl. Ross. III, 597; Shmal'g., Fl. II, 431; Turcz. Fl. baic.-dah. II, 1, 98; Kryl., Fl. Zap. Sib. IV, 727; Kom. Fl. Kamch. II, 7 No. 296; Boiss. Fl. Or. IV, 1184. — Ic.: Rchb. Ic. Fl. Germ. XI, tab. 612, f. 1267 — 1269. — Exs.: HFR No. 486, 1088, 2256 — 2258.—Vernacular names: chernotal, chernoloz [from "chernyi," black].

A handsome tree to 16 m high, up to 75 years of age, or a shrub 3 — 5 m high with a spreading top; in peatbogs and near upper forest zone a procumbent shrub; bark of old trunks gray or dark brown, cracked, lustrous; annotinous branches dark gray or olivaceous-yellow, glabrous, lustrous, readily flaking; young branchlets and expanding leaves viscous, leaving a yellow imprint on paper, aromatic; buds ovoid, curved at the tip, 2-angled, brown, lustrous; stipules oblong or broadly ovate, glandular-dentate, soon caducous; petioles 0.2 — 1.4 cm long, with numerous large glands, glabrous, often colored; leaf blades coriaceous, dark green and lustrous above, paler beneath, ovate-oblong to broadly lanceolate, 5—13 cm long and 2—4 cm broad, broadest about the middle, obtuse or cuneate at base, acuminate at apex, closely glandular-serrate on the margin, quite glabrous, blackening on drying, bitter-tasting, 2.5—3 times as long as broad (f. latifolia Hrtm.) or to 3—5 times (f. angustifolia Anderss.); rarely leaves of vegetative shoots elliptic-lanceolate to lanceolate, those of upper flowering shoots strongly narrowed toward base, obtusish and enlarged at apex (v. imminuta E. Wolf); catkins serotinous, very fragrant; staminate subcoaetaneous, cylindric, 2—7 cm long, 1—1.5 cm thick, very dense, stalked, with serrulate glabrous bracts (in v. imminuta the staminate catkins stalkless, small, ca. 1.2 cm long, with 2—5—7 stamens occurring in the same catkin); pistillate 1—6 cm long and 0.8 cm thick, pendulous, with rather long glabrous stalks; rachis white-hairy; catkins after fruiting retained on the tree well into the winter; scales yellowish-green, ciliate at base, serrulate or in pistillate flowers dentate, caducous, mostly shorter than ovary, rarely (f. bracteata E. Wolf) twice the length of the ovary; stamens normally 5, rarely 2—3—4 (f. tetrandra (L.) Willd.) or more than 5, up to 12—24 (f. polyandra Bray), with crisp hairs at base; glands 2, anterior and posterior, in both sexes, often 2-parted and forming a 4—6-parted disk, or rarely in pistillate flower gland solitary; ovary green, glabrous, ovoid-conical, the stipe as long as the posterior gland; style short, thick, bifid at apex, or to 1.5 mm long and cleft to base (f. macrodistyla Görz); stigmas with short divergent lobes; capsule to 7 mm long, glabrous, lustrous, ripening at the end of summer. Fl. June—July; fr. July—August. (Plate IX, Figure 8).

206 Grassy bogs, wet-meadow and swampy valleys, inundated areas, and damp woods. Forest-tundra, forest and steppe belt; in the mountains up to the forest zone, 2,000—2,500 m.— Throughout the USSR, except the Arctic Region, Crimea, and the S. part of Central Asia. **Gen. distr.:** all of Europe, N. and W. Asia, Mongolia, Jap.-Ch., Dzu.-Kash. Replaced in North America by the closely related S. lucida Marsch. Described from Europe. Type in London.

Hybridizing with S. alba, australior, fragilis, and triandra.

In the mountains reduced in all parts, with paler bark and leaves; leaves with obsolescent glands and without stipules; such a mountainous form of S. pentandra L. grows in the Urals (in Kryl., Fl. Zap. Sib.: f. nana Bolle), in E. Sayan Mountains and in Kamchatka; it has been recently separated by Dr. Floderus as subsp. pseudopentandra Flod.; it corresponds apparently to v. microphylla Anderss.

Economic importance. A good late honey-plant. The bark is used for salicyl extraction and for tanning (tannin content 5.67—10.33%). The leaves are used for the production of a yellow dye. The twigs are suitable for coarse wickerwork; young twigs are made into cordage and are used as brushwood. The wood is firmer than in other species, is suitable as fuel and is used for odd jobs. Ornamental; planted in streets and used in planting at barrages. Suitable for pollarding. As compared with other species, more difficult to propagate by seed and cuttings, also grows more slowly. Frost-hardy.

Section 29. **URBANIANAE** O. v. Seem. Salic. Japon. (1903) 15.— Staminate catkins declinate, falling after flowering; pistillate pendulous; bracts glandular at base; scales 3—5-nerved, in pistillate flowers caducous, in staminate with 5 (—8) adnate stamens, of these 1 posterior very long opposite the posterior gland, 2 lateral very short, 2 anterior somewhat shorter than or about equaling the inner ones; filaments elongated, hairy at base; in staminate flowers 1 anterior and 2 posterior glands united into 1; in pistillate flowers anterior gland none, the lateral absent or united with the posterior, these distinct or united to above the middle; ovary short-stipitate; style deeply cleft; stigmas parted, elongated, narrow, breaking off after flowering.

1. Leaves ovate-elliptic, mostly cordate or almost round at base, abruptly tapering at apex to a short point, not more than 3 times as long as broad, thin, not blackening; stipules broad, orbicular, reniform, or oblong; pistillate catkins in fruit to 5—10 cm long; ovary short-stipitate . . . . 166. **S. cardiophylla** Trautv. et Mey.

207 + Leaves ovate-oblong, more or less rounded at base, long-tapering toward apex, more than twice as long as broad, firm, blackening on drying; stipules lanceolate; catkins in fruit 10—15 cm long; ovary rather long-stipitate . . . . .167. S. Maximowiczii Kom.

166. S. cardiophylla Trautv. et Mey. in Middend. Sibir. Reise, I, 3 Abth. 2, 3 (1856) 77, tab. 19 and 20; Anderss. Mon. Salic. 37 et tab. III; Ej. in DC. Prodr. XVI, 2, 207; Kom. and Alis., Opred. rast. Dal'nevost. kr. I, 424. — T o i s u s u c a r d i o p h y l l a Kimura in Tok. Bot. Mag. XLII (1928) 289. — Ic.: Trautv. Pl. imagin. (1844) 16.

A tree to 12—17 (—35) m high and to 1 m (commonly ca. 50 cm) in trunk diameter; bark deeply fissured, with a thick cork layer; branches gnarled, faintly purple, lustrous; annotinous shoots brown or reddish-brown, glabrous, lustrous; buds oblong, hooked at the tip, reddish-brown, glabrous, lustrous; stipules oblong, orbicular, or reniform, glandular-serrate, commonly half the length of the petiole; petioles 2—6 mm long, mostly eglandular, pubescent; leaf blades ovate-elliptic, cordate or rounded at base, abruptly attenuate to a point, glabrous on both sides, with showy venation, green above, paler or rarely glaucescent beneath, turning black on drying, 3—9 cm long and 2—4.5 cm broad, one-third longer than broad; midrib slightly hairy beneath; lateral veins 7—10—14 pairs, very prominent, the lower at nearly a right angle, the upper at an acute angle; catkins sub-coaetaneous, on a rather long leafy-bracted stalk; narrowly cylindric, to 9—10 cm long and 5—7 cm thick; scales caducous, ovate or broadly obovate, obtuse or truncate, mostly glabrous on the back, rough-ciliate on the margin, less so in pistillate flowers; pale, translucid, 3—5-nerved, almost completely covering the ovary; stamens normally 5, rarely 4—8, at first concealed by the scale, finally exserted; filaments hairy at base; anthers yellow; glands 2 or 3; ovary ovoid-conical, glabrous, ca. 3.5 mm long, covered by the scale; style very short to fairly long, cleft down to base; stigmas with 4 narrow divergent lobes; stipe shorter than to nearly twice as long as the paired narrow glands; capsule to 4.5 mm long. Fl. May—June; fr. July.

Valleys of mountain streams; solitary trees. — E. Siberia: Dau., Lena-Kol.; Far East: Ze.-Bu., Okh., Uda, Sakh., Uss. **Gen. distr.:** Jap.-Ch. Described from the Okhotsk area. Type in Leningrad.

Economic importance. Used for building, miscellaneous carpentry jobs, fuel, and ornament; suitable for matchwood and for dug-out boats.

167. **S. Maximowiczii** Kom. in A. H. P. XVIII (1901) 442; Fl. Manchzh. II, 1 (1903) 25; Kom. and Alis., Opred. rast. Dal'nevost. kr. I, 424, Plate 129; Nakai, Fl. Sylv. Kor. XVIII, 72 et tab. VI, VII; C. K. Schn. Pl. Wilson. in
208 Arn. Arboret. IX, 1,100. — Toisusu cardiophylla var. Maximowiczii Kim. in Tok. Bot. Magaz. XLII (1928) 289.

A handsome tree to 12—20 m high and to 35—80 cm in trunk diameter; bark dark brownish-gray or cinereous-brown, irregularly fissured lengthwise; branches upright or slightly pendulous, greenish or olivaceous-tawny in summer, yellowish in winter; buds ovoid-oblong, lustrous; stipules ear-shaped or ovate, unilaterally truncate at base, acute, dentate; petioles 0.5—1.6 cm long, at first hairy, becoming glabrous; leaf blades firm, 3—12 cm long and 1.2—3.5 cm broad, ovate-oblong to ovate-lanceolate, broadest at base, mostly obtuse, rarely cordate or acute, terminating in a point, more than twice as long as broad, sharply serrate, dark green above, intensely glaucous beneath, glabrous, turning black on drying; midrib dark-hairy, at length glabrate; lateral veins to 15 pairs, very prominent beneath, evident above; leaves of vigorous shoots to 15 × 5 cm; catkins borne on a long (2—6 cm) stalk, loose, pendulous, with 2—4 small bracts at base, up to 11—14 cm long, the rachis glabrous; staminate scales obovate, concave, 3—5-nerved, to 2.5 cm long, membranous, white-hairy on the margin; pistillate scales elliptic, acute, caducous, 3-nerved; stamens 5, the inner one inserted opposite the posterior gland, to 6—7 mm long, the 2 lateral ones ca. 3—4 mm long, the 2 anterior about equaling or exceeding the posterior gland; filaments of all stamens hairy at base; anthers round, yellow; staminate flowers with one posterior oblong gland and one broadly subulate anterior gland; pistillate flowers with 2 or 3 posterior glands, these often united at base; ovary glabrous, ovoid-lanceolate, yellowish, the stipe to 1 mm long; style and stigmas 2-parted; stigmas caducous. Fl. May—June; fr. June—July. (Plate IX, Figure 9).

Forming compact thickets on pebbles in wooded mountain valleys. — Far East: Uss. **Gen. distr.:** Manchuria, Korea. Described from Manchuria. Type in Leningrad.

Economic importance. Used for construction, wooden articles, fuel, and ornament. Also used for matchwood. More slender and elegant than the preceding species. Characterized by rapid growth.

DUBIOUS SPECIES AND UNKNOWN AUTHORS
(Species dubiae, auctori ignotae).

1. **S. salviaefoliae** affinis Bge. Reliqu. Lehmann. 321; Borshch., Bot. geogr. Aralo-Casp. kr. (1865) 50; Trautv. Incr. No. 4801. Recorded for Centr. Asia (left tributaries of upper Zeravshan).

Branches outwardly resembling S. cinerea, villous-tomentose; leaves oboval-oblong, obliquely acuminate or obtuse, densely silky-
209 tomentose beneath, densely covered above with soft gray tomentum, flat, the margins quite entire; stipules oval-lanceolate, acute, erect, entire, more than twice the length of the petiole; buds gray-silky.

Apparently synonymous with one of the species described above.

2. **S. anomala** E. Wolf in Fedde, Repert. sp. nov. V (1908) 22.

(Section Incubaceae ?). A rather tall shrub; young branches silky; petioles 0.5—0.8 cm long; leaf blades lanceolate or oblong to oblanceolate, 4—5.5 cm long, 5—6 (rarely 7) times as long as broad, entire or subentire with few shallow dentations, sericeous above, silvery-silky beneath; lower leaves often green, glabrous or slightly hairy beneath, smaller, borne on lateral branchlets, 2.3—4.5 cm long; lateral veins at an acute angle, prominent above; pistillate catkins unknown; staminate catkins small, 5—7 mm long, sessile, with bracts at base; scales covered with short hairs, brown, "richly tinted" at apex; stamens 2; filaments completely connate, rarely distinct, hairy below the middle; anthers yellow, glands 2, often growing out into a lobed cup. "Native in S. Russia."

We conclude that this description by Vol'f probably refers to S. rosmarinifolia L. with abnormal connate stamens — a phenomenon often encountered in this species (!!) — f. cladostemma.

3. **S. macropoda** Stschegl. in Bull. Soc. Nat. Mosc. (1854) I, 197, No. 304; Trautv. Incr. No. 4765. — S. livida Kar. et Kir. (1841) No. 1977. — S. Starkeana Willd. sec. Anderss. in DC. Prodr. XVI, 2, 289.

(Section Lividae). A shrub; branches buff, densely tomentose when young, glabrous in age; stipules much shorter than the petiole, small, semicordate-ovate or ovate-lanceolate, the margin glandular-dentate; stipules short, tomentose; leaf blades elliptic-oblong or obovate-oblong, to 4 cm long and ca. 1 cm broad, narrowed toward base, terminating in a short straight point, with entire subrevolute margins, in age glabrate and lustrous above, densely hairy-pubescent beneath; catkins lateral; pistillate to 4 cm long, borne on a very short leafy-bracted stalk, cylindric, plump; scales brown, dark at apex, densely covered with long hairs; ovary ovoid-conical, to 3 mm long, glabrous; style rather long, mostly cleft down to base; stigmas with 2 small lobes; stipe 4 mm long or longer, 3—4 times the length of the scale, as long as or longer than the ovary.— Centr. Asia: Dzu.-Tarb. Described from Ayaguz. Type in Kharkov (not seen).

Note. Undoubtedly belonging to the section Lividae Nym., but, according to Shchegleev (l. c., not.), differing from all the other species
210 in the densely long-hairy scales, the stipe exceeding the ovary, and the deeply cleft style.

Following the submission of the manuscript, a study of willows from Hokkaido and from Sakhalin by Dr. Arika Kimura appeared in the Journal of the Faculty of Agriculture, Hokkaido University, Sapporo XXVI, 4 (1934). Five species, as described below, are also indicated there for the Soviet part of Sakhalin.* As we do not dispose of the specimens, we are not in a position either to confirm or to reject them. We have not so far found in the USSR anything that would conform to these descriptions.

4. **S. taraikensis** Kimura in Journ. of the Facult. of Agricult. Hokkaido Univers., Sapporo XXVI, 4 (1934) 419—422.

(Section Capreae). A shrub to 5 m high and 11 cm in trunk diameter, with a round much-branched top, rarely arborescent, with grayish-brown

* [i.e., the northern part.]

bark; summer shoots greenish or brownish, glabrous or initially pubescent; exposed wood with scattered striation to 8 mm long; buds oblong, obtuse, glabrous; stipules (sometimes absent) semicordate, dentate, with often glandular teeth, glabrous, 1.3—6 cm long and 0.9—3 mm broad or larger; petioles to 1.6 cm long, channeled, glabrous; young leaves often covered above with appressed white silky hairs, silky beneath on the midrib, often reddish-brown with a green tip; grown leaves 6.6 —10.4 cm long and 2.1—3.6 cm broad, oblanceolate to obovate or subrhombic-elliptic (in var. latifolia Kim. broadly elliptic), rarely lance-ovate, attenuate toward apex, obtuse or pointed at base, coriaceous, lustrous dark green above, glaucous beneath, glabrous on both sides, dentate, dentate-serrate, or erose on the margin; veins 10—13, somewhat impressed above, prominent beneath, at an angle of 40—60 (80)°; leaves on the lower part of branches small, sometimes subentire; catkins coaetaneous or subprecocious with a silky rachis; pistillate ovoid, 2.3 —3.8 cm long and 1.8—2.7 cm broad, on a stalk to 0.7 cm long, densely flowered, at anthesis 2—2.7 cm long and 0.7—1.4 cm broad, on a stalk to 1.2 cm long; bracts 3 or 4 (—6), sometimes with revolute margins; scales oblong, obtusish, brown or black at apex, yellowish-green beneath or at the middle sometimes reddish, hairy on both sides; ovary oblong-conical, densely white-silky, to 3 mm long; style glabrous, yellowish-green, to 0.7 mm long, with 2 stigmas, their lobes oblong, 0.5—0.8 mm long; stipe silky, 1.9—2.3 mm long; gland 1, inner, yellowish-green, truncate, 0.4—1 mm long; capsule to 0.9 cm long, clothed with scattered hairs or sometimes glabrate, to 11 mm long; anthers yellow, ovate, ca. 1 mm long and 0.9 mm broad. Fl. June; fr. July.

Far East: Sakh. **Gen. distr.:** Hokkaido, Sakh. Described from S. Sakhalin. Type in Japan.

211

5. **S. poronaica** Kimura in Journ. of the Faculty of Agricult. Hokkaido Univers., Sapporo XXVI, 4 (1934) 426—427. — S. phlebophylla apud Kudo in Journ. Coll. Agr. Hokk. Univ. XII, 1 (1923) 30 No. 6589.

(Section Myrtilloides sec. Kimura (?— M. H.)). A shrub 0.5—1 m high; branches ascending, grayish-brown, densely pubescent at the ends, yellowish-green when young; annotinous branches purplish-castaneous, glabrous, lustrous, or slightly pubescent; buds ovoid, obtuse, to 9 mm long and 4 mm broad; stipules rarely developed, semicordate, glaucous beneath, tomentose on both sides, 2 × 1.2 mm; petioles channeled, puberulous, to 9 mm long; leaf blades stiff, obovate, terminating in a short point or almost rounded at apex, narrowed toward base, subentire or slightly serrulate toward apex, revolute at the margins or flat, green and glossy above, glaucescent beneath, the lower glabrous on both sides, the upper white-hairy or silky-tomentose on both sides, 2.7—6.5 cm long and 1.6—3.2 cm broad; midrib of upper leaves tomentose on both sides, that of lower leaves commonly glabrous; lateral veins 12—18 pairs, at an angle of 50—70°, anastomosing at the margin, the lower surface prominently and strikingly reticulate-veined; catkins coaetaneous; staminate cylindric, 2—3 cm long and 1.2 cm thick; pistillate to 2.2—3.5 cm long and 6—9 mm thick; rachis silky; peduncle sometimes up to 1.3—3 cm long; bracts 4—6, sometimes covered with glandular hairs or white-hairy; scales oblong, obtuse and black at apex, yellowish-green at base, hairy outside, with scattered hairs within, to 2.5 mm long and 0.9 mm broad; gland 1, inner, ovate, truncate, to 0.7 — 1 mm long; filaments of stamens free at base, to 7.5 mm long; anthers broadly elliptic, to 0.8 mm long, finally turning brown; ovary long-conical, 2.7 — 3.6 mm long, violet-colored, silky-pubescent; stipe pubescent, 1.5 — 2 mm long; style glabrous, yellowish-green, 0.4 — 0.6 mm long, with brownish 2-parted stigmas. Fl. June; fr. July.

Location not indicated. — Far East: Sakh. Endemic (?). Described from S. Sakhalin. Type in Japan.

Note. To judge by the description, very close to S. myrsinites L. and S. fuscescens Anderss., but the latter has commonly a glabrous ovary.

6. **S. nyiwensis** Kimura in Journ. of the Facult. of Agricult. Hokk. Univers., Sapporo XXVI, 4 (1934) 442. — S. minutiflora var. pubescens (non E. Wolf) Kudo in Journ. Coll. Agric. Hokk. Univ. XII, 1, 30 (1923) No. 143.

(Section not indicated). A shrub with ascending grayish-brown slender branches; young branches densely gray-silky; annotinous pubescent; stipules elliptic-oblong, glabrous or pubescent on both sides, to 3 × 1.3 mm;
212 petioles to 4.5 mm long, channeled above, silky-pubescent, at length glabrous beneath; leaf blades coriaceous, stiff, elliptic, attenuate at apex, rounded or obtuse or broadly cuneate at base, mostly entire, revolute, at first densely appressed-silky, later glabrescent, sometimes almost glabrous, glaucescent beneath, 1.5 — 2.7 cm long and 0.9 — 1.6 cm broad; midrib and lateral veins prominent beneath; lateral veins 6 — 9 pairs, at an angle of 50 — 70° (becoming more acute toward apex), divergent; catkins in fruit to 3.5 cm long and 1 cm thick, densely flowered, with a silky rachis, stalk 0.2 — 1 cm long, and 3 or 4 small bracts; scales ovate, very obtuse, to 1 mm long and 0.8 mm broad, brownish-tipped, pale below, long-hairy on both sides; gland 1, inner, ovate-oblong, truncate, to 1 mm long and 0.4 mm broad; ovary elongate-conical, obtuse, to 2.5 mm long including the short stipe (0.2 mm), silky, glabrous at the top; style somewhat compressed, 0.7 mm long, with straight divergent 2-parted stigmas and oblong lobes. Fr. August.

Woods. — Far East: Sakh. Endemic (?). Described from N. Sakhalin. Type in Japan.

Note. The author adds to this description: "at first sight the leaves resemble in shape and color those of S. myrtilloides L., but distinguishing characters are the silky pubescence of branchlets, leaves, and ovary; also the densely flowered catkins, the short stipe, and the longer and more deeply cleft style."

7. **S. orotchonorum** Kimura in Journ. of the Facult. of Agricult. Hokkaido Univers., Sapporo, XXVI, 4 (1934) 444 — 445. — S. cinerea (non L.) Kudo in Journ. Coll. Agr. Hokk. Univ. XII, 1, 30 (1923) 137; Ej. Rep. Veg. North. Saghal. 99 No. 193 (1924); Makino et Nemoto, Fl. Japon., ed. 1 (1925) 1122; ed. 2 (1931) 162.

(Section not indicated). A shrub with slender yellow branches; stipules semiovate, acute, on the outer margin remotely serrate, to 4.5 — 8 mm long and 2.5 — 4.5 mm broad; petioles 3.5 — 6 mm long, channeled above, woolly-tomentose; leaf blades coriaceous, stiff, elliptic to ovate-elliptic, terminating

in an oblique point, attenuate or obtuse at base, the lower entire, the upper dentate-serrate, green above, glaucous beneath, densely or sparsely downy-sericeous, 3.5—6.5 cm long, 1.2—3.9 cm broad; midrib tomentose on both sides; lateral veins 7—10 pairs, at an angle of 40—60°, forking near the margin and confluent with the peripheral vein, forming loops; catkins more or less crowded at the ends of branches, in fruit to 4.7 cm long and 1 cm thick, with small silky bracts; scales subulate (? probably liguliform — M. H.), 2.5 mm long and 0.6 mm broad, glabrate within, slightly hairy at
213 the margin on the back (color not indicated); capsule silky, to 6 mm long; stipe silky, 3.2 mm long; style very short; stigmas 2-lobed. Fr. August.

Location not indicated. — Far East: Sakh. Endemic (?). Described from N. Sakhalin. Type in Japan.

Note. To judge by the description, the plant belongs to the section Lividae. Dr. A. Kimura maintains that this willow deviates insignificantly from S. xerophilae Floder.

8. **S. stoloniferoides** Kimura in Journ. of the Facult. of Agric. Hokkaido Univers. Sapporo XXVI, 4 (1934) 448. — S. phlebophylla (non Anderss.) Kudo, Rep. Veg. North. Saghal. (1924) 98 No. 192 (ex parte.).

(Section not indicated). A shrub, apparently trailing, with prostrate buff glabrous lustrous branches; young branches to 1 mm thick, yellowish, sparsely covered with short flexuous hairs; annotinous buff, lustrous, glabrate; buds to 3 mm long, ovoid, obtuse, ribbed, dark yellow or fulvous, puberulous, becoming glabrous; stipules small, elliptic or obliquely ovate, entire or lacerate-serrate, gray beneath, pubescent or glabrate, 1—2 mm long and 0.4—0.8 mm broad; petioles channeled above, finally glabrous beneath, to 3.5 mm long; grown leaves coriaceous, stiff, obovate or elliptic, rarely narrowly ovate, acuminate at apex, obtuse at base, entire, 1—2.8 cm long and 0.5—1 cm broad, the lower ones smaller; upper surface green, lustrous, without stomata; lower surface glaucous; both surfaces at first slightly pubescent, finally glabrous; veins prominent beneath, the lateral 6—9 pairs, at an angle of 40—50°, the subsidiary not evident beneath; catkins in fruit 1.5—2.5 cm long and to 1 cm thick, the rachis silky, the stalk to 7 mm long, with 3 or 4 small oblong-elliptic bracts; scales obovate, obtuse, hairy-sericeous, densely within, more sparsely outside, to 0.8 mm long and 0.5—0.6 mm broad (color not indicated); gland 1, inner, oblong, to 0.7 mm long; ovary glabrous, lance-conical, to 1.7 mm long; stipe glabrous, in fruit to 0.3 mm long; style to 0.4 mm long, longer than the small 2-parted or emarginate stigmas; capsule ca. 5 mm long. Fr. August.

Location not indicated. — Far East: N. Sakh. Described from the vicinity of Okha. Endemic (?). Type in Japan.

Note. Dr. Kimura is inclined to relate this species to S. stolonifera Coville with the reservation that the latter is known to him only from description and illustration. We do not discern in this willow anything in common with S. phlebophylla.

9. **S. chilkoana** Sukacz. in "Selektsiya i introduktsiya bystrorastushch. drevesn. porod" (Selection and Introduction of Fast-Growing Species), Izd. Narkomzema SSSR (1934), pp. 65 and 71—72.

A shrub. "Cultivated in an experimental nursery from cuttings which I collected by the Khilok River (near Khilok station) in Transbaikalia, in 1925. It cannot be identified with any known willow. Its characters (only staminate samples are available) resemble to some extent those of S. dahurica (in leaves), but it has 2 stamens. My first assumption was that it represents a hybrid in which may have participated either S. rorida, or S. Sjusevii, or S. dahurica ... All the investigated male samples were found to have 2 distinct stamens, which excludes the possibility that any willow of the Purpureae may have had a part in the origin of S. chilkoana ... Leaving aside for the time being the problem of origin and specific distinctiveness of this willow, we shall continue to refer to it as S. chilkoana. Rather fast-growing, attaining the dimensions of a tall shrub" (V. N. Sukachev).

Note. The botanical description of this willow and its specimens are unknown to us.

Willow Hybrids Known Also Under a Distinct Name
(Not Recorded Above)
(Figures refer to serial numbers of the parental species)

S. acuminata Koch.	=	*S. caprea* × *viminalis*. 55, 94.
„ *alopecuroides* Tausch.		„ *fragilis* × *triandra*. 163, 148.
„ *ambigua* Ehrh.		„ *aurita* × *repens* 65, 88.
„ *americana* hort.		„ *purpurea* × *triandra*. 111, 148.
„ *bicolor* Sm.		„ *caprea* × *phylicifolia*. 55, 38.
„ *Brownei* (Anderss.) Lundstr.		„ *arctica* × *reptans*. 11, 29.
„ *calliantha* Kern.		„ *daphnoides* × *purpurea*. 145, 111.
„ *canescens* Fr.		„ *caprea* × *lapponum*. 55, 34.
„ *capreaeformis* Wimm.		„ *caprĕa* × *viminalis*. 55, 94.
„ *cuspidata* Schultz.		„ *fragilis* × *pentandra*. 163, 165.
„ *Doniana* Sm.		„ *purpurea* × *repens*. 111, 88.
„ *elaeagnifolia* Tausch.		„ *purpurea* × *viminalis*. 111, 94.
„ *excelsior* Host.		„ *alba* × *fragilis*. 151, 163.
„ *ferruginea* Forb.		„ *cinerea* × *viminalis*. 62, 94.
„ *finmarkika* Willd., Whlnb, Fr.		„ *aurita* × *myrtilloides*. 65, 76.
„ *fissa* Ehrh.		„ *purpurea* × *viminalis*. 111, 94.
„ *Forbiana* Sm.		„ „ „ „ „ „ „
„ *Friesii* Kern.		„ *fragilis* × *pentandra*. 163, 165.
„ *geminata* Forb.		„ *cinerea* × *viminalis*. 62, 94.
„ *gracilenta* Tausch.		„ *alba* × *fragilis*. 151, 163.
„ *Hartmanniana* Anderss.		„ *hastata* × *lanata*. 79, 30.
„ *hexandra* Erhr.		„ *alba* × *pentandra*. 151, 165.
„ *hippophaëfolia* Thuill.		„ *triandra* × *viminalis*. 148, 94.
„ *hirtula* Anderss.		„ *livida* × *myrtilloides*. 71, 76.
„ *holosericea* Gaud.		„ *caprea* × *viminalis* sec. Wimm. 55, 94.
„ „ Koch et Ziz.		„ *cinerea* × *viminalis* Wimm. 62, 94.
„ „ Willd.		„ „ „ „ „ „ „
„ *Hostii* Kern.		„ *caprea* × *viminalis*. 55, 94

215 *S. Koernickei* Anderss.	= *S. daphnoides* × *phylicifolia*. 145, 38.
„ *Krausei* Anderss.	„ *cinerea* × *triandra*. 62, 148.
„ *Laestadiana* Hrtm.	„ *cinerea* × *lapponum*. 62, 34.
„ *lanceolata* DC.	„ *caprea* × *viminalis* sec. Wimm. 55, 94.
„ „ Fr.	„ „ „ „ sec. Anderss. „ „
„ *lancifolia* Doell.	„ *cinerea* × *viminalis*. 62, 94.
„ *latifolia* Forb.	„ *caprea* × *nigricans*. 55, 53.
„ *laurina* Sm.	„ *caprea* × *phylicifolia*. 55, 38.
„ *lispoclados* Dode	„ *alba* × *pentandra*. 151, 165.
„ *livescens* Doell.	„ *aurita* × *livida*. 65, 71.
„ *lutescens* Doell.	„ *aurita cinerea*. 65, 62.
„ *macrorhyncha* Anderss.	„ *daphnoides* × *phylicifolia*. 145, 38.
„ *macrostipulcea* Forb.	„ *caprea* × *viminalis*. 55, 94,
„ *Massalskyi* Gorz. in sched.	„ *micans* × *triandra*? 152, 148.
„ *mollissima* Ehrh.	„ *triandra* × *viminalis*. 148, 94.
„ „ Sm.	„ *caprea* × *viminalis*? (sec. Wimm.) 55, 94.
„ „ Whlnb.	„ *purpurea* × *viminalis*. 111, 94.
„ *multiformis* Doell.	„ *triandra* × *viminalis*, 148, 94.
„ *multinervis* Doell.	„ *aurita* × *cinerea*. 65, 62.
„ *myrtoides* Doell.	„ *livida* × *nigricans*? (sec. Anderss.) 71, 53.
„ *neglecta* Gorski in sched.	„ *lapponum* × *myrtilloides*. 34, 76.
„ *Neisseana* Kern.	„ *caprea* × *viminalis*. 55, 94.
„ *obtusifolia* Willd.	„ *aurita* × *lapponum*. 65, 34.
„ *onusta* Bess.	„ *aurita* × *myrtilloides*. 65, 76.
„ *paludosa* Hartm.	„ „ „ „ „ „
„ *palustris* Host.	„ *alba* × *fragilis*. 151, 163.
„ *paradaphnoides* E. Wolf.	„ *pyrolaefolia* × *rorida*. 78, 147.
„ *parviflora* Host.	„ *purpurea* × *repens*. 111, 88.
„ *Patzeana* Anderss.	„ *daphnoides* × *repens*. 145, 88.
„ *Patzei* Wimm.	„ *aurita* × *livida*. 65, 71.
„ *pischpekensis* E. Wolf.	„ *cinerea* × *pyrpurea*? 62, 111.
„ *plicata* Fr.	„ *aurita* × *repens*. 65, 88.
„ *podolica* Blocki.	„ ?
„ *Pokornyi* Kern.	„ *fragilis* × *pentandra*. 163, 165.
„ *puberula* Doell.	„ *cinerea* × *nigricans*. 62, 53.
„ „ Lasch. (non Koch).	„ *purpurea* × *viminalis*. 111, 94.
„ *Reichardtii* A. Kern.	„ *caprea* × *cinerea*. 55, 62.
„ *reticuloides* Anderss.	„ *hastata* × *reticulata*. 79, 1.
„ *rubra* Huds.	„ *purpurea* × *viminalis* 111, 94.
„ *rugulosa* Anderss.	„ *aurita* × *myrtilloides*. 65, 76.
„ *Russeliana* Koch	„ *alba* × *fragilis*. 151, 163.
„ *Schatilowi* Schroed.	„ *lapponum* × *purpurea*. 34, 111.
„ *Schrenkiana* Andrs. (non Rgl.)	„ *hastata*×*myrtilloides* (sec. Anderss.) 79, 76.
„ *semihelix* Laksch.	„ *purpurea viminalis*. 111, 94.
„ *sericans* Tausch.	„ *caprea* × *viminalis*. 55, 94.
„ *Schumanniana* O. v. Seem.	„ *pentandra* × *triandra*. 165, 148.
„ *Smithiana* Hart.	„ *caprea* × *viminalis*. 55, 94.
„ *spathulata* Willd.	„ *aurita* × *repens*. 65, 88.
„ *stipularis* Sm.	„ *dasyclados* × *viminalis*. 109, 94.
„ *subsericea* Doell.	„ *cinerea* × *repens*. 62, 88.
„ *Trevirani* Spr.	„ *triandra* × *viminalis*. 148, 94.
„ *undulata* Ehrh.	„ *alba* × *triandra*. 151, 148.
„ *turgaiskiensis* E. Wolf.	„ *caspica* ×*rosmarinifolia*. 115, 88.

216 S. *versifolia* Ser. = *S. aurita* × *repens*. 65, 88.
" " Whlnb. = " *lapponum* × *myrtilloides*. 34, 76.
" *vilnensis* Gorski. = " *aurita* × *myrtilloides*. 65, 76.
" *viridis* Fr. = " *alba* × *fragilis*. 151, 163.
" *Vratislaviana* Kern. = " *caprea* × *viminalis*. 55, 94.
" *Wichurae* Anderss. = " *glauca* × *phylicifolia*. 25, 38.
" *Zacharowi* Schroed. = " *acutifolia* × *dasyclados*. 146, 109.
" *Zedlitziana* Kern. = " *cinerea* × *viminalis*. 62, 94.

Genus 357. **POPULUS*** L. **

L. Gen. pl. (1737) 317.

A fast growing dioecious tree; terminal and lateral buds protected by numerous scales, often profusely resinous; shoots long with larger leaves or short and partly fertile; leaves petiolate, deltoid, rhombic, elliptic, ovate, or cordate, rarely lanceolate or linear, often varying on the same tree; inflorescence a catkin (spikelike raceme), developing as a rule before the leaves, cylindric, pendulous, not leafy at base; bracts subtending the flowers toothed, dissected or laciniate, falling off before fruit ripening; pistillate flowers short-pediceled, with a cup-shaped or patelliform disk (torus) at base, the disk corresponding to the glands of willow; staminate flowers with actinomorphic androecium; flowers in various species exceptionally bisexual; stamens 3 — 30; stigmas commonly 2, rarely 3 or 4; capsule mostly 2-, rarely 3- or 4-valvular; seeds many, comose; flowers pollinated by wind, without nectaries and with dry pollen, but nevertheless visited by bees.

The poplar occurs in mountain and riverside woods; particularly widespread in E. Asia and in the southeastern seaboard of the United States.

The wood of all poplars is soft, white, suitable for carpentry and to some extent for building.

Fossils of the genus Populus occur in the USSR in geological formations from Upper Cretaceous to Quaternary.

P. arctica Heer in Upper Cretaceous layers of Sakhalin; reported for Novosibirskie Islands and Ze.-Bu. (Tsagayan), also in Upper Cretaceous; reports for Tertiary formations have not been confirmed. — P. balsamoides Goepp. in Tertiary formations, in the Sarmatian of Bl. (Krynka), in the Paleocene of Ob (Tomsk), Uss. and Sakh.; in the Neocene of Ob (near Tara on the Irtysh). — P. Gaudinii Heer in Sakh. (Mgachi). — P. glandulifera Heer in Lower Tertiary formations of Sakh. (Dui, Mgachi). — P. Heliadum Unger in Tertiary formations (Pliocene ?)
217 of Balkh. (Chingistai). — P. latior A. Braun in the Oligocene of M. Dnp. (Tim), in the Paleocene of Uss. (Pos'et, Novokievskoe, Fatashi), and Sakh. (Dui, etc.); in Maiotis formations of Bl. (Khadzhibei). — P. mutabilis Heer in the Miocene of Cisc. (Groznyi District); in the Oligocene of Ar.-Casp. (Kara-Sandyk). — P. Richardsonii Heer in early Paleocene or in Upper Cretaceous of An. (Anadyr R.); Novosibirskie Islands and in Cretaceous layers of Ze.-Bu. (Tsagayan on Bureya R.). — P. tremula L. in the Pliocene of E. Transc. (Shiraki). — P. Zaddachii Heer in the Paleocene of L. V. (Kamyshin, Ushi); in the Paleocene of Uss. (Rechnoi Peninsula, Nikol'skii mine), Sakh. (near Aleksandrovsk), Okh. (Taui Bay),

* From Latin "populus," people, signifying a tree grown around public squares and other places of public assembly. Name for poplar used by Horace.

** Treatment by V. L. Komarov.

and Kamch. (according to Natthorst); reported for the Upper Cretaceous of Ze.-Bu. (Tsagayan on Bureya R.).

There are also numerous records referring to Populus sp., without designation of species, for various formations from the Paleocene to Postpliocene.

Key to Subgenera and Series of the Genus Populus L.

1. Disk deciduous, deeply parted, with sharp teeth; leaf glands rather inconspicuous, warty, often concave at the top and then cup-shaped; leaves at the ends of branchlets sublinear, the lower ones broader than long . . . . Subgenus 1. **Turanga** Bge.
\+ Disk persistent in fruit to maturity . . . . 2.
2. Disk obliquely truncate; leaf glands prominent, spherical, often extending to the petiole . . . . Subgenus 2. **Leuce** Duby, see below, 3.
\+ Disk evenly truncate, rather deeply emarginate or toothed; leaf glands irregularly spherical, rounded or elongated to subpectinate, see below stage 4, p. 173 . . . . Subgenus 3. **Eupopulus** Dode.
3. Grown leaves of long shoots white-tomentose beneath, more or less palmate; leaves of old branches more or less tomentose beneath; bracts more or less dissected; reproducing chiefly by suckers . . . . Series **Albidae** Dode.
\+ Leaves of shoots glabrous or with rather sparse short pubescence, more or less serrate-dentate; leaves of old branches stiff, glabrous or appressed-puberulous; bracts deeply divided; petioles laterally flattened near the blade . . . . Series **Trepidae** Dode.
4. Petioles more or less flattened laterally; blades dark green beneath; leaf glands elongate-spherical; petioles on shoots as long as on fertile branches . . . . Series **Aegirus** Asch.
\+ Petioles quadrangular; leaves gray or whitish beneath; leaf glands cup-shaped; petioles on shoots shorter than on old branches; buds viscid, with a strong balsamic odor . . . Series **Tacamahaca** Spach.

218 Key to Species

1. Leaves coriaceous, glaucous, concolor, on the shoots*, entire, on old branches broadly deltoid or reniform; stigmas particularly large, carmine-red; capsule coriaceous, slender, elongate . . . . 2.
\+ Lower surface of leaves tomentose or smooth, paler than the upper; leaves varying in shape, the margins crenate, serrate, or dentate; disk persistent in fruit; stigmas green or pink; capsules elongate to globose . . . . . . 6.
2. Disk cut merely to the middle; leaves of old branches with distinct rather large teeth . . . . 3.
\+ Disk cut nearly to the base; leaves of old branches reniform, subentire; capsule densely pubescent, cinereous . . . 5. **P. pruinosa** Schr.

* The term "shoots" comprises suckers as well as long branches arising from the bark of the trunk or produced at the ends of old branches.

3. Stipe of ovary short; rachis of the catkin and pedicels covered with few long straight hairs . . . . 4.

\+ Stipe of ovary longer, ca. 2 mm; ovary covered with short velvety pubescence . . . . 5.

4. Leaves of shoots linear, with straight margins; young branches greenish, smooth, even at apex; leaves of mature branches rounded-deltoid, with 5 teeth on each side . . . . 2. **P. ariana** Dode.

\+ Leaves of shoots lanceolate, with wavy margins; young branchlets red, at the tips glaucous-pubescent; leaves of mature branches dilated, rounded at apex, with 6 — 9 teeth on each margin . . . . 4. **P. euphratica** Olivier.

5. Leaves of long and intermediate shoots linear, generally broader, with 7 teeth on each margin; rachis of the catkin, pedicels, and ovary densely pubescent, gray . . . . 3. **P. Litwinowiana** Dode.

\+ Leaves of intermediate shoots narrowly lanceolate, those of short shoots smaller on the average, with 4 teeth on each margin; catkin and ovary velutinous, rather soon glabrescent . . . . 1. **P. diversifolia** Schrenk.

6. Leaves, especially when young, white or gray beneath, tomentose or arachnoid, commonly more or less lobed; bracts hairy, toothed . . . . 7.

\+ Leaves glabrous or inconspicuously pubescent, crenate or serrate or dentate; bracts deeply cut . . . . 10.

219

7. Leaves of long shoots more or less lobed, densely white-tomentose beneath; leaves of short shoots heavily white-felted beneath . . . . . . 8.

\+ Leaves of long shoots not lobed or but slightly lobed, white or gray beneath with flocculent pubescence; leaves lobed, toward fall green beneath . . . . 9.

8. Leaves of long shoots straight at base, with 3 — 5 rather deep lobes, toward fall very densely tomentose beneath, the margins dentate; leaves of short shoots sparingly toothed . . . . 6. **P. nivea** L.

\+ Leaves of long shoots large, cuneately truncate at base, densely puberulous, bright green and lustrous above, deeply lobed, the lobes often lobulate; leaves of short shoots elliptic-orbicular, irregularly angular, with sharp teeth . . . . 7. **P. Bolleana** Lauche.

9. Leaves of short shoots deltoid, with 2 basal lobes, dentate, white beneath with flocculent pubescence; leaves of short branches irregularly orbicular, sharply dentate, retaining into fall their gray pubescence . . . 8. **P. alba** L.

\+ Leaves of short shoots deltoid-elliptic, cordate at base, covered beneath with gray flocculent pubescence; leaves of short branches orbicular, with undulate twice-serrate margins . . . . 9. **P. canescens** Sm.

10. Leaf petioles compressed laterally in upper part . . . . 11.

\+ Petioles rounded-quadrangular or terete . . . . 16.

11. Catkin-scales deeply fringed, dark brown, covered with gray villous pubescence; stamens 6—8; leaves of short branches orbicular or ovate, with sinuately toothed margin . . . . 12.

\+ Catkin-scales glabrous; stamens 12 — 30; leaves of short branches more or less rhombic, acuminate, cuneate or blunt at base and crenate-serrate in upper part, rarely orbicular with crenate but not sinuate margins . . . . 13.

12. Leaves of short branches orbicular, minutely point-tipped, rather coarsely dentate; expanding leaves ciliate, covered with scattered hairs . . . . 10. **P. tremula** L.

\+ Leaves of short branches orbicular, minutely point-tipped, serrulate; expanding leaves glabrate . . . . P. Davidiana Dode.

13\. Branches upright, appressed to the stem; head pyramidal . . . . 12. **P. pyramidalis** Roz.

\+ Branches spreading; head compoundly branched . . . . 14.

220 14. Rachis of catkins and pedicels densely clothed with raised hairs; petiole elongated; leaf blade long-tipped . . . 15. **P. tadshikistanica** Kom.

\+ Rachis of catkin and pedicels quite smooth; petiole not elongated . . . 15.

++ Rachis of catkin and pedicels quite smooth; petiole strongly elongated; capsule valves broad . . . . 13. **P. usbekistanica**

15\. Leaves deltoid-orbicular, stiff, grayish . . . . 14. **P. cataracti** Kom.

\+ Leaves deltoid, clearly cuneate at base, green . . . . . . 11. **P. nigra** L.

16\. Young shoots angular, prominently lined along the angles; their narrow leaves often lanceolate . . . . 17.

\+ Young shoots more or less terete, straight or knotted . . . . 20.

17\. Leaves of branches broadly ovate, very large, cordate or straight at base . . . . ***P. candicans** Ait.

\+ Leaves of branches orbicular, slightly cordate at base; fruit small . . . . 17. **P. pamirica** Kom.

++ Leaves of branches narrowly or broadly ovate, gradually tapering to a point . . . . 18.

18\. Ovary and fruit glabrous; branches olivaceous-brown or reddish-brown; leaves rhomboid-ovate or rhomboid-elliptic, prominently point-tipped, at base narrowly rounded or broadly cuneate . . . . **P. Simonii** Carr.

\+ Young branches yellow; leaves of mature branchlets ovate-lanceolate, at base broadly cuneate . . . . 19.

19\. Fruit of medium size . . . . 16. **P. laurifolia** Ldb.

\+ Fruit large, to 1 cm long . . . . 18. **P. densa** Kom.

20\. Ovary and fruit hairy . . . . 21.

\+ Ovary and fruit glabrous . . . . 22.

21\. Bark of young branches yellowish . . . . **P. pilosa** Rehder.

\+ Bark of young branches light grayish . . . . 25. **P. amurensis** Kom.

22\. Leaf petioles on short branchlets glabrous, sometimes puberulous . . . . 20. **P. suaveolens** Fisch.

\+ Leaf petioles more or less hairy, sometimes almost tomentose . . . . 23.

23\. Pubescence of petioles very dense, the hairs raised . . . . 24.

\+ Pubescence of petioles less dense, the hairs not raised . . . . 25.

24\. Leaves large, coriaceous; catkins 10—18 cm long . . . . 23. **P. ussuriensis** Kom.

\+ Leaves of medium size; catkins 4—5 cm long . . . 24. **P. baikalensis** Kom.

221 25. Shoots terete; leaves strongly ribbed; buds and young leaves aromatic, very viscous . . . . 22. **P. koreana** Rehder.

\+ Shoots angled; leaves flat; buds and young leaves almost odorless; leaves on the average larger and broader than in the preceding, more distinctly apiculate . . . . 21. **P. Maximowiczii** Henry*

* The species P. talassica, which I have established (see p.188) cannot be included in this key, since the collected material is incomplete and the systematic position of this poplar cannot as yet be determined.

Subgenus 1. **TURANGA** Bge. in Mém. Sav. Etr. Acad. Pétersb. VII (1851). 498. — Leaves coriaceous, glaucous, on long shoots entire, the distribution of stomata similar on the two concolor surfaces; heterophyly pronounced; buds very small, short, obtuse; disk deciduous, deeply divided, with acute teeth; stigmas large, bright red; stamens ca. 12; anthers elongated, the connective apiculate; capsule elongated, conical, stalked. Trees of saline soils. Cannot be propagated by cuttings or grafts.

Series 1. **Euphraticae** (Dode) Kom. — Disk divided to the middle or somewhat lower down; leaves of short branches strongly dentate.

1. **P. diversifolia** Schrenk. in Bull. Acad. Sc. Pétersb. X (1842) 253 et in Fisch. et Mey. Enum. pl. nov. Schrenk II (1842) 15; Ldb. Fl. Ross. III, 627; Dode in Mém. Soc. Hist. Nat. Autun XVIII (1905) 15. — P. euphratica B. Fedtsch. (non Olivier) Rast. Turkest. (1915) 292. — Ic.: Trautv. Pl. imag. Fl. Ross (1844) 23, tab. 16.

A tree of medium size, with a spreading head; branches terete, with sparse pubescence visible with the aid of a magnifier; leaves of short shoots rounded-ovate, 2.4 — 4 cm long, cordate or truncate at base, often cuspidate, the cusp with 4 diverging teeth on each side, the 2 glands at the base of the blade very small and concave; petiole long, compressed laterally, at first very sparsely pubescent; leaves of long shoots (often in the top part of the tree) elongate-lanceolate, acute, entire; catkins lateral, oblong, pendulous, 2.5 — 5 cm long; rachis, pedicels, ovaries, and young fruits clothed with dense velvety pubescence; stalk about half as long as the capsule, to 6 mm long, rugose, glabrous; seeds small. April — May.

Riverbanks and sands; solitary or forming groves. — Centr. Asia: Ar.-Casp., Balkh., Kyz. K., Kara K. Spreading northward as far as the Irgiz and Emba rivers and the Bol'shie Barsuki Sands. An uninterrupted strip of this poplar extends westward along the Syr Darya River to the Karmakcha River, with occasional trees reaching as far as Kara-tugai. Endemic. Described from Lake Dzhalanashkul'. Type in Leningrad.

Economic importance. Provides fuel and material for small wooden articles.

222

2. **P. ariana** Dode in Mém. Soc. Hist. Nat. Autun XVIII (1905) 16. — P. euphratica Lipskii, Lesn. rast. v Turk. (1911) 17 (non Olivier). — Ic.: Dode l. c. tab. XI, 1a.

A tree; leaves of long shoots linear, entire; leaves of short shoots rounded-deltoid, almost straight at base, acute, with 5 teeth on each margin; buds pubescent; staminate catkins 3 — 4 cm long, the rachis densely clothed with appressed erect teeth; pedicels hairy, 1 — 1.5 mm long; anthers blackish-violet; rachis of pistillate catkins glabrous or with few scattered rather long hairs; ovary and young fruit quite glabrous; leaves greener than those of P. diversifolia. March — April.

Riverbanks and often saline soils in proximity of rivers. Particularly dense thickets occur along the Tedzhen River (a narrow strip more than 100 km long) and the Murgab River. — Centr. Asia: Kara K., Mtn. Turkm. and Amu D. (e. g., at Faraba). **Gen. distr.:** E. Iran. Type not recorded.

Economic importance. As in the preceding.

3. **P. Litwinowiana** Dode in Mém. Soc. Hist. Nat. Autun (1905) 17. — Ic.: Dode l. c. tab. XI, lc.

A tree, bigger than the preceding; leaves of long shoots linear, entire; leaves of short shoots broad, with up to 7 teeth on each margin; buds strongly pubescent, viscid in winter; fruit-stalk longer than in related species; top of ovaries and fruits directed toward the base rather than the apex of the catkin; a conspicuous feature are the broad leaves, especially on the middle branchlets; rachis of catkin and pedicels relatively hairy; fruit larger than in the two preceding species. April.

Riverbanks; forming whole forests. — Centr. Asia: Balkh., in the Ili Valley. Gen. distr.: Kuldja and the whole of Kashgaria as far as Kunlun.

Economic importance. One of the few trees growing in this area; hence its practical value.

4. **P. euphratica** Olivier, Voy. Emp. Ottoman III (1807) 449 fig. 45—46; Dode in Mém. Soc. Hist. Nat. Autun XVIII (1905) 16; Grossg., Fl. Kavk. II (1930) 2. — Ic.: Dode l. c., tab. XI, l. c.— Iranian name: patta.

A tree; buds pubescent; leaves of long shoots linear or lanceolate; leaves of short shoots ovate-rhombic or elliptic-orbicular, with shallow dentation, all pale green; buds pubescent; rachis of young catkins pubescent; pedicels with scattered hairs; capsule smooth, glaucescent, 3-valvular. April. (Plate X, Figure 4, a—d).

River valleys of the Araks system (e. g., Alindzha-chai) and in Zangezur. Caucasus: S. Transc. **Gen. distr.**: Arm.-Kurd., W. Iran. Described from Mesopotamia. Type in Paris.

223 Note. A low tree, forming small groves.

Economic importance. The resin of the bark, under the name "buri-armeni," is considered a remedy for syphilis. Wood reddish, coarse-grained, light, suitable for miscellaneous carpentry use.

Series 2. **Pruinosae** (Dode) Kom. — Disk divided down to base; leaves of short branches entire or slightly toothed, glaucous.

5. **P. pruinosa** Schrenk in Bull. Acad. Sc. Pétersb. III (1845) 210; Ldb. Fl. Ross. III, 628; O. and B. Fedchenko, Conspectus Fl. Turk. VI (1916) 331; Dode in Mém. Soc. Hist. nat. Autun, XVIII, 18. — Ic.: Dode l. c., XI, 2a. — Turkmenian: petta; Kashgarian: tograk; Kazakh: turangá, turangyl.

A tree of medium or below medium size, 4—7 m high, with a whitish uneven trunk; leaves of short shoots coriaceous, reniform, emarginate or minutely apiculate, with entire or wavy or slightly sinuate margins, heavily pruinose on both sides, often pubescent at first, at length somewhat velvety; petiole terete; rachis of catkin, pedicels, and capsules densely cinereous-pubescent. March—May. (Plate X, Figure 5).

Riverside sands and pebbles in desert areas, solitary or in groves, often together with other species of the subgenus Turanga; rarely near springs or wells. — Centr. Asia: Balkh., Syr D., Kyz. K., Amu D., Mtn. Turkm., Pam.-Al. **Gen. distr.**: Iran., Dzu.-Kash. Described from the banks of the Ili River. Type in Leningrad.

The northernmost point attained by this species is to be found beside the Syr Darya River, just below 46°N. lat., where it grows together with P. diversifolia Schrenk.

P. glaucicomans Dode in Mém. Soc. Hist. nat. Autun XVIII (1905) 18 = P. ariana Dode × P. pruinosa Schrenk. — Ic.: Dode l. c. tab. XI, 2.
Similar to the preceding, but leaves in the lower part serrate-crenate and heavily pubescent to densely velvety.
Centr. Asia: Mtn. Turkm. Described from Turkestan.

Subgenus 2. **LEUCE*** Duby in DC. Bot. Gall. I (1828) 427. — Leaves darker above than beneath, often lobed, variable in shape; petiole strongly compressed laterally in upper part, often with rather large cup-shaped glands; disk persistent in fruit, obliquely truncate, with wavy margin; stigmas 2, pale pink to purple; stamens 5 — 20; capsule small, elongate; catkins rather compact; bark very light-colored, remaining smooth for a long time.

224

Series 1. Albidae Dode. — Leaves more or less white-tomentose beneath, more or less lobed; disk more or less cut; propagated chiefly by suckers.

6. P. nivea Willd. Berl. Baumzucht (1796) 227; Dode in Mém. Soc. Hist. Nat. Autun XVIII (1905) 21. — P. alba f. nivea C. K. Schn. Handb. Laubholzk. I (1904) 22. — Ic.: Dode l. c. tab. XI, 7. — Tadzhik name: mirza safidor; Turkish: ak-terek; German: weisse Pappel.
A tree; leaves of shoots clearly lobed, straight at base, dull dark green above, densely tomentose snow-white beneath, 3—5-lobed and toothed; leaves of short branches elliptic, with few teeth, the subparallel secondary veins slightly flexuous; petiole unevenly hairy. April—May.
As compared with the common silver poplar, this form is more westerly, common in Central Europe (Austria) and Bal.-As. Min. The origin is to be sought in the Caucasus: S. and E. Transc., and in Centr. Asia: Ar.-Casp., Syr D., Amu D., Pam.-Al. For this last region Dode also establishes a distinct "species" P. Paletzkyana Dode l. c. 21 et tab. XI, 8, distinguishing it from P. nivea by the slightly undulate leaf margins, the less dull and more sparsely tomentose lower leaf surface, and other equally insignificant characters. In V. A. Paletzky's words, this poplar was collected in Tashkent where it is cultivated under the name P. turkestanica and represents a pyramidal form of silver poplar.
Note. According to the material of the Paris Herbarium, P. Paletzkyana Dode is none other than the hybrid P. alba × P. tremula = P. hybrida. It would probably be preferable to use this appellation. P. Morisetiana Dode l. c., p. 22 et tab. XI, 11 (Des Balkans à l'Himalaya) appears to be P. nivea with pinnatilobed leaves, a form not infrequent in the gardens of Central Asia (number of lobes 3—7).

* From Greek "leuke," the name for silver-poplar used by Aristophanes.

7. **P. Bolleana** Lauche, Hütt. d. deut. Garten (1878) ex Dode in Mém. Soc. Hist. Nat. Autun XVIII (1905) 23. — P. Bachofeni Wierzb. in Rchb. Icon. V, II, tab. 616; B. Fedtsch. Consp. Fl. Turk. VI, 329. — Exs.: HFR No. 1436, 1437. — Ic.: Dode l. c. tab. XI, 14. — Known also as Samarkand poplar; Uzbek name: shor-teryak, chini-teryak (pyramidal form); Persian: sofidar.

A big tree, spreading or often pyramidal, with upright appressed branches; buds very large; leaves of shoots, large, broadly subcuneate at base, green lustrous above, rather sparsely white-tomentose beneath, deeply 5—7-lobed; leaves of short branchlets elliptic-orbicular, irregularly angular, with sharp teeth; anthers bright red. Leaf variations include marked leathery consistency, cordate shape, sharp or blunt teeth, and flat or wavy margins. Pubescence diminishes greatly toward fall. March.

225 Cultivated in gardens and growing wild along streams in foothills. — Caucasus: Tal.; Centr. Asia: Kyz. K., Amu D., Syr. D., Pam.-Al., Mtn. Turkm. **Gen. distr.**: Iran., Kashg.

Yields lumber and material for woodworking.

8. **P. alba** L. Sp. pl. (1753) 1034; Dode in Mém. Soc. Nat. Autun XVIII, 25; Shmal'g., Fl. II, 439; Ldb. Fl. Ross. III, 626; Kryl., Fl. Zap. Sib. IV, 780. — P. triloba Dode, P. treyviana Dode, l. c. (1905). — Ic.: Dode tab. XI, 19. -- Vernacular names: belolistka [white-leaved; German: Silber-pappel; French: peuplier blanc, ypréau; English: white or silver leaf poplar; Italian: pioppo bianco, alberello, pobbia; Georgian: alis-khe, verkhvi; Armenian: kagamakh-spitak; Tataric: komakh, ak-soukh; Turkish: ak-terek; Chinese: yangu.

A tree 18—35 m high, with a wide head; bark very light-colored, smooth when young; that of old trees greenish-gray, shallowly fissured; young branches white-tomentose; winter buds pubescent, rarely glabrous with ciliate margins; leaves of shoots deltoid, straight at base, with 2 very small lobes at the angles, toothed above, the lower surface white, with profuse flocculent pubescence; leaves of branches orbicular, slightly truncate at base, sharp-toothed, very irregular, the gray flocculent pubescence persistent till fall; staminate catkins stout, 3—7 cm long; scales ferruginous, elliptic to cuneate-elliptic, irregularly toothed or subentire, moderately villous-ciliate; stamens 6—8 (3—30); anthers initially light purple, becoming yellow; fruiting catkins 10—12 cm long, the rachis sparingly covered with soft hairs; ovary oblong-conical, short-stipitate; disk membranous, oblique; stigmas 2, pinkish-red, the narrow lobes cruciately spreading; capsule glabrous. March—May (flowers appearing with the leaves).

Growing wild in riverside deciduous woods (inundated woods) and in water meadows, solitary or forming woods; also in barchan [crescent-shaped] sand dunes. — European part: U. and M. Dnp., Bl., V.-Don, Transv., Urals (S.), V.-Kama, V.-Don, L. Don, L. V., Crim.; Caucasus: Cisc.; W. Siberia: Ob, U. Tob., Irt., Alt.; Centr. Asia: Ar.-Casp., Balkh., Dzu.-Tarb., Mtn. Turkm., Amu D., Syr D., the lower-lying part of Pam.-Al. **Gen. distr.**: S. part of Centr. Eur., Atl. Eur., Med., Bal.-As. Min., Dzu.-Kash. Described from Central Europe. Type in London.

Note. P. triloba Dode (Monogr. p. 21) was established for silver poplars of Altai (Ledebour's collections) and Dzungaria (Karelin prope Ajagus), i. e., for the eastern race of this species with a simpler leaf outline. P. Trevyiana Dode (Monogr. p. 21) was established for Chernyaev's specimen from Kharkov (1859) which undoubtedly belongs to common silver poplar.

Cultivated in nearly all regions for ornament in gardens and grounds.
226 The N. limit for cultivation is 68°N. lat. in Norway; for the wild-growing tree 58°15', near Novo-Ilinskii on the Ob River, somewhat to the south of the Ker River mouth. The E. limit is 58° E. long., on the Chulym River; the S. limit is to be found in Chinese Dzungaria (Urungu R., etc.).

Economic importance. The wood is coarse-grained, shiny, with yellowish hardwood and white alburnum, very soft, cleaving readily, lightweight. Used for paneling, doors, windows, parquet floors, and for various carpentry and turning jobs. The foliage is fed to livestock; the bark is suitable for tanning. Produces suckers abundantly. Propagated by cuttings and suckers (Medvedev, Der. i kust. Kavk. 285).

9. **P. canescens** Sm. Fl. Brit. III (1805) 1080; Dode in Mém. Soc. Nat. Autun XVIII, 26. — P. hybrida M. B. Fl. taur.-cauc. II (1808) 422; Grossg., Fl. Kavk. II, 2. — P. alba × P. tremula Shmal'g., Fl. II, 439. — Ic.: Dode l. c. tab. XI, 22.

A tall tree with whitish-gray bark; leaves of shoots subcordate, slightly lobed, irregularly serrate; head rather lean, almost weeping, light green; leaves of branches orbicular, irregularly emarginate, with toothed margin, sparsely flocculate-pubescent; stigmas carmine-red, rising parallel to the long axis of ovary. March—April.

Growing wild in inundated woods. — European part: M. Dnp., Bl., L. V. (N. edge; southward down to Aksu on the Ilek River, 50°30' N. lat.), Crimea; Caucasus: Cisc., Dag., W. and E. Transc., S. Transc., Tal. **Gen. distr.**: Bal.-As. Min. Described from cultivated specimens.

Note. In Dode's opinion, this is not by any means a hybrid. He collected it abundantly in Bulgaria, near Varna. According to other evidence, this is a hybrid of P. alba × P. tremula.

Series 2. **Trepidae** Dode. — Leaves mostly glabrous or with short appressed pubescence; bracts deeply divided; leaf petiole compressed laterally near the blade.

10. **P. tremula** L. Sp. pl. (1753) 1034; Shmal'g., Fl. II, 439; Kryl., Fl. Zap. Sib. IV, 781; Turcz. Fl. baic.-dah. II, 2, 125; Ldb. Fl. Ross. III, 627; Fedch., Rast. Turk. 292; O. and B. Fedchenko, Conspectus Fl. Turk. VI, 330. — Russian — osina; English — asp, trembling poplar; French: tremble; German: Espe, Zitterpappel, Flatteraspe; Karelian: gabu; Tataric: uzyak-agach, etc.; Georgian: khavalo, verkh vi; Armenian: korani; Polish: osa, osina.

A big tree, to 25—50 m high and to 5 m in circumference; bark smooth, light greenish-gray; young branches glabrous; leaves of shoots large, deltoid or deltoid-elliptic, slightly cordate or straight at base, distinctly apiculate, the margins denticulate; leaves of branches glabrous or obscurely

pubescent beneath (rarely more densely pubescent — var. villosa), rounded-deltoid or orbicular, minutely apiculate, with clearly dentate
227 margin; occasional specimens have leaves cuneate at base; stipules soon caducous, whitish, linear; petiole about as long as the leaf blade, strongly compressed in upper part; catkins 4—14 cm long and 2 cm thick, villous; bracts palmatifid, dark, villous-ciliate on the margin; anthers purplish-red, becoming paler; ovary pale green, conical, with 2 purple stigmas; disk pale, glabrous or more or less ciliate on the margin; pedicel very short. March—May (flowering before the leaves). Nectariferous glands present at the base of young leaf blades (extranuptial nectaries). (Plate X, Figure 8).

Woods of all types, most often at wood margins, clearings, and in places where trees have been broken by the wind, often as part of birch forests or in small pure stands, especially on felling sites, etc., as forerunners of natural reforestation; also gullies, shores, and swamp margins. Generally a species requiring light, but the seedlings prefer a certain amount of shade. — European part: Kar.-Lap. (almost the whole of the Kola Peninsula, including the N. coast; in the mountains up to the timberline, Dv.-Pech., Lad.-Ilm., U. V., Urals, V.-Kama., U. Dnp., M. Dnp., V.-Don, Transv., Bl., Crim.; Caucasus: Cisc., Dag., W., E., and S. Transc.; W. Siberia: Ob, U. Tob., Irt., Alt.; E. Siberia: Yen., Lena.-Kol., Ang.-Say., Dau.; Far East: Kamch., Okh., Ze.-Bu., Uda, Uss., Sakh.; Centr. Asia: Ar.-Casp., Balkh., Dzu.-Tarb., T. Sh. **Gen. distr.**: Scand., Centr. Eur., Med., Bal.-As. Min., Arm.-Kurd., Dzung., Mong., Manch. Described from N. Europe. Type in London.

Economic importance. Wood white, without heartwood, soft, easily cleft, in dry condition rather durable; suitable for paper manufacture (cellulose) for match production (exported in blocks to Japan), for turning roof shingles, hoops, bobbins, rural dishware, etc. The trunks are made in some places into canoes and small boats. The shavings provide good packing material. The bitter bark is used in tanning and in medicine. The wood contains, in addition to cellulose (ca. 50%) lignin, galactan, and pentosan; yields on distillation 4.17% acetic acid. The bark contains the glucosides salicin and populin, and the enzyme salicase (Wehmer). Easily propagated by seed and within twenty years reaching mature condition at 12 m height. The fast turnover represents a considerable economic advantage.

* **P. Davidiana** Dode in Mém. Soc. Hist. Nat. Autun XVIII (1905) 31. — P. pellostachys Dode, ibid. — P. wutaica Mayr, Fremdl. Wald-und Park-bäume (1906) 474. — P. tremula var. Davidiana C.K. Schn. in Sarg. Plantae Wilsonianae. Vol. III, part. I (1916) 24. — Also referred to as Chinese aspen.

A tree to 20 m high, with rather thin oval or round head; bark
228 smooth, greenish or pale gray, that of old trees dark gray and fissured in the lower part of the trunk; the leaves differing from those of common aspen in the finer and shallower dentation; they are often orbicular, terminating in a small point; catkins so far not studied. April.

Schneider assumes that here belong the collections from Amur and Ussuri. Kudo also refers here the aspen of North Sakhalin, but the specimens of our herbarium do not confirm this. — **Gen. distr.**: mountains of China, Manchuria, and Korea.

*P. Sieboldii Miquel in Ann. Mus. Lugd. Bat. III (1867) 20; C. K. Schn. in Sarb. Plantae Wilsonianae Vol. III, part 1, 38. — P. tremula var. villosa Franchlet et Sav. Enum. Pl. Jap. I (1875) 463. — Also referred to as Japanese aspen. Japanese: hago-yanagi, khako-yanagi.

A tree; differing from aspen in the tomentose young branches; leaves at first white-hairy on both sides, finally with gray hairs only on the lower surface; anthers turning black on drying, and some other characters.

Grows in South Sakhalin (Korsakovo); may possibly be found also in North Sakhalin. — Gen. distr.: only in N. and Centr. Japan.

Subgenus 3. **EUPOPULUS** Dode. — Characters in the key.

Section 1. AEGIRI* Dode. — Aegirus Aschers. Pl. Prov. Brand. I (1864) 645. — Petioles more or less compressed laterally, their length and shape the same on shoots and branches; leaves more or less glaucous beneath, the glands pectinate; branches and leaves glabrous.

11. **P. nigra** L. Sp. pl. (1753) 1034; Asch. und Gr. Synopsis IV, 36; Dode Mém. Soc. Hist. Nat. Autun XVIII 47; Shmal'g., Fl. II, 439; Ldb. Fl. Ross. III, 628; Kryl., Fl. Zap. Sib. IV, 783; Fedchenko, Rast. Turk. 294. — P. hypomelae Dode. — P. euxylon Dode l. c. — Russian: osokor', chernyi topol' [black poplar]; English: black-barked or common black poplar, water poplar; French: liardier, bouillarg [?], peuplier noir, peuplier franç.; German: Schwarz Pappel; Italian: pioppo commune, p. nero; Polish: sokora, topola plodna; Turkish: teryak, kara-terek; Georgian: opi; Armenian: kagamakh seav; Abkhazian: amyshva.

A handsome tree, to 30 m high, with a broad head; bark thick, splitting, dark gray; leaves of shoots broadly deltoid, 5—12 cm long and 4—15 cm broad, almost straight at base, shortly acuminate; leaves of branches subdeltoid, broadly rounded at base, terminating in a slender point, the naked translucent margin shallowly serrate; all leaves lustrous, the lower surface but slightly paler than the upper; staminate catkins 6—9 cm
229 long and to 1 cm thick, at first compact, becoming looser; pistillate catkins more slender, to 12 cm long, later elongating; bracts palmatifid, glabrous, soon caducous, brownish to red, with convex margins; stamens 6—30, with white filaments and purplish-red anthers; fruit ovoid, 7—9 mm long, slightly warty, distinctly stalked; stigmas 2, yellow, recurved. March—April (flowering before the leaves). (Plate X, Figure 7).

Inundated river valleys, pebbly and sandy banks, water meadows, riverside sands, and rarely terraces; also brooks, oxbow lakes, and lake shores. Often cultivated and readily naturalized due to seed dispersal. — European part: Lad.-Ilm., U. V., V.-Kama U. Dnp., M. Dnp., V.-Don, Urals, Transv., Bl., Crim., L. Don, L. V.; Caucasus: Cisc., Dag., W., E., and S. Transc., Tal.; W. Siberia: Ob, U. Tob, Irt., Alt.; E. Siberia: Yen.; Centr. Asia: Ar.-Casp., Balkh., Dzu.-Tarb., Amu D, Pam.-Al. Gen. distr.: Centr. Eur. (E. part), Bal.-As. Min., Arm.-Kurd., Dzu.-Kash. Cultivated forms of this poplar are widespread in all civilized countries of the Temperate Zone. It also occurs in the towns of the Far East. It rises along the rivers in the Caucasus to 1,500 m. The northernmost outpost of the wild tree lies at 64°N. lat., on the Yenisei; the eastern limit is the

* From Greek "aigeiros" — black poplar.

Kana River (Yen.), at 96° E. long. Described from Central Europe. Type in the Linnaeus Herbarium.

Economic importance. At the tips of the leaf teeth there are nectariferous glands. The wood is soft and light, with brownish heartwood, yellow sapwood, and tawny streaks; used for turning, carpentry, and carving; employed in the making of spoons, scoops, troughs, bowls, etc. The bark is used for dressing leather and dying it yellow. An ointment prepared from the buds is used in popular medicine (the buds, "Gemmae seu oculi populi," contain, according to Wehmer (Die Pflanzstoffe I, 250), 0.5% essential oil, malic acid, mannite, salicin, populin, chrysin, etc.). The bark contains 3% of tannic acid. Easily propagated by cuttings; produces suckers in profusion. Growing well on loose soils with a high moisture content. Suitable for town and country planting, and for afforestation of steppe locations. This species requires plenty of light.

Note. According to the material of the Paris Herbarium, P. hypomeleana Dode is but P. nigra L. of Altai; Dode maintains that it stretches down to Kansu Province in China. The description of P. euxylon Dode was based on specimens from Ashkhabad in Turkmenia and Hermannstadt in Hungary*; these specimens vary so much from each other that they provide no basis for the separation of a distinct species from P. nigra L.

230 12. **P. pyramidalis** Rozier, Cours d'agric. VII (1786) 619; Boiss. Fl. Or. IV, 1194; Dode Mém. Soc. Hist. Nat. Autun XVIII 50. — P. italica Moench, Baüme Weissenstein (1785) 79. — P. pyramidata Moench, Meth. (1794) 339. — P. fastigiata Desf. Tabl. écol. Paris (1804) 213. — P. nigra var. italica Duroi Harbk. Baumz. II (1772) 141; Asch. und Gräbn. Synops. IV, 41. — P. nigra var. pyramidalis Spach in Ann. Sc. Nat. (1841) 31; Shmal'g., Fl. II, 440. — P. dilatata Ait. Hort. Kew. ed. I, 3 (1789) 46. — P. scytica Dode, l. c. — Referred to in Russian as pyramidal, Italian or Ukrainian poplar; other Russian names: raina, rakita italianskaya; English: fastigiate or Lombardy poplar, cypress poplar, poplar pine; French: peuplier d'Italie, p. pyramidal, p. de Lombardie; German: Pyramiden, Chaussee, Italienische, Lombardische Pappel; Italian: pioppo cipressino; Polish: topola wloska; Czech: turecki topol; Tataric (in Crimea): sal'vi; Persian: safidar; Turkish: teryak, mirza-teryak.

A handsome erect tree to 30 m high; branches upright from base, imparting a pyramidal shape to the head; 2-year-old branches yellow or lurid; young shoots and leaves at first slightly pubescent; leaves of shoots broadly triangular, broadly cuneate at base, prominently tipped; leaves of branches rhombic, the base broadly cuneate, rounded or slightly cordate, the margin serrate; flowers and fruits as in P. nigra. Staminate trees predominate.

Wild in Centr. Asia, along rivers; cultivated in European part: U. Dnp., Bl., Crim., V.-Don; Caucasus: Cisc., Dag., W., E., and S. Transc., Tal.; Centr. Asia: Amu D., Syr D., Pam.-Al. **Gen. distr.:** Iran., Bal.-As. Min., Med. The country of origin is apparently Afghanistan, whence gradually spread to the Mediterranean area.

P. nigra L. var. afghanica Aitchison et Helmsley in Aitchison Flora of the Kurram Valley etc., Afghanistan (1880) No. 161.

A tree to 30 m high, with extremely slender ascending terminal branches; leaves small, membranous, ovate-rhomboid, markedly cuneate at base,

* [Now Sibiu in Rumania.]

PLATE X. 1. Populus suaveolens Fisch. – 2. P. amurensis Kom. – 3. P. ussuriensis Kom. – 4. P. euphratica Oliv.; 4, a–d) leaves. – 5. P. pruinosa Schrenk. – 6. P. laurifolia Ldb.; 6, a – 7, a) P. nigra L. – 8. P. tremula L.; a) catkin.

short-acuminate, the margin crenate-dentate; fruiting catkins slender, loose; disk cup-shaped, obscurely lobed; immature capsules ovaloid, as long
233 as their stalks.

We have no reliable reports concerning the occurrence of this beautiful form of pyramidal poplar in Soviet Central Asia. Since, however, these poplars have not as yet been studied, the search for the Afghan poplar, especially in Tadzhikistan, may prove successful upon further investigation. All this provided that it is not identical with P. Kanjilaliana Dode.

* **P. kanjilaliana** Dode, Mém. Soc. Hist. Nat. Autun XVIII (1905) 60. — Ic.: id. Tab. XII, f. 96.

A tree; leaves of shoots elliptic, narrowly rounded or broadly cuneate at base, shortly point-tipped; middle leaves lance-obovate, shortly apiculate, the cuneate base somewhat rounded at the petiole base; leaves of branches suborbicular, somewhat cordately rounded at base, short-acuminate; young petioles pubescent, the young leaves ciliate on the margin, becoming smooth; dentations of the leaf margin broad, irregular; lower surface of leaf whitish; shoots slightly pubescent; network of veins prominent, less so beneath. This poplar represents a transitional stage toward the group Pseudobalsamifera. Fruit smooth.

"Middle Asia, Turkestan."

This description does not enable identification of P. kanjilaliana Dode with any of the specimens that are to be found in our herbaria. Even though this poplar resembles the preceding, it is not possible to consider the two as synonymous. It was found in the Paris herbarium that P. kanjilaliana is based on specimens gathered by Capus on the Yagnob and Voru rivers (Zeravshan River system) and represents "mirza terek," i. e., the pyramidal poplar of Zeravshan villages. Upon naturalization it retains its characteristic shape (Dalmatia and the Adriatic coast).

Very closely related but more frost-hardy forms (cultivated in Mosvow) are known under the name P. pannonica Besser in Flora XV, Beil. 2 (1821) 14; Fedchenko in 6th ed. of Maevskii's Fl. Sr. Ross. (1933) 246. — Exs.: HFR No. 1186. — Referred to as Hungarian poplar.

The leaves are triangular as in P. nigra and not rhombic as in P. pyramidalis; branches less appressed to trunk; dentation of leaves more pronounced.

Economic importance. The pyramidal poplar is easily propagated by slips from the roots; grows rapidly but senescence sets in early. Because of the shallow root system, the tree is easily uprooted by wind. It is also very subject to diseases. The age of 20 years is therefore most suitable for felling. In Central Asia, the pyramidal poplar "mirza teryak" (P. kanjilaliana Dode) is often the main source of timber (girders of bridges over rivers, etc.). The wood is soft and light, rot-resistant. It is considered in W. Europe as good material for carving, furniture, and
234 boxes.

Note. P. scytica was established for Chernyaev's specimens from Kharkov (1859) that belong to the common pyramidal poplar of the Ukraine.

* **P. deltoides** Marshall, Arb. amer. (1785) 106. — P. canadensis Moench, Bäume Weissenst. (1785) 81. — P. marilandica Poiret in Lam. Dict. Suppl. IV (1816) 378. — P. carolinensis Moench l. c. (1785) 81. — Designated in Russian as Canadian poplar. English: Carolina poplar, river poplar, big or yellow cottonwood, cotton-tree.

A tree to 45 m high and to 2.25 m diameter; bark grayish-green, in age becoming very rough; leaves of branches broadly deltoid-ovate, at base broadly cuneate and decurrent onto the petiole, prominently short-tipped, the margin crenulate, the petiole the length of the blade, 4 —7 cm long; catkins plump, sessile, 6 — 10 cm long; bracts smooth, doubly fringed; stigmas sessile, angular; fruit 2 — 4-valvular, as long as or longer than the stalk.

Native in North America where it grows beside the rivers and lakes of Canada and of the Atlantic States. The biggest of all poplars and favored for cultivation in W. Europe and in the USSR (Moscow; here and there in the Ukraine, near Kuibyshev).

Economic importance. Wood soft, dark brown, highly valued for carpentry and constituting a rather significant part of Canadian export.

* **P. berolinensis** Dippel, Laubholzkunde II (1892) 210.

A cultivated tree, used for city planting; distinguishable by the large bright green leaves and the tawny, angular branches; leaf shape variable, mostly rhombic. Most probably a hybrid between P. pyramidalis and P. laurifolia (see below, p. 186) which first appeared in the Berlin Botanical Garden. A similar hybrid, P. nigra × P. laurifolia is P. Rasumowskyana C. K. Schn.

* **P. Krauseana** Dode, Mém. Soc. Hist. Nat. Autun XVIII (1905) 46.

According to Dode this is a hybrid of uncertain origin. Herbarium material shows it to be P. nigra × P. pruinosa, e. g., Michelson No. 2832, Tashkent, 3 X 1909 (in garden).

13. **P. usbekistanica** Kom. in Journ. botan. de l'URSS (1934) No. 5, 501. — Ic.: Kom. ibid. Tab. 1.

A tree; glabrous in all parts; young branchlets yellowish, becoming light gray; buds very small, ovoid, dull when dry; petioles 2 —6.8 cm long, flexible; leaf blades rounded-ovate to broadly ovate, gradually tapering to a short point, straight or subcordate at base, about as broad as or broader than long, 3 — 7.5 cm broad and 3 — 7 cm long, firm, prominently and finely reticulate-veined on both sides; fruiting catkins dense, 5 —7 cm
235 long; flowers borne on very short pedicels, in our specimens hermaphrodite and actinomorphic; disk spreading horizontally, slightly angular; stamens 10, persistent in fruit, in a regular circle around the ovary base; fruit ovoid, ca. 7 mm long; hairs of the tuft rufescent.

This species differs from P. nigra not only in leaf shape but also in the short fruit-stalks, the broad lobes of the open capsule, and the color of the tuft-hairs.

Stony slopes and beside water in valleys of mountain streams. — Centr. Asia: Pam.-Al. (N. slope of the Turkestan Range in its N. part). Endemic. Described from Iordan in the Shakhimardan River system. Type in Leningrad.

14. **P. cataracti** Kom. in Journ. botan. de l'URSS (1934) No. 5, 501.

A tree of medium size, with a dense head; bark of branches light gray; all parts quite smooth; leaves relatively small; petioles slender, whitish, scarcely compressed, 7.5—2.5 [?] long; blades almost concolor on the two sides, grayish-green, dull, suborbicular, commonly broader than long, 2—4 cm by 2.5—5 cm, almost straight at base, terminating in a very short straight or oblique point, the margin crenulate-denticulate, the network very prominent; buds short, conical, viscous.

Centr. Asia: Pam.-Al. (collected 7 July 1913 by B. A. Fedchenko on the Voru River, Zeravshan channel, in Osie-sabs natural boundary area, at waterfall). Endemic. Described from the Voru River. Type in Leningrad.

Note. Related to P. nigra from which it differs in the firmer round leaves. Catkins not collected.

15. **P. tadshikistanica** Kom. in Journ. botan. de l'URSS (1934) No. 5, 509.— Ic.- Kom. l. c. tab. 2.

A tree of medium size; branches nodding, yellowish, their youngest parts (tips) densely velvety; petioles slender, compressed, densely pubescent, 2—3 cm long; leaves of branches 2.5—5 cm long and 1.5—4 cm broad, broadly lanceolate through ovate to deltoid-ovate, gradually tapering to a long point, the shallowly crenate margins ciliate, the base rounded or broadly cuneate; fruiting catkins 7—8 cm long, the rachis and the pedicels densely pubescent; disk wavy-margined; fruit strongly tuberculate covered with raised hairs.

Banks of mountain streams.— Centr. Asia: Pam.-Al. (Bal'dzhuan, between villages of Khazret-Sultan and Aspringon (1913) No. 3536, A. I. Michelson). Endemic. Type in Leningrad.

Note. Differs from all the other poplars in the raised hairs.

236

Section 2. **TACAMAHACAE*** Dode.— Tacamahaca (1841) l. c.— Bark rugose; buds broad, resinous, aromatic; leaves whitish beneath, devoid of a pellucid margin, cuneate to subcordate at base; petioles terete or square in section, commonly channeled above; stamens 18—60; anthers oblong to round; stigmas 2—4, broad, borne on a short style or subsessile; capsule 2—4-valvular.

Series 1. **Laurifoliae** Kom.— Young shoots quadrangular, with winged corky protrusions on the ribs.

16. **P. laurifolia** Ldb. Fl. Alt. IV (1833) 297; Ldb. Fl. Ross. III, 629; Kryl., Fl. Zap. Sib. 784.— Ic.: Pallas, Fl. Ross. I (1784) 67, ex parte; Ldb. Ic. pl. Fl. Ross. V, t. 479.

A tree 10—20 m high; bark gray; shoots and young branchlets yellow, ribbed, sometimes almost wiηged; leaves of shoots narrow, lance-deltoid; leaves of branches ovate to oblong-ovate, rounded or broadly cuneate at base, gradually acuminate, with close small obtuse teeth on the margin, glabrous or sparsely ciliate, dark green above, pale or grayish-green beneath, 6—15 cm long and 2—7 cm broad, 2.5—9 times as long as the petiole, this subterete, channeled above, often covered with spreading hairs;

* English designation for balsam poplars in North America; at one time it denoted aromatic resins imported into England from tropical colonies.

stipules soon caducous, ovate-lanceolate to lanceolate, acute; catkins rather loose; bracts suborbicular, 3—5 mm long, laciniate, glabrous or sparsely ciliate; stamens numerous (up to 65); anthers purple; disk covering the pistil up to the middle; stigmas broadly lobed; ovary and fruit bearded or glabrous, to 6 mm long. April—May. (Plate X, Figure 6, 6a).

In river valleys, on pebbles, in riverside deciduous woods, on waterside sands and gravelly mountain slopes. — W. Siberia: U. Tob., Ob, Irt., Alt.; E. Siberia: Yen., Ang.-Say.; Centr. Asia: Dzu.-Tarb. **Gen. distr.**: Dzu.-Kash., Mong. (Tuva). The most northerly outpost at 64°N. lat. (Fatyanova on the Yenisei River), further east along the Angara River. Described from banks of the Uba and Ul'ba rivers in the Altai foothills. Type in Leningrad.

Economic importance. As for other poplars. Grown in towns and near railroad stations of S. Siberia and the [former] Far Eastern Territory.

17. **P. pamirica** Kom. in Journ. botan. de l'URSS (1934) No. 5, 510. — Ic.: Kom. ibid. tab. 3.

A tree of medium size; shoots brownish, sulcate, quadrangular; their leaves quite glabrous, ovate, broadly cuneate at base, acuminate, doubly sharp-serrate on the margin, the sharp teeth more deep-cut than in other
237 poplars; branchlets whitish, with a very soft cork layer; leaves of branches 4—6.5 cm long, 3—6 cm broad, orbicular, subcordate at base, short-pointed at apex, crenate and sparingly pubescent on the margin; petioles 2—5 cm long, terete, like the principal veins slightly pubescent; young branches slightly velvety; catkins loose in fruit, to 6 cm long, with densely pubescent rachis; fruit subsessile, 4 mm long, sparingly hairy; disk sinuate-lobed, membranous. May—June.

Banks of mountain rivers. — Centr. Asia: Pam. Al. (Shugnan, Shakhdara River, 26 July 1897, S. I. Korzhinskii). Endemic. Described from the Shakhdara River. Type in Leningrad.

18. **P. densa** Kom. in Journ. botan. de l'URSS (1934) No. 5, 510. — P. suaveolens β macrocarpa Schrenk, Enumeratio altera pl. nov. a cl. Shrenk lectarum (1842) 16. — P. balsamifera Kar. et Kir. Enum. pl. (1841) No. 821, 747. — P. suaveolens in Vvedenskii, Popov, et al., Opredelitel' r. okr. Tashkenta (1923) 76 (non Fisch.). — Ic.: Kom. ibid. tab. 4.

A tree of medium size, with rather loose spreading head; shoots square in cross section, sandy, with thin corky wings; their leaves broadly lanceolate with transitions toward elliptic, uniformly attenuate toward apex, acute, cuneate at base, the margin sharply serrate, both blade and petiole glabrous; leaves of branches ovate to broadly elliptic, acuminate, broadly cuneate at base, crenate and pubescent on the margin, 5—7 cm long and 2—2.5 cm broad, the almost flat and very sparingly pubescent petiole 1—5.5 cm long; catkins ca. 10 cm long, with sparsely hairy rachis; capsule glabrous, rugose, ca. 1 cm long and 6 mm broad, borne on a very short stalk. April.

Banks of rivers and lakes, often forming small woods among meadows. — Centr. Asia: Dzu.-Tarb., Pam.-Al., T. Sh. Endemic. Described from specimens of Schrenk, from Dzhagarak-akso tract, and specimens of Komarov from Lake Iskander-kul'. Type in Leningrad.

19. **P. talassica** Kom. in Journ. botan. de l'URSS (1934) No. 5, 509. — Ic.: Kom. ibid.

A small tree with a dense head; bark very light-colored; leaves of branches ovate, acute, shallowly crenate, 2—4 cm broad; petioles 1.5—4 cm long, sparsely pubescent, terete, channeled above; fruiting catkins ca. 8 cm long, with pubescent rachis; fruit subsessile, 1 cm long, glabrous, slightly rugose.

Centr. Asia: Syr D. (N. V. Pavlov, 25 July 1931, No. 1176, Talass Ala Tau Mountains, shrub thicket on the banks of the Aksu River, alt. 1,500 m). Endemic. Type in Leningrad.

Note. Differing from all other poplars in the distinctive shape (oval base and gradual transition to pointed apex) and tint of the leaves.

238

Series 2. **Suaveolentes** Kom. — Young shoots terete, smooth or knotty, without corky outgrowths.

20. **P. suaveolens** Fisch. in Bull. Sc. Acad. Sc. Pétersb. IX (1842) 348; Ldb. Fl. Ross. III, 629; Turcz. Fl. baic.-dah. II, 125; Kom. Fl. Kamtsch. II, 2. — P. foliis ovatis, acutis, serratis Gmel. I (1747) 152. — P. balsamifera Pall. Fl. Ross. I, 67 (non L.). — Ic.: Gmel. l. c., tab. XXXIII; Pall. l. c., tab. 41 (pl. alt. exclusa). — Russian: topol'nik; German: Pappel, Pappelbaum; Kamchatkan: koetgal; Buryat-Mongol: khai-ulya; Tungus: ulan.

A big tree, to 25 m high; bark greenish-gray, rugose on the lower part of the trunk, smooth above; shoots terete; their leaves broadly lanceolate or elliptic, acuminate; leaves of branches dark green above, whitish beneath, 7—11 cm long and 5—7 cm broad, the margin serrate, the petiole sparsely pubescent; catkins to 10 cm long; bracts broadly obovate, whitish, deeply fringed; pedicels very short; disk cup-shaped, as long as the conical ovary; style 3-fid; stigmas broadly ovate or reniform, with wavy margins; capsule subsessile, 3-parted, with rugose lobes; leaves of young seedlings narrow, with much elongated cuneate base. (Plate X, Figure 1).

Grows in rows and in small woods along the banks of rivers and lakes, especially on pebbly or sandy ground. — E. Siberia: Lena-Kol., Ang.-Say., Dau.; Far East: Kamch., Okh., Ze.-Bu., Uda. Spreading to the north along the Penzhina and Anadyr rivers, throughout the Lena River basin, along the Khatanga River almost to 72° N. lat. The exact boundary between P. laurifolia Ldb. and P. suaveolens on the Yenisei has not been determined. Described from Transbaikalia.

Economic importance. Dugouts ("baty") are made from the trunks. An infusion of buds has medicinal use for its diuretic action.

21. **P. Maximowiczii** A. Henry in Gard. Chron. 53 (1913) 198; Elwes and Henry, Trees of Great Brit. and Ir. VII (1913); Nakai Fl. Silv. Kor. XVIII, 199. — Japanese: doro-noki.

A big tree to 30 m high; bark gray, lighter and greenish in upper part, smooth; young branches pubescent; leaves 10—12 cm long, 7—9 cm broad, firm, with an impressed network of veins, broadly ovate, broadly elliptic, or orbicular, slightly cordate at base, terminating in a distinct short point, the veins pubescent beneath; catkins 10—12 cm long, glabrous; capsule smooth. April—May; fr. June.

Riverside pebbles or mountain slopes, in mixed or broad-leaved woods, in the latter case solitary. Seedlings and young individuals characterized by the elongated cuneate base of their leaves. Petioles often red. — Far East: Ze.-Bu. (SE part), Uss., Sakh. **Gen. distr.**: Jap.-Ch. (N.). Described from cultivated specimens. Type in London.

239 22. **P. koreana** Rehder in Journ. Arnold Arboretum III (1922) 226; Man. cult. Trees (1927) 89; Mitteil. d. deutsch. Dendrol. Ges. XXXVIII (1927) 37; Nakai Fl. Silv. Koreana XVIII (1930) 196. — Ic.: Nakai ib. tab. XLIX.

A tree to 20 m high, the trunk 40—80 cm in diameter; bark dark gray; shoots terete, rather dark, aromatic, exceedingly viscous; shoot-leaves with petioles 4 — 15 mm long, oblong to ovate-elliptic, green above with an impressed network of veins, whitish beneath, glabrous on both sides, short-acuminate; leaves of short branches crowded at the ends; petioles reddish, smooth or appressed-hairy, terete, 1—3.5 cm long; leaf blades obovate-oblong or orbicular, 4—12 cm long, 2.4—8.8 cm broad, dull green above with an impressed network of veins, acute or short-cuspidate, the margin slightly crenate-serrate or subentire, the smooth veins often red below the middle; staminate catkins pendulous, ca. 3.5 cm long; bracts round or reniform, 3—4 mm long, laciniate; disk cup-shaped, whitish; stamens 10—30; anthers blackish-purple, to 1 mm long; pistillate catkins 3—5 cm long, smooth; flowers sessile; bracts and disk as in P. Maximowiczii A. Henry; ovary green, spherical, with 2—4 stigmas. April—May.

Beside water and on slopes; needs more sunshine than the preceding species. Nakai stresses particularly the red petioles of this poplar. — Far East: Uss. **Gen. distr.**: Korea. Described from Korea. Type in the Arnold Arboretum.

Economic importance. The most ornamental species of the group.

23. **P. ussuriensis** Kom. in Journ. botan. de l''URSS (1934) No. 5, 510. — Ic.: Kom. ibid., tab. 5.

A rather small tree, to 12 m high; bark gray; head round, dense; young branches pubescent; shoots square in cross section; buds dark, resinous, viscous, lance-conical, long-acuminate; leaves of short fertile branches crowded at the ends, on young long branches alternate; petioles 2—4 cm long, velutinous with dense raised white or yellowish hairs; leaf blades broadly elliptic, rounded-ovate or orbicular, narrowly cordate at base, round-tipped, minutely mucronulate, the slightly crenate margin obscurely ciliolate; displaying pronounced anisophylly, 6—12 cm long and 3—10 cm broad, firm, dark green above with an impressed network of veins, with pubescence of the petiole spreading onto the principal veins, the white lower surface with 4—8 pairs of large lateral veins; catkins 12—18 cm long; the rachis densely pubescent, especially in lower part; capsule subsessile, smooth, to 7 mm long. May. (Plate X, Figure 3).

Sandy river valleys or tall crests; in groups. — Far East: Uss. Endemic. Described from islands in the Ussuri River estuary. Type in Leningrad.

240 24. **P. baikalensis** Kom. in Journ. Bot. de l'URSS (1934) No. 5, 511.

A tree of medium size, with light gray bark; young branchlets pubescent; leaves of branches rounded-ovate; petiole 12—16 mm long, densely velutinous; blade 6—7 cm long, 4—5—6 cm broad, short-acuminate, the flat upper surface dark green, the lower surface densely pubescent on the veins and sparsely so between them, the margin dentate-crenate and hairy, the base semicordate or rounded; catkins 4— 5 cm long, the rachis pubescent; capsule 4—5 cm long, smooth. May.

Sandy riverside plains (terraces ?) with thickets of Siberian stone pine, near Lake Baikal. — E. Siberia: Dau. Endemic. Described from V. N. Sukachev's specimen from the vicinity of Sosnovka village on the western shore of Lake Baikal. Type in Leningrad.

25. **P. amurensis** Kom. in Journ. Bot. de l'URSS (1934) No. 5, 510. — Ic.: Kom. ibid. tab. 6.

A tree; leaves of fertile branches broadly ovate, short-pointed at apex, smooth, rounded at base; petioles ca. 3 cm long, densely pubescent, as are the principal veins; catkins 6—7 cm long, with a densely pubescent rachis; capsule hairy, ca. 0.5 cm long. (Plate X, Figure 2).

Forest margins. — Far East: Ze.-Bu. Description based on K. I. Maximovich's specimens from the right bank of the Amur River, slightly above the Pikhtsa River mouth, at the Goldi village of Ukhsumi. 24 July 1855. Type in Leningrad.

***P. pilosa** Rehder in American Museum Novitates No. 292 (1927) 1.

A tree 5—12 m high; trunk diameter 30—75 cm; bark whitish-gray, deeply fissured; young branchlets densely hairy, yellowish; buds viscid, pubescent on the outside; leaves of fertile branches ovate to broadly ovate, 4.5—8 cm long, 4—6 cm broad, short-acuminate; base rounded, straight, or shallowly cordate; margin crenulate, broadly ciliate; upper surface pubescent on the prinicpal veins, the paler lower surface also on subsidiary veins, the hairs yellowish; petiole subterete, 1—2.5 cm long, densely clothed with yellowish hairs; fruiting catkins subsessile, 5—8 cm long, compact, with hairy rachis; bracts broader than long, fimbriate, smooth; capsule ca. 4.5 cm long, sessile, spherical-ovoid, pubescent on the outside; disk pubescent, crenate-margined, 4—5 mm in diameter.

On alluvial formations, solitary or in groups. South of the distribution area of P. laurifolia, representing, in a way, its southern extension. Morphologically, P. pilosa and P. amurensis are similar, but the former is genetically connected with P. laurifolia and the latter with
241 P. Maximowiczii. — **Gen. distr.**: Dzu.-Kash., Mong. Described from the valleys of the Baga-bogdo Range in the Mongolian Altai system. Type in Washington D. C., in the National Museum of Natural History.

***P. Simonii** Carrière in Revue Horticole (1867) 360; C. K. Schn. in Sarg. Plantae Wilsonianae III, I (1916) 21; Rehder in Journ. Arnold Arbor. XII (1931) 63. — Referred to as Chinese poplar.

A tree to 16 m high; leaves of shoots short-petioled, ovate, rounded at the ends, often minutely mucronulate; shoots square in cross section, with prominent or even winged corky ribs; leaves of mature short branchlets ovate, cuneate at base; catkins narrow; capsule smaller than in other species of this group, mostly 2-valvular, smooth.

Schneider referred here S. I. Korzhinskii's specimen from the upper Amur. This young plant seems rather to be akin to P. suaveolens. It is possible that some of the still young trees planted in the streets of Khabarovsk belong here; more likely, however, it is P. laurifolia of which the young branches also have winged shoots. — **Gen. distr.:** China, Mong. (E. border), Manch., and Korea. Described from the vicinity of Peking. Type in France.

It has not so far been possible to determine what exactly is signified by P. Mandschurica Nakai, J. Pharm. Soc. Japan (1924) 4, 513; it may be synonymous with P. Simoni Carrière.

***P. moskoviensis** Schroeder in Pl. Inst. Rur. Mosc. colunt. (1899) 43; Russkii ogorod (1918) 512; Syreishch., Mosk. Fl. II, 43. — P. suaveolens × P. laurifolia.

An upright tree of medium size; bark whitish, on the young somewhat ribbed branches tawny; buds strongly aromatic, viscous; petioles 1—2 cm long, puberulous, sometimes with 2 glands; leaf blades ovate, rounded or slightly cordate at base, often inequilateral, gradually tapering at the ends, dark green above, grayish-green beneath, ciliolate on the irregularly denticulate margin; catkins as in P. suaveolens Fisch.

Used for city planting in U. V. and V.-Don, owing to its rapid growth. Described from Petrovsko-Razumovskoe near Moscow. Type not preserved.

***P. balsamifera** L. Sp. pl. (1753) 1034; Aschers. und Gr. Synops. Mittel ur. Fl. IV, 50; Syreishch., Fl. Mosk. gub. II, 42; Shmal'g., Fl. II, 440. — English: balsam or Carolina poplar, Tacamahac, cottonwood; German: Balsampappel; French: peuplier baumier.

A tree to 24 m high, 4—5 m in diameter; bark smooth, gray; head spreading; shoots terete or slightly angled; branches glabrous; petioles terete, long, glabrous; leaf blades ovate or elliptic, always longer than
242 broad, rounded or shallowly cordate at base, 5—12 cm long, 2.5—7 cm broad, whitish beneath; petioles rather long; catkins to 15 cm long, borne on a long glabrous stalk; bracts round, smooth, fimbriate; capsule 2-valvular, subovoid, borne on a very short stalk. April; fr. May—June.

Cultivated here and there almost throughout Europe in gardens and in streets. — **Gen. distr.:** N. Am., where use is made of the wood and bark.

Economic importance. In the USSR use is made of the buds, "gemmae populi," which contain essential oil, resin, salicin, tannic acid, and chrysin. Growth rapid. The wood provides material for carpentry work. Described from North America. Type in the Linnaeus Herbarium.

***P. candicans** Ait. Hortus Kew. III (1789) 406; Shmal'g., Fl. II, 440; Syreishch., Fl. Mosk. II, 41. — P. ontariensis auct. — Vernacular : sedoi; English: balm of Gilead; German: Ontario Balsampappel; Dutch: balsemperboom.

A big tree to 30 m high, with a broad head and splitting bark; branches at first pubescent, becoming smooth, olivaceous to reddish-brown, with pale yellow lenticels; buds to 1 cm long, acute, viscous; shoot-leaves very large, 7—22 cm long, 7.5—18 cm broad, triangular-elliptic, point-tipped, slightly cordate at base; leaves of old branches 6—15 cm long, 5—12 cm broad, ovate-triangular or broadly ovate, broadly cordate at base, the irregularly toothed margin more or less ciliate; catkins to 17 cm long; capsule 2-valvular, stalked, narrowly ovoid, pointed, glabrous. April—May.

Naturalized in the E. States of N. America; in Europe cultivated since 1752 in gardens and along roadsides. Origin not established. In the USSR cultivated: European part: M. Dnp., V.-Don, Bl.; Centr. Asia: Syr D. Described from New England. Type in London.

Order 12. **Myricales** ENGL.

Flowers achlamydeous, dioecious or monoecious, bracteate; stamens 2—16, mostly 4; carpels 2, unilocular; stigmas 2, filiform; fruit a drupe, with a waxy coat; seeds exalbuminous; shrub or undershrubs with simple or rarely pinnatifid leaves; flowers in aments.

Family XLI. **MYRICACEAE** LINDL. *

Flowers unisexual, without a perianth; staminate flowers often with
243 2 or more very small scalelike bracts; stamens mostly 4 (rarely more or fewer); filaments short, distinct or connate at base, the erect ovate anthers opening by 2 longitudinal slits; pistillate flowers with 2—4 scalelike bracts at base; ovary of 2 united carpels; fruit a small dry drupe. Shrubs with spirally arranged leaves.

Myricaceae were widespread in the Tertiary period, and the genera Myrica, Comptonia, and Comptoniophyllum were components of the Arcto-Tertiary flora. Of these, the genus Comptonia has survived only in North America.

Species of Myrica: M. acuminata in the Paleocene (Eocene ?) of L. Don (Osinovka in former Kharkov Province) — M. banksiaefolia Ung. in Sarmatian formations of E. Transc. (Khvteeba). — M. Brylkiniana Heer in the Lower Tertiary series of Sakh. (Dui). — M. (Comptonia) dryandroides Ing. in Paleocene formations of Ob (Tomsk). — M. M. deperdita Ung. in Sarmatian formations of E. Transc. (Iora R., Khvteeba). — M. hakeaefolia (Ung.) Staub in Lower Miocene (Mediterranean) of Dag. (Kemakh) and in the Miocene (Chokrak series) of Dag. (Khaivo-dere). — M. lignitum Ung. in Sarmatian formations of E. Transc. (Khvteeba), in the Upper Dui series of Sakh. (Mgachi), and the Oligocene (Tongrian stage) of Bl. (Kremennaya in Mariupol' [Zhdanov] district). — M. salicina Heer in the Sarmatian of E. Transc. (Khvteeba) and Upper Dui series of Sakh. (Mgachi). — M. tenuifolia Heer in the Upper Dui series of Sakh. (Mgachi). — M. vindobonensis Ettings. in the Sarmatian of Bl. (Orekhov), in the Lower Dui series of Sakh. (Brodyazh., etc.), and the Paleocene of Uss. (Amagu).

* Treatment by O. I. Kuzenev.

Comptonia cf. acutiloba (Brongn.) Heer in Upper Dui series of Sakh. (Rogatyi, Kirpichnaya, Tokova, etc.). — C. cf. rotundata Watelet in the Oligocene (Tongrian stage) of Bl. (Kremennaya in Mariupol' district). — C. Ungerii (Ett.) in the Aquitanian series of Ar.-Casp. (Karasandyk, Yar-kue). — C. oeningensis Heer in the Lower Miocene of Transv. (Sterlitamak).

Comptoniphyllum Naumanni Nath. in Upper Dui series of Sakh. — C. japonicum Nath. in Upper Dui series of Sakh.

Genus 358. **MYRICA** L.
L. Sp. pl. (1753) 1024.

Flowers monoecious or dioecious, in terminal or axillary upright aments; each bract subtending a single flower; in staminate flowers 4 stamens united at base; in pistillate flowers the bract coalescent with the base of the pistil, forming a winged fruit; ovary unilocular; drupe 1-seeded.

1. Leaves on both sides densely tomentose . . . . 2. **M. tomentosa** (DC.) Asch. et Graebn.
+. Leaves glabrous above, with sparse hairs beneath . . . . . . 1. **M. gale** L.

1. **M. gale** L. Sp. pl. (1753) 1024; Ldb. Fl. Ross. III, 661, excl. pl. kamtschat. et sitchens. — Exs.: Pl. Finl. exs. No. 588. — Ic.: C. K. Schn. Laubholzk. I (1906) fig. 30.

A shrub 1—1.5 m high, much branched, profusely covered with golden-yellow glands; branches dark brown, densely leafy; leaves elongate-obovate to oblanceolate, 2—4 (5) cm long, (0.5) 0.8—1.5 cm broad, cuneate
244 at base, acuminate or rounded and mucronulate at apex, short-petioled, with scattered teeth in upper part, dull dark green above, paler and slightly tomentose beneath, covered with small glands; staminate aments to 1—1.6 cm long, light brown; bracts densely glandular, tawny, more or less whitish on the margin; pistillate aments 5 mm long, in fruit 0.7—1.5 cm long, greenish, racemiform. April—May.

Seacoasts, wood margins, wet meadows, grassy bogs, and sands. — European part: Lad.-Ilm. **Gen. distr.**: Scand., Centr. Eur., Atl. Eur. Described from Europe. Type in London.

2. **M. tomentosa** (DC.) Asch. et Graebn. Synops. IV (1910) 353; Kom., Fl. Kamch. II, 36. — M. gale β tomentosa DC. Prodr. XVI, 2 (1868) 148. — M. gale Ldb. Fl. Ross. III, 2 (1849—51) 661, quoad pl. kamtsch. et sitchens. — Ic.: Kom. and Alis., Opred. rast. Dal'nevost. kr. I (1931), Plate 130.

A shrub to 1 m high; branches dense, spreading, dark gray; young branches pubescent and profusely covered with glands; leaves oblong-obovate with transitions to cuneate-lanceolate, (2) 3—5.5 cm long, 1—1.5 cm broad, with 2 or 3 teeth on each margin in upper part, rarely toothless, round-tipped, cuneate at base, tomentose, heavily so and glandular beneath; staminate aments numerous, sessile, gathered at the ends of branches, 0.8—1.2 cm long; bracts purplish-brown, light-colored on the margin, broad, distinctly point-tipped, covered with numerous glands; pistillate aments shorter, developing a little later than the staminate, compact,

(245)

PLATE XI. 1. *Myrica tomentosa* (DC.) Asch. et Gr. – 2. *Pterocarya fraxinifolia* (Lam.) Spach.

0.3 — 0.5 cm long in flower, 0.6 — 1.5 cm long in fruit, ovaloid-lanceolate, acute; stigmas filiform, purple. April — May. (Plate XI, Figure 1).

Lake shores, mossy or rarely grassy bogs, and sands exposed to the action of the sea. — Far East: Kamch., Okh., Uda, Uss., Sakh. **Gen. distr.**: N. Japan, N. Am. Described from Kamchatka. Type in Paris.

Order 13. **Juglandales** ENGLER

Flowers achlamydeous or haplochlamydeous, diclinous, monoecious; stamens 3 — 40; carpels 1; ovary inferior, 1-locular; fruit a 1-seeded drupe or nut; seeds without endosperm; trees with alternate, mostly pinnate, exstipulate leaves.

247 Family XLII. **JUGLANDACEAE** LINDL.*

Flowers in aments, subtended by a bract and 2 bracteoles, mostly with a perianth; staminate flowers with 8 — 20 stamens in 2 or more series; pistillate mostly with perianth, more or less covered in lower part by a cupuliform involucre formed by the bracts united with the ovary and transformed in fruit into a fleshy coat or a flight apparatus; ovary unilocular with a solitary ovule; style short, with a 2-lobed stigma; fruit a drupe or rarely a nut, incompletely 2 — 4-septate. Monoecious trees with alternate imparipinnate leaves; the unisexual flowers wind-pollinated.

Genera of this family, Juglans, Pterocarya, Hicoria, and Engelhardtia, spread far to the north in the Tertiary period.

Genus Juglans: J. acuminata A. Br. was very widespread in the Old World, especially in early Tertiary floras of Asia and Neogene floras of Europe; in the Oligocene of M. Dnp. (Tim); Ar.-Casp. (Kara-sandyk, Chegan), Ob (Tomsk), Alt. (Ashutas), Uss. (Rechnoi), Sakh. (Upper Dui series of Dui, Mgachi, Berez. Polyana, etc.), and Kamch.; in the Sarmatian formations of Bl. (Krynka) and of Cisc. (Khadyzhensk). — J. cinerea L. fossilis Bronn. in the Pliocene (?) of Lena-Kol. (Aldan, Mount Mamontova) and W. Transc. (Goderskii Pass). — J. crenulata Schmalh. in the Oligocene (?) of Alt. (Chingistai). — J. densinervis Schmalh. in the Oligocene of Ob (Tomsk) and in Tertiary layers of Bukhtarma in the Altai. — J. nigella Heer in Lower Dui series of Sakh. (Dui, Mgachi); in Tertiary layers of Kamch. (Korf) and Uss. (Rechnoi). — J. attica Ung. in the Pliocene of the Goderskii Pass. — J. sp. in Paleocene layers of Uss. (Amagu), etc.

The genus Pterocarya was widespread in the Tertiary period on the territory of USSR, reaching far beyond its present distribution limits.

P. castaneifolia (Goepp.) Menzel in the Miocene of Ob (Tara). — P. cf. denticulata Heer in the Paleocene of Uss. (Friz [De Vries] Strait). — P. Massalongii G. et Str. in Sarmatian formations of Bess. (Lipkany).

* Treatment by V. L. Nekrasova.

The genus Hicoria (Carya) now occurring only in North America, was widely distributed throughout the Palearctic Region and was represented by the following species: H. bilinica Ung. in the Sarmatian of Bl. (Grebeniki, Krynka), in Maiotis layers of Bl. (Grebeniki), in the Sarmation of E. Transc. (Khvteeba), and the Paleocene of W. Transc. (Goderskii Pass). — H. Heeri (Ett.) Krysht. in the Paleocene of M. Dnp. (Ekaterinopol'e [Katerinopol']). — H. (Carya) sp. in the Oligocene of U. Don (Tim) and in the Paleocene of Uss. (Rechnoi).

The genus Engelhardtia, now confined to North America, occurred widely in Europe in the Tertiary period, although finds in the USSR have been few up to the present. — A fruit of E. sp. was found in the Paleocene of Uss. (Novokievskoe).

Key to Genera

1. Pistillate flowers in many-flowered aments; fruit with a thin exocarp and a hard endocarp; bracts in fruit developing into 2 wings; perianth scarious . . . . 359. **Pterocarya** Kunth.
\+ Pistillate flowers solitary or in groups of 2 or 3; fruit a drupe with fleshy exocarp and a hard endocarp, wingless . . . . 360. **Juglans** L.

248

Genus 359. **PTEROCARYA** KUNTH.

Kunth, Ann. Sc. Nat. II (1842) 345.

Staminate flowers in dense many-flowered axillary aments, with a 3—6-lobed perianth and 9—18 stamens; pistillate flowers in long terminal aments, the bract and 2 bractlets coalescent with ovary and developing in fruit into wings, the 4-lobed tubular perianth united at base with ovary; lobes of stigma linear; fruit a nutlike drupe, with 2 wings; fruit coat of 2 layers: an indehiscent dry membranous exocarp, and a smooth hard indehiscent endocarp, incompletely 4-locular; cotyledons thick, folded.

1. **P. fraxinifolia** (Lam.) Spach, Hist. Veg. Phan. II (1834) 180. — Juglans fraxinifolia Lam. Encycl. meth. IV (1797) 502. — Pterocarya caucasica C. A. M. Verz. Pfl. cauc. (1831) 134. — Ic.: Ann. Sc. Natur. XVIII (1862) pl. 5, fig. 47, 48, 49; C. K. Schn. Handb. Laubholzk. I (1906) fig. 49 — 51.

A tree to 25 — 30 m high, 1 — 1.5 m in diameter, with whitish non-fissured bark; leaves pinnate; leaflets 8 — 12 pairs, sessile or short-petiolulate, oblong or elongate-ovate, serrate-margined, acuminate, glabrous and lustrous above, paler beneath with hairs at the axils of veins; staminate aments 5 — 7 cm long, solitary at the ends of branchlets; flowers borne on very short pedicels; pistillate aments terminal, with numerous sessile flowers; fruit irregularly turbinate, the wings broader than the body. April—May. Fr. September — October. (Plate XI, Figure 2).

Boggy or inundated places, in mixed woods near rivers, rarely forming small pure stands. — Caucasus: W. and E. Transc., Tal. **Gen. distr.**: Iran. Described from the Caucasus. Type in Paris.

Economic importance. Wood reddish-tinged, light, soft; used for making troughs, vats, bowls, etc. The bast is used for tying grapevines and for weaving bats. All part of this tree have diaphoretic properties.

Genus 360. **JUGLANS*** L.

L. Gen. pl. ed. 1 (1737) 291.

Staminate flowers in solitary pendulous many-flowered aments; perianth 5- or 6-lobed; stamens 8—40, adnate to perianth; pistillate flowers
249 terminal, solitary or 2 or 3 together, the double perianth united with the ovary; stigma lobes 2, linear or lanceolate; fruit a large wingless drupe; exocarp fleshy; endocarp (nutshell) hard, the nut ovoid or spherical, wrinkled, incompletely 2—4-locular; cotyledons thick, folded.

1. Nut abruptly pointed; nut with 6—8 rather sharp longitudinal ridges (Section Cardiocaryon Dode) . . . . 1. **J. manshurica** Maxim.
+ Nut not pointed, rugose, without longitudinal ridges (Section Dioscaryon Dode) . . . . 2.
2. Leaflets 3—5 pairs, relatively narrow, acute; fruit large, ovoid or ovaloid, with rather thin shell . . . . 2. **J. regia** L.
+ Leaflets 2—4 pairs, relatively broad, obtuse; fruit smaller, spherical or subspherical, with harder shell . . . . 3. **J. fallax** Dode.

1. **J. manshurica** Maxim. in Bull. Phys. Mat. Acad. Pétersb. XV (1857) 127; Prim. fl. Amur. (1859) 76; B. V. Skvortsov, Manchzh. lesnoi orekh [The Manchuria Forest Nut], Harbin (1928) 1—11; B. W. Skvorzow in Linguan Sc. Journ. Canton vol. 6, No. 3 (1928) 206. — Ic.: Maxim. Mél. biol. VII (1872) 631; Skvortsov l. c., Plates I—VI; Kom. and Alis., Opred. rast. Dal'nevost. kr. I (1931), Plate 128.

A big tree with a straight upright trunk, dark gray wrinkled bark, and relatively small loose head; leaflets 6—7—9, elongate-elliptic, serrulate, rounded at base, tapering toward a rather long cuneate point; glabrous above, grayish beneath and loosely hairy only on the veins, the young leaves with heavier vesture; exocarp brownish-green, readily separating from the nut; nut dusky brown, ovoid or ovaloid, abruptly terminating in a short point, with 6—8 oblong rather sharp ridges and coarsely rugose between the ridges. The shape of the nut is very variable, ranging from round — var. rotunda Skv., var. triquetra Skv. and var. gracilis Skv., to elongate — var. genuina Skv., var. oblonga Skv., var. mifunensis Skv., var. Komarovi Skv., var. girinensis Skv., var. Dode Skv., and var. depressa Skv. The size of the nut is also extremely variable, fluctuating between 2.7 and 5 cm in length. Fl. May—June; fr. September.

Mixed woods, in soil with a high humus content in valleys and inundated forests; also upon pebbles beside mountain streams. — Far East: Ze.-Bu., Uda, Uss. **Gen. distr.**: Manch., N. Korea. Described from the Amur River. Type in Leningrad.

* A name for walnut used by Pliny and other ancient writers.

250 Note. J. stenocarpa described by Maximovich (Prim. Fl. Amur. (1859) 78; Icon. Mél. biol. VIII (1874) 62) from a single fruit, is probably not an independent species but merely one of the varieties of J. manshurica with an elongated nut. Observations conducted by E. L. Vol'f on specimens of J. manshurica and J. stenocarpa grown in the dendrological garden of the Leningrad Forest Institute, do not resolve the doubts concerning J. stenocarpa since the origin of the seeds from which it was grown is most uncertain.

Economic importance. The wood of the Manchurian nut is hard and durable, easily worked, polishing well, and endowed with beautiful texture. It is used for furniture veneer and wall trimming. It is also employed for turning jobs. The bark of young trees is woven into thin ropes. The nuts, though very thick-shelled, have a high fat content (52%) and yield edible and industrial oil. Tanning agents are sometimes obtained from the exocarp. The tree is of value for ornament and is occasionally cultivated in the USSR and in W. Europe.

2. **J. regia** L. Sp. pl. (1753) 997. — Ic.: Duhamel, Traité des arbres et arbustes, éd. nouv. IV (1809) pl. 47; C. K. Schn. Handb. Laubholzk. I (1906) figs. 42—44; Bull. of appl. bot. XXII, No. 3 (1929) figs. 16, 17, 18, 21; VIII series No. 1 (1932) figs. 4—13. Russian: orekh gretskii [Greek].

A big tree, to 20 m high, with a straight upright trunk to 1.5—2 m and even up to 4—7 m in diameter, dark gray fissured bark, and a large spreading head; leaflets commonly 3—5 pairs, elongate-ovate, slightly crenate, acuminate, glabrous above, hairy beneath in the axils of veins; staminate aments with lanceolate bracts; flowers 6-lobed; stamens 12—18, equaling the perianth lobes; staminate aments 2—5-flowered; flowers sessile with a small-toothed outer perianth and a glabrous inner perianth; exocarp greenish, glabrous; nut yellowish, ovoid, slightly pointed, reticulately rugose, sometimes obtusely ridged, usually with rather thin shell and a large kernel. Fl. May—June; fr. September.

Banks of streams, bottoms of ravines and valleys, and mixed broad-leaved woods on mountain slopes. — Caucasus: W., S., and E. Transc., Tal.; Centr. Asia: Mtn. Turkm. **Gen. distr.:** Bal., Iran.

Note. In the Caucasus apparently naturalized in most cases from remnants of old groves. The shape of nuts is most variable; elongated and rounded, thick-shelled, and thin-shelled nuts occur, but this variability has not so far been studied.

As regards the walnut from Mtn. Turkm. (Kopet Dagh Mountains), M. G. Popov (Tr. po prikl. bot, gen. i selekts. XXII, No. 3 (1929) 294) separates
251 it as ssp. turcomanica Popov, while Gurskii (Tr. po prikl. bot., gen. i selekts., series VIII, No. 1 (1932), page 173) reports for this species various varieties, mostly based on nut shape, such as var. globosa Gursky with round flat-based nuts; var. pyriformis Gursky, with nuts strongly elongated at the top, flat at base and compressed laterally; var. compressa Gursky with a very prominent seam and cuneate base; and var. globosa-angulosa Gursky, with nut depressed at the top on the side of the seam; there are, in addition, a great number of transitions between these varieties.

Economic importance. The wood is hard, somewhat shiny, polishing beautifully especially in basal part of tree, with an exceptionally beautiful texture. Owing to these properties it is very valuable for furniture and finishing veneers. Particularly appreciated are burls or excrescences in the lower part of the trunk, reaching up to 600—800 kg in weight. The leaves, containing the alkaloid juglandin, have medicinal use. The exocarp yields a dark dye, while the nuts are used for food and for extraction of edible and industrial nut oil. Widely cultivated in the Caucasus, Crimea, European part of the USSR, Central Asia, W. Europe, and N. America.

3. **J. fallax** Dode in Bull. Soc. Dendr. France (1906) 89. — J. regia ssp. fallax Popov in Bull. of appl. bot. XXII, No. 3 (1929) 204. — Ic.: Bull. Soc. Dendr. France (1906) 83, 96; Bull. of appl. bot. XXII, No. 3 (1929), figs. 19, 20, 22.

A tree; similar to the preceding species; leaflets 2—4 pairs, elongate-ovate, slightly crenate, subobtuse or obtuse; nut mostly spherical, brownish, with a thick strong shell and a small kernel. Fl. May—June; fl. September.

Forming woods in valleys and on mountain slopes of Syr Darya District and scattered small groves in the mountains of Pam.-Al. Rising in the mountains to 1,800—1,900 m, in Darvaz up to 2,300 m. — Centr. Asia: Syr D., Pam.-Al. **Gen. distr.**: Iran, Afghanistan, Tibet, and India.

Note. Nut shape varies greatly; the variations have not yet been studied.

Economic importance. As for J. regia L., occasionally cultivated in Soviet Central Asia.

All the three species of USSR nuts appear in different forms, providing rich material for selection. In each species the nuts can be classified according to their shape and appearance, e. g., elongate and spherical, thick- and thin-shelled, relatively smooth and variously rugose, short- or long-pointed.

252 Often cultivated for ornament in the European part of the USSR are the species J. cinerea L. of the section Trachycaryon, with elongated nuts, and J. nigra L. of the section Rhysocaryon, with spherical nuts, both from North America. A report claims cultivation along the Zeravshan River of the Himalayan species J. kamaonia Dode, related to J. fallax and J. regia, with large edible nuts strongly emarginate at the top.

In addition, several representatives of the American genus Carya Nutt. (Hicoria Raf.), with smooth angular fruits, have recently appeared in cultivation in the European part of the Soviet Union.

Order 14. **Fagales** ENGLER

Flowers cyclic, monochlamydeous or less often achlamydeous, rarely bisexual, usually diclinous, monoecious; stamens opposite the perianth lobes; ovary inferior; carpels 2—6, each containing 1 or 2 ovules; fruit nutlike, 1-seeded; endosperm absent; ovule chalazogamous; pollen 2-nuclear. Trees with alternate stipulate leaves; flowers in simple spikes, dichasia, or coils.

1. Stamens 2—12; stigmas 2; pistillate flowers in aments or surrounded by an involucre . . . . Family XLIII. **Betulaceae** C. A. Agardh.
\+ Stamens 4—7 or 8—14; stigmas 3; fruits solitary or in groups, surrounded by an accrescent cupule; trees with stipulate undivided or deeply pinnate leaves . . . . Family XLIV. **Fagaceae** A. Br.

Family XLIII. BETULACEAE C. A. AGARDH.

Trees, shrubs, or undershrubs; flowers unisexual, monoecious; staminate flowers commonly in aments; pistillate in aments or in clusters; perianth simple, scarious, adnate or free; sometimes perianth none; stamens 2 or many; filaments distinct, often 2-parted together with anthers; perianth united with upper part of ovary or absent; ovary inferior, 2-locular, more or less compressed laterally, with 2 distinct filiform styles; fruit a nut or a nutlet, with a scarious or coriaceous wing, or wingless, sometimes crowned with vestiges of perianth and style or covered in various ways by the often accrescent bract and bracteoles forming an involucre; seeds commonly solitary, exalbuminous, with a membranous coat; leaves alternate, petiolate, pinnately veined, nearly always toothed; stipules free, mostly deciduous; aments terminal or lateral, initiated in fall and coming into flower before leaf emergence.

253 This family was exceedingly widespread in the Tertiary period throughout the Old World, especially as a component of Arcto-Tertiary flora which consisted of deciduous trees and developed in Asia since the Paleocene and in the European part of the USSR since the Eocene.

Key to Tribes

1. Staminate flowers solitary in the axils of bracts; perianth absent; stamens 3 to indefinite, inserted at the base of the calyx; perianth of pistillate flowers united with the upper part of the ovary; bracteoles forming an involucre coalescent with the fruit . . . . Tribe I. **Coryleae** Meissn.
\+ Staminate flowers with perianth, in the axils of bracts, in dichasia; stamens 2—4, opposite the perianth lobes; pistillate flowers without perianth; bracteoles united with the calyx . . . . Tribe II. **Betuleae** Döll.

Tribe I. CORYLEAE MEISSN. *
Meissn. Gen. (1842) 846.

Characters in the key

Key to Genera

1. Staminate flowers with 2 bracteoles; expanding leaves folded along the midrib; nutlet large . . . . 363. **Corylus** L.

* Treatment of the tribe Coryleae by E. G. Bobrov.

\+ Staminate flowers without bracteoles; expanding leaves folded along the lateral veins; nutlet small . . . . 2.

2\. Fruiting involucre flat, sometimes but slightly curved at base, 3-lobed, or more or less rounded and toothed on the margin . . . . 362. **Carpinus** L.

\+ Involucre bladderlike, cup-shaped, closed, embracing the nutlet . . . . 361. **Ostrya** Scop.

Genus 361. **OSTRYA*** SCOP.

Scop. Fl. carniol. (1760) 414.

Monoecious trees; staminate aments cylindric, with imbricated scalelike bracts; perianth none; stamens 6—14, inserted at the base of the bract; filaments forked; anthers hairy at the top; pistillate flowers in dense strobilaceous cones; a pair to each deciduous bract, each enclosed
254 in a membranous urceolate initially open involucre, the 2-locular 2-ovuled ovary united with the perianth and terminating in a short ciliate limb; style short, with 2 filiform stigmas; fruit a 1-locular 1-seeded smooth nutlet enveloped in an accrescent scarious closed cupuliform involucre; seeds exalbuminous, with fleshy epigeous cotyledons; leaves alternate, short-petioled, with deciduous stipules, simple, deciduous, pinnately veined, irregularly or doubly dentate, acuminate; buds alternate, sessile, scaly.

Ostrya kieviensis Schmalh. in Lower Oligocene layers of M. Dnp. (Ekaterinopol'e in former Kiev Province). — O. palaeocarpinoides Bugaczev. in Middle Miocene layers of Transc. (Nakhichevan).

1\. **O. carpinifolia** Scop. Fl. carniol. ed. II (1772) 224; Boiss. Fl. Or. IV, 1178. — Ic.: Rchb. Ic. Fl. Germ. XII (1850) f. 1299; H. Winkl. in Pflanzenr. 19 H. (IV, 61) 1904, 21. — Exs.: HFR No. 1741.

A tree to 15—20 m high and 0.5 m in trunk diameter; young branches more or less covered with long hairs, brown; bark of the trunk dark brown, coming off in flakes; buds ovoid, with obtusish scales; leaves ovate-oblong, subcordate or truncate at base, acuminate at apex, sharply double-toothed, at first appressed-hairy, finally with hairs on the veins, these 15—18, the short petiole densely hairy; staminate aments pendulous, with brownish-red bracts, developing in fall; pistillate aments strobilaceous, ovoid-cylindric; bracts soon caducous, ovate, acute; involucre oblong-ovoid, acuminate, compressed, pale reddish, hairy at base, with a tuft of hairs at apex, many-nerved; nutlet ovoid, pointed, slightly compressed, yellowish-brown, smooth, glabrous, obscurely nerved, crowned with remnants of the ciliate perianth limb; involucre several times as long as the nutlet. April—July. (Plate XII, Figure 1).

A rather rare species in forests of the lower mountain levels, up to 1,000—1,500 m. — Caucasus: Cisc. (Khodza Valley, Baksan), W. and E. Transc. **Gen. distr.**: Med., Bal.-As. Min. Described from S. Europe.

Economic importance. A beautiful tree deserving cultivation in garden and pleasure grounds. The very firm wood is suitable for production of various small articles. The bark, with a tannin content of 6.5%, is useful for tanning.

* Name used by Theophrastus.

Genus 362. **CARPINUS*** L.

L. Gen. pl. (1737) 292 (excl. Ostrya).

Monoecious trees and large shrubs; staminate flowers in cylindric lateral aments, with bracts, destitute of perianth; stamens 4—12, at the
255 base of the calyx; filaments forked; anthers hairy at the top; pistillate flowers in terminal aments, a pair subtended by each of the small early deciduous bracts; perianth, with a short limb toothed on the margin, united with the ovary and furnished with an accrescent leafy involucre; ovary 2-locular, inferior; ovules solitary in each locule; style short, with 2 filiform stigmas; fruit a 1-locular 1-seeded almost woody ridged nutlet enclosed in a leafy involucre; seeds exalbuminous, with fleshy epigeous cotyledons; leaves alternate simple, deciduous, ovate, pinnately veined, doubly toothed, with early deciduous leaflets, folded in bud along the lateral veins; buds sessile, scaly.

The genus is extremely widespread in fossil condition. It appears in Asia from the earliest phases of the Tertiary period; in Europe later, attaining wide distribution in the Miocene. Sometimes erroneously named Carpinus or appearing with variants, e. g., Carpinophyllum.

Carpinophyllum pyramidalis Goepp. in Upper Dui series of Sakh. (Akhengy) and in the Paleocene of Uss. (Rechnoi).

Carpinus betuloides Unger in Tertiary (Pliocene?) layers of Alt. (Bukhtarma, Chingistai). — C. betulus L. in Postpliocene layers of U. Dnp. (Timoshkovichi) and in interglacial formations (Kletsova) of U. V. (Likhvin), Cisc. (Zheleznovodsk), and Dag. (Tarku-tau). — C. grandis Unger; reports for early Paleocene of the European part of the USSR need confirmation; in Eocene (?) layers of V.-Don (Lava in Kursk Region); in the Oligocene of M. Dnp. (Mogil'no ?, Volyanshchina, Ryzhany); in Sarmatian formations of M. Dnp. (Bondarevka in Podolia), Bl. (Orekhov, Krynka), and Crim. (Yarglyki); Maiotis formations of Bl. (Khadzhibei); Sarmatian of Cisc. (Apsheronskaya), and Paleogene formations of Asia: Ob (Tomsk), Ar.-Casp. (Kara-Sandyk, Dzhilan), Uss. (De-Friz, Sikhote-Alin, Nikol'skii mine, Amagu), Okh. (Taui Bay), Sakh. (Mgachi, Dui, etc.), Kamch., Ang.-Say. (Mishikha, SE Lake Baikal, etc.). Reports for Upper Cretaceous layers of Bureya lowlands (Ze.-Bu.) need confirmation as regards both species and genus. — C. pyramidalis Heer in Sarmatian formations of Cisc. (Adagum), in the Pliocene of W. Transc., and the Miocene of E. Transc. (Ossetia). — C. sp. in Tertiary layers of Ang.-Say. (Baikal). Balkh. (Ashutas), and Sakh. (various locations).

1. Involucral bracts in fruit ovate, with toothed margin . . . . 2.
\+ Involucral bracts in fruit lobed or obliquely triangular, with a small lobe at base . . . . 3.
2. Leaves 5—11 cm long, cordate at base, the teeth point-tipped; bracts with a basal auricle covering the nutlet . . . . 1. **C. cordata** Blume.
\+ Leaves 2—6 cm long, doubly toothed on the margin; bracts without auricle or lobes at base . . . . 2. **C. orientalis** Mill.
3. Bract 3-lobed, the lateral lobes about equal, one-third to half as long as the middle lobe . . . . 4. **C. betulus** L.

* The name for hornbeam used by Pliny and other ancient authors.

\+ Bract obliquely triangular, the outer (broad) margin coarsely toothed, the inner margin with a small lobe at base . . . . 3. **C. schuschaensis** H. Winkl.

256 Section 1. **DISTEGOCARPUS** (Sieb. et Zucc.) Sargent, Sylva North. Am. IX (1896) 60. — Distegocarpus Sieb. et Zucc. Fl. jap. fam. nat. II (1846) No. 798, pro genere. — Bracts of staminate flowers ovate-lanceolate, stalked; fruiting aments densely imbricated, with thin bracts.

1. **C. cordata** Blume, Mus. bot. Lugd.-bat. I (1849—51) 309. — Distegocarpus cordatus DC. Prodr. XVI, 2 (1868) 128. — Ostrya mandschurica Bud. ex Trautv. in A.H.P. IX(1884) 166. — Ic.: Pflanzenr. H. 19 (IV, 61) 1904, 27; Kom. and Alis., Opred. rast. Dal'nevost. kr. I (1931), Plate 132.

A tree to 15 m high, with spreading head and grayish-brown fissured bark; young branchlets hairy, becoming glabrous; leaves ovate, acuminate, narrowly cordate at base, to 11 cm long and 6 cm broad, commonly double-toothed on the margin; the lateral veins terminating in the large teeth, the intermediate teeth smaller, all teeth long-pointed; veins 15—25, impressed above, prominent beneath, the 2—4 lower ones giving rise to subsidiary veins, the lower surface hairy on the veins, the upper surface only on the midrib, the young leaf hairy also between the veins, the hairy petioles 1—3 cm long; stipules soon deciduous, lanceolate, to 3 cm long, 0.4 cm broad, clothed with long hairs; staminate aments 4—8 cm long, loose, borne on a densely hairy stalk to 1 cm long; bracts scarious, oblong, acuminate, tawny, long-ciliate on the margin and at apex, 4—7 mm long, about twice the length of the stamens; pistillate aments cylindric, dense, in fruit 6—10 cm long and to 5 cm broad, with imbricated accrescent bracts, borne on a stalk 2—3 cm long; bracts of fruiting aments ovate, acuminate, 2—3 cm long, 0.7—1.5 cm broad, reticulate-veined, on the margin and at apex irregularly and coarsely sharp-toothed, at base with a broad toothed auricle covering the nutlet, covered with bristles on the outside at base; nutlet ellipsoid, 0.8—1.3 cm long, 2—3 cm broad. May; fr. July—August.

A tree of second growth in shady mixed forests. — Far East: Uss. (extreme SW). **Gen. distr.**: Jap.-Ch. (E.). Described from Siebold's Japanese collections.

Economic importance. The firm white wood of this hornbeam is used for turning and for preparation of various small articles. The nutlets are sometimes pressed to provide an edible oil.

Section 2. **EUCARPINUS** Sargent, Sylva North. Am. IX (1896) 40. — Bracts of staminate aments broadly ovate, subsessile; fruiting aments loose, with coriaceous scales.

257 2. **C. orientalis** Mill. Gard. dict. ed. 7 (1759) No. 3. — C. duinensis Scop. Scop. Fl. carniol. II (1772) 243, t. 60; Ldb. Fl. Ross. III, 2 (1849—51) 586. — C. minor Pall. Bemerk. auf einer Reise II (1801) 95. — Ic.- Radde-Fom. in Mém. des Sc. Phys. et Math. Acad. de l'Ukraine XV, livr. I (1929) tab. VIIa. — Exs.- HFR No. 1740.

(259)

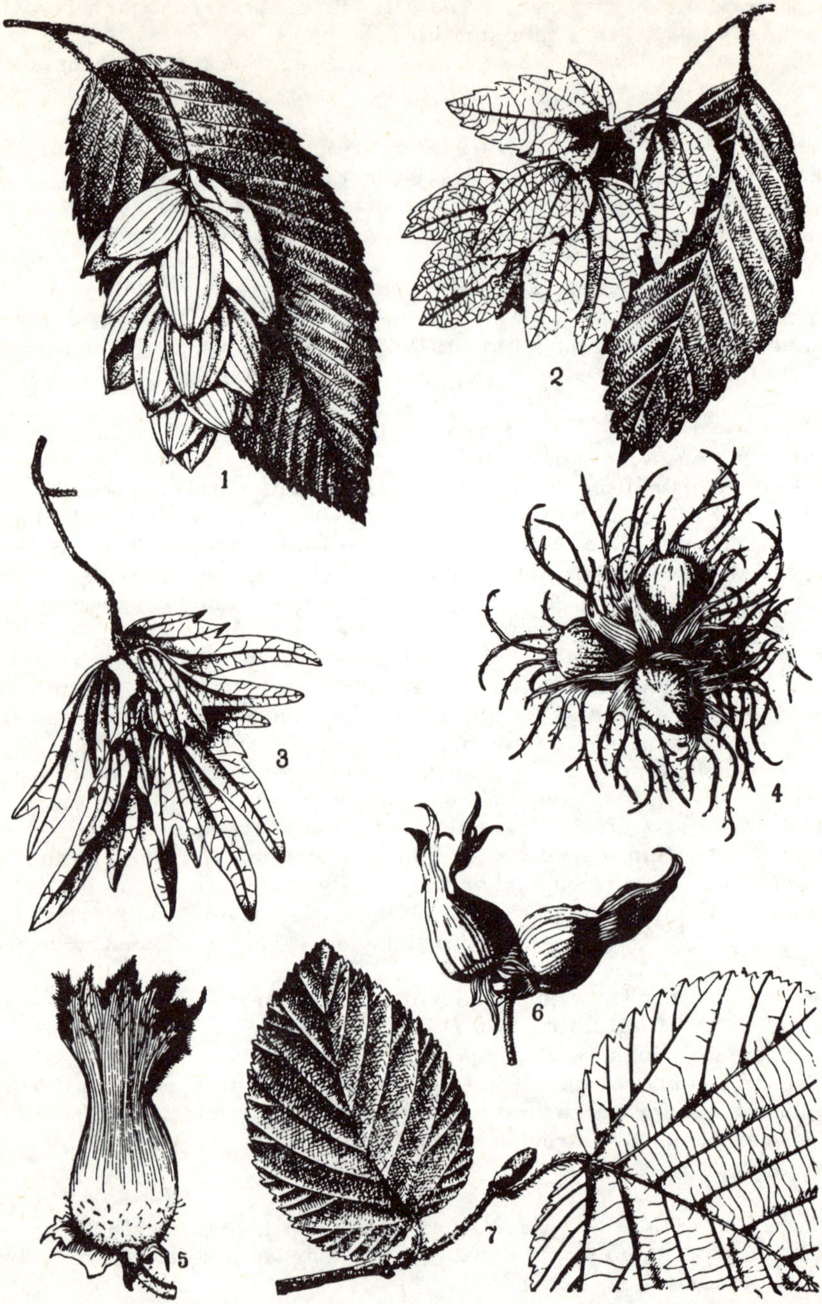

PLATE XII. 1. Ostrya carpinifolia Scop. – 2. Carpinus schuschaensis H. Winkl. – 3. C. oxycarpa H. Winkl. – 4. Corylus colurna L. – 5. C. pontica C. Koch. – 6. C. maxima Mill. – 7. C. colchica Alb.

A tree or shrub, with gray bark; young branchlets and leaf petioles villous; leaf blades ovate-lanceolate to ovate, obtusish or slightly cordate at base, acuminate or rarely rounded at apex, doubly toothed on the margin, with 10 — 15 veins impressed above, pubescent on midrib and on lateral veins beneath, densely hairy at vein axils, 2 —5 cm long, 1.2 —2.5 cm broad, the petiole to 8 mm long; stipules lanceolate, hairy on the outside; staminate aments dense, 1.5 cm long; bracts obtusish, almost flat, ciliate on the margin, twice the length of the hairy-tipped anthers; pistillate aments dense, ovaloid or oblong-ovaloid, 3 — 8 cm long, 2 —3.5 cm broad, borne on a stalk 12 — 18 mm long; bracts of fruiting aments ovate, flat, without lobes, acuminate, 1.4 — 2.2 cm long and 7 — 13 mm broad, irregularly toothed on the margin, the 5 — 8 nerves glabrous above, pubescent beneath; nutlet ovaloid, slightly compressed, brown, somewhat lustrous, with 8 — 12 rather faint ridges, hairy at the top. Fl. April; fr. June — July.

The lower mountain zone, mainly wood margins, very often on dry stony slopes. — European part: Crim.; Caucasus: Cisc. (south of the line Anapa — Batalpashinsk — Makhach-Kala), Dag., W., S., and E. Transc. (S. slopes of the Greater Caucasus in Azerbaijan — Nukha, Shemakha). **Gen. distr.**: E. Med. (E.), Balkans. Described from S. Europe.

Economic importance. An outstandingly hardy species with modest requirements, much branched, profusely leafy, very suited for green fencing. Wood very firm, suitable for making small articles. The leaves and young branches provide winter feed for livestock. Almost unknown in cultivation, but undoubtedly of interest.

3. **C. schuschaensis** H. Winkl. in Pflanzenr. 19 Heft (IV, 61) (1904) 32. — Ic.: Mém. des Sc. Phys. et Math. Acad. de l'Ukraine XV, livr. 1 (1929) tab. VIIa.

A tree (?); young branches reddish-brown, covered with soft hairs; leaves ovate, acuminate, 5 — 7 cm long, scarcely cordate at base, irregularly double-toothed on the margin, with 10 — 14 veins impressed above and hairy beneath, the pubescent petioles 6 — 10 mm long; fruiting aments loose, 4 — 6 cm long, 3 — 3.5 cm broad; bracts of fruiting aments 2 — 3 cm long and 1 — 1.5 cm broad, obliquely triangular, more or less acuminate, obtusish
258 at apex, the broad (outer) margin doubly toothed, the narrow (inner) margin simple-toothed or subentire, with a small pointed lobe embracing the nutlet; nutlet ellipsoid, flattened, prominently ridged in upper part, puberulous and glandular-dotted. Fl. March — April; fr. June — July. (Plate XII, Figure 2).

Distributed as undergrowth at lower mountain levels; in the forests of Talysh mountains occupying higher levels. — Caucasus: E. Transc. (Shemakha, Nukha, Geokchai), S. Trans. (Karabakh), Tal. **Gen. distr.**: Iran. (NW). Described from Azerbaijan.

4. **C. betulus** L. Sp. pl. (1753) 998; Ldb. Fl. Ross. III, 2, 587; Shmal'g., Fl. II, 428. — Ic.: Rchb. Ic. Fl. Germ. XII (1850) t. 632, 633. — Exs.: HFR No. 77.

A tree to 25 m high, with a spreading head, sometimes slightly twisted trunk; bark brown, smooth, slightly fissured; young branches brown, lustrous, sometimes hairy; annotinous branchlets sericeous; buds acuminate, with reddish-brown glabrous ciliate-margined scales; leaves ovate, acuminate, slightly and asymmetrically cordate or rounded at base, doubly toothed on the margin, with 10—15 pairs of veins, to 15 cm long and 6 cm broad, the petiole 15 mm long; young leaves sericeous, finally glabrous above, the lower surface with scattered hairs, densely hairy at the vein axils; veins impressed above; staminate aments loose, to 6 cm long and 1 cm broad; bracts yellowish, reddish-brown and ciliate on the margin, acuminate, with 5—7 stamens at base; anthers pubescent at the top; pistillate aments to 15 cm long and to 6 cm broad; bracts of fruiting aments coriaceous, 3-lobed, to 6 cm long, with entire or more or less toothed margin, 3-nerved; middle lobe pointed, 2—3 times as long as the lateral lobes, these rather deeply toothed inflexed and embracing the nutlet; nutlet ovaloid, slightly flattened, lustrous, brown, with 7—11 ridges, crowned with remnants of the perianth. Fl. March—April; fr. June—July.

In mixed broad-leaved forests, sometimes forming pure stands; in E. Transc. rising in the mountains to 2,000 m. — European part: U. Dnp., M. Dnp., Bl., L. Don (W. upper reaches of the Mius River), Crim.; Caucasus: Cisc. (S. of Stavropol), Dag., W., S., and E. Transc., Tal. **Gen. distr.**: Scand. (S.), Centr. Eur., Bal.-As. Min., Iran (SW). Described from Europe.
261 Type in London.

Economic importance. Lumber of little value for construction purposes because of the crookedness of the trunk, but provides material for the making of small articles and for fuel. The durability of the wood renders it suitable wherever great resistance to friction is required, e. g., propellers, vanes of mill wheels, cobblers' wooden rivets, agricultural tools, carts, wheel flanges, etc. Valued in forestry as a profusely suckering coppice tree. Young shoots and foliage are used as feed for livestock. The bark is employed in tanning, while its deeper layers are sometimes used to provide a yellow dye for wool.

Note. Winkler described as distinct species C. oxycarpa, C. hybrida, and C. grosseserrata from Karabakh and Talysh, while Radde-Fomin described C. geokczaica from Geokchai in Azerbaijan. These species differ in the shape of the involucre and the extent to which it is dissected, the shape of the nutlet, its hairiness, and ridging. In our opinion, all these forms are hybrids between C. schuschaensis and C. betulus. Their identification is rendered particularly difficult by the fact that they hybridize in turn among one another.

a. **C. oxycarpa** H. Winkl. l. c. — Ic.: Radde-Fom. l. c. t. III.

Bracts subfalcately curved, with deeply cut outer margin and a small basal lobe on the inner side; nutlet taper-pointed at the top. (Plate XII, Figure 3).

Described from Shusha; cotype in Leningrad.

There is a close similarity between var. betuloides established by Winkler and the typical C. betulus. Various other specimens known from the Caucasus differ from the latter in some features associated with C. oxycarpa.

b. **C. hybrida** H. Winkl. l. c.

Bracts similar to those of C. schuschaensis, but the outer margin is long-hairy at base, the midnerve sending out 2 or 3 lateral nerves; nutlet pointed or rarely somewhat blunted, with short hairs at the top.

Described from Karabakh; cotype in Leningrad.

Winkler proposes that this plant is a hybrid between C. orientalis and C. schuschaensis; in our view it hardly differs from the latter.

c. **C. grosseserrata** H. Winkl. l. c. — Ic.: Radde-Fom. l. c. tab. VIII.

Bracts almost 3-lobed, large, deeply dissected, coriaceous. Described from Radkan (Iran); cotype in Leningrad.

This form, occurring in the USSR in Talysh, is manifestly of hybrid origin.

262

d. **C. geokczaica** Radde-Fom. l. c. — Ic.: ibid. tab. VI, a.

Bracts narrow, not lobed; nutlet ovaloid-cylindric.

Described from Geokchai. Type in Leningrad.

Hybrid origin of this form is possible; it does not appear to us, however, to differ from C. schuschaensis.

Genus 363. **CORYLUS** * L.

L. Gen. pl. (1737) 730.

Monoecious, often big shrubs, or rarely trees; staminate flowers in pendulous cylindric aments, consisting of 4 forked stamens united with the bract and a pair of bracteoles, without perianth; anthers 1-locular, hairy at apex; aments developing in fall; pistillate flowers in particolored dichasia (a third flower abortive), subtended by a bract, the dichasia enclosed in an imbricated globose bud from which the reddish stigmas protrude at anthesis; ovary inferior, 2-locular, ovule development confined to one of the locules; style bearing 2 filiform stigmas; perianth inconspicuous, adherent to ovary, with a limb of 4—8 unequal teeth, each flower with 2 bracteoles; fruit a nut, enclosed in a leafy tubular or much dissected involucre; nut ligneous, 1-locular, 1-seeded; seed without endosperm, the fleshy plano-convex cotyledons raised to the surface in germination; leaves alternate, angulinerved, simple, deciduous, ovate, short-acuminate, doubly toothed; stipules lanceolate, early deciduous.

The genus Corylus was widely distributed in the Tertiary, or possibly even in the Upper Cretaceous period, throughout the deciduous forest belt with temperate climate, reaching beyond the limits of the Arctic Ocean. At a later stage it appears in the region of subtropical and tropical floras of European Paleocene.

C. avellana L. in Tertiary layers (Pliocene ?) of Alt. (Chingistai); in Postpliocene layers of Lad.-Ilm. (Luga) U. Dnp. (Mikulino, Drozhzhino, Dubrova), U. V. (interglacial formations of Likhvin, U. V. (interglacial, Troitskoe), U. V. (interglacial, Galich), U. Dnp. (Murava, in Minsk Region), M. Dnp. (tuffs, Mushkutinets), Cisc. (tuffs, Mashuka), E. Transc. (Khevsha R. in Georgia); also forms oblonga Anderss. in the Postpliocene of

* Name for hazel used by Virgil.

U. V. (interglacial, Likhvin), Pliocene of E. Transc. (Shiraki); f. silvestris Anderss. in the Postpliocene of U. V. (Belolipki, Tula Region). — C. colurna L. in the Pliocene of E. Transc. (Shiraki). — C. insignis Heer in Ob (Tomsk Region), and Paleogene formations of Uss. (Rechnoi) and Sakh. (near Aleksandrovsk). — C. Macquarrii (Forbes) Heer in Sarmatian formations of Bl. (Krynka), in the Upper Cretaceous of Ze.-Bu. (Tsagayan; the taxonomic identity needs confirmation), and in Paleogene formations of Uss. (Sikhote-Alin, Amagu, Novokievsk, Fatashi), Sakh. (Mgachi, Dui, etc.), Kamch., Balkh. (Ashutas), and Okh. (vicinity of Taui Bay); also reported as var. macrophylla Heer from Sakh. (near Aleksandrovsk). — C. turgaica Pojark. in Oligocene layers of Ar.-Casp. (Dzhilan, Kara-sandyk, Yar-kue).

Corylus, without determination of species, has been recorded in Cretaceous layers of Ze.-Bu. (Sagib. Boguchan), in Tertiary layers of Uss. (Amagu) and Balkh. (Ashutas), and in the Postpliocene of U. V. (Troitskoe).

263
1. Involucre leafy . . . . 2.
\+ Involucre contracted above the nut, setose . . . . 7.
2. Involucre of 2 bracts . . . . 3.
\+ Involucre of 1 bract, more or less deeply cut on one side . . . . 5.
3. Bracts of the involucre deeply cut into unequal lobes, these irregularly toothed, lanceolate or linear-lanceolate, recurved, or palmate . . . . . 4.
\+ Bracts of the involucre shallowly cut (to 1/8 — 1/6) into subequal lobes, these triangular, sometimes subentire; leaves truncate, almost 2-lobed, short-acuminate . . . . 3. **C. heterophylla** Fisch.
4. A tree to 25 m high; involucre several times as long as the nut, dissected into long often recurved lobes . . . . 1. **C. colurna** L.
\+ A shrub to 4 — 6 m high; involucre not exceeding (mostly shorter than) the nut, cut into unequal lobes . . . . 2. **C. avellana** L.
5. Lower part of the involucre, leaf petioles, and young branchlets covered with stalked glands . . . . 6.
\+ Plant devoid of stalked glands; leaf petioles and young branchlets covered with whitish bristles . . . . 5. **C. colchica** Alb.
6. Involucre campanulate; nut broad, subglobose; shrub to 5 m high . . . . 4. **C. pontica** C. Koch.
\+ Involucre fleshy below, firmly embracing the nut, gradually somewhat contracted above, lobed at the top; nut ovoid to subcylindric; shrub to 10 m high . . . . **C. maxima* Mill.
7. Involucre to 6 cm long, thrice (rarely twice) the length of the nut . . . . 6. **C. manshurica** Maxim.
\+ Involucre shorter, twice the length of the nut . . . . 7. **C. brevituba**.

1. **C. colurna** L. Sp. pl. (1753) 999; Boiss. Fl. Or. IV, 1176. — Ic.: H. Winkl. in Pflanzenr. H. 19 (IV, 61) 1904, 49. — Russian: leshchina drevovidnaya [arboraceous hazel], medvezhii orekh [bear-nut].

A tree to 25 m high; trunk to 30 cm in diameter; young branches gray; annotinous tawny, commonly covered with bristles or with stalked glands; older covered with gray corky bark; buds ovoid, with reddish pubescent scales; leaves 7 — 12 cm long and 5 — 8 cm broad, cordate at base, obovate to suborbicular, acuminate, doubly toothed, when young pubescent above, pilose or stipitate-glandular beneath, especially on the veins; petiole 1 — 3 cm long; hairy or rarely with stalked glands; stipules lanceolate, acuminate;

264 staminate aments to 8 cm long, with ovate acuminate bracts; fruits 3—7 together; involucre considerably surpassing the fruit, much dissected into narrow lanceolate recurved often toothed lobes; nut flattened laterally, sparsely hairy at apex. Fl. March—April. (Plate XII, Figure 4).

Shady mixed forests of the middle mountain zone (1,000—1,800 m). — European part: Crim. (cultivated); Caucasus: Cisc. (upper course of the Kuban River), W. Transc. (Krasnaya Polyana), E. Transc. (Belyi Klyuch, Borzhomi, Gori, former Nukha County), Tal. **Gen. distr.**: Bal.-As. Min., Iran., Ind.-Him. Described from Greece. Type in London.

Economic importance. Firm, durable, beautiful wood, with rosy tint, used for making furniture and turning small articles. The nutshell is very hard, but the kernels are used for food like the ordinary hazelnuts and are sometimes pressed for oil.

2. **C. avellana** L. Sp. pl. (1753) 998; Ldb. Fl. Ross. III, 588; Shmal'g., Fl. II, 427. — Ic.: Vol'f and Palibin, Opred. derev'ev i kustarn. II (1904) 176.

A shrub; young branches gray, glabrous; annotinous branches yellowish-gray, hairy or glandular-setose; buds subspherical or ovoid, slightly compressed; bracts rounded, glabrous or sparingly pubescent, ciliate on the margin; leaves orbicular or ovate, asymmetrically cordate at base, acuminate at apex, irregularly double-toothed on the margin, glabrous, with 8—12 pubescent veins, darker above, 6—12 cm long, 5—9 cm broad; petiole 1—2 cm long, glandular-hispid; stipules oblong-ovate, obtusish, hairy; staminate aments to 5 cm long, the bracts densely pubescent with ciliate margin; anthers glabrous, with a tuft of hairs at the apex; fruits 1—5 together; involucre campanulate, open, consisting of 2 bracts cut into broad toothed lobes, stipitate-glandular below, about as long as the nut; nut subglobose or ovoid. Fl. February—April; fr. September.

In open mixed or broad-leaved forests, mostly as undergrowth, and often at the margins. — European part: Lad.-Ilm. (N. border: Leningrad — Tikhvin—Belozersk), Dv.-Pech. (extreme SW), U. Dnp., U. V., V.-Kama (SW of the line Osa—Krasnoufimsk—Kananikol'skii Zavod; in the south of the Nakas Range), M. Dnp., Transv. (N.), Bl., V.-Don (SW, in the NE along the Volga, as far south as Khvalynsk), Crim.; Caucasus: Cisc., Dag., W., S. (Karabakh), and E. Transc., Tal. **Gen. distr.**: Scand. (except the north), Centr. Eur., Atl. Eur., Med., Bal.-As. Min. Described from Europe. Type in London.

265 **Economic importance.** Until recent times the wild hazelnut was used in the USSR almost exclusively as a delicacy in roasted form. It may, however, be of considerable importance as a food article. Up to twenty varieties of nuts are known in W. Europe and in America, and the ancestor of most of them is the wild hazelnut. Most of these varieties have relatively large fruit; they are cultivated commercially over thousands of hectares. Certain varieties are of purely ornamental value (the forms pendula, quercifolia, laciniata, aurea, atropurpurea).

The nuts yield one of the best vegetable oils. The branches are used for the making of pipe stems, hoops, and canes; they are also woven into baskets, etc. Hazelnut cultivation is confined in the USSR to certain districts of the Crimea, Transcaucasia, and the Black Sea coast of the

Caucasus. On account of its high profitability, it could be considerably extended northward, and for this purpose suitable European varieties would have to be selected.

Note. The cultivated forms include some rare hybrids between common hazel and C. *colurna*, such as C. *intermedia*, C. *colurnoides*, and C. *arborescens*, which display intermediate characteristics.

3. **C. heterophylla** Fisch. ex Besser in Beibl. zur Flora I (1834) 24; Turcz. Fl. baic.-dah. II, 1 (1854) 134. — C. *Avellana* β *dahurica* Ldb. Fl. Ross. III (1851) 558. — Ic.: Trautv., Russk. Fl. (1844) I, Plate 4; Kom. and Alis., Opred. rast. Dal'nevost. kr. I (1931) 433. — Goldi: atsieng-kura; Chinese: chendza.

A shrub not more than 2 m high; young branches pubescent; leaves broadly obovate, slightly cordate at base, terminating in point not usually surpassing the broad lateral lobes, irregularly toothed on the margin, green on both sides, glabrous above, pubescent on the veins beneath, 6—10 cm long and about as broad; petiole to 3 cm long, hairy and sparsely stipitate-glandular; fruits 2 or 3 together at the ends of branchlets, on a stalk to 3 cm long; involucre campanulate, overtopping the nut, the rounded involucral bracts cut into 6—9 subequal toothed lobes, with stalked glands below; nut globose, flattened at the top, slightly broader than long. Fl. May; fr. August.

Wood margins and mountain slopes, forming dense thickets. — E. Siberia: Dau. (Argun Valley), Far East: Ze.-Bu. (S. of the line Albazin—Zeya-pristan), Uss. **Gen. distr.**: E. Mongolia, Manchuria, N. China. Described from the Argun River. Type in Leningrad.

266 **Economic importance.** The nuts of this hazel are harvested commercially. The modest requirements greatly facilitate cultivation. The most elementary care of wild stands results in a considerable improvement in yield and quality of the nuts. Cultivated in Japan.

4. **C. pontica** C. Koch in Linnaea XXII (1849) 329; Ej. Dendrol. II, 2 (1873) 11. — C. *Avellana* var. *pontica* H. Winkl. in Pflanzenr. H. 19 (IV, 61) (1904) 47, 51. — C. *colurna* Boiss. Fl. Or. IV (1879) 1176, p. p., non L. — C. *imeretica* Kem.-Nath. in Travaux Inst. Bot. Tiflis (1934) 111. — C. *byzantina* hort.

A shrub; young branches and leaf petioles stipitate-glandular and densely pubescent; leaves rounded, short-acuminate, cordate at base, irregularly double-toothed on the margin, at first pubescent above, finally with soft hairs only on the lower surface; stipules oblong-lanceolate; fruits crowded; involucre 1-bracted, pubescent, glandular below, greatly exceeding the fruit, wide-open at the top, campanulate, cut at apex into toothed lobes, deeply dissected on one side; nut broad, compressed at base, grayish-pubescent at the top. Fl. April; fr. August. (Plate XII, Figure 5).

Lower mountain zone. — Caucasus: W. Transc. (Adzharia); European part: Crim. (cultivated). Described from the Pontic Range. Type in Berlin.

Note. Widely cultivated in S. Europe under the name Constantinople or Byzantine nut since very remote times. The species was described in

1849, but its existence has been doubted by many taxonomists up to the present time. As a matter of fact, the species exists in a natural state and specimens from the former Artvin District and the vicinity of Batumi are preserved in the Herbarium of the Botanical Institute of the Academy of Sciences. The illustration (Plate XII, Figure 5), which we have borrowed from Koehne, Deutsche Dendrol., conforms to the type and the nuts which we have seen resemble it closely.

The new species C. *imeretica*, described by Kemularia-Nathadze (l. c.), scarcely differs from C. *pontica* C. Koch, at least insofar as can be deduced from the description. We have not seen the specimens.

In the southern part of W. Transcaucasia there are under cultivation some twenty cobnut varieties of C. *pontica* ancestry.

***C. maxima** Mill. Gard. Doct. ed. 7 (1759) No. 3. — C. *tubulosa* Boiss. l. Or. IV (1879) 1176. — Exs.: Fl. exs. Austro-Hung. No. 3872. — Also referred to as Lombardy nut.

A big shrub or a tree, to 10 m high; annotinous branches sometimes glabrous, mostly stipitate-glandular; leaves orbicular, short-acuminate, very often red, pubescent beneath, 7 — 12 cm long and 6 — 10 cm broad;
267 petiole 1 — 2.5 cm long; staminate aments to 10 cm long and to 1 cm in diameter; fruits (solitary) 3 — 6 together on a stalk to 3 cm long; involucre in lower part somewhat fleshy, velutinous, stipitate-glandular, enveloping the nut, splitting in maturity, somewhat contracted above and from here cut into broad toothed lobes; nut ovoid, sometimes subcylindric, acuminate. Fl. March; fr. September. (Plate XII, Figure 6).

European part: Crim. (cultivated); Caucasus: W. Transc. (cultivated along the Black Sea coast). **Gen. distr.**: S. Europe (Istria, Banat), Bal.-As. Min. Described from cultivated specimens.

Economic importance. Cultivated in the Crimea and on the Black Sea coast for more than a century. It may be assumed that cultivation could be extended into the S. part of the Ukraine and into many other parts of the Caucasus. The possibility of cultivation is limited by a minimum temperature requirement, the trees dying off at temperatures below 18°C. Of some interest for ornamental use is the red-leaved form (f. *atropurpurea*) that occurs in the parks of the Caucasus, Kharkov, and the Chernigov area.

5. **C. colchica** Albov, Prodr. Fl. Colch. (1895) 219. — Ic.: H. Winkl. in Pflanzenr. H. 19 (IV, 61) (1904) 53.

A shrub not more than 1 m high; young branches gray, rather densely covered with silky hairs, those of the preceding year glabrous; buds oblong-ovoid, large, reddish-brown, pubescent; leaves ovate, rounded or sometimes subcordate at base, short-acuminate, doubly sharp-toothed on the margin, with scattered hairs above, densely pubescent on the veins beneath, 5 — 7 cm long, 3.5 — 4.5 cm broad; veins 11 or 12, impressed above; petiole densely pubescent, ca. 1 cm long; stipules lanceolate, acuminate; staminate aments reddish-brown, solitary or in pairs in the leaf axils, with ovate acuminate ciliate bracts; fruits solitary or several together; involucre 1-bracted, appressed to the nut, hairy, contracted above the nut into a short beak crowned by a lacerate sericeous limb, often not completely covering the top of the nut; nut short-ovoid, 13 mm long and 11 mm in diameter. Fl. April; fr. September. (Plate XII; Figure 7).

Forming thickets at the upper timberline. — Caucasus: W. Transc. (Mingrelia: Migaria, Dzhvari; Abkhazia: Bzybskii Range). Endemic. Described from Mingrelia. Type in Geneva; cotype in Leningrad.

Economic importance. The nuts are small and of little interest as such. However, this species, growing in the rather severe conditions of the upper timberline, could undoubtedly be of value for hybridization with large-fruited species and varieties.

268

6. **C. manshurica** Maxim. in Bull. Acad. Pétersb. XV (1857) 137; Primit. Fl. amur. (1859) 241. — C. r o s t r a t a var. m a n s h u r i c a Rgl. in Mém. Acad. Pétersb. VII, 4 (1862) 129; H. Winkl. in Pflanzenr. H. 19 (IV, 61) (1904) 52. — Ic.: H. Winkl. ibid. 49; Kom. in Bull. Jard. Bot. Princ. XXVIII (1929) 217.

A shrub 3—4 m high, with several stems from the ground, with fissured bark; leaves rounded-ovate or elliptic, in upper part cut into toothed lobes, taper-pointed, sharp-serrate on the margin, more densely pubescent at the axils of veins beneath, the upper surface pubescent only in young leaves; staminate aments rarely 2, mostly 3 or 4 together; involucre narrow, long-cylindric, tubular, to 6 cm long, 2—3 times as long as the nut, fimbriate on the margin, densely clothed on the outside with rufous bristles; nut sharp-pointed. Fl. May; fr. September.

As underwood in mixed and coniferous forests, sometimes forming woods. — Far East: Ze.-Bu. (E. of Bureya), Uda, Uss. **Gen. distr.**: Ch.-Jap. Described from Amur. Type in Leningrad.

Economic importance. The bristly vesture of the involucre prevents extensive use of this and the next species for food. They are nevertheless of interest for nut production.

7. **C. brevituba** Kom. in Bull. Jard. Bot. Princ. XXXIII (1929) 219. — Ic.: ibidem, 217.

A big shrub or a tree to 3 m high, with dark gray fissured bark and pubescent young branches; leaves often cordate at base, rounded-ovate, in upper part sometimes almost lobed, toothed on the margin, the lobes mostly little in evidence and thus the leaves apparently doubly toothed, paler beneath and hairy on the veins, to 10 cm long and to 9 cm broad; petiole pubescent, to 2.5 cm long; fruits 2 or 3 together, crowded, 1 or 2 abortive; involucre tubular, the broad tube 3—4 cm long, 1.5 times as long as the nut, densely setose outside, lobed on the margin; nut puberulous, obtuse-tipped. Fl. May; fr. August.

In sparse mixed forests. — Far East: Ze.-Bu. Endemic. Described from the Zeya-Bureya watershed, upper Tom River, upper Bysa River, and the Bysa-Selemdzha watershed. Type in Leningrad.

N o t e. A species scarcely different from the preceding.

269 Tribe II. **BETULEAE** DÖLL.

Döll. Zur. Erklär. des Laubknosp. der Ament. (1848) 10; Aschers., Fl. Prov. Brandenburg (1864) 619.

Characters in the key

Key to Genera

1. Scales of fruiting aments thin, 3-parted, falling in maturity; stamens bifid . . . . 364. **Betula** L.

\+ Scales of fruiting aments woody, 5-parted, persistent in maturity; stamens scarcely parted . . . . 365. **Alnus** L.

Genus 364. **BETULA*** L.**

L. Sp. pl. (1753) 982; Spach in Ann. Sc. Nat. 2 sér., XV (1841) 182; Regel in DC. Prodr. XVI, 2 (1868) 161; H. Winkl. in Engl. Pflzr. H. 19 (IV, 61) (1904) 56. — Betulaster Spach, c., p. 198.

Staminate and pistillate flowers in aments; staminate flowers in very short-stalked dichasia, subtended by an ovate-rhombic thickish ciliate-margined reddish-brown scale; bracts 2, thickish, ciliate-margined; perianth 4-lobed, usually only the anterior lobe developed, the lateral reduced, the posterior rudimentary; stamens mostly 2, rarely 3, the anthers 2-parted, hence apparently 4 or 6; pistillate flowers in 3's, or by reduction of the middle flower in pairs, subtended by a scale; perianth absent; bracts 2, united with the calyx and forming a rigid 3-lobed caducous scale; ovary 2-locular, with 1 ovule in each locule; stigmas 2, included in the calyx; fruit (by abortion of one of the ovules) 1-seeded, flattened, with 2 membranous wings, crowned by the withered stigmas. Trees or shrubs with alternate leaves; staminate aments in clusters of 2—4, terminal or axillary; pistillate solitary, upright or spreading.

Fossilized Betula is reported already in Cretaceous formations of North America, although identity has not been confirmed. There is no doubt concerning its occurrence in the Eocene of Europe and Asia.

Betula alba L. in the Postpliocene of Lad.-Ilm. (Vyaz, Luga District) and U. Dnp. (Vyshegory station); in interglacial formations of U. Dnp. (Prechistaya, Smolensk Region); in the Postpliocene of U. V. (Troitskoe), Lena-Kol. (Berezovka), and Arc. Sib. (Novosibirskie Islands, Bol'shoi Lyakhovskii Island). — B. Brongniartii Etting. in Lower Tertiary layers of Sakh. (Dui, Mgachi, etc.), Uss. (Pos'et), and Okh. (Taui Bay area). — B. caudata Goepp. in the Pliocene of W. Transc. (Goderskii Pass). — B. elliptica Saporta in Lower Tertiary formations of Uss. (Sikhote-Alin) and in Upper Dui series of Sakh. (Dui). — B. gypsicola Saporta in
270 the Paleocene of L. Don (Ushi). — B. lenta Wilden. in Tertiary (Pliocene ?) formations of Alt. (Chingistai). — B. macrophylla Goepp. in Sarmatian formations of Bl. (Krynka). — B. nana L. in the Postpliocene of Lad.-Ilm. (Late Ice Age layers at Tosno River in the Leningrad area, the Postpliocene

* Name used by Pliny.

** Treatment by O. I. Kuzeneva.

formations of ancient Vyaz); in interglacial layers of U. V. (Galich); interglacial formations of U. Dnp. (Prechistaya, Vyshegory); and in the Postpliocene of Dv.-Pech. (near Vologda), U. V. (village of Il'ya-prorok), Ob (Demyanskoe), and Yen. (Lower Yenisei River). — B. prisca Etting. in Lower Tertiary formations of Uss. (Sikhote-Alin — Botchi); in Lower Tertiary formations (Sakh.: Dui and Mgachi). — B. odorata Bechst. in the Postpliocene of U. Dnp. (Kletsova). — B. pubescens Ehrh. in the second interglacial period of U. Dnp. (Kletzova); in the Postpliocene of Lad.-Ilm. (Petrozavodsk, Vyaz) and E. Transc. (Khevsha River, travertine deposits). — B. sachalinensis Heer in the Postpliocene of Uss. (Botchi) and Sakh. (Dui, Agnevo). — B. Sokolowii Schmalh. in Tertiary layers of Alt. (Chingistai, Pliocene ?). — B. verrucosa Ehrh. in the Postpliocene of Lad.-Ilm. (Vyaz) and U. V. (Troitskoe). — B. sp., unspecified, often recorded, beginning with Upper Cretaceous formations: in Tsagayan formations of Ze.-Bu. (Belogor'e); in Lower Tertiary formations of Balkh. (Ashutas), etc.; in the Postpliocene of U. Dnp. (Timoshkovichi, Tuganovichi, Murava, Mikulino, Drozhzhino), U. V. (Lyalovo, Troitskoe, Borok, former Tver Province*), Dv.-Pech. (Vologda), L. Don (Archedskaya woodland), and Yen. (Lower Yenisei).

Fossil birch wood has also been frequently recorded: in the Cretaceous system of V.-Don (Komarovo district) and in the Paleocene of L. Don (Osinovo).

1. Trees or rarely very big shrubs . . . . 3.
\+ Shrubs . . . . 2.
2. Leaves rounded, often broader than long, rounded or obtusish at apex, blunt-toothed, with 2—4 (5) pairs of veins (Section Nanae Rgl.) . . . 11.
\+ Leaves oblong, always longer than broad, acute or subacute, sharp-toothed, with 4—7 pairs of veins (Section Fruticosae Rgl.) . . . . . 14.
3. Fruiting aments subspherical or ovoid or elongate-ovoid or semicylindric; scales often elongated; leaves on both sides more or less ribbed, the veins impressed above . . . . 4.
\+ Fruiting aments more or less cylindric; scales rather short; leaves on both sides 5—7 (8)-ribbed . . . . 5.
4. Nutlets with membranous wings (Section Costatae Rgl.) . . . . 6.
\+ Nutlets marginate but not winged (Section Asperae Nakai) . . . . 7. B. Schmidtii Rgl.
5. Wing 2—3 times as broad as the nutlet, rarely about as broad (Section Albae Rgl.) . . . . 17.
\+ Wing at most half as broad as the nutlet (Section Dahuricae Rgl.). . . . . 16. B. dahurica Pall.
6. Fruiting aments retaining the scales until the following summer . . . 7.
\+ Scales shed in late fall or at the beginning of winter . . . . 8.
271 7. Leaves obovate, ovate, rounded, or elliptic, at base cuneate, (3.5) 5—8—10 cm long . . . . 1. B. Medwedewii Rgl.
\+ Leaves ovate, rounded or slightly cordate at base, 3—7 cm long . . . . 2. B. megrelica D. Sosn.
8. Leaves 4—14 cm long . . . . 9.
\+ Leaves 2.5—4.5 cm long . . . . 10.

* [Now Kalinin Region.]

9. Leaves with 7—10 pairs of veins, oval or ovate, short-acuminate . . . . 4. B. Ermani Cham.

\+ Leaves with (10) 12 — 14 pairs of veins, elongate-oval, long-acuminate . . . . 3. B. costata Trautv.

10. Wing much narrower than the nutlet; fruiting ament 20—25 mm long . . . . 6. B. Raddeana Trautv.

\+ Wing about as broad as the nutlet; fruiting ament to 15 mm long . . . . 5. B. Prochorowii Kuzen. et Litw.

11. Wings broader than the nutlet; bracts and nutlets large . . . . 11. B. Middendorffii Trautv. et Mey.

\+ Wings narrower than the nutlet; bracts and nutlets small . . . . 12.

12. Wings usually much narrower than the nutlet; leaves mostly flabellate . . . . 10. B. rotundifolia Spach.

\+ Wings one-fourth to one-half as broad as the nutlet . . . . 13.

13. Branches glandular; wings one-fourth to one-half the breadth of the nutlet . . . . 9. B. exilis Sukacz.

\+ Branches eglandular; wings one-fourth to one-third the breadth of the nutlet . . . . 8. B. nana L.

14. Wings broader than or rarely as broad as the nutlet; leaves cuneate at base . . . . 15. B. Gmelini Bge.

\+ Wings narrower than the nutlet; leaves rounded or rarely broad-cuneate at base . . . . 15.

15. Wings slightly narrower than or half as broad as the nutlet; bracts with broad lobes; leaves resinous and densely glandular-dotted beneath . . . . 14. B. ovalifolia Rupr.

\+ Wings one-third to one-half as broad as the nutlet; bracts very small, with narrow lobes; leaves not resinous or with scattered resinous spots beneath . . . . 16.

16. Leaves 6—7- or rarely 5-nerved, finely crenate-serrate; bark pale . . . . 13. B. fruticosa Pall.

\+ Leaves 4—5- or rarely 6-nerved, coarsely crenate-serrate; bark brown . . . . 12. B. humilis Schrank.

17. Young branches not warty; grown leaves subcoriaceous, ovate or rhombic, acute or acuminate, subcordately rounded or cuneate at base . . . . 25.

272 + Young branches rather heavily warty . . . . 18.

18. Grown leaves thin, doubly serrate or lobed-serrate, acuminate, rhombic, truncately or broadly cuneate at base . . . . 19.

\+ Leaves firm or subcoriaceous, simply or doubly crenate-serrate, acute, more or less cuneate at base . . . . 29.

19. Leaves very large, 5—12 cm long . . . . 20.

\+ Leaves of medium size, 3.5—7 cm long . . . . 21.

20. Leaves suborbicular to ovate, taper-pointed, simply or doubly serrate, firm . . . . 21. B. grandifolia Litw.

\+ Leaves ovate, long-acuminate, subsinuate on the margin with slender double teeth, thin, chartaceous . . . . 22. B. ajanensis Kom.

21. Leaves ovate-rhombic, broadly cuneate at base . . . . 22.

\+ Leaves ovate-deltoid . . . . 23.

22. Wings 2—3 times as broad as the nutlet; leaf petiole glabrous . . . . 17. B. verrucosa Ehrh.

\+ Wings slightly broader than to twice as broad as the nutlet; leaf petiole pubescent . . . . 20. **B. japonica** Sieb.

23\. Leaf base truncate; bracts long-cuneate . . . .18. **B. platyphylla** Sukacz.

\+ Leaf base broad-cuneate; bracts short-cuneate . . . . 24.

24\. Lateral lobes of the bract broad and spreading . . . . .19. **B. mandshurica** (Rgl.) Nakai.

\+ Lateral lobes of the bract ascending . . . . 23. **B. Cajanderi** Sukacz.

25\. Low, almost shrublike trees with a straight trunk; leaves broadly ovate; wing 1.5 times as broad as the nutlet. . . 24. **B. pubescens** Ehrh.

\+ Low trees with mostly crooked stem .26.

26\. Wing about half as broad as the nutlet . . . . . .28. **B. ircutensis** Sukacz.

\+ Wing about as broad as the nutlet . . . . .27.

27\. Leaf petiole glabrous; wing about as broad as or slightly broader than the nutlet . . . . 27. **B. baicalensis** Sukacz.

\+ Leaf petiole pubescent; wing about as broad as or slightly narrower than the nutlet . . . . .28.

273 28. Leaves finely toothed, the network of veins prominent. . . . . 25. **B. tortuosa** Ldb.

\+ Leaves coarsely toothed, the network of veins inconspicuous . . . . 26. **B. Kusmisscheffii** (Rgl.) Sukacz.

29\. Leaf petioles glabrous . . . . .30.

\+ Leaf petioles pubescent. . . . . .33.

30\. Peduncles of aments 3—4 mm long . . . . .31.

\+ Peduncles of aments 8—12 mm long . . . . .32.

31\. Leaves rhombic, ca. 1.5 cm long, 1.2 cm broad; wing half as broad as the nutlet . . . . 34. **B. Saposhnikowi** Sukacz.

\+ Leaves narrow-ovate, ca. 4.5 cm long and 2.5 cm broad; wing about as broad as the nutlet . . . . .30. **B. tianschanica** Rupr.

32\. Peduncles of aments glabrous; leaves broadly ovate, to 40 mm long . . . . 39. **B. Korshinskyi** Litw.

\+ Peduncles of aments densely pubescent; leaves cuneate . . . . 37. **B. procurva** Litw.

33\. Leaves large, 50—65 mm long . . . . .34.

\+ Leaves small, 20—40 mm long . . . . .35.

34\. Peduncles of aments up to 17 mm long; lateral lobes of fruiting bracts rounded at apex . . . . 35. **B. turkestanica** Litw.

\+ Peduncles of aments short, stout; lateral lobes of fruiting bracts obliquely truncate at apex . . . . 29. **B. kirghisorum** Sav.-Ryczg.

35\. Leaf blade spatulate or hastate. . . . . 40. **B. pamirica** Litw.

\+ Leaf blade unlike the above . . . . .36.

36\. Leaf blade ovate . . . . .37.

\+ Leaf blade obovate or rhombic-ovate. . . . . .39.

37\. Leaf base cuneate . . . . .38.

\+ Leaf base truncate; wing as broad as or broader than the nutlet . . . . 36. **B. alaica** Litw.

38\. Wings about half as broad as the nutlet . . . . .33. **B. Kelleriana** Sukacz.

\+ Wings 1.5 times as broad as the nutlet or about equally broad . . . . 38. **B. schugnanica** Litw.

39\. Lateral lobes of the bracts upright, mostly about as broad as the middle lobe, obtuse or acutish . . . . .31. **B. microphylla** Bge.

274 + Lateral lobes of the bracts subhorizontally spreading, about twice as broad as the middle lobe, emarginate or obliquely truncate . . . . 32. **B. Rezniczenkoana** (Litw.) Schischk.

Subgenus 1. **EUBETULA** Rgl. Monogr. Betul. (1861) 74. — Fruiting inflorescences subspherical, ovoid or short-cylindric, solitary; wings of the nutlet completely or almost completely concealed under the bract.

Section 1. COSTATAE Rgl. in DC. Prodr. XVI, 2 (1868) 162. — Fruiting aments subspherical or ovoid, elongate-ovoid, or semicylindric; bracts more or less elongated; nutlet with membranous wings; leaves with 6 or more pairs of veins on both sides, the veins impressed above.

1. **B. Medwedewii** Rgl. in A. H. P. XI (1887) 375; H. Winkler in Engler's Pflanzenr. Heft 19 (1904) 64; Medv., Der. i kust. Kavkaza, 3rd edition (1919) 333. — Ic.: Rgl. in Gartenflora XXXVI (1887) 384, tab. 95, f. 1—4; C. K. Schn. Handb. Laubholzk. I (1904) 98, 100. — Exs.: HFR No. 1742; Herb. Fl. Cauc. No. 210.

A small tree to 3 m in height, with whitish bark and upright branches; older branches dark brown, with scattered lenticels; annotinous branched furrowed, hairy, becoming glabrous; leaves membranaceous, petiolate, obovate, ovate, suborbicular, or elliptic, broadest about the middle of the blade, cuneate or rounded or sometimes subcordate at base, mostly short-pointed from a rounded apex, rarely tapering to a point, unequally and doubly denticulate on the margin, (3.5) 5—8 (10) cm long, 3—6 (7) cm broad, with 8—11 pairs of veins on each side, the upper surface dark green, dull or slightly lustrous, glabrous or with few scattered hairs to 2 mm long, the lower surface paler, with appressed hairs on the veins, especially on the midrib; petiole hairy, (4) 8—20 mm long; staminate aments 2 or 3 together at the ends of branchlets, cylindric, spreading or pendulous, 4—5 cm long, 0.5—1 cm broad; fruiting aments 2.5—4 cm long, 1.5—1.8 cm broad, oblong-cylindric, straight or slightly curved, the peduncles pubescent; bracts persistent throughout winter until the following summer, elongate-cuneate, 3-lobed, strongly lignified at base, hairy-ciliate on the margin, the linear-oblong lobes obtuse, the middle lobe somewhat enlarged upward and much longer than the spreading lateral lobes; nutlet obovoid, ca. 3 mm long, the wing one-fourth the breadth of the nutlet. May—June. (Plate XIII, Figure 1).

In mountains, close to the timberline. — Caucasus: W. Transc. Endemic. Described from the Adzhar-Imeretian Range. Type in Leningrad.

275 2. **B. megrelica** D. Sosn. in Trav. de l'Inst. Bot. de Tiflis (1934) 42. — Ic.: ib. 37, f. 1.

A tree of medium size, with upright branches; annotinous branchlets brown, angled, covered with black glands, large lenticels, and scattered long white hairs; bark of 3—4-year-old branches cinereous to dark gray, that of older branches castaneous; leaf petioles 4—16 mm long, with scattered pubescence and black glands; blades ovate, rounded or slightly cordate at base, gradually attenuate to the acutish apex, 3—7 cm long, 1.7—5.7 cm broad; leaves of short branchlets suborbicular, abruptly acuminate, glabrous or sparingly hairy above, paler beneath, more or less pubescent on the veins and rather profusely glandular, obtusely double-serrate, with 6—10 pairs of veins; staminate aments forming a short cluster at the ends of branches; fruiting aments sessile, persistent till the following summer; bracts hairy on the margin, cuneate, 3-lobed, the linear-oblong lobes obtuse, the middle lobe twice the length of the divaricate lateral lobes; nutlet sublanceolate, with very narrow wings. May—June.

In forests of the subalpine zone. — Caucasus: W. Transc. Endemic. Described from Mingrelia. Type in Tiflis.

3. **B. costata** Trautv. in Maxim. Primit. Fl. Amur. (1859) 253; Maak, Putesh. po dol. r. Ussuri II (1861) 299; Kom., Fl. Manchzh. II (1903) 43. — B. Ermani var. costata Rgl. Monogr. (1861) 129. — B. ulmifolia var. costata Rgl. in DC. Prodr. XVI, 2 (1868) 176; H. Winkler in Engl. Pflzr. Heft 19 (1904) 64. — Ic.: Rgl. Monogr. (1861) tab. XIII, f. 1—6, tab. VI, f. 36; Kom. and Alis., Opred. rast. Dal'nevost. kr. I (1931), Plate 131, Figure 4.

An elegant tree to 20 m high, with smooth or slightly fissured pale yellow or buff bark; bark of old branches darker, flaking; young branches dark tawny, glabrous, with white lenticels or rarely at the tips of branchlets glandular; leaves petiolate, ovate or elongate-oval, strongly tapering to a characteristic point, rounded or rarely broadly cuneate at base, conspicuously and sharply long-toothed on the margin, with 12—14 pairs of approximate veins, 4.5—8 cm long, 2—4 cm broad; upper surface dark green, sparsely pubescent especially on the veins; lower surface paler, with scattered glands, pubescent on the veins, with axillary tufts of hairs; petiole pubescent, 0.7 cm long; staminate aments commonly 3, rarely 8 or many; fruiting aments solitary, rounded-ovoid, 1.2—1.5 cm long, 1—1.2 cm broad, short-pedunculate, upright or slightly spreading; bracts woody; widely imbricate, long-cuneate, 6—9 mm long, glabrous or sometimes with short hairs on the margin, the middle lanceolate lobe 3—4 mm long, 1.5—3 times as long as the spatulate somewhat spreading lateral lobes; nutlet
276 2—2.5 mm long, the wings half as broad as the nutlet, broadest in upper part. May.

Isolated trees in primeval mountain forests. — Far East: Uss. **Gen. dist.:** Manchuria. Described from the Khekhtsir Mountains near Khabarovsk. Type in Leningrad.

4. **B. Ermani** Cham. in Linnaea VI (1831) 537; Ldb. Fl. Ross. III, 653; Trautv. in Maxim. Primit. Fl. Amur. 242; Rgl. Monogr. (1861) 122; Rgl. in DC. Prodr. XVI, 2, 176; Kom., Fl. Manchzh. II, 49; H. Winkler in Engl. Pflanzenr. Heft 19 (1904) 66; Sukachev, Tr. Bot. Muz. Akad. Nauk VIII (1911) 233; Kuzeneva, Tr. Bot. Muz. Akad. Nauk XII (1914) 70; Hultén, Fl. of Kamtch. II, 26; Kom., Fl. Kamch. II, 38. — Ic.: Cham. ib. tab. VI, f. D. — Also referred to in Russian as stone birch.

A big or moderate tree; branches in the forest upright, at the seacoast and near the upper distribution limit spreading; bark on the trunk strongly exfoliating, on old trunks fissured, yellowish, yellowish-gray, pinkish-gray,

(277)

PLATE XIII. 1. Betula Medwedewii Rgl. — 2. B. Ermani Cham. — 3. B. Schmidtii Rgl. — 4. B. rotundifolia Spach. — 5. B. nana L.

dark gray, or brownish; young branchlets glandular-verrucose, more or less pubescent; fertile branchlets reddish-brown, with white lenticels; leaves ovate, oval, or broadly oval, truncate or short-cuneate or rounded, rarely cordate, short-acuminate, doubly and sharply dentate-serrate, with 6—9 (10) pairs of veins, 4—14 cm long and 3—10 cm broad (in var. litoralis Kom. the leaves more or less cordate at base, 6—14 cm long and 5—10 cm broad; in var. parvifolia Koidz. ca. 4 cm long and 3 cm broad), hairy on the veins and on the petiole, dark green above, paler beneath and bearded at the axils of the veins; staminate aments pendulous, to 8 cm long; fruiting aments subsessile, erect, 2—4 cm long, 1—2 cm broad (in var. litoralis Kom. 2—4 cm long); bracts large, to 1 cm long, ciliate on the margin, with 3 lanceolate lobes, the middle lobe surpassing the lateral ones. May—June. (Plate XIII, Figure 2).

In coniferous and mixed mountain forests; at higher altitudes on mountain slopes, nearer to the subalpine zone, solitary or in groups, occasionally forming the upper timberline; isolated specimens on stony screes and on the banks of mountain streams. In Kamchatka forming
279 open stands of parkland character, without participation of other species; nearly everywhere it constitutes there the timberline, near the sea it displays a series of windblown forms. — E. Siberia: Lena-Kol. (S. and E.). Dau.; Far East: throughout. **Gen. distr.:** Korea, Japan. Described from Kamchatka. Type in Leningrad.

Note. Cultivated and occurring in botanical gardens of the USSR and Europe.

Gives rise to the following hybrids:

1) B. Ermani Cham. × B. Middendorffii Tr. et Mey., Sukach., ib. 224; Kuz., ib. 72. — Ic.: Kuz., ib., Plate VII, Figures 7—11. — A big spreading shrub; petioles short; wings of nutlets and leaves broader than those of B. Ermani.

2) B. Ermani Cham. × B. japonica Sieb. — B. avatschensis (= avaczensis) Kom. in Fedde Repert. sp. nov. No. 355 (1914) 166; Kom., Fl. Kamch. II (1929) 43. — B. platyphylla Sukacz. × B. Ermani Cham., Hultén Fl. of Kamtch. II (1927) 33. — Ic.: Hultén ib. f. 5a, b. — A small tree with dark castaneous to almost black bark and bright white lenticels; the bark at the base of the trunk dark gray and fissured; young branches verrucose-glandular; leaves cuneate-oval, acuminate, doubly serrate, smooth; fruiting aments compact, cylindric, 2 cm long and 0.8 cm broad, short-peduncled, nodding; scale minutely stalked, ciliolate, the narrowly oboval middle lobe but slightly longer than the lateral ones, these oboval-spatulate and free nearly down to base; nutlet obovaloid, smooth, uniformly winged; catkins persistent after leaf-fall.

Far East: Kamch.

5. **B. Prochorowii** Kuzen. et Litw. in Trav. Mus. Bot. Ac. Petrogr. XII (1914) 70. — Ic.: Ib. tab. III, f. 1—5.

A small tree; branches straight, glandular; leaves ovate-lanceolate, acuminate or acute, at base cuneate or rounded or subcordate, 2.5—3 cm long and 2—2.5 cm broad, unequally and almost doubly toothed, the upper surface glabrous or on the midrib and at base pubescent, more or less glandular-dotted, the lower surface paler, hairy on the veins, more or less resinous-glandular; petioles 3—5 mm long, hairy; fruiting aments borne on a peduncle 2—3 mm long, erect, subcylindric, ca. 15 mm long and 10 mm thick; scales long-cuneate at base, ca. 5 mm long, cut to the middle into 3 pubescent-margined lobes, the somewhat projecting median lobe oblong, the lateral ones elongate-oval; nutlet ovoid, pubescent at apex, the entire wings about as broad as or slightly narrower than the nutlet. May.

Stony slopes. — Far East: Ze.-Bu., Uss. Endemic. Described from the upper reaches of the Urkan and Gilyui rivers in the Tukuringra Range. Type in Leningrad.

280 6. **B. Raddeana** Trautv. in A. H. P. X (1887) 129; H. Winkler in Engl. Pflanzenr. 19 (1904) 69; E. Bush, Tr. Bot. Muz. Ak. Nauk XII (1914) 333. — Ic.: Rgl. in Gartenflora XXXVI (1887) 384, tab. 95, f. 5—11. — Exs.: Herb. Fl. Cauc. No. 414; Pl. orient. exs. No. 34.

A small tree, with pinkish, white, or brownish bark; the bark of branches dark brownish, with lenticels; annotinous branchlets yellowish-brown, velutinous, with scattered glands; leaves petiolate, ovate to ovate-oblong, cuneate or rounded at base, acuminate, unequally sharp-toothed on the margin, with 6 or 7 pairs of veins, green and glabrous above, paler and pubescent in the axils of veins beneath, 3—4.5 cm long, 2—3.5 cm broad, the densely puberulous petiole 10—15 cm long; staminate aments 2 or 3 together, slender, elongate-cylindric, 2—2.5 cm long and 10—14 mm in diameter, the pubescent peduncle 5—6 mm long or rarely longer; bracts caducous, long-cuneate, ca. 8 mm long, puberulous on the margin, the straight oblong or ovate median lobe narrowed toward base, the lateral lobes half to two-thirds as long, oval, acutish to obtusish, slightly divergent; nutlet obovoid, ca. 3 mm long, pubescent at apex, often pointed at base, the wings about as broad as the nutlet. June.

Subalpine birch groves. — Caucasus: Cisc. (central part of the Greater Caucasus), Dag., E. Transc. (Gr. Cauc.). Endemic. Described from Dagestan. Type in Tiflis; cotype in Leningrad.

Note. Hybridizing with B. pubescens Ehrh.

Section 2. **ASPERAE** Nakai, Botan. Mag. Tokyo XXIX (1915) 41. — Fruiting aments upright, oblong; lateral lobes of the bract narrow; nutlet marginate but without membranous wings; leaves on both sides with 3—5 pairs of more or less impressed veins. Medium-sized or tall trees.

7. **B. Schmidtii** Rgl. in Bull. Soc. Natur. de Moscou XXXVIII 3 (1865) 412 et in DC. Prodr. XVI, 2 (1868) 175; Kom., Fl. Manchzh. II (1903) 52. — Ic.: Rgl. l. c. (1865) tab. 6, f. 14—20.

A sparsely branched tree, 20—25 m high, with brownish-gray smooth or fissured bark; the bark of the branches smooth, dark cherry-colored, dark brown or almost black, with white lenticels; leaves short-petioled, oval or elongate-ovate, gradually acuminate or rarely elongate-acuminate sharp-toothed on the margin, glabrous or rarely very sparsely hairy or
281 glandular, slightly pubescent on the veins and sparingly glandular beneath, with 5—7 pairs of very prominent veins, 4—8 cm long and 2—6 cm broad, the pubescent petiole 0.5—1 cm long; fruiting aments elongated, straight or scarcely curved, 2—3 cm long and 5—8 mm broad, the peduncle one-fourth

to one-third the length of the ament, 0.5—1 cm long; bracts 5—7 mm long, divided into 3 oblong to sublinear lobes, the median lobe elongate, 2—3 mm long, the lateral ones 1.5—2 mm long, somewhat divergent to almost spreading, acuminate; nutlet ovaloid or ovoid, 1.5—2 mm long, wingless, narrowly marginate. May. (Plate XIII, Figure 3).

Rocky crests and stony slopes, in deciduous or mixed forests; also isolated trees in sparse woods and coppices.— Far East: Uss. (only S.). **Gen. distr.**: N. Korea. Described from Slavyanka Bay area in Pos'et. Type in Leningrad.

Economic importance. Recently conducted investigations disclosed some outstanding properties of the wood of B. Schmidtii, such as durability and resilience. These properties have been recognized by the local population, which uses it for instrument making.

Section 3. **NANAE** Rgl. in DC. Prodr. XVI, 2 (1868) 162.— Fruiting aments small, upright; leaves suborbicular, often as broad as long, obtuse or rounded at apex, with 2—4 (5) pairs of veins and a conspicuous network of veinlets, the margin obtusely toothed. Shrubs, often small.

8. **B. nana** L. Sp. pl. (1753) 983; Rgl. in DC. Prodr. XVI, 2 (1868) 171; H. Winkler in Engl. Pflzr. Heft 19 (1904) 69; Sukach., Tr. Bot. Muz. Akad. Nauk VIII (1911) 212; Kryl., Fl. Zap. Sib. IV (1930) 796.— B. nana α europaea Ldb. Fl. Ross. III (1851) 654.— B. nana var. genuina Rgl. Monogr. (1861) 101.— Ic.: Rgl. Monogr. (1861) tab. IX, f. 1—8, 9, 11; Hegi Fl. Mitt. Eur. (1891) tab. 84, f. 4.— Exs.: HFR No. 587.

A low, much branched shrub, 20—70 (120) cm high, with ascending or prostrate branches; young branches heavily velutinous or pubescent, often becoming glabrous, without resin-glands, their bark dark tawny or dark reddish-brown; leaf petioles 1—2 (4) mm long; blades orbicular, often broader than long, rarely rounded-oval, rounded or often broadly cuneate or subcordate at base, sometimes truncate at apex, 5—12—15 mm long, 10—20 mm broad, obtusely toothed on the margin, with 2—4 pairs of veins, at first glutinous and with scattered pubescence beneath, finally glabrous or with isolated hairs in the axils of veins, dark green and shiny above, paler green beneath; staminate aments sessile, upright, 0.5—1.5 (2) cm long, 1.5—2 mm broad, with yellow anthers; pistillate aments borne on a short pubescent peduncle, ovaloid or elongate-ovoid, light tawny, 5—8 mm long (in fruit to 12 mm) and 3—5 mm thick (in fruit to 6 mm); bracts 2.5
282 2.5—3 mm long, with 3 upright linear-oblong slightly ciliate lobes; nutlets ellipsoid or ovoid, ca. 2 mm long and 1 mm broad, the wings one-fifth to one-third as broad as the nutlet and not projecting beyond it at apex. Fr. April—June. (Plate XIII, Figure 5).

Arctic tundra, the alpine zone, as well as sphagnum and hypnum moss bogs of the forest belt.— Arctic: Arc. Eur., Nov. Z., Arc. Sib. (W.); European part: Kar.-Lap., Dv.-Pech., Lad.-Ilm., U. V., V.-Kama, U. Dnp.; W. Siberia: Ob; E. Siberia: Yen. **Gen. distr.**: Arctic, Scand., Centr. Eur. Described from Lapland. Type in London.

Forming the following hybrids:

1. B. nana L. × B. tortuosa Ldb. —B. Sukaczewii Soczava in Ocherki po fitosots. i fitogeogr. (1929) 393.— Ic.: ib. 391.

A shrub 10—70 cm high, more or less procumbent; bark of branches brown; branches glabrous, very rarely slightly glandular; leaves obovate or flabellate, cuneate or rarely almost round at base, crenate from the very base; lower teeth smaller, somewhat pointed, the upper obtuse and parted at the middle, the blade 2—3 cm long and 1 1/3—2 1/2 cm broad, rarely as broad as long, the petiole relatively long; fruiting aments 5—10 mm long, 3—4 mm broad, the pubescent peduncle one-fourth to half as long as the ament; nutlet oblong-ovaloid, the wings one-third to one-half as broad as the nutlet and not projecting beyond its apex which forms a small tubercle. May.

Stony and stone-and-lichen tundras occasionally corroded by wind. — N. Urals, basin of the Lyapin River, a tributary of the Sosva River; basin of the Kosyu River, a tributary of the Usa River.

2. B. nana L. × B. exilis Sukacz. — Characterized by the presence of pubescence together with resin-glands on the young branchlets; most widespread in the Eniseisk District. — Arctic: Arc. Sib. (chiefly between the Yenisei and Lena rivers); E. Siberia: Yen., Lena-Kol.

3. B. nana L. × B. pubescens Ehrh. = B. pubescens Ehrh. × B. nana f. alpestris (Fries) H. Winkler in Engl. Pflzr. H. 19 (1904) 93. — B. alpestris Fries, Summa veg. Scand. I (1846) 212.

Economic importance. In the summer B. nana provides one of the sources of food for deer. The local population in the Arctic Region stores the shrubs for winter fuel.

9. **B. exilis** Sukacz. Trav. Mus. Bot. Acad. Sc. Pétersb. VIII (1911) 213; Kom. Fl. Kamch. II, 44. — B. nana β sibirica Ldb. Fl. Ross. III (1849) 654 (ex parte et excl. syn.); Trautv. et Mey. Flor. Ochot. 85. — B. nana var. sibirica et var. intermedia Rgl., Monogr. Betul. (1861) 43 (ex parte). — B. glandulosa H. Winkler in Engl. Pflzr. H. 19 (1904) 73 (ex parte). — B. glandulosa β rotundifolia Rgl. in DC. Prodr. XVI, 2 (1868) 172, ex parte. — Ic.: Sukacz. ib. tab. I, f. 4, 5 and (sub B. fruticosa) f. 16.

283 A much branched shrub with procumbent or ascending dark branches; young branchlets densely glandular; petioles 1—3 mm long; leaf blades suborbicular, often broader than long, 1—2 cm long and 1—2.5 cm broad (in var. subtilis Sukacz. 6—10 mm long and 8—15 mm broad) rounded at apex, mostly rounded or rarely subcordate or broadly subcuneate, obtusely dentate, dark green and smooth above, pale beneath, with 2—4 principal veins and a dense network of veinlets, often glandular about the veins; fruiting aments upright, pedunculate, green, finally brown, 7—12 mm long and ca. 5 mm broad (smaller in var. subtilis); bracts with 3 sublinear parallel lobes, the middle lobe slightly surpassing the lateral ones; nutlet ovaloid, ca. 2 mm long, the wings half as broad as the nutlet or narrower, slightly projecting beyond the apex of the nutlet. April—June.

Arctic tundra, sphagnum bogs, and barren heights of Transbaikalia. — Arctic: Arc. Sib. (E.), Chuk., An.; E. Siberia: Yen., Lena-Kol., Ang.-Say., Dau.; Far East: Kamch., Okh., Ze.-Bu. (N.), Sakh. Described from the surroundings of Eniseisk. Type in Leningrad.

Note. B. sessilis Kom. (Not. Syst. ex Herb. H. B. P. II (1921) 131), described from the Ayan area, closely resembles this species and is possibly a hybrid. Even more problematic are B. Abolini Sukacz. and

B. wiluica Sukacz. (nom. nuda in Travaux Com. pour l'étude rep. Jakoute X (1929) 367), very likely also of hybrid nature.

10. **B. rotundifolia** Spach. in Ann. Sc. Nat., 2 sér. XV (1841) 194; Sukach., Tr. Bot. Muz. Akad. Nauk VIII (1911) 215; Kryl., Fl. Zap. Sib. IV (1930) 797. — B. nana β sibirica Ldb. Fl. Ross. III (1842) 654, ex parte; Rgl. Monogr., 43, ex parte. — B. glandulosa H. Winkler in Engl. Pflanzenr. Heft 19 (1904) 73, ex parte. — B. glandulosa var. rotundifolia Rgl. in DC. Prodr. XVI, 2 (1868) 172. — Ic.: Sukach., ib. tab. I, fig. 2 and 15 (the latter sub B. ovalifolia); Litv., Tr. Bot. Muz. Akad. Nauk XII (1914) tab. XI, f. 13, tab. XII, f. 14.

A shrub from 20 to 100 cm high (rarely higher), erect or prostrate, with blackish-brown or buff bark; branches densely closed with resinous glands, only the youngest puberulous; leaf petioles 2 — 5 mm long; blades orbicular or often flabellate (larger than in B. nana, 1 — 2.5 cm long and 0.8 — 2 cm broad) slightly narrowed and rarely rounded or cordate at base, coarsely toothed on the margin; grown leaves glabrous, only at the axils of veins sparingly pubescent; pistillate aments ovaloid, (8) 10 — 15 (18) mm long and 6 — 8 mm thick; bracts with upright linear-oblong lobes, 3.5 — 5 mm long; nutlet ellipsoid or obovoid, ca. 2.5 mm long and 1.5 mm broad, the relatively broad wings half the breadth of the nutlet, slightly projecting beyond the apex of the nutlet. June — August. (Plate XIII, Figure 4).

284 Forming compact thickets, chiefly in the alpine zone and the moss-lichen subzone of the tundra. The shrubs reach greatest height in the subalpine zone; their size decreases near the upper distribution limit where the shrubs trail upon the ground. — W. Siberia: Alt.; E. Siberia: Ang.-Say. **Gen. distr.**: N. Mong. Described from Siberia. Type in Paris.

11. **B. Middendorffii** Trautv. et May. Fl. ochot. in Middend. Reise 1, 2 (1856) 293; Trautv. in Maxim. Prim. Fl. Amur. (1859) 257; Rgl. Monogr. Bet. (1861) 41 et in DC. Prodr. XVI, 2, 169; Kom. Fl. Manch. II, 53; H. Winkl. in Engl. Pflzr. H. 19 (1904) 87; Sukach., Tr. Bot. Muz. Akad. Nauk VIII (1911) 216; Kuzen., Tr. Bot. Muz. Akad. Nauk XII (1914) 66; Kom., Fl. Kamch. II, 46 (ex parte). — Ic.: Trautv. et Mey. ib. 21.

A spreading shrub, averaging 0.5 — 2 m in height; young branches glandular and pubescent, at length fairly smooth, lustrous; leaves obovate to suborbicular or rounded-rhombic, at base rounded or cuneately rounded, at apex rounded-obtuse, on the margin bluntly dentate-crenate, with 3 — 6 pairs of veins, at first with scattered hairs finally on both sides glabrous, firm, dark green above, lustrous, 15 — 40 mm long, 10 — 30 mm broad, the petiole 2 — 3 mm long; fruiting aments pedunculate, elongate-cylindric or subspherical or ovoid-spherical, mostly nodding, rarely upright, 1 — 2 cm long, 0.5 — 0.8 cm thick; bracts obovate-cuneate, 3-lobed, ciliolate, the erect or somewhat divergent rounded lateral lobes slightly shorter than the middle lobe; nutlet ellipsoid, the wings 1.5 times as broad or rarely as broad as the nutlet, projecting beyond its apex but not reaching the tips of stigmas. Fl. May — June.

Forming undergrowth of deciduous or mixed, often waterlogged woods; in mossy bogs; participating in the composition of subalpine shrub thickets (together with Pinus pumila and Alnus fruticosa), replacing Pinus in the mountains; on mountain passes and on top of barren heights. — Arctic: An.; E. Siberia: Lena-Kol., Dau.; Far East: Kamch., Okh., Ze.-Bu., Uda, Uss. (N. part), Sakh. **Gen. distr.**: N. Manchuria (rare). Described from the Shantar Islands. Type in Leningrad.

Note. Appearing in a number of forms and apt to produce hybrids with other species.

1) var. vallensis Sukacz. — a semispherical shrub, 0.5 — 1 m high; leaves 15 — 2 5 mm long and 10 — 20 mm broad; fruiting aments and bracts mostly small.

2) var. subalpina Sukacz. — a big shrub, 1.25—2 m high; leaves large, ca. 40 mm long and 30 mm broad; fruiting aments and scales mostly large.

3) var. alpina Sukacz. — a low shrub, 0.25 — 0.5 m in height, with procumbent or ascending branches; leaves small. Outwardly resembling
285 B. exilis, but differing from that species in the broad wing of the fruit. There are transitional forms between these varieties.

Forming a number of hybrids: 1) B. Middendorffii × B. platyphylla, Kuzen. ib. 72, tab. IV, tab. VI, f. 19, 24; 2) B. Middendorffii × B. Ermani, Kuzen. ib. 72, tab. VII, f. 12 — 15; 3) B. Middendorffii × B. exilis, Kuzen. ib. 73, tab. V, f. 2; tab. VI, f. 25 — 27.

Economic importance. The foliage of B. Middendorffii serves as food for deer. The shrubs are gathered by the population of the Arctic Region and stored for winter fuel.

Section 4. **FRUTICOSAE** Rgl. in DC. Prodr. XVI, 2 (1868) 162. — Leaves oblong, always longer than broad, acute or subacute, with 4 — 7 pairs of veins, finely toothed on the margin. Shrubs.

12. **B. humilis** Schrank, Baier. Fl. (1789) 420; Ldb. Fl. Ross. III, 653; Rgl. Monogr. 104 (ex parte) et in DC. Prodr. XVI, 2 (1868) 173 (ex parte); Shmal'g., Fl. II, 430; H. Winkler in Engl. Pflzr. H. 19 (1904) 73 (ex parte); Sukach., Tr. Bot. Muz. Akad. Nauk VIII (1911) 208; Kryl., Fl. Zap. Sib. IV, 794. — Ic.: Hegi III. Fl. Mittel-Eur. II, tab. 480, fig. a, b, c, d, e. — Exs.: HFR No. 339.

A shrub 1 — 2.5 m high, with upright branches; young branchlets beset with resinous warts and sparsely puberulous; old branches blackish-brown; leaves ovate or rounded-ovate, rounded or bluntly angled at base, short-acuminate or subobtuse, rather coarsely toothed on the margin, at first pubescent, finally glabrous above, paler and sparingly pubescent on the veins or glabrous beneath, with 4 or 5 (6) pairs of veins, 1 — 3.5 (4) cm long and 0.7 — 2.5 cm broad, the pubescent petiole 0.2 — 0.6 (1) cm long; fruiting aments borne on a short pubescent peduncle, subtended by 2 leaves, erect, oblong-ovoid, ovaloid, or cylindric, 1 — 1.5 cm long, 0.5 — 0.8 cm thick; bracts cuneate, 3-lobed, 3 — 5 mm long, puberulous on the upper margin, the linear-oblong middle lobe exceeding the obliquely ascending broader lateral lobes; nutlet broadly ellipsoid, the wings one-third to one-half as broad as and not surpassing the nutlet. May — June.

Moss and sedge bogs of the forest zone; below the alpine zone and the timberline, wettish meadows and banks of mountain streams. — European part: Dv.-Pech., Lad.-Ilm., U. V., V.-Kama, U. Dnp., M. Dnp., V.-Don; W. Siberia: Ob, Alt.; E. Siberia: Yen., Ang.-Say., Dau. (W. part). **Gen. distr.**: Scand. (S. part), Centr. Eur., N. Mong. Described from Bavaria. Type in Munich.

Note. Varieties established: 1) var. vulgaris Perf. in Journ. Bot. URSS 20, 6 (1935) 639. — base of leaves rounded; leaves ovate, broadly
286 ovate, elliptic, or orbicular, at apex acuminate, obtusish, or round-tipped; lateral veins at an angle of 34—50°, the network of veinlets prominent beneath; 2) var. cuneata Perf. ib., Plate III, Figure 7a, b; Plate I, Figure 1. — leaf base narrowly to broadly cuneate; leaves oval-lanceolate, rhombic, elliptic, or obovate, at apex acute, obtusish, or round-tipped; lateral veins at an angle of 28—30°; network of veins inconspicuous or obscure; 3) var. subcordata Perf. ib. Plate 1, Figure 3. — leaf base subcordate; leaves ovate, elliptic, or broadly elliptic, at apex acuminate, obtuse, or round-tipped; lateral veins at an angle of 50—70°; network of veinlets prominent beneath. The first two varieties are widespread, the third was only once encountered (Vologda).

Of interest is var. cretacea Litw. (Tr. Bot. Muz. XI (1913) 10), found on the chalk of the Voronezh Region; it is distinguished by the small size of leaves and aments.

13. **B. fruticosa** Pall. Reise III, Anhang (1776) 758; Fl. Ross. I (1784) 62; Spach, Rev. Bet. (1841) 193; Ldb. Fl. Ross. III, 653 (ex parte); Turcz. Fl. baic.-dah. II, 130 (ex parte); Kom., Fl. Manchzh. II, 50 (ex parte); Sukach., Tr. Bot. Muz. Akad. Nauk VIII (1911) 209. — Ic.: Pall. Fl. Ross. tab. XL, f. A. B. C. — Vernacular name: ernik.

A shrub, 0.75—2.5 m high, with upright branches and white bark; young branchlets glandular; leaves ovate to elliptic, rounded or rarely broadly cuneate at base, irregularly serrulate, pale green, glabrous or sparingly pubescent above, pubescent on the veins and sometimes diffusely and obscurely glandular beneath, with 5 or 6 pairs of veins, 1.5—2.5 (4) cm long, 1—2 (2.5) cm broad, the petiole 0.3—0.5 (0.8) cm long; fruiting aments upright or slightly divergent, cylindric, 1—2 cm long and ca. 0.5 cm thick, the peduncle 3—6 mm long; bracts cuneate, with a short or somewhat elongated basal portion, 2.5—3.5—5 mm long, 1—1.5 mm broad, 3-lobed, the somewhat divergent lateral lobes short and slightly broader than the median; nutlet ovaloid, 2—2.5 mm long, the wing one-fourth to one-half the breadth of the nutlet, 0.3—0.5—1 mm broad. (Plate XIV, Figure 4).

River valleys and waterlogged areas; forming thickets (the so-called "erniki") which are widespread in the southern part of the distribution area. Also occurring on riverbanks and penetrating along the riverbeds far to the north. — E. Siberia: Lena-Kol., Dau.; Far East: Okh., Ze.-Bu., Uss. (W.). **Gen. distr.**: China. Described from Lake Baikal. Type in London.

289 Note. On the banks of small rivers attaining considerable size, and arborescent specimens occur 2.5—3 m in height, with stems 2—3—5 cm in diameter.

Hybridizing with B. platyphylla Sukacz. and B. Gmelini Rupr.

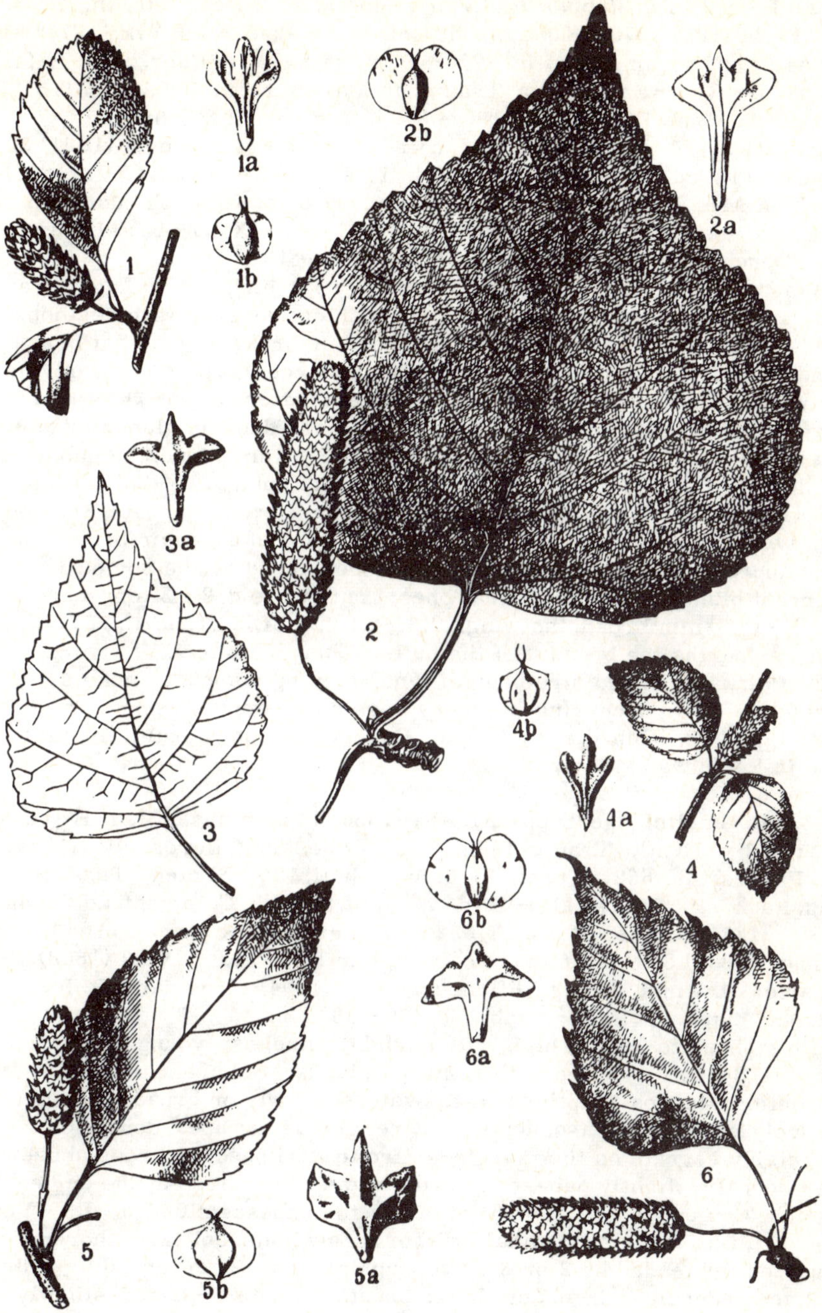

PLATE XIV. 1. Betula ovalifolia Rupr. – 2. B. grandifolia Litw. – 3. B. platyphylla Sukacz. – 4. B. fruticosa Pall. – 5. B. dahurica Pall. – 6. B. verrucosa Ehrh.

14. **B. ovalifolia** Rupr. in Bull. phys.-math. Ac. Sc. St. Pétersb. XV No. 23—24 (1857) 378; Sukach., Tr. Bot. Muz., Akad. Nauk VIII (1911) 210. — B. palustris var. Rupr. ib. 377. — B. reticulata Rupr. ib. — B. fruticosa var. Ruprechtiana Trautv. in Maxim. Prim. Fl. Amur. (1859) 258; H. Winkl. in Engl. Pflzr. H. 19 (1904) 87. — B. humilis Schrank ε Ruprechtii, ζ reticulata, η ovalifolia Rgl. Monogr. (1861) 52. — B. humilis Schrank var. Ruprechtii Rgl. in DC. Prodr. XVI (1868) 174; Korsh. in A. H. P. XII (1892) 389. — B. fruticosa Kom., Fl. Manchzh. II, 50 (ex parte). — Ic.: Rgl. ib. tab. XIII, 40—47; tab. IX, 56—67; Kom. and Alis., Opred. rast. Dal'nevost. kr. I (1931) Plate 131.

A medium-sized or large shrub, (0.75) 1—2 m high, with upright branches; old branches grayish-brown; young branchlets covered with resinous glands; leaves oval or ovate-oval, rounded or broadly angled at base, round-tipped or rarely obtusely acuminate at apex, rather regularly dentate, fairly firm, dark green and sometimes with scattered long white hairs above, paler beneath with hairs on the veins and densely glandular-dotted, with 5—6—7 pairs of veins, 2—4 cm long, 1.2—3 cm broad, the pubescent petiole 0.3—0.5 (0.8) cm long; fruiting aments upright, 1—2.8 cm long, 7—8 (10) cm broad, the peduncle 2—4 cm long; bracts cuneate, 3-lobed, slightly pubescent on the margin, 4—5 mm long and 3—4 mm broad, the lobes approximate, the obtusish middle lobe exceeding the somewhat divergent bluntly rhombic lateral ones; nutlet ovoid, 2—3 mm long, 1—1.5—(1.8) mm broad, the wings 0.3—0.8 mm broad, (one-third) one-half to three-fourths the breadth of the nutlet. May. (Plate XIV, Figure 1).

On flat waterlogged stretches of land, forming compact thickets ("erniki"). Also occurring in mossy bogs on swampy river banks. — Far East: Uda, Uss. **Gen. distr.**: China. Described from Ussuri Territory. Type in Leningrad.

15. **B. Gmelini** Bge. Suppl. Fl. Alt. (1836) 113 (in observ. ad Betulam microphyllam); Trautv. Plant. imag. et descript. fl. ross. ill. (1844) 10; Ldb. Fl. Ross. III, 652; Turcz. Fl. baic.-dah. II, 128; Sukach., Tr. Bot. Muz. Akad. Nauk VIII (1911) 211. — B. divaricata Ldb. Comment. ad Gmel. Fl. Sib. (1841) 59; Fl. Ross. III, 652. — B. fruticosa var. Gmelini Rgl. Monogr. (1861) 34. — B. fruticosa Rgl. in DC. Prodr. XVI, 2 (1868) 169 ex parte; H. Winkl. in Engl. Pflanzenr. H. 19 (1904) 87 ex parte. — Ic.: Trautv. l. c. tab. 5; Rgl. l. c. tab. VI, f. 39—53.

290 A big shrub to 2—3 m high, with upright branches; young branchlets glabrous, densely covered with resinous glands; old branches dark, smooth, not peeling; leaves elliptic-oval to ovate, narrowly or broadly cuneate; or rarely rounded at base, acute or rarely obtuse at apex, doubly or irregularly serrate on the margin, entire toward base, glabrous above, paler beneath, slightly pubescent on the midrib, beardless at the angles of veins, 2—4 cm long, 1.2—2.5 cm broad, the pubescent petiole 0.6—1 cm long; fruiting aments terminal on short branchlets, solitary, short-peduncled, subtended by 2 or 3 leaves, upright, 1.5—2 cm long, 0.6—0.8 cm thick, the peduncle 0.3—0.7 cm long; bracts cuneate, 3-parted, slightly pubescent on the margin, 4—6 mm long, the acutish median lobe surpassing the obtuse-tipped lateral lobes; nutlet ellipsoid, ca. 3 mm long. May.

Thickets on dry slopes; rarely sandy valley crests, and bottoms of ravines. — E. Siberia: Ang.-Say., Dau. Gen. distr.: N. Mongolia. Described from Dauria. Type in Leningrad.

Forming hybrids: B. Gmelini × B. rotundifolia, B. Gmelini × B. fruticosa.

Section 5. DAHURICAE Rgl. in DC. Prodr. XVI, 2 (1868) 162. — Fruiting aments rather short; wing half the breadth of the nutlet or much narrower. Trees, often tall.

16. **B. dahurica** Pall. Fl. Ross. I (1784) 60; Ldb. Fl. Ross. III, 651; Turcz. Fl. baic.-dahur. II, 2, 128; Rgl. Monogr. 55; Rgl. in DC. Prodr. XVI, 2 (1868) 174; Trautv. in Maxim. Primit. Fl. Amur. (1859) 250; Kom., Fl. Manch. II, 45; H. Winkl. in Engl. Pflzr. H. 19 (1904) 86. — B. dioica Pall. Reise III, 224, 321, 421, nom. nud. — B. Maximowiczii Rupr. in Bull. Acad. Pétersb. XV (1857) 139. — B. Maackii Rupr. in Bull. Acad. Pétersb. XVI (1857) 564. — Ic.: Pall. Fl. Ross. I (1784) tab. 39, f. A — B.

A tree of medium size, 6 — 18 m high; trunk 30 — 60 cm in diameter, in the south erect, in the northern part of the distribution area forking at about the middle of its height; top broad and spreading; bark of old trees dark gray to blackish, fissured mostly longitudinally; young branchlets pinkish to reddish, tawny-brown, or dark brown; the bark of branches light or dark brown, with white lenticels; leaves oval or ovate, broadly or rarely narrowly cuneate or rounded at base, acuminate, irregularly or doubly toothed on the margin, with 7 — 9 pairs of veins, at first pubescent, finally with scattered hairs on the veins above, pubescent on the veins and with tufts of hairs in the vein axils beneath, (3) 4 — 6 (7) cm long, (2) 3 — 4 (5) cm
291 broad, the more or less pubescent or rarely glabrous petiole 6 — 11 (13) mm long; staminate aments 3.8 — 9 cm long; fruiting aments elongate-cylindric, upright or nodding, 2 — 3 cm long and 7 — 9 mm thick, the peduncle ca. 1 cm long; bracts 4 — 5 mm long, short-cuneate or rarely long-cuneate at base, glabrous or glandular, 3-angled; middle lobe broadly lanceolate, narrower and longer than the lateral, rarely about as long, acutish or obtuse; lateral lobes divergent, short, elongate-oval or rounded; nutlet ovaloid or obovoid, 1.5 — 2.5 mm long, pubescent at apex, the wing one-fourth to one-third as broad as the nutlet and projecting beyond its apex. April — May. (Plate XIV, Figure 5).

In pure stands or mixed with oak in the southern part and with white birch in the northern part of the distribution area. Growing on dry well drained ridges, slopes, terraces, riverbanks, sandy crests and river valleys. It requires much light and does not grow in coniferous mountain woods. Solitary trees occur on open mountain slopes among shrub thickets. — E. Siberia: Dau.; Far East: Ze.-Bu., Uda (S.), Uss. **Gen. distr.**: Mong. (E. part), China. Described from Dauria. Type in London.

Economic importance. As early as 1774 Pallas noted that the wood of B. dahurica is stronger and yellower than that of white birch and, being tougher, it is more suitable for carriage construction than the wood of common birch. Recent studies have indicated that the wood is characterized by hardness and tenacity. Besides being used for the making of miscellaneous articles and for fuel, the tree is useful as an indicator of good drainage for purposes of land reclamation. In the northern part of its territory, the presence of B. dahurica may be taken as first evidence that the land is in no need of amelioration.

Section 6. ALBAE Rgl. Bull. Soc. Natur. Mosc. XXXVIII (1865) 396 et in DC. Prodr. XVI, 2 (1878) 162. — Fruiting aments mostly regularly cylindric; bracts rather short; leaves on both sides with 5—7 (8) pairs of prominent veins, without furrows along the veins; wing one-third to half as broad or rarely as broad as the nutlet. Tall or rarely low trees or big shrubs.

17. **B. verrucosa** Ehrh. Beitr. Naturk. VI (1791) 98; H. Winkl. in Engl. Pflanzenr. H. 19 (1904) 75; Kryl., Fl. Zap. Sib. IV, 786. — B. alba L. Sp. pl. (1753) 982 (ex parte); Ldb. Fl. Alt. IV, 244 (ex parte); Ldb. Fl. Ross. III, 650 (ex parte). — B. alba α vulgaris γ verrucosa Rgl. Monogr. Betul. (1861) 75, 77. — B. alba subsp. verrucosa var. vulgaris Rgl. in DC. Prodr. XVI, 2 (1868) 163. — B. alba var. verrucosa Wallr. Sched. crit. (1822) 495; Shmal'g., Fl. II, 430. — Ic.: H. Winkl. ib. f. 21; Syreishch., Il. Fl. Mosk. gub. II (1907) 47.

A tree to 20 m high; bark smooth, white, in old trees blackish-gray at the base of the trunk, deeply fissured, firm; branches often pendulous; annotinous branches reddish-brown; young branches glabrous, rather
292 densely studded with resinous warts (young sucker growth usually pubescent); leaves triangular-ovate or rhombic-ovate, broadly cuneate or almost truncate at base, mostly with lateral angles not rounded, on suckers sometimes cordate, acuminate at apex, doubly sharp-toothed on the margin, rather thin, smooth on both sides, 3.5—7 cm long, 2.5—5.5 cm broad; petiole 2—3 cm long, one-third to one-half, rarely one-fourth, the length of the blade; staminate aments 2 or 3 together at the ends of branches, pendulous, 5—6 cm long; pistillate aments solitary on short lateral branchlets, subtended by 2 or 3 leaves, 2.5—3 cm long and 9—10 mm thick; bracts of fruiting aments greenish or greenish-brown, ca. 5 mm long and broad, ciliate, in upper part often covered with short hairs; lateral lobes broad, subovate, obtuse, somewhat reflexed, the smaller median lobe ligulate; nutlet oblong-elliptic, the wings 2—3 times as broad as the nutlet, produced as far as the tips of stigmas and forming a narrow sinus. May. (Plate XIV, Figure 6).

In pure stands and together with other species in the forest belt and to some extent in the forest-steppe belt. — European part: Kar.-Lap., Lad.-Ilm., U. V., V.-Kama, U. Dnp., M. Dnp., V.-Don., Transv., Crim.; W. Siberia: Ob, U. Tob. (Mugodzhary), Irt., Alt. **Gen. distr.**: Scand., Centr. Eur., Med., Bal. Described from Europe. Type in Paris.

Note. Giving rise to hybrids, especially with B. pubescens, which often grow close to B. verrucosa, or else to B. pubescens and B. humilis.

Economic importance. Owing to its quality, the wood provides valuable material for various jobs, especially in carriage construction. Also furnishes valuable fuel.

18. **B. platyphylla** Sukacz. in Trav. Mus. Bot. Ac. Sc. Pétersb. VIII (1911) 220. — B. alba Ldb. Fl. Ross. III, 650 (ex parte); Turcz. Fl. baic.-dah. II, 2, 127. — B. alba var. typica Trautv. in Maxim. Prim. Fl. Amur. (1859) 249 (ex parte); Korsh. A. H. P. XII (1893) 389 (ex parte). — B. alba subsp. latifolia α Tauschii Rgl. Bull. Soc. Nat. Mosc. XXXVIII (1865) 399 et in DC. Prodr. XVI, 2 (1868) 165 (ex parte). — B. latifolia Kom., Fl. Manchzh. (1904) 38, non Tausch. — B. japonica β Tauschii H. Winkl. in Engl. Pflzr. H. 19 (1904) 78 (ex parte). — Ic.: Sukacz. ib. tab. III.

A tree to 20 m high, with white bark; branches smooth, partly glandular, rarely heavily glandular or eglandular; leaves squarely truncate, rounded-truncate, or broadly cuneate at base, acuminate, simply or rarely doubly serrate, sparingly or scarcely resin-dotted and without pubescence in the axils of veins beneath, glabrous, on the margin sparsely hairy, 5—7 cm long, 3.5—6 cm broad, the smooth petiole 1.5—2.5 cm long; fruiting aments cylindric, 2.5—3 cm long, 7—9 mm broad, the peduncle 1 cm long; lateral
293 lobes of the bract divergent, rounded or truncate, the short middle lobe lanceolate; wings as broad as or slightly broader than the nutlet, projecting beyond it and nearly reaching the stigmas. May. (Plate XIV, Figure 3).

In pure and mixed stands; profiting from forest fires to increase its distribution area. — E. Siberia: Lena-Kol., Dau.; Far East: Ze.-Bu. Described from Dauria.

19. **B. mandshurica** (Rgl.) Nakai in Botan. Mag. Tokyo XXIX (1915) 42. — B. alba subsp. mandshurica Rgl. Bull. Soc. Nat. Mosc. XXXVIII (1865) 399 et in DC. Prodr. XVI, 2 (1868) 165. — B. latifolia Kom., Fl. Manchzh. II, 38, ex parte, non Tausch. — B. japonica var. mandshurica H. Winkl. in Engl. Pflzr. H. 19 (1904) 78. — Ic.: Rgl. Bull. t. 7, f. 15; Kom. and Alis., Opred. rast. Dal'nevost. kr. I (1931) Plate 131, 1.

A tree to 15 m high, with white bark; young shoots reddish-brown, glabrous, finally covered with round white lenticels; leaves deltoid, broadly cuneate at base, long-acuminate, coarsely toothed on margin, glabrous, dark green above, pale beneath, with 5—7 (8) pairs of veins, 5—6 cm long, the petiole 1.5—2 cm long; fruiting aments broadly cylindric, 3—4.5 cm long, 0.8—1 cm thick; bracts cuneate, the oblong median lobe slightly longer than the broad divergent lateral lobes; wing as broad as or somewhat broader than the nutlet. April—May.

Forests, forest margins. — Far East: Uss. **Gen. distr.**: Manchuria. Described from the shores of Ol'ga Gulf. Type in Leningrad.

20. **B. japonica** Sieb. in Verh. Batav. Gen. XII (1830) 25. — B. japonica var. kamtschatica H. Winkl. in Engl. Pflzr. H. 19 (1904) 61; Kom., Fl. Kamch. II (1929) 41. — B. alba subsp. latifolia var. kamtschatica Rgl. in DC. Prodr. XVI, 2 (1868) 165. — B. platyphylla Hultén, Fl. of Kamtch. II, 32, non Sukacz. — Vernacular name: presnets.

A big tree, with white or in age gray bark and a slim or rarely spreading top, young branches verrucose-glandular; leaves oval or broadly oval, cuneately truncate or rounded or rarely shallow-cordate at base, acuminate, dentate or doubly dentate-serrate, glabrous or rarely in the axils of veins verrucose or puberulous, with 4 or 5 pairs of veins, dull above, paler beneath,

3—7 cm long; fruiting aments cylindric, 6—7 mm long, pedunculate; bracts with a short-cuneate basal portion, the divergent rounded or angular lateral lobes as long as or shorter than the middle lobe; nutlet oblong, ca. 3 mm long; wings of the nutlet undulate or rarely toothed on the margin. Fl. May; fr. July.

In dense mixed forest stands or forming small groves in river valleys. — Far East: Kamch., Okh. **Gen. distr.:** N. Japan. Described from Japan.

294 21. **B. grandifolia** Litw. in Trav. Mus. Bot. Ac. Sc. Pétersb. II (1905) 98. — B. grandifolia var. pubescens Kuzen., Tr. Bot. Muz. Akad. Nauk XII (1914) 64. — Ic.: Litw. l. c.; Kuzen. ib. tab. II, f. 1; tab. VI, f. 1—4.

A small or medium-sized tree, with light whitish or pinkish smooth bark; branches blackish-brown, smooth, partly dark with lenticels; leaves suborbicular or ovate, broadly cuneate or truncate or rarely subcordate at base, tapering to a point, simply or doubly serrate, coriaceous, dark green smooth and glandular-dotted on the veins above, paler and glabrous beneath (in var. pubescens Kuzen. with tufts of hairs in the axils of veins), 8—12 cm long and 7—11 cm broad; a small proportion of the leaves smaller, 4—6 cm long and 3—5 cm broad; petiole glabrous, 3—3.5 cm long; fruiting aments cylindric, 4—4.5 cm long and ca. 1 cm broad, the slender smooth peduncle 1—1.5 cm long; bracts 6—7 mm long and ca. 5 mm broad, hairy, with a stalk ca. 3 mm long, the angular horizontally spreading lateral lobes about as long as or slightly longer than the lance-elongate obtusish middle lobe; nutlet ovoid or ovaloid, 2—2.5 mm long and ca. 1.5 mm broad; wings 1.5 times as broad or rarely as broad as the nutlet and greatly surpassing it at apex. April—May. (Plate XIV, Figure 2).

Deciduous or mixed woods on mountain slopes and in valleys of mountain streams. — E. Siberia: Lena-Kol.; Far East: Ze.-Bu. (NW). Endemic. Described from the Lena-Kolyma area, from the Amga River basin. Type in Leningrad.

22. **B. ajanensis** Kom. in Not. system. Herb. Horti Bot. Petrop. II, 33—34 (1921) 130.

A medium-sized tree; branches blackish-purple, smooth; young branchlets glandular, puberulous, brown, dingy; buds glutinous; leaves ovate, broadly cuneate or subtruncate at base, long-acuminate, almost sinuately double-dentate with slender teeth, chartaceous, thin, with 5 or 6 pairs of veins, 5—9 cm long and 5—7 cm broad, the petiole 2—3 cm long; fruiting aments 3.5—4 cm long, 10—12 mm broad; bracts slightly hairy, 5—7 mm long and 4—5 mm broad, attenuate-tapering toward base, the middle lobe acuminate, the lateral ones falcate; nutlet 2.5—3 mm long and ca. 1.2—1.4 mm broad, the wing much broader than the nutlet and almost reaching the stigmas. May.

Valleys. — Far East: **Okh.** Endemic. Described from the Aldoma River basin. Type in Leningrad.

23. **B. Cajanderi** Sukacz. in Acta Forest. Fennica 34/13 (1929) 1. — Ic.: ib. f. 11.

A tall tree; young branches smooth, with scattered glands and warts;
295 grown leaves rhombic or broadly rhombic, with broad-cuneate or rounded

base, long-acuminate, doubly dentate-serrate, smooth, nerveless, 4—5 (6) cm long, 3—4 (5) cm broad; petiole smooth, eglandular, 1 (2) cm long; fruiting aments cylindric, slightly nodding, 2—2.5 cm long, 0.7—0.8 cm thick, the peduncle 10—12 cm long; bracts rather indurated, pubescent on the margins at the back, somewhat tapering at base, the middle lobe more or less elongated, the rounded-rhombic lateral lobes erect; nutlet elongate or ovaloid, the wings as broad as or slightly narrower than the nutlet, surpassing its pubescent apex. May.

Forests. — E. Siberia: Lena-Kol. Endemic. Described from the Lena River. Type in Leningrad.

24. B. pubescens Ehrh. in Beitr. Naturk. V (1789—90) 160; Ldb. Fl. Ross. III, 651; H. Winkler in Engl. Pflanzr. H. 19 (1904) 81; Kryl., Fl. Zap. Sib. IV, 789. — B. alba L. Sp. pl. (1753) 982, ex parte.; Ldb. Fl. Alt. IV, 244, ex parte. — B. alba var. pubescens Rgl. Monogr. (1861) 82 et in DC. Prodr. XVI, 2 (1868) 166.

A medium-sized or low tree, sometimes frutescent and sterile; bark of the trunk white; branches ascending or spreading (not pendulous); young branchlets reddish-brown, pubescent, without resinous warts; leaves ovate or rarely rhombic-ovate, at base rounded or rarely cordate, acuminate, rather sharply toothed on the margin, firm, lustrous above, paler beneath, 4—6 cm long, 2.5—5 cm broad; young leaves and shoot-leaves glutinous, rather heavily pubescent, finally pubescent only beneath, with warts in the axils of veins; petiole pubescent, one-fourth to one-half the length of the blade; staminate aments 2 or 3 together at the ends of branches, pendulous, 5—6 cm long; fruiting aments solitary on short 2- or 3-leaved lateral branchlets, 2.5—3 cm long and 9—10 mm thick, borne on a pubescent peduncle; bracts 3—5 mm broad, covered in upper part with short hairs, ciliate on the margin; median lobe elongate-oblong; lateral lobes spreading, broad, round-tipped or obtuse to subtruncate; nutlet oblong-ellipsoid, ca. 2 mm long, the wings 1.5 times as broad as the nutlet and reaching tips of stigmas. April — May. (Plate XV, Figure 3).

Damp forests and forest margins, bogs, edges of swamps, lakeshores, and burned sites. In the Caucasus in the high-mountain zone. — European part; all regions except the Crimea, sporadically in the S.; Caucasus: all regions except Talysh; W. Siberia: all regions; E. Siberia: Yen., Ang.-Say. Gen. distr.: Scand., Centr. and Atl. Eur. Described from W. Europe.

The following varieties are recorded.

1) var. ovalifolia Sukacz. in Bull. de l'Acad. des Sc. de St. Pétersb. VIII, No. 3 (1914) 233. — B. pubescens var. alapica Baran., Izv. Zap. Sib. Otd. Russ. Geogr. obshch. IV, I (1924) 50. — B. alba var. vulgaris
296 subvar. typica f. ovalis C. K. Schn. Handb. Laubholzk. I (1904) 117. — Leaves ovate, short-acuminate, at base rounded or subcordate;

2) var. rhombifolia (Tausch) Sukacz. in Bull. de l'Acad. des Sc. de St. Pétersb. VIII, No. 3 (1914) 233. — B. rhombifolia Tausch in Fl. Ratisb. 2 (1838) 752. — B. alba subsp. pubescens var. rhombifolia Rgl. in DC. Prodr. XVI, 2 (1868) 168. — B. alba var. vulgaris subvar. typica f. rhomboidalis C. K. Schn. l. c. 117. — Leaves rhombic, acuminate, at base broadly cuneate;

(297)

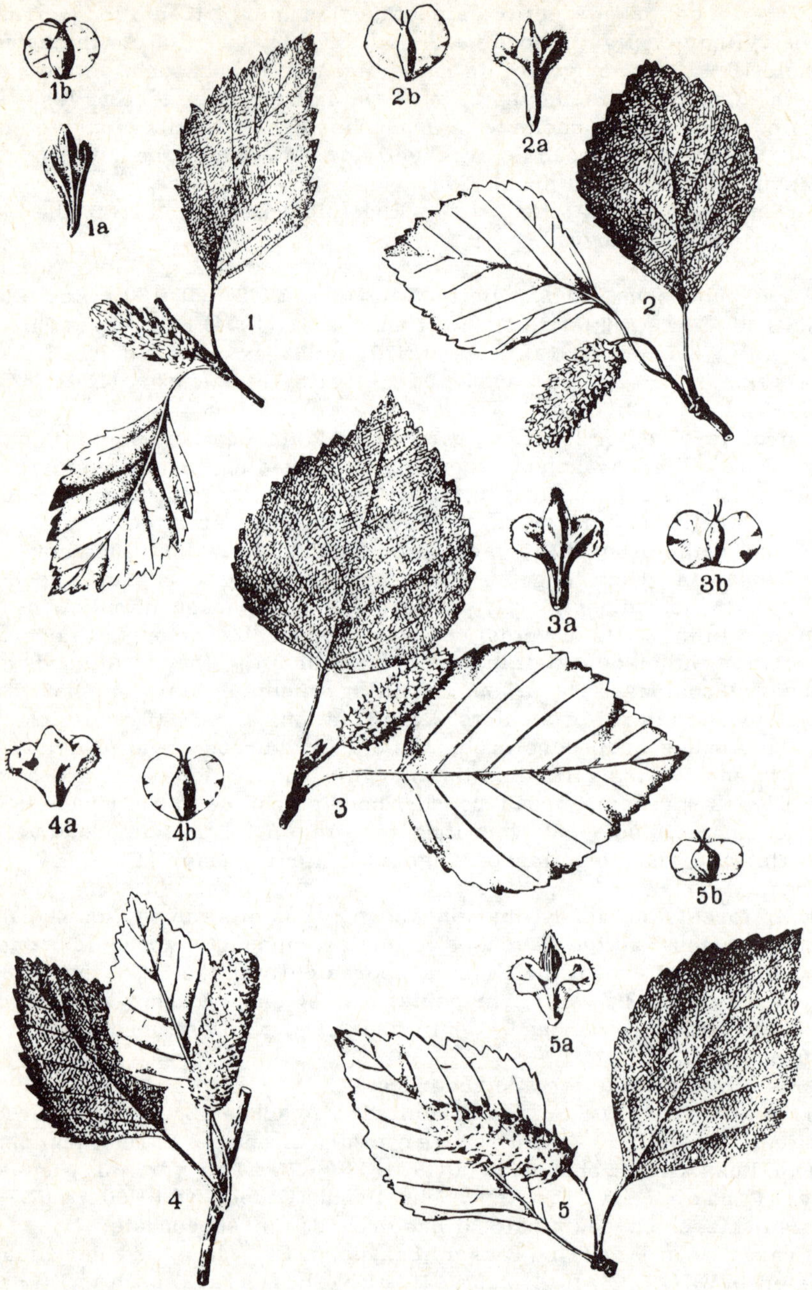

PLATE XV. 1. Betula microphylla Rgl. — 2. B. tortuosa Ldb. — 3. B. pubescens Ehrh. — 4. B. schugnanica Litw. — 5. B. procurva Litw.

3) var. glabra Fick. ex C. K. Schn. Handb. Laubholzk. I (1904) 117; Asch. et Graebn. Synops. Mitt. Eur. Fl. IV (1908—13) 400; Andreev, Tr. po prikl. bot, gen. i sel., XXI, 2 (1928—29) 194; Pachos., Opis. rast. Khers. gub. Lesa (1915) 185. — Leaves and shoots glabrous or scarcely pubescent even when young. Sands in the S. part of the distribution area;

4) var. sibakademica Baran., Izv. Zap. Sib. otd. russk. Geogr. obshch. IV, I (1924) 52. — Differs from the type in the pronounced brownish-black coloration of the bark.

The following hybrids occur:

B. pubescens × B. nana, H. Winkl. in Engl. Pflzr. H. 19 (1904) 93. — B. intermedia Thomas in Gaud. Fl. Helv. VI (1830) 176. — B. hybrida Rgl. Monogr. (1861) 97. — In the area of occurrence of both species.

B. pubescens × B. verrucosa (with predominating characteristics of B. pubescens grex pseudopubescens Morgent. in Vierteljahr. der Naturf. Gesell. in Zürich. 60 (1915). — B. subpubescens H. Winkl. Mitteil. d. deutsch. Dendr. Gesell. 52 (1930) 40.

Economic importance. The wood of B. pubescens differs little in quality from that of B. verrucosa and is similarly used in cart construction, etc. Valuable carving material is furnished by the "birch burls," accrescences formed about the root or more rarely on the trunk or on sucker growth. The wood represents a variant of normal birch wood in having irregular twisted annual rings and medullary rays. This wood is highly valued for carpentry and turning jobs and also for veneers. Considerable use is also made of birch bark, especially in the North: the Zyrian-Ostyak peoples, sometimes stitching together two or three layers, use it as waterproof cover for their tents, etc.; to increase its flexibility they sometimes scour it with fat. It is also used for canoes and certain small articles (wickerwork, bast shoes, etc.), and for tar production. A beverage is made out of the spring sap (as in the case of B. verrucosa).

299 25. **B. tortuosa** Ldb. Fl. Ross. III (1851) 652; Kryl., Fl. Zap. Sib. IV, 791. — B. davurica Ldb. Fl. Alt. IV (1833) 245, non Pall. — B. tortuosa α genuina Rgl. Monogr. (1861) 90. — B. alba subsp. tortuosa α genuina Rgl. in DC. Prodr. XVI, 2 (1868) 169. — B. pubescens var. tortuosa Koehne ex H. Winkl. in Engl. Pflzr. H. 19 (1904) 82.

A rather low rugged tree with grayish-white or yellowish bark and flexuous branches; young branchlets pubescent; annotinous grayish-brown, resinous-verrucose; leaves of fertile shoots rhombic-ovate to suborbicular-rhombic, obtusely angular or broad-cuneate at base (leaves of sterile shoots with rounded or shallowly cordate base), short-acuminate or rarely subobtuse, unequally dentate, at first pubescent, becoming glabrous, 2.5—5 cm long and 2—3.5 cm broad (leaves of sterile shoots to 7 cm long and 6 cm broad); petiole one-third to one-half the length of the blade, with scattered short hairs; fruiting aments 15—25 mm long and 6—8 mm broad; bracts ciliate, the oblong obtuse lateral lobes obliquely ascending; wings about as broad or somewhat narrower than the broadly ellipsoid nutlet, their upper margin surpassing the nutlet but overtopped by the slender stigmas. May. (Plate XV, Figure 2).

Subalpine subzone, approaching the timberline in the Arctic Region. — Arctic: Arc. Eur., Arc. Sib. (W.); European part: Kar.-Lap. (N.), Dv.-Pech. (N.); W. Siberia: Alt. Described from Altai. Type in Leningrad.

26. **B. Kusmisscheffii** (Rgl.) Sukacz. in Bull. Acad. Sc. Pétersb. VIII (1914) 233; Kryl., Fl. Zap. Sib. IV, 791. — B. alba var. Kusmisscheffii Rgl. in Gartenfl. IX (1860) 303. — B. pubescens var. Kusmisscheffii H. Winkl. in Engl. Pflanzenr. H. 19 (1904) 82. — B. tortuosa var. Kusmisscheffii Rgl. Monogr. (1861) 90; DC. Prodr. XVI, 2 (1868) 169. — Ic.: Rgl. Gartenfl. IX (1860) tab. 303; Monogr. tab. VII, f. 1—2.

A tall spreading shrub, with somewhat crooked stems, or a small tree 2—4 m high, with twisted trunk, stout crooked branches, and short tawny lustrous branches; young branchlets pubescent; leaves suborbicular or broadly suborbicular, rounded-truncate or subcordate at base, obtuse or slightly acuminate at apex, coarsely and unequally dentate (leaves of sterile shoots somewhat narrower, acute, more deeply incised between the teeth or double-toothed); young leaves puberulous, becoming glabrous, not verrucose in the vein axils, thick, coriaceous, prominently reticulate-veined, 3—4 cm long, 2—4 cm broad; petiole one-third to half as long as the blade; fruiting aments 0.5—2 cm long; bracts 3-lobed with short cuneate basal portion; lobes subequal or the middle lobe slightly narrower and longer than the lateral, these erect or slightly divergent, rounded or acuminate; wing as broad as or slightly narrower than the ovaloid nutlet. May—June.

300 Arctic: Arc. Eur., Arc. Sib. (W.). Endemic. Described from the White Sea coast. Type in Leningrad.

27. **B. baicalensis** Sukacz. in Bull. Acad. Sc. Pétersb. VIII (1914) 233.

A tree; trunk low, tortuous; branches recurved, ramose; bark more or less brown, exfoliating into slightly separating sheets; young branches glabrous or puberulent, glandular; leaves rhombic-ovate, rather long-cuneate at base, broadest at or somewhat above the middle, acutish or obtusish, coarsely and unequally crenate-serrate, entire near the base, coriaceous, glabrous and lustrous above, paler and glabrous or with scattered hairs on the veins beneath, with (4) 5 (6) pairs of veins, 3—4 (7) cm long, 2—3 (5.5) cm broad, the petiole 2.5—3 times shorter than the blade; fruiting aments cylindric, slightly nodding or rarely almost upright, 2—3 cm long and 5—9 mm broad, the glabrous or puberulous peduncle one-third to two-fifths the length of the ament; bracts cuneate at base, with middle lobe slightly pubescent on the margin, ciliate, and erect roundish or rounded-rhombic lateral lobes half the length of the middle one; nutlet oblong, rarely obovoid, pubescent at apex, the wings about equaling or slightly surpassing the nutlet. June.

Sandy shores of Lake Baikal. — E. Siberia: Dau. (mouths of the Upper Angara and Turka rivers). Endemic. Described from the shores of Lake Baikal. Type in Leningrad.

28. **B. irkutensis** Sukacz. in Trav. Mus. Bot. Acad. Sc. Pétersb. VIII (1911) 226. — Ic.: ib. tab. IV.

A tree to 2 m high, with crooked stem; bark dark brown below, light-colored with yellowish tint higher up; young branchlets sparingly glandular, puberulous, becoming glabrous; leaves ovate-oblong or ovate-rhombic, with cuneate or rounded-cuneate base, acute or more or less acuminate, coarsely and unequally serrate, on both sides with 7—9 pairs of veins, pubescent on the veins above, with minute axillary tufts of hairs beneath, 5—7 cm long and 3—4 cm broad, the rather densely pubescent petiole 1.2—1.5 cm long; fruiting aments elongate-cylindric, suberect or slightly nodding, 2.5—3 cm long and ca. 1 cm broad, borne on a peduncle 7—9 mm long; bracts more or less long-cuneate at base, 3-lobed to one-third the length or less, ciliate, the straight lateral lobes shorter than the elongate middle lobe; nutlet ovaloid or obovoid, pubescent at apex, the wing half as broad as the nutlet. May—June (?).

Subalpine zone of barren Siberian heights. — E. Siberia: Ang.-Say. Endemic. Described from the Podgolechnaya River between the Lena and Kirenga rivers. Type in Leningrad.

301 29. **B. kirghisorum** Sav.-Ryczg. in Monit. du Jardin bot. de Tiflis 29 (1912) 11. — Vernacular name = "kyzyl kain."

A rather small tree, with dull white bark and a loose top; branches upright or obliquely ascending; young branchlets reddish-brown with or without a grayish tint, densely beset with whitish resinous warts; leaves ovate to ovate-oval, enlarged at or somewhat below the middle, cuneate at base, acuminate, equally double-serrate, with prominent brown veins beneath, verrucose in the vein axils, 2.7—6.5 cm long, 1.3—4.4 cm broad; petiole pubescent, 0.7—1.7 cm long, (2/11) 2/7 (1/2) the length of the blade; fruiting aments 1.2—2.3 cm long, 0.4—0.8 cm broad, obliquely ascending; bracts stiff, pubescent; wings narrower than or rarely as broad as the nutlet. May (?).

Amid pine woods. — Centr. Asia: Ar.-Casp. Endemic. Described from the vicinity of Turgai. Type in Leningrad.

30. **B. tianschanica** Rupr. in Mém. Acad. Sc. Pétersb. VII ser. XIV, No. 4 (1869) 72; Litv., Tr. Bot. Muz. Akad. Nauk XII (1914) 96. — B. alba ssp. songarica var. microphylla Rgl. in Bull. Soc. Nat. Moscou LI (1868) 223. — Ic.: Litv. ib., Plate X, 10, Plate XII, 10.

A tree more than 2 m high; bark pinkish; leaves ovate or narrow-ovate to subrhombic-oval, acuminate, broadly cuneate, more than 4 cm long, 2.5 cm broad, entire at the cuneate base, unequally serrate on the margin; petiole 1 cm long, glabrous like the blade; fruiting aments upright, 1.8 cm long, 0.7 cm thick, the slightly pubescent stalk 3—4 mm long; bracts ca. 5.5 mm long, sparingly ciliate, the ascending rounded lateral lobes much shorter than the linear middle lobe; nutlet 2 mm long, ovoid, slightly pubescent at the base of the style, the wings about as broad as the nutlet, their upper margin somewhat surpassing it. May.

Valleys of mountain streams. — Centr. Asia: T. Sh. **Gen. distr.**: Dzu.-Kash. Described from the Tien Shan Mountains. Type in Leningrad.

31. **B. microphylla** Bge. in Mém. prés. à l'Acad. Pétersb. par div. sav. II (1835) 606; Ldb. Fl. Ross. III, 652; Rgl. in DC. Prodr. XVI, 2, 169; H. Winkl. in Engl. Pflzr. H. 19 (1904) 88; Kryl., Fl. Zap. Sib. IV, 792. — B. fruticosa var. cuneifolia Rgl. Monogr. (1861) 35. — Ic.: Tr. Bot. Muz. Akad. Nauk XII (1914) I, 11, 12.

A tree 4 — 5 m high; bark yellowish-gray, sometimes pinkish; annotinous branches brownish-gray, densely beset with resinous warts; young branchlets also pubescent; leaves rhombic-ovate or rhombic-obovate,
302 entire at the broad-cuneate or obtusely angular base, short-acuminate or rarely obtuse at apex, 1.5 — 3 (4) cm long and 10 — 15 cm broad, at first slightly pilose on the veins and on the margin, finally glabrous, somewhat paler beneath; petiole puberulous or glabrous, one-fifth to one-third the length of the blade; fruiting aments borne on short lateral usually 2-leaved spurs, oblong-ovaloid, 10 — 22 mm long and 7 — 8 mm thick, the stalk 2 — 9 mm long; bracts obcuneate, pubescent on both sides, ciliate, 3-lobed, 4.5 — 6 mm long and 2 — 3 mm broad, the ascending slightly divergent lateral lobes slightly shorter than or rarely as long as the oblong or sublanceolate middle lobe; nutlet broadly ellipsoid, 2.5 — 3 mm long, the wings slightly broader than, or to 1.5 times as broad as the nutlet and slightly surpassing it. May. (Plate XV, Figure 1).

Desert-steppe valleys of mountain streams. — W. Siberia: Alt. **Gen. distr.**: N. Mongolia and the Tannu-Tuva Republic.* Described from Altai. Type in Leningrad.

32. **B. Rezniczenkoana** (Litw.) B. Schischk. in Kryl., Fl. Zap. Sib. IV (1930) 793. — B. *microphylla* var. *Rezniczenkoana* Litw., Trudy Bot. Muz. Akad. Nauk XII (1914) 97. — B. *kirghisorum* Andreev, Tr. Biuro po prikl. bot., gen. i sel. XXI, 2 (1929) 190, non Sav.-Ryczg. — Ic.: Andreev, ibid. Fig. 3.

A rather small tree with yellowish-gray bark; branches brownish or brownish-gray, covered with greenish and finally whitish resinous warts; young branchlets also pubescent; leaves small, 15 — 40 mm long and 10 — 30 mm broad, rhombic-ovate, entire at the broad-cuneate base, elsewhere unequally dentate, short-acuminate or obtuse, paler beneath, often pilose on the veins or glabrous; fruiting aments borne on a short pubescent peduncle, oblong-ovaloid, 10 — 25 mm long, 5 — 7 mm thick; bracts 3 — 5 mm long, glabrous or scarcely hairy, the subhorizontally spreading hairy-margined truncate or emarginate lateral lobes nearly twice as broad as the middle lobe; nutlet 2 — 3 mm long and 3 — 4 mm broad; the wings slightly broader than, or to 1.5 times as broad as the nutlet. April—May.

Banks of mountain streams and ravines. — W. Siberia: Alt. (western part and Kolbinskii Range). Described from the upper reaches of the Bol'shaya Bukon River, a tributary of the Irtysh. Type in Leningrad.

33. **B. Kelleriana** Sukacz. in Bull. de l'Acad. des Sc. St. Pétersb. VIII (1914) 235; Kryl., Fl. Zap. Sib. IV, 795.

A tree 4 — 6 m high, with yellowish-gray bark; young branches densely
303 covered with resinous warts, slightly pubescent, becoming glabrous; leaves ovate or elliptic, acuminate, unequally obtusely dentate, glabrous, (1.5) 3 — 3.5 (4.5) cm long and 1.5 — 3.5 cm broad, the petiole pubescent; fruiting aments upright, ovaloid-cylindric, 17 — 20 mm long and 6 — 9 mm thick, borne on a peduncle 1 — 10 mm long; bracts ca. 5 mm long, cuneate at base, ciliate, the angular-rhombic or rounded lateral lobes about as long as the narrower middle lobe; wings about half as broad as the obovoid nutlet. April (?).

* [Now Tuva Autonomous Region of the USSR.]

Steppe valleys of mountain streams. — W. Siberia: Alt. Endemic. Described from the Chu River. Type in Leningrad.

34. **B. Saposhnikovii** Sukacz. in Bull. de l'Acad. des Sc. de St. Pétersb. VIII (1914) 235.

A tree 3—5 m high; young branches densely covered with dark and finally white warts, slightly pubescent, becoming glabrous; leaves rhombic or ovate-rhombic, cuneate at base, acute at apex, crenate-serrate, entire at base, glabrous or with scattered short hairs on the veins beneath, (11) 13—16 (22) mm long, (8) 10—11 (114) mm broad, with 4 or 5 pairs of veins; petioles glabrous, one-third the length of the blade; fruiting aments upright, oblong, 10—12 mm long, 5—6 mm thick, borne on a peduncle 4 mm long; bracts 3—3.5 mm long, with a short cuneate basal portion, 3-lobed nearly to the middle; lobes subequal, scarcely recurved at apex, oblong, obtusish, erect; nutlet oblong, 2—2.5 mm long, with very narrow wings. May.

Subalpine zone. — Centr. Asia: T. Sh. Endemic. Described from the Indyl'chek River natural boundary area. Type in Leningrad.

35. **B. turkestanica** Litw. in Trav. Mus. Bot. Ac. Sc. Petrogr. XII (1914) 90. — Ic.: Ib. tab. XII, f. 5.

A tree; bark peeling off, dull lurid; young branches glandular, more or less pubescent; leaves ovate, with a cuneate, broad-cuneate, or obtusish base, acute or acuminate, at first more or less pubescent on both sides, finally glabrescent, more or less pubescent on the veins beneath but glabrous in the axils, entire at base, sharply double-serrate elsewhere, ca. 5 cm long, 3.5 cm broad; petiole ca. 1.75 cm long, finely pubescent and covered with long hairs; fruiting aments cylindric, 2.5 cm long, 0.7—0.8 cm broad, nodding, the slender pubescent peduncle covered with scattered long hairs; bracts pubescent on the margin; cuneate, ca. 5 mm long; lateral lobes shorter than the triangular middle lobe, spreading or somewhat ascending, more or less angular; nutlet ovoid or almost obovoid, 2.5 mm long, the wings about as broad as or slightly narrower than the nutlet, their upper margin not surpassing the nutlet. May.

Along mountain streams. — Centr. Asia: Pam.-Al., T. Sh. Endemic. Described from the Aleksandrovskii [Kirghiz] Range in the Frunze area. Type in Leningrad.

304 36. **B. alajica** Litw. in Trav. Mus. Bot. Ac. Sc. Petrogr. XII (1914) 89. — Ic.: tab. IX, f. 6, tab. XI, f. 7.

A tree; bark of the trunk yellow, peeling off; old branches brownish-black, glandular; young branchlets slightly pubescent and pilose, resinous-glandular; leaves ovate or broad-ovate, truncate or broadly rounded at base, acute, 3.5 cm long and 2.7—3 cm broad, entire at base, elsewhere simply and sharply serrate, pubescent on the veins beneath, on both sides rather densely resinous-glandular; petiole pubescent, 1 cm long; fruiting aments cylindric, ca. 2 cm long and 1 cm thick, upright (at the ends of branches in USSR specimens), the peduncle puberulous; bracts ca. 5 mm long, cuneate at base, slightly pubescent on the margin, the middle lobe narrow, the ascending broad lateral lobes truncate or entire at apex, nutlet 2.5 mm long, ovoid, the wing barely as broad as the nutlet.

Along mountain streams. — Centr. Asia: Pam.-Al. Described from the Alai Range, at an altitude of 3,300 m (Khorzinskii, No. 4906). Type in Leningrad.

37. **B. procurva** Litw. in Trav. Mus. Bot. Ac. Sc. Petrogr. XII (1914) 93. — Ic.: ib, tab. X, f. 9, tab. XII, f. 9. — B. alba ssp. songorica Rgl. in A. H. P. VI (1879) 476.

A tree of medium size; trunk branched, more or less twisted, with yellowish or orange-yellow bark; young branches slightly pubescent, resinous-glandular; leaves elongate-subrhombic or broadly ovate, cuneate or broad-cuneate at base, acute or slightly acuminate, entire at base, elsewhere coarsely and often unequally dentate, ca. 6.5 cm long and 3.5 cm broad, sometimes slightly pubescent on the veins beneath; petiole 1—1.5 cm long, glabrous; fruiting aments ca. 2.8 cm long and 1 cm broad, cylindric, rounded at apex, upright or divergent, the rather densely pubescent peduncle 1—1.5 cm long; bracts ca. 5 mm long, slightly ciliate, short-cuneate at base, the prominent middle lobe ligulate, the ascending lateral lobes rounded; nutlet ca. 2.5 mm long, ovoid, the wings about as broad as or somewhat broader than the nutlet and slightly surpassing it at apex. (Plate XV, Figure 5).

Valleys of mountain streams. — Centr. Asia: Pam.-Al. Endemic. Described from the vicinity of Arslanbob (Andizhan area, D. I. Litvinov, 1899). Type in Leningrad.

38. **B. schugnanica** (B. Fedtsch.) Litw. in Trav. Mus. Bot. Ac. Sc. Petrogr. XII (1914) 93. — B. alba var. schugnanica B. Fedtsch. Trav. Mus. Bot. Ac. Sc. Pétersb. I (1902) 163. — Ic.: Litw., tab. X, f. 8, tab. XII, f. 8.

A low tree; branches reddish; young branchlets densely pubescent; leaves subcoriaceous, elongate to subovate, ca. 3.5 cm long and 2.25 cm broad, acute or acuminate, at base cuneate, unequally and almost doubly serrate, on both sides densely pubescent on the veins, slightly resinous-
305 glandular beneath; petiole pubescent, 1.5 cm long; fruiting aments upright, cylindric, ca. 2.5 cm long and 0.7 cm thick, the villous peduncle ca. 1.2 cm long; bracts ca. 4 mm long, cuneate at base, the margin somewhat pubescent, the short triangular middle lobe obtusish, the spreading subfalcate lateral lobes round-tipped or obliquely truncate; wings about as broad to 1.5 times as broad as the elongate-ovoid nutlet. (Plate XV, Figure 4).

Banks of mountain streams. — Centr. Asia: Pam.-Al. Endemic. Described from the Gunt River valley. Type in Leningrad.

39. **B. Korshinskyi** Litw. in Trav. Mus. Bot. Ac. Sc. Petrogr. XII (1914) 89. — Ic.: ib. tab. IX, f. 4; tab. XII, f. 4.

A shrub (?); young branches glandular, sparsely long-pilose or glabrous; leaves broad-ovate or rarely narrow-ovate, acuminate, at the cuneate or broad-cuneate base entire, elsewhere simply serrate, more than 4 cm long and 3 cm broad, glabrous or with scattered hairs on the veins beneath and on the margin; petiole to 12 mm long, glabrous or sparingly hairy; leaves and petioles of young branchlets densely hairy; fruiting aments very small, 1—1.5 cm long, 0.8—1 cm broad, upright, the erect rigid glabrous peduncle ca. 8 mm long; bracts 4 mm long, glabrous, broad-cuneate, the narrow acute middle lobe much longer than the lateral ones, these ascending, round-tipped, about as broad as the middle lobe; nutlet 3 mm long, obovoid, the wing about as broad as or slightly broader than the nutlet and surpassing it at apex.

Mountain slopes. — Centr. Asia: Pam.-Al. Endemic. Described from the Kugart Pass, at an altitude of 1,500 — 2,000 m, in the Fergana Range. Type in Leningrad.

40. **B. pamirica** Litw. in Trav. Mus. Bot. Ac. Sc. Petrogr. XII (1914) 91. — Ic.: ib. tab. IX, f. 5, tab. XII, f. 6.

A shrub (?), much-branched; branches reddish, densely glandular; young branchlets heavily pubescent and sparsely long-pilose; leaves subcoriaceous, spatulate or lanceolate, acute to obtusish, at the broad-cuneate base entire, elsewhere unequally dentate, ca. 3.5 cm long and 2.7 — 3 cm broad, glabrous above, rather heavily pubescent on the veins beneath, the heavily pubescent petiole ca. 1 cm long; fruiting aments upright, cylindric, ca. 2 cm long, 1 cm thick, the pubescent peduncle 0.5 — 0.7 cm long; bracts ca. 5 mm long, evenly cuneate, somewhat hairy on the margin, the slightly projecting middle lobe ligulate, the shorter ascending lateral lobes round-tipped or slightly angular at apex; nutlet ca. 3 mm long, ovoid, the wings about as broad as or slightly broader than the nutlet and somewhat surpassing it at apex.

Banks of mountain streams. — Centr. Asia: Pam.-Al. Endemic. Described from the Kizil-su River (S. I. Korzhinskii, 19 August 1897, No. 5530). Type in Leningrad.

306

Genus 365. **ALNUS*** GAERTN. **

Gaertn. Fruct. et sem. II (1791) 54; H. Winkl. in Engl. Pflzr. 19 Heft IV (1904) 61.

Flowers unisexual, monoecious; staminate flowers in 3-flowered dichasia forming cylindric aments with scalelike bracts; bractlets 4; perianth of 4 or rarely fewer segments; stamens 4, opposite, their filaments extremely short, the ovaloid anthers with parallel locules, without hairs at the apex; pistillate flowers in 2-flowered dichasia, forming upright cylindric or mostly oblong aments; bracts becoming lignified, persistent in maturity; perianth none; ovary bilocular; styles short, with long stigmas; ovule solitary in each locule, pendulous, anatropous; nutlet flat, mostly winged, 1-seeded. Trees or shrubs, with alternate deciduous leaves and early caducous stipules. Venation of leaves pinnate; staminate aments developing from terminal buds of the previous year, well developed by fall, emerging in early spring; pistillate aments solitary or clustered, appearing in the leaf axils toward fall or rarely with the leaves, on short branchlets.

Up to 20 species in Europe, in the mountains of Algeria, in N. and E. Asia, in N. America, and in S. American Andes.

Fossils of the genus *Alnus* were reported for Cretaceous formations, but these reports have not been fully verified. In the Tertiary *Alnus* occurs in the Eocene and acquires wide distribution everywhere toward the end of that period when species of the Quaternary already make an appearance.

* The name for this plant used by Roman writers Vitruvius, Pliny, etc.

** Treatment by V. L. Komarov.

— Alnus cordifolia Tenore (mutatio) in Tertiary (Pliocene ?) formations of Alt. (Chingistai). — A. corylifolia Lesq. in Lower Tertiary layers of Uss. (Amagu). — A. fruticosa Rupr. in the Postpliocene of Lena-Kol. (Berezovka) and in interglacial formations of Arc. Sib. (Bol'shoi Lyakhovskii Island). — A. glutinosa Gaertn. in Tertiary (Pliocene ?) formations of Alt. (Chingistai), in the Postpliocene of Lad.-Ilm. (Petrozavodsk area; Vyaz village), U. Dnp. (Kletsovo), U. V. (Troitskoe, Likhvin, Krushma River), V.-Kama (Galich), V.-Don (near Krivobor'e, etc.), and E. Transc. (Makart in Georgia); in Pliocene (Shiraki). — A. incana Willd., as v. sibirica and mut. rotundifolia in Tertiary (Pliocene ?) formations of Alt. (Chingistai); the typical form in the Postpliocene of U. V. (glacial formations of Troitskoe and interglacial formations of Troitskoe and Likhvin). — A. Keferst einii (Goepp.) Unger., widely distributed among Tertiary formations of the Palaearctic Region; in Sarmatian formations of Bl. (Krynka); in Maiotis formations of Bl. (Odessa area: Khadzhibei, Lanzheron); in Upper Cretaceous formations of Ze.-Bu. (Tsagayan in Bureya area, the identity in need of confirmation); in Lower Tertiary formations of Uss. (Botchi), Sakh. (Korf Bay), An. (between Omocha and Cape Telegraficheskii); in addition, two varieties reported for these formations: var. longifolia Heer. and var. latifolia Heer. — A. errulata (Michaux) Willd. in Tertiary (Pliocene ?) formations of Alt. (Chingistai). — A. speciosa Eichw. reported for the Paleocene of L. Don (Kamyshin-Ushi). — A. subcordata C. A. M. in the Pliocene of
307 W. Transc. (Shiraki). — A. sp., not specifically identified, reported from a large number of locations in Tertiary and Postpliocene layers: in the Pliocene of V.-Don (Uryv: Ivnitsy, Voronezh, Krivobor'e); in the Eocene (?) of the former Kursk Province (Radubichi); in the Miocene (or Oligocene ?) of V.-Kama (Sterlitamak area); in Upper Cretaceous formations of Ze.-Bu. (Belogor'e on the Bureya River and Boguchan ?); in Lower Tertiary formations of Balkh. (Ashutas); in Lower Tertiary of Uss. (Novokievsk, Fatashi, Rechnoi, Amagu, Sikhote-Alin), Sakh. (Vaskina, Nikolaevka, etc.), Kamch. (Korf, etc.), Lena-Kol. (Tastakh); in the Postpliocene of U. Dnp. (Mikulino, Drozhzhino, Dubrova, Murava, Timoshkovichi), V.-Don (Likhvin, Krushma River), U. V. (Troitskoe), V.-Kama (Galich), L. Don (Archedinskaya woodland), Lena-Kol. (Berezovka), and Sakh. (various locations).

1. Pistillate aments clustered, developing in spring at the ends of short sparingly leafy branchlets . . . . 2.

\+ Pistillate aments solitary or clustered, developing in fall in the leaf axils and persistent through winter . . . . 5.

2. Perianth longer than stamens; anther-cells connate throughout their length; central staminate bract more or less distinctly inflated at center on the outer side . . . . 3.

\+ Perianth shorter than stamens; anther-cells free at both ends . . . . 4.

3. Leaves mostly broad-ovate, uniformly attenuate at both ends, acute, glabrous beneath; lateral veins 8—10 pairs . . . . . 1. **A. fruticosa** Rupr.

\+ Leaves mostly broad-elliptic, obtusely short-tipped, the petiole twice as long as in the preceding species . . . . 2. **A. manshurica** (Call.) Hand.-Mazz.

4. Central staminate bract not inflated at center; leaves broad-ovate or rounded-ovate, with broad subcordate base, glabrous beneath; teeth of the serrate margin very slender and longer than in related species; lateral veins 7—10 pairs . . . . 3. A. **Maximowiczii** Call.

\+ Central staminate bract more or less inflated at center; leaves broad-ovate, obtusely short-tipped; lateral veins 8 or 9 pairs . . . . 4. **A. kamtschatica** (Call.) Kom.

5. Pistillate aments solitary or rarely 2 or 3 in the bud . . . . 6.

\+ Pistillate aments several in the bud, forming a compound raceme . . . 7.

6. Leaves rounded or cordate at base, oval or oval-oblong, the teeth of the margin subequal . . . . 5. **A. subcordata** C. A. M.

\+ Leaves cuneately narrowed toward base, oblong-lanceolate or oval-oblong, the margin sharply serrate . . . . . . 6. **A. japonica** Sieb. et Zucc.

7. All strobiles borne on a rather long peduncle . . . . 8.

308 + Terminal or often all strobiles sessile or nearly so . . . . 9.

8. Leaves even in maturity more or less hairy in the axils of veins, with yellowish tomentose tufts beneath; strobiles oblong, to 20 mm long . . . . 8. **A. barbata** C. A. M.

\+ Leaves glabrous except for small axillary tufts of hairs; strobiles broadly ovaloid, to 14 mm long . . . . 7. **A. glutinosa** L.

9. Strobiles large, massive; leaves larger, thin, broad-cuneate at base, glabrous above, densely pilulose beneath . . . . . . 12. **A. tinctoria** Sarg.

\+ Strobiles smaller; leaves relatively small but firm, broad-cuneate or rounded or slightly cordate at base . . . . 10.

10. Leaves densely velutinous beneath . . . . 11.

\+ Leaves quite glabrous beneath or hairy on the veins but glabrous in the axils . . . . 10. **A. sibirica** Fisch.

11. Leaves ovate or elliptic, acute, grayish-pubescent beneath . . . . 9. **A. incana** (L.) Moench.

\+ Leaves broad-ovate or rounded-ovate or suborbicular, obtuse, densely velutinous-pubescent with rufous hairs . . . . 11. **A. hirsuta** Turcz.

Note. Hybrids between the various alder species have not been taken into consideration in the composition of this key because of the marked inconsistency of their characteristics. Guidance for their identification will have to be sought in the text and in notes referring to individual species.

Section 1. **ALNOBETULA** W. D. Koch, Synops. I (1838) 660. — Alnaster Spach, Ann. Sc. Nat. 2 sér. XV (1841) 200. — Strobiles clustered, on short leafy branchlets.

1. **A. fruticosa** Rupr. Beitr. z. Pflanzenk. Russ. Reich. II (1845) 53; Kryl., Fl. Zap. Sib. 799; Trautv. in Midd. Reise I (1847) 8, 146, 152, 168, 172. — Alnaster fruticosus Ldb. Fl. Ross. III (1851) 656. — Alnaster viridis Spach in Ann. Sc. Nat. 2 sér. XV (1841) 184 (ex parte); Turcz. Fl. baic.-dah. II, 2, 131. — Alnobetula fruticosa Rupr. in Bull. Acad. Pétersb. (1856) 434. — Alnus viridis var. sibirica Rgl. Monogr. Betul. in Mém. Soc. Nat. Mosc. III (1861) 137. — Ic.: Rgl. l. c. tab. XIV, f. 16—20. — Yakutian: abaga.

A dwarf shrub in the northern USSR, a tall shrub in the south where often arborescent with a trunk to 6 m high; bark dark gray; young branches reddish-brown, with scattered subsessile lenticels; leaves 5—10 cm long, 3—7 cm broad, ovate to broad-ovate, acute, rounded and often asymmetric or rarely broad-cuneate at base, sharply denticulate with unequal teeth
309 and often doubly dentate, the upper surface dark green, dull or lustrous, glabrous; lower surface paler, often resinous from deliquescent glands, with short rufous hairs on the principal veins, especially at the axils; staminate aments developing with the leaves, 3.5—6 cm long, with violet-brown bracts; anthers bright yellow, pistillate aments ovaloid, all borne on peduncles representing ramifications of a common stalk subtended by 1—3 leaves; nutlet winged, elliptic, the wing as broad as or slightly narrower than the nutlet; stigmas bright red. End of April, May, June, and even July (in the tundras). (Plate XVI, Figure 2).

In mountain valleys on riverbanks, pebbles and gravelly slopes, stony screes, in coniferous and mixed woods, often reaching the height of Alnus incana; in the northern USSR on riverside sands, in deciduous forests and forest borders. Greatly reduced in size in the tundras where it becomes dwarfed and procumbent and produces small-leaved forms with abbreviated tortuous branches (var. microphylla Scheutz 1888). — Arctic: Arc. Eur. (east of the Mezen River, 45°W. long., 66°N. lat.), Arctic Urals (to 68°N. lat.), Arc. Sib. (penetrating beyond the 72nd parallel); European part: Dv.-Pech., V.-Kama (Malmyzh on the Vyatka River), Urals (S. limit at 60°30'N. lat.); W. Siberia: Ob, Alt.; E. Siberia: Yen., Lena-Kol., Ang.-Say., Dau.; Far East: Ze.-Bu. **Gen. distr.**: Tannu-Tuva and N. Mongolia. Described from the Mezen River. Type in Leningrad.

2. **A. manshurica** (Call.) Hand.-Mazz. in Oesterr. Bot. Zeit. LXXXI (1932) 306. — A. fruticosa var. mandschurica Call. in C. K. Schn. Laubholzk. I (1904) 121 et in Fedde Repert. Sp. nov. X (1911) 226; Kom. in A. H. P. XXII, 59. — Ic.: Nakai, Fl. Sylv. Kor. II (1914) tab. 22.

A tree to 8 (10) m, with dark gray smooth bark, or a shrub about 3 m high; leaves 7—8 (up to 15 cm) long, 2.5—8 cm broad, with 7—9 and up to 11 pairs of lateral veins, broad-elliptic, obtusely short-tipped, broad-cuneate or shallowly cordate at base; aments as in the preceding species; wing as broad as or narrower than the nutlet and either of the same length as or surpassing the nutlet and then forming a sinus at the apex. May.

Banks of mountain streams; rarely slopes. — Far East: Uss., Uda. **Gen. distr.**: Jap.-Ch. (Manchuria, Korea). Described from Amur. Type in Leningrad (Maksimovich's specimen).

Economic importance. Along the Amur River a bark decoction yields a yellow dye for hides.

310

3. **A. Maximowiczii** Call. in C. K. Schn. Laubholzk. I (1904) 122; Kom. et Al. Key pl. Far. East. Reg. I, 428. — Gilyak: kheungi.

A shrub or tree to 10 m; bark gray, with suborbicular lenticels; young branches pale brown, with numerous very narrow lenticels; buds 1—1.3 cm long; leaves broad-ovate or rounded-ovate, with broad often cordate base, 7—10 cm long, 7—8 cm broad, the even margins serrulate with very slender and rather long teeth; lateral veins 7—10 pairs; strobiles 1.5—2 cm long, their peduncles 0.8 — 2 cm or even 3 cm long; nutlet winged, with 2 distinct lateral wings, these somewhat narrower than the nutlet. May — June.

Banks of wooded mountain rivulets, in shade. — Far East: Uss. (N. part near the sea), Uda, Sakh. **Gen. distr.**: Jap.-Ch. (N. and Centr. Japan). Described from N. Japan (Hakodate).

4. **A. kamtschatica** (Call.) Kom. comb. nova. — A. sitchensis var. b. kamtschatica Call. in C. K. Schn. Laubholzk. I (1904) 123. — A. fruticosa var. kamtschatica Kom. in Fl. penins. Kamtsch. I (1929) 47. — A. Maximowiczii Hultén (non Callier) Fl. Kamtch. II (1928) 34. — Ic.: Kom. l. c. tab. 7, 8.

A shrub with a stout main stem almost completely appressed to the ground; branches ascending forming a dense head 1 — 3 m high; bark dark gray, with lighter lenticels; leaves oval, short-acuminate, with a rounded asymmetric or symmetric base, 5 — 10 cm long and 3 — 7 cm broad, with petiole 1 — 2 cm long, sharply denticulate on the even or undulate-sinuate margin, lighter beneath, glandular; lateral veins 8 or 9 pairs; buds pointed, strongly resinous; staminate aments ca. 5 —7 cm long, rather loose, from the axils of upper leaves; pistillate clustered 3 — 6 together, with slender often branched peduncles; strobiles ovaloid, dark brown, 12 mm long; nutlet winged, the wings narrowed toward base. May — June.

Mountain slopes, on rocks and stony taluses, amid undergrowth of birch woods, and on the crests of river valleys; in the mountains, a distinct alder zone producing compact and almost impenetrable thickets; where it approaches its upper limit, the plant becomes dwarfed and small-leaved. Flowering precocious, the staminate aments persistent on the branches for about 15 days and falling after leaf emergence. — Arctic: An., Chuk. (S. border, rare); Far East: Kamch., Okh., Uda, Sakh. (N.). Endemic. Described from Kamchatka. Type in Leningrad.

Economic importance. Provides refuge for animals of commercial value. Bark and foliage yield a dye for hides.

311 Note. The profusion with which this species occurs in Kamchatka naturally gives rise to variability. This is particularly pronounced in leaf shape: strongly sinuate, approaching A. sitchensis Sarg. (1861) = A. sinuata Rydberg (1897); or with completely plain margins, resembling A. Maximowiczii Call. but nevertheless distinctive, with a distribution area in NE Asia, including Kurile Islands.

Section 2. GYMNOTHYRSUS Spach in Ann. Sc. Nat. 2 sér. XV (1841) 204. — Pistillate aments axillary, their individual peduncles shorter than aments; inflorescences mostly leafless; leaves distinctly serrate.

5. **A. subcordata** C. A. M. Verz. der Pfl. in d. Jahr. 1829 u. 1830 in Cauc. u. am Casp. Meer ges. (1831) 43; H. Winkl. in Engl. Pflzr. 19 Heft (IV, 61) (1904) 113; Grossg., Fl. Kavk. (1930) 19. Talishi: razdar.

A tree 10 — 18 m high; branches olivaceous-brown or reddish-brown, with thin light-colored lenticels; buds pubescent like young branchlets, ovoid, obtuse; leaves with petioles 1 — 3 cm long, cordate at base, orbicular, ovate-orbicular, oval, or ovate-oblong, acute or acuminate or rarely rounded at apex, with subequal teeth on the margin, somewhat glutinous, glabrous above, with 10 — 12 pairs of veins and axillary tufts of hairs beneath, 7 — 14 cm long, 4.5 — 7 cm broad, rarely 5 — 16 cm long and 3.5 — 11 cm broad; staminate aments in a terminal cluster of 3 — 5; strobiles axillary, solitary or in pairs, rarely several together; nutlet broadly oval, with a very narrow wing. April. (Plate XVI, Figure 1).

Broad-leaved woods of the lower mountain zone, banks of rivulets, up to 1,000 m above sea level. — Caucasus: Tal. **Gen. distr.**: Iran (Gilan). Described from the vicinity of Astrabad [Gurgan]. Type in Leningrad.

Economic importance. Wood reddish, streaked, firm, stands immersion, suitable for carpentry and turning.

Note. Hybrids with A. incana W. have been described as A. Regelii Call. (from Tal.); in W. Med. replaced by the related species A. cordata Desf.

6. **A. japonica** Sieb. et Zucc. Fl. Jap. fam. nat. II (1846) 230; Rgl. Mon., Betul. in Mém. Soc. Nat. Moscou XXXIV (1861) 143; Kom. in A. H. P. XXII (1903) 60. — A. maritima var. arguta Rgl. in Bull. Soc. Nat. Moscou XXXVIII (1865) 428. — Japanese: gannoki, harinoki.

A tree 6 — 10 m high, with trunk circumference of 0.7 m; young branches more or less pubescent; annotinous branches glabrous, light olivaceous-brown or reddish-brown, with distinct lenticels; buds glabrous, reddish-brown, resinous; leaves oval or elliptic-oblong, narrowed or rounded at
312 base, acuminate, 3 — 10 cm long and 2 — 5 cm broad, glabrous above, with axillary tufts of hairs or slightly pubescent on both sides, sharply serrate; petiole pubescent or glabrous, 2 — 3.5 cm long; strobiles ovaloid or ovaloid-oblong, 1.2 — 2 cm long, 1 — 1.5 cm in diameter; nutlet with a very narrow wing. April — May. (Plate XVI, Figure 3).

Coastal strip, among coastal ridges, lake shores, boggy places, etc. — Far East: Jap.-Ch. Described from the coasts of S. Japan. Type in Munich.

Economic importance. Provides good solid wood, suitable for woodworking.

Note. A closely related species, A. maritima (Marsh) Nutt. grows along the Atlantic coast of N. America. The USSR plant produces a great number of germinable seeds and has long been in cultivation for ornamental purposes. It is endowed with considerable winter hardiness.

7. **A. glutinosa** (L.) Gärtn. Fruct. et sem. II (1791) 54; Ldb. Fl. Ross. III, 657; H. Winkl. in Engl. Pflzr. 19 Heft (IV, 61) 115; Shmal'g., Fl. II, 428; Kuzeneva in Fl. Yugo-Vost. IV, 70; Kryl., Fl. Zap. Sib. 800; Grossg., Fl. Kavk. II, 20. — Ic.: Gärtn. l. c. tab. 90; Rchb. Ic. Fl. Germ. XII (1850) tab. 631; Rgl. Monogr. Betul. in Mém. Soc. Natur. Moscou XIII (1861) tab. II. — Exs.: HFR No. 1090 et 1090a. — Also referred to in Russian as black alder; Finnish: terva-leppa; Lettish: melnais alksnis; Tataric in the Crimea: charykh agach; Azerbaijani: kizil agach; Georgian: murkgani; Armenian: barti, asnateneval, lasteni; Ossetian: sauferv; Kazakh: kandalach; Imeretian: tkhchela; Avar: makar; Lezghian: nukkhne.

PLATE XVI. 1. Alnus subcordata C.A.M. – 2. A. fruticosa Rupr. – 3. A. japonica Sieb. et Zucc. – 4. A. hirsuta Turcz.

A tree to 20 m, occasionally up to 35 m; bark of the trunk dark brown, fissured; young branches smooth, often glutinous, rarely rather sparsely pubescent, reddish-brown, with whitish transversal lenticels; buds obovaloid, obtuse or acutish, glutinous, stalked; leaves oboval, orbicular, or oval-elliptic, round-tipped or often emarginate at apex, broad-cuneate at base, in lower part entire, denticulate elsewhere with cartilaginous wide-angled teeth, rarely coarsely double-toothed (A. *denticulata* C. A. M. in the Caucasus); young leaves glutinous; grown leaves 4 — 9 cm long, 3 — 7 cm broad, with 5 — 7 pairs of veins; upper surface dark green, quite glabrous, lustrous; lower surface glabrous except for yellowish axillary tufts of hairs or pubescent between subsidiary veins, paler, resin-dotted; petiole 1 — 2 cm long; staminate aments terminal, in clusters of 3 — 5, 315 4 — 7 cm long, on peduncles 8 — 12 mm long; strobiles 3 — 5 together, ovaloid, to 15 mm long, the peduncle usually longer than strobile; nutlet with a very narrow coriaceous wing, reddish-brown, flattened, 2 — 2.5 mm across. April — May.

Riverbanks and lakeshores, swamps and grassy bogs, often near springs. — European part: Kar.-Lap. (S. border), Lad.-Ilm., Dv.-Pech. (S. part), U. Dnp., U. V., V.-Kama, M. Dnp., V.-Don, Bl., L. Don, Transv., L. V., Crim., Urals; Caucasus: Cisc. (in the mountains up to 1,500 above sea level); W. Siberia: Ob (SW corner), U. Tob., Irt.; Centr. Asia: Balkh. (Karkaraly). The N. limit follows on the whole the 62nd parallel, the E. limit approximates 66° E. lat. **Gen. distr.**: Scand. (northward to 63 — 64° N. lat.), Atl. and Centr. Eur., Med. (in Spain as far south as 38° N. lat.), Bal.-As. Min. Described from Centr. Europe. Type in London.

Economic importance. Wood light red, yellowing upon exposure to moisture, large-fibered, soft and light, brittle, more durable in water, hence used for piles, gutters, well frames, mine props, etc. Also employed in carpentry and turning. Alder charcoal was reputed for gunpowder. The bark and strobiles are employed for tanning. The bark yields a dye for hides, the resulting color being black, red, or yellow, depending on the mordant. A bark tincture is also used in the Caucasus to dye cloth and silk. A foliage infusion is of medicinal value as a laxative and, for its astringent action, against throat ailments. The bark contains 5 — 9% tannin, a red pigment "alnein," emodin, alnulin $C_{30}H_{50}O$, etc. The leaves contain salicin, populin, the alkaloids glutanol and glutinel $C_{14}H_{28}O_1$, etc. Black alder is irreplaceable for afforestation of wet places, mires, and swampy shores. According to some indications, the foliage is harmful to animals.

Particularly recommended for ornamental purposes is the mutant f. *incisa* Willd. Sp. pl. IV (1805) 335, with 5 — 7-lobed leaves cut right to the midrib, with deeply crenate-serrate lobes. Another similar form, f. *quercifolia* Willd. Berl. Baumz. (1796) 44, has leaves with 3 — 5 lobes on each side, toothed margins, and red petiole. Similar forms occur in nature on very rare occasions.

8. **A. barbata** C. A. M. Verz. Pfl. im Cauc. u. am west. Ufer des Kasp. Meeres ges. (1831) 43, No. 331; Grossg., Fl. Kavk. II, 20. — *Alnus glutinosa* var. *barbata* Ldb. Fl. Ross. III (1846 — 51) 657; H. Winkl. in Engl. Pflzr. — 19 Heft, 1, 61, 118.

A tree resembling on the whole the preceding; differing chiefly in oval or oboval-oblong obtuse or acute doubly serrate leaves, with dense yellowish tomentose axillary tufts beneath, both surfaces densely clothed with short hairs; strobiles oblong, 16—20 mm long, 6—8 mm broad, usually as long as or slightly longer than their peduncles. April.

316 Boggy places and shores. — Caucasus: Cisc., W. Transc., Tal. **Gen. distr.**: Asia Minor. Described from Talysh. Type in Leningrad.

9. **A. incana** (L.) Moench, Meth. (1794) 124; Willd. Spec. pl. IV (1805) 335; Ldb. Fl. Ross. III, 656 (var. exclusis); Shmal'g., Fl. II, 429; Grossg., Fl. Kavk. II, 19; Kryl., Fl. Zap. Sib. IV, 800. — A. *lanuginosa* Gilib. Exerc. Phytol. II (1792) 402. — A. *februaria* var. *incana* O. Ktze. Taschenb. Leipz. (1867) 239. — *Betula alnus* β *incana* L. Sp. pl. (1753) 983. — B. *incana* L. fil. Suppl. (1781) 417. — Russian names: ol'kha seraya [gray], ol'kha belaya [white], vil'kha, vol'kha, elokha, eleshina, leshinnik, etc.; Finnish: harman leppa; Lettish: baltais alksnis; Estonian: izalep; Tataric: irye; Ossetian: ferv; Svanetian: beklesh; Chechen: maà, muut dechik,

A medium-sized tree, 5—15 m high, or a shrub; bark fairly smooth, gray; young branches pubescent, not glutinous; buds pubescent, stalked, ovoid to ovoid-globose; leaves oval to oval-lanceolate or ovate-orbicular, rarely elliptic, acute or acuminate, rarely obtusish, rounded or slightly cordate at base, sharply double-serrate, at first densely pubescent, finally glabrous or more or less appressed-hairy above, rather sparsely pubescent or glabrous on the glaucous lower surface, more densely pubescent or even gray-tomentose on the veins, in the northern USSR (Murman) green and glabrate beneath with 8—10 pairs of veins (var. *virescens* Whlb. Fl. Lapp. 250, identical with A. *borealis* Norrlin); petiole 1—2 cm long, softly hairy or tomentose; staminate aments smaller and lighter-colored than those of A. *glutinosa*; pistillate 3—8 together on a common leafless stalk, subsessile except for the terminal aments, in maturity ca. 1.5 cm long; nutlet with narrow membranous wings. March — April.

Forest borders; playing a major part in early stages of forest regeneration and then often forming considerable thickets. Suckering profusely. Banks of rivers, streams and rivulets, hillock bogs, hill slopes, and riverside terraces. — European part: Kar.-Lap. (northward to the Kola Gulf estuary and the village of Penei, the typical form only as far as 66°19'N. lat., more northerly parts being occupied exclusively by var. *virescens* Whlb., otherwise A. *borealis* Norrlin), Dv.-Pech. (to the Mezen, Leza and Tsyl'ma rivers), Lad.-Ilm., U. V., V.-Kama, Transv., V.-Don, U. Dnp., M. Dnp., Urals (from 62°N. lat. to 52° in the S.); Caucasus: Cisc., Dag., W., E. and S. Transc. (the middle mountain zone); W. Siberia: Ob (rare) to 35° E. long. **Gen. distr.**: Scand., Centr. Eur., Med. (only mountains of N. Italy), Bal.-As. Min. (only along the Danube), Atlantic N. Am. Described from Europe.

Economic importance. Wood lighter-colored and firmer than that of black alder, used for carpentry and turning. Provides the best charcoal for drawing. The bark is rich in tannin; it also contains alnulin and alniresinol. The wood contains the resin "xylan." The gum and bee bread are used in beekeeping. The tree is often cultivated in gardens;

317 particularly valued are variants with cut leaves. — f. *pinnata* Lundmark in Vetensk. Akad. Nya Handl. (1790) 130, with incised-lobed or pinnately dissected leaves.

Note. The species, most widespread in Scandinavia and NE Europe, becomes less frequent further south, where it concentrates in mountains, but it does not rise above 1,800 m in Europe and the Caucasus (Greater Caucasus). Occurring fairly frequently on the W. slopes of the Ural Mountains in the latitudes indicated above; disappearing rapidly to the east where it is known only in some isolated locations, e. g., Samarovo on the Irtysh River and Kyshtymskii Zavod. Reports for Central Asia are more than dubious.

The genus *Alnus* does not occur at all in the mountains and deserts or the oases of Central Asia. It was reported for Pam.-Al. by Litvinov (Trudy Bot. Muzeya Akad. Nauk XI (1913) 69), as collected in "the Kugart subdistrict of Andizhan County" on the Karatal River and was named by him *A. incana* DC. var. *turkestanica*, without description. Also, *A. glutinosa* was reported in the work: Osten-Sacken u. Ruprecht, Sertum tianschanicum in Mém. Acad. Petrop. VII, ser. vol. XIV No. 4 (1869) 71, as collected in one of the ravines on the S. slopes of Tien Shan on the road to Kashgar. Both specimens are preserved in the herbarium of the Komarov Botanical Institute. It has been found that the specimen referred to by Ruprecht is merely a deformed cauline offshoot of *Populus laurifolia* Ldb., while the specimen mentioned by Litvinov belongs to the genus *Corylus*, as admitted by himself on an earlier occasion (Trudy Bot. Muzeya II (1905) 50). It may thus be confidently stated that alders are not to be found at all in the republics of Soviet Central Asia, if one disregards the northern borders of Kazakhstan.

A. glutinosa × *A. incana*, Krause, Schles. Ges. vaterl. Cult. (1845) 58; Aschers. und Gr. Synopsis IV, 430; Hjelt in Acta Soc. Fauna et Flora Fennica XXI (1902) No. I, 52. — *A. pubescens* Tausch. Flora XVII (1834) 520; Syreishchikov, Mosk. Fl. II, 51, with illustration.

More closely related to *A. incana* than to *A. glutinosa*. Bark gray; young branchlets, leaf petioles and stipules pubescent; leaves partly round-tipped or emarginate at apex and partly acuminate, the upper surface rather sparsely pubescent, the glaucous lower surface glabrous or slightly pubescent, velutinous on the veins and slightly tufted in the vein axils; strobiles stalked.

Scattered occurrence, together with *A. incana*. — European part: Kar.-Lap., Lad.-Ilm., U. V., V.-Don, L. Don. **Gen. distr.:** Scand., Centr. Eur., Bal. Described from Germany.

10. **A. sibirica** Fisch. in Tourcz. Cat. Baical. (1838) No. 1063 et in Fl. baic.-dah. II, 2 (1854) 132. — *A. incana* var. *sibirica* Ldb. Fl. Ross. III (1851) 656.

A shrub or a small tree; leaves orbicular, mostly obtuse, occasionally
318 acute or mucronulate, dark green above, intensely glaucous and quite glabrous or hairy on the veins beneath, 4—7 cm long, 3—5.5 cm broad, slightly lobed, serrate or dentate, with 7 or 8 pairs of veins; aments as in *A. incana* Moench. May.

Forest borders and undergrowth of coniferous forests. — E. Siberia: Dau. (only to the E. of the Yablonovyi Range); Far East: Ze.-Bu. **Gen. distr.:** reported for Korea and W. Manchuria. Described from E. Transbaikalia. Type in Leningrad.

11. **A. hirsuta** Turcz. Cat. Baical. (1838) No. 1064 et apud Rupr. in Bull. Phys.-Math. Acad. Petersb. XV (1857) 376; Kryl., Fl. Zap. Sib. IV, 801; Kom. in A. H. P. XXII, 54 et in Fl. pen. Kamtsch. II, 51. — A. incana var. hirsuta Spach in Ann. Sc. Nat. 2-e sér. (1841) 207; Turcz. Fl. baic.-dah. II, 2, 133; Ldb. Fl. Ross. III, 656. — A. Alissoviana Mandl. in Botanikai Közlemények XIX (1921) 89 et in Oesterr. bot. Zeitschr. 1922, No. 7—9, 172. — Ic.- Mandl. ibid. 1922, 173.

A tree or big shrub, 4—16 m high, with a dense top and smooth brownish-tawny bark; young branches gray, pubescent; buds ovoid, ca. 9 mm long; leaves to 12 cm long and 11 cm broad, orbicular or broad-oval, obtuse or short-acuminate (var. Cajanderi Call.), dark green and more or less pubescent or glabrous above (var. glabrescens Call.), glaucescent and densely soft-pubescent, often velvety to the touch, the pubescence rufescent throughout, the margin erose dentate-crenate or doubly dentate, the base rounded or cuneate or rarely (in var. Cajanderi Call.) subcordate; lateral veins 5—10 pairs; staminate aments on peduncles ca. 10 mm long, pendulous, cylindric, 10—16 cm long, dark purplish-brown; staminate aments cherry-red, 8 mm long; strobiles ovoid to subglobose, in clusters of 3—10, to 1.5 cm long, dark brown, the scales yellowish-brown on the inside; nutlet rufescent-brown, oboval, with a narrow thickened wing. April—May. (Plate XVI, Figure 4).

Banks of rivulets and small rivers, rarely large rivers, grassy bogs, near springs, and wet places at the foot of elevations. — W. Siberia: Ob (Ket River, 58°45' N. lat. and 87°30' E. long., according to Krylov); E. Siberia: Yen. (northward to 61°30' N. lat.), Lena-Kol., Dau.; Far East: Ze.-Bu., Uss., Sakh., Okh., Kamch. **Gen. distr.:** Jap.-Ch. (N. Japan, Korea, Manchuria). Described from the Upper Angara River. Type in Leningrad.

Note. Displaying great variability in size and shape of leaves which may be glaucous or green beneath, even on the same tree; the scales also vary in size, firmness, and thickness.

Economic importance. The wood is excellent for working. Yields planks to 30 or even 40 cm broad.

319 12. **A. tinctoria** Sarg. in Garden and Forest X (1897) 473; Call. in C. K. Schn. Laubholzk. I, 134; Kom. in A. H. P. XXII, 56. — A. hirsuta C. K. Schn. in Plantae Wilsonianae II, 2 (1916) 497 (ex parte). — Ic.: Sarg. l. c., f. 59.

A tree or a big shrub, 6 — 20 m high, with smooth dark bark; leaves orbicular, fairly firm, ca. 10—15 cm long, 8—10 cm broad, round-tipped or obtuse, rounded at base, bluntly and shallowly lobed, double-serrate, dark green and somewhat lustrous above, sparsely but uniformly pubescent beneath on and between the veins, the petiole to 4 cm long; strobiles compact, fleshy, much larger than in any other species native to the USSR, nearly twice the size of the strobiles of A. hirsuta. March—May.

Banks of small rivers and rivulets in the taiga. — Far East: Uss. (S. part). **Gen. distr.:** N. Japan, Korea? Described from N. Japan. Type in the Arnold Arboretum (N. America).

A. hirsuta Turcz. × A. japonica Sieb. et Zucc. — A. nikolskensis Mandl. in Oesterr. Bot. Zeitschr. No. 7—9 (1922) 175.

Hybrids between these two alder species are very variable and often differ markedly from both parents. There are forms with small, orbicular, almost glabrous leaves, and also forms with elongated leaves. — Far East: Uss. (banks of the Suifun River, below the city of Nikol'sk).

Family XLIV. **FAGACEAE** A. BR. *

Flowers unisexual, or rarely the pistillate with rudimentary stamens, monoecious, solitary or in 2's or 3's, the staminate in capitate or cylindric aments, the pistillate in short spikelike inflorescences or few-flowered clusters; perianth 4—8-lobed; stamens 5—20; ovary inferior, 2—6-locular, with 1 or 2 ovules in each locule; styles 2—6, united at base; fruit a 1- or rarely 2-seeded nut, partly or wholly enclosed by an indurescent subligneous involucre or cupule formed by union of bracteoles, the cupule covered outside with appendages, scales or spines, containing 1 or several nuts; seed exalbuminous, with a straight embryo; leaves simple, alternate, pinnately veined; stipules soon deciduous; flowers developing with or after the leaves. Trees.

Remnants of Dryophyllum Debey, a genus with leaves resembling those of chestnut and oak, occur in Upper Cretaceous and Lower Tertiary formations, to the Lower Oligocene. This genus is especially typical for the Volga Paleogene — D. Dewalquei Sap. et Mar. in the Paleocene of L. Don (Ushi; near Kamyshin) and U. Dnp. (Volyanshchina, Ryzhany). — D. subcretaceum Sap. et Mar. in L. Don (Ushi). — D. furcinerve (Rossm.) Schmalh. in the Oligocene of M. Dnp. (Ekaterinopol'e), Bl. (Adzhamka), and V.-Don (Tim); in the Pliocene of W. Transc. (Goderskii Pass). — D. rossicum Kretschet. in the Paleocene of V.-Don (village of Shovskoe).

320

Key to Genera

1. Cup cupuliform, not splitting, enclosing the lower part of a separate nut . . . . 367. **Quercus** L.
\+ Cup completely enclosing 2 or 3 nuts, splitting in maturity into 4 valves . . . . 2.
2. Staminate aments cylindric, erect, with pistillate flowers at base; cup enveloping 1—3 nuts, prickly outside . . . . 366. **Castanea** Mill.
\+ Staminate aments capitate, pendulous; cup covered with bristlelike or scalelike appendages, enclosing 2 triquetrous nuts . . . . . 368. **Fagus** L.

* Family characters, keys, and description of Castanea supplied by E. G. Bobrov.

Genus 366. **CASTANEA*** MILL.

Mill. Gard. Dict. ed. VII (1759). — Fagus L. Gen. pl. ed. 1 (1737) 292, p. p.

Flowers monoecious, clustered few together, in long erect cylindric aments, the staminate composing most of the ament, the pistillate separated near the base; staminate flowers consisting of a 5- or 6-lobed perianth and 10—20 stamens inserted at base around a glandular disk, the filaments long, the subglobose anthers 2-locular; pistillate flowers solitary or 3 (rarely more) together in a 4-valved involucre; perianth of 5—8 lobes united below with the ovary; ovary inferior, lageniform, commonly 6-locular, with 2 ovules in each locule; the pistillate flowers sometimes containing rudiments of stamens; fruit with a coriaceous prickly cup dehiscing in maturity into 2—4 valves and enclosing 1—3 nuts with coriaceous shell; seed commonly solitary; embryo without albumen, straight; cotyledons thick; germination hypogeal; leaves undivided, arranged in 2 ranks or spirally, deciduous, convolute in the bud; buds alternate, with numerous scales. Trees.

The chestnut appears in Cretaceous formations and is widely distributed through all Tertiary layers. — Castanea Kubinyi Kov. in Sarmatian formations of Bl. (Orekhov, Krynka) and V.-Kama (Sterlitamak District); in Lower Tertiary formations of Uss. (Novokievskoe [Kraskina], Rechnoi, De-Friz, Aleksinskii mine) and Sakh. (near Aleksandrovsk on the coast north of the settlements of Kamennaya and Takinosav). — C. sativa Ung. in the Sarmatian series of W. Transc. (Khvteeba). — C. Ungeri Heer in Lower Tertiary formations of Uss. (Rechnoi, De-Friz) and Sakh. (Mgachi). — C. sp. reported without determination of species for Lower Tertiary formations of Balkh. (Ashutas) and Sakh. (Agnevo R.).

321 1. **C. sativa** Mill. Gard. Dict. ed. VIII (1768) No. 1. — Fagus castanea L. Sp. pl. (1753) 997. — Castanea vulgaris Lam. Encycl. I (1783) 708. — C. vesca Gärtn. De Fruct. et Sem. I (1788) 181, tab. 37. — Ic.: Vol'f and Palib., Opr. (1904) 185. — Exs.: Fl. Cauc. exs. No. 83. — Russian: kashtan posevnyi; Georgian: tsabli; Tataric: shabalut; Abkhazian: akhe.

A tree to 30 m high, with a trunk 1.5—2 m in diameter and a wide spreading top; branches spreading, the young pubescent, reddish-brown, with lenticels; buds ovoid, covered with 2 outer scales; leaves borne on long shoots, 10—25 cm long, lanceolate, tapering toward apex, subulate-tipped, narrowed and entire at base, serrate dentate elsewhere, the subulate teeth often curved toward the margin; leaves at first covered above with short hairs and densely hairy above, at length densely clothed beneath with stellate hairs and glabrous above, the lower surface prominently nerved; lateral veins 18—30, terminating in teeth; stipules soon caducous, linear or lanceolate; staminate flowers yellowish, 3 or more together on a long axis, the catkins not surpassing the adjoining leaves, falling after flowering; pistillate flowers greenish, clustered 3 or rarely 4—7 together; stigmas long, filiform, reddish; fruit a 1-seeded (rarely 2-seeded) nut, with coriaceous castaneous shell; involucre accrescent, coriaceous, spherical, covered with subulate prickles, enclosing 1—3 nuts. Fl. June; fr. Septembe (Plate XIX, Figure 6).

* Name already used by the Roman writers Columella and Pliny.

In broad-leaved woods, mostly at 800—1,800 m above sea level. — European part: M. Dnp. (cultivated); Crimea (cultivated); Caucasus: Cisc. (Caucasian Reserve, Maikop District), W., E. and S. Transc. (slopes of Greater Caucasus and Karabakh), also cultivated in these regions. **Gen. distr.**: Centr. Eur. (S. part), Med., Bal.-As. Min., Arm.-Kurd. Described from Italy. Type in London.

Economic importance. Chestnut is one of the most valuable arboreous species with a wide variety of uses. It is of greatest interest as a fruit tree. The nuts are consumed raw, roasted, or cooked; they are also ground into a meal and are of considerable importance as a substitute for coffee. The bark has a high tannin content. The leaves, fruit involucres, and the bark provide a black or brown dye. Oil derived from the fruits as well as extracts of the foliage and the bark have medicinal uses. The flowers are nectariferous. The wood of chestnut is light brown,* with white sapwood; yields highly valuable material for carpentry. Particularly suitable for wine casks.

Chestnuts are propagated by seed. They bear fruit within 15 —20 years and yield profusely after 2—3 years. They attain an age of a thousand
322 years and more. Suckering profusely, they are excellent for coppices.

Chestnut cultivation was known in ancient Greece and Rome. In our times it is widespread in southern Europe where it covers hundreds of thousands of hectares. It has also been introduced into America. Apart from the need for great extension of Castanea sativa cultivation, of considerable interest for the USSR is C. dentata Marsh., a species closely related to the USSR chestnut and cultivated (according to Prof. Kichunov) in the Mtsensk District ["raion"] of the Orel District ["okrug"]. This species is characterized by relatively small, very sweet fruit, but is very susceptible to infection by the ascous fungus Endothia.

Dozens of varieties of cultivated chestnut are known in Europe and undoubtedly many of them are adaptable to the climatic conditions of the southern regions of the Soviet Union. As regards the American chestnut, it may prove quite satisfactory for regions with a climate resembling that of Kursk Region.

Genus 367. QUERCUS ** L.†

L. Sp. pl. ed. 1 (1753) 994.

Staminate flowers in long pendulous slender-stalked aments; perianth 4—7-parted (mostly 6-parted) to the middle or almost down to base; stamens 4—12, mostly 6; pistillate flowers solitary or 2 or several together on a very short or fairly elongated peduncle; ovary 3- or sometimes 4-locular, with 2 ovules in each locule, surrounded by a cup-shaped cupule covered outside with numerous scales; styles 3; stigma large, completely covering the upper face of the style; fruit a 1-celled 1-seeded acorn, with rudiments of abortive seeds in upper or lower part; shell of the acorn thin, glabrous within, or else thick 3-layered

* Acquiring in time a reddish-brown color.

** The name used by Cicero and other ancient authors.

† Treatment by V.P. Maleev.

and then pubescent within; cupule accrescent in fruit and enveloping the acorn rather high up, its scales appressed, spreading or recurved, sometimes united into concentric rings surrounding the cupule. — Trees or rarely shrubs; leaves evergreen, persistent through winter, or deciduous, entire, toothed or more or less deeply lobed, with persistent or caducous stipules.

Economic importance. Most oaks are important and widely distributed species, yielding durable hard lumber used for a variety of purposes. The bark contains tannin and is extensively used for leather tanning; of similar use are the cups of certain species, notably those of Q. *aegilops* L., and galls formed by the puncture of insects. The bark of the cork oak and
323 of some other species is the principal source of cork. The acorns are used as a coffee substitute and as feed for swine; in some species they are sweet and edible in raw condition. Many oak species are eminently suited for pleasure-ground, street and avenue planting, etc. and are recommended for extensive ornamental use. The drought resistance of some of the species and their modest soil requirements enable their use for afforestation of arid regions.

Fossils of *Quercus* occur in the earliest Cretaceous layers of America which contain remnants of dicotyledonous plants. Oaks were widely distributed through the Upper Cretaceous and all the Tertiary floras, although the geographical distribution of the sections was quite different than it is at the present time. Oak fossils have also been found in Tertiary formations of Australia. The Cretaceous oaks of Europe resemble most closely the present-day tropical species of Asia and Central America with which North American species show affiliation. Gradual transition to European forms took place in the course of the Tertiary period. Representatives of the sections *Tozza* and *Infectoria* appear in Europe since the Miocene. — Q. *aizoon* Heer in the Lower Tertiary formations of Sakh. (Mgachi). — Q. *Alexeevi* Pojark. in the Oligocene of Ar.-Casp. (Sarry-bulak). — Q. *chlorophylla* Unger in the Oligocene of M. Dnp. (Volyanshchina) and V.-Don (Molotychi). — Q. cf. *crispula* Bl. in Lower Tertiary formations of Uss. (Rechnoi). — Q. *deuterogona* Unger in the Sarmatian formations of Bl. (Krynka). — Q. *diplodon* Sap. in the Paleogene of L. Don (Ushi). — Q. *platania* (Heer) in L. Don (Ushi). — Q. *drymeja* Unger in the Sarmatian formations of Bes. (? Lipkany) and in the Oligocene of Ar.-Casp. (Kara-bulak, Yar-koe), also reported from ? for Sakh. (Dui). — Q. *etymodrys* Unger in Tertiary (Pliocene ?) formations of Alt. (Chingistai). — Q. *Gmelini* A. Br. in the Oligocene of V.-Don (Tim) and Ar.-Casp. (Dzhilan). — Q. *Gmelini* A. Br. ? in the Eocene of M. Dnp. (Lava, Kursk Region) and V.-Don (Tim). — Q. *groenlandica* Heer in the Oligocene of Ob (Tomsk). — Q. *Heeri* A. Braun in the Oligocene of M. Dnp. (Volyanshchina). — Q. *kamyschiensis* Goepp. in the Paleocene of L. Don (Ushi) and in Eocene-Oligocene formations of L. Don (Privol'e in Izyum District). — Q. *lonchitis* Unger in the Oligocene of Cisc. (Maikop District) and in Lower Tertiary formations of Uss. (Pos'et, Novokievskoe, Fatashi, Rechnoi Peninsula). — Q. *macranthera* F. et M. in Postpliocene formations of W. Transc. (Sukhumi). — Q. *magnoliaefolia* Goepp. in the Oligocene of V.-Don (near Fatezh;

Molotychi). — Q. neriifolia A. Br. in the Oligocene of V.-Don (Tim) in the Tongrian stage (Oligocene) of Bl. (Kremennaya Ravine, Mariupol' [Zhdanov] area), and in Eocene-Oligocene formations of L. Don (Osinovaya). — Q. Nimrodi Unger in the Oligocene of Ob (Tomsk) and Ar.-Casp. (from ? in Yar-kue). — Q. odontophylla Sap. in the Paleocene of L. Don (Ushi). — Q. Olufsenii Heer in the Paleocene of L. Don (Ushi) and in Lower Tertiary formations of Sakh. (Mgachi). — Q. palaeovirens Schmalh. in the Paleocene of L. Don (from ? in Ushi) and in the Oligocene of M. Dnp. (Ekaterinopol'e). — Q. pedunculata Ehrh. in Postpliocene formations of U. V. (Likhvin, Krushma R.) and Cisc. (in tuffs at Zheleznovodsk). — Q. platania Heer in Lower Tertiary formations of Uss. (Amagu). — Q. parceserrata Sap. et Marion in the Paleocene of L. Don (Ushi). — Q. pseudocastanea Goepp. in Sarmatian formations of Bl. (Krynka). — Q. robur L. in Postpliocene tuffs of Bes. (Daraban), M. Dnp. (Mishkutinets), and Cisc. (Mashuka); in the Postpliocene of Ar.-Casp. (Karadzhar on the Dzhilanchika River). — Q. spathulata Eichw. in the Oligocene of V.-Don (Kursk Province). — Q. Steenstrupii Heer in the Paleocene of L. Don (Ushi). — Q. timensis Palib. in the Oligocene of V.-Don (Tim). — Q. venulosa Eichw. in the Oligocene of V.-Don (Molotychi, Tim). —
324 Q. sp., unidentified, reported for many distribution areas in Tertiary and Postpliocene formations, e. g., in the Paleocene of L. Don (Ushi; vicinity of Karavaev), Pliocene of E. Transc. (Karamal-Naftalan, Kisatib), Lower Tertiary formations of Uss. (Rechnoi), postglacial formations of U. V. (Krushma), and the Postpliocene of M. Dnp. (Timoshkovichi, Tuganovichi).

Fossilized wood of Quercinium kamyschinense Goepp., Q. magnoliaefolium, and Q. rossicum Merckl. with var. montanum Merckl. was found in the Paleocene of L. Don (Ushi near Stalingrad [Volgograd]). — Quercinium reticulatum Eichw. in the Oligocene of V.-Don (Tim). — Q. rossicum Merckl. in Tertiary formations of Zaporozh'e, Bl. — Q. rossicum Merckl. var. montanum Merckl. in Sarmatian formations of Bl. (Zaporozh'e-Aleksandrovsk area, Orekhov — same age ?). — Q. venulosum Eichw. in the Oligocene of V.-Don (Tim).

1. Leaves evergreen . . . . 2.
\+ Leaves deciduous or persistent on the tree in withered condition . . . 4.
2. Bark of the trunk and of the branches with thick cork layer; scales of the cup long and spreading; lateral veins not forking . . . . 3.
\+ Bark of the trunk and of the branches without a cork layer; scales of the cup small and appressed; some of the lateral veins forking . . . . ***Q. ilex** L.
3. Acorns ripening during the first year . . . . ***Q. suber** L.
\+ Acorns ripening in the following year . . . . . . ***Q. occidentalis** J. Gay.
4. Scales of the cup long, narrowly lanceolate, spreading and recurved . .5.
\+ Scales of the cup small, imbricate, appressed or sometimes slightly spreading . . . . 6.
5. Leaves with very short obtuse lobes, large, to 20—30 or even up to 50 cm long, very short-petioled heavily appressed-pubescent beneath; stipules deciduous . . . . 19. **Q. dentata** Thnb.
\+ Leaves with large almost triangular acuminate teeth, with grayish-white pubescence beneath, or shallowly and obtusely lobed and then glabrous or nearly so, smaller, to 15—18 cm long, with petioles ca. 2 cm long; stipules persistent . . . . 1. **Q. castaneifolia** C. A. M.

6. The terminal lobe and lateral lobes terminated by a long slender point; acorn shell thick, densely pubescent within . . . . . . *Q. rubra L.

\+ Lobes of leaves obtuse or acuminate, always without a long slender point; acorn shell thinner, glabrous within . . . . 7.

7. Leaves very large, to 20 and sometimes to 30 cm long, undivided, the margin with small upcurved teeth . . . . 3. Q. pontica C. Koch.

325 + Leaves commonly smaller, more or less deeply lobed or coarsely toothed . . . . 8.

8. Pistillate flowers and fruits long-stalked, the stalks greatly surpassing the leaf petioles . . . . 9.

\+ Pistillate flowers and fruits sessile or short-stalked, the stalks not surpassing the leaf petioles . . . . 14.

9. Leaves shallowly and equally lobed; the lobes 9—12 pairs, broadly semioval or obtusely angular, the sinuses between the lobes not more than one-sixth to one-tenth the breadth of the blade; lateral veins parallel and ending in the lobes, without intermediate veins, the axils of veins beneath with persistent rufous pubescence . . . . 6. **Q. Hartwissiana** Stev.

\+ Leaves more deeply pubescent; the lobes fewer, unequal, rather long, straight or curved; lateral veins not quite parallel with pronounced intermediate veins; pubescence fugacious or persistent, never rufous . . . . 10.

10. Leaves firm, subcoriaceous, heavily pubescent beneath; fruit-stalks stout; cup thick, firmly woody, the rather large scales arranged in more or less regular rows around the cupule . . . . 11. **Q. longipes** Stev.

\+ Leaves thinner, usually glabrous or slightly pubescent beneath; cup less thick, its scales smaller, not forming regular rows around the cup . . . . 11.

11. Leaves deeply lobed; lobes 6—8, long and narrow, rarely relatively short and broad, always with 1 or several large teeth on the margin . . . . 10. **Q. erucifolia** Stev.

\+ Leaf lobes fewer, entire or rarely with 1 or 2 small teeth . . . . 12.

12. Leaves subsessile, strongly auriculate, the auricles completely or almost completely concealing the petioles . . . . 13.

\+ Petioles longer, clearly evident . . . . 7. **Q. robur** L.

13. Leaves pale green and glabrous beneath . . . . 9. **Q. imeretina** Stev.

\+ Leaves glaucescent, puberulous . . . . 8. **Q. pedunculiflora** C. Koch.

14. Stipules large, to 1.5 cm long, persistent; shoots, annotinous branchlets and the lower surface of leaves densely shaggy-tomentose; scales of cup narrowly lanceolate, all or at least those on the lower part of the cup persistent . . . . 2. **Q. macranthera** F. et M.

326 + Stipules caducous or sometimes persistent around the terminal buds and then much smaller; shoots and leaves glabrous beneath or covered with short grayish hairs; scales of the cup appressed or merely their tips spreading . . . . 15.

15. Leaves varying greatly on the same plant, at the base of shoots often entire, elsewhere with few large teeth or short lobes . . . . 16.

\+ All leaves more or less alike, deeply or shallowly lobed or rather coarsely toothed, the teeth or lobes numerous . . . . 17.

16. Shoots and at least the young leaves pubescent beneath and on the petioles, the lower surface of leaves pale green . . . . 4. **Q. araxina** (Trautv.) Grossh.

\+ Shoots and leaves beneath glabrous; lower surface of leaves brownish . . . . 5. **Q. Woronowii** Maleev.

17. Leaves with short broad obtuse or rarely acuminate lobes or rather coarsely toothed, with parallel veins ending in the lobes; intermediate veins absent . . . . 18.

\+ Leaves more deeply lobed with subequal lobes; lateral veins commonly curved or flexuous, not quite parallel; at least 1 or 2 pairs of intermediate veins present . . . . 21.

18. Leaves subsessile, with petioles 0.3 – 0.4 cm long, the lower surface of young leaves clothed mainly with simple hairs . . . . 18. **Q. mongolica** Fisch.

\+ Leaves long-petioled, clothed with fugacious or persistent mainly stellate hairs . . . . 19.

19. Leaves crisped on the margin, 6 – 9 cm long; scales of cup dark brown, slightly pubescent, terminating in a spreading brownish-orange tip . . . . 14. **Q. longifolia** C. Koch.

\+ Leaves not crisped, commonly larger; scales of cup heavily pubescent, with an appressed brownish tip . . . . 20.

20. Leaves with 8 – 11 pairs of lobes, firm, subcoriaceous, pale green beneath, 7 – 14 cm long . . . . 12. **Q. iberica** Stev.

\+ Leaves with fewer lobes, commonly thinner, yellowish-green beneath, usually larger, 12 – 18 cm long . . . . 13. **Q. hypochrysa** Stev.

21. Leaves large, 15 – 18 cm long, very firmly coriaceous, with few very unequal and more or less curved lobes; fruits several together on a stalk to 3 cm long . . . . 17. **Q. Kozlowskyi** Woron.

\+ Leaves smaller, less firm, more equally lobed; fruits solitary or several together, sessile . . . . 22.

327 22. Leaves 8 – 12 cm long; mature, leaves and shoots glabrous or nearly so . . . . 15. **Q. petraea** Liebl.

\+ Leaves smaller, heavily clothed beneath, like the shoots, with gray pubescence . . . . 16. **Q. pubescens** Willd.

Subgenus 1. **LEPIDOBALANUS** Endl. Gen. Suppl. IV, II (1847) 24; Örsted in Vidensk. Meddel. (1866) 65, emend. — Styles recurved or upright, enlarged or slender at apex, pointed; fruit ripening the first or second year; shell of acorn thin, naked within; rudimentary ovules basal; scales of the cup not connate, small and appressed or larger and spreading to almost recurved.

Section A. CERRIS Spach, Hist. Nat. Veg. Phan. XI (1842) 166; Örsted in Vidensk. Meddel. 74 (pro subg.). — Styles linear, slender at apex, pointed, upright or somewhat spreading; scales of cup elongated, linear or narrowly lanceolate, spreading or recurved; fruits ripening the second year, rarely the first year; perianth of staminate flowers parted at most to the middle into broad-lanceolate or oval lobes; leaves evergreen, wintering or deciduous, entire or more or less coarsely toothed, rarely lobed with pointed teeth or lobes; stipules persistent.

Subsection 1. SUBER Spach, Hist. Nat. (1842) 171; Örsted in Vidensk. Meddel. 76 (pro sect.). — Leaves evergreen, rarely wintering, entire or finely toothed; fruits ripening sometimes the first year; scales of cup sometimes short and upright.

*Q. **suber** L. Sp. Pl. (1753) 995; Boiss. Fl. Or. IV, 1168; Wenzig in Jahrb. d. Bot. Gart. zu Berlin IV (1886) 201; C. K. Schn. Laubholzk. I (1907) 186. — Ic.: Rchb. Ic. Fl. Germ. XII (1850) t. 641; Kotschy, Die Eichen Europa's u. d. Orients (1862) t. XXXIII; C. K. Schn. l. c., f. 117 b—f, 1.

A tree to 20 m; bark of the trunk and of the branches with a thick cork layer; young shoots and buds densely clothed with yellowish-gray tomentum; leaf petioles 1—1.5 cm long; blades oval or elongate-oval, 3—7 cm long and 1.5—3.5 cm broad, entire or with few small sharp teeth, dark green and lustrous above, heavily gray-pubescent or rarely glabrate beneath, evergreen, coriaceous; acorns 1—3 on a peduncle 1—2 cm long, ripening the first year, elongate-ovaloid, to 3 cm long and 1.5 cm in diameter, enclosed by the cupule to between one-half and one-third; scales of cup pubescent, the upper ones elongate-lanceolate, ca. 0.5 cm long, upright or spreading, the lower ones shorter. May.

Cultivated in S. Crimea, in W. (S. of Sochi) and E. Transc. (Tiflis [Tbilisi], Baku). **Gen. distr.:** W. Med. to Istria. Described from S. Europe. Type in London.

328

Economic importance. The bark of the cork oak and of the next species is used for cork production. The exploitation is largely confined to natural stands within the distribution area. The principal cork-producing countries are Portugal, up to 93,000 tons of cork annually; Spain, to 80,000 tons; Algeria, to 75,000; also Tunisia, etc. The average annual output in Mediterranean cork-producing countries amounts to 250,000 tons, to the value of 10 million francs. To satisfy the cork requirements of the USSR and to eliminate the need for imports of this commodity, it has been planned to plant up to 80,000 hectares of cork oak in the Crimea and in the Caucasus in the course of 30 years. It will however take some 30 to 50 years before the need for imports may be eliminated, since the first usable cork can be removed from the trees only when they have reached the age of 25—30 years. Cork oak was first introduced to the territory of the USSR in the 1820's. The largest stands are in Agudzeri near Sukhumi, about 500 trees; near Gagra, about 500 trees; and in the Nikitskii Botanical Garden, some 150 trees, including those of the next species. Individual trees and small groups of cork oak occur throughout the Crimea as well as in western, and to extent eastern Transcaucasia. The wood is heavy with specific gravity 0.8—1.0, easily warping, splitting and rotting; chiefly used for fuel and charcoal. The bast has a tannin content of up to 20% and is used for leather tanning.

***Q. occidentalis** J. Gay in Ann. Sc. nat. sér. 4, VI (1856) 243; C. K. Schn. Laubholzk. I, 187. — Ic.: C. K. Schn. l. c. f. 117 g—k; Kern in Bull. appl. bot. XVIII, 2 (1927—1928) p. 457—8, f. 2—3.

This species differs from the preceding in that its leaves are less heavily pubescent beneath, persistent through winter and falling the next spring after the expansion of new leaves. Another distinguishing feature is the biennial ripening of acorns. May.

Cultivated in S. Crimea and W. Transc. **Gen. distr.**: W. Med. (Portugal, S. France, Corsica). Described from Portugal. Yields cork of better quality than the preceding species and provides most of the cork produced in Portugal. This species was introduced into the USSR before Q. suber by the Nikitskii Botanical Garden, in 1819, from acorns of cork oak obtained from Lisbon. An oak grove derived from these acorns exists to the present time in the Nikitskii Botanical Garden. Together with the preceding species it has been distributed in cultivation, chiefly in S. Crimea. In the new plantations of cork oak in the USSR it will be included together with Q. suber and in certain areas it will be given preference.

Subsection 2. EUCERRIS Örsted in Vid. Meddel. (1866) 75 (pro sect.). — Leaves deciduous or persistent through winter, entire or rarely coarsely toothed or more or less deeply lobed; fruits maturing the second year; scales of cup long, spreading, rarely upright.

329 1. **Q. castaneifolia** C. A. M. Verzeichn. d. Pflanz. Cauc. (1831) 44; Eichwald, Plant. nov. fasc. 2 (1831—33) 9; Ldb. Fl. Ross. III, 591; Boiss. Fl. Or. IV, 1174; Wenzig in Jahrb. d. Bot. Gart. Berlin, IV, 212; Freyn in Bull. de l'Herb. Boiss. II sér., 2 (1902) 904; C. K. Schn. Laubholzk. I, 178; Medvedev in Moniteur du Jard. Bot. de Tiflis X (1908) 38; Grossg., ibidem 46—47 (1919) 55. — Q. aegilops β castaneaefolia C. Koch in Linnaea XXII (1849) 319. — Ic.: Eichwald l. c. tab. I; Kotschy Die Eichen tab. XL; C. K. Schn. l. c. f. 108, 1, 109 c—e; Grossg., l. c. tab. IV, f. 1—2. — Exs.: Herb. Fl. Cauc. No. 211.

A tree to 25 m, with smooth gray bark; young shoots gray with heavy pubescence; annotinous shoots gray, with fugacious pubescence; buds small, silky-pubescent or glabrate; stipules to 1.5 cm long, linear, heavily pubescent like the leaf petioles, these to 2 cm long; leaves cuneate, rounded or slightly cordate at base, narrowly elongate-ovate or elongate-elliptic, 10—18 cm long and 4—8 cm broad, with a triangular acuminate terminal lobe and (7) 10—12 (15) large triangular acute teeth, the sinuses one-fifth to one-fourth the breadth of the blade, firm, subcoriaceous, dark green and at first sparingly pubescent, becoming glabrate above, grayish with dense stellate pubescence beneath; or else leaves broader, with broad obtuse mucronulate teeth or lobes, the sinuses to one-third or more of the leaf breadth, and then thinner, pale green and glabrate beneath (f. obtusiloba Freyn); lateral veins as many as and ending in the teeth, parallel, straight or (in f. obtusiloba) more or less arched; staminate inflorescence 7—10 cm long, with a densely pilose peduncle; pistillate flowers and fruits solitary or 2 or 3 together, borne on a very short stout stalk; cup hemispherical, to 2 cm long and 2.5 cm in diameter; its scales gray-pubescent, pointed and brownish at the tips, the lower ones narrowly oval and spreading, the middle and upper ones longer (to 1.5 cm), narrowly lanceolate, recurved; acorns 2.5—3.5 cm long, 2—3 times the length of the cup. April—May. (Plate XVII, Figure 1, 1a).

Forming forests in plains and in mountains up to 1,800 m above sea level. — Caucasus: Tal., E. Transc. (Azerbaijan, former Geokchai County). **Gen. distr.**: Iran. (N. Iran — Gilan, Mazanderan). Described from Talysh. Type in Leningrad.

Note. The form *obtusiloba* Freyn grows only in lowland forests of Talysh, while the typical form with sharp-toothed leaves grows mainly in mountains. The fruits of the two forms are identical, whereas the
330 leaves are joined by gradual transition. Schmidt pointed out recently (zhurn. Subtropiki [Subtropics Journal], 1931, No. 3) that peculiar leaves dissected nearly to the midrib, with obtuse lobes are formed on sucker shoots of Q. *castaneifolia* trees nibbled by livestock. Thus f. *obtusifolia* Freyn is merely a modification due to ecological conditions, devoid of taxonomic significance. The species Q. *Sintenisiana*, described in 1935 by O. Schwarz (Notizblatt d. Bot. Gart. u. Mus. Berlin-Dahlem, XII, 114 (1935) 468) from N. Iran, hardly differs in fact from the typical Q. *castaneifolia* C. A. M.

Economic importance. One of the most widely distributed forest species of Talysh, yielding valuable wood for construction and various other purposes, including clapboards for wine casks. The striking appearance and beauty of this tree recommend it for planting in the south of the USSR where it is to be found occasionally in cultivation (Crimea, Ukraine); it deserves considerable extension.

Section B. **CERRIDOPSIS** Maleev in Journ. Bot. URSS 20, 2 (1935) 163. — Styles upright, elongated, scarcely enlarged at the end, not pointed; scales of cup narrowly lanceolate, elongated but much shorter than the cup, in unripe fruits somewhat spreading, in ripe ones subappressed; fruits ripening the first year; perianth of staminate flowers parted to between one-half and three-fourths or nearly to the base into suboval or narrowly lanceolate lobes; leaves deciduous before winter, coarsely toothed or rather deeply lobed with large-toothed lobes; stipules, especially around the terminal buds, persistent.

Subsection 1. MACRANTHERAE Stefanoff in Ann. de l'Univ. de Sofia ser. V, vol. VIII (1929 — 30) 53 (pro sect.). — Perianth of staminate flowers parted merely to one-half to three-fourths into broad-oval lobes.

2. **Q. macranthera** F. et M. ex Hohenacker in Bull. Soc. Nat. Mosc. VI (1838) 260; Ldb. Fl. Ross. III, 591; C. Koch in Linnaea XXII, 319; Steven in Bull. Soc. Natur. Mosc. XXX (1857) 388; Boiss. Fl. Or. IV, 1165; Wenzig in Jahrb. d. Bot. Gart. zu Berlin IV, 198; C. K. Schn. Laubholzk. I, 102; Medvedev in Monit. du Jard. Bot. de Tiflis XI, 28. — Ic.: Radde in Museum Caucas. II (1901) tab. 4; C. K. Schn. l. c., f. 108 b.

A tree to 20 m, but usually lower, the short stout trunk covered with thick fissured bark; shoots densely shaggy with long yellowish-gray hairs, the 1- and 2-year old branchlets gradually glabrescent; buds 5 — 6 mm long, obtuse, ovaloid, with few heavily gray-pubescent scales; stipules to 1.5 cm long, densely villous; leaf petioles 1 — 2 cm long, subcoriaceous, 6 — 18 cm, on the average ca. 10 cm, long and 3 — 12 cm broad, dark green and at first sparingly puberulous above, becoming glabrate with pubescence confined to the veins, yellowish-gray beneath with dense sometimes partly disappearing pubescence, oboval or elongate, commonly cuneate or rarely

331 truncate or subcordate at base, with a short obtuse terminal lobe and 8 — 12 pairs of short, obtuse entire or coarsely toothed lateral lobes, the sinuses one-fifth to one-sixth the breadth of the blade; the lobes sometimes longer, the sinuses thus attaining one-third to one-fourth the breadth of the blade, and then they are commonly narrowed and rather pointed (f. pinnatifida Medw.); lateral veins as many as and directed only toward the lobes, straight or arched, parallel, intermediate veins none or 1 or 2 faint ones in the lower part of the blade; staminate inflorescence to 10 — 15 cm long, with a densely pubescent peduncle; anthers large, to 1.5 mm long; pistillate flowers and fruits 1 — 5, sessile or borne on a stalk to 2 cm long; cupule hemispherical, to 1.5 cm long and 1.5 — 2 cm in diameter, enclosing the acorn to between one-half and one-third; lower scales and those of young fruits somewhat spreading, the middle and upper ones appressed, narrowly lanceolate, gray-pubescent, with a brown glabrous tip; acorns 2 — 2.5 cm long, initially appressed-pubescent, finally glabrous. May. (Plate XVII, Figure 2, 2a).

Forming forests in the mountains, on dry, mostly southern slopes, at altitudes between 800 and 2,400 m above sea level. — Caucasus: S. and E. Transc., Tal., Dag. **Gen. distr.**: Iran. (N. Iran: Gilan, Mazanderan), Arm.-Kurd. (Olty-chai River basin). — Described from Talysh. Type in Leningrad.

Economic importance. A hardy and, above all, drought-resistant oak that constitutes the only high-mountain forest-forming species in arid areas of E. and S. Transcaucasia, such as Armenia. Deserves attention as a valuable tree for afforestation and ornamental planting in the dry parts of Transcaucasia and Central Asia and, possibly, also for the south of the European part of the USSR. The wood is used like that of other oaks. The leaves often display very large spherical long-villous galls that need study as to their tannin content.

Section C. EULEPIDOBALANUS Örsted in Vidensk. Meddel. (1866) 65 (emend.). — Styles enlarged and rounded at the ends; scales of the cup small, tightly appressed with only the tips sometimes slightly spreading; fruits ripening the first year; perianth of staminate flowers parted to the base into narrowly lanceolate lobes; leaves evergreen, persistent through winter or deciduous, entire or denticulate or mostly coarsely toothed or rather deeply lobed; stipules soon deciduous, sometimes persistent especially near the terminal buds.

Subsection 1. ILEX Örsted in Vidensk. Meddel. (1866) 69 (pro sect.). — Perianth of staminate flowers parted to the middle; leaves evergreen, entire or shallowly sharp-toothed; stipules persistent.

332 ✶Q. ilex L. Sp. pl. (1753) 995; Boiss. Fl. Or. IV, 1167; Wenzig in Jahrb. Bot. Gart. zu Berlin IV, 199; C. K. Schn. Laubholzk. 1, 188. — Ic.: Rchb. Ic. Fl. Germ. tab. 642; Kotschy Die Eichen, tab. XXXVIII; C. K. Schn. l. c. f. 114 g — l, f. 118; Hegi III. Fl. v. Mitteleur. III, 115, f. 433 i — r.

A tree to 20 m; trunk covered with smooth dark gray bark; buds small, ovaloid, with heavily pubescent scales; shoots densely gray-tomentose; leaf petioles heavily pubescent, 0.5 — 1.5 cm long; blades elliptic, oval, narrowly oval, or sometimes broad-lanceolate, coriaceous, lustrous green and glabrous or nearly so above, densely covered beneath with persistent or on fully developed leaves sometimes with partly or wholly transient pubescence, rounded or cuneate at base, entire or with few sharp teeth, 2.5—7.5 cm long and 1—4 cm broad; lateral veins 8—10 pairs, straight or curved, some forked; pistillate flowers and fruits 1—3 on a short peduncle or sessile; cup enclosing the acorn to between one-half and one-third; scales narrow-lanceolate, almost flat, densely covered with short hairs; acorns 2—3.5 cm long. April—May.

Frequently and extensively cultivated in S. Crimea, W. Transcaucasia, and more rarely in E. Transc. — **Gen. distr.**: W. and E. Med. Described from S. Europe. Type in London.

Note. Very variable in respect of leaf size and shape as well as the amount of vesture.

Economic importance. Widely distributed in cultivation in the southern USSR, chiefly in S. Crimea; fully adapted to the conditions of the south and growing luxuriously. Very undemanding, drought-resistant and hardiest of all evergreen oaks. Highly ornamental and, compared with other oaks, fast growing and thus suitable not only for ornamental use but also for afforestation purposes in southern countries with Mediterranean climate. The qualities of this tree are particularly appreciated in Italy. It is recommended for extensive cultivation in S. Crimea and in the drier parts of W. Transcaucasia. The wood is very durable, hard and heavy, with specific weight 1.04; the heartwood is brownish. The wood is highly valued for construction, submerged installations, and miscellaneous purposes. Of particular value is the wood of the roots for carpentry. The bark has a tannin content of 7.25%.

Subsection 2. PONTICAE Stefanoff in Ann. de l'Univ. de Sofia VIII (1929—30) 53 (pro sect.). — Perianth of staminate flowers parted to the middle into triangular lobes; leaves deciduous, undivided, with small teeth; stipules deciduous.

3. **Q. pontica** C. Koch in Linnaea XXII (1849) 319; Boiss. Fl. Or. IV, 1166; Wenzig in Jahrb. Bot. Gart. zu Berlin IV, 210; C. K. Schn. Laubholzk.
333 I, 192; Medvedev in Monit. du Jard. Bot. de Tiflis XI, 33. — Ic.: Gartenflora 40 (1891) 510, tab. 2; C. K. Schn. l. c. f. 102 i. — Exs.: Fl. Cauc. exs. No. 155.

A tall shrub or a small tree; annotinous branchlets glabrous, reddish-brown, with small light-colored lenticels; buds broad-ovaloid to subglobose, the middle much bigger than the lateral ones, to 1.5 cm long, the dorsal scales cuneate, brownish-yellow, the narrow membranous margin dark-rimmed; leaf petioles 1—1.5 cm long; blades firm, very large, 10—20 and up to 35 cm long, 4—10 (to 15) cm broad, undivided, elliptic or broadly oboval, attenuate at apex, acuminate or more or less round-tipped and mucronate, cuneately narrowed toward base, with 15—30 obtusish or pointed upturned teeth on each margin, lustrous dark green above, paler beneath with a dense very thin coating of stellate hairs or glabrous; lateral veins 15—30 pairs, straight, parallel, directed toward the teeth, very prominent beneath; staminate inflorescence to 10 cm long, with an almost black hairy stalk; pistillate flowers 2—5 at the ends of shoots; acorns borne on short stout peduncles; cup patelliform, to 1.5—2 cm in diameter, enclosing the acorn to one-fourth; scales rather large, inflated, with appressed tips, heavily gray-pubescent; acorns broadly ovaloid, to 3 cm long and 2 cm broad, light brown, at first sparsely pubescent, becoming glabrous. May—June. (Plate XVII, Figures 3, 3a).

The upper forest and subalpine zones, up to the timberline, at altitudes between 1,300 and 2,100 m, on clayey or rarely calcareous soil. — Caucasus: W. Transc. (Abkhazia, W. Georgia, Adzharistan). Gen. distr.: Bal.-As. Min. (NW Anatolia—Pontic Range, westward to Trebizond). — Described from the N. slopes of the Pontic Range in the [former] Rize District. Type in Berlin.

Economic importance. A very striking and beautiful oak, with leaves resembling those of the chestnut, especially effective in fall on account of the bright colors of the foliage. Cultivated in pleasure grounds of W. Europe. Recommended for extensive planting in the USSR for ornamental purposes. The plant is apparently rather hardy and could be grown in the Ukraine.

Subsection 3. GALLIFERAE Spach, Hist. Nat. XI (1842) 170 (pro sect.). — Perianth of staminate flowers parted nearly to the base into narrow-lanceolate lobes; leaves subcoriaceous, winter-deciduous, coarsely toothed or unequally short-lobed; stipules deciduous or persistent only at the terminal buds.

4. **Q. araxina** (Trautv.) Grossh., Fl. Kavk. II (1930) 23. — Q. r o b u r var. a r a x i n a Trautv. A. H. P. II (1873) 587. — Q. i n f e c t o r i a b p e t i o l a r i s Medw. in Moniteur Jard. Bot. Tiflis, XI (1908) 37, non DC.

334 A small tree or shrub; shoots initially heavily gray-pubescent, the annotinous glabrate, lustrous, reddish-brown; buds small, globose or ovaloid, with glabrous ciliolate-margined scales; terminal buds surrounded by long-persistent small linear bracts; leaf petioles 1 — 1.5 cm long; blades coriaceous, lustrous green above, pale green and at first densely velutinous with short hairs, becoming glabrous or nearly so, very variable in shape, 6—10 cm long and 2—5 cm broad, elongate-obovate or elongate-elliptic, at the base of shoots often entire or with 1 or 2 large teeth, elsewhere coarsely toothed to broadly short-lobed, on each margin with 3 or 4 broad obtuse teeth or lobes, divided by shallow and sometimes barely perceptible broad sinuses; midrib and lateral veins very prominent beneath; lateral veins 6—8 pairs, more or less flexuous and curved, not quite parallel, ending in the lobes, their margins or, in the upper and lower part of the leaf, in imperceptible lobes; tertiary veins forming a dense network with irregular alveoles; peduncle ca. 1 cm long, as long as or longer than the petioles, with 1 — 3 fruits; cup scales more or less convex, gray-pubescent, tapering into a short brownish appressed point. May. (Plate XVII, Figure 4).

Forming thickets on dry stony slopes together with junipers and other xerophytic plants. — Caucasus: S. Transc. (S. Armenia, on slopes to the Araks in the Megri and Zangezur districts). Gen. distr.: Iran. (N. Iran ?). Described from Karchevan. Type in Leningrad.

(335)

PLATE XVII. 1. Quercus castaneifolia C.A.M., 1a) staminate flower. – 2. Q. macranthera F. et M., 2a) staminate flower. – 3. Q. pontica C.Koch, 3a) staminate flower. – 4. Q. araxina (Trautv.) Grossh. – 5. Q. Hartwissiana Stev., 5a) scale of cupule. – 6. Q. imeretina Stev., 6a) scale of cupule.

Economic importance. A very drought-resistant oak that ought to be tested with a view to cultivation in barren dry soils, and possibly also in saline soils, of the southern USSR. Studies are needed concerning the contents of tanning agents and their quality since other oaks of the same subsection, such as Q. infectoria Oliv., yield much high-grade tannin.

5. **Q. Woronowii** Maleev sp. nova in Addenda IV, p. 546. — Q. amblyoprion Woron. in schedis. — Q. dschorochensis Maleev in Bot. Zhurn. SSSR No. 2 (1935) 164 and 174, non C. Koch.

A shrub or a small tree (?); shoots glabrous; annotinous branchlets
337 reddish-brown, lustrous; buds globose, small; petioles 1—2 cm long; leaves subcoriaceous, quite glabrous, lustrous green above, brownish beneath, at the base of shoots broadly lanceolate, oval or elliptic, entire or with 1 or 2 large teeth, 2—5 cm long and 1—3 cm broad; other leaves broadly obovate, 5—8 cm long and 3—4 cm broad, with 3—5 broad obtuse sometimes barely perceptible lobes on each margin; lateral veins 8—10 pairs, more or less curved, not quite parallel, ending in the lobes, their margins, in the undeveloped lobes; tertiary veins strongly flexuous, forming a dense network with small very irregular alveoles; young fruits several together on a stalk 0.5—1 cm long; cup scales of young fruits subtriangular, almost flat, lustrous brown, slightly pubescent, tapering into an obtuse appressed pubescent tip; ripe fruits unknown. May. (Plate XVIII, Figure 5).

Dry stony slopes, up to 1,000 m above sea level. — Caucasus: W. Transc. (Adzharistan — along Chvana River Gorge). **Gen. distr.**: Arm.-Kurd. (Artvin District). Described from Adzharistan near the village of Tskhemlisi. Type in Leningrad.

Note. A species closely related to the preceding; also varying widely in leaf shape, but differing in that the leaves are quite glabrous, brownish-green beneath. Distinguishable from Q. dschorochensis C. Koch by the broader leaves with short broad obtuse lobes. The true Q. dschorochensis C. Koch does not occur in the USSR.

Economic importance. As for the preceding species.

Subsection 4. ROBUR Rchb. Fl. germ. exs. (1831) 176 (pro sect.). — Perianth of staminate flowers parted nearly to base into narrow-lanceolate lobes; leaves winter-deciduous (sometimes persistent on the tree in withered condition until spring), coarsely toothed or often more or less deeply lobed; stipules deciduous, rarely persistent around the terminal buds; bark of the trunk thick, splitting.

a. PEDUNCULATAE Maleev. — Pistillate flowers and fruits borne on long peduncles greatly exceeding the leaf petioles.

Series 1. **Hartwissianae** Maleev. — Leaves regularly short-lobed or rather large-toothed, with 9—12 pairs of lobes; lateral veins straight or somewhat arched, parallel; intermediate veins none or in lower part of the leaf barely perceptible; tertiary veins almost parallel, forming a network with large almost regular alveoles.

6. **Q. Hartwissiana*** Stev. in Bull. Soc. Nat. Mosc. XXX (1857) 387. — 338 Q. armeniaca Kotschy, Die Eichen (1862) 25, p. p.; Boiss. Fl. Or. IV, 1164; Wenzig in Jahrb. Bot. Gart. Berlin IV, 186; Medvedev in Moniteur Jard. Bot. Tiflis XI (1908) 10. — Ic.: Kotschy l. c. tab. XXV; Medvedev l. c. tab. 3.

A tree to 10—15 m; shoots glabrous; annotinous branchlets dark reddish-brown, sometimes almost black; buds large, to 0.6—0.7 cm long, broadly ovaloid to subglobose; scales ciliolate; terminal buds with long-persistent linear stipules; leaf petioles 1.5—2 and up to 4 cm long, yellowish-brown; blades oboval or elongate, rather firm, lustrous green and glabrous above, paler and brownish-tinted and minutely grayish-pubescent beneath with longer rufescent hairs persisting only on the veins and in their axils or sometimes glabrate, 8—12 and up to 15—20 cm long and 4—7 or sometimes 9—12 cm broad, somewhat obliquely subcordate to subcuneate at base, with a short obtuse terminal lobe and 9—12 pairs of very regular semiovate or triangular obtuse or subacuminate lobes, the sinuses one-sixth to one-tenth the breadth of the blade; midrib and lateral veins very prominent beneath, brownish; lateral veins as many as lobes, frequent, longer than the blade; tertiary veins slightly flexuous, subparallel, forming a network with almost rectangular alveoles; fruiting peduncles 4 or 5, rarely 2—4 or up to 10 cm long, upright, bearing 1—4 fruits, the terminal often abortive; cup hemispherical, short-cylindric, to 1.5 cm high and 2 cm in diameter; scales rather large, dark gray, with almost rectangular base and triangular upper portion, terminating in a small brownish tip, gray-pubescent, flat or sometimes the upper ones slightly inflated; acorn long-cylindric, strongly exserted from the cup, to 2.5—3 cm long and 1.2—1.5 cm in diameter, sometimes larger and to 4 cm long and then usually solitary (f. macrocarpa Medw.). (Plate XVII, Figures 5, 5a).

Forests of the maritime and the lower mountain zone, up to 1,000—1,200 m. — Caucasus: W. Transc. (northward to Dzhubga and Arkhipo-Osipovka), Cisc. rare along the upper course of the Bol'shaya Laba and Belaya rivers and their tributaries). **Gen. distr.**: Bal.-As. Min. (NE Lazistan). Described from W. Transc. (Kutaisi). Type in Tiflis.

Note. In the Nikitskii Botanical Garden there is a specimen of this oak, grown in Steven's time from seed brought by Hartwiss from W. Transcaucasia (Steven's description).

Economic importance. Together with the Iberian Oak (Q. iberica Stev.), distributed in the same parts of the Caucasus, it is used for fuel and other purposes, though to a limited extent since it hardly appears in any compact stands and only solitary trees occur in the woods.

339

Series 2. **Eurobora** Maleev. — Leaves very short-petioled, often almost sessile, auriculate at base, deeply and irregularly lobed with (3) 5—7 pairs of very unequal lobes; secondary veins more or less curved, not parallel; intermediate veins pronounced; tertiary veins strongly flexuous, not parallel, forming a dense network with irregular alveoles.

* Named for Hartwiss (1824—1860), late director of the Nikitskii Botanical Garden.

7. **Q. robur** L. Sp. pl. (1753) 996; Wenzig in Jahrb. Bot. Gart. Berlin IV, 184; C. K. Schn. Laubholzk. I, 197. — Q. *pedunculata* Ehrh. Arb. (1789) No. 77 (nomen nudum); M. B. Fl. taur.-cauc. II (1808) 402; III (1819) 621; Ldb. Fl. Ross. 590; Steven in Bull. Soc. Nat. Mosc. XXX, 385; Boiss. Fl. Or. IV, 1163; Shmal'g., Fl. II, 425; Zelenetskii, Mat. dlya Fl. Kryma (1906) 375; Medvedev in Monit. Jard. Bot. Tiflis XI, 1, pp. — Ic.: Rchb. Ic. Fl. Germ. XII (1856) tab. 648; C. K. Schn. l. c. I, f. 124, 125 and 126 a—f; Medvedev, l. c. tab. I, f. 1—2; Hegi III. Fl. Mitteleur. III, tab. 86, f. 1. — Exs.: Callier, Iter taur. No. 730.— Russian: dub chereshchatyi [pedunculate].

A tree to 40—50 m; trunk forming several stout branches below the top; shoots glabrous; annotinous branchlets reddish-brown; buds ovoid to subglobose; leaves crowded at the ends of shoots; petioles 0.5—1 cm long; leaves 7—15 and sometimes 20—30 cm long, 4—7 cm broad, more or less elongate-obovate, cordate and distinctly auricled at base, with an obtuse and rather elongate terminal lobe and 4—6 (8) rather long obtuse or rarely subacuminate unequal straight or curved lateral lobes, the sinuses between them to 1/3 — 1/5 the breadth of the blade; the lobes entire or rarely with 1 — 3 large teeth; upper surface lustrous green, glabrous; lower surface paler, at first pubescent, at length glabrous with few hairs along the veins; lateral veins ending in the lobes remote, not very prominent beneath; intermediate veins less pronounced, sometimes reaching the margin of the blade; pistillate flowers and fruits solitary or 2 or 3 together on a peduncle to 6—8 cm long; cup cupuliform, ca. 1 cm tall, sometimes small saucer-shaped and ca. 0.5 cm tall, 1.5—2 cm in diameter; scales of cup slightly convex or almost flat, rather broadly ovate, sometimes transversely extended, gray-pubescent, short-tipped; acorn 1.5—3.5 cm long, enclosed to between one-third and one-half its length, at first sparingly tomentose, becoming glabrous, brownish-yellow, with more or less distinct brown striation. April—May.

Forming woods on sandy-loam, loam, podzolic, and chernozem soils in the steppe zone, in ravines and gullies of mountain slopes. — European part: M.Dnp., Bl., Crim., V.-Don, L. Don, L. V., Transv., U. Dnp., U. V., V.-Kama, Lad.-Ilm., Dv.-Pech.; Caucasus: Cisc., Dag. **Gen. Distr.**:
340 Atl. Eur., Scand. (to 63° N. lat.); N. Bal.-As. Min. (N. part of the Balkan Peninsula). Described from Europe. Type in London.

Note. A species very variable in all its characters. Particularly inconsistent are the leaves, as regards size, shape, degree of dissection, and vesture. The variability of leaves is observable on different shoots of the same tree as well as on different trees, and is apparently due, in the first place, to ecological conditions. It has generally been noted that, under the drier conditions of the southern USSR, those forms predominate which have smaller, relatively firm, more deeply cut, and more hairy leaves. With lower light intensity and an ample moisture supply, the leaves tend to be larger, thinner, and scantily lobed. The length of the fruiting peduncle also varies: there are forms with peduncle ca. 2 cm and less in length and hence almost sessile acorns (f. *brevipes* Beck; f. *pseudosessilis* Asch. et Gr.). Variability also involves other characters. Most interesting are two races that differ ecologically while being indistinguishable morphologically. One of them, var. *tardiflora* Czern., sometimes referred to as winter oak, is distributed through the Ukraine and the SE part of Central Europe; its spring sprouting and leaf expansion occur 2—3 weeks later than in the early race, var. praecox Czern. Moreover, var. tardiflora usually retains its withered leaves until the following spring. Its wood is heavier and tougher; it is hardier and stands up to spring frosts; it is therefore valuable for afforestation purposes near the distribution limit of the oak. A distinctive form of pedunculate oak grows on the sandy loams of Volhynia, the so-called pink or soft oak, characterized by its softer and less durable wood of initially pink and finally yellow color. There are also numerous climatic races of Q. robur, distinguished by their adaptation to various climatic conditions and the duration of the growing period.

Economic importance. This is one of the most important forest trees of Europe and of the European part of the USSR, covering extensive areas in the forest and forest-steppe belts, in gullies and ravines. One of the hardiest species for afforestation in the steppe belt. As indicated by the studies of Cieslar, Kobranov, and others, the provenance of the seed of this oak is of great silvicultural importance. The seed should be taken from areas with similar climate, preferably from the race most adapted to given climatic conditions. The wood of Q. robur is very hard, durable, heavy (specific weight 0.54—0.74). It is valued and widely used for construction and for the production of miscellaneous articles. The bark has a tannin content of 8—20% and it is extensively put to use directly or as tannin extract for leather tanning. The acorns are fed to swine and are used as a substitute for coffee. Q. robur is also a very valuable ornamental tree; its shortcoming is the rather slow growth.

341 8. **Q. pedunculiflora** C. Koch in Linnaea XXII (1849) 320. — Q. pedunculata d. glaucescens Medw. in Monit. Jard. Tiflis XI (1908) 6.

Tree; allied to the preceding; petioles ca. 0.5 cm long; leaves cordate at base, with auricles concealing the petiole; lower surface glaucescent, slightly pubescent or glabrate with hairs confined to the veins; lobes 3—5, very unequal, in lower part of the leaf very short, commonly narrow elongate obtuse entire or with 1 or 2 large teeth; sinuses separating the lobes to between one-fifth and one-fourth the breadth of the blade, rarely deeper and reaching nearly to the middle of the blade; secondary veins strongly curved; intermediate veins always pronounced; peduncles ca. 4 cm long; scales of cup dark brown, sparsely clothed with short rarely falcate hairs, quite flat, broadly triangular, terminating in a short spreading brown tip. April—May. (Plate XVIII, Figures 1, 1a).

Forests in the lower mountain zone, up to 1,000 above sea level. — Caucasus: Dag., E. Transc. (former Kuba County). Type in Berlin.

Note. Lack of sufficient material makes it impossible to determine the precise northern distribution limit of this species in Dagestan where it is replaced in the N. part by Q. robur. It is apparently associated with the maritime and the lower mountain zones, while Q. robur L. grows higher up in the mountains. Q. pedunculiflora C. Koch is also reported for the Balkan Peninsula, but there it is replaced by a related race described by Koch as β virens (Q. brutia Griseb. in Spicil. Fl. Rum. et Bit., 338).

Economic importance. As for the preceding.

9. **Q. imeretina** Stev. ex Voronov in Ezhegodnik Ekzotlesa I (1930) 91; Addenda IV, p. 545. —Q. sessiliflora ε Tchorochensis Albov, Prodr. Florae Colch. (1895) 218, non DC.

Tree; annotinous branchlets glabrous, reddish-brown, turning gray; buds very small, 0.2 — 0.3 cm long, ovoid, obtuse; petioles very short; leaves subsessile, 5 — 14 cm long and 2.5 — 5 cm broad, often curved, thin, pale green beneath, glabrous or sometimes with scattered hairs along the veins, dull green above, strongly cordate at base, the large auricles completely concealing the petiole; lobes 4 — 6 on each side, markedly unequal, in lower part of the leaf barely perceptible, straight or curved, narrow, obtuse, the sinuses between them one-third or more of the breadth of the blade; lateral veins ending in the lobes commonly strongly flexuous or curved, markedly longer than the leaf, remote; intermediate veins several on each side, always distinct; peduncles to 10 cm long, equaling or exceeding or rarely shorter than the leaves, with 1 or 2 developed acorns
342 and commonly with 1 or 2 abortive ovules at the very slender upper end of the peduncle; cup small, low, to 1.5 cm in diameter, enclosing only the base of the acorn; scales small, closely appressed, gray-pubescent, almost flat except for the slightly inflated lowest ones, lanceolate, prolonged into an appressed slightly pubescent brown tip; acorn narrow-cylindric, to 3 cm long. April. (Plate XVII, Figures 6, 6a).

Forming forests in the lower forest zone, up to 200 m above sea level, chiefly on Upper Quaternary terraces. — Caucasus: W. Transc. (Rion and South Abkhazian lowlands, eastward to Kutaisi and Kviril, northward to the Kodor River). Endemic. Type in Tiflis.

Series 3. **Haas** Maleev. — Petioles longer; leaves exauriculate, firm, rather densely pubescent beneath; cup deeply cupuliform, thick, woody.

10. **Q. erucifolia** Stev. in Bull. Soc. Nat. Mosc. XXX (1857) 388. — Q. pedunculata f. erucaefolia Medw. in Monit. Jard. Bot. Tiflis XI (1908) 8.

Tree; shoots and annotinous branchlets glabrous; buds obtuse, broad-ovoid, ca. 0.5 cm long; petioles ca. 0.5 — 1 cm long; leaves bright green and glabrous above; densely covered beneath with short fine pubescence and with fugacious longer hairs, 6 — 9 cm long and 4 — 6 cm broad, elongate-obovate, scarcely cordate at base; terminal lobe short, narrow, obtuse; lateral lobes 6 — 8 pairs, long and narrow, rarely shorter and broad, with crisped margin and commonly with several teeth; sinuses deep, to one-third or more of the breadth of the blade; lateral veins very prominent beneath, usually curved and flexuous; intermediate veins 1 or several on each side; fruiting peduncles shorter than leaves, 2.5 — 3 or rarely 5 cm long; cup cupuliform, ca. 1 cm high and 1 — 1.5 cm in diameter; scales of cup small, flat, gray-pubescent, rounded-triangular, with a small spreading tip. May. (Plate XVIII, Figure 2).

Forests in the lower mountain zone, at 150 — 750 m above sea level.— Caucasus: E. Transc. (foothills of the Greater Caucasus from South Ossetia to Geokchai County in [Soviet] Azerbaijan). Endemic. Described from the Nukha District. Type in Tiflis.

11. **Q. longipes** Stev. in Bull. Soc. Nat. Mosc. XXX (1857) 387. — Q. pedunculata var. longipetiolata Medw. in Monit. Jard. Bot. Tiflis XI (1908) 4. — Exs.: Fl. Cauc. exs. No. 335; Pl. orient. exs. No. 380.

Tree; shoots grayish-brown, at first densely pubescent, becoming glabrous; buds ovoid, obtuse, 0.3 — 0.4 cm long, finely silky-pubescent, rarely glabrous; petioles 0.5 — 1.5 cm long; leaves 7 — 12 cm long and 5 — 7 cm broad, firm, subcoriaceous, dull green and glabrous above, paler
343 beneath and densely covered with fine pubescence, sometimes glabrescent, with longer hairs along the veins, shallowly cordate at base, with small subcuneate auricles; terminal lobe short, broad, obtuse; lateral lobes 3 or 4 pairs, dissimilar in size and shape; sinuses one-fourth the breadth of the blade; rarely lobes 5 or 6 pairs, longer, narrow, the sinuses between them deeper, and then the base rather long-cuneate; midrib and lateral veins very prominent beneath, the lateral veins remote; intermediate veins pronounced, ending also at the margins of the lobes; fruiting peduncles stout, 3 — 7 cm long, shorter than leaves, bearing 1 — 3 acorns; cup large, thick, deeply cupuliform, to 1.5 cm high and 1.5 — 2 cm in diameter; scales of cup rather large, disposed in distinct parallel rings surrounding the cup, densely pubescent, broadly triangular or rounded, with a short obtuse spreading brownish-pubescent tip; acorns narrowly cylindric, strongly exserted from the cup, to 3.5 cm long. April. (Plate XVIII, Figures 3, 3a).

Forests in lowlands, on alluvial soil. — Caucasus: E. Transc. (in the plain of the Kura River and its tributaries). Endemic. Described from the Nukha District and from Kirovabad. Type in Tiflis.

Economic importance. Worthy of note, since it is apparently a very drought-resistant oak that could be employed for afforestation purposes in the arid regions of the southern USSR.

b. SESSILIFLORAE Maleev. — Pistillate flowers and fruits sessile or borne on peduncles shorter than or equaling the leaf petioles.

Series 1. Ibericae Maleev. — Leaves short-lobed, with rather shallow sinuses; veins straight or arched, parallel; intermediate veins absent or sometimes faintly visible in the lower part of some leaves; stipules usually long-persistent around the terminal buds.

12. Q. iberica Stev. ex M. B. Fl. taur.-cauc. II (1808) 402; III (1819) 620; Stev. Bull. Soc. Nat. Mosc. XXX, 388. — Q. sessiliflora auct. plur. fl. Cauc. — Q. sessiliflora β iberica Ldb. Fl. Ross. III (1851) 590; Medvedev in Monit. Jard. Bot. Tiflis XI (1908) 17. — Q. pubescens γ iberica Wenzig in Jahrb. Bot. Gart. Berlin IV (1886) 190. — Ic.: Eichw. Plant. nov. (1831 — 33) fasc. 2, tab. 3.

A tree of medium size, with reddish-brown branchlets and shoots; buds with long-persistent stipules, ovoid, acuminate, to 0.7 cm long, or sometimes globose; petioles 1 — 3 cm long; leaves rather firm, subcoriaceous, retained in withered condition until the following spring, 7 — 14 or sometimes up to 20 cm long and 3 — 8 sometimes to 10 cm broad, lustrous green above, paler, glabrous or finely gray-pubescent beneath,

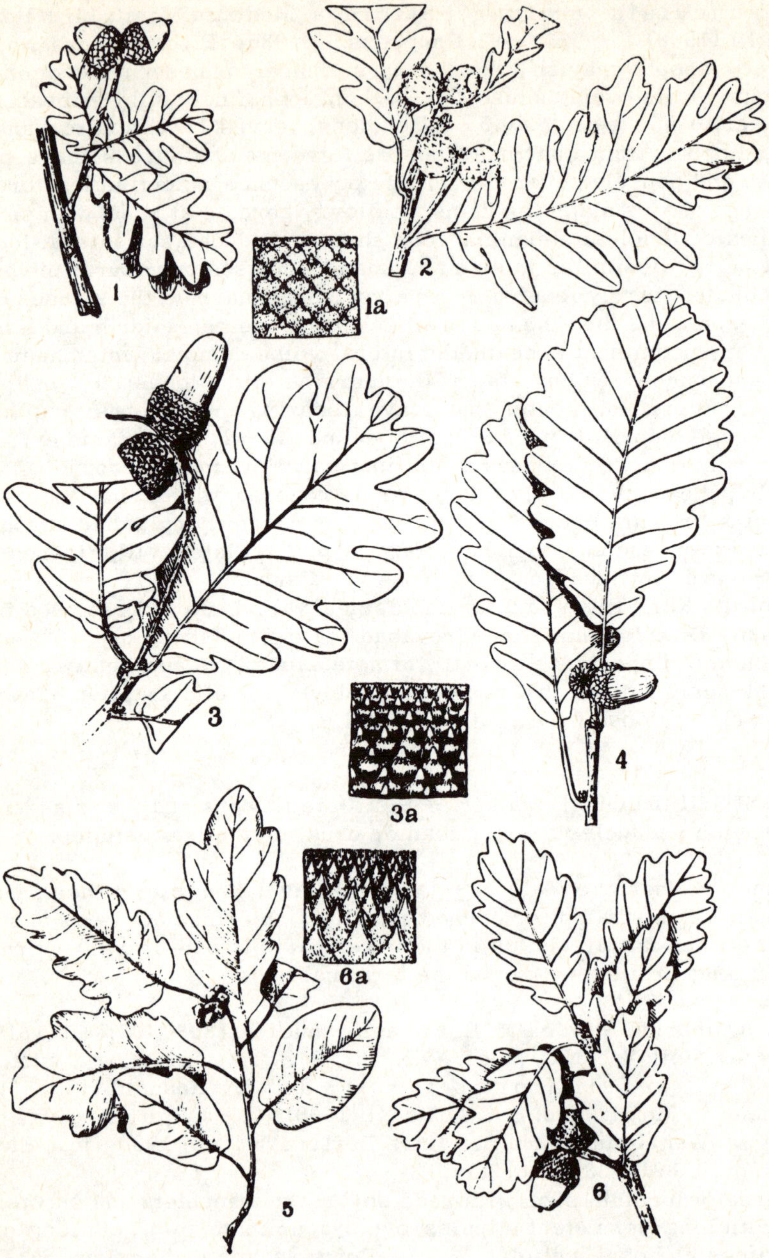

PLATE XVIII. 1. Quercus pedunculiflora C.Koch; 1a) portion of cup.— 2. Q. erucifolia Stev.— 3. Q. longipes Stev.; 3a) portion of cup.— 4. Q. iberica Stev.— 5. Q. Woronowii Maleev.— 5. Q. longifolia C.Koch; 6a) portion of cup.

344 more densely so and with scattered hairs along the veins, obovate or narrower and elongate; terminal lobe short, obtuse; base somewhat cuneate, rarely subcordate; lateral lobes 8—11 pairs, short, sometimes scarcely more than large teeth, in upper and lower part of the leaf barely perceptible; sinuses one-fifth to one-tenth the breadth of the blade; veins prominent beneath, commonly brownish; lateral veins as many as lobes, straight or slightly arched and, in relation to the length of the blade, rather frequent; occasionally 1 or 2 faint intermediate veins visible on the lower part of the blade; tertiary veins but slightly flexuous, subparallel, forming a rather loose network with regular alveoles; pistillate flowers and fruits sessile, solitary or in pairs, or borne on a short peduncle and then usually several crowded together; cup cupuliform, to 1.5 cm high and 1.5 cm in diameter enclosing the acorn to between one-half and one-third; scales of cup narrow-ovate or triangular-lanceolate, densely gray-pubescent, the lower ones commonly strongly convex, the upper ones flat, with a short appressed brown tip; acorn to 3.5 or sometimes to 4.5 cm long and 1.5 cm in diameter. April—May. (Plate XVIII, Figure 4).

Forming forests chiefly on dry southern slopes. — Caucasus: W. Transc. (northward to Sochi, sporadically to Novorossiisk), E. and S. Transc., Tal., Dag., Cisc. (sporadically further W. and rarely on the slopes of the Greater Caucasus and at the upper reaches of the Belaya and Bol'shaya Laba rivers). Endemic. Described from W. Georgia and from the middle course of the Aragva River. Type in Tiflis.

Note. Q. iberica displays considerable variability in respect of various characters. Of rather frequent occurrence are larger leaves with mostly cuneate base and very short lobes; such leaves usually characterize samples derived from low-lying shaded locations. On the other hand, samples of trees growing under dry conditions and exposed to light usually have smaller firm leaves with rather long and narrow lobes and more numerous veins. Larger leaves with a reduced number of veins often display a deviation from typical venation consisting in a shift of lateral veins of a pair with each arising at a different level; such leaves usually have 1 or 2 intermediate veins. The length of peduncle also varies, the number of fruits varying correspondingly from 1 to several. Thus in Abkhazia predominate trees with elongated peduncles and crowded
347 fruits (var. sorocarpa Woron. — Q. sorocarpa Woron. in schedis), but side by side with these there are others with solitary sessile acorns. Similar variability of peduncle length has been described by Yaroshenko for North America. In spite of this lack of constancy, Q. iberica Stev. is a species clearly distinguishable from other related species.

Economic importance. Q. iberica is the most widely and profusely distributed oak of the Caucasus. It forms extensive forests and is therefore of greater economic importance for the Caucasus than any other oak. The wood does not differ in quality and durability from that of European oaks, but the tree is inferior to the summer and winter oaks in its size.

13. **Q. hypochrysa** Stev. in Bull. Soc. Nat. Mosc. XXX (1857) 338. — Q. sessiliflora γ mannifera Medw. in Monit. Jard. Bot. Tiflis (1908) 20. — Q. Szowitsii Wenzig in Jahrb. Bot. Gart. Berlin IV (1886) 20. — Russian: dub zolotistyi [golden].

Tree, related to the preceding. Differing in its thinner leaves, golden-yellow beneath, with persistent axillary pubescence of longer stellate hairs, relatively long-obovate, 10—18 cm long and 5—8 cm broad; terminal lobe short and broad; lateral lobes fewer, 5—7 pairs, very short broad and obtuse; lateral veins considerably longer than the leaf, more remote than in the preceding; pistillate flower and fruits 3 or 4 together; peduncles ca. 2 cm long, about equaling the petioles.

Caucasus: E. Transc. (mountainous Karabakh, SE and coastal part of Azerbaijan), Dag. Endemic. Described from Kirovabad. Type in Tiflis.

14. **Q. longifolia** C. Koch in Linnaea XXII (1849) 328. — Exs.: Herb. Fl. Cauc. No. 416.

A shrub or a small tree; annotinous branchlets glabrous, lustrous, reddish-brown; buds elongate-ovoid, ca. 0.5 cm long, acuminate; petioles 0.5—1 cm long; leaves coriaceous, quite glabrous on both sides or with few isolated hairs along the veins, pale green beneath, elongate-obovate, 6—9 cm long and 3—4 cm broad, with a short narrow obtuse terminal lobe, narrowed toward the rounded or subcordate base, crisp-margined; lateral lobes 3—7 pairs, short, narrow, obtusish; sinuses one-sixth to one-eighth the breadth of the blade; lateral veins arising at an angle of 30—50° and then strongly recurved toward the base; pistillate flowers and fruits 2—4 together, rarely solitary, on a peduncle 0.5—1 cm long; cup cupuliform, ca. 1 cm high and 2 cm in diameter, sometimes subglobose and enclosing the acorn; scales of acorn slightly gray-pubescent, dark brown, almost flat or slightly convex, lanceolate or ovate, the elongated narrow spreading brownish-orange tips conspicuously contrasting against
348 the dark surface of the cup; acorn stoutly cylindric, to 1.5—2 cm long and ca. 1 cm in diameter, rarely up to 3 cm long. (Plate XVIII, Figures 6, 6a).

Dry stony slopes. — Caucasus: W. Transc. (Adzharistan in mountains, northward to the Adzhar-Imeretian Range). **Gen. distr.**: Arm.-Kurd. (Chorokh River basin). Described from the Chorokh River valley. Type in Berlin.

Note. A species related to Q. iberica but quite distinct in its smaller leaves with crisped margin, a smaller number of lobes and of veins, the cup which is characteristically less pubescent and darker than in other species, and the bright spreading tips of cup scales.

Series 2. **Eusessiles** Maleev. — Leaves less deeply unequally lobed, the more or less curved lateral veins not parallel; intermediate veins commonly present although on some leaves faint; all stipules soon caducous.

15. **Q. petraea** Liebl. Flora fuldensis (1784) 403. — Q. sessiliflora Salisb. Prodr. Stirp. (1796) 392; Ldb. Fl. Ross. III, 589; Steven in Bull. Soc. Nat. Mosc. XXX, 385; Boiss. Fl. Or. III, 1164; Wenzig in Jahrb. Bot. Gart. Berlin IV, 186; Shmal'g., Fl. II, 5; Zelenetskii, Mat. Fl. Kryma, 377; Medvedev in Monit. Jard. Bot. Tiflis XI, 13. — Q. robur β L. Fl. suecica ed. II (1755) 340. — Q. robur M. B. Fl. taur.-cauc. II (1808) 417. — Q. sessilis Ehrh. Arb. XX (1789) 87 (nomen nudum); C. K. Schn. Laubholzk. I, 196. — Q. calcarea Troitzky in Zhurn. Russk. Bot. O-va XVI, 4 (1931) 349. — Q. iberica Troitzky ibidem 350. — Ic.: Loud. Arb. VII (1838) tab. 691; C. K. Schn. l. c. f. 123; Hegi III. Fl. Mitteleur. III, tab. 86, f. 2. — Russian: dub skal'nyi, dub sidyachetsvetnyi [rock or sessile-flowered].

Tree to 20—30, sometimes up to 40 m; branchlets and shoots glabrous or sometimes sparingly pubescent, reddish-brown; petioles 1—2.5 cm long; leaves 8—12 cm long and 3.5—7 cm broad, retained on the tree in withered condition until the following spring, glabrous bright green above, paler beneath and glabrate or finely pubescent with longer hairs along the veins, rounded or more or less cuneate at base; terminal lobe elongate, obtuse; lateral lobes 5—7 pairs, entire or coarsely toothed, elongate, sometimes curved, obtuse, unequal, longest in the middle of the blade; sinuses one-fifth to one-fourth or more of the breadth of the blade; lateral veins more or less curved, sometimes flexuous, not quite parallel, in relation to leaf length remote; intermediate veins 1 or 2, mainly in the lower part of the blade; pistillate flowers and fruits solitary or up to 5 together, mostly 2 or 3, sessile or rarely on a very short peduncle; cup cupuliform, ca. 1 cm high and 1.5 cm in diameter; scales of cup almost flat or slightly inflated, triangular-lanceolate, gray-pubescent, brownish-tipped; acorn 1.5—2.5 cm long, enclosed to between one-half and one-third by the cup. April—May.

349 Forming forests on mountain slopes, more rarely in plains. — European part: M. Dnp. (W.), Bl. (W.), Crim.; Caucasus: Cisc., Dag., W. Transc. (N. part to Tuapse). **Gen. distr.:** S. Scand. (to 60° N. lat.), Atl. Eur., Bal.-As. Min. (N. part of Balkan Peninsula), Med. (W. part, in mountains). Described from England. Type in London.

Note. Varying greatly, chiefly in leaf size and shape. Thus, of frequent occurrence are leaves with long-cuneate base (f. decipiens Bechst., rather often in the Crimea). Sometimes the leaves are entire (f. mespilifolia A. et Gr.). Particularly widespread in Ciscaucasia are trees with deeply cut leaves (f. pinnatifida Medw.). Amount of vesture and other characters also vary. Hybrids sometimes occur of this species with Q. robur L., rather inconsistent in their characteristics which are intermediate between those of the parental species. Hybrids occur where the distribution areas of the two species come into contact. In the USSR they are known from the Crimea.

Economic importance. Of same economic value as Q. robur, but with a much more restricted distribution in the USSR. The bark has a tannin content of 7—10%. The wood has a specific weight of 0.65—0.75; it is softer than the wood of Q. robur, more easily worked, and therefore preferred in carpentry. As the trunk is straight and less branched than in Q. robur, it yields a greater amount of serviceable lumber, with less waste. These features also contribute to the ornamental value of this tree.

16. **Q. pubescens** Willd. Berl. Baumzucht (1796) 279; M. B. Fl. taur. cauc. II, 402; III, 621; Ldb. Fl. Ross. 590; C. Koch in Linnaea XXII; Steven, Bull. Soc. Nat. Mosc. XXX, 385; Wenzig Jarhrb. Bot. Gart. Berlin IV, 189; Medw. Monit. Jard. Bot. Tiflis (1908) 22; C. K. Schn. Laubholzk. I, 199. — Q. lanuginosa Thuil. Fl. Paris (1799) 502. — Q. sessiliflora II lanuginosa DC. Prodr. XVI, 2 (1868) 10; Zelenetskii, Mat. Fl. Kryma, 382. — Q. pubescens Boiss. Fl. Or. IV (1867) 1165. — Q. sessiliflora β pubescens Shmal'g., Fl. II (1897) 425. — Q. crispata Stev. Bull. Soc. Nat. Mosc. XXX (1857) 386. — Ic.: Rchb. Ic. Fl. Germ, tab. 647; Kotschy, Die Eichen, tab. XXXIV; C. K. Schn. l. c. I, f. 122; Hegi III. Fl. Mitteleur. III, f. 491 a—3. — Exs.: Callier Iter taur. 732 No. (sub Q. pinnatifida).

A small tree, 5—10 m, rarely higher, usually with an uneven crooked trunk; shoots densely gray-tomentose, the tomentum gradually disappearing on annotinous branchlets; buds ca. 0.5 cm long, light brown, with densely pubescent scales; expanding leaves and ends of shoots reddish; petioles 0.5—1 and up to 2 cm long, densely pubescent; leaves retained on the tree in withered condition until the spring, firm, at first densely tomentose above, at length glabrescent with scattered hairs, dark green, on the lower surface densely clothed with spreading and finally completely or partly disappearing hairs, varying considerably in shape and size, 4—7 and up
350 to 10 cm long, 2—6 cm broad, slightly cordate or subcuneate at base, elongate-obovate; terminal lobe short, obtuse: lateral lobes 4 or 5, sometimes 3 or 7 pairs, rather broadly rounded, entire or coarsely toothed, the sinuses between them one-third to one-fifth the breadth of the blade; or leaves narrower, with narrow acuminate crisp-margined lobes — f. crispata Beck. (—Q. crispata Stev.); lateral veins more or less curved; intermediate veins 1 or several in lower part of the leaf; staminate inflorescence 3—4 cm long, with heavily pubescent stalk; fruits 1—4, sessile or on a peduncle to 0.5 cm long; cup to 1 cm high, hemispherical; scales of cup gray-tomentose, slightly convex or almost flat, lanceolate, with elongate appressed light brown tips; acorn 1.5—2.5 cm long. April—May.

Dry places, on calcareous, schistose, or argillaceous soil, up to 400—500 m above sea level. — European part: M. Dnp. (SW, rare), Crimea; Caucasus: W. Transc. (N. part, southward to Dzhubga), Dag., E. Transc. (NW part to the Caspian coast). **Gen. distr.:** Atl. Eur. (N. part), Centr. Eur. (S. part), E. Med., Bal.-As. Min. Described from the vicinity of Paris. Type in Paris.

Note. As pointed out above, the leaves vary greatly in size and shape. Specimens growing in more shaded and moist situations sometimes have larger leaves, the pubescence disappearing almost completely on the lower surface of the leaves and on the shoots, thus easily confounded with Q. petrea. They can, however, be distinguished by remnants of the tomentum which persist on the shoots and on the cup. Most distinctive is f. crispata, described by Steven as a separate species, but merely representing sucker shoots of pubescent oak. Q. pubescens sometimes hybridizes with Q. robur and much more often with Q. petrea. In the USSR these hybrids are known from the Crimea.

Economic importance. The uses are like those of other species, but they are restricted by the small size of the tree and the crookedness of the trunk. Of interest for propagation in arid areas of the southern USSR, especially on calcareous soils. Extensively used for afforestation of Austrian karst.

17. **Q. Kozlowskyi*** Woron. ex Grossg., Fl. Kavk. II (1930) 24.

Tree; annotinous branchlets reddish-brown, glabrous, lustrous; buds 0.5 cm long, ovoid, with ciliate scales; petioles stout, reddish-brown, 2.5—3 cm long; leaves firmly coriaceous, large, to 15—18 cm long and 10 cm broad, lustrous dark green and glabrous above, paler prominently reticulate-veined and very sparingly pubescent beneath, cuneate or

* Named for V. Kozlovskii, famous collector of Caucasian flora.

351 subcordate at base; lobes 4—6 pairs, very unequal, straight or curved, acuminate, the sinuses between them 1/4—1/5 the breadth of the blade; lateral veins curved; intermediate veins present in lower part of the blade; tertiary veins forming a dense network with irregular alveoles; fruits several on a peduncle ca. 3.5 cm long; cup hemispherical; scales of cup small, lanceolate, gray-pubescent, with glabrous brown tips. May.

Caucasus: E. Transc. (Belyi Klyuch natural boundary area). Type in Leningrad.

Note. Oak known only from a single herbarium leaf from the location indicated; differing conspicuously from leaves of other Caucasian oaks. As regards the leaves and the cup of immature fruits, it is most closely related to Q. pubescens. Further study of this oak and its distribution is needed.

Subsection DIVERSIPILOSAE C. K. Schn. Laubholzk. I (1907) 208. — Leaves shortly and not deeply lobed, with straight or somewhat arched lateral veins parallel to each other; intermediate veins absent. The pubescence on lower surface of the leaf consists of simple and stellate hairs, with a predominance of the former.

18. **Q. mongolica** Fisch. ex Turcz. in Bull. Soc. Nat. Moscou XI (1838) 101 (nomen); Turcz. Fl. baic.-dah. II (1854) No. 1065; Ldb. Fl. Ross. III, 589; Maxim. Primitiae Fl. Amur. (1859) 241, 390; Kom., Fl. Manchzh. II, 68; C. K Schn. Laubholzk. I, 209. — Ic.: C. K. Schn. l. c. f. 126 i; Nakai Fl. Sylv. Koreana VI (1917) tab. 11; Kom. and Alis., Opred. rast. Dal'nevost. kr. I (1931), Plate 133, p. 1, 2, 4, 5.

Tree to 10 m; young shoots glabrous; annotinous branchlets lustrous, reddish-brown; buds elongate-ovoid, large, 0.7—1 cm long; petioles 0.3—0.4 cm long; leaves firm, subcoriaceous, obovate or elongate, 8—15 and sometimes up to 20 cm long, 7—10 and to 15 cm broad, strongly narrowed toward the somewhat cordate base; terminal lobe short, obtuse; lateral lobes 7—13 pairs, broad, obtuse or rarely acuminate, the upper ones very short and often barely perceptible; sinuses 1/2—1/10 the breadth of the blade; upper surface quite glabrous, lustrous, green; lower surface at first pubescent, finally glabrous or with few hairs along the veins; lateral veins straight or slightly curved, in relation to blade length approximate, as many as lobes; tertiary veins subparallel, forming a rather loose network with regular alveoles; pistillate flower and fruits sessile, 2—6 together; cup thick, hemispherical, enclosing the acorn to 1/2—1/3, ca. 1.5 cm in
352 diameter; scales of cup thick, convex, densely pubescent, with appressed tips, the lower rounded-ovate, the upper linear; acorn broad-cylindric, 1.5—2 cm long and 1—1.5 cm in diameter. May.

Hills, foothills, river valleys, sometimes crags; forming extensive forests, especially on southern slopes. — Far East: Ze.-Bu.*, Uss., Uda, Okh. (S.), Sakh. **Gen. distr.**: Jap.-Ch. (Manchuria, N. China, Korea). Described from the right bank of the Argun River. Type in Leningrad.

Economic importance. An important forest species of the Far East, yielding valuable lumber for construction, miscellaneous woodwork and plywood. Specific weight 0.78.

* As far west as Albazin on the Amur River, to Tygda station on the railroad line, and to the Tukuringra Range in the northwest. — Editors.

Section D. DENTATAE C. K. Schn. Laubholzk. I (1907) 209. — Styles enlarged and rounded at the ends, spreading; scales of cup long, narrow-lanceolate, reflexed, rarely upright; fruit ripening in the first year; perianth of staminate flowers rather shallowly parted into triangular lobes; leaves deciduous, with short broad obtuse lobes; stipules caducous.

19. **Q. dentata** Thnbg. Fl. Jap. (1784) 177; Kom., Fl. Manchzh. II, 76; C. K. Schn. Laubholzk. I, 209. — Q. o b o v a t a Bge. in Mém. Acad. Sc. Pétersb. (1830) 135. — Ic.: Thnbg. Ic. Pl. Jap. V (1805) tab. 6; C. K. Schn. l. c. f. 133; Nakai Fl. Sylv. Koreana tab. 20; Kom. and Alis., Opred. rast. Dal'nevost. Kr. I, Plate 133, p. 3.

Tree to 15—20 m, with fissured bark; shoots densely shaggy with rufescent stellate hairs; annotinous and 2-year-old branchlets with partially persistent tomentum disappearing only in the second year, gray, with large raised light gray lenticels; buds large, ovoid, densely pubescent; petioles 0.5 cm long; leaves firm, 10—20 cm long and 7—12 cm broad, sometimes up to 30 or even 50 cm long and to 20—30 cm broad, obovate, bright orange-red in fall, tardily deciduous; basal auricles small and often indistinct; terminal lobe short and broad; lateral lobes 8—13 pairs, very short and broad; upper surface dark green, glabrate; lower surface densely clothed with stellate hairs; veins very prominent; lateral veins 8—13 pairs, directed into the lobes, parallel; intermediate veins none; pistillate flowers and fruits 2 or 3 (to 6) together, sessile; cup large, hemispherical; cup scales numerous, narrow-lanceolate, to 1.5 cm long, free, at first upright,
353 at length reflexed, densely pubescent outside, glabrous within; acorn subglobose, to 2 cm in diameter, enclosed by the cup to about the middle. May.

Dry slopes. — Far East: Uss. (along the seacoast from Tumyngan to the mouth of the Suchan River). **Gen. distr.:** Jap.-Ch. Described from Japan. Type in Uppsala.

Economic importance. The foliage of this oak provides food for silkworms. The tree is very ornamental owing to the large leaves which display very effectively bright autumn colors.

Subgenus 2. **ERYTHROBALANUS** Spach, Hist. Nat. XI (1842) 160; Örsted Vidensk. Meddel. 70. — Styles spreading, capitate; fruit maturing in the second year, rarely in the first year; abortive ovules apical; shell of acorn thick, 3-layered, densely tomentose within; scales of acorn appressed; leaf lobes and teeth bristle-pointed.

Section RUBRAE Loud. Arb. III (1838) 1877. — Leaves commonly deep-lobed, with acute sharp-toothed lobes, deciduous, in fall bright red or orange.

***Q. rubra** L. Sp. pl. (1753) 996; Sarg. Silva VIII (1895) 125; C. K. Schn. Laubholzk. I, 175. — Ic.: Sarg. l. c. tab. 409—410; C. K. Schn. l. c. f. 105; Elwes a. Henry Trees of Gr. Brit. V (1910) tab. 314, 333, f. 3.

Tree to 20 m; shoots densely rufescent-tomentose; annotinous branchlets glabrous, dark reddish-brown; buds ovoid, acute, to 0.8 cm long; petioles 3—5 cm long; leaves thin, dark green lustrous and glabrous above, finely appressed-pubescent beneath, to 15—20 cm long and 10—18 cm broad, cuneate at base; terminal lobe narrow, long-pointed; lateral lobes 2—4 pairs, elongated, narrow; all lobes sharply toothed, rarely entire; in addition to lateral veins, several subsidiary veins ending in the sinuses and at their margins, less pronounced and sometimes failing to reach the margin; pistillate flowers and fruits sessile or short-peduncled; cup enclosing the acorn to 1/3; scales of cup coriaceous, lustrous, brown, finely pubescent, ovate, appressed; acorn subspherical, ca. 2 cm in diameter, lustrous brown, finely pubescent. May.

North America (eastern states). Often cultivated in the USSR and introduced into forest stands of the Ukraine. Described from Virginia and Carolina. Type in London.

Economic importance. An important attribute of this oak is its ability to flourish on poor soils where other oaks grow less well. The wood, although inferior in quality to European oaks, and therefore less durable, is extensively used for construction and other purposes. The bark is widely employed in North America for leather tanning.

354 Genus 368. **FAGUS** L.*

L. Gen. pl. ed. 1 (1737) 292.

Flowers unisexual, monoecious, with a simple inconspicuous perianth; staminate flowers in many-flowered heads, the perianth 5-lobed; stamens 8—12, at anthesis greatly exceeding the perianth; anthers elongated, obtuse or pointed at base; pistillate flowers 2—4, surrounded by an involucre (cupule); perianth adnate to ovary, with a short 4- or 5-lobed limb; ovary inferior, 3-locular; styles 3, elongated, hairy; ovules anatropous, 2 in each locule; involucre (cupule) becoming woody in fruit, covered outside with leaflike or subulate appendages; nuts 2—4 in each involucre, sharp-edged; seeds 1 or rarely 2 in each nut; cotyledons reniform. Trees with smooth gray bark; leaves alternate, coriaceous, lustrous, entire or sometimes sparingly toothed, pubescent on the margin and along the veins beneath.

Economic importance. The value of beech consists chiefly in its wood. This takes one of the first places as regards caloric value. On dry distillation it yields alcohol and creosote. It is widely used for a variety of woodwork, such as bent (so-called Viennese) cabinet work, oars, shoe lasts, small articles for domestic use, spoons, trays, etc. It is also employed in boat construction (keels) and carriage construction (wheel spokes, etc.). Beech staves are made into containers for oil. For this purpose only so-called "white" beech is used, since "red" beech imparts to the oil a bitter taste and a dark color. These differences in the wood are due to the formation of so-called false heartwood as a result of fungal infection. Such modified wood (normally no heartwood is produced) becomes permeated with tannic substances that bring about the darkening.

* Treatment by E. V. Vul'f.

In recent times beech wood has been used for the production of railroad ties, which are impregnated with creosote or other chemical compounds. Wood with strongly developed false heartwood impedes the penetration of chemicals, as cellular pores become blocked by tyloses.

Beech nuts contain a considerable amount of valuable oil which is extracted by pressing. Apart from its applications for food and burning, this oil is used as adulterant for walnut, poppy, or olive oil. Large yields of seed are not obtainable annually but rather in 5—10-year cycles. The foliage provides feed for goats and cattle. It is also used to replace straw as bedding for livestock.

355 The earliest representatives of the genus have been described from Cretaceous formations of North America and from the Saxonian Cenomanian. The genus attained extensive development in Tertiary formations of both hemispheres, especially in regions of temperate forest flora.

F. Antipovii Heer in V.-Kama (Sterlitamak District), Ar.-Casp. (Karasandyk, Dzhilan, Chegan, Kenkous, Barsuki, Kop-mulla, Yar-kue, etc.), Ob (Tomsk), Uss. (Pos'et, De-Friz, Aleksinskii mine, Fatashi, Novokievskoe), and Sakh. (Gemonai, Agnevo, the coast near Orgomnaya Canyon and further south along the coast; Kamennaya and Takinosava). — F. Deucalionis Ung. in Sarmatian formations of Bl. (Krynka); in the Paleocene of L. Don (Ushi); Tertiary (Pliocene ?) formations of Alt.; Lower Tertiary layers of Uss. (Amagu and possibly Rechnoi). — F. Feroniae Ung. in Sarmatian formations of Cisc. (Apsheronskaya). — F. ferruginea Ait. mut. altaica Schmalh. in Tertiary (Pliocene ?) layers of Alt. (Chingistai). — F. orientalis Lipsky in Postpliocene tuffs of N. Cisc. (Zheleznovodsk) and in E. Transc. (Keramal-Naftalan). — F. silvatica L. in Postpliocene formation of V.-Don (Likhvin) and Cisc. (Mashuka — tuffs). — F. sp. in Tertiary formations of E. Transc. (Naftalan) and Lower Tertiary of Uss. (Amagu).

1. Appendages of involucre of two kinds: the upper ones subulate, the lower ones leaflike, often green, and nerved; perianth of staminate flowers broadly campanulate, the ovate round-tipped lobes often overlapping and usually shorter than the perianth tube; exine of pollen grains somewhat rough . . . . 1. **F. orientalis** Lipsky.

\+ Appendages of involucre all alike, subulate; perianth of staminate flowers infundibular-campanulate, its elongate linear-lanceolate lobes commonly longer than the perianth tube; exine of pollen grains smooth . . . . 2. **F. silvatica** L. (See also transitional forms between the two species).

1. **F. orientalis** Lipsky in A. H. P. XIV, 2 (1898) 300. — F. silvatica auct. Fl. caucas.; Ldb. Fl. Ross. III (1847—49) 593 (p. p.); Shmal'g., Fl. II, 424 (p. p.). — F. silvatica a) typica Medvedev in Mon. Jard. bot. 17 (1910) 3. — F. silvatica β macrophylla Hohenack. in Bull. Soc. Nat. Mosc. XI, 3 (1838) 259; Ldb. l. c. — F. silvatica γ asiatica DC. Prodr. XVI, 2 (1868) 119 ex parte. — F. asiatica H. Winkl. Pflanzengeogr. Stud. üb. Buchenwäld. (1901) 2, 5. — F. Hohenackeriana Palib. Bull. Herb. Boiss. 2 ser. VIII (1908) 378. — F. Hohenackeri Palib. (in sched.); Grossg., Fl. Kavk. II (1930) 21. — F. Sieboldi Koehne, Dendrol. (1893) 121, ex parte. — Ic.: W. Turill in Hooker's Ic. plant. 5 ser. II (1930) tab. 3137 (illustration of the flower incorrect). — Exs.: HFR No. 1739; Fl. cauc. exsicc. No. 84.

Tree to 30—50 m, with smooth light gray bark; leaves in 2 ranks, elliptic or ovate-elliptic, acuminate, rounded or more or less cuneate at base, (1.7) 5 —15 (20) cm long, (1.3) 2.4—8 (11.6) cm broad, appressed-
356 pubescent beneath especially along the veins; perianth of staminate flowers broadly campanulate, its lobes broad-ovate or almost round (very rarely elongate lanceolate), (0.6) 0.8—3 (3.6) mm long, about as long as or shorter than the perianth tube, this (1) 1.4—4 (4.4) mm long, often with a black spot at apex, the margin beset with black or white hairs; stamens to 12, (0.3) —1—6.8—(7.6) mm, anthers (0.8)—1 —2 mm long; appendages of involucre of two kinds, the lower ones leaflike, green, many-nerved, (2.6)—4.2—15—(17) mm long and 0.4—2 —(4) mm broad, the upper ones subulate, (1.2)—2—6.8—(8.2) mm long; stalk of involucre (0.9) 1.4—4.5—(9—7 [?]) cm long, appressed-pubescent; nut triquetrous, (1.2) 1.3—1.9 (2.2) cm long, (0.5) 0.6—1 (1.15 [?]) broad, not exceeding the involucre. Fl. May. (Plate XIX, Figures 1, 4).

Fossils have been found in Azerbaijan, near the village of Kasum-begli, the Keramal-Naftalan oil industry plant, and in the N. Caucasus in the vicinity of Zheleznovodsk (Krasnov, Palibin) in layers dating from the end of the Tertiary and the beginning of the Quaternary period.

Forests. — European part: Crim.; Caucasus: all regions. **Gen. distr.**: Iran, Syria, Asia Minor, Greece (?), Macedonia, Bulgaria, Dobruja. Described from the Caucasus. Type in Leningrad.

2. **F. silvatica** L. Sp. pl. (1753) 998; Ldb. Fl. Ross. III, 593 (p. p.); Shmal'g., Fl. II, 424 (p. p.). — Ic.: Fl. Dan. VIII (1810) tab. 1283; Rchb. Ic. Fl. Germ. XII (1850) tab. 639.

Tree to 30—50 m, with smooth light gray bark; leaves alternate, distichous, mostly ovate, rounded at base, mostly broad-acuminate at apex, entire, slightly undulate and remotely denticulate, sometimes lobed, 4—10 (—15) cm long, 2.5—7 (10) cm broad, with 5—8 lateral veins, the margin and the veins beneath appressed-pubescent; petiole 0.8—13 (1.8) cm long; perianth of staminate flowers infundibulate-campanulate, deeply cut, the elongate-linear or linear-lanceolate lobes 1.5—2.5 mm long, commonly longer than the tube, this 1—2.25 mm long, usually with white hairs on the margin; stamens 4—15, 3—7 mm long; anthers 1—1.5 mm long; appendages of involucre alike, subulate, the upper and the lower ones of equal length and hence terminating at different levels; stalk of involucre 0.8—1.3 (1.8) cm long, appressed-pubescent; nut triquetrous, 1— 1.6 cm long, equaling or shorter than the involucre. Fl. May. (Plate XIX, Figures 3, 5).

Forests. — European part: Crim., U. Dnp. **Gen. distr.**: W. and S. Europe; in the north — Scotland between 56° and 57° N. lat., SW Norway to 60°31', W. Sweden; in the east to E. Prussia, Lithuania, Volhynia, Podolia; in the south — Pyrenees, Cévennes, Alps, Apennines, the mountains of Sicily and the Balkan Peninsula. Described from W. Europe. Type in London.

359 F. orientalis Lipsky × F. silvatica L. = F. taurica Popl., pro parte, Mat. Krymsk. Zapovedn. (1925) 84.

Transitional forms, linking the species, are characterized by leafiness of the involucre, as in F. orientalis, but the perianth of staminate flowers

(357)

PLATE XIX. 1,4. Fagus orientalis Lipsky.— 3,5. F.silvatica L.— 2. F.orientalis × F.silvatica.— 6. Castanea sativa Mill.

resembles or approaches that of F. silvatica. Reversed combination occurs more rarely. (Plate XIX, Figure 2).

Forests. — European part: Crim. Described from the Crimean National Forest. Type in the Herbarium of the National Forest.

Note. Both species display great variability in their characters, especially as regards the vegetative parts. In the Crimea, side by side with specimens having all the characteristic features of F. orientalis, there are others that display typical characteristics of F. silvatica. There are also trees that combine the characteristics of both species and constitute transitional forms. These forms may be of a hybrid nature or else they manifest here, where the distribution areas of the two species converge, traces of an as yet unaccomplished transformation of the Tertiary F. orientalis into the Quaternary F. silvatica. Attempts at disentangling the complicated systematics of beech in the Crimea by giving the entire aggregate another name — as attempted by G. I. Poplavskaya who separated the Crimean beech as a new species F. taurica Popl. — do not provide a solution to the problem and actually obscure the picture. The proposed approach is all the more inacceptable since analogical occurrence of both species and their intermediate forms applies also to the Balkan Peninsula.

Order 15. **Urticales** LINDL.*

Flowers cyclic, monochlamydeous, rarely achlamydeous; stamens opposite the perianth segments; carpel nearly always 1; ovary superior, 1-seeded; fruit a drupe or nut; pollen usually binuclear; herbs, shrubs, or trees; leaves alternate or opposite, stipulate.

Key to Families

Flowers bisexual or unisexual; perianth segments 4 or 5 (rarely 3—8); stamens as many or twice as many as perianth segments; stigmas 2; ovary unilocular; fruit a nut or drupe; trees with distichous simple asymmetric stipulate leaves; flowers small in axillary cymes or pistillate flowers solitary . . . . Family XLV. **Ulmaceae** Mirb.

Flowers unisexual, monoecious or dioecious; perianth of 2+2 or rarely 2+6 segments, often becoming fleshy in fruit; stamens as many as perianth segments, rarely 1; stigmas 2 or 1; ovary unilocular; fruit a nut or drupe; trees or rarely herbs; leaves stipulate; flowers small, often in heads or aments; plants containing milky juice . . . . .Family XLVI. **Moraceae** Lindl.

360 Flowers rarely bisexual, mostly unisexual; perianth segments 4 or 5 (rarely 2 or 3); stamens as many as perianth segments; stigma 1, simple or divided; ovary unilocular; fruit a nut or drupe; seeds containing endosperm; herbs or rarely shrubs, with opposite or spiral stipulate leaves; flowers in aments, cymes, spikes, or heads; plants without milky juice . . . . Family XLVII. **Urticaceae** Endl.

* Treatment by A. V. Yarmolenko.

Family XLV. **ULMACEAE** MIRB.

Perianth simple, of 4—6 segments; stamens 4—6; ovary superior, free, 1- or 2-locular, the locules 1-seeded; stigmas 2; flowers bisexual or unisexual, clustered or solitary, axillary; leaves with mostly caducous stipules, distichous, alternate, petiolate.

Representatives of the genera Ulmus, Celtis, and Zelkova were widely distributed in the Tertiary era, particularly in the Northern Hemisphere where they were a component of deciduous forests.

Key to Genera

1.	Fruit a drupe	370. **Celtis** L.
+	Fruit dry, nutlike	2.
2.	Fruit a samara	369. **Ulmus** L.
+	Fruit a wingless wrinkled nutlet	371. **Zelkova** Spach.

Subfamily **Ulmoideae** ENGL.

in Engl. et Prantl. Pflanzenfam. III, 1 (1896) 61.

Flowers clustered; fruit not becoming indurated; embryo straight.

Genus 369. **ULMUS** L.*

L. Gen. pl. ed. V (1754) 83.

Flowers bisexual; perianth persistent, 5- or 6-parted or cleft, the segments imbricated in bud; stamens opposite and mostly as many as perianth segments; ovary sessile or stipitate, without a basal disk; stigma 2-lobed, persistent, united with the winglike appendages, the ovary, and the fruit; fruit a samara, i. e., a nutlet surrounded by a membranous rim [wing]. Trees or rarely shrubs; leaves distichous, undivided, with caducous stipules; inflorescence a fascicle, sometimes elongated; pedicels bracteolate at base. Russian: "vyaz" or "berest."

The genus was widespread in the Tertiary era in Europe, Asia, and North America, to the Arctic Region.

U. appendiculata Heer in Upper Dui (Oligocene ?) formations of Sakh. (Dui). — U. Braunii Heer in Sarmatian formations of Cisc. (along the Armavir-Tuapse railroad); in Maiotis formations of Bes. (Seimeny); in Lower Tertiary formations of Uss. (Rechnoi); in Upper Dui (Oligocene ?) series of Sakh. (Dui). — U. Braunii Unger in Tertiary formations of Cisc. (Adagum). — U. campestris L. in Postpliocene formations of
361 U. V. (Grigorovo in Moscow Region) and Cisc. (tuffs of Mashuka). — U. effusa Willd. in the Postpliocene of Cisc. (tuffs of Mashuka). — U. longifolia Unger in Lower Tertiary formations of Uss. (Rechnoi). — U. planoides Lesq. in the Lower Dui series of Sakh. (Pil'vo). —

* Name of elm used by Roman authors.

U. plurinervia Unger in Lower Tertiary formations of Sakh. (on the coast of Sakhalin between Kamennaya Canyon and Takin-Sav; also at Mgachi). — U. sp., not identified, in the Upper Dui series of Sakh. (Cape Rogatyi); in the Postpliocene of U. V. (Zvenigorod, Belolipki on the Krushma River) and Ze.-Bu., Astashikha (at the mouth of the Bureya River).

Key to Species Based on Fruit and Leaf Characters

1. Samara with densely ciliate margin, long-pediceled (Section Blepharocarpus Dum.) (Russian: vyaz) . . . . 2.
+ Samara glabrous or remotely hairy, short-pediceled (Section Madocarpus Dum.) . . . . 3.
2. Leaves unequally cordate at base; mature samara 12 — 16 mm long . . . . 1. **U. laevis** Pall.
+ Leaves equal at base, cuneate or orbicular; mature samara 8 — 12 mm long . . . . **U. celtidea** (Rogov.) Litw.
3. Nutlet near the apical notch of the wing (Subsection Foliaceae C. K. Schn.) . . . . 4.
+ Nutlet separated from the apical notch by the suture (Subsection Glabrae (Moss.) C. K. Schn.) (Russian: il'm) . . . . 9.
4. Nutlet in upper part of the elongated samara (Series Nitentes Moss.) (Russian: berest) . . . . 5.
+ Nutlet in the center of a round samara (Series Pumilae C. K. Schn.) (Russian: il'movnik) . . . . 8.
5. Samara gradually cuneately narrowed toward base, oblong-ovate . . . . 5. **U. propinqua** Koidz.
+ Samara rounded or rounded-cuneate at base, short-stalked, rounded-elliptic or obovate . . . . 6.
6. Lower surface of leaves hairy and along all the veins covered with minute red glands . . . . 3. **U. foliacea** Gilib.
+ Leaves hairy beneath, not glandular . . . . 7.
7. Grown leaves thin, papery; branches brownish or reddish-brown . . . . 4. **U. suberosa** Moench.
+ Grown leaves coriaceous; branches gray or yellowish-gray . . . . 6. **U. densa** Litw.
8. Leaves ovate-lanceolate, the subsidiary veins remotely dichotomizing, the teeth simple or rarely double . . . . 7. **U. pumila** L.
+ Leaves ovate, the subsidiary veins forked, the teeth double or triple . . .
362 . . . . 8. **U. Androssowii** Litw.
9 Samara glabrous (Series Euglabrae C. K. Schn.) . . . . 10.
+ Samara pubescent over the locule or over the whole surface . . . . . . 11.
10. Mature leaves of fertile branchlets acuminate, without lobes . . . . 9. **U. scabra** Mill.
+ Mature leaves of fertile branchlets 3 — 5-lobed at the summit . . . . 10. **U. laciniata** (Trautv.) Mayr.
11. Samara with fine silky hairs merely over the locule . . . . 11. **U. elliptica** C. Koch.
+ Samara hispidulous over the whole surface . . . . 12. **U. macrocarpa** Hance.

Key to Species Based on Leaf Characters

1. Leaves relatively equal at base . . . . 2.
\+ Leaves distinctly unequal at base . . . . 3.
2. Leaves thin, papery, the tapering teeth upcurved . . . . 2. **U. celtidea** (Rogov.) Litw.
\+ Leaves firm, coriaceous, the straight teeth obtusish . . . .7. **U. pumila** L.
3. Subsidiary veins at most once or twice branched . . . . 1. **U. laevis** Pall.
\+ Subsidiary veins branched 2 or 3 (4) times or the leaves very firm, coriaceous, lustrous . . . . 4.
4. Petioles 4—5 times the length of the adjacent buds, glabrous or pubescent . . . . 5.
\+ Petioles at most 3 times the length of adjacent buds, densely woolly with spreading hairs . . . . 10.
5. Leaves dotted beneath along the veins with red glands . . . . 3. **U. foliacea** Gilib.
\+ Leaves hairy or glabrous, eglandular . . . . 6.
6. Leaves enlarged above the middle, tapering toward base, acuminate at apex . . . . 7.
\+ Leaves enlarged about the middle, asymmetrically rounded toward base, acute or acuminate at apex . . . . 8.
7. Leaves unequally trebly serrate with upcurved tapering teeth . . . . 5. **U. propinqua** Koidz.
\+ Leaves doubly serrate with short teeth . . . . 12. **U. macrocarpa** Hance.
8. Leaves relatively thin, papery . . . . 4. **U. suberosa** Moench.
\+ Leaves thick, coriaceous . . . . 9.
9. Leaves sparingly pubescent . . . . 6. **U. densa** Litw.
\+ Leaves densely pubescent . . . . 8. **U. Androssowii** Litw.
363 10. Leaves of fertile branchlets with 3—5 lobes at apex . . . . 10. **U. laciniata** (Trautv.) Mayr.
\+ Leaves of fertile branchlets acuminate, not lobed . . . . 11.
11. Lower surface of leaves and petioles soft-hairy . . . . 11. **U. elliptica** C. Koch.
\+ Lower surface of leaves and petioles scabrous . . . . 9. **U. scabra** Mill.

Section 1. **BLEPHAROCARPUS** Dumort. Fl. Belg. Prodr. (1827) 25.— The section referred to in Russian as "vyaz." Samara with glabrous surface and densely ciliate margin; pedicels slender, 2—5 times the length of the mature samara; perianth with unequal lobes.

1. **U. laevis** Pall. Fl. Ross. I (1784) 75, tab. 48, f. F.; C. Koch, Dendr. (1872) 419; C. K. Schn. in Oesterr. Bot. Zeitschr. 64 (1916) 66.— U. pedunculata Foug. in Mém. Acad. Sc. Paris 1784 (1787) 215; Ldb. Fl. Ross. III, 48; Shmal'g., Fl. II, 423.— U. effusa Willd. Fl. Berol. Prodr. (1787) 94; M. B. Fl. taur.-cauc. 1, 194.— Ic.: Elwes et Henry, The Trees of Gr. Brit. and Irel. VII (1913) tab. 411, f. 7.

Big trees; young branches glabrous or pubescent, brown; older branches lustrous, with a hoary bloom; leaf buds acute; leaves rounded-obovate, with a strongly unequal cordate base, acuminate at apex, doubly serrate, with 12—19 pairs of veins, glabrous above, glabrous or with axillary tufts of hairs or pubescent beneath; subsidiary veins at most once or twice branched; perianth with 5 — 7 segments, very asymmetric; samara with the nutlet at center, 12 — 16 mm long, the margins of the apical notch overlapping. April — June. (Plate XX, Figure 1).

Commonly a tree of deciduous and mixed woods, often on river floodplains. — European part: Lad.-Ilm., V.-Kama, U. V., V.-Don, Transv., M. Dnp., Bl., L. Don, L. V., Crim.; Caucasus: Cisc. **Gen. distr.**: Scand., Centr. Eur., Atl. Eur. Described from the USSR.

Note. The regions enumerated above delimit the natural distribution area of U. laevis. The tree adapts itself successfully to planting in the cities of the Caucasus, Central Asia, and Siberia. The species U. simplicidens Wolf, described by E. L. Vol'f from a specimen grown in the grounds of the Forest Institute of Leningrad and described in Not. syst. H. Petrop. IV (1923) 118, is not included in the Flora, even though it is claimed by Vol'f to have grown from seed brought from the Caucasus. It may be merely a mutation arisen under cultivation.

364 **Economic importance.** The tree yields valuable lumber for carpentry and bast fiber of rather poor quality.

2. **U. celtidea** (Rogov.) Litw. Schedae ad HFR VI (1908) 167. — U. pedunculata var. glabra Trautv. Bull. phys.-math. Ac. Sc. Pétersb. XII (1857) 375. — U. pedunculata var. celtidea Rogov., Izv. Kievsk. Univ. (1869); Shmal'g., Fl. II, 424. — Exs.: HFR No. 1995.

A small tree; branches light brown or reddish-brown; leaves oblong-lanceolate or rounded-cuneate, gradually attenuate toward apex, unequally dentate-serrate, glabrous or pubescent; subsidiary veins branching at most once or twice; perianth with 5 — 7 obtuse somewhat unequal lobes; 8 — 12 mm long, the nutlet central, the margins of the notch divergent at an acute angle. May. (Plate XX, Figure 2).

Forests. — European part: U. Dnp. Endemic. Described from the vicinity of Chernigov. Type in Leningrad.

Section 2. **MADOCARPUS** Dumort. Fl. Belg. Prodr. (1827) 25. — Samara glabrous throughout or with lanulose cell, rather sparingly ciliate; pedicel shorter than the samara.

Subsection 1. FOLIACEAE C. K. Schn. in Oesterr. Bot. Zeitschr. 64 (1916) 26. — Nutlet in the upper part of the samara, touching the apical notch, or central in a round samara.

Series 1. **Nitentes** Moss. Cambridge Brit. Fl. II (1914) 89. — Series referred to in Russian as "berest." — Samara obovate or rather narrowly elliptic, the nutlet in its upper part.

3. **U. foliacea** Gilib. Exercit. Phytol. II (1792) 395; C. K. Schn. in Oesterr. Bot. Zeitschr. 64 (1916) 75. — U. campestris L. Sp. pl. (1753) 225, pro parte (cum U. scabra confusa); M. B. Fl. taur.-cauc. I, 193. — U. campestris var. vulgaris Ldb. Fl. Ross. III, 2 (1851) 646. — U. glabra Mill. Dict. ed. 8, (1768) No. 4. — U. campestris var. glabra Shmal'g.,

Fl. II (1897) 423. — Ic.: Elwes et Henry The Trees of Gr. Brit. and Irel. VII (1909) tab. 412, f. 28. — Exs.: HFR No. 1986, 1989 (sub U. corylifolia Host), 1990 (sub U. tiliaefolia var. Akinfiewi Litw.).

Tree; bark of old branches brownish-gray with ashy bloom, smooth; annual shoots yellowish-brown, glabrous or covered with spreading pubescence and dotted with red glands; leaf buds obtuse; stipules oblong-linear, 5—7 mm long and to 1 mm broad; leaves oblong-obovate, narrowed
365 toward the unequally rounded base, doubly serrate with sharp upcurved teeth, papery, glabrous above, with tufts of white hairs in the axils of subsidiary veins and minutely dotted with red glands beneath, to 12 cm long and 6 cm broad, the petioles 0.7—1.5 cm long; perianth with 4 or 5 obtuse ciliate lobes; stamens 3 or 4; samara obovate, slender-pediceled, with scattered red fugacious glands, 15—20 mm long and 10—14 mm broad. March—April. (Plate XX, Figure 3).

Mixed and broad-leaved woods, in mountain areas on floodplains, in the steppe zone in gully woods. — European part: U. Dnp., M. Dnp., V.-Don, Bl., L. Don, L. V., Crim.; Caucasus: all regions. **Gen. distr.**: Centr. Eur., Atl. Eur., Med., Bal.-As. Min., Arm.-Kurd. Described from Grodno. Type in Kiev.

Note. A species habitually mixed with other related forms. The name U. campestris ought to be discarded as it embraces samples of different species even in the Linnaeus Herbarium. The name U. glabra Mill., as nomen praeoccupatum (c. f. U. glabra Huds.) is also eliminated. As compared with U. suberosa and U. propinqua, the species is more mesophilic, typically arboraceous and very rarely frutescent.

Economic importance. Extensively cultivated in pleasure grounds and in towns. Widely used for park and avenue planting in the southern USSR, to some extent in the Ukraine. The wood is very firm and is used for miscellaneous woodwork.

4. **U. suberosa** Moench, Bäume Weissenstein (1785) 136; M. B. Fl. taur.-cauc. 1, 77. — U. campestris var. suberosa Ldb. Fl. Ross. III, 2 (1851) 647. — U. campestris var. suberosa Shmal'g., Fl. II (1897) 123. — Exs.: HFR No. 1987, 1988.

A tree or shrub; bark of old branches ashy-black, often with winglike corky outgrowths; annotinous shoots reddish-brown or brown, mostly glabrous; leaf buds obtuse; stipules linear, to 12 mm long and 1—3 mm broad; leaves obovate, unequally rounded at base, acute or rarely apiculate, doubly or trebly serrate with short obtuse teeth, papery, glabrous or diffusely hispid eglandular beneath, smooth or scabrous above, to 10 cm long; perianth with 4 or 5 obtuse ciliate segments; stamens 3—5; samara obovate, stalked, glabrous, 15—20 mm long and 10—14 mm broad. March—April.

Mixed woods, slopes and gullies, wood margins and coppices. — European part: V.-Kama, U. V., U. Dnp., V.-Don, Transv., Bl., L. Don, L. V., Crim.; Caucasus: all regions. **Gen. distr.**: Centr. Eur., Atl. Eur., Med., Bal.-As.
366 Min., Arm.-Kurd., Iran. Described from Europe. Type in Berlin.

Note. A relatively xerophytic race of this series, often producing shrubby forms, with corky excrescences. Readily distinguishable from the preceding species by the absence of red glandular hairs on the leaves and fruits.

Economic importance. Wood used for production of various small articles. Cultivated with a view to consolidating slopes of gullies and ravines. A species with relatively modest moisture requirements. Some of the town planting of Samarkand and Tashkent in Soviet Central Asia involves this species together with U. *densa*, whereas U. *foliacea* is rarely encountered there. U. *suberosa* occurs as a cultivated form in the ravines of Mtn. Turkm., Amu D., and Syr D.

An interesting experiment in large-scale cultivation was carried out in Turkmenia (Chuli natural boundary area near Firyuza) but the plantations have been severely damaged by insects. The corky outgrowth may be of interest only for manufacture of pressed products.

5. **U. propinqua** Koidz. in Tokyo Bot. Mag. XLIV (1930) 95. — U. *japonica* Sarg. Trees et Shrubs II (1907) 1, tab. 101, non Sieb. (1830). — U. *campestris* var. *japonica* Sarg. ex Rehder in Bailey, Cycl. Amer. Hort. IV (1902) 1852. — U. *campestris* Kom. Fl. Mansh. II (1903) 82, non L. — U. *Davidiana* var. *japonica* Nakai Fl. Sylv. Kor. 19 (1932) 26, tab. IX. — U. *campestris* Ldb. Fl. Ross. III, 2 (1851) 646 (pro parte). — Ic. - Kom. and Alis., Opred. rast. Dal'nevost. kr. I, Plate 134.

Tall trees or shrubs; trunks whitish; bark of branches grayish-brown, often with corky outgrowths; leaves short-petioled, glabrous or rough-pubescent, obovate, acuminate toward apex, cuneately and unequally narrowed toward base, doubly serrate, 2—8 cm long and 1.5 cm broad; samara obovate, cuneately narrowed toward base, not stalked, 10—16 mm long, 4—8 mm broad, glabrous. April—June. (Plate XX, Figure 4).

Mixed woods in river valleys and on mountain slopes. — E. Siberia: Dau.; Far East: Ze.-Bu., Uss., Sakh. **Gen. distr.**: Mong., Jap.-Ch. Described from Japan. Cotype in Leningrad.

Note. This species represents a Far Eastern race of the series. Komarov (Fl. Mansh. II (1903) 84) describes for it a number of variations: 1) *laevis denudata* — with smooth leaves and corkless branches, 2) *laevis suberosa* — with smooth leaves and corky outgrowths on the branches, 3) *scabra suberosa* — with very rough leaves and corky outgrowths on the branches, 4) *scabra denudata* — with very rough
369 leaves and corkless branches, and 5) *pumila* — a dwarf depressed form with very rough leaves and winglike corky outgrowths on the branches.

These variations are not geographically strictly delimited but ecologically they are clearly differentiated.

Economic importance. Suitable for city planting.

6. **U. densa** Litw. in Schedis ad HFR VI (1908) 163. — Exs.: HFR No. 1991, 2414, 2787. — Russian: vyaz gustoi [dense], berest gustoi, sadovyi [garden] karagach, narvan.

A big tree with a straight dark trunk and an exceptionally dense top; bark of branches gray or yellowish-brown; leaf buds obtuse; mature leaves firm, coriaceous, oblongly lance-ovate, rounded-cuneate at base, acute at apex, doubly serrate, glabrous or pubescent, eglandular, 5—7 cm long and 2—3 cm broad, the petiole diffusely pubescent; stipules oblong or linear, often bearded at apex; stamens 4; samara oblong-obovate, ca. 2 cm long, 1.2 cm broad, rounded-cuneate at base, commonly stalked. March—April.

(367)

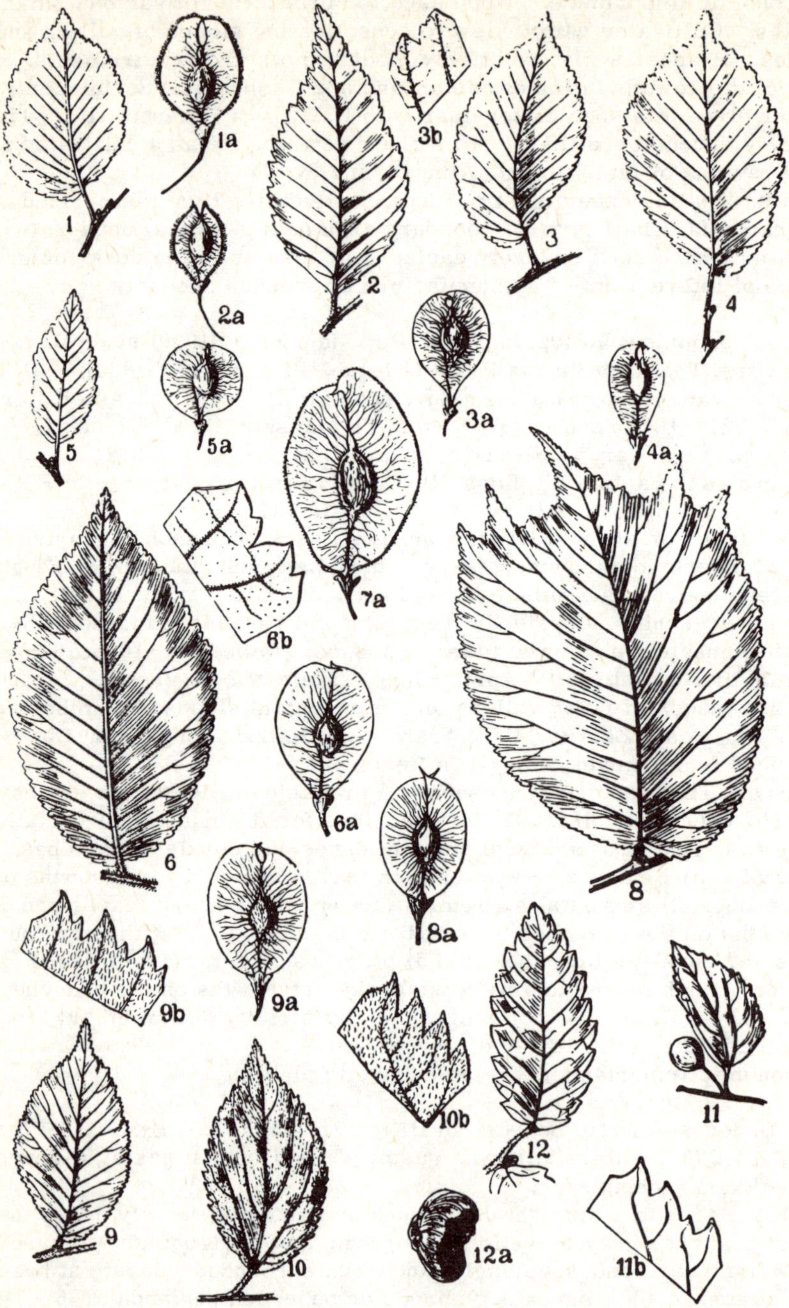

PLATE XX. 1. Ulmus laevis Pall.– 2. U.celtideca (Rogov.) Litw.– 3. U.foliacea Gilib.– U.propinqua Koidz.– 5. U.pumila L.– 6. U.scabra Mill.– 7. U.macrocarpa Hance.– 8. U.laciniata (Trautv.) Mayr.– 9. U.elliptica C.Koch.– 10. Celtis caucasica Willd.– 11. C.glabrata Stev.– 12. Zelcova carpinifolia Pall.– Designations signify throughout: a – fruit, b – leaf margin.

Ravines, and cultivated stands near irrigation ditches and in gardens. — Centr. Asia: Mtn. Turkm., Amu D., Syr D., Pam.-Al., T. Sh. Endemic. Described from Farab. Type in Leningrad.

Note. As often happens in the case of Central Asian arboreous species, cultivated and wild forms of this elm are often confused. Because of the special liking of the Uzbeks for this tree, it is cultivated wherever possible, including places with a very poor water supply. It would be interesting to know whether U. *densa* occurs in Iran. In any case, in Transcaucasia, U. *campestris* var. *umbraculifera* Trautv. (A. H. P. II (1872) 590), "Nal'band" or "Narband" (Araks River valley) gravitates toward U. *foliacea* and U. *suberosa* rather than U. *densa*, as assumed by Litvinov (Schedae HFR l. cit.). Here also belongs, as a cultivated form, U. *Bubyriana* Litw. = *Ulmus densa* var. *Bubyriana* Litw. in Schedis ad HFR l. c. 23 No. 2444. U. *turkestanica* Rgl. in Gartenflora (1884) 76 and 396, undescribed.

Series 2. **Pumilae** C. K. Schn. in Oesterr. Bot. Zeitschr. 66 (1916) 32. — Russian: "il'movnik." Samara rounded-elliptic or broad-elliptic, often broader than long, hence nutlet central.

7. **U. pumila** L. Sp. pl. (1753) 327; Turcz. Fl. baic.-dah. II, 95; Kom. Fl. Mansh. II, 85. — U. *pumila* var. *arborea* Litw. Schedae ad HFR VI (1908) 400. — U. *manshurica* Nakai, Fl. Sylv. Kor. XIX (1932) 22. — Ic.: Nakai l. c. tab. VII; Kom. and Alis., Opr. rast. Dal'nevost. kr. I, Plate 13, Figure 2. — Exs.: HFR No. 1992. — Russian: vyaz prizemistyi [dwarf], il'movnik.

Big trees or shrubs; bark of young branches yellow, that of older
370 ones yellowish-cinereous; leaf buds obtusely ovoid; leaves coriaceous, small, broad-lanceolate, acute or acuminate, equally double-serrate; stipules foliaceous, green; samara broad-elliptic, slightly oblique or distinctly unequal, with a central nutlet, to 1.5 — 2 cm in diameter. February — April. (Plate XX, Figure 5).

Dry sandy or stony soils, pebbles of river valleys, slopes, sometimes on rocks. — E. Siberia: Dau.; Far East: Ze.-Bu., Uss.; Centr. Asia: Dzu.-Tarb., Mtn. Turkm., Amu D., Syr D., T. Sh. **Gen. distr.**: Dzu.-Kash., Mong., Jap.-Ch. Described from Transbaikalia. Type in Leningrad.

8. **U. Androssowii** Litw. in Sched. HFR VIII (1922) 23. — Exs.: HFR No. 2445, 2788.

Trees with a round head; bark of young branchlets brownish-cinereous or yellowish, that of older branches gray; leaf buds ovoid; leaves rounded-ovate or ovate, unequal at base, short-acuminate, doubly serrate, 5 — 6 cm long, 3 — 4 cm broad, the long petiole heavily pubescent; stipules oblong, densely ciliate; samara angular-orbicular, with a central nutlet, ca. 2.5 cm in diameter. March — April.

Only cultivated. — Centr. Asia: Amu D., Syr D., Mtn. Turkm. Endemic. Described from Samarkand. Type in Leningrad.

Note. Not a single specimen is known from anything approaching natural conditions. All the same, U. *Androssowii* displays some most unusual features, as the fruits are typical for the series *Pumilae*, whereas the leaves would rather place this species in the series *Nitentes*.

This outstanding combination of characters persists in specimens from a great number of locations and testifies to the stability and consistency of this form. The cultivated tree population of local villages in Fergana is often composed exclusively of U. Androssowii, while U. suberosa, U. densa, and U. pumila occur there very rarely. Might not this species be a hybrid between U. densa and U. pumila?

The distribution areas of U. pumila L. in Central Asia and in the East are isolated from each other. It is therefore natural to assume that these are two different vicariant species. E. Koehne described a tree raised in the Späth and Sievers nurseries from West Siberian and Turkestan seeds, which is distributed by these nurseries under the name U. pinnato-ramosa Dieck (Cat. 1895, nomen nudum):

U. pinnato-ramosa Dieck ex Koehne in Fedde Repert. spec. nov. VIII (1910) 74; Nakai, Flora silv. Kor., XIX (1932) 24.

Branches slender; branchlets regularly distichous, grayish-pubescent,
371 with internodes one-third to half the length of the leaves; buds ovoid, quite smooth, barely 1.5 cm long; stipules 5—7 mm long, obliquely oblong from cordate base, caducous; petioles 3—11 mm long, pubescent at first, becoming glabrous; leaf base straight or but slightly oblique at base, rarely broad-oblong, mostly oblong-lanceolate, 2.5 — 6.5 cm long, 1—2.3 or rarely to 3 cm broad, coarsely and closely, simply or almost doubly serrate with 9—14 blunt teeth on each margin, quite smooth; veins parallel, as many as primary teeth; fascicles of subsessile flowers 4—6 mm long, 4—7 mm broad; calyx 4-lobed, ca. 2 mm long; stamens barely exceeding the calyx lobes; fruit obovate-orbicular, 9—14 mm long, the quite glabrous wings with apical incision reaching the cell.

The species belongs to the section Madocarpus Dumort. of the series Pumilae C. K. Schn. and is related to U. pumila L., but the latter has glabrous branches, longer petioles, 3—11 cm. The distichous arrangement of branchlets is a less conspicuous feature.

It is thus perfectly possible to separate the Central Asian small-leaved elm from that of the Far East. Even if the characters indicated by Koehne are not accurate, others can be established.

Editors

Subsection 2. GLABRAE C. K. Schn. in Oesterr. Bot. Zeitschr. 66 (1916) 29. — Nutlet at some distance from the notch of the wing, separated from it by a fairly long suture.

Series 1. **Euglabrae** C. K. Schn. l. c. 26. — Russian: "il'm." Both surfaces and margin of samara completely glabrous.

9. **U. scabra** Mill. Gard. Dict. ed. VIII (1768) No. 2. — U. campestris L. Sp. pl. (1753) 225 (pro parte, cum U. suberosa confusa); Ldb. Fl. Ross. III, 2, 646 (pro parte). — U. glabra Huds. Fl. Angl. ed. 1 (1762) 95. — U. montana With. Bot. arrang. veg. Gr. Brit. I (1787) 250; Shmal'g., Fl. II, 423. — Exs.: HFR No. 1993. — Russian: vyaz shershavyi, vyaz gornyi [scabrous or mountain elm].

Big trees; bark of young branches dark brown, warty, pubescent, that of mature branches gray or yellowish-brown; leaf buds small, conical,

densely woolly, slightly shorter than adjoining petioles; leaves large, obovate, strongly unequal at base, acuminate at apex, coarsely serrate with twice or thrice cut sharp long-pointed teeth, scabrous above, tough, diffusely pubescent beneath, 18—20 cm long and 10—12 cm broad, the petiole 1—4 mm long; samara obovate, to 3 cm long and 1.8 mm broad, the central nutlet separated from the apical notch by suture 5—8 mm long, quite
372 glabrous. March—May. (Plate XX, Figure 6).

Broad-leaved forests on inundated land and on flat interfluve areas. European part: Lad.-Ilm., U. V., V.-Kama, U. Dnp., M. Dnp., V.-Don, Transv., Bl., L. Don, L. V., Crim.; Caucasus: all regions. **Gen. distr.**: Scand., Centr. Eur., Atl. Eur., Med., Bal.-As. Min. Described from Europe. Type in London.

Note. U. glabra Huds. is the priority name of this species (Fl. angl. (1762) 95). Considering, however, that the name is homonymous, U. glabra Miller (Gard. Dict. (1868) No. 4) being often used to designate U. foliosa, it may be discarded, in conformity with the Cambridge Congress recommendations, as liable to lead to confusion and carrying different connotations. On such grounds the name U. campestris L. was discarded, since under this Linnaean species two species were confounded. The same approach should be adopted in the case of U. glabra which embraces in the literature two entirely different species.

10. **U. laciniata** (Trautv.) Mayr, Fremdl. Wald. u. Parkbäume (1906) 523. — U. major var. heterophylla Maxim. et Rupr. in Bull. Acad. Sc. Pétersb. XV (1857) 139. — U. montana var. laciniata Trautv. in Mém. Sav. étr. Ac. Sc. Pétersb. IX (1859) 246; Kom. Fl. Mansh. II, 88. — U. campestris Ldb. Fl. Ross. III, 2, 646, pro parte. — Ic.: Kom. and Alis., Opred. rast. Dal'nevost. kr., Plate 134.

A tree similar to the preceding, differing in annotinous branchlets being slightly pubescent or glabrous, gray or yellowish; leaves of fertile branches mostly 3—(5)-lobed at apex. March—May. (Plate XX, Figure 8).

Broad-leaved mountain woods. — Far East: Uss., Uda, Sakh. **Gen. distr.**: Jap.-Ch. Described from the Lower Amur. Type in Leningrad.

Note. The species represents an East Asian race replacing the European U. scabra after a considerable geographical gap. Features characteristic of U. laciniata may often be observed in Ural representatives of U. scabra.

Series 2. **Fulvae** C. K. Schn. in Oesterr. Bot. Zeitschr. 66 (1916) 26. — Wing quite glabrous; only the cell covered with slender brittle silky hairs.

11. **U. elliptica** C. Koch in Linnaea XXII (1849) 599. — U. scabra var. elliptica Yarmolenko in Zhurn. Russk. Bot. Obshch. 12 (1927) 17. — Exs.: HFR No. 1039, 1994; Pl. Or. Exs. No. 131; Fl. Cauc. exs. No. 336 a, b.

Big trees; young branchlets grayish-brown or reddish-brown, glabrous or pubescent; leaf buds acute, clothed with rusty-brown hairs; leaves large, elliptic, unequal at base, acuminate at apex, doubly dentate with
373 curved teeth, scabrous above, soft-hairy beneath, to 15 cm long and 5—6 cm broad; samara obovate, 3 cm long and 1.5—1.8 cm broad, the central nutlet

separated from the apical notch by a suture 5—8 mm long, the cell silky-pubescent. April— May. (Plate XX, Figure 9).

Broad-leaved woods. — European part: V.-Don, M. Dnp., L. Don, Transv., Crim.; Caucasus: Cisc., W., E., and S. Transc., Dag. Endemic. Described from the Caucasus. Type in Berlin; cotype in Leningrad.

Note. A very interesting form in the systematics of the genus Ulmus, gravitating, according to Schneider (l. cit.), toward the North American U. fulva Michx. Noteworthy is its sporadic occurrence in the forest-steppe belt of the European part of the USSR and in the Crimea and the more or less contiguous distribution in the Caucasus where additional study is needed.

Series 3. **Wallichianae** C. K. Schn. in Oesterr. Bot. Zeitschr. 66 (1916) 26. — Samara pubescent over the entire surface of wing and cell.

12. **U. macrocarpa** Hance in Journ. of Bot. (1868) 332; Kom. Fl. Mansh. II, 81. — Ic.: Kom. and Alis., Opr. rast. Dal'nevost. kr. I, Plate 134, Figure 4.

A small tree; branches gray, or the younger ones yellow or brownish; leaf buds conical; leaves 7—9 cm long and 5 cm broad, obovate, unequally cuneate at base, acuminate at apex, doubly serrate with short teeth; 3—3.5 cm long and 2.2—2.5 cm broad, with a central nutlet separated from the apical notch by a suture 9—10 mm long, lanulose over the entire surface of wing and cell. April—June. (Plate XX, Figure 7).

Stony slopes, rocky screes, and river valleys. — E. Siberia: Dau.; Far East: Ze.-Bu., Uss. **Gen. distr.**: Mong., Jap.-Ch. Described from Peking. Type in London.

Note. A rare relict plant belonging to the Japano-Sino-Himalayan series of species, reaching on USSR territory the northern limit of its distribution.

Subfamily **Celtoideae** ENGL.

Engl. in Engl. u. Prantl, Pflanzenfam. III, 1 (1896) 63.

Flowers bisexual, very rarely unisexual, commonly solitary, axillary, more rarely fasciculate; fruit indurated; embryo curved.

Genus 370. **CELTIS*** L.

L. Gen. pl. ed. V (1754) 382.

Flowers staminate or hermaphrodite; perianth with 5 (4) caducous lobes;
374 stamens 4, the filaments inserted below the disk; ovary 1-locular, 1-seeded; stigma 2-lobed with unequal linear lobes; fruit a drupe with sweetish fleshy pulp and a woody reticulate kernel. Trees or shrubs with alternate serrate leaves and caducous stipules; leaf veins 3, divergent. Russian: "karkas."

C. australis L. in the Postpliocene of Cisc. (tuffs of Mashuka). — C. trachytica Kov. in the Sarmatian series of Bl. (Krynka).

* A Latin name for the genus Zizyphus, transferred by Tournefort to the nettle-tree.

1. Leaves dark green and scabrous above, pubescent beneath . . . . 1. **C. caucasica** Willd.
\+ Leaves yellowish-green, glabrous on both sides . . . . 2. **C. glabrata** Stev.

1. **C. caucasica** Willd. Sp. pl. IV (1805) 994; Ldb. Fl. Ross. III, 2, 632; Boiss. Fl. Or. IV, 1156. — C. australis M. B. Fl. taur.-cauc. II, 440 non L.; Ldb. Fl. Ross. III, 632.

A tree or shrub; young branches brown or reddish-brown; leaf buds acute, flattened; leaves unequal, ovate or ovate-lanceolate, serrate with sharp curved teeth, rounded or cuneate at base, acute or acuminate, pubescent beneath, rough above, coriaceous, 4 — 10 cm long and 2.5 — 5 cm broad; fruiting peduncles glabrous or at base pubescent, 1.5 — 3 cm long; fruit globose, reddish-yellow or dark, pruinose; stone slightly rugose, somewhat flattened at the top. Fl. March — May; fr. September — October. (Plate XX, Figure 10).

Dry stony bluffs, rocks, ravines, and more rarely as undergrowth in clearings. — Caucasus: all regions; Centr. Asia: Mtn. Turkm., T. Sh., Syr D., Amu D., Pam.-Al. **Gen. distr.**: Arm.-Kurd., Iran. Described from the Caucasus. Type in Berlin; cotype in Leningrad.

Note. This species is often confounded with the typical Mediterranean species C. australis L. from which it differs in the more ovate leaves without a long-acuminate tip. The coloring of the ripe fruit, a character proposed to distinguish these two species by Medvedev (Vest. Tifl. Bot. Sada 14, 1904), black for C. australis (p. 2) and reddish-yellow for C. caucasica (p. 3), does not provide a reliable criterion. Both in Central Asia and in the Caucasus light- and dark-colored forms occur side by side.

Economic importance. The wood is of considerable potential value, being unusually firm and of high quality; hence the plant is often referred to as "iron-tree." The tree is exceptionally modest as regards water requirements and may be employed in mountain forestation, especially for consolidation of mountain slopes and screes. The fruits are mealy, with a pleasant taste, and are used for food both fresh and ground for
375 preparation of gruel or pressed cakes. It is highly desirable to study the various forms of this tree concerning both wood quality and fruit palatability, as considerable variability may be expected in both regards.

2. **C. glabrata** Stev. ex Planch. Ann. Sc. Nat. VIII (1848) 285. — C. Tournefortii M. B. Fl. taur.-cauc. II (1808) 448; Ldb. Fl. Ross. III, 2, 632 (non Lam.). — C. Tournefortii var. glabrata Boiss. Fl. Or. IV (1879) 1157. — Ic.: Elwes et Henry, The Trees of Gr. Brit. and Irel. IV (1909) t. 267, f. 10. — Exs.: Pl. Or. Exs. No. 109.

A tree or shrub; young branches light brown; leaf buds less flattened than in the preceding species; leaves firm, unequal, ovate or rhombic, rounded-cuneate at base (on short shoots of shrubby forms subcordate), sharp-tipped at apex, unequally denticulate-serrulate on the margin, yellowish-green on both sides, glabrous, 4 — 6 cm long and 4 cm broad, the pedicels always glabrous; fruiting peduncles glabrous, 1 — 2 cm long; fruit globose, orange or reddish, pruinose; stone slightly rugose, round. Fl. March — April; fr. September — October. (Plate XX, Figure 11).

Stony slopes and ravines. — European part: Bl., Crim.; Caucasus: Cisc., E., W. and S. Transc. Endemic. Described from the Caucasus. Type in Helsinki.

Note. All statements in the note to the preceding species, concerning the color of ripe fruits, apply equally to C. glabrata. This species is often confounded with C. Tournefortii Lam. from which it differs in the pale yellow color of the glabrous and less coriaceous leaves.

Economic importance. The least studied form, somewhat inferior in economic value to C. caucasica. The wood and fruit resemble the preceding species in quality.

Genus 371. **ZELKOVA*** SPACH.

Spach in Ann. Sc. Nat. 2 ser. XV (1841) 355.

Perianth 4- or 5-lobed, campanulate, persistent; stamens 4 or 5; ovary solitary, 1-seeded, with 2 toothlike styles; nutlet dry, gibbous-ovoid, 1-seeded; flowers bisexual and staminate, axillary. Subarboreous or arboreous plants with alternate dentate pinnate-veined leaves.

Fossils of the genus are widely distributed in the Tertiary formations of Europe and Asia.

Z. keaki Sieb. in Tertiary (Pliocene) formations of Alt. (Chingistai). — Z. Ungeri Kov. in M. Dnp. (Bondarevka), Bl. (Orekhov, Krynka); in the Pliocene of E. Transc. (Shiraki); in Lower Tertiary formations of Uss. (Novokievskoe, Fatashi, Rechnoi, Khanka); Sakh. (Mgachi). — Z. sp. in Postpliocene formations of Ze.-Bu. (Astashikha). — Z. Richardii Mich. in Tertiary (Pliocene ?) formations of Alt. (Chingistai). — Z. sp. in Lower Tertiary formations of Balkh. (Ashutas) and in Postpliocene formations of Ze.-Bu. (Astashikha).

376 1. **Z. carpinifolia** (Pall.) Dipp. Handb. Laubholzk. II (1892) 32. — Rhamnus carpinifolia Pall. Fl. Ross. II (1788) 24, tab. 60. — Planera Richardii Michx. Fl. bor.-am. II (1803) 248. — P. crenata Desf. Cat. Hort. Paris (1829) 476. — Zelkova crenata Spach in Ann. Sc. Nat. 2 sér. XV (1841) 356; Ldb. Fl. Ross. III, 2, 645; Boiss. Fl. Or. IV, 1159. — Ic.: Pall. l. cit., tab. 60. — Exs.: Herb. Fl. cauc. No. 219; Pl. or. exs. No. 381.

A tree; young branchlets dark brown, covered with spreading hairs; leaves ovate-oblong, acute, unequally subcordate at base, acute, remotely crenate-serrate with round-pointed or bluntish teeth; perianth of pistillate flowers campanulate; staminate flowers in axillary fascicles or pairs; nutlet glabrous, rugose. Fl. March; fr. August—September. (Plate XX, Figure 12).

Forests. — Caucasus: W., E., and S. Transc. **Gen. distr.:** Arm.-Kurd., Iran. Described from the Caucasus.

Family XLVI. **MORACEAE** LINDL.

Trees, shrubs, or herbs; leaves alternate or opposite, simple or dissected; stipules present; flowers unisexual, monoecious or dioecious;

* A local Imeretian name of the plant.

perianth 4- or 5-parted, 4-lobed, or undivided, accrescent; ovary superior or inferior, 1-locular; ovule solitary, suspended; stigmas 2, filiform, or style 2-parted; fruit a nutlet or drupe; embryo curved or spiral.

Key to Genera

1. Trees or shrubs . . . . 2.
\+ Herbs, erect or climbing and then the stem sometimes woody . . . . . 5.
2. Axis of inflorescence accrescent into a fleshy globular-pyriform syncarp containing the flowers . . . . 373. **Ficus** L.
\+ Axis of inflorescence not accrescent . . . . 3.
3. Pistillate flowers in heads . . . . 4.
\+ Pistillate flowers in spikelike inflorescences . . . . 372. **Morus** L.
4. Branches with slender sharp thorns . . . . *__Maclura__ Nutt.
\+ Branches unarmed . . . . *__Broussonetia__ L'Hérit.
5. Climbing plants; leaves opposite, lobed . . . . 374. **Humulus** L.
\+ Erect plants; leaves dissected into narrow lanceolate segments; lower leaves opposite, the upper ones alternate . . . . . 375. **Cannabis** L.

377 Genus 372. **MORUS*** L.

L. Gen. pl. ed. V (1754) 424.

Flowers unisexual, in catkins; perianth 4-parted, in male flowers unchanging, in pistillate flowers accrescent and becoming fleshy in fruit; colored; fruits aggregated in a berrylike syncarp; stamens 4; stigmas 2, filiform. Monoecious or dioecious trees, with alternate undivided or lobed or toothed leaves; stipules present.

1. Stigma distinctly hairy; the accrescent perianths of the syncarp villous outside; leaves deeply cordate at base . . . . 1. **M. nigra** L.
\+ Stigma minutely papillose, glabrous; the accrescent perianths of the syncarp glabrous outside; leaves truncate or scarcely cordate at base . . . . 2. **M. alba** L.

1. **M. nigra** L. Sp. pl. (1753) 986; Ldb. Fl. Ross. III, 2, 644. — Ic.: Elwes et Henry, Trees Gr. Br. and Irel. IV (1909) t. 267. — Russian: tut chernyi [black].

A tree; branches reddish-brown; leaves ovate, undivided or often lobed, deeply cordate at base, obtuse at apex, obtusely crenate-dentate, stiff, scabrous, 6—10 cm long and broad; fruiting perianth dark or light-colored, always villous outside; stigma distinctly shaggy; syncarp large, 2—2.5 cm long, blackish-violet or black. May—June.

Cultivated in gardens and often naturalized. — European part: M. Dnp., V.-Don, Bl., L. Don, L. V., Crim.; Caucasus: Centr. Asia. **Gen. distr.**: Med., Bal.-As. Min., Centr. Eur., Iran. Described from the coasts of Italy. Type in London.

* The name used for this tree by the Greek writer Aeschylus; the fruit called "morum" in Latin by Ovid. French: mûrier; English: mulberry tree; German: Maulbeerbaum; Polish: morwa.

Note. This species is sometimes reported as growing wild in Soviet Central Asia (O. and B. Fedchenko). Some authors (C. K. Schneider, Rehder) consider W. Asia as its native area, but this assumption has not so far been verified. In any case, the present-day distribution of this species is associated with man to such an extent that it is more correct to consider it as a cultivated though readily naturalized tree.

2. **M. alba** L. Sp. pl. (1753) 986; Ldb. Fl. Ross. III, 2, 643; M. B. Fl. taur.-cauc. II, 307; Kom. in A. H. P. XXII (1904) 91. — M. taurica M. B. Fl. taur.-cauc. II (1808) 398. — Ic.: Elwes et Henry Trees Gr. Brit. a. Irel. IV (1909) tab. 267; Kom. and Alis., Opr. rast. Dal'nevost. kr. I (1931) 135. — Russian: tut belyi [white]; Ukrainian: bila shovkovitsa, shovkun; tuta.

378 A tree; branches grayish-brown; leaves ovate, truncate or rounded or subcordate at base, acute at apex, on young shoots mostly undivided, on fruiting and annotinous branches lobed or sinuate, crenate-dentate with rounded teeth, thin, soft, mostly glabrous; fruiting perianth light-colored, glabrous outside; stigma papillose, not villous; fruit white (var. vulgaris Bureau) or purple-black (var. tatarica Seringe = M. tatarica Pall. Fl. Ross. I (1784), tab. 52). April — June. (Plate XXI, Figure 2).

Extensively cultivated and in the southern USSR often naturalized. — European part: U. Dnp., M. Dnp., V.-Don, Bl., L. Don, L. V., Transv., Crim.; Caucasus: all regions; Centr. Asia: all regions; Far East: Uss. **Gen. distr.**: Jap.-Ch., As. Min. (cultivated), Centr. Eur., Med. Described from China. Type in London.

Economic importance. The wood is used for odd jobs. The tree is of primary importance for sericulture, the foliage providing food for silkworm. The fruit is eaten fresh or dried. In Darvaz and other parts of Pam.-Al., a meal made of dried mulberry fruits constitutes an important food article. One-year-old stems yield mulberry fiber used in the cotton and paper industry.

Silk manufacturing in Georgia makes use chiefly of male trees, called locally "purtseli," as they produce more foliage and their growth is faster. As sprout leaves may serve as food for silkworm, freezing of a part of the tree's crown is of no serious consequence and thus cultivation may extend right into the taiga region. The mulberry is a beautiful tree, with a dense round head, and is used in the southern USSR for town and country planting. It is suitable for ravine forestation and for shelterbelts. The wood, with yellowish-brown heartwood and narrow yellow sapwood, is very hard and is used for odd jobs, musical instruments, and clapboards; in many instances it replaces oak. The base, an excellent source of fiber, is employed for the production of yarn, cardboard, paper, cordage, ropes, and twine for tying up grapevines. A yellow dye is extracted from the leaves.

Mulberry and its cultivation were unknown in Europe before the 12th century. In China and India the tree had already been grown in remote antiquity. In Transcaucasia, especially in the eastern part, mulberry is often found in naturalized condition and rises to 1,000 m above sea level. It is also widely distributed through Central Asia where it reaches an altitude of 1,600 m (Gissar Range). Old trees, 200—300 years of age, may be found, attaining a height of more than 20 m in a trunk diameter up to

1 m. The tree bears fruit in great profusion. The fruit has a high sugar content suitable for alcohol production, and it is indeed used for this purpose in Transcaucasia (mulberry vodka). The fruit is also made into liqueur, vinegar, jam, jelly, syrup, molasses ("bekmaz" or "dushab"), or candy ("lavash"). The bark of the roots and the leaves as well as fruit syrup have medicinal uses. — Editors.

379 Genus* **MACLURA*** NUTT.

Nutt. The Genera of North Amer. Plants II (1818) 233 (nom. conserv.) — Ioxylon Raf. Amer. Month. Mag. II (1817) 118.

Tree with spiral phyllotaxy; leaves large, undivided, entire; flowers dioecious; staminate aments cylindric; perianth of male flowers 4-parted; pistillate inflorescences capitate, the 4-parted perianth of female flowers fleshy in fruit; style 1; stigma short-lobed; fruit dry, embedded together with perianth in the strongly accrescent fleshy inflorescence axis; seeds exalbuminous.

M. aurantiaca Nutt. Gen. N. Am. II (1818) 233; Asch. und Gr. Syn. IV, 582. — Ioxylon pomiferum Raf. Amer. Month. Mag. II (1817) 118.

A tree to 20 m (mostly not more than 10 m); bark deeply fissured, dark; young branches pubescent, becoming glabrous, at first green, becoming olivaceous, and finally brown; winter-buds short, globose, each bud subtended by a slender sharp spine; leaves ovate, acute, rarely broad-oval, mostly 7—16 cm long, 3.5—10 cm broad, at first pubescent, finally glabrous, the petiole 1.5—4 cm long; syncarp large, globose, orangelike but strongly wrinkled, yellowish-green; fruitlets [achenes] small, light brown, buried in the pulp of the syncarp. Fl. May, June; fr. August, September.

Growing wild in North America from Missouri to Kansas and Texas. Introduced into Europe in 1818. Cultivated on the S. coast of the Crimea and along the Black Sea coast of the Caucasus, in Transcaucasia, and in Central Asia. Used in gardens as a component of thorny hedges. According to Zelenetskii, occasionally naturalized in the Crimea.

Economic importance. The tree yields valuable wood of golden-yellow color. It produces impenetrable hedges. The fruit is edible.

Genus* **BROUSSONETIA**** L'HÉRIT.

L'Hérit. in Vent., Tabl. III (1799) 547.

A shrub or tree, dioecious; leaves opposite; staminate flowers in cylindric aments, the perianth 4-parted; pistillate flowers in globose heads, the perianth tubular or 4-toothed; fruits with a fleshy outer layer, borne on thick fleshy gynophores; syncarp consisting of the fleshy receptacle, with the orange-red fruits protruding on their gynophores.

B. papyrifera L'Hérit. in Vent. Tabl. règn. végét. II (1799) 547; Asch. und Gr. Synops. IV, 583. — Ic.: C. K. Schn. Laubholzk. I, 240, fig. 155.

A shrub, branching from base; branches upright; bark reddish-gray; annotinous branches olivaceous or grayish-green; leaves all petiolate,

* Named for the geologist W. Maclure.

** Named for the naturalist P. Broussonet [Initial should be A. for Auguste.]

380 broadly rhombic-ovate or 3-lobed, 10—20 cm long, 5—15 cm broad, the lower surface densely pubescent throughout or merely on the veins; flower characters as described for the genus. Fl. May; fr. September.

Native in Japan and China. Cultivated. Suitable for avenue planting, etc. In the USSR, cultivation trials are conducted in W. Transcaucasia. The tree tends to become naturalized.

Economic importance. The plant yields tough bast which is used for the production of superior grades of paper and related products.

Genus 373. **FICUS** L.

L. Gen. Pl. ed. V (1754) 482.

Monoecious or rarely dioecious; flowers unisexual, borne inside a globose or pyriform accrescent receptacle with a small opening at apex; male flowers with a 2—6-parted perianth and 2—6 stamens; female flowers with a 5-lobed perianth; ovary 1-locular; style lateral, slender; fruit a nutlet. Trees or often shrubs with alternate or rarely opposite leaves and connate caducous stipules.

Ficus was widely distributed especially in the evergreen area of the Tertiary flora of the USSR. There are also reports concerning its occurrence outside this area but they are not always reliable.

Ficus densifolia Knowlt.? in the Lower Dui series of Sakh. (Pil'vo, tunnels at Pyatyi Klyuch). — F. Giebelii Heer in the Eocene (spondylous layer) of M. Dnp. (Kiev) and in the Oligocene of V.-Don (Tim). — F. junx Unger in the Oligocene of U. Dnp. (Volyanshchina). — F. kiewiensis Schmalh. in the spondylous Eocene layers of M. Dnp. (Kiev). — F. lanceolata Heer in Sarmatian formations of Cisc. (Adagum) and E. Transc. (Khvteeba). — F. multinervis Heer in the Miocene of W. Transc. (Maidany, Goderskii Pass), in Sarmatian formations of E. Transc. (Khvteeba), and in Lower Tertiary formations of Sakh. (Akhengy). — F. populina Heer in the Pliocene of W. Transc. (Goderskii Pass) and in the Oligocene Aquitanian of Ar.-Casp. (Yar-kue). — F. procarica Krysht. in the Maiotis formations of Bes. (Seimeny). — F. Rogowiczii Schmalh. in the Oligocene of M. Dnp. (Ekaterinopol'e) and U. Dnp. (Ryzhany).— F. tiliaefolia (A. Br.) Heer in Lower Tertiary formations of Uss. (Rechnoi; De-Friz; Aleksinskii mine; Khabarovsk base) and Sakh. (Berezovaya Polyana, Cape Rogatyi, Mgachi, Ogorodnaya Canyon, Nikolaevka, etc.). Certain finds of similar leaves have been referred to the genus Buettneria and others to Alangium. — F. ucrainica Persidsky in the Eocene of M. Dnp. (Lavy-Pugivi') and in Lower Tertiary (or uppermost Cretaceous ?) formations of Ob (Lozva River). — F. uralica Krysht. in Lower Tertiary (or uppermost Cretaceous ?) formations of Ob (Lozva River).

1. **F. carica** L. Sp. pl. (1753) 1059; Pall. Fl. Ross. II, 44; M. B. Fl. taur.-cauc. II, 452; Ldb. Fl. Ross. III, 644, Shmal'g., Fl. II, 422; Rollov, Dikorast. r. Kavkaza 204; Medvedev, Der. i kust. Kavk. 226; Fedch., Konsp. Fl. Turk., part 6, 342. — Ic.: C. K. Schn. III. Laubholzk. I, ff. 151, 156, 158. — Russian: smokovitsa obyknovennaya [common]; vinnaya yagoda [wine berry], figa;

Tataric: nizhir; Georgian: legvi; Mingrelian: lugi; Armenian: dzeni, tuz; Abkhazian: allakhá; Turkmenian: andzhir; Moldavian: smokiny, nizhir; Serbian: smokvina, smokva.

381 A shrub or tree to 7 — 10 m, sparsely branched; bark light gray; branches upcurved, the young ones pubescent; leaves deciduous, rough above with stiff hairs; more or less pubescent beneath, large, borne at the ends of branches on petioles 2 — 5 cm long; leaf blade orbicular or broad-ovate, 8 — 15 cm long and 6 — 18 cm broad, with a cordate sinus at base, 3 — 5-lobed, sinuate, rarely undivided; inflorescences and the compound fruits axillary, solitary on short stalks, pyriform, 5 — 8 cm long, pale yellow to violet-brown; flowers borne inside the receptacle, the staminate with a 3 — 5-parted perianth and 3 — 5 stamens, the pistillate with a 5-parted perianth and a 1-seeded ovary; style lateral, with 1 or 2 stigmas; winter-buds glabrous, the lateral globose or ovoid with numerous scales, the terminal oblong-ovoid and long-acuminate with 1 or 2 scales. Fl. April — May; fr. June — November (depending on variety). (Plate XXI, Figure 1).

Wild or naturalized on exposed stony slopes and among rocks, also on walls of old dilapidated buildings; growing wild in the undergrowth of broad-leaved woods. — European part: Crim. (Sudak, Kacha River, etc.); Caucasus: W., S. and E. Transc., Tal. (wild); Centr. Asia: Mtn. Turkm. (ravines in the Kopet Dagh Range); Pam.-Al. (Panj River). **Gen. distr.**: Med., As. Min., Iran-Afghanistan to NW India. May be found in cultivation to the north of the areas indicated, but information concerning the potential northward extent is incomplete. The fruit ripens at room temperature even in the north of the European part of the USSR. Described from the Mediterranean area. Type in London.

Note. Both unisexual and hermaphrodite specimens of *Ficus carica* L. are known to occur. The male plants contain in their receptacles, in addition to staminate flowers, galls produced by a minute wasp *Blastophaga psenes* (L.) Loew that develops in the flower. Upon hatching, the wasp leaves the host receptacle loaded with pollen from the staminate flowers. In the course of its flight it creeps into receptacles of female plants and brings about their fertilization. In this context the following varieties have been distinguished: var. *orinosyce* Tschirch und Ravasini, the always infertile var. *caprificus* of the same authors, and the fruiting var. *domestica* Tschirch und Ravasini. Only the first of these can be properly considered as wild-growing. Like all plants of ancient culture, the fig has given rise to a large number of cultivated varieties. A description of these varieties would outgrow the framework of our Flora. M. G. Popov in Tr. Prikl. Bot., Sel. i Gen. XXII, 3 (1929) 307, indicated the existence of various geographic races which differ rather consistently in the shape of leaves and fruits. It would, however, be risky to catalog them before a monographic treatment of this form cycle. It is possible that the Kopet Dagh and the Tadzhikistan figs belong to distinct though closely related species.

Economic importance. The fruit is eaten both fresh and dried. In many parts of Transcaucasia the fig is a favorite delicacy and is also made in a coffeelike beverage. The fruit has lenitive properties for which it is used in popular medicine and as a home remedy as an infusion for gargling

382 or as a poultice in the case of pectoral and laryngeal ailments; also applied as emollient for abscesses and tumors. According to Wehmer, the fruit contains 20—40% dry matter: sugar, and citric, malic, acetic, and boracic acids. The firm wood is suitable for turning. The plant is propagated by layers, suckers, and cuttings. Fast-growing and high-yielding.

Genus 374. **HUMULUS*** L.

L. Gen. pl. ed. V (1754) 453.

Flowers unisexual, dioecious, in axillary inflorescences; perianth lobes 5, opposite the stamens; staminate flowers in panicles, the perianth unaltered after flowering; pistillate flowers in heads, the entire perianth 1-sidedly accrescent in fruit; style bifid. Perennial or annual climbing plants with opposite leaf blades, the leaves stipulate. Russian: "khmel'."

1. Fruiting perianth 1-sidedly accrescent, scarious, puberulous and glandular; perennial . . . . 1. **H. lupulus** L.

\+ Fruiting perianth not scarious, only slightly accrescent, hispid; annual . . . . 2. **H. japonicus** Sieb. et Zucc.

1. **H. lupulus** L. Sp. pl. (1753) 1028; Ldb. Fl. Ross. III, 635; M. B. Fl. taur.-cauc. II, 419; Kryl., Fl. Zap. Sib. IV, 805. — Ic.: Syreishch., Ill. Fl. Mosk. gub. II (1907) 55.

Perennial; rootstock long creeping; stems climbing, angled, rough-hairy; leaves entire or 3—5-lobed, deeply cordate, acuminate, coarsely serrate, with large lanceolate approximate stipules; perianth of pistillate flowers one-sidedly enlarging and becoming scaly, winglike, obtuse, investing the achene at base, covered with small glands and diffusely puberulous outside. June—July. (Plate XXI, Figure 7).

River valleys, gullies, damp broad-leaved woods, coppices, and osier-beds. Cultivated in ornamental and truck gardens. — European part: U. Dnp., M. Dnp., Bl., U. V., V.-Don, L. Don, L. V., Crim.; Caucasus: W. Siberia: U. Tob., Irt., Alt.; Centr. Asia: T. Sh. **Gen. distr.**: Centr. Eur., Med., Bal.-As. Min., N. Am. Described from Europe. Type in London.

Note. Being widely cultivated, hops may have spread by the agency of man and thus the natural distribution area of this plant is rather
383 uncertain. — In the countries of the Far East it is known only in cultivation and it is possible that it is partly introduced from the west and partly a cultivated Japanese race of H. cordifolius Miq.

Economic importance. Hop cones are used in beer production. The glands excrete lupuline which contains a bitter hop acid (humulone, $C_{21}H_{30}O$, and lupulone, $C_{18}H_{26}O_4$) and resinous substances, as well as a distinctive camphor $C_{10}H_{16}$, etc. (Wehmer).

2. **H. japonicus** Sieb. et Zucc. Fl. Jap. fam. Nat. II (1846) 213; Kom. Fl. Mansh. II (1904) 92. — Ic.: Kom. and Alis., Opr. rast. Dal'nevost. kr. (1931) 136.

Annual; stem climbing, angled, rough with spreading bristles; leaves (3) 5—7-lobed, cordate at base, with ovate-oblong lobes, coarsely and

* A Latin word, apparently derived from a Slavic name of the plant.

sharply serrate, pubescent, the petioles very long; stipules small, subulate-triangular, distant; perianth of pistillate flowers not enlarging in fruit, acute, coriaceous, hispid outside and on the margin, vary sparsely glandular. June—September. (Plate XXI, Figure 6).

Sandy soils in river valleys; often close to human habitations, as a weed. — Far East: Uss., Ze.-Bu. Gen. distr.: Jap.-Ch. Described from Japan. Type in Leningrad.

Genus 375. **CANNABIS** L.

L. Gen. pl. ed. V (1754) 453.

Dioecious plants; pistillate flowers in axillary spikes; staminate flowers in panicles, the perianth 5-parted; stamens 5; perianth of pistillate flowers undivided, frill-like, slightly enlarging in fruit; style bifid. Herbs with erect stem; lower leaves opposite, upper ones alternate; stipules present. Russian: "konoplya."

1. Fruit with perianth persistent on the achene in the form of variously shaped dark shreds, marbled, jointed at base . . . . 2. **C. ruderalis** Janischewsky.
+ Fruit smooth, narrowed toward base or globose, not jointed; perianth a rudimentary frill . . . . 1. **C. sativa** L.

1. **C. sativa** L. Sp. pl. (1753) 1027; M. B. Fl. taur.-cauc. II, 419; Ldb. Fl. Ross. III, 634. — C. *foetens* Gilib. Exerc. phytol. II (1792) 450. — Ic.: Hegi III. Fl. Mittel. Eur. III (1909) tab. 88. — Turkish: kinder; Tadzhik: bang, anasha; staminate plants: poskon'.

Annual; stems 50—150 cm, covered with appressed hairs; leaves long-petioled, with 3—5—7 elongated-lanceolate lobes, coarsely serrate; inflorescences in the axils of upper leaves, spreading, the staminate
384 flowers in panicles, the pistillate in spikes; fruit a gray achene, not jointed at base, strongly adherent. June—July.

Growing wild in fluvial deposits, on crags and stony slopes of hills and mountains; often escaped from cultivation. — European part: Transv., L. V. (delta islands, etc.); W. Siberia: Alt.

Economic importance. Cultivated since antiquity for its strong fiber, the oil content of the seeds, and essential oils with delirifacient and analgesic properties (hashish or anasha). The fiber ("pen'ka") provides material for packing, strings, ropes, sweep nets, canvas, lacework, etc. The seeds contain 30—35% of fatty oil used for food, soap manufacture, production of soft soap, drying oil, and oil paints. Oil cake provides concentrated feed for cattle. The leaves of hemp inflorescences excrete, especially at flowering time, drops of liquid with a characteristic potent odor. Outstanding in this respect is the so-called Indian hemp, *Cannabis indica* Lam. These drops, containing a resinous cannabin oil $C_{21}H_{30}O_2$ and the nicotine-like base cannabinine, constitute the active principles of hashish which also has medicinal importance.

2. **C. ruderalis** Janishewsky, Uchenye Zap. Gos. Sar. Univ. II, 2 (1924) 14. — Ic.: Idem., Fig. 1—4.

Annual. Differing from the preceding in the smaller size of stems and leaves, and the marbled articulate achene. May—July.

In the steppe region, very often as a field weed. — European part: M. V., L. V.; W. Siberia: U. Tob., Irt., Alt.; Centr. Asia: Balkh., T. Sh. Endemic. Described from Saratov. Type in Saratov.

Note. This hemp has not so far been sufficiently explored and it is possible that t ere are other as yet undescribed forms beside that described by Yanishevskii. As regards distribution, only regions of mass occurrence have been listed, but sporadic occurrence is much wider.

Economic importance. The seeds are smaller than those of *Cannabis sativa* and cultivation is hampered by the proneness of the seeds to shed. The plant is of interest for its high-quality fiber which has already been put to industrial use. Notwithstanding its value, the plant is a noxious weed.

Family XLVII. **URTICACEAE** ENDL.

Perianth simple of 2—5 connate or distinct segments; stamens 4 or 5; ovary superior, free, 1-locular (unicarpellate), sessile or short-stipitate; fruit an achene, slightly or sometimes strongly compressed, straight or oblique; stigma very variable in shape, from tufted or capitate to linear and vermicular, commonly sessile, more rarely on a long style; flowers
385 bisexual or staminate (actually bixesual, but with reduced or rudimentary ovary) or pistillate; perianth of pistillate flowers accrescent or rarely not enlarging in fruit; inflorescence a dichasium and forms derived from it, often with reduced axis and enlarging or reduced bracts, or seemingly racemose or spiciform, mostly axillary, rarely terminal, bearing staminate or pistillate or bisexual flowers or all these together. Plants monoecious or dioecious; annual or perennial herbs, rarely undershrubs or shrubs, very rarely trees; leaves alternate or opposite, with or without stipules, simple, entire or toothed or dissected; vesture consisting of simple or glandular-capitate, or long geniculately articulate stinging hairs, the latter with caustic contents which is extruded upon pricking of the skin by the hairs and causes painful and — in the case of certain plants (*Laportea*, *Girardinia*) — sometimes rather dangerous stings.

No authenticated fossils of this family have so far been found in the USSR although fossil seeds will undoubtedly be uncovered among Quaternary floras.

Macclintockia — a genus recorded up to the present only in the lowest Tertiary or Upper Cretaceous layers of the Ural area (Ob) and the Dnieper area; its systematic position is dubious. — *M. Lyellii* Heer in Lower Tertiary (or Upper Cretaceous ?) formations of Ob (Lozva) and in Upper Cretaceous formations of Dv.-Pech. (Lemva). — *M. trinervia* Heer in Lower Tertiary and Upper Cretaceous formations of Ob (Lozva). — *M. ucrainica* in Lower Tertiary formations of U. Dnp. (Volhynia).

Key to Genera

1. Leaves opposite . . . . 2.
+ Leaves alternate . . . . 4.
2. Plants vested with stinging hairs in addition to simple and glandular ones . . . . 376. **Urtica** L.
+ Plants without stinging hairs . . . . 3.
3. Perianth of pistillate flower 4-parted . . . . 379. **Pilea** Lindl.
+ Perianth of pistillate flowers 5-parted . . . . 380. **Achudemia** Blume.
4. Plants with stinging hairs . . . . 5.
+ Plants without stinging hairs . . . . 381. **Parietaria** L.
5. Perianth of pistillate flowers 4-parted; achene compressed . . . . 377. **Laportea** Gaudich.
+ Perianth of pistillate flowers 2-parted; achene biconvex, lenticular . . . . 378. **Girardinia** Gaudich.

386 Genus 376. **URTICA*** L.

L. Gen. pl. ed. V (1754) 423.

Flowers bisexual with rudimentary ovary (staminate) or pistillate; perianth of staminate flowers 4-parted, not enlarging; perianth of pistillate flowers with 2 inner (dorsal) segments enlarging in fruit and 2 outer (lateral) unchanging segments; stamens 4; ovary with sessile stigma or with a very short style; achene compressed, flat or lenticularly biconvex; inflorescence apparently spiciform, branched, commonly (in the USSR, except U. urens L.) bearing only staminate or pistillate flowers; monoecious or dioecious forms occur. Annual or perennial herbs with stinging and simple hairs; leaves undivided, toothed or deeply 3—5-fid; stipules in pairs, sometimes connate. Russian: "krapiva."

Of more than 40 species of the genus which is distributed through the Temperate and Torrid zones of both hemispheres, two (U. dioica and U. urens) are cosmopolitan, associated with man and occurring whenever the natural vegetation is disturbed through man's interference.

Economic importance. Young plants of U. dioica and U. urens are sometimes used for food seasoning instead of sorrel. The stems contain contain fibers which finally become lignified. These fibers deserve investigation with a view to utilization in the textile industry. Particular attention should be paid to U. platyphylla Wedd., since the Ainu of Sakhalin have long been using it as a source of fiber; similarly, U. dioica calls for investigation in this respect.

1. Leaves deeply dissected . . . . 3. **U. cannabina** L.
+ Leaves dentate or serrate, not dissected . . . . 2.
2. Inflorescence containing both staminate and pistillate flowers . . . . 1. **U. urens** L.
+ Inflorescence containing either staminate or pistillate flowers . . . . . 3.
3. Plants monoecious . . . . 4.
+ Plants dioecious . . . . 7.

* Name used for nettle by Horace.

4. Pistillate flowers gathered at the end of the inflorescence into a spherical head; dorsal segments of the female perianth strongly swelling . . . . 2. **U. pilulifera** L.
\+ Pistillate flowers in clusters scattered along the axis of the branched inflorescence; dorsal segments of the perianth not swelling . . . . 5.
5. Cystoliths in the leaves rod-shaped; plants pale green . . . . 4. **U. laetevirens** Maxim.
387 + Cystoliths in leaves punctate; plants dark green . . . . 6.
6. All stipules united in pairs . . . . 5. **U. platyphylla** Wedd.
\+ Stipules of the lower leaf pairs distinct, those of the upper united . . . . 6. **U. kioviensis** Rogov.
7. Cystoliths in leaf tissue botuliform; leaves broad, with very large sharp teeth, often turning blue on drying . . . . . . 7. **U. cyanescens** Kom.
\+ Cystoliths in leaf tissue round . . . . 8.
8. Stems and leaves very sparingly covered with stinging hairs . . . . 9. **U. angustifolia** Fisch.
\+ Stems and leaves profusely covered with stinging hairs . . . . 9.
9. Beside stinging hairs, plants densely woolly-pubescent with short simple hairs . . . . 10. **U. pubescens** Ldb.
\+ Beside stinging hairs, plants sparsely covered with short simple hairs or hairless . . . . 8. **U. dioica** L.

1. **U. urens** L. Sp. pl. (1753) 984; Ldb. Fl. Ross. III, 637; Turcz. Fl. baic.-dah. II, 91; Shmal'g., Fl. II, 419; Kom. Fl. Mansh. II, 94; Kryl., Fl. Zap. Sib. IV, 807. — Ic.: Rchb. Ic. Fl. Germ. XII (1850) t. 652. — Exs.: Pl. Finland. exs. No. 182; Fl. Pol. exs. No. 876.

Perennial; stems erect, 15—40 cm long; leaves elliptic or ovate, deeply incised-dentate, rounded-cuneate at base, acuminate, the long petioles usually exceeding the inflorescences; stipules small, green, free; inflorescence containing both staminate and pistillate flowers; the lateral perianth segments short, the dorsal ones enlarging in fruit, ovate, investing and as long as the achene; achene to 1.5 mm long. May—October.

Weed-infested places, near habitations, and at roadsides. — European part: throughout; Caucasus: all regions; W. Siberia (sporadically, adventive); E. Siberia: (Ang.-Say. and Dau., adventive); Far East: Ze.-Bu., Uss. (adventive). **Gen. distr.**: Scand., Centr. Eur., Atl. Eur., Med., Bal.-As. Min. Described from Europe. Type in London.

2. **U. pilulifera** L. Sp. pl. (1753) 983; Ldb. Fl. Ross. III, 636; Shmal'g., Fl. II, 420. — Ic.: Rchb. Ic. Fl. Germ. 12 (1850) t. 653. — Exs.: Rchb. Fl. Germ. exs. No. 22; Dörfler Herb. Norm. No. 5172; Sennen, Pl. d'Esp. No. 6364.

Perennial; stems erect or ascending, 20—75 cm long, angled, smooth, covered with bluish bloom and scattered stinging hairs; leaves broad-ovate, rounded or cordate at base, acuminate, coarsely serrate-dentate, to 10 cm long and 9 cm broad, long-petioled; stipules free, triangular,
388 herbaceous; inflorescence containing either pistillate flowers in a globular terminal head or scattered staminate flowers; dorsal segments of fruiting perianth enlarging cucullately to 3—4 mm, the small lateral ones persisting on the inflorescence after shedding of seeds; achene to 3 mm long, cordiform. (Plate XXI, Figure 3).

In weedy places and as a weed among crops. — European part: Crim.; Caucasus: E. Transc. Gen. distr: Med., Bal.-As. Min. Described from S. Europe. Type in London.

3. **U. cannabina** L. Sp. pl. (1753) 984; Ldb. Fl. Ross. III, 638; Kom. Fl. Mansh. II, 95; Kryl., Fl. Zap. Sib. IV (1930) 807. — Ic.: Kom. and Alis., Opr. rast. Dal'nevost. kr. I, tab. 137.

Perennial; stems 70—150 cm long, angular; rootstock creeping; leaves deeply 3—5-parted with pinnatisect and sometimes doubly pinnatisect lobes, to 15 cm long, with a long slender petiole; stipules free, oblong-linear; inflorescence branched, long, densely beset with flower clusters; plants monoecious or dioecious; segments of fruiting perianth connate to one-third, the dorsal accrescent to equal the achene, the linear lateral ones with a short free portion; achene 2—2.5 mm long. (Plate XXI, Figure 4).

Slopes, roadsides, and weed-infested places. — European part: occasionally adventive; W. Siberia: U. Tob., Irt., Alt.; E. Siberia: Ang.-Say., Dau.; Far East: Ze.-Bu., Uss.; Centr. Asia: Balkh., T. Sh. **Gen. distr.**: Dzu.-Kash., Mong., Jap.-Ch. Described from Siberia. Type in London.

4. **U. laetevirens** Maxim. Bull. Ac. Sc. Pétersb. XXII (1877) 236; Kom. Fl. Mansh. II, 96.

Perennial; stems 40—100 cm long, diffusely pubescent or glabrous, with scattered stinging hairs, simple or short axillary shoots, rarely branched; leaves broad-ovate or ovate, acute or rarely acuminate, cuneate or subcordate at base, coarsely serrate with sharp distant teeth; cystoliths in leaves rod-shaped; stipules free, linear-lanceolate; lower inflorescences short, interrupted, with pistillate flowers; upper inflorescences long, bearing staminate flowers; the lateral segments of fruiting perianth short, the dorsal segments enlarging to fit the shape and size of the achene; achene ovate, to 1.5 mm long. June—August.

In shady woods, and at the foot of and in fissures of rocks. — Far East: Uss. **Gen. distr.**: Jap.-Ch. Described from Japan. Type in Leningrad.

391 5. **U. platyphylla** Wedd. Arch. Musc. Hist. Nat. IX (1857) 98; Kom. Fl. penins. Kamtsch. II, 53; Hultén Fl. of Kamtch. II, 40. — U. dioica var. platyphylla Wedd. in DC. Prodr. XVI, 1 (1869) 51; Maxim. Bull. Ac. Sc. St. Pétersb. XXII (1877) 236.

Perennial; stem 50—150 cm long, erect, with slender axillary shoots, puberulous or glabrous, with scattered stinging hairs; leaves (4) 5—20 (27) cm long and to 12 cm broad, ovate-elliptic or ovate, acute or acuminate, cuneate or rounded or cordate at base, coarsely and sharply serrate with incurved teeth; stipules of the opposite leaves united, scarious, to 1.5 cm long; plants dioecious; staminate inflorescence branched, with fleshy axes and with scarious triangular bracts to 2 mm long; pistillate inflorescences with flowers gathered in clusters, long, branched; fruiting perianth with a pair of small lateral segments and large dorsal segments surpassing the achene; achene ovate, 1—1.25 mm long. June—September. (Plate XXI, Figure 8).

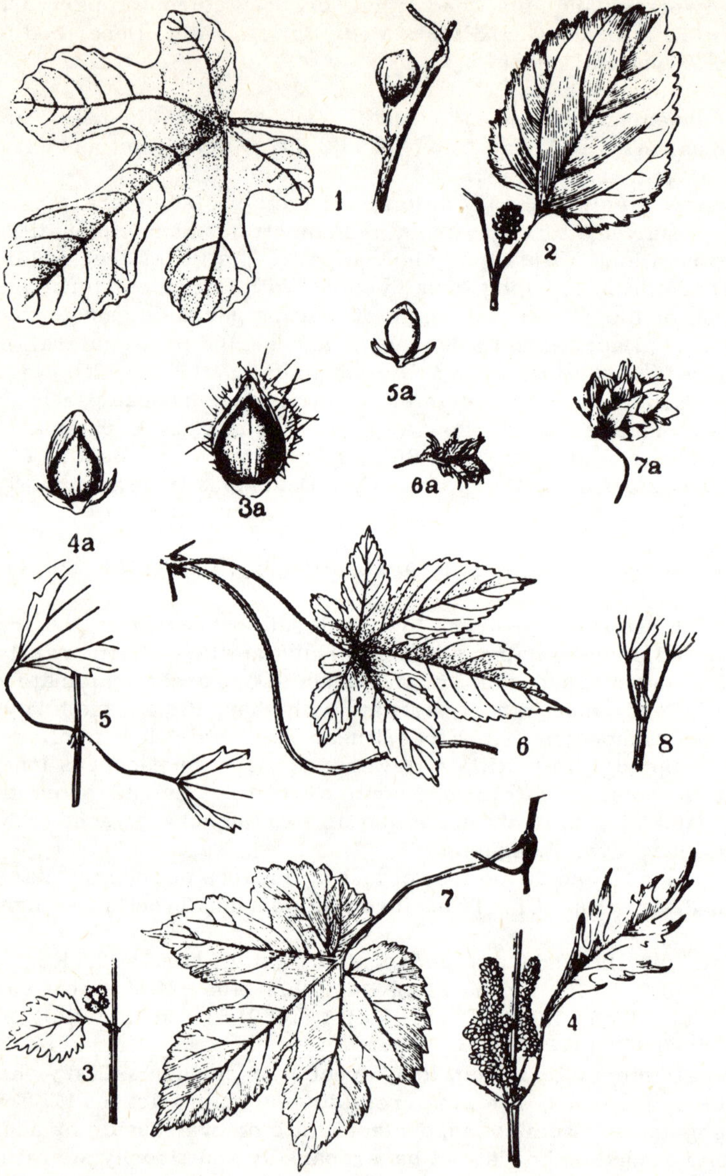

PLATE XXI. 1. *Ficus carica* L., leaf and fruit. – 2. *Morus alba* L., leaf and fruit. – 3. *Urtica pilulifera* L., inflorescence and leaf, 4a) fruit with perianth. – 4. *U. cannabina* L., inflorescence and leaf. – 5. *U. angustifolia* Fisch., stipules, fruit and perianth. – 6. *Humulus japonicus* Sieb. et Zucc., a) inflorescence. – 7. *H. lupulus* L., leaf, a) inflorescence. – 8. *Urtica platyphylla* Wedd., stipules.

Riverside woods and osier-beds, crags, and bluffs. — Far East: Kamch., Sakh. **Gen. distr.**: Jap.-Ch. (Kurile and Commander Islands). Described from Avacha in Kamchatka. Type in Kew.

Note. This species was also reported for Uss., but herbarium investigations have indicated its complete absence in this region. Of interest is a dwarf tundra form collected by P. T. Novograblenov near Kichiga (No. 3006), but the single specimen is insufficient to justify separation into a distinct species. Additional collections are desirable. The name U. Takedana Ohwi (Acta phytotax. et Geob.) has been proposed for replacement of the homonym U. platyphylla Wedd. (non Hamilt.).

6. **U. kiovensis** Rogov. in Bull. Soc. Nat. Mosc. XVI (1843) 324. — U. dioica var. monoica Trautv. Bull. phys.-math. St. Pétersb. XIII (1855) 189. — U. dioica var. kioviensis Wedd. in Arch. Mus. Hist. Nat. IX (1857) 88 et in DC. Prodr. XVI, I, 51; Shmal'g., Fl. II, 420.

Perennial; rootstock creeping; stems numerous, 80 — 120 cm long, abundantly covered with recurved stinging hairs; leaves ovate-lanceolate to oblong-lanceolate, coarsely serrate-dentate with incurved teeth; stipules large, herbaceous, the lower free, the upper united throughout or merely at base; lower inflorescences exclusively staminate, the upper ones exclusively pistillate; bracts of staminate flowers small, scarious, to 0.75 mm long; fruiting perianth with a pair of lateral segments to 1 mm long and a pair of ovate dorsal segments equaling the ovate achene, 1.5 mm long. August.

392 Wet osier-beds and ravines. — European part: U. Dnp., M. Dnp. Described from Kiev. Type in Leningrad.

Note: This species, described from the vicinity of Gadyuchii Yar near Kiev, is reported by Schmalhausen for Chernigov and Podolia. It is very likely that this species is identical with U. radicans Botta (Verhandl. Verein Naturk. Presburg I (1856) 6 — 7) and U. major var. krakoviensis Kanitz. Bot. Zeit. (1863) 51, but this could only be confirmed by thorough treatment of the entire cycle U. dioica s. lato. We separate U. kiovensis on account of two circumstances: firstly, the morphological break consisting in the united stipules and the monoecious disposition of sexes; secondly, the geographic isolation which would maintain the distinctiveness of the distribution area even if identity of the USSR form with the Hungarian were to be admitted.

7. **U. cyanescens** Kom. sp. nova in Addenda IV, p. 546. — U. platyphylla Kom. et Alis., Opred. rast. Dal'nevost. kr. I (1931) 446, non Wedd.

Perennial; rootstock horizontally creeping; stems several, prominently 4-angled, to 80 cm long and ca. 4 cm in diameter; stipules extraaxillary, quite free, paired, sublinear, acute, membranous, ca. 12 mm long and slightly more than 1 mm broad; petioles slender, 1 — 3 cm long; leaves oval-elliptic, rounded broad-cuneate or slightly sinuate at base, 8 — 14 cm long, 6 — 8 cm broad, turning blue on drying, puberulous on the veins beneath, long-acuminate, coarsely and sharply serrate with 7 — 12 teeth on each margin, the teeth to 2 cm long, the leaf tissue containing numerous rodlike cystoliths with rounded ends; flowers dioecious; staminate racemes axillary, in pairs, upright, rather compact, with pubescent axis; bracts very small, oblong or linear, acute, partly caducous; flowers borne on very short pedicels or sessile, 1.2 mm in diameter; perianth 4-parted, the strongly concave segments with unequal margins and stinging hairs; anthers large, yellow; the rudimentary style globular; pistillate flowers not aggregated. June — August.

Growing in stony soil on crests of mixed forests, in shade; also along streams in dense thickets of herbaceous vegetation. — Far East: Uss. (S.). Endemic. Described from the pass leading from the Suputinka Valley into the Maikhe Valley, near the village of Khotunichi. Type in Leningrad.

Note. Distinguishable by the blue color in drying, larger flowers, elongated deeply incised-toothed leaves, the short rodlike cystoliths.

8. **U. dioica** L. Sp. pl. (1753) 984; Ldb. Fl. Ross. III, 637 excl. var. — U. dioca var. vulgaris Wedd. in DC. Prodr. XVI, 1 (1869) 50; Shmal'g., Fl. II, 419, excl. var.; Boiss. Fl. Or. IV, 1146, excl. var.; Kryl., Fl. Zap. Sib. IV, 808, pro parte. — Ic.: Rchb. Ic. Fl. Germ. XII (1850) t. 654. — Exs.: Pl. Finl. exs. No. 598.

393 Perennial; rootstock creeping underground; stems erect, 60 — 100 cm long, glabrous or sparingly pubescent, with a dense coating of stinging hairs especially at the nodes; leaves ovate-cordate, coarsely serrate with incurved teeth, mostly cordate at base, acuminate at apex; stipules free, oblong, scarious, to 12 mm long; plants dioecious; inflorescences long, branched, covered with dense appressed simple hairs and scattered stinging hairs; bracts of staminate flowers very small, scarious; perianth of pistillate flowers with a pair of short lateral segments and a pair of dorsal segments accrescent in fruit to a length of 1.25 — 1.5 mm and exceeding the achene. May — September.

Roadsides, as weed among crops and near human habitations, and in damp woods. — European part: throughout; Caucasus; W. Siberia: throughout; E. Siberia: Ang.-Say. (adventive in Dau. and Lena-Kol.); Far East: (adventive in Ze.-Bu., Uss., Okh., and Kamch.); Centr. Asia: throughout adventive as anthropochore. **Gen. distr.**: Scand., Centr. and Atl. Eur., Med., Bal.-As. Min., Arm.-Kurd., Iran. Spread by man into North and South America and into Australasia. Described from Europe. Type in London.

Note. Widely distributed by the agency of man, U. dioica is now to be found everywhere in the Temperate Zone of both hemispheres. It is therefore very difficult to distinguish such species as U. angustifolia Fisch. and U. pubescens Ldb., their areas being intersected by the distribution of ruderal U. dioica and differences of morphological and ecological nature liable to be masked by manifestations of hybridization. In the Caspian Lowland the more common species is U. pubescens, while in Transbaikalia U. dioica is displaced by U. angustifolia; this is largely the position in woods, whereas in ruderal habitats occurs the adventive U. dioica.

9. **U. angustifolia** Fisch. ex Hornem. Suppl. Hort. Hafn. (1819) 107; Kom. Fl. Mansh. II, 95; Hultén, Fl. of Kamtch. II, 40; Kom. Fl. penins. Kamtsch. II, 153. — U. dioica var. angustifolia Ldb. Fl. Alt. IV (1833)

240 et Fl. Ross. III, 637; Turcz. Fl. baic.-dahur. II, 90. — Ic.: Kom. and Alis., Opr. rast. Dal'nevost. kr. I (1931) 137.

Perennial; rootstock creeping; stem erect, 50—150 cm long, sparingly covered with stinging hairs; leaves oblong-lanceolate to lanceolate or rarely ovate-lanceolate, to 15 cm long, petiolate, rounded or cordate at base, acute at apex, the margin remotely serrate, the teeth sometimes taper-tipped; stipules scarious, linear-oblong, to 1 cm long; plants dioecious; inflorescences long, branched, covered with appressed hairs interspersed with few stinging hairs; bracts of staminate flowers scarious, to 1 mm long; dorsal perianth segments of the fruiting perianth slightly exceeding to half as long again as the mature achene, the lateral segments half the length of the dorsal ones; achene round, ca. 1 mm long. June—October. (Plate XXI, Figure 5).

Riverside and mixed mountain woods, springs, rapids, and rocky screes,
394 more rarely near dwellings and in weedy places. — E. Siberia: Dau., Lena-Kol.; Far East: Ze.-Bu., Uss., Okh., Kamch. **Gen. distr.**: Jap.-Ch., Mong. Described from Siberia. Type in Leningrad.

Note. The species represents an eastern geographical vicariad of U. dioica. Komarov's indication (Fl. Mansh. II (1904) 96) concerning the absence of the latter in the Far East refers to the past; by now U. dioica has also been spread to this area and occurs here and there in ruderal habitats. The narrow leaves, characteristic of U. angustifolia, are also to be found in U. dioica of Europe and W. Siberia. The same may be noted for the shape and size of seeds which differ clearly in the typical forms. The most consistent character appears to be the paucity of stinging hairs in U. angustifolia as against their abundance in the case of U. dioica.

10. **U. pubescens** Ldb. Fl. Alt. IV (1833) 240; Fl. Ross. III, 638. — U. dioica var. pubescens Trautv. in Bull. phys.-math. XIII (1855) 188; Weddel in Arch. Mus. Hist. Nat. IX, 78 et in DC. Prodr. XVI, 1, 51; Boiss. Fl. Or. IV, 1146; Shmal'g., Fl. II, 420.

Perennial; rootstock short; stems erect, stout, 30—100 cm long, with a dense woolly coating of simple short hairs interspersed with numerous stinging hairs; leaves cordate-lanceolate, cordate at base, long-tapering at apex, coarsely and closely serrate with sharp teeth, to 12 cm long and 4—5 cm broad, the lower surface and petiole densely woolly-pubescent; stipules free, herbaceous, oblong, to 75 [7.5 ?] mm long; plants monoecious; inflorescences short, branched, densely woolly; bracts of staminate flowers obsolescent; lateral lobes of the fruiting perianth one-fourth the length of the ovate dorsal segments, these as long as the achene; achene ovate-elliptic, 1.25 mm long. May—August.

Coppices, riverside deciduous woods, and weed-infested places. — European part: L. V., L. Don; Caucasus: Cisc., Dag. Endemic. Described from Astrakhan. Type in Leningrad.

Note. A geographical race of U. dioica, distinguished by the dense woolly pubescence of the leaf undersurface, of the petioles, and of the stem. The limited distribution of this race may be the result of reduced adaptability to ruderal conditions, as in the case of U. angustifolia.

The shape and serration of the leaves are very consistent and these characters distinguish this race from others of its cycle. It is noteworthy that the collections of S. I. Korzhinskii from the Volga Delta consist solely of this race, without any admixture of U. dioica.

Genus 377. **LAPORTEA*** GAUD.
Gaud. in Freyc. Voy. Uran. Bot. (1826) 498.

Flowers either staminate (bisexual with rudimentary ovary) or pistillate;
395 perianth simple, or 4- or 5-merous; stamens 5; perianth with a sessile oblong persistent stigma; achene flat, inequilateral, obliquely ovate; inflorescences corymbs or cymes, axillary and terminal, containing either staminate or pistillate flowers. Perennial herbs, shrubs, or trees, covered with stinging hairs; leaves alternate, toothed or entire; stipules 2-nerved, caducous.

1. **L. bulbifera** (Sieb. et Zucc.) Wedd. Arch. Mus. Hist. Nat. IX (1857) 139; Kom. Fl. Mansh. II, 97. — Urtica bulbifera Sieb. et Zucc. Fl. Jap. famil. nat. II (1846) 214.

Perennial; stems herbaceous, angled, 30—100 cm long; leaves ovate, palmately 3-veined, crenate-serrate, rounded at base, acuminate at apex, to 15 mm long, on petioles to 5 cm long, usually with 1—3 reniform fleshy tubers in the axils of the petioles; pistillate inflorescences terminal; staminate axillary; peduncles compressed, slightly winged; perianth of staminate flowers 4-parted, with 4 acute segments; fruiting perianth 4-fid, the accrescent lateral segments exceeding the dorsal; achene compressed. August—September.

Shady mixed woods. — Far East: Uss. (right tributaries of the Suifun River). Gen. distr.: Jap.-Ch. Described from Japan. Type in Leningrad.

Genus 378. **GIRARDINIA** GAUD.
Gaud. in Freyc. Voy. Uran. Bot. (1826) 498.

Flowers either staminate or pistillate; perianth simple, in staminate flowers 4-or 5-parted, in pistillate flowers 2-lobed with one lobe greatly exceeding the other; stamens 4 or 5; ovary with an elongated filiform caducous stigma; achene ovate-lenticular, biconvex; flowers bracteolate; inflorescence a racemiform dichasium bearing either staminate or pistillate flowers. Perennial or annual herbs with alternate toothed leaves and persistent stipules.

1. **G. cuspidata** Wedd. in DC. Prodr. XVI, 1 (1869) 103; Maxim. Bull. Ac. Sc. Pétersburg, 22 (1877) 240; Kom. Fl. Mansh. II, 98.

Perennial; stems to 1 m long, angular, covered with appressed simple hairs and scattered strongly stinging hairs to 6 mm long; leaves orbicular-ovate, rounded at base, coarsely sinuate-toothed on the margin, acuminate, to 14 cm long and 11 cm broad, the slender pedicels to 8 cm long; cystoliths rather small, spherical; lower inflorescences staminate, upper pistillate; perianth of pistillate flowers 2-parted; achene ovate, lenticularly biconvex. July—October.

* Named for the orientalist de Laporte or the zoologist de Laporte de Castelnau.

396 Slopes and mountain woods, sometimes at the foot of crags. — Far East: Uss. (left tributaries of the Suifun River). **Gen. distr.**: Jap.-Ch. Described from China. Type in Paris.

Genus *BOEHMERIA JACQ.

Jacq. Enum. pl. Carib. (1760) 9.

Flowers in unisexual clusters gathered in racemes of corymbs; staminate flowers with 2—4-toothed perianth; stigma long, filiform; fruiting perianth thin, firm, dry; seed containing endosperm. Monoecious herbs, shrubs, or small trees; leaves opposite or alternate, undivided, toothed, subtended by stipules.

B. nivea (L.) Gaud. in Freyc. Bot. voy. (1826) 499. — Urtica nivea I Sp. pl. (1753) 985. — Also known as "rami."

Perennial; stems erect or ascending, to 1 m long; leaves large, broad-ovate or rounded-elliptic, more or less cordate or rarely cuneate at base, 15 cm long and 10 cm broad, sparingly pubescent above, snow-white from dense tomentum beneath; flower clusters greenish, gathered in loose racemes. Fl. June—September.

Native in China. In the USSR cultivated in W. Transc. (Gagra, etc.) and in Centr. Asia (Tashkent, Vakhsh) where it requires irrigation. Described from China.

Economic importance. Cultivated for the base which yields excellent fiber.

Genus 379. PILEA* LINDL.

Lindl. Collect. (1821) t. 4.

Flowers staminate or pistillate; perianth simple, in pistillate flowers 2- or 3-lobed, in staminate 4- or rarely 2- or 3-parted; stamens (2 or 3) 4; stigma sessile, tufted; inflorescence a dichasium, usually shortened and cluster-shaped; achene ovate, slightly flattened. Annual or very rarely [?] perennial herbs with decussate leaves and connate stipules.

1. Leaves entire, 0.5—1 cm long; perianth of pistillate flowers with unequal segments . . . . 1. **P. peploides** Hook. et Arn.

\+ Leaves dentate-serrate, 2—7 (12) cm long; perianth of pistillate flowers with 3 equal segments . . . . 2. **P. viridissima** Makino.

1. **P. peploides** Hook. et Arn. Bot. Voy. Beech. (1831) 96; Kom. Fl. Mansh. II, 98.

Small perennial plants with slender stems 1—7 cm long; leaves paired below, toward the ends of stems apparently verticillate, acute, broad-cuneate at base, entire or rarely with 1 or 2 small teeth on the margin, the lower
397 surface dotted with rusty glands, the upper side with rod-shaped cystoliths; inflorescence an abbreviated dichasium, mostly shorter than the leaf petiole; perianth of pistillate flowers with 1 strongly developed lateral segment, the 1 or 2 other segments barely perceptible. July—September.

* From Greek "pilos," hair, referring to the hairlike perianth segments.

Hummocks among lowland meadows, and at seacoasts, and rock crevices. — Far East: Uss. **Gen. distr.**: Jap.-Ch., Sandwich Islands, Malay Peninsula. Described from the Sandwich Islands. Type in London.

2. **P. viridissima** Makino in Tokyo Bot. Mag. (1909) 87. — P. pumila Maxim. Prim. Fl. Amur. (1859) 246, non A. Gray; Kom. Fl. Manchzh. II, 99. — Urtica pumila L. Sp. pl. (1753) 984.

Perennial; stems 5—80 cm long, angular, glassy-green; leaves ovate, cuneate at base, acute or apiculate, 2—7 (12) cm long, eglandular, with short rod-shaped cystoliths; inflorescences axillary, usually shorter than petioles; perianth of pistillate flowers 3-parted, the segments triangular-subulate; scarious staminodes adnate to the perianth segments within, 1.5—2 mm long. July—September.

Wet shaded places; woods and coppices, inundated meadows, and shores. — E. Siberia: Dau.; Far East: Ze.-Bu., Uda, Uss. **Gen. distr.**: Jap.-Ch. Described from Japan. Type in Tokyo.

Note. Worthy of note is the isolated occurrence of P. viridissima near the warm springs at the village of Verkhne-Azharskoe in Transbaikalia, well to the north of the general limit of the species. The true P. pumila Asa Gray, Man. Bot. U. S. ed. 1, 437, grows in North America and has been described from Canada.

Genus 380. **ACHUDEMIA** BLUME

Blume Mus. Bot. Lugd.-batav. II (1852) 57.

Flowers unisexual; clusters containing staminate, pistillate and hermaphroditic flowers, gathered in loose paniculate cymes borne singly in the leaf axils; perianth of staminate flowers 5-parted, with unequal segments, 5 stamens, and a barely visible rudimentary style, the peduncles jointed; perianth of pistillate flowers 5-parted; rudimentary stamens 6, opposite the perianth segments, scalelike, incurved; ovary free, slightly oblique, the stigma sessile; hermaphroditic flowers with 5 stamens and a pistil as in pistillate flowers; fruit lenticularly flattened, enveloped by persistent perianth segments; seeds erect; leaves decussate, petiolate, equilateral, 3-nerved; stipules axillary. (Two species, in Java and in Japan).

1. **A. japonica** Maxim. in Bull. Acad. Pétersb. XXII, 241 et in Mélanges Biol. IX (1876) 627; Kom. in A. H. P. XXII (1903) 100.

Annual; stem slender, erect, 4—30 cm long, simple or rarely branched;
398 lower leaves often entire, 15 mm long, 10 mm broad; other leaves coarsely serrate with 3—7 teeth on each margin, rather long-petioled, cuneate at base, distinctly apiculate, more or less rhombic, varying in size, the largest 30 mm long and 20 mm broad, more crowded at the end of the stem; stipules axillary, short, deltoid, the lower ones soon caducous; peduncles solitary from the leaf axils or arising from an aggregate of short clusters, these many-flowered at the ends of peduncles; some flowers on the longer peduncles with oval 1-nerved infertile stamens half the length of the perianth and fruit somewhat larger and finely granular; flowers on the shorter

peduncles with obsolescent infertile stamens, the fruit smaller and smooth; the larger fruits falling early, the small ones retained among the perianth segments. August—October.

Moss-covered rocks and stone heaps, in the shade of broad-leaved woods.— Far East: Uss. (Pos'et area, Kedrovaya Canyon). **Gen. distr.:** Jap.-Ch. Described from Kyushu Island in Japan. Type in Leningrad.

Genus 381. **PARIETARIA** L.

L. Gen. pl. ed. V (1754) 471.

Flowers polygamous, in clusters or cymes, bracteolate; perianth of hermaphrodite flowers of 4 subequal segments, these connate at base, concave, spreading on flowering; stamens 4, inserted at the base of the perianth opposite the segments, folded transversely, inflexed, straightening at anthesis; ovary ellipsoid, sessile or borne on a style, tufted or capitate; perianth of pistillate flowers tubular, 4-toothed, persistent in fruit and not accrescent; style exserted from the perianth; stigma tufted; achene ovate, slightly compressed, enclosed in the persistent perianth. Perennial or annual herbs with alternate leaves.

1. Style of pistillate flowers long; the lower flowers and inflorescences bisexual . . . . 2.
\+ Style of pistillate flower short or stigma almost sessile; the lower flowers and inflorescences always bisexual, the others pistillate . . . 4.
2. All flowers and inflorescences bisexual . . . . 3. **P. judaica** Strand.
\+ The lower flowers and inflorescences pistillate, the others bisexual . 2.
3. Inflorescence compact; bracts connate at base . . . . 1. **P. officinalis** L.
\+ Inflorescence loose; bracts free at base . . . 2. **P. erecta** Mart. et Koch.
4. Bracts of pistillate flowers scarious, enlarging after flowering; perianth of pistillate flowers covered with capitate hairs . . . . 6. **P. alsinaefolia** Delile.
\+ All bracts herbaceous, linear or oblong; perianth of pistillate flowers covered with simple hairs or glabrous . . . . 5.
399 5. Bracts subulate-linear; perianth of hermaphrodite and pistillate flowers scarious, not changing after flowering . . . . 5. **P. micrantha** Ldb.
\+ Bracts lanceolate or oblong; perianth of hermaphrodite flowers scarious and not changing after flowering, that of pistillate flowers becoming firm and brown . . . . 4. **P. chersonensis** Dorfler.

Subgenus **EUPARIETARIA** Kom. comb. nova. — Flowers hermaphrodite or pistillate; staminate none; perianth simple, 4-toothed or 4-parted, accrescent in fruit; stamens 4; ovary superior, free, 1-locular, in pistillate flowers the stigma borne on a long style; inflorescence a dichasium, often abbreviated; bracts connate at base or free; basal flowers of inflorescence pistillate, the upper hermaphrodite, or else all flowers hermaphrodite; achene of hermaphrodite flowers falling together with the perianth, that of pistillate flowers falling out of it. Perennial herbs, woody at base, very rarely annuals, covered with simple or glandular-capitate hairs; leaves alternate, entire, without stipules.

1. **P. officinalis** L. Sp. pl. (1753) 1052. — P. officinalis diffusa Wedd. in DC. Prodr. 16 (1869) 235, 42. — P. ramiflora Moench, Meth. pl. (1794) 327; Shmal'g., Fl. II (1897) 420. — P. diffusa Mert. et Koch in Röhlings Deutschl. Fl. ed. III (1823) I, p. 827; Ldb. Fl. Ross. III (1851) 639. — P. judaica Willd. Sp. pl. (1805) 954; M. B. Fl. taur.-cauc. II (1808) 440 (non Strand et L.). — Ic.: Curtis, Fl. Lond. II, IV (1798) t. 63. — Exs.: Hayek, Fl. Stiriaca exs. No. 946 (sub nom. P. judaica).

Stems tufted from a woody rootstock, widely branching, densely covered with spreading hairs, 30—50 cm long; leaves lanceolate or oblong-elliptic, cuneate at base, acute or acuminate, sparingly hairy beneath, ciliate on the margin, glabrous and densely minutely papillose above; petiole as long as or shorter than the blade, heavily coated with spreading hairs; inflorescences dense, subtended by bracts connate at base, the winged lateral axes commonly surrounding the basal flower; basal flower pistillate and, when (rarely) the inflorescences monochromatic (only on the lower part of the stem), the others hermaphrodite, or occasionally lower pistillate flowers 2 or 3; perianth of pistillate flowers tightly enclosing the achene, the oblong-lanceolate segments covered on the margins with brown hairs and beaked at apex; perianth of hermaphrodite flowers tubular, to 3—4 mm long, the tips of the segments folded into the tube; achene dark olivaceous, ca. 1 mm long; style of pistillate flowers long, with tufted stigma. May — August. (Plate XXII, Figure 2, a, b).

Rocks and fissures, dry stony places; rarely a weed. — European part:
400 Crim.; Caucasus: Cisc., Dag., W. Transc. Gen. distr.: Centr. Eur., Atl. Eur., Med., Bal.-As. Min. Described from Centr. Europe. Type in London.

Note. The species has given rise to a number of geographic races in the West Mediterranean region. The plants growing in the USSR represent the principal race which is characteristic of Central Europe. In Dagestan it is replaced by another, distinguished by less hoary appressed pubescence, rather stunted prostrate growth, and smaller lanceolate leaves. Pending treatment of the entire form-cycle we record it provisionally as P. caespitosa Jarm. (ad int.). A form with more rounded leaves, occurring in Abkhazia, approaches the South European races rather than the typical P. officinalis, even though it gravitates toward the latter in respect of the spreading pubescence. The Abkhazian form was published by Yu. N. Voronov (Pl. Abchas. exs. No. 231) but coining a name must be considered premature before completion of treatment of the whole cycle.

2. **P. erecta** Mert. et Koch, Röhling's Deutschl. Flora ed. III, 1 (1823) 492; Ldb. Fl. Ross. III, 2, 639. — P. officinalis γ erecta Wedd. in DC. Prodr. 16, 1, 235[43]. — P. officinalis Willd. Sp. pl. (1805) 953 (et auct. plur. non L.); M. B. Fl. Taur.-cauc. II (1808) 439. ; Shmal'g., Fl. II, 420. — Ic.: Rchb. Ic. fl. Germ. XII (1850) 651. — Exs.: Billot, Fl. Gall. et Germ. exs. No. 644 et 722; Kmet Fl. Austro-Hung. exs. No. 3868; Hayek Fl. stir. exsic. No. 735; Fiori, Béguinot et Pampanini Fl. Ital. exs. No. 781.

Perennial; stems erect, 30—100 cm long, simple or branched, hoary-pubescent with short hairs; leaves oblong-lanceolate, to 12 cm long and 4.5 cm broad, elongate-cuneate or rarely rounded-cuneate at base, long-acuminate or rarely acute at apex, sparingly hairy beneath, glabrous and minutely papillose above, rather long-petioled; inflorescence an open many-flowered dichasium with nonelongating axes and distinct bracts; basal flowers pistillate, borne freely in the axils of inflorescence branches, 1-bracted; upper flowers hermaphrodite, terminal, with 2 or rarely 3 bracts; perianth of pistillate flowers tightly enclosing the achene, the oblong-lanceolate segments connivent into a beak above the achene; perianth of hermaphrodite flowers campanulate, scarious, ca. 3 mm long, falling with the achene; achene black, lustrous, ca. 1 mm long; style of pistillate flowers long, with tufted stigma. April—September. (Plate XXII, Figure 1a—f).

Shady thickets, wooded gullies, and banks of streams. — European part: M. Dnp., Bl., Crim.; Caucasus; possibly occurring also in Centr. Asia (Mtn. Turkm.). **Gen. distr.**: Centr. Eur., Atl. Eur., Med., Bal.-As. Min. Described from Germany (Wetterau). Type in Leningrad.

3. **P. judaica** Strand. Linnaei Am. Acad. IV (1759) 464; L. Sp. pl. (1763) 1492 (non Willd., Weddel, Boiss. et auct. plur.).

Perennial; stems many from branching rootstock, glabrous or pubescent, to 50 cm long; leaves lanceolate or ovate; inflorescences compact; bracts
401 connate at base; all flowers hermaphrodite; achene olivaceous. April—September.

Rock crevices in mountains, caves, and near springs. — Caucasus; Centr. Asia: T. Sh., Pam.-Al., Amur D., Syr D., Mtn. Turkm. **Gen. distr.**: Bal.-As. Min., E. Med., Arm.-Kurd., Iran., Dzu.-Kash. Described from Palestine. Type in London.

Note. A composite East Mediterranean species, associated with mountainous habitats. Described by Strand and accepted by Linnaeus and later authors (Moench, Meth. Pl. (1794) 327; Willd. Sp. pl. (1805) 954; Weddel in DC. Prodr. 16, 1 (1869) 235, 43; Boiss. Fl. Or. IV (1879) 114), this species has been confused with P. *officinalis* although it can be easily distinguished by its exclusively hermaphrodite flowers. A number of species of this cycle were described by Blume Mus. bot. lugd.-bat. II (1856) 250—251 and C. Koch, Linnaea 22 (1849) (605—607), but the bulk of them are concentrated in the Caucasus and in Soviet Central Asia, and have not so far been described.

Subgenus **FREIREA** (Gaudich.) Kom. comb. nova. — *Freirea* Gaudich. Bot. in Freyc. Voy. Uran. (1826) 502, pro gen. Flowers bisexual or pistillate; none staminate; perianth simple, 4-parted, accrescent or unchanging after flowering; stamens 4; ovary superior, free, 1-locular, in pistillate flowers with sessile capitate stigma; inflorescence a dichasium, abbreviated or growing out into an apparent bostyx; basal flowers hermaphrodite, the upper pistillate, or occasionally some inflorescences with exclusively pistillate flowers; achene commonly falling together with the surrounding perianth. Annuals with alternate entire exstipulate leaves, covered with simple or glandular capitate hairs, but without stinging hairs.

4. **P. chersonensis** (Lang et Szov.) Dörfler, Herb. Norm. (1898) No. 3581. — *Parietaria lusitanica* β *chersonensis* Lang. et Szovits, Dritte Beibl. z. Flora 1 (1827) 67. — *Freirea chersonensis* (Dörfler) Jarm.

h. l. — Exs.: Lang. et Szovits, Herb. ruth. Cent. I, No. 67 (typus); Dörfler l. cit.; Adamovič Pl. Balc. exs. (sub nom. Parietaria serbica); Degen, Pl. Banatus exs. a. 1897; Degen, Fl. exs. Austro-Hung., No. 3867.

Annual; stems slender, 5—30 cm long, erect, pubescent, sometimes branched from the middle; leaves ovate to ovate-lanceolate, cuneate at base, acute or acuminate or obtusish; inflorescences many-flowered; bracts oblong or oblong-lanceolate, usually longer than perianth; perianth of hermaphrodite flowers scarious, not changing after flowering; perianth of pistillate flowers becoming firm, brown, with navicularly valvate oblong-lanceolate segments, scabrous, 1.5—2 mm long; achene ovate, ca. 1 mm long. May—August. (Plate XXII, Figure 3a—d).

Limestone and chalk rocks, clayey and sandy taluses, and bluffs; in the
402 shade of stones, on damp gravel, in gullies and ravines. European part: U. Dnp., M. Dnp., Bl., L. Don, Crim.; Caucasus: Centr. Asia: Mtn. Turkm., T. Sh., Balkh., Syr D., Amu D., Pam.-Al. **Gen. distr.**: Bal.-As. Min., Iran.

Note. P. chersonensis is a geographical vicariad of the West Mediterranean P. lusitanica L. from which it differs in the erect (not bushy-spreading) stems, the oblong-lanceolate perianth segments, the bracts commonly exceeding the perianth (especially in Caucasian specimens), and the more ovate-lanceolate shape of the leaves.

5. **P. micrantha** Ldb. Ic. pl. alt. 1 (1829) 7; Fl. Alt. IV (1833) 303. — Ic.: Ldb. Ic. pl. alt. (1829) I, tab. 22. — Freirea micrantha (Ldb.) Jarm., h. l.

Annual; stems slender, delicate, glabrous or sparingly pubescent; leaves thin, filmy, ovate to broad-ovate or triangular-cordate, rounded or cordate or rarely rounded-cuneate at base, obtuse or subacute at apex, sparingly hairy, the petioles long and slender; inflorescences 3—5-flowered, with slender branches; bracts linear-subulate, equaling or almost equaling the flowers, partly covered with capitate hairs; perianth of all flowers scarious and unchanging, in pistillate flowers campanulate, commonly shorter than achene, 1 mm in length. June—September. (Plate XXII, figure 4).

Granite cliffs, stone fissures, damp and shaded crevices, near springs, and in caves. — European part: S. Urals in V.-Kama; Caucasus; W. Siberia: U. Tob., Irt., Alt.; E. Siberia: Ang.-Say., Dau., Ze.-Bu. (S.), Uss.; Centr. Asia: throughout, except desert. **Gen. distr.**: Bal.-As. Min., Arm.-Kurd., Dzu.-Kash., Ind.-Him., Iran., Mong., Jap.-Ch. Described from Altai. Type in Leningrad.

6. **P. alsinaefolia** Delile, Bot. Egypt. (1817) 137. — Freirea alsinaefolia (Delile) Gaudich. Bot. in Freyc. Voy. Uran. (1826) 502. — Ic.: Delile, l. cit. tab. 50, f. 2.

Annual; stems erect or ascending, 15—20 cm long, simple or rarely
405 branched; leaves orbicular-ovate or broad-ovate, rounded-cuneate at base, pubescent, the petioles long and slender; inflorescence with 1 oblong basal bract and large scarious accrescent lateral bracts, in 3's around each flower, covered with simple hairs; perianth of hermaphrodite flowers becoming firm and brown after flowering, elongate-pyriform, half as long again as the achene; perianth of pistillate flowers also turning brown, pyriform, equaling the achene, covered with capitate hairs. June—July.

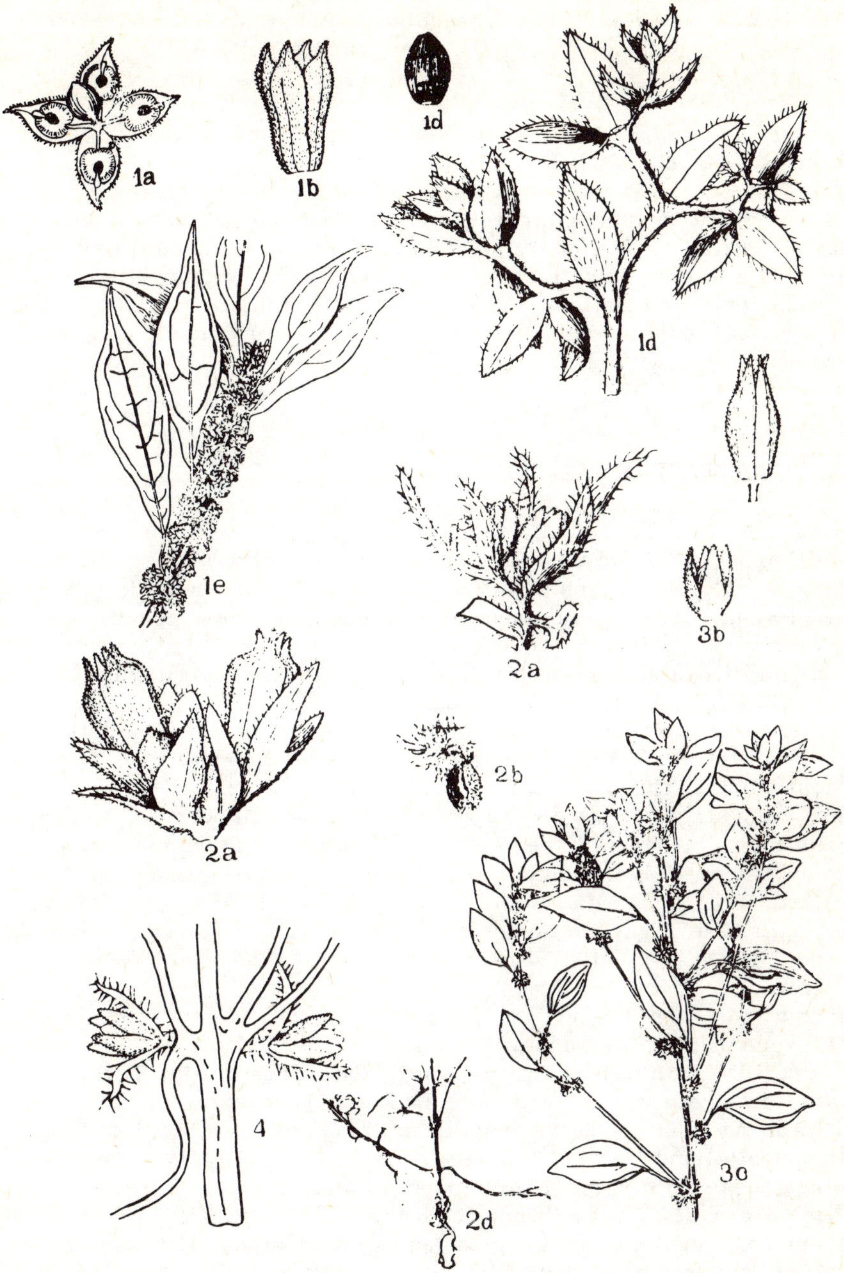

PLATE XXII. 1. Parietaria erecta et Koch., a) hermaphrodite flower at the onset of flowering; b) fruiting perianth of a hermaphrodite flower; c) achene; d) inflorescence; e) portion of stem with leaves and inflorescence; f) perianth of staminate flower after flowering. — 2. P. officinalis L., a) inflorescence; b) pistillate flower. — 3. P. chersoniensis Dörfl., a) inflorescence; b) fruiting perianth of a pistillate flower; c–d) plant aspect. — 4. P. micrantha Ldb., axillary inflorescence with hermaphrodite flowers.

Ravines. — Centr. Asia: Mtn. Turkm., found by D. I. Litvinov in the Saandak Ravine near Ashkhabad. **Gen. distr.**: Egypt, Med., Arm.-Kurd., Arabia, Iran. Described from Egypt. Type in Montpellier.

Order *Proteales ENGLER.

Flowers cyclic, monochlamydeous, mostly 4-merous, bisexual or unisexual, regular or irregular; perianth corolloid; stamens rarely free, mostly adnate to the perianth segments with only the anthers distinct; ovary superior, unicarpellate. Trees or shrubs with alternate undivided or pinnately cut leaves, without stipules; flowers in racemes or spikes; embryo with 3 — 8 cotyledons.

Family *PROTEACEAE J. ST. HIL.

This family was undoubtedly widely distributed in the Tertiary era over the parts of Europe occupied by evergreen flora (the Poltavian region of A. Kristofovich). Representatives of the family have been definitely identified by the presence of seeds in Pliocene formations of the Netherlands. The genera recorded for the USSR are: Banksia, Dryandra, Lomatia, Grevillea, Persoonia, Palibinia, Protea, Hakea, and Embothrium.

Banksia agastachoides Schmalh. in the Oligocene of M. Dnp. (Ekaterinopol'e). — B. Deickeana? Heer in the Oligocene of U. Dnp. (Volyanshchina, Ryzhany). — B. helvetica Heer in the Oligocene of U. Dnp. (Volhynia). — B. rossica Schmalh. in the Oligocene of M. Dnp. (Ekaterinopol'e). — Banksites integer Heer in the Oligocene of U. Dnp. (Ryzhany, Volyanshchina). — Dryandra Brongniartii Etting. in the Eocene of M. Dnp. (Lavy in Kursk Region). — D. Ungeri Etting. in the Aquitanian Oligocene layers of Ar.-Casp. (Kara-sandyk; Yar-kue).

Note. The last-mentioned plant more likely belongs to the family Myricaceae.

Dryandroides lignitum Unger? in the Oligocene (Tongrian stage) layers of Bl. (Kremennaya).

Dryandra Schrankii Etting. in the Oligocene of Mtn. Turkm. (Er-oilan-duz); similarly species of Palibinia Korov.

Hakea myrtilloides Schmalh. in the Oligocene of M. Dnp. (Ekaterinopol'e) and U. Dnp. (Ryzhany). — H. spathulatha Schmalh. in the Oligocene of M. Dnp. (Ekaterinopol'e) and U. Dnp. (Volhynia).

Lomatia ucrainica Schmalh. in the Oligocene of M. Dnp. (Ekaterinopol'e) and U. Dnp. (Volyanshchina, Ryzhany). — L. sp. in the Oligocene-Eocene formations of Irt. (south of Pavlodar).

Palibinia laxifolia E. Korov, P. densifolia E. Korov., P. acutifolia E. Korov., P. lanceolata E. Korov., and P. latifolia E. Korov in the Oligocene of Mtn. Turkm. (Er-oilan-duz).

Note. Described previously as Dryandra Schrankii Etting.

406 Order 16. **Santalales** LINDL.

Flowers cyclic, monochlamydeous, rarely dichlamydeous; ovary inferior, 2- or 3-carpellate, rarely monocarpellate; fruit 1-seeded.

1. Flowers monochlamydeous with a cup-shaped receptacle; ovary inferior, unilocular, with central bearing 1—3 (rarely 4 or 5) ovules without integuments; fruit a nut; seeds solitary, containing endosperm; herbs or shrubs with spiral or opposite leaves; mostly green hemiparasites, with roots and small flowers. . . . . Family XLIX. **Santalaceae** R. Br.

\+ Flowers monochlamydeous with a cup-shaped receptacle fully united with the carpels, the margin of the receptacle often projecting above the base of the perianth as an irregular or toothed rim; the inner layer of the receptacle becomes slimy and viscid in fruit; endosperm more or less developed; shrubby or rarely herbaceous plants becoming attached to branches of trees by means of haustoria; green hemiparasites, mostly with fully developed leaves . . . . Family XLVIII. **Loranthaceae** D. Don.

Family XLVIII. **LORANTHACEAE** D. DON.*

Flowers regular or slightly zygomorphic, hermaphrodite or unisexual; flower axis cupuliformly enlarged, in pistillate flowers united with ovary; perianth simple, of 2—2, 2—3, or 3—3 distinct or connate lobes; stamens as many as perianth lobes, opposite and sometimes adnate to them; ovary embedded in the receptacle, placenta and ovules mostly not differentiated, the ovary containing 1—3 embryo sacs; fruit united with receptacle into a false berry, its inner layer becoming slimy and viscid; seeds mostly with endosperm, the embryo with 2 or rarely 3—6 cotyledons. Commonly (USSR) parasitic plants with green, sometimes scalelike leaves, living on trees or shrubs.

Key to Genera

1. Leaves large, elliptic-lanceolate . . . . 2.
\+ Leaves scalelike . . . . 383. **Razoumowskia** Hoffm.
2. Leaves persistent; perianth lobes and stamens 4 . . . . 384. **Viscum** L.
\+ Leaves deciduous; perianth lobes and stamens 6 . . . . 382. **Loranthus** L.

407 Genus 382. **LORANTHUS**** L.

Gen. pl. ed. 5 (1854) 154.

Flowers small or fairly large, in a paniculate or spikelike inflorescence, commonly hermaphrodite, rarely unisexual; the short enlarged cup-shaped receptacle truncate or toothed; perianth lobes 4—6, valvate, free or connate at base; filaments of stamens adnate at base or higher up to the

* Treatment by B. A. Fedchenko.

** From "loron," strap, and "anthos," flower.

base of perianth lobes; anthers immobile, dorsobasifixed; style filiform or twisted, with obtuse capitate terminal stigma; fruit a berrylike pseudocarp with a juicy middle layer; embryo surrounded by copious endosperm. Green parasites on broad-leaved or coniferous trees; leaves opposite or alternate, the usually distinct nervature pinnate or consisting of 3—5 longitudinal veins.

1. L. **europeus** Jacq. Enum. Stirp. 55 (1762) 230; Ldb. Fl. Ross. II, 81 (excl. loco natali). — Ic.: Jacq. Fl. Austr. I, 30. — Exs.: Fl. Hung. exs. No. 31.

Shrubby, 10—30 cm, glabrous, dichotomously branched; leaves deciduous, oblong-oval obtuse, distinctly pinnatinerved; flowers small in small spikelike inflorescences, yellowish-green, with 6 perianth lobes and 6 stamens; fruit berrylike, oblong-ovaloid, yellowish. April—May. (Table XXIII, Figure 2a, b).

European part: M. Dnp., Crim. **Gen. distr.**: Med., Eur., Bal.-As. Min. Described from Austria. Type in Vienna.

Note. Parasitic exclusively on oaks.

Pallas's report (Puteshestvie, pt. I, 1809, p. 546), repeated by Ledebour (l. c.) concerning the occurrence of this species on the Ural River near Mergenev is clearly due to a misunderstanding. Apparently Pallas mistook for Loranthus the so-called "witches'-broom" on species of Salix L.

Genus 383. **RAZOUMOWSKIA*** HOFFM.

Hoffm. Herb. Mosq. Introd. No. 1 (1808) fig. 1a—i. — Arceutobium M. B. Fl. taur.-cauc. II (1819) 629.

Flowers solitary in the leaf axils, unisexual, dioecious; staminate flowers with a 2—5-parted perianth; anthers sessile; pistillate flowers with 2-parted perianth; pseudo-ovary ovoid; style short, conical; stigma obtuse; pseudocarp berrylike, ovoid, crowned by remnants of the perianth; limb of perianth separating in maturity by a transversal split, causing the fruit to fall out; embryo embedded in copious endosperm. Small, much branched parasites; branches becoming lignified; leaves small, scalelike, in pairs.

408

1. R. **oxycedri** (DC.) F. Schultz, Herb. Norm. (1853) No. 1219. — Arceutobium oxycedri (DC.) M. B. Fl. taur.-cauc. III (1819) 629; Ldb. Fl. Ross. II, 380; Shmal'g., Fl. I, 430. — R. caucasica Hoffm. l. c. — Viscum oxycedri DC. in Lam. et DC. Fl. Fr. ed. 3, vol. IV (1815) 274. — Ic.: Rchb. Ic. Fl. Germ. XXIV, tab. 141. — Exs.: Herb. Fl. Cauc. No. 213; Fl. Hungar. exs. No. 32.

Shrubby, 2—20 cm long, glabrous, densely branched, articulate; leaves small, slightly sheathing; pseudocarp azure-blue.

Parasitic on junipers (Juniperus oxycedrus L., J. excelsa M. B., J. turcomanica B. Fedtsch., J. semiglobosa Rgl., and J. seravschanica Kom.). (Plate XXIII, Figures 4a—d).

* Named for Count Rasoumovsky (1748—1822), founder of the Gorenki Garden.

(409)

PLATE XXIII. 1. Cytinus rubra (Fourr) Kom., a) staminate flower; b) pistillate flower. — 2. Loranthus europaeus Jacq., a) flower; dissected fruit. — 3. Viscum album L., a) inflorescence; b) flower; c) dissected fruit; d) site of attachment to host plant; e) leaves of mistletoe from an almond tree; f) same from spruce; g) same from pine. — 4. Arceutobium oxycedri (DC.) M.B.; a) staminate flower; b) same in longitudinal section; c) female flower; d) same in longitudinal section.

Genus 384. **VISCUM*** L.
L. Gen. pl. (1737) 193; ed. 5 (1754) 448.

Green parasites penetrating with their branched roots under the bark of tree branches and then sending out haustoria into the wood; branches opposite, becoming lignified; flowers crowded in the axils of ramifications, monoecious or dioecious; perianth of staminate flowers with a short tube, the lobes connate with the anthers, the anther cells opening on the proximal side by numerous pores; pistillate flowers with a 3- or 4-parted limb; stigma thick, pulvinate; pseudocarp juicy, containing 1 or several seeds embedded in nutritive tissue; inner part of receptacle and fruit become transformed into a slimy viscid mass called viscin. Russian: "omela."

1. Fruits white . . . . 1. **V. album** L.
\+ Fruits yellow or orange . . . . 2. **V. coloratum** (Kom.) Nakai.

411 1. **V. album** L. Sp. pl. (1753) 1029; Ldb. Fl. Ross. II, 360. — V. nervosum Andrz., sec. Rogovich, Obzor rast. Kievsk. uch. okr. (1869) 252. — Ic.: Rchb. Ic. Fl. Germ. XXIV, 139, 140.

Perennial, spherically branched, 20 — 120 cm in diameter, with lignified branches; leaves thick, commonly indistinctly nerved, oblong-oval to oblong, obtusish; flowers in 5's or 6's, yellowish-green; fruit globular, to 9 — 10 mm in diameter, white; seeds flat-angled (var. platyspermum Keller) or convex-angled (var. austriacum Beck. = V. austriacum Wiesb.). March — April. (Plate XXIII, Figure 3a — g).

Parasitizing on broad-leaved trees (poplar, maple, birch, lime, elm, pear, apple, oak), rarely on conifers (pine), sometimes causing considerable damage, especially to fruit trees.

European part: U. Dnp., M. Dnp., V.-Don, Bl., Crim.; Caucasus: Cisc., S. and E. Transc. **Gen. distr.**: Scand., Atl. and Centr. Eur., Med., Bal.-As. Min., Ind.-Him., Jap.-Ch. Described from Europe. Type in London.

Note. European mistletoe lives parasitically on trees of different genera and, in this connection, a number of physiological races can be distinguished. Thus mistletoe of broad-leaved trees usually has broader leaves, translucent white and often somewhat abbreviated fruits, while the number of seeds fluctuates from 1 to 4. Mistletoe on fir mostly has elongated fruits; on pine, its fruits are often somewhat yellowish, smaller than on fir, with a tendency to oblongness, and the leaves are narrower. Both races associated with the conifers mentioned were separated by Wiesbauer into a distinct species — V. austriacum Wiesb. in Gener. Doublett. Verz. Schles. Tauschvereins (1882 — 83), the same in Deutsche Bot. Monatschr. (1884) 60 and Oesterr. Bot. Zeitschr. (1888) 183.

Economic importance. As a parasite of valuable trees, mistletoe fairly often causes considerable damage, especially in orchards. On the other hand, it has value as a medicinal plant and as feed for domestic animals and birds. The fruits are used for extraction of so-called bird glue (Vogelleim in German).

2. **V. coloratum** (Kom.) Nakai in Mori Enum. pl. Kor. (1922) 128.— V. album subsp. coloratum Kom. Fl. Mandsh. II (1903) 107. — V. album

* Ancient Romans used the name "viscum" both for mistletoe and for the glue produced from it.

Maxim. in Bull. Acad. Pétersb. XV, 1135, non L. — Ic.: Kom. and Alis., Opr. rast. Dal'nevost. kr. I, Plate 140. Japanese: yadorighi; Chinese: dun-tsin; Goldi: poktsolya.

Perennial; very closely related to the preceding species, but readily distinguishable by the color of its mature fruits, viz., yellow (var. lutescens Mak.) or reddish-orange (var. rubroaurantiacum Mak.). — Far East: Uss. Gen. distr.: Jap.-Ch. Described from the Amur. Type in Leningrad.

412 Family XLIX. **SANTALACEAE** R. BR.*

Flowers bisexual, regular; perianth united with the disk, simple, more or less tubular, 4- or 5-lobed; stamens as many as anteposed, and inserted at the base of the perianth lobes; ovary inferior, unilocular; flowers mostly solitary in racemiform or paniculate inflorescences; fruit nutlike, nerved, with persistent perianth; leaves sessile, undivided, exstipulate. USSR herbs perennial.

Genus 385. **THESIUM**** L.
L. Gen. pl. ed. V (1754) 97.

Flowers hermaphrodite; perianth greenish outside, funnel-shaped, cup-shaped, or rarely claviform (with long tube and limb), 4- or 5-lobed; disk indistinct; stamens inserted at the base of perianth lobes; anthers oblong, dehiscent by a longitudinal slit; ovary inferior; style filiform; stigma capitate or often 3-parted; nutlet globular or ellipsoid, ribbed with longitudinal nerves or reticulate; flowers sessile or short-pediceled bearing 3 bracteoles or rarely 1 bracteole; the paired lateral bracteoles short, the odd lower one usually much longer; leaves linear, lance-linear, rarely broad-lanceolate, the upper leaves transformed into unpaired bracteoles. Perennial mostly hemiparasitic herbs. Russian "lenets."

1. Flowers 15—18 mm long; perianth claviform, with long tube and 5 short lobes; nutlet 8—10 mm long; leaves broad-lanceolate . . . . 1. T. Minkvitzianum B. Fedtsch.
\+ Flowers not more than 10 mm long; perianth campanulate, infundibular, or cupuliform, 4- or 5-lobed; nutlet at most 7 mm long; leaves narrow-lanceolate or linear . . . . 2.
2. Perianth 4-lobed, the persistent portion of the perianth as long as or slightly longer than the nutlet; leaves linear . . . . . . 21. T. alpinum L.
\+ Perianth 5-lobed . . . . 3.
3. Nutlet longitudinally nerved, lateral nerves none or ascending but not transversal . . . . 8.
\+ Nutlet longitudinally nerved, reticulate with ascending and transversal lateral nerves connecting the longitudinal ones . . . . 4.

* Treatment by E. G. Bobrov.

** A plant name used by Pliny, probably derived from the name of the mythological hero Theseus.

4. Persistent portion of the perianth one-fourth to one-third the length of nutlet; flowers ca. 2 mm long; perianth rather flat, hemispherically bowl-shaped . . . . 5. **T. procumbens** C. A. M.

413 + Persistent portion of the perianth about as long as the nutlet; perianth deeply lobed, more or less funnel-shaped . . . . 5.

5. Nutlet globular, 2 — 2.4 mm long; stems glabrous, erect, not prostrate; leaves thin, not fleshy . . . . 6. **T. chinense** Turcz.

\+ Nutlet ca. 4.5 mm long; leaves, peduncles and bracteoles more or less fleshy . . . . 6.

6. Glaucous, glabrous plants; stem erect; inflorescence paniculate, its branches straight, spreading, dichotomously branched; peduncles bearing 2 or 3 (5) bracteoles . . . . .4. **T. Szovitsii** DC.

\+ Dark green plants, with only the stems sometimes light-colored; stems prostrate or ascending, very rarely suberect; plants scaberulous in upper part . . . . 7.

7. Perianth lobed to the middle; nutlet short-stalked; one peduncle often bearing 2 or 3 flowers; plants usually suffrutescent in appearance . . . . 2. **T. Kotschyanum** Boiss.

\+ Perianth lobed to one-third; nutlet sessile; peduncles commonly bearing a single flower . . . . 3. **T. maritimum** C. A. M.

8. Plants with creeping rhizomes or with a long underground portion of stem . . . . 9.

\+ Taproot straight, the crown producing numerous stems, these more or less crowded at base . . . . .12.

9. Bracteoles 3, the lateral pair usually shorter than the lower unpaired bracteole . . . . .10.

\+ Bracteole 1, unpaired, large; lateral bracteoles . . . . 8. **T. ebracteatum** Hayne.

10. Bracteoles rough on the margin and on the back, the lateral ones about as long as the flower, the unpaired somewhat longer; inflorescence paniculate . . . . 7. **T. linifolium** Schrank.

\+ Inflorescence a simple raceme (very rarely apparently branched because of stem ramifications); unpaired bracteole several times as long as the flower . . . . 11.

11. Rhizome slenderly branched; fruit-stem fleshy, rugose, 2 — 3 mm long . . . . 9. **T. repens** Ldb.

\+ Underground portion of the stem to 15 cm long; rhizome none; nutlets subsessile; leaves usually borne 1-sidedly . . . . .10. **T. alatavicum** Kar. et Kir.

12. Stems prostrate to ascending, simple; lower leaves scalelike; flowers

414 broadly infundibular, ca. 3 mm long and 3 mm broad; inflorescence a 1-sided raceme; peduncles not longer than the flowers . . . . 20. **T. brachyphyllum** Boiss.

\+ Stems erect or declinate (erect only in one form of T. ramosum); lower leaves sometimes shorter than the middle ones but not scalelike; inflorescences not 1-sided . . . . .13.

13. Peduncles not longer than flowers; fruiting peduncles appressed to rachis; lobes of the persistent perianth portion somewhat inflexed only at the tips; the linear leaves ascending and more or less appressed to the stem . . . . .14. **T. longifolium** Turcz.

\+ Peduncles on the middle part of the rachis considerably (sometimes several times) longer than the flowers; fruiting peduncles spreading or even recurved . . . . .14.

14. Perianth tube slightly accrescent and becoming indurated in fruit, hence the oblong-ellipsoid nutlet (ca. 7 mm long) apparently constricted at the base of the indurated part of the tube; stems simple . . . . 12. **T. rupestre** Ldb.

\+ Nutlet ellipsoid, not constricted; perianth tube not accrescent and not indurated; stems branched . . . . 15.

15. Fruiting peduncles strongly spreading or even recurved; inflorescence branches flexuous; stems branched only in upper part; plants glabrous (except the form hirtulum), only the upper leaves and bracteoles scabrous; flowers 3—4 mm long . . . . 11. **T. refractum** C. A. M.

\+ Fruiting peduncles ascending, not reflexed; inflorescence branches not flexuous . . . . 16.

16. Persistent portion of the perianth about as long as the nutlet; stems branched in middle part; flowers 5—7 mm long . 13. **T. saxatile** Turcz.

\+ Persistent portion of the perianth one-fifth to one-third the length of the nutlet . . . . 17.

17. Tufted plants; stems stiffly erect, glabrous, branched in upper part; branches upright, diverging but little from the stem, bearing simple racemes; leaves not distinctly veined; nutlet oblong-ellipsoid, ca. 7 mm long and 2 mm in diameter . . . . 15. **T. multicaule** Ldb.

\+ Plants many-stemmed, not tufted; stems erect or slightly declinate, very rarely almost prostrate; branches not appressed to stem . . . . 18.

18. Lateral bracteoles scarcely half the length of the flower, the middle unpaired bracteole twice as long and about equaling the flower; very strongly branched plants; fruit-stalk 2—3 mm long . . . . 19. **T. ramosissimum** Bobr.

415 + Lateral bracteoles about as long as the flower; the unpaired middle bracteole 2—3 times the length of the flower; nutlet borne on a stalk ca. 1 mm long or almost sessile . . . . 19.

19. Plants glaucescent; stems branched at the ends; inflorescence paniculate; leaves 2—6 cm long and 3—7 mm broad, rather distinctly 3—5-nerved (especially beneath); nutlet 3 mm long . . . . 17. **T. ferganense** Bobr.

\+ Plants dark green, sometimes brownish; stems branched from base and in upper part; inflorescences simple, racemiform; leaves shorter and narrower, 1—3-nerved, not always distinctly; nutlet large . . . . . 20.

20. Perianth 2—3 mm long; nutlet ca. 3.5 mm long; pedicels 0.4—1 cm long; upper part of stem, leaves, and peduncles rather densely scabrous; leaves linear, to 3 cm long and to 2.5 mm broad . . . . 16. **T. ramosum** Hayne.

\+ Perianth ca. 4 mm long; nutlet 4.5 — 5 mm long; peduncles 1.5—4 cm long; bracteoles very finely scaberulous; lower leaves to 4 (5) cm long and to 4 mm broad . . . . 18. **T. Gontscharovii** Bobr.

Section 1. CHRYSOTHESIUM Jaub. et Sp. III. pl. or. (1842) t. 104. — Flowers with a long tube and a short 5-lobed limb (claviform), solitary in the axils of bracteoles, in a simple racemiform inflorescence.

1. **T. Minkwitzianum** B. Fedtsch. in Not. Syst. ex Herb. H. B. P. IV (1923) 113.

Perennial; root rodlike, woody (?), many-headed; stems numerous, more or less erect, 20—30 cm long, to 5 mm in diameter, ribbed, glabrous, densely leafy, few-branched from the middle; leaves thick, coriaceous; the cauline broadly lanceolate, 15 mm long, 5—10 mm broad, rather distinctly 5—7-nerved, especially beneath; leaves at the end of the stem and those of the branches lanceolate or linear-lanceolate, to 20 mm long and 2—5 mm broad, 1—3-nerved; inflorescence a raceme, essentially unbranched, few-flowered, crowned with sterile bracts; peduncles 5—10 mm long, ascending, more or less incurved; bracteoles 3, about equaling the flower, 15—18 mm long, the lateral linear-lanceolate, 3-nerved, the middle one broadly lanceolate, 5—7 nerved; flowers yellowish, ca. 18 mm long, almost sessile or borne on a very short pedicel; perianth long-tubular, cut at the end to 3 mm into 5 lobes; nutlet ellipsoid, subsessile, 8—10 mm long and 6 mm in diameter, alveolate by a dense network of branched subsidiary nerves arising from 5 longitudinal ribs; persistent portion of the perianth as long as the nutlet. Fr. end of May. (Plate XXIV, Figure 14).

416 Slopes of low foothills. — Centr. Asia: T. Sh. (low foothills along the Chimkent highway, between Dzhilga and Ak-Dzhar stations, vicinity of Kel'te-Mashat, slopes of Mount Myng-Chupur, 1,000 m).

Described from the foothills near Dzhilga station. Type in Leningrad.

Note. Besides the USSR species, this section includes two other species distributed through the eastern part of Hither Asia: T. stelleroides Jaub. et Sp. and T. aureum Jaub. et Sp. (both from the Euphrates).

Section 2. EUTHESIUM DC. Prodr. XIV (1857) 639. — Perianth campanulate, funnel-shaped, or cup-shaped, 5- or rarely 4-lobed; inflorescence rarely racemiform, mostly paniculate.

Cycle A. **Reticulata** Bobr. — Nutlet longitudinally nerved, reticulate by transverse and oblique subsidiary nerves.

Series 1. Macrantha Bobr. — Peduncles 1—3 (5)-flowered; perianth deeply lobed; persistent portion of the perianth as long as or even exceeding the nutlet. Beside USSR species, the section contains the following species of Hither Asia: T. macranthum Fenzl, T. crassifolium Hauskn., and T. impressum Steud.

2. **T. Kotschyanum** Boiss. Diagn. ser. 1, VII (1846) 86. — T. impressum β Kotschyanum Boiss. Fl. Or. IV (1879) 1066.

Perennial; root straight, woody, branched at the top, hence sometimes apparently rhizomatose; stems many, to 20 cm long, prostrate and ascending, rarely erect, sparingly branched, curved, scaberulous in upper part; leaves somewhat fleshy, scabrous especially on the margin, the lower scalelike and to 5 mm long, the linear middle ones acuminate and 2.5 cm long, the upper ones transformed into unpaired bracteoles; leaves of prostrate stems more or less 1-sided; inflorescence a slender scantily branched few-flowered raceme; peduncles about half the length of the flowers, these borne singly or up to 3 on the peduncle; bracteoles commonly 3, the lateral one-third to one-half the length of the flower, the middle one as long as or slightly longer than the flower; on 3-flowered peduncles the middle flower sometimes ebracteolate; flowers 8—10 mm long, pedicellate, cut to the middle, the lanceolate lobes inflexed at the tips, scaberulous outside; nutlet 4—5 mm long, short-stalked, globular-ellipsoid, reticulate with subsidiary branched nerves connecting the longitudinal ribs; persistent portion of the perianth as long as or slightly longer than the nutlet. Fl. May; fr. June. (Plate XXIV, Figure 3).

Gravelly mountain slopes in the steppe region. — Centr. Asia: Mtn. Turkm. (central Kopet Dagh). **Gen. distr.**: Iran. Described from S. Iran (Shiraz Mountains). Cotype in Leningrad.

417 3. **T. maritimum** C. A. M. Verzeichn. Pfl. Cauc. (1831) 40; Ldb. Fl. Ross. III, 2, 542; Boiss. Fl. Or. IV, 1066.

Perennial; root straight, sparingly branched, many-headed; stems 10—20 cm long, prostrate and ascending, rarely erect, branched in middle part, angled, sometimes minutely scaberulous; leaves 1—2 cm long and to 3 mm broad, fleshy, sometimes almost triquetrous, rarely with a more or less distinct single nerve, acuminate, rough-margined; inflorescence a 2-sided few-flowered raceme; peduncles fleshy, half as long as flowers; bracteoles 3, fleshy, in upper flowers subequal, in lower ones the ovate scabrous acuminate middle bracteole much longer, the lateral bracteoles about as long as the nutlet; flowers sessile, to 8 mm long, scaberulous outside, cut to one-third or more into lanceolate lobes; nutlet ca. 4.5 mm long, sessile, scabrous, prominently and coarsely reticulate with subsidiary nerves connecting the longitudinal ribs; persistent portion of the perianth as long as or even exceeding the nutlet, the perianth lobes inflexed at the tips. Fl. April— May; fr. June.

Coastal sands. — Caucasus: E. Transc. (Baku district — Sumgait, Kalyazi). Endemic. Described from the village of Kalyazi. Type in Leningrad.

4. **T. Szovitsii** DC. Prodr. XIV (1857) 649; Boiss. Fl. Or. IV, 1066.

Perennial; root straight, giving rise to 2 or 3 stems; stems 20—45 cm long, divaricately branched from the middle, sulcate, glabrous, erect; leaves linear, 2—6 cm long, somewhat fleshy, obscurely 1-nerved, glabrous, glaucescent, acuminate; inflorescences paniculate, the branches straight, divergent from rachis, at the ends often dichotomously branching; peduncles 2- or 3- (5)-flowered, as long as or slightly longer than the flowers; lateral bracteoles shorter than the flower; middle bracteole as long as or slightly longer than the flower; on several-flowered peduncles the number of bracteoles sometimes less than 3; flowers to 8 mm long, pedicellate; perianth cut nearly to base, the 5 narrowly lanceolate lobes inflexed, overtopping the anthers; nutlet short-stalked, ellipsoid, ca. 4.5 mm long, reticulate; persistent portion of the perianth promptly caducous, about as long as the nutlet, with inflexed lobes. Fl. June; fr. July. (Plate XXIV, Figure 11).

Saline argillaceous deserts. — Caucasus: S. Transc. (Nakhichevan), E. Transc. (Baku district, Kazakh). Endemic. Described from Nakhichevan. Type in Leningrad.

Series 2. **Procumbentia** Bobr. — Flowers solitary, small, mostly cup-shaped; persistent portion of the perianth one-fourth to one-third the length of the nutlet. Beside USSR species, this section contains T. B e r g e r i Zucc. of the Balkans and Asia Minor.

418 5. **T. procumbens** C. A. M. Verzeichn. Pfl. Cauc. (1831) 40; Ldb. Fl. Ross. III, 2, 542; Bordz. in Protok. Kievsk. o-va for 1913 (1914) 24. — T. d i f f u s u m Andrz. ex Besser in DC. Prodr. XIV (1857) 644. — T. d i f f u s u m Andrz. Kievsk. Univers. Izv. No. 7 (1862) 137. — T. r a m o s u m Ldb. Fl. Ross. III, 2, 541, quoad pl. Andrz. e Podolia.

Perennial, with a fusiform taproot, in gravelly soils sometimes rhizomatose; stems numerous, 10—25 cm long, prostrate, 1-sidedly leafy, often branched, sulcate, minutely scaberulous at the ends; leaves linear, to 3 cm long, obscurely 1-nerved, sometimes crowded on short shoots; inflorescence rarely simple, more or less 1-sided, mostly branched, with slender branches; peduncles commonly not longer than the flowers, only in vigorous specimens 2—3 times as long; lateral bracteoles about the length of the flower, the middle one somewhat longer, acuminate, curved, minutely scaberulous; flowers subsessile, ca. 2 mm long, broader than long, 5-lobed, the perianth applanately hemispherical-cupuliform; nutlet subsessile, ovoid, ca. 2.5 mm long, longitudinally nerved, somewhat reticulate with transverse and oblique subsidiary nerves; persistent portion of the perianth one-fourth to one-third the length of the nutlet. Fl. July; fr. August. (Plate XXIV, Figure 13).

Chalks, dry steppe meadows, and gravelly slopes at the upper levels of mountains. — European part: U. Dnp. (SE — former Krolevets County), M. Dnp. (Podolia, Lubny, Belgorod), V.-Don (Voronezh); Caucasus: Cisc. (Greater Caucasus), W., S. (Nor-Bayazet) and E. Transc. (Shemakha). Endemic. Described from W. Caucasus. Type in Leningrad.

Series 3. **Decurrentia** Bobr. — A monotypic group, with a very isolated distribution area. The only known species resembles outwardly T. a l p i n u m L. of the cycle N e r v o s a.

6. **T. chinense** Turcz. in Bull. Soc. Nat. Moscou (1837) No. VII, 157. — Th. B a s n i n i a n u m Turcz. Fl. baic.-dah. II, 2 (1856) 74. — T. d e c u r r e n s Blume ex DC. Prodr. XIV (1857) 652. — T. r u g u l o s u m Bge. in sched. — T. r e t i c u l a t u m Fisch. in sched. — Ic.: Kom. and Alis., Opr. rast. Dal'nevost. kr. I, Plate 138 (habit).

Perennial; with a taproot; stems 2—8, 20—40 cm long, erect, only the lateral ones somewhat declinate, sulcate, glabrous, branched especially in upper part; leaves linear, acuminate, to 3.5 cm long and to 3 mm broad, more or less distinctly 1-nerved; inflorescences mostly paniculate; peduncles commonly longer than flowers; bracteoles 3, scabrous, the middle ones 2—3 times the length of the flower, the lateral ones about equaling the flower; bracteoles and the upper leaves terminating in a

419 cartilaginous point; pedicels 3—5 mm; style not exceeding the stamens; nutlet globular, 2—2.5 mm in diameter; remnant of perianth about as long as or slightly shorter than the nutlet; nutlet ribbed, with distinct fine reticulation. Fl. June; fr. July. (Plate XXIV, Figure 10).

Dry grassy slopes, small meadows, and sparse coppices. — E. Siberia: Dau. (E. — Nerchinskii Zavod); Far East: Ze.-Bu. (Blagoveshchensk), Uss. Gen. distr.: Jap.-Ch. (E. Mongolia, Manchuria, N. China, Japan). Described from N. China. Type in Leningrad.

Cycle B. **Nervosa** Bobr. — Nutlet longitudinally nerved; subsidiary nerves oblique, not transverse, hence the nutlet reticulate-rugose.

Series 1. **Linophylla** Bobr. — Leaves lanceolate or linear-lanceolate; flowers distinctly pedicellate; bracteoles shorter than the flower. Besides USSR species, the series contains the species T. Bavarum Schrank.

7. **T. linifolium** Schrank, Baier. Reise (1786) 129. — T. intermedium Schrad. Spicil. Fl. Germ. (1794) 27; Ldb. Fl. Ross, III, 2, 540, Shmal'g., Fl. II, 403. — T. linophyllon L. Sp. pl. (1762) 301, p. p. — Ic.: Rchb. Ic. Fl. Germ. XI, 546, f. 1160. — Exs.: HFR No. 738.

Perennial, with creeping rhizomes; stems to 40 cm long, arising singly from each rhizome, rarely 2 or 3 together, erect or somewhat declinate, sulcate, glabrous, commonly branched only at the ends; leaves lance-linear or narrowly linear-lanceolate, to 4 mm broad and up to 4 cm long, not always distinctly 3—5-nerved, acuminate, glabrous; inflorescences paniculate, with declinate branches, the lower branches bearing 3—5 flowers, the upper monochromatic and essentially representing peduncles; peduncles commonly 2—3 times the length of the flower, only the upper ones shorter; lateral bracteoles about as long as the flower, the unpaired middle bracteole somewhat longer and surpassing the flower; the bracteoles scaberulous on the margin and on the back; flowers campanulate, to 3 mm long, lobed to the middle or lower down, the lobes sometimes lobulate at base; nutlet ellipsoid, twice the length of the stalk and 3 times as long as the remainder of the perianth; nutlet longitudinally nerved, with 5 of the nerves rather distinct. Fl. May; fr. June. (Plate XXIV, Figure 8).

Forest margins and coppices. — European part: U. Dnp. (S.), M. Dnp., Bl. (N.). Gen. distr.: N. and S. Eur. Described from Centr. Europe.

Series 2. **Repentia** Bobr. — Shade plants of the forest and forest-steppe belts (in the mountains in the forest and scrub zone), with long often creeping rhizomes and usually erect stems.

8. **T. ebracteatum** Hayne in Schrad. Journ. I (1800) 33; Ldb. Fl. Ross. III, 2, 543; Shmal'g., Fl. II, 403. — ? T. monophyllum Gilib. Exerc.
420 phyt. II (1792) 428. — T. comosum Roth, Catal. II (1800) 29. — T. bracteatum Gromov, Enum. Chark. in Tr. Obshch. Nauk Khar'k. Un. I, 143. — Ic.: Schrad. Journ. l. c. t. VII. — Exs.: HFR No. 184; Fl. Polon. No. 254, a, b.

Perennial, with creeping rhizomes; stems ascending, erect at the top, 10—30 cm long, commonly unbranched, glabrous, sparsely leafy, with crowded sterile bracts at the ends; leaves linear, obscurely 3-nerved, glabrous; inflorescence a simple unbranched raceme; flowers short-pediceled, with 1 bracteole; perianth tubular, strongly inflexed at apex; nutlet stalked, ellipsoid, with branched longitudinal nerves, 3 times as long as the persistent perianth portion. Fl. May; fr. June.

Grassy forest glades, forest margins, and coppices. — European part: U. Dnp., U. V., V.-Kama, V.-Don (except SW), Transv.; W. Siberia: U. Tob. (Ob). The northern limit runs along the line Minsk — Moscow — Kirov — Perm. **Gen. distr.**: N. Europe. Described from the vicinity of Berlin. Cotype in Leningrad.

9. **T. repens** Ldb. Fl. Alt. I (1829) 274; Ldb. Fl. Ross. III, 2, 539; Kryl., Fl. Zap. Sib., No. 4 (1930) 812. — T. nanum Patrin ex DC. Prodr. XIV, 2 (1857) 648. — T. pumilum Patrin ex DC. l. c. — Ic.: Ldb. Ic. Fl. Ross. t. 233.

Perennial; rootstock creeping, branched, long, slender, ca. 1 mm in diameter; stems few, solitary, angled, unbranched, 10 — 25 cm long; leaves linear, acuminate, narrowed toward base, obscurely 1-nerved, 1.5 — 4 cm long and to 2 mm broad; inflorescence a simple raceme; peduncles 3 — 4 times the length of the flowers, obliquely ascending, bearing 3 foliaceous bracteoles below the flower; middle bracteole long, sometimes surpassing the peduncle, the lateral bracteoles twice as long as the flower; perianth yellowish-white within, ca. 4 mm long, cut to the middle or lower down into 5 acuminate lobes; stamens two-thirds as long as the perianth lobes; style scarcely exceeding the stamens; nutlet rounded-ellipsoid, to 4 mm long and 2 mm broad, sometimes obscurely 5-ribbed, glabrous, 2 — 3 times as long as the persistent perianth, this with inflexed tips; nutlet borne on a fleshy wrinkled stalk 2 — 3 mm long. Fl. June; fr. August.

Forest belt, in grassy forest glades, forest margins, open woods, rarely inundated meadows; small meadows in the alpine zone; river valleys in mountainous steppe regions. — W. Siberia: Ob (S., to Tara in W.), Irt. (N. — Tyukalinsk, and E.), Alt.; E. Siberia: Yen. (extreme S.), Ang.-Say., Dau. (Baikal, Selenga, Turkinskie Vody). **Gen. distr.**: Mongolia. Described from the vicinity of Ridder [Leninogorsk]. Type in Leningrad.

421

10. **T. alatavicum** Kar. et Kir. in Bull. Soc. Nat. Moscou XV (1842) 445; Ldb. Fl. Ross. III, 2 (1849 — 51). — T. Petrovii N. Pavl. in sched.

Perennial; underground part of the stem pale yellow, very long (to 15 cm) and to 1.5 mm in diameter, hence root apparently creeping; stems 3 or 4 from one root, glabrous, sulcate, commonly simple, rarely branched at the ends, 10 — 40 cm long; leaves linear-lanceolate, 2 — 5 cm long and to 3 mm broad, obscurely 3-nerved, the upper ones acuminate; inflorescence slender, unbranched (very rarely apparently branched due to stem ramifications); peduncles mostly not longer than or at most 2 — 2.5 times as long as the flowers; lateral bracteoles commonly not surpassing the flower, the middle one (sometimes leaflike) 2 — 8 times as long, on the upper peduncles not more than twice the length of the flower; flowers 3 — 4 mm long, the perianth lobes often auriculate on one side and sometimes on both sides; stamens not surpassing the perianth lobes; nutlet appressed to rachis, subsessile, glabrous, ovaloid, to 5 mm long and to 2.5 mm in diameter, the longitudinal nerves not always evident, the persistent portion of the perianth one-fifth to one-fourth the length of the nutlet; leaves and flowers often in apparently 1-sided arrangement. Fl. July; fr. August. (Plate XXIV, Figure 1).

Woods, especially spruce, ravines, more rarely grassy slopes. — Centr. Asia: T. Sh., W. Pam.-Al. Endemic. Described from the alpine zone of the Ala Tau Range, valley of the Baskan River. Type in Leningrad.

Series 3. **Longifolia** Bobr. — A group of the mountain-Siberia speciation center. In E. Siberia, some of them are not adequately delimited. These species grow in the forest-steppe and the steppe (E. part) zones, in sparsely wooded areas, and in mountain meadows. Morphologically they are difficult to separate from the series Ramosa.

11. **T. refractum** C. A. M. in Bong. et Mey. Verzeichn. (1841) 58; Ldb. Fl. Ross. III, 2, 539. — T. ramosum Ldb. Fl. Alt. I (1829) 275, non Hayne. — T. ramosum Turcz. Fl. Baic.-Dah. II, f. 2, 78, non Hayne. — R. Rytschkowii Fisch. ex Trautv. A. H. P. IX (1884) 153 (pl. dahur.). — T. longifolium auct. fl. Or. Extr., non Turcz. — Ic.: Bong. et Mey. l. c. tab. XIII. — Also referred to as Siberian toadflax.

Perennial, with a straight rather stout many-headed root giving rise to several stems; stems 10—45 cm long, sulcate, more or less curved, branched at the ends; leaves linear-lanceolate, narrowed toward base, commonly obtusish, to 5 cm long and to 4 mm broad, the 3 nerves not always evident; inflorescence paniculate; peduncles borne on inflorescence branches, 0.5—1 cm long, commonly much longer than the flowers and fruits; fruiting peduncles spreading or even recurved; rachis often
422 flexuous; bracteoles 3, the middle ones about as long as or slightly longer than the flower, the lateral ones much shorter; flowers ca. 4 mm long; perianth yellowish-white within, ca. 3 mm long, broadly funnel-shaped, cut into 5 acuminate lobes; nutlet ellipsoid, ca. 4 mm long and 2 mm in diameter, 5-ribbed, with longitudinal nerves between the ribs; persistent portion of the perianth one-fourth to one-third the length of the nutlet, borne on a stalk ca. 1 mm long. Fl. July; fr. August.

f. hirtulum Kryl., Fl. Alt. (1909) 1174. — Dwarf plants, the whole plant or merely the stem sparsely setose; peduncles commonly not longer than the flowers; occurring on gravelly slopes.

Lowland steppes, forest margins, mountainous areas, steppes and meadows. — W. Siberia: U. Tob., Irt., Alt.; E. Siberia: Ang.-Say., Dau., Lena-Kol.; Far East: Ze.-Bu., Uda, Sakh., Uss.; Centr. Asia: Dzu.-Tarb., T. Sh. (E.). **Gen. distr.:** Mongolia and Manchuria. Described from Batovskii outpost (Baty in the Zaisan area). Type in Leningrad.

Note. Plants from E. Siberia and from the Far East differ from the type in their larger flowers.

12. **T. rupestre** Ldb. Fl. Alt. I (1829) 277; Ldb. Fl. Ross. III, 2, 542; Kryl., Fl. Zap. Sib., No. 4, 815. — Ic.: Ldb. Ic. Pl. Fl. Ross. III, t. 241.

Perennial, with a stout woody many-headed root; stems 15—20, crowded, sulcate, 10—25 cm long, the central ones erect, the peripheral somewhat declinate, unbranched except for sterile branchlets sometimes arising in lower part; leaves linear, acuminate, narrowed toward base, 1.5 — 3 cm long, 1 — 2 mm broad, with a narrowly scarious-rimmed margin, distinctly 1-nerved, the lateral nerves usually imperceptible; flowers on the upper part of stem in simple racemes; peduncles obliquely ascending, 2 — 3 times as long as the flowers and fruits; bracteoles 3, the middle one about as long as the fruit, the lateral pair about half the length of the middle bracteole; perianth 3.5 — 4 mm long, yellowish-white within, campanulate, cut to about the middle into 5 lanceolate lobes, the perianth tube somewhat accrescent and indurated in fruit; stamens somewhat shorter than the perianth segments; style not surpassing the anthers; fruit oblong-ellipsoid, 7 mm long, 2.5 mm broad, rather distinctly 5-ribbed, rugose between the ribs, slightly narrowed in middle part, with a constriction at the base of the indurated perianth tube, the tube narrowed at apex, the tips of the perianth lobes inflexed; the fruit borne on a stalk not more than 2 mm long. Fl. June; fr. July.

Rocks and gravelly slopes. — W. Siberia: Alt. (near the Chernyi Anui River in the Charysh River valley, near Ridder). Endemic. Described from the Charysh River valley. Type in Leningrad.

13. T. **saxatile** Turcz. ex DC. Prodr. XIV (1857) 640; Turcz. Cat. Baic.-Dah. No. 1001. — T. pratense Turcz. Fl. baic.-dah. II, 2 (1856)
423 79, non Ehrh. — T. refractum auct. Fl. Dah., p. p., non C. A. M.

Perennial, with a straight many-headed root; stems several, 10 — 20 (40) cm long, more or less erect, ribbed, in the middle part divaricately branched; leaves linear, to 3 cm long, rather distinctly 1-nerved, obtusish, only the upper ones acuminate; inflorescence branched in lower part; fruiting peduncles obliquely ascending, about twice the length of the flowers; bracteoles 3, the lateral pair about as long as or sometimes shorter than the flower, the unpaired middle bracteoles slightly exceeding the flower; peduncles and the margin of bracteoles scabrous; flowers 5 — 7 mm long, pedicellate, narrowly funnel-shaped, the perianth lobed to one-third; nutlet ellipsoid-globular, ca. 3 mm long, borne on a stalk 1 — 1.5 mm long, with longitudinal ribs and longitudinal subsidiary nerves between them; persistent portion of the perianth about as long as or slightly shorter than the nutlet. Fl. May — June; fr. July — August. (Plate XXIV, Figure 5).

Plants growing on sands and sandy soils are characterized by particularly vigorous growth; their flowers are ca. 7 mm long; the large prominently ribbed nutlets ca. 6 mm long.

Gravelly mountain slopes and sands. — E. Siberia: Ang.-Say. (SE), Dau. (SW — Selenginskaya Dauriya*); Far East: Ze.-Bu. (Blagoveshchensk), Uss. (W.). **Gen. distr.:** N. Mongolia and Manchuria. Described from the upper reaches of the Oka River (Eastern Sayan Mountains). Type in Leningrad.

Note. This species is fully distinguishable from T. refractum C. A. M. only in its typical forms by the larger flowers and the size of the persistent perianth portion. In some cases it is difficult to distinguish this species from large-flowered T. refractum of the Far East.

* [A mountainous area in Dauria, in the Buryat ASSR.]

14. **T. longifolium** Turcz. Cat. Baic.-Dah. (1838) No. 999; Fl. Baic.-Dah. II, 2 (1856) 78; Ldb. Fl. Ross. III, 2, 541. — T. multicaule Turcz. Cat. l. c., non Ldb. — T. Vlassovianum Fisch. ex Trautv. in A. H. P. IX, 1 (1884) 153.

Perennial with a descending root; stems 2—7, 10—40 cm long, suberect, branched at the ends; leaves linear, to 4 cm long and to 2.5 mm broad, rather distinctly 1-nerved; inflorescence apparently paniculate owing to the ramifications at the end of the stem; peduncles obliquely ascending, short, usually not longer than the flowers and fruits; lateral bracteoles commonly about as long as the fruit, the middle bracteole 2—3 (4) times as long; fruiting peduncles more or less appressed to the rachis; persistent portion of the perianth one-fourth to one-third the length of the nutlet, the perianth lobes inflexed only at the tips. Fl. May—June; fr. end of June. (Plate XXIV Figure 12).

Steppes, dry grassy slopes, and small meadows among coppices and sparse woods. — E. Siberia: Dau. (Shilka River Valley, Verkhneudinsk,
424 Aginskaya Steppe, Chita District, Nerchinsk District, Lena-Kol. (Yakutsk District). **Gen. distr.:** N. Mongolia. Described from the Shilka River valley. Type in Leningrad.

Note. This species is readily distinguishable from T. refractum C. A. M. only at maturity when its short peduncles are appressed to rachis, as opposed to the long spreading or even recurved peduncles of T. refractum.

Series 4. **Caespitosa** Bobr. — Rather compactly cespitose plants, the numerous items with slightly divergent branches only at the ends. The section consists of a single species which stands apart from the rest.

15. **T. multicaule** Ldb. Fl. Ait. I (1829) 276; Ldb. Fl. Ross. III, 2, 541; Kryl., Fl. Zap. Sib. IV, 814. — Ic.: Ldb. Ic. pl. Fl. Ross. III, t. 237.

Perennial with a taproot branched in lower part; stems tufted, 30—60 cm long, strict, angled, glabrous, leafy, in upper part branched, the branches diverging but little, the stem and branches terminating in unbranched racemes; leaves linear, without distinct veins, 2—4 cm long and to 3 mm broad, terminated by a cartilaginous point, the lower ones short and scalelike; flowering and fruiting peduncles 10—15 mm long, obliquely ascending; bracteoles 3, the lateral pair about half the length of the fruit, the middle one usually lower and 2—3 times as long as the lateral; perianth yellow within; nutlet oblong-ellipsoid, ca. 7 mm long and 2 mm in diameter, very short-stalked, with 5 ribs and network of longitudinal nerves; the persistent perianth one-fourth to one-third the length of the nutlet. Fl. June; fr. July.

Steppes, gritty soils, gravelly slopes, and rarely sands. — W. Siberia: U. Tob.?, Irt., Alt. (S.); Centr. Asia: Ar.-Casp. (N.), Balkh. (N.), Dzu.-Tarb. **Gen. distr.:** Mongolia. Described from Bukhtarminsk. Type in Leningrad.

Series 5. **Ramosa** Bobr. — Abundantly branched plants, often scabrous at the top; nutlet prominently nerved, the persistent portion of the perianth one-fifth the length of the nutlet. A group of species of the Mediterranean

(425)

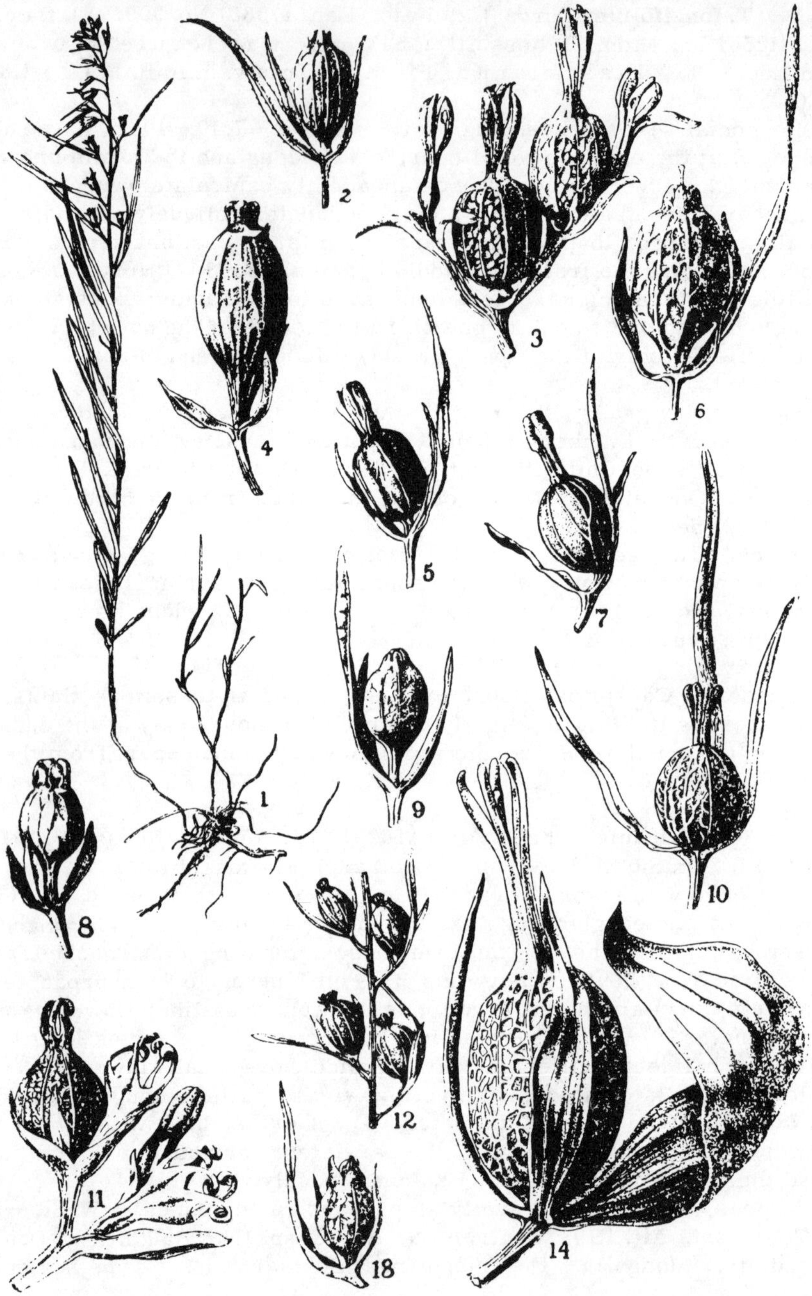

PLATE XXIV. 1. Thesium alatavicum Kar. et Kir. – 2. T. ramosum Hayne. – 3. T. Kotschyanum Boiss. – 4. T. ramosissimum Bobr. – 5. T. saxatile Turcz. – 6. T. Gontscharovii Bobr. – 7. T. alpinum L. – 8. T. linifolium Schrank. – 9. T. brachyphyllum Boiss. – 10. T. chinense Turcz. – 11. T. Szovitsii DC. – 12. T. longifolium Turcz. – 13. T. procumbens C.A.M. – 14. T. Minkvitzianum B. Fedtsch.

(in a wide sense), of steppes and mountain-steppes (except extreme E.). The series apparently includes also the West European T. humifusum DC. and T. divaricatum Jan.

427 16. **T. ramosum** Hayne in Schrad. Journ. I (1800) 30; Ldb. Fl. Ross. III, 2, 540; Shmal'g., Fl. II, 404. — T. divaricatum Ldb. l. c., non Jan.; Shmal'g., l. c., non Jan. — Ic.: in Schrad. Journ. l. c. t. 7.

Perennial with a woody many-headed vertical root; stems numerous, erect or slightly declinate, very rarely almost decumbent, 15 — 30 cm long, sulcate, often ribbed and scabrous in upper part, branching from base; leaves linear, narrowed at base, acuminate, to 3 cm long, 1 — 2.5 mm broad, obscurely 1 — 3-nerved, the upper ones scabrous-margined; inflorescence racemiform, branched at base; peduncles obliquely ascending, scabrous, monochromatic, 4 — 10 mm long, 2 — 4 times as long as the fruit; bracteoles 3, the lateral about as long as the flower and the fruit, the middle one longer; all bracteoles scabrous, with a prominent midrib on the back and hence apparently 3-angled; perianth 2.5 — 3 mm long, campanulate, cut into lanceolate lobes, white within; flowers borne on a short pedicel or almost sessile; nutlet subsessile or short-stalked, ovaloid, ca. 3.5 mm long, 3 — 4 times the length of the persistent portion of the perianth; perianth lobes inflexed; nutlet rather densely reticulate with branched longitudinal nerves, the 5 principal nerves more prominent than the rest. Fl. June; fr. July. (Plate XXIV, Figure 2).

Steppes, steppe meadows, wood margins; meadows and open woods in mountains. — European part: U. Dnp. (except N.); V.-Don (N.: Eletz and Livny districts), V.-Kama (S. — steppe regions of the Urals), M. Dnp., L. Don, Transv., Bl., Crim., L. V., U. Tob. (except N.); Caucasus: Cisc., Dag., W., S. and E. Transc., Tal.; W. Siberia: Irt. (S.); Centr. Asia: Ar.-Casp. (N.), Balkh. (N.), Dzu.-Tarb., T. Sh. (Aulie-Ata, Karatau, Dzhebogly); Mtn. Turkm. (W. Kopet Dagh).

Note. In the S. regions of the European part of the USSR and of Central Asia, the characters of this species are rather constant; on the other hand, in the Caucasus, where it ranges through different habitats from deserts to the alpine zone, it displays great variability. A number of described species, varieties, and forms of T. ramosum affinity are in fact ecological forms of this species; the more important of these are:

f. asperulum DC. Prodr. XIV (1857) 644. — T. asperulum Boiss. et Buhse in Nouv. Mém. Soc. Nat. Moscou XII (1860) 194.

The upper part of the stem, branches, peduncles, leaves, and bracteoles strongly scabrous or setose-scabrous; plants of a xerophylous aspect.

Occurring in Dag., S. Transc., and Tal. — Plants hardly distinguishable from this form are known from the Crimea and S. Ukraine. Described from Nakhichevan. Type in Leningrad.

428 f. laxiflorum (Trautv.) Bordz. in Bull. Jard. de Kieff, livr. VII—VIII (1928) 7. — T. laxiflorum Trautv. A. H. P. IX (1884) 152. — T. pratense auct. fl. Cauc.; Ldb. Fl. Ross. III, 542, non Ehrh.

Sparsely flowered plants; flowers smaller; persistent portion of the perianth about half the length of the nutlet; a mesophytic shade plant. — Occurring in W. Transc. Described from Imeretia; cotype in Leningrad.

f. brevibracteatum Bordz. in Bull. Jard. de Kieff, livr. VII—VIII (1928) 9.

The unpaired middle bracteole equaling or even smaller than the fruit; the lateral bracteoles much smaller; peduncles short, about equaling or at most twice the length of the fruit.

Occurring in the subalpine zone of the Greater Caucasus. Described from the upper reaches of the Kuban River.

f. caespitans Ldb. Fl. Ross. III, 2, 541. — T. Dollineri Murb. ex Velen. Fl. Bulg. (1891) 498. — Plant with prostrate stem; occurring on rock outcrops in S. Ukraine. West European authors consider it to be a distinct species.

17. **T. ferganense** Bobr. in Acta Inst. Bot. Ac. Sc. URSS ser. 1, II (1936).

Perennial, with a straight many-headed taproot; stems 5—15, erect, 25—30 cm long, sulcate, branched at the ends; leaves lanceolate, 2—6 cm long and 3—7 mm broad, acuminate, narrowed at base, rather distinctly 3- or rarely 5-nerved, especially beneath; leaves and stems glaucescent, the upper surface of leaves and the bracteoles scaberulous; inflorescence paniculate, the lower branches 7—12 cm long; peduncles 1—2 cm long, slender, obliquely ascending, in fruit more declinate; bracteoles 3, the lateral linear and about as long as the flower, the middle ones lance-linear and twice the length of the flower; flowers narrowly funnel-shaped, 4—5 mm long, with a very short pedicel; perianth lobed to one-third; nutlet ca. 3 mm long, short-stalked, ellipsoid, with branched longitudinal nerves; persistent portion of the perianth one-fifth to one-third the length of the nutlet. Fl. July; fr. August.

Coppices and woods on mountain slopes. — Centr. Asia: T. Sh. (Andizhan District: Arslanbob, Karacha-Bulak, Kugart; Osh District, Gul'cha). Endemic. Described from Arslanbob. Type in Leningrad.

18. **T. Gontscharovii** Bobr. in Acta Inst. Bot. Ac. Sc. URSS ser. 1, II (1936).

Perennial, with a woody vertical many-headed root; stems numerous, 15—40 cm long, ascending to erect, rarely almost prostrate, smooth, sulcate, branched in middle part; leaves linear, glabrous, narrowed at base,
429 terminating in a short point, the lower to 4 (5) cm long and 3—4 mm broad, the upper to 2 cm long and 2 mm broad; rather distinctly 1-nerved; inflorescence racemiform, very loose, with 3—7 loosely few-flowered clusters at base; peduncles 1.5—4 cm long, shorter at the top, declinate, each bearing a solitary flower; bracteoles 3, minutely scaberulous, the lateral about as long as or shorter than the flower, the middle one 2—4 times the length of the flower; flowers subsessile; perianth campanulate, ca. 4 mm long, lobed to one-third; nutlet ellipsoid, subsessile or borne on a very short stalk formed by the scarious base of its ribs, rather densely reticulate by branched longitudinal nerves; 4.5—5 mm long and ca. 3—5 mm in diameter; persistent portion of the perianth one-fifth to one-fourth the length of the nutlet; style very often persistent in fruit. Plants growing on gypseous soils have abbreviated vegetative parts, somewhat fleshy leaves, a heavily knotted aerial portion of the strongly developed woody taproot, and a generally suffruticose aspect. Fl. June; fr. July. (Plate XXIV, Figure 6).

Stony slopes of middle mountain altitudes (1,500—2,200 m). — Centr. Asia: Pam.-Al. (Vakhsh Range; Gissar Range, Kandara; Sangalak Range, on gypsiferous slopes; Darvaz). Endemic. Described from the Tundak Pass. Type in Leningrad.

19. **T. ramosissimum** Bobr. in Acta Inst. Bot. Ac. Sc. URSS ser. 1, II (1936).

Perennial, with a many-headed woody taproot; stems numerous, erect, to 60 cm long, ribbed, abundantly branched from the base, slender, sparingly leafy; leaves linear; those at the base of lower branches larger, to 4 cm long and to 3 mm broad, narrowed toward base, acuminate at apex; upper leaves smaller, 1 — 2 cm long, rarely distinctly 1 — 3-nerved; peduncles divergent from rachis, the upper ones about as long as the flowers, the lower twice as long; bracteoles 3, the lateral 3 — 4 mm long, barely reaching the middle of the flower, the middle bracteole usually inserted at the same level, twice as long as the lateral and about the length of the flower; all bracteoles lanceolate, keeled, acuminate; flowers ca. 5 mm long, funnel-shaped, cut to about the middle into obtusish lobes; pedicels to 3 mm long; nutlet ovaloid, ca. 4.5 mm long and 2.5 — 3 mm in diameter; persistent portion of the perianth one-fifth to one-fourth the length of fruit; stalk 2 — 3 mm long, in maturity yellowish-brown, wrinkled, readily breaking off; nutlet rugose with sometimes closed ramifications of the longitudinal nerves. Fl. and fr. July. (Plate XXIV, Figure 4).

Gravelly mountain slopes. — Centr. Asia: Pam.-Al. (water divide between Kizyl-su and Mazarsultan, W. slopes of the Karimonok Mountains on gypsiferous calcareous slopes at the upper reaches of the Daraz-Nitau River; Kulyab, near Kala-i-Kunya). Endemic. Described from the Karimonok Mountains. Type in Leningrad.

430 Series 6. **Micrantha** Bobr. — Plants with short prostrate unbranched stems and small flowers. This series contains T. parnassi DC. of the Balkan Peninsula and T. libanoticum Ehrenb. of Syria.

20. **T. brachyphyllum** Boiss. Diagn. ser. I, V (1844) 48; Fl. Or. IV, 1062; Transhel', Protok. SPB. Obshch. Estestvoisp. XXXIV (1904) 1, 227; Bordz. in Bull. Jard. de Kieff VII—VIII (1928) 14.

Perennial, with a fusiform root; stems numerous, 5 — 15 cm long, prostrate to ascending, unbranched, slightly ribbed, densely almost 1-sidedly leafy; lower leaves scalelike, lanceolate, 2 — 3 mm long; middle leaves linear, to 10 mm long, acuminate, scabrous-margined, rather distinctly 1-nerved; upper leaves transformed into unpaired bracteoles to 15 mm long; inflorescence a slender simple raceme, more or less 1-sided; peduncles 2 — 3 mm long, about the length of the flower, obliquely ascending; bracteoles 3, the lateral longer than the flower, narrowly linear, the middle one 5 — 8 times as long as the flower, leaflike, to 2 mm broad; flowers short-pediceled, broadly funnel-shaped, ca. 3 mm long and to 3 mm in diameter; nutlet ellipsoid, ca. 2.5 mm long, borne on a stalk 1 — 1.5 mm long, the inconspicuous nerves but slightly branched; persistent portion of the perianth one-third the length of the nutlet. May. (Plate XXIV, Figure 9).

Gravelly mountain slopes and stony places in steppes with sparse vegetation. — European part: Crim. (Yaila, vicinity of Simferopol'); Caucasus: S. Transc. (Ararat, Akhalkalaki). **Gen. distr.:** Bal.-As. Min., E. Med. Described from Philadelphia* in Palestine*. Type in Geneva.

Series 7. **Rostrata** Bobr. — Persistent portion of the perianth tubular, as long as or longer than the nutlet; perianth lobes inflexed only at the very tips. Beside USSR species, this series contains the West European T. rostratum Mert. et Koch and T. pratense Ehrh.

21. **T. alpinum** L. Sp. pl. (1762) 301; Ldb. Fl. Ross. III, 2, 542; Shmal'g., Fl. II, 404. — L. Litwinowii V. Petr. in Maevskii Fl. Sr. Ross. (1933) 266. — Exs.: Fl. exs. Austro-Hung. No. 3829.

Perennial, with a straight slender descending root; stems several, 15 — 35 cm long, erect, often declinate, very rarely few-branched, sulcate, glabrous; leaves linear, quite distinctly 1-nerved, acuminate, narrowed at base, glabrous; lower leaves 3 — 4 cm long and 2 — 3 mm broad; upper leaves transformed into unpaired bracteoles, 1 — 2 cm long; inflorescence a slender spikelike simple raceme, usually 1-sided; peduncles commonly not longer than the flowers, in fruit appressed to rachis; bracteoles 3, the lateral not longer than the flower and
431 fruit, the middle one 2 — 4 times as long as the flower, lanceolate, leaflike, acuminate; flowers to 6 mm long, funnel-shaped, short-pediceled, cut to the middle into 4 obtusish lobes (very rarely individual flowers 5-lobed); nutlet ca. 3 mm long, globular-ellipsoid, borne on a yellowish-brown stalk 1.5 mm long, inconspicuously ribbed; persistent portion of the perianth tubular, as long as or slightly longer than the nutlet, the lobes inflexed at the tips. Fl. June; fr. July. (Plate XXIV, Figure 7).

Wood margins and lakeshores; alpine meadows, stony slopes, and open woods at higher mountain altitudes, 1,500 — 2,500 m above sea level. — European part: Lad.-Ilm. (Valdai District — Lake Bologoe, Glybokoe, Berezaika station; Opochka District — village of Andreitsevo), U. Dnp. (Sebezh District, Zargor'e), U. V. Vyshnii Volochek District, between Lakes Beloe and Pyavochnoe); Caucasus: Cisc., Greater Caucasus (Teberda, Baksan, Tskhenis-Tskhali, Kobi, upper Kuban River Balkaria**, South Ossetia). **Gen. distr.:** mountains of Centr. and S. Europe. Described from S. Europe. Type in London.

Note. Plants from the Valdai Hills and from adjacent areas are identical with those from S. Sweden and there are no sufficient grounds for separating them as a species. There would be at least as much justification for separating the Caucasian specimens which we have referred to this species.

Order 17. **Aristolochiales** LINDL.

Flowers monochlamydeous; ovary mostly inferior, 1 — 6-locular, rarely 5-locular; ovules many; fruit a many-seeded capsule or berry.

* [Amman in Jordan.]

** [Part of the present-day Kabardinian-Balkar ASSR.]

1. Perianth corolloid, mostly of 3 united lobes; stamens 6—36, rarely 5, free or adnate to styles; fruit a capsule; seeds containing endosperm and a minute embryo; herbs or woody climbers; leaves cordate or reniform at base . . . . Family L. **Aristolochiaceae** Blume.
\+ Perianth of 4- or 5-merous; stamens many, inserted around the stylar column below the stigmatic ring; ovary unilocular; fruit a many-seeded berry; embryo undifferentiated, submerged in the endosperm tissue . . . . Family LI. **Rafflesiaceae** Dumort.

Family L. **ARISTOLOCHIACEAE** BLUME.*

Flowers actinomorphic or zygomorphic, hermaphrodite; perianth simple, corolloid, united, 3-lobed; stamens 6, united with the style, or 12, free; ovary inferior or half-inferior, 6-locular, rarely 4- or 5-locular; styles united into a column; stigma mostly lobed; ovules anatropous; fruit a many-seeded capsule; seeds containing endosperm and a minute embryo.

432 Key to Genera

1. Herbs with erect or climbing stems, more rarely woody climbers; leaves numerous, alternate; perianth zygomorphic; stamens 6 . . . . 387. **Aristolochia** L.
\+ Herbs without aerial stems; leaves 2, developing at the top of a sometimes superficial rootstock; Perianth regular; stamens 12 . . . . 386. **Asarum** Tourn.

Genus 386. **ASARUM**** L.

L. Gen. pl. (1756) 133.

Flowers solitary, axillary, short-pediceled, regular, with inferior or half-inferior ovary; perianth campanulate, 3-lobed; stamens 12, in a circle around the styles; filaments very short, as long as 2—3 times the length of the anthers, the connectives continued beyond the anthers; styles 6, united at base into a furrowed column; fruit an irregularly dehiscent hemispherical capsule crowned with remnants of the perianth; seeds grayish-brown, elongate-ovoid or navicular, the furrowlike raphe with a fleshy outgrowth. Russian: "kopyten'."

Herbs with branched rootstock, the branches terminating in 2 or 3 scarious scales and 2 long-petioled leaves, the slender petiole 2—3 times the length of the blade; blades reniform or cordate, entire.

1. Flowers outside and peduncles glabrous; limb of perianth spreading, the constricted throat surrounded by a narrow annular ridge . . . . 1. **A. Sieboldi** Miq.
\+ Flowers outside and peduncles pubescent; limb of perianth straight or slightly curved, not constricted at throat . . . . 2.

* Treatment by N. A. Ivanova.

** The name of this plant used by the Greek writer Dioscorides.

2. Leaves rounded-reniform, always covered beneath with short hairs . . . . 2. **A. europaeum** L.
\+ Leaves rounded-cordate, at flowering time glabrous beneath . . . . 3. **A. ibericum** Stev.

1. **A. Sieboldi** Miq. in Ann. Mus. Bot. Lugd.-Bat., II (1865 —1866) 134; Maxim. Bull. de l'Acad. Sc. St. Pétersb. XVII (1872) 163; Kom. and Alis., Opr. rast. Dal'nevost. kr. I (1931) 453. — A. heterotropoides Schmidt., Reis. im Amurl. (1868) 171.

Perennial, with an underground rhizome; scales 2, orbicular, to 1—1.5 cm long, obtuse, conduplicate; leaves coriaceous, cordate or triangular-cordate, acute, 4—9 cm long and 5—10 cm broad, green, the upper surface at flowering and subsequently pubescent only on the veins, the lower surface paler, in plants from Sakhalin glabrous (var. sachalinensis Maxim.), in plants from [former] Ussuri Territory covered with
433 very short appressed hairs (var. mandshurica Maxim.); petiole 5—15 cm long, glabrous or puberulous; peduncles at first short and curved, at length straightening and elongating to 3—4 cm; perianth mostly green with dingy purple margin, rarely green or purple throughout; perianth tube initially hemispherical, finally short-campanulate, with very prominent longitudinal nerves within; lobes of the limb triangular-oval, acutish; throat constricted, surrounded by a ringlike ridge but not pectinate; anther-connectives obsolescent; ovary half-inferior, the free portion conical; styles united at base into a column, prolonged into radially spreading longitudinally grooved appendages with round villous stigmas at base; capsule 10—15 mm long and 15—20 mm broad; seeds 3—3.5 mm long and ca. 1.5 mm broad, not flattened, almost smooth, with slightly impressed raphe. May. (Plate XXV, Figure 5).

Coniferous and mixed woods and dense shrub thickets. — Far East: Uss., Sakh. **Gen. distr.**: Japan, Korea, Manchuria, China. Described from Japan. Type in Leyden.

Note. The varieties of A. Sieboldi established by Maksimovich, var. mandshurica and var. sachalinensis, are distinguished by their leaf shape. Later collections of V. L. Komarov from Ussuri Territory and from Manchuria have shown that these differences are not consistent. However, the absence of pubescence on the lower leaf surface of Sakhalin plants and its presence in Ussuri plants distinguish these geographic races. It is probable that more ample material from Sakhalin will disclose additional differences and will thus enable their separation as species.

Economic importance. A medicinal plant, particularly valued in China. In Mukden Province it is known as wild or mountain "zhen'-shen'."

2. **A. europaeum** L. Sp. pl. ed. I (1753) 442; Ldb. Fl. Ross. III, 553; Shmal'g., Fl. II, 400; Kryl., Fl. Zap. Sib. 817. — Ic.: Thomé Fl. Deutsch. III (1888) t. 456; Rchb. Ic. Fl. Germ. XII, t. 668. — Exs.: Meinsh. Herb. Fl. Ingr. No. 549.

Perennial; rhizome with superficial annual branches; scales 3, ovate, acutish, conduplicate, 1—2 cm long; leaves dark green, in live condition lustrous, coriaceous, becoming brownish or spotted and in this condition persistent through winter till the following spring; blade rounded-reniform,

4—6 cm long and 5—8 cm broad, covered on both sides with short appressed hairs; petiole 5—10 cm long, channeled above, clothed with longer spreading hairs; flowers on a nodding peduncle to 1—2 cm long; perianth cinnamon-brown; lobes triangular-ovate, 7—9 mm long and 4—6 mm broad, terminated
434 by a lanceolate inflexed point; filaments at first as long as the anthers, elongating to 2—3 times their length; appendages of connectives slightly longer than to 1.5 times as long as the anthers; ovary inferior, its upper part subconical-applanate; stylar column 6-angled, enlarged at the end into a 6-rayed disk bearing the villous stigma; seeds trigonous-ovoid, slightly flattened, 3—3.5 mm long and ca. 2 mm broad, grayish-brown, finely rugose, with a broad and deep raphe. End of May, beginning of June.

Woods, mostly broad-leaved or mixed. — European part: Dv.-Pech. (S.), Lad.-Ilm., U. V., V.-Kama, U. Dnp., M. Dnp., V.-Don, Transv., Bl. (N.), Crim., L. Don; W. Siberia: Ob (W.), Alt. **Gen. distr.**: W. Europe (except extreme N. and S. parts). Described from W. Europe. Type in London.

Economic importance. Used in veterinary and human medicine (Radix et Herba Asari). The root was used in the past in medicine as an emetic. Applied in popular medicine as antipyretic and for treatment of mangy horses. It can be used for the production of a light brown dye. Contains camphorlike "azorone."

3. **A. ibericum** Stev. ex Ldb. Fl. Ross. III (1850) 553 (nom. nud.); Woron. in Sched. HFR, VI (1908) 70. — A. intermedium (C. A. M.) Grossh., Fl. Kavk. II (1930) 36. — A. europaeum var. intermedium C. A. M. Ind. cauc. (1831) 46; Ldb. Fl. Ross. III, 553. — A. europaeum var. caucasicum Duchartre in DC. Prodr. XV (1864) 423. — Exs.: HFR No. No. 1786.

Perennial; rhizome creeping, its annual branches superficial, pubescent; scales 3, ovate, acutish, conduplicate, 1—2 cm long; leaves dark green, in live condition lustrous, coriaceous, apparently not persistent through winter; blades rounded-cordate, 4—8 cm long and 6—12 cm broad, at flowering glabrous beneath, mostly covered on the veins above with short appressed hairs and with few hairs between the veins near the margin; petioles 5—12 cm long, channeled above, at the lower and upper end mostly pubescent especially along the groove, glabrous or almost so elsewhere; peduncles nodding, 1—2 cm long; perianth olivaceous-brown; perianth lobes triangular-ovate, 7—9 mm long and 4—6 mm broad, terminated by a lanceolate inflexed point; connective tips subulate, slightly longer than to 1.5 times as long as the anthers; ovary inferior, its upper part conical; stylar column 6-angled, enlarged at the end into a 6-rayed disk bearing the villous stigma; seeds trigonous-ovoid, slightly flattened, 3—3.5 mm long and ca. 2 mm broad, yellowish-brown, finely rugose, with a broad and deep raphe. Fl. May. (Plate XXV, Figure 6).

Woods, mainly beech, and shady coppices. — Caucasus: Cisc., W. Transc.; European part: V.-Don (Vesenmeyer ex gubern. Simbirsk?). **Gen. distr.**: An endemic Caucasian species. Described from the Caucasus. Type in Leningrad.

435 Note. A record indicating A. ibericum in former Simbirsk Province raises doubts as to the correctness of the herbarium label, although some specimens of A. europaeum, collected further south in the Volga Hills and the Donets Ridge, do show a varying degree of transition towards A. ibericum.

Genus 387. **ARISTOLOCHIA*** L.
L. Gen. pl. (1756) 133.

Herbs with erect or climbing stem, rarely woody climbers; leaves many, alternate, entire, cordate at base, petiolate; perianth zygomorphic, the oblique limb deciduous, the tube inflated at base around the gynostemium; ovary inferior, 6-locular, rarely 4- or 5-locular; stamens 6, arranged around the styles and united with them into a columnar gynostemium; capsule dehiscent by 6 longitudinal slits; seeds flat, 3-angled, with rounded angles. Russian: "kirkazon."

1. Flowers in axillary fascicles; perianth tube almost straight, the limb ligulate . . . . 2.
+ Flowers solitary, rarely in pairs, axillary; perianth tube strongly curved; limb not ligulate . . . . 3.
2. Stem erect; perianth limb obtuse . . . . 5. **A. clematis** L.
+ Stem climbing; perianth limb acute, prolonged into a bristlelike point . . . . 7. **A. contorta** Bge.
3. A woody climber; limb of perianth almost regular, 3-lobed, the throat surrounded by a ringlike ridge; capsule to 11 cm long and 3 cm broad . . . . 1. **A. manshuriensis** Kom.
+ Herbs with erect stems and underground tubers; limb of perianth oblong-cordiform with obtuse auricles or elliptic exauriculate; throat not surrounded by a ring; capsule 2—2.5 cm long; leaves ovate or orbicular, the lower ones obtuse . . . . 4.
++ Leaves triangular in outline, rarely oblong, the lower ones acute . . . . 5. **A. Bottae** Jaub. et Sp.
4. Flowers large; limb acute, 3—5 cm long; the lower inflated part of the tube 5—8 mm long . . . . 2. **A. pontica** Lam.
+ Flowers smaller, limb mostly obtusish, 2—2.5 cm long, the lower inflated part of the tube 5—8 mm long . . . . 5.
5. All leaves obtuse or slightly retuse; limb of perianth spreading, golden-yellow with appressed purplish-violet hairs, purplish-brown or dark violet on the margin . . . . 4. **A. Steupii** Woron.
436 + Lower leaves obtuse, the upper ones more or less acute; limb of perianth straight or more or less declinate, distinctly auriculate, glabrous within, varying in color but not golden-yellow at center . . . . 3. **A. iberica** Fisch. et Mey.

1. **A. manshuriensis** Kom., Fl. Manchzh. II (1904) 112; Kom. and Alis., Opr. rast. Dal'nevost. kr. I (1931) 454.

Perennial; leaves rounded-cordate, 11—29 cm long, acutish or acuminate, at first pubescent beneath and sparsely pilose above, at length sparsely covered with short hairs; blade several times as long as the petiole; flowers solitary, rarely in pairs, on short axillary branchlets; peduncles 1.5—3 cm long, slightly pubescent below, glabrous above, slightly curved, with 1 or 2 dry brownish densely pubescent scales at base; bracts 2, inserted on the lower part of the pedicel, to 1 cm long, green, cordate-ovate or cordate, subsessile, mostly not clasping, or bracts absent; tube of perianth curved horseshoelike, inflated around the gynostemium (to 15 mm

* From the Greek words "aristos" and "locheia," referring to supposed value in aiding childbirth.

broad), narrowed above and then again enlarged at the curvature (to 16 — 18 mm), greenish glabrous and bluntly ribbed outside, with purple rings and speckles within, inflated in upper part, hairy around the gynostemium; limb brownish or greenish-yellow, ca. 22 mm in diameter, rather shallowly 3-lobed, almost regular, when fully expanded almost flat; lobes broadly triangular, sharply or obtusely long-tipped, short-hairy on the margin; throat surrounded by a ringlike sparsely papillose ridge; ovary cylindric, ribbed; gynostemium 3-angled; stigma 3-lobed; stamens in pairs on the outer face of the stigmatic lobes; capsule hexagonal-cylindric; seeds grayish-brown, cordate-triangular, 6 — 7 mm in diameter, their dorsal side convex and minutely tuberculate, the ventral side glabrous and smooth. (Plate XXV, Figure 2).

Mixed mountain woods and wood margins. — Far East: Uss. **Gen. distr.**: Manchuria and N. Korea. Described from the Suifun River. Type in Leningrad.

Note. Differing from the related cultivated A. macrophylla Lam. in larger flowers, relatively short peduncles and nonamplexicaul bracts.

2. **A. pontica** Lam. Encycl. meth. (1789) 255; Ldb. Fl. Ross. III, 555; Boiss. Fl. Or. IV, 1081. — Ic.: Gard. Chron. Ser. III, XXXI (1902). — Exs.: Fl. cauc. exs. No. 33.

Perennial; stems simple, slightly flexuous, sparsely pubescent, mostly 30 — 50 cm long; leaves 6 — 10 cm long and broad, rarely larger, cordate-ovate, the lower mostly obtuse, the upper subacute, glabrous, covered
437 beneath especially on the veins with very short hairs or glabrate; petioles sparsely pubescent, 4 — 6 cm long, the lower about as long as the blade, the upper one-third to one-half its length; peduncles 1.5 — 2.5 cm long, mostly curved below the flower; bracts in bud densely pubescent, after flower expansion mostly with but few hairs on the nerves; perianth tube mostly greenish-purple, the straight limb darker purple, glabrous within; ovary covered with spreading hairs; capsule ovaloid, 2 — 2.5 cm long. End of April, beginning of May.

Shady woods and coppices. — European part: V.-Don (Kharkov, adventive); Caucasus: W. Transc. **Gen. distr.**: Turkish Armenia. Described from "the Levant" (probably from Transcaucasia or Turkish Armenia). Type in Paris.

3. **A. iberica** Fisch. et Mey. in Szovits herb.; Boiss. Fl. Or. IV (1879) 1081. — A. pontica β parviflora Duchartre in DC. Prodr. XV (1864) 493. — A. pontica var. iberica (Fisch. et Mey.) Woron. in Fl. cauc. exsicc. (1907) No. 85. — Exs.: Pl. Orient. exs. No. 382; Fl. cauc. exs. No. 85.

Perennial; stems simple, slightly flexuous, sparsely pubescent, mostly 20 — 30 cm long; leaves 6 — 8 cm long and broad, cordate to cordate-ovate, the lower mostly obtuse, the upper subacute, the upper surface glabrous or sparingly covered with very short hairs, the lower surface initially pubescent, sometimes heavily so, finally sparsely pubescent or glabrate; petiole slightly pubescent, 2 — 5 cm long, in lower leaves two-thirds, in upper one-third to one-half the length of the blade; peduncles 1 — 2 cm long, mostly curved below the flower; bracts in bud densely pubescent, after expansion of flowers more sparsely pubescent and sometimes merely on the nerves; limb of perianth glabrous within, purple or greenish-purple, sometimes dark at the throat, the tube and limb concolor outside; ovary covered with spreading hairs; capsule ovaloid, 2—2.5 cm long; seeds ca. 6 mm in diameter, black, finely tuberculate, the dorsal face flat, the ventral concave. End of March—April.

Shady woods and coppices. — Caucasus: W. Transc. (central part). Endemic. Described from the Caucasus. Type in Leningrad.

4. **A. Steupii** Woron in Grossg. Fl. Kavk. II (1930) 37; Woron. in Acta Inst. Bot. Ac. Sc. URSS, ser. I, 1 (1933) 214.

Perennial; stems simple, slightly flexuous, sparsely pubescent, mostly 10—25 cm long; leaves varying greatly in size, 3—8 cm long and broad, cordate-orbicular to cordate-oval, retuse or obtuse, only the uppermost sometimes slightly acuminate, the upper surface mostly covered with very short hairs, the lower surface at first rather heavily pubescent, finally sparsely pubescent or glabrate; petioles sparsely pubescent, 1—4 cm long, the lower about as long as or shorter than the blade, the upper one-third to one-half the length of the blade; pedicels 5—10 mm long, mostly curved below the flower; lower flowers often rudimentary, cleistogamic; perianth in bud densely pubescent outside, after expansion
438 more sparsely pubescent; tube greenish with purple veins outside, golden-yellow within, variegated with small purplish-violet stripes and appressed hairs; limb elliptic; ovary covered with spreading hairs; capsule pyriform, ca. 3 cm long and 1.5 cm broad. April—May. (Plate XXV, Figure 1).

Woods. — Caucasus: Cisc. (Maikop) and W. Transc. (NW part). Endemic. Described from the Caucasus. Type in Leningrad.

5. **A. Bottae** Jaub. et Sp. Illustr. I (1842—43) 173, tab. 98.

Perennial; stems branched, scabrous; leaves broad-cordate and deeply sinuate at base, scabrous; petiole 0.5—1 (2) cm long; perianth 3—5 cm long, dark purple, more or less pubescent outside; limb 2.5—3.5 cm long, auriculate, pointed. — Caucasus: S. Transc. (Nakhkrai). Gen. distr.: Arm.-Kurd., Iran. Described from the vicinity of the city of Diyarbakir. Type in Paris.

6. **A. clematitis** L. Sp. pl. (1753) 962; Ldb. Fl. Ross. III (1847—1849) 554; Shmal'g., Fl. II (1897) 401. — A. B r a n i c i a n a Andrz. ex Trautv. in A. H. P. IX (1884) 155 (nom. nud.). — Ic.: Thomé, Fl. Deutsch. III (1888) t. 457; Rchb. Ic. Fl. Germ. XII (1850) t. 669. — Exs.: HFR No. 286; Fl. polon. exs. No. 568.

Perennial, glabrous throughout; rhizome creeping; stem to 70 cm long, simple, erect or slightly flexuous; leaves obtuse or scarcely retuse, very rarely acutish, slightly rough on the margin; petiole one-third to one-half the length of the blade; peduncles 10—12 mm long; perianth pale yellow, the tube ca. 12 mm long, the limb as long as or 1.5 times as long as the tube; capsule pendulous, pyriform; seeds ca. 1 cm in diameter, brownish, glabrous, finely rugose. Fl. May—September. (Plate XXV, Figure 4).

PLATE XXV. 1. Aristolochia Steupii Woron., aspect, a) longitudinal section of a flower; b) gynostemium; c) capsule.— 2. A. manshuriensis Kom., fl., a) gynostemium.— 3. A. contorta Bge., flower.— 4. A. clematis L., flower.— 5. Asarum Sieboldi Miq., a) longitudinal section of a flower; b) stamens and pistils.— 6. A. ibericum Stev., a) longitudinal section of a flower; b) seed.

Inundated woods, coppices, and meadows; steep riverbanks and seashores, in gullies; sometimes synanthropic in towns, gardens, etc. — European part: U. V. (S. part), V.-Kama (S.), U. Dnp., M. Dnp., V.-Don, Transv., Bl., Crim., L. Don, L. V.; Caucasus: Cisc., W. Transc., Dag. **Gen. distr.**: W. Eur., Asia Minor. Described from Europe. Type in London.

7. **A. contorta** Bge. Enum. plant. Chin. bor. (1831) 58; Maxim. Primit. Fl. Amur. (1859) 238; Kom., Fl. Manchzh. II (1904) 111; Kom. and Alis., Opr. rast. Dal'nevost. kr. I (1931) 453. — Ic.: Bot. Mag. Tokyo, XXVI (1912) 183;
441 Jabe, Ic. Fl. Manchur. I (1914) t. 8. — Also referred to as riverside birthwort.

Perennial, glabrous throughout; leaves cordate, obtuse or acute; petiole one-fourth to one-half the length of the blade; peduncles ca. 1.5 cm long, often branched; perianth pale yellow, the tube ca. 1.5 cm long, the limb the same length; capsule ovaloid, 35—50 mm long and 25—32 mm broad; seeds 5—6 mm in diameter, dark castaneous at center, light-colored at the periphery. Fl. July—August.

Osier-beds. — Far East: Uss. **Gen. distr.**: Jap.-Ch. Described from N. China. Type in Leningrad.

***A. macrophylla** Lam. Encycl. I (1783) 252; Asch. und Gr. Syn. IV, 681; Syreishchikov, Mosk. Fl. II, 59. — A. s i p h o L'Herit. Stirp. nov. I (1784) 13. — S i p h i s i a g l a u c a Rafin. Med. Fl. I (1828) 65. — S. s i p h o Klotzsch. Monatsb. Berl. Acad. (1859) 602.

A woody climber, 4—10 m tall, with gray rugose bark; young branches green; leaves long-petioled, ovate, cordate at base, 10—23 cm long, with scattered hairs; flowers solitary or in pairs, axillary; the long peduncles bearing at base and about the middle a leaflike clasping bract; perianth glabrous, brown, greenish-brown outside, dark purple at throat, the tube curved, the limb regularly 3-lobed; fruit oblong, dark gray, rather large, suspended on a long stalk. Fl. May—July; fr. September—October.

Only cultivated; valued as a tall climbing ornamental with very dense foliage. Native in N. America. Winter-hardy in Moscow and Leningrad; flourishing even better in the south.

Family LI. **RAFFLESIACEAE** DUMORT.*

Flowers mostly unisexual, rarely hermaphrodite, with a simple perianth, regular; perianth 4- or 5-merous; stamens many; ovary mostly inferior, 1-locular, with 4 or 6—8 parietal placentae, or with numerous contorted cells; ovules many; fruit a many-seeded berry. Parasites destitute of chlorophyll, with characteristic haustoria; shoots very short, with scalelike leaves and terminal inflorescence.

Genus 388. **CYTINUS** L.

L. Gen. pl. ed. V (1754) 567.

Monoecious plants; flowers unisexual in a terminal inflorescence, the lower pistillate, the upper staminate; perianth tubular or tubular-campanulate,

* Treatment by V. L. Komarov.

4 — 6-(rarely 9)-lobed; stamens twice the length of perianth lobes; fruit fleshy; cells of ovary with 8 — 14 ovules borne on parietal placenta; stigma capitate, 8 — 10-furcate. Russian: "podladannik."

442 1. **C. rubra** (Fourr.) Kom. comb. nova. — C. hypocistis L. Syst. nat. ed. 12, 11 (1767) 602 (ex parte); Grossg., Fl. Kavk. II, 37. — Asarum hypocistis L. Sp. pl. (1753) 442 (ex parte). — Hypocistis rubra Fourr. Ann. Soc. Linn. XVII (1869) 148. — Cytinus Clusii Nyman, Consp. (1881) 645.

Perennial, juicy, fleshy plants; stems short, solitary or in tufts, erect, up to 10 cm long; leaves scalelike, rather bright carmine-red (var. kermesinus Guss. Fl. Sic. Syn. II (1844) 619), ovate to oblong, often acuminate, sometimes fringed; clusters 5 — 10-flowered; flowers lurid to pinkish-red, almost sessile and crowded at the end of the stem, sometimes umbellately disposed with the uppermost flowers barely as long as the bracteoles; perianth with a constriction at the throat, the incisions oval obtuse, the middle part glabrous; stamens 8. April — June. (Plate XXIII, Figures 1, a, b).

Parasitic on the roots of Cistus tauricus Presl., hence appearing under these shrubs. — Caucasus: W. Transc. (according to Grossgeim, found so far only in Pitsundskaya woodland). Gen. distr.: Canary Islands, Med., Bal.-As. Min. Described from Sicily. Type in London.

Note. The parasite, developing in root tissues, has thalloid or myceliumlike form. The flowering shoots are initiated within the root and then, upon bursting through the bark, emerge outside. The headlike inflorescence is surrounded by the foliar scales, the flowers are all more or less at the same level. This plant, while biologically interesting, has not so far found any practical application. The juice is blood-red.

Order 18. **Polygonales** LINDL.

Flowers cyclic, haplochlamydeous or heterochlamydeous, perianth segments 3 — 6, stamens 6 — 9, rarely more or fewer; superior ovary unilocular 1-seeded; herbs or shrubs; leaves simple over their greater part, provided at base with dry or herbaceous ocreae; fruit an achene.

Family LII. **POLYGONACEAE** LINDL.

Flowers cyclic, haplochlamydeous to heterochlamydeous, regular, hermaphrodite or unisexual; perianth segments 3 — 6; stamens 6 — 9, rarely more or fewer, sometimes partly doubled; carpels 3 or 2, rarely 4; styles 2 — 4; fruit an achene; seeds with copious mealy endosperm; embryo straight or curved. Herbs or rarely shrubs; leaves alternate, simple or rarely pedate or dissected; stipules scarious or herbaceous, sheathing, referred to as ocreae; flowers rather small, crowded in inflorescences.

443 Fossils known in the USSR:

Polygonum amphibium L. in the Pliocene of E. Transc. (Ol'tu District). — P. bistorta L. in the Postpliocene of V.-Don (Aleksin). — P. persicaria L. in the Postpliocene of V.-Don (Likhvin, Aleksin).

Key to Genera

1. Shrubs or undershrubs with clearly lignified branches . . . . 2.
\+ Herbaceous plants, perennial or annual . . . . 4.
2. Fruits 3-angled, not united with perianth even when shorter than the perianth . . . . 394. **Polygonum** L. (some species).
\+ Fruit tightly enveloped by accrescent perianth, or winged, or bearing simple or branched hard outgrowths or bristles; strongly branched shrubs . . . . 3.
3. Fruit apparently broad-winged by the surrounding firm and strongly accrescent often pink or red perianth; branches often spiny-tipped . . . . 392. **Atraphaxis** L.
\+ Fruit winged or bearing variously shaped outgrowths, often shaggy; branches always unarmed . . . . 393. **Calligonum** L.
4. Perianth green, herbaceous, 6-parted, the outer segments not accrescent spreading or reflexed in fruit, the inner accrescent and enveloping the fruit . . . . 5.
\+ Perianth corolloid; all segments alike in fruit . . . . 7.
5. Flowers 2-merous; a perennial plant with entire orbicular, cordate, or reniform leaves . . . . 389. **Oxyria** Hill.
\+ Flowers 3-merous; leaves of different form from the above . . . . . . 6.
6. Fruit 3-winged, 3-angled; stigmas capitate; large plants with mainly radical leaves . . . . 391. **Rheum** L.
\+ Fruit 3-angled; stigmas fringed . . . . 390. **Rumex** L.
7. Flowers 3-merous, disposed at the ends of the stem and of its branches in small groups surrounded by bracts; fruit flat; a very small annual . . . . .396. **Koenigia** L.
\+ Perianth mostly 5-merous; achene 3-angled or planoconvex . . . . . . .8.
8. Flowers in racemes gathered into a corymbose panicle; leaves triangular-cordate, somewhat fleshy; fruit much exserted from the perianth . . . . 395. **Fagopyrum** Moench.
\+ Fruit little exserted . . . . 394. **Polygonum** L.

444 Genus 389. **OXYRIA*** HILL.**

Hill. Veg. Syst., X (1765) t. 24, Engl. Nat. Pfl. III, 1a (1893) 19. — Donia R. Br. in Ross. Voy. ed I App. XLI (1819) 41. — Oxylapathon St. Lager, Ann. Soc. Bot. Lyon, VIII (1861) 159 pro parte.

Flowers hermaphrodite; perianth 4-parted, the outer segments spreading, the larger inner segments erect; stamens 6, with short filaments; ovary compressed laterally; styles 2; stigmas tufted; fruit a laterally compressed winged achene; leaves long-petioled; stems branched in upper part, bearing loose paniculate inflorescences. Russian: "kislichnik" ["kislyi"= sour].

* From Greek "oxys," sour.

** Treatment by A. S. Lozina-Lozinskaya.

1. **O. digyna** (L.) Hill. Hort. Kew. (1769) 158; Kryl. Fl. Zap. Sib. V, 837. — Rumex digynus L. Sp. pl. (1753) 337. — Oxyria reniformis Hook. Fl. Scot. III (1821) 111; Ldb. Fl. Alt. II, 56; Ej. Fl. Ross. III, 498. — Ic.: Rchb. Ic. Fl. Germ. XXIV, 60. — Exs.: H. F. A. M. No. 101a.

Perennial; rhizome stout, 0.5 — 1 cm in diameter; stems solitary or several together, glabrous, erect or ascending, in inflorescence branched and somewhat angled; leaves all radical, rarely 1 or 2 on the stem, long-petioled, reniform or rounded-reniform, 2 — 4 cm in diameter; flowers hermaphrodite, 2 — 6 at the ends of stems and branches and forming a slender racemiform inflorescence, the slender pedicels jointed below; perianth of 4 segments, of these 2 obovate, almost flat, 1.5 mm long and 1 mm broad, the 2 others navicularly folded and about half as broad; stamens 6, the length of the perianth; pistils shorter than stamens, with compressed ovary, 2 short divergent styles, and tufted stigmas; fruit compressed laterally, winged, the colored wings as broad as the achene. June — July.

Alpine zone of mountains and the Arctic Region; tundras and banks of streams, damp rocks, and pebbles. — Arctic: Arc. Eur., Nov. Z., Arc. Sib., Chuk., An.; Caucasus: the alpine zone throughout the Caucasus; W. Siberia: Alt.; E. Siberia: Ang.-Say.; Far East: Kamch., Ze.-Bu.; Centr. Asia: Pam.-Al., T. Sh., Dzu.-Tarb. **Gen. distr.:** Scandinavia, Atl. Eur., Centr. Eur. (alpine zone), Bal.-As. Min., Iran., Arm.-Kurd., Mong., China. Described from Lapland. Type in London.

Economic importance. The fleshy sour leaves are used for food and have antiscorbutic properties.

Genus 390. **RUMEX*** L.**

Gen. Pl., ed. I (1737) 105, ed. 5 (1754) 156. — Campd. Monogr. du genre Rumex (1819).

Flowers unisexual or hermaphrodite; perianth of 6 almost free segments disposed in 2 whorls, the outer wilting in fruit, the inner [valves]
445 accrescent, enveloping the fruit, often one or all of them with a lumpy tubercle on the back; stamens 6, in pairs between the inner perianth segments; pistil with 3 filiform styles and tufted stigmas; fruits a trigonous achene. Perennial or annual plants, from 10 cm to 1.5 m tall, with large or small entire or subentire leaves. Russian: "shchavel'."

1. Fruiting perianth scarcely accrescent, slightly surpassing the achene . . . . 2.
\+ Fruiting perianth strongly accrescent, winged . . . . 4.
2. Stigmas pink; leaves hastate, oblong-lanceolate, or linear-lanceolate . . . . 3.
\+ Stigmas golden-yellow, surpassing the perianth; leaves spatulate or cochlear . . . . 3. **R. aureostigmaticus** Kom.
3. Valves united with the achene . . . . 2. **R. acetoselloides** Bal.
\+ Valves not united with the achene . . . . 1. **R. acetosella** L.
4. Leaves sagittate or hastate at base, or leaves linear . . . . 5.
\+ Leaves rounded, cordate, or cuneate at base . . . . 11.

* Name used by Pliny.

** Treatment by A. S. Lozina-Lozinskaya.

5\. Leaves sagittate, long- or broad-triangular, large, with large lobes; flowers hermaphrodite . . . . 6.

\+ Leaves rather small, sagittate, oblong-lanceolate, ovate-triangular, ovate, lyrate, or linear; flowers unisexual, the plants dioecious . . . . 7.

6\. Leaves glabrous beneath except for the scabrous veins . . . . 10. **R. Gmelinii** Turcz.

\+ Leaves white-tomentose beneath . . . . 11. **R. jakutensis** Kom.

7\. Leaves lyrate, with hastate lateral lobes; valves unappendaged, to 9 mm long . . . . 9. **R. acutatus** L.

\+ Leaves linear, or filiform-linear, glaucous, sometimes with 1 or 2 teeth below . . . . 4. **R. graminifolius** Lam.

++ Leaves sagittate, lanceolate, oval-lanceolate, or triangular-ovate; valves with a deflexed fleshy appendage . . . . 8.

8\. Roots tuberiferous; valves to 5 mm long, reniform-orbicular, cordate at base . . . . 7. **R. tuberosus** L.

\+ Roots not tuberous; valves smaller . . . . 9.

446 9. Plants with fibrous roots; panicle cylindric or ovaloid . . . . 10.

\+ Plants with a stout taproot; panicle pyramidal . . . . 6. **R. thyrsiflorus** Fingerh.

10\. Ocreae toothed; leaves ovate-oblong, with a narrow sinus . . . . 5. **R. acetosa** L.

\+ Ocreae entire, acuminate; leaves ovate-triangular, with a broad sinus . . . . 8. **R. arifolius** All.

11\. Valves entire or obscurely toothed . . . . 12.

\+ Valves toothed or spiny . . . . 35.

12\. Radical leaves cuneate at base . . . . 13.

\+ Radical leaves rounded, subtruncate or cordate at base . . . . 20.

13\. All valves or one of them with a lumpy tubercle . . . . 14.

\+ Valves without a tubercle . . . . 17.

14\. Plants of medium size, to 50 cm tall . . . . 15.

\+ Plants big, to 120 cm tall . . . . 16.

15\. All tepals tubercled . . . . 15. **R. sibiricus** Hult.

\+ Only one tepal tubercled . . . . 16. **R. Komarovii** Schischk. et Serg.

16\. Leaves slightly hairy on the margin, to 60 cm long; tepals to 7 mm long, ovate-triangular; inflorescence broad-paniculate . . . . 20. **R. hydrolapathum** Huds.

\+ Leaves heavily hairy on the margin, to 25 cm long; tepals rounded-ovate, to 5 mm long; inflorescence narrowly cylindric . . . . 17. **R. crispus** L.

17\. Inflorescence branched; tepals all alike in fruit . . . . 18.

\+ Inflorescence simple, unbranched; tepals unequal in fruit . . . . 33. **R. angustifolius** Campd.

18\. Leaves hairy on the margin; plants green . . . . 19.

\+ Leaves smooth on the margin; plants reddish . . . . 14. **R. kamtschadalus** Kom.

19\. Inflorescence narrow-cylindric . . . . . 19. **R. pseudonatronatus** Borbas.

\+ Inflorescence broad-ovaloid . . . . 18. **R. Fauriei** Rech. f.

20\. Valves oblongly triangular-ovate, with elongated ligulate tips . . . . . . 21.

\+ Valves orbicular, ovate, or triangular-ovate . . . . 22.

447 21. All tepals tubercled; panicle leafy ...... 34. **R. conglomeratus** Murr.
\+ Only one tepal tubercled; panicle leafless ...... 35. **R. sanguineus** L.
22. Valves rounded-reniform or rounded-cordate ...... 23.
\+ Valves ovate-oblong, triangular-ovate, or cordate, acute ...... 29.
23. All tepals without tubercles ...... 24.
\+ All tepals or one of them tubercled ...... 25.
24. Plants green; leaves oblong-ovate or ovate-lanceolate; tepals rounded at apex ...... 12. **R. domesticus** Hartm.
\+ Plants reddish; leaves long-lanceolate, linear-lanceolate, or oval-lanceolate; tepals slightly tapering at apex ... 13. **R. pamiricus** Rech. f.
25. Leaves ovate-triangular or broad-ovate, deeply cordate ...... 26.
\+ Leaves oblong-ovate or ovate-lanceolate ...... 27.
26. Tepals 4 mm long ...... 22. **R. Fischeri** Rchb.
\+ Tepals 6–7 mm long ...... 21. **R. confertus** Willd.
27. Tepals 6–8 mm long; leaves oblong-ovate, ovate, or ovate-lanceolate ...... 23. **R. patientia** L.
\+ Tepals 4–6 mm long; leaves oblong-oval or oblong-lanceolate .... 28.
28. Tepals thin, reticulate; leaves glaucous, with rather long hairs on the margin ...... 25. **R. aschabadensis** A. Los.
\+ Tepals firm; leaves pale green, smooth or with very short hairs on the margin ...... 24. **R. Rechingeri** A. Los.
29. Leaves 3–7 cm long; plants to 50 cm high ...... 30.
\+ Leaves much longer ...... 31.
30. Plants with a taproot; tepals ovate ...... 31. **R. Schischkinii** A. Los.
\+ Plants with fibrous roots; tepals oblong ..... 32. **R. arcticus** Trautv.
31. Inflorescence diffuse ...... 31.
\+ Inflorescence slender, with appressed branches ...... 32.
32. Tepals not tubercled ...... 27. **R. Paulsenianus** Rech. f.
\+ One of the tepals tubercled ...... 26. **R. thianschanicus** A. Los.
448 33. Leaves oval-oblong; tepals to 1 cm long .... 28. **R. armenus** C. Koch.
\+ Leaves ovate or triangular-ovate; tepals to 6 mm long ...... 34.
34. Leaves acuminate ...... 29. **R. aquaticus** L.
\+ Leaves obtuse ...... 30. **R. alpinus** L.
35. Tepals unequally toothed on the margin ...... 35.
\+ Tepals spiny on the margin ...... 44.
36. Inflorescence paniculate, ovaloid or cylindric, compact ...... 37.
\+ Inflorescence divaricately branched, remotely verticillate ...... 39.
37. Tepals 6 mm long, 8 mm broad, broadly reniform-cordate ...... 38. **R. Regelii** F. Schmidt.
\+ Tepals 4 mm long, ovate-triangular ...... 38.
38. Tepals acuminate ...... 36. **R. stenophyllus** Ldb.
\+ Tepals obtusish ...... 37. **R. ussuriensis** A. Los.
39. Leaves small, to 10 cm long ...... 40.
\+ Leaves large ...... 42.
40. Perennial; tepals and their teeth tough; pedicels rather stout ...... 41. **R. pulcher** L.
\+ Annual or biennial plants; pedicels and teeth of tepals slender ..... 41.
41. Tepals triangular, acute, subtruncate at base, with 4 or 5 setiform teeth ...... 40. **R. Halaczi** Rech.

\+ Tepals oblong-triangular, with 3—5 spreading slender teeth . . . . 39. **R. reticulatus** Bess.
42. Inflorescence with upright branches; valves oval-triangular, 2—3 mm long . . . . 43. **R. obtusifolius** L.
\+ Inflorescence with nodding branches arising at an angle of 45°; valves broadly triangular, ovate, or rounded-cordate, 7—8 mm long . . . . . . 43.
43. Valves rounded-cordate, equal . . . . 42. **R. syriacus** Meish.
\+ Valves broadly triangular-ovate, one of them to 8 mm long, the two others shorter . . . . 44. **R. foveolatus** A. Los.
44. All tepals spiny-margined . . . . 45.
\+ Only one of the tepals with 2 symmetric spines 3—4 times the breadth of the tepal . . . . 45. **R. amurensis** F. Schmidt.
449 45. All tepals tubercled . . . . 46.
\+ Only one of the tepals tubercled . . . . 46. **R. Marschallianus** Rchb.
46. Stem branched from base, the branches spreading . . . . 47. **R. ucranicus** Fisch.
\+ Stem not branched at base . . . . 47.
47. Valves with 1 or 2 spines equaling the breadth of the valve . . . . 48. **R. rossicus** Murb.
\+ Valves with 2—5 spines exceeding the breadth of the valve . . . . 49. **R. maritimus** L.

Section 1. ACETOSELLA Meisn. in Mart. Fl. Brasil. XIV (1855) 10; DC. Prodr. XIV (1856) 63. — Flowers unisexual; tepals not accrescent in fruit; leaves sagittate at base or linear.

1. **R. acetosella** L. Sp. Pl. (1753) 338; Ldb. Fl. Ross. III, 2, 511; Kryl., Fl. Zap. Sib. IV, 833. — Acetosa parva Gilib. Exerc. phytol. II (1792) 446. — Ic.: Rchb. Ic. Fl. Germ. XXIV, tab. 192; Kom. and Alis., Opred. rast. Dal'nevost. kr. I, tab. 1402. Russian: shchavel' vorob'inyi [sparrow-sorrel], shchavelek [diminutive of "shchavel'"].

Perennial; rhizome creeping; stems numerous, erect, branched, 15 —55 cm long; lower leaves petiolate, hastate, 1—5 cm long, 2—1.5 mm [?] broad, the terminal lobe lanceolate or ovate-lanceolate, the two smaller narrow lateral lobes perpendicularly spreading or upcurved; upper leaves sessile, lanceolate or linear-lanceolate; flowers unisexual, dioecious, in rather loose clusters disposed in a loose panicle; pedicels not jointed; all tepals, in both pistillate and staminate flowers, erect, the outer shorter and narrower than the inner; valves of fruiting pistillate flowers about equaling the fruit, 1.5 mm long, 1 mm broad, ovoid, acuminate, unappendaged and not tubercled at base; achene lustrous, trigonous, light brown, less than 1 mm long. May — August.

Wood margins, slopes, pine woods, sands, bluffs, fallows, and roadsides. — European part: throughout, except L. V. and the Arctic Region; Siberia and Far East: throughout; Caucasus: Cisc., Dag. Described from Europe. Type in London.

Note. Recorded:

var. multifidus (L.) DC. in Lam. et DC. Fl. Fr. III (1815) 378. — R. multifidus L. Sp. pl. ed. II (1762) 482. — Leaves with 2—4 pairs of lateral lobes;

var. integrifolius Wallr. Sched. crit. (1822) 186. — All leaves without lateral lobes. A widespread weed, with a multilayered root system, infesting crops, wastes, and fallows.

450 2. **R. acetoselloides** Baill. Bull. Soc. Bot. Fr. I (1854) 282. — Exs.: Balansa Pl. or. exs. No. 381 (1854).

Perennial; root slender, rather long; stem erect, glabrous, sulcate, glaucescent, branched in upper part; leaves petiolate, glabrous, glaucous, lanceolate, acuminate, cuneate at base, 3—4 cm long, the dissected or simple lateral lobes to 1 cm long; ocreae white, scarious, lacerated into a number of pointed divisions; inflorescence terminal, pyramidal, loosely flowered; flowers diclinous, borne on a peduncle branched in upper part; tepals of staminate flowers subequal, the outer oblong-lanceolate, the inner ovate; tepals of pistillate flowers smaller, erect; valves 1.5 mm long, united with the achene, ovate; fruit as that of R. acetosella L. June—July.

Dry sunny slopes, open woods, roadsides, and weedy places. — Caucasus: W. Transc. (Abkhazia, Black Sea coast). **Gen. distr.**: Bal.-As. Min., Arm.-Kurd. Described from the vicinity of Smyrna [Izmir].

Note. A common weed of the Caucasus, infesting crops.

3. **R. aureostigmaticus** Kom. in Fedde Rep. sp. nov. XIII (1914) 166; Kom., Fl. Kamch. II (1929) 56. — R. graminifolius var. sublanceolatus Scheutz ex E. Hulten Fl. of Kamtch. II (1928) 47. — Ic.: Kom., Fl. Kamch. II, tab. IX.

Perennial; stem simple, rarely with 1 or 2 branches, sulcate, 10—20 cm long, slender, glabrous; radical leaves rosulate, long-petioled, spatulate, cochlear, oblong, or lanceolate, sometimes sagittate with lateral lobes; ocreae few, linear-oblong, fleshy; panicle rather diffuse, with straight slender branches; pedicels slender, shorter than the flower; flowers small, unisexual, the plants dioecious; valves twice as broad as the fruit, purple, oboval, mucronulate; stigmas broad, tufted, golden-yellow, with slender linear divisions; achene small, lustrous, trigonous, pale. July.

Slopes near the timberline. — Far East: Kamch. Described from Mount Shiveluch. Type in Leningrad.

4. **R. graminifolius** Lamb. Transact. Linn. Soc. X (1811) 264, t. 10; Ldb. Fl. Ross. III, 512. — R. acetosella var. graminifolia Schrenk ex Herder in A. H. P. XI, 2 (1892) 204. — R. angustissimus Ldb. in Mém. Acad. Pétersb. V (1812) 536.

Perennial; rhizome vertical, ropelike, branched at the top; stems several, erect, glabrous, sulcate, straight or curved, 7 — 25 cm long, slightly thickened at the nodes; ocreae large, white-hyaline, lacerate; leaves
451 filiform-linear, 2—6 cm long, 0.5— 1.5 mm broad, undivided or with filiform lateral lobes; inflorescence branched, paniculate, elongated, 3—10 mm long [?]; flowers in loosely flowered whorls; ocreae white-hyaline; pedicels jointless; staminate flowers with small oval dingy-purple tepals; pistillate flowers purple or bright yellow; valves ovate-rhombic, membranous, unappendaged, slightly surpassing the fruit, to 2 mm long; achene yellow, lustrous, trigonous-ovoid, with rounded angles. July—August

Sandy places, dunes, riverside terraces, gravels, in the Arctic Region and the alpine zone. — W. Siberia: U. Tob.; E. Siberia: Yen., Ang.-Say., Lena-Kol.; Far East: Kamch., Sakh., Chuk. **Gen. distr.**: Ber. Described from Kamchatka and the Kurile Islands.

Section 2. **ACETOSA** Meissn. in DC. Prodr. XIV (1856) 64. — Flower unisexual or hermaphrodite; tepals strongly accrescent in fruit; leaves hastate or sagittate at base.

5. **R. acetosa** L. Sp. pl. (1753) 337; Ldb. Fl. Ross. III, 2, 510; Kryl., Fl. Zap. Sib. IV, 835. — Acetosa pratensis Mill. Gard. Dict. ed. 8 (1768) No. 1. — Lapathum acetosum Scop. Fl. Carn. ed. 2, 1 (1772) 260. — Lapathum pratense Lam. Fl. Franc. III (1778) 8. — Rumex micranthus Campd. ex Meisn. in DC. Prodr. XIV (1856) 65. — R. pratensis Dulac, Fl. Hautes Pyren. (1867) 165. — Ic.: Rchb. Ic. Fl. Germ. XXIV (1909) 194. — Russian: shchavel' obyknovennyi, kislyi [common or sour].

Perennial; root short, fibrous; stem erect, sulcate, to 1 m long, with small toothed ocreae in the inflorescence, branched; leaves somewhat fleshy; the lower and radical long-petioled, 2.5—13 cm long, 1.5—5 cm broad, ovate-oblong, short-acuminate or obtuse, sagittate at base with descending triangular acute lobes, sometimes subhastate; upper stem leaves sessile, narrower than the lower ones; inflorescence a slender rather loose panicle; flowers unisexual (plants dioecious) rose-colored, red, yellowish, or multicolored, disposed in rather loose clusters within the inflorescence; pedicels jointed at the middle; tepals of staminate flowers oblong-oval, the inner somewhat larger than the outer, deciduous, all erect; outer tepals of pistillate flowers reflexed and appressed to pedicel; valves erect, accrescent in fruit to 3.5—4 mm, suborbicular, entire, cordate at base, with a fleshy recurved appendage in the sinus; achene trigonous, dark brown, lustrous, pointed, 1.5—2 mm long, less than 1 mm broad. June—July. (Plate XXVI, Figures 3, 7).

Meadows, sparse woods, grassy slopes, in the alpine zone of mountains. — European part: all regions, except the Crimea; Caucasus: Cisc., Dag., Transc.; W. Siberia: U. Tob., Alt.; E. Siberia: Yen., Ang.-Say., Dau.;
452 Centr. Asia: Pam.-Al., T. Sh.; Far East: Uss., Ze.-Bu., Sakh. **Gen. distr.**: Arctic, Scand., Centr. Eur., Atl. Eur., S. Eur., Bal.-As. Min., Med., Mong., Jap.-Ch., Dzu.-Kash., Ind.-Him., N. Am. Described from Europe. Type in London.

Note. The following varieties have been distinguished:

var. pratensis (Mill.) Wallr. Sched. critic. (1882) 182. — R. acetosa var. vulgaris Koch, Syn. ed. 2 (1845) 709. — Lower leaves oblong-elliptic, with almost parallel descending lobes;

var. velutinus K. Johansson, Sv. Vet.-Akad. Handl. XXIX (1897) 108. — Plants densely white-pubescent throughout. Lad.-Ilm. (Oredezh River);

var. fissus Koch, Syn. ed. I (1836) 616. — Lower leaves long, with incised lobes.

Economic importance. The sap contains 1.36% of potassium oxalate and free oxalic acid (Wehmer). The leaves are used as a vegetable. Introduced into cultivation.

(453)

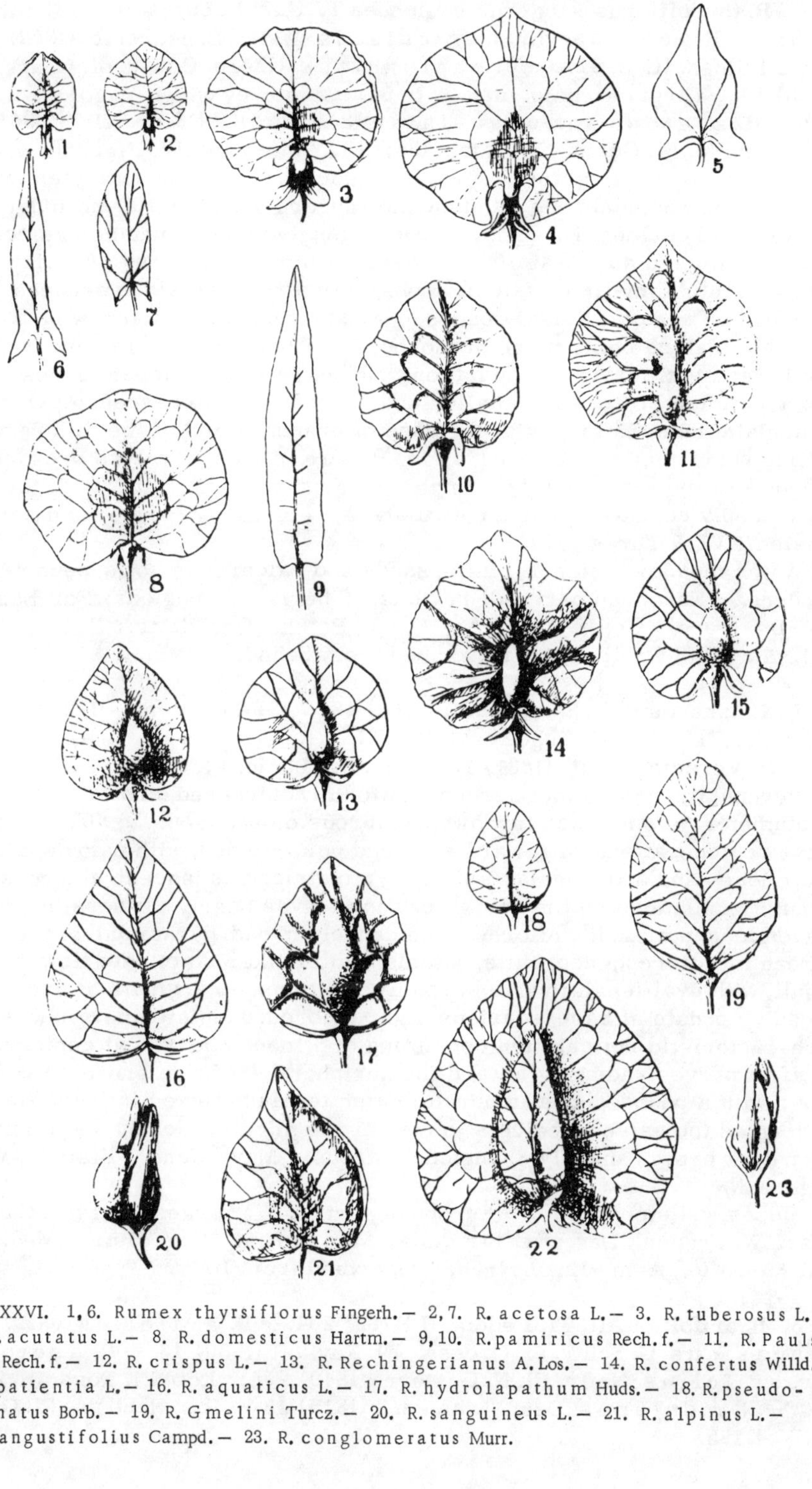

PLATE XXVI. 1, 6. Rumex thyrsiflorus Fingerh. — 2, 7. R. acetosa L. — 3. R. tuberosus L. — 4, 5. R. acutatus L. — 8. R. domesticus Hartm. — 9, 10. R. pamiricus Rech. f. — 11. R. Paulsenianus Rech. f. — 12. R. crispus L. — 13. R. Rechingerianus A. Los. — 14. R. confertus Willd. — 15. R. patientia L. — 16. R. aquaticus L. — 17. R. hydrolapathum Huds. — 18. R. pseudonatronatus Borb. — 19. R. Gmelini Turcz. — 20. R. sanguineus L. — 21. R. alpinus L. — 22. R. angustifolius Campd. — 23. R. conglomeratus Murr.

6. **R. thyrsiflorus** Fingerh. in Linnaea IV (1829) 380; Kryl., Fl. Zap. Sib. IV, 837. — R. *acetosa* var. *auriculatus* Wallr. Sched. crit. (1822) 182; Ldb. Fl. Ross. III, 2, 511. — R. *haplorhizus* Czern. Consp. pl. Chark. (1859) 53; Turcz. Fl. baic.-dahur. II, 54; Grigor'ev in Fl. Yugo-Vost. IV, 122. — R. *acetosa* var. *angustata* Meisn. in DC. Prodr. XIV (1856) 65. — Ic.: Rchb. Ic. Fl. Germ. XXIV, tab. 196; Fl. Yugo-Vost. IV, Fig. 250.

Perennial; root vertical, stout, long, sometimes branched; stem erect, deeply furrowed, glabrous or glandular-puberulous, 50—100 cm long; leaves 3—13 cm long, 1.5—4 cm broad, oblongly ovate-lanceolate, strongly attenuate toward apex, sagittate at base, with more or less spreading narrow acute sometimes falcate lobes, with a rounded sinus between them, glabrous or rarely diffusely pubescent, surrounded by a narrow undulate rim, this feature being more pronounced in Central Asian forms; radical and lower stem leaves borne on long ribbed petioles; upper leaves short-petioled or sessile, smaller and narrower; inflorescence pyramidal-paniculate, dense; bracts at the base of branches and of the inflorescence whorls scarious, 1—3-toothed, as in R. *acetosa* L.; valves broadly rounded-ovate, with slightly crenate margins, somewhat attenuate toward
455 apex, deeply cordate at base, reticulate, 3—4 mm long. June—July. (Plate XXVI, Figures 1, 6).

Water meadows, slopes, bluffs, sandy and calcareous soils, pebbles, and borders. — European part: all regions; Siberia: all regions; Far East: Ze.-Bu. (rare); Centr. Asia: Ar.-Casp., Balkh. **Gen. distr.**: Scand., Bal.-As. Min., Centr. Eur. Described from Scandinavia.

7. **R. tuberosus** L. Sp. pl. ed. II (1762) 481; Ldb. Fl. Ross. III, 2 (1849) 511; Grigor'ev in Fl. Yugo-Vost. IV, 123. — R. *acetosa* α *tuberosus* C. A. M. Verzeichn. Pfl. (1831) 156. — Ic.: Rchb. Ic. Fl. Germ. XXIV, t. 200.

Perennial; roots tuberiferous, fusiformly thickened, brownish, to 0.7 cm in diameter; stem erect, furrowed, glabrous or scabrous, to 60 cm long; leaves oblong, attenuate toward apex and point-tipped, with narrow and acute lobes and a triangular sinus, asymmetric, glaucescent, glabrous, minutely spinulose on the veins; cauline leaves nearly all sessile; radical leaves long-petioled; inflorescence densely broad-pyramidal, branched; ocreae in inflorescence white, obtusish, collarlike; staminate flower very small, with oval tepals; valves broadly reniform, emarginate at apex, broadly cordate at base, scarious, light-colored, undulate-margined, with a characteristic reticulation consisting of 6 looped nerves at center and straight nerves radiating toward the periphery, 4—5 mm broad, 3—5 mm long; with a rounded-oval tubercle; outer tepals reflexed in fruit, almost as long as the valves, pedicels jointed at the middle; achene ca. 1 mm long, dark brown, sharply 3-angled, lustrous. May—June. (Plate XXVI, Figure 3).

Slopes and stony soils. — European part: L. V. (Ergeni Hills); Caucasus: Cisc., W. Transc., Dag.; Centr. Asia: Mtn. Turkm. **Gen. distr.**: Med., Bal.-As. Min., Arm.-Kurd., Iran. Described from Italy. Type in London.

8. **R. arifolius** All. Fl. Pedem. II (1785) 204. — R. *arifolius* var. *caucasicus* G. Schir. ex Grossh., Fl. Kavk. II (1930) 43. — R. *acetosa* var. *arifolius* Neilr. Fl. N. Oesterr. (1849) 294; Ldb. Fl. Ross. III, 2, 510. — R. *montanus* Desf. Tabl. ed. 2 (1815) 48. — Ic.: Rchb. Ic. Fl. Germ. XXIV, t. 195.

Perennial; root fibrous as in R. acetosa L.; stem erect, straight, coarsely sulcate, to 100 cm long; ocreae yellowish-hyaline, cuspidate;
456 leaves ovate-triangular, acute, broadly sagittate at base with spreading broad-triangular lobes and a broad obtuse-angled sinus, wavy-margined, the radical long-petioled, the middle and upper sessile; inflorescence paniculate, with a number of racemiform branches arising from a common peduncle; pedicels jointed at the middle; flowers unisexual, in rather loose few-flowered whorls; tepals of staminate flowers narrow; those of pistillate flowers oblong, undulate marginally, reticulate, surpassing the fruit, minutely appendaged; achene sharp-angled, darkish. June—July.

Tundra, subalpine and alpine mountain zones. — Arctic: Arc. Sib; European part: Kar.-Lap., Dv.-Pech., V.-Kama (Urals); W. Siberia: Alt., U. Tob., Ob (W. part); Far East: Kamch., Sakh.; Caucasus: Cisc., Dag., E. and W. Transc. **Gen. distr.**: S. Eur., Med., Bal.-As. Min., Scand. Described from Piedmont.

Note. Caucasian forms were designated by Shiryaev as var. caucasicum without, however, any statement of distinguishing characters. The Altai specimens differ from some Caucasian and European plants in more rounded leaves but, as the leaf shape is very variable, it is difficult to set the form apart. Pohle marked herbarium plants from northern areas of the European part of the USSR, characterized by their diminutive size and small leaves, as var. alpinus Pohle. Kamchatkan forms, designated by Komarov and Hultén as R. acetosa var. alpinus but rightly referable to the species R. arifolius All., differ from the European plants in their leaves having more rounded basal lobes and a rounded (not pointed) sinus.

The hybrid R. aplinum × R. arifolius stands closer to R. arifolius but the plants are larger (R. Brüggeri Gurke, P. Eur. II (1897) 109).

9. **R. acutatus** L. Sp. pl. (1753) 337; Boiss. Fl. Or. IV, 1015. — R. hastifolius M. B. Fl. taur.-cauc. I (1808) 290. — R. pubescens C. Koch in Linnaea XXII (1849) 211. — Ic.: Rchb. Ic. Fl. Germ. XXIV, tab. 195.

Perennial; root slender, branched; stems many, branched, erect or ascending, flexuous, glabrous or minutely papillose, slightly sulcate, glaucescent, 30—60 cm long; leaves fleshy; the radical long-petioled, broad-oval or triangular, with oval lobes or auricles, cordate at base, obtuse at apex; the upper hastate, broadly ovate-rhombic, with broad-oval lobes, hastate-cordate at base, acute; flowers in terminal racemes consisting of rather distant 2—4-flowered whorls; pedicels short, filiform, jointed at the middle; staminate flowers terminal, with reddish oval tepals and strongly exserted yellowish-red anthers; outer tepals of pistillate flowers
457 very small; valves round, reniform-cordate at base, thin yellow or pink, with darker nerves and an uneven margin, 7—8 mm in diameter; achene sharp-angled, pointed at apex and at base, yellow. Fl. May—June. (Plate XXVI, Figures 4, 5).

Rocks, embankments, limestone, and slopes. — European part: Crim.; Caucasus: W. and E. Transc., Dag. **Gen. distr.**: Centr. Eur., Med., Bal.-As. Min., Arm.-Kurd., Iran. Described from Switzerland. Type in London.

Varieties recorded:

var. typicus Beck. — Stems simple or sparsely branched, with a solitary inflorescence.

var. hastatus Schult. — Leaves triangular, acute, with indistinct triangular lobes (E. Transc.).

var. pubescens Beck. — Plants pubescent (Cisc.).

var. glaucus Jacq. — Plants glaucous, strongly branched, with numerous inflorescences (Cisc., Dag., Transc.); the most frequent form.

Section 3. **HOLOLAPATHUM** A. Los. sect. nova. — Lapathum Meisn. in DC. Prodr. XVI (1856) 42, p. p. — Flowers hermaphrodite; leaf base cuneate, rounded, or cordate; valves entire or undulate-margined.

10. **R. Gmelini** Turcz. Cat. baic. (1838) 986, nom. nud.; Ldb. Fl. Ross. III, 2 (1851) 508. — R. alpinus Georgi, It. I (1775) 287 non L.

Perennial; rhizome fibrous; stem 25 — 40 cm long, erect, sulcate, reddish or yellowish, in inflorescence scabrous; leaves long-petioled, broadly triangular-sagittate, rounded at apex, with rounded lobes, deeply sinuate at the broad-cordate base, to 15 cm long, 10 cm broad, glabrous above, appressed-hairy on the veins beneath; upper leaves narrower, oblong; flowers in loose whorls forming a racemiform cylindric inflorescence; pedicels about twice the length of the perianth, jointed at base; outer tepals ovate, to 2 mm long; valves to 7 mm long and 5 mm broad, oval-ovate, rounded at apex, subtruncate at base, entire, all without tubercles. July. (Plate XXVI, Figure 19).

River banks and seacoasts. — Arctic: An.; E. Siberia: Dau.; Far East: Uda, Okh., Sakh. Endemic. Described from Transcaucasia. Type in Leningrad.

11. **R. jakutensis** Kom. in Not. Syst. Herb. H. B. P. II (1921) 131.

Perennial; root fusiform, stout; stem erect, simple, to 50 cm long, smooth, terete; radical leaves 3, appressed to stem, on petioles to 8 cm long; blade 8 — 11 cm long, 2 cm broad, oblong-triangular, rounded-obtuse at apex, with
458 rounded-oblong basal auricles 0.5 — 2.5 cm long and 1 — 1.5 cm broad, white-tomentose beneath with imperceptible lateral veins, dark green and glabrous above; raceme erect, 25 cm long, with few appressed branches, the lower whorls 20-flowered, the upper 3 — 10-flowered; valves without tubercles, entire. Differing from the closely related R. Gmelini Turcz. in having pubescent leaves. May.

Grass-covered marshy places. — E. Siberia: Lena-Kol., Aldan River area. Endemic. Described from the Aldan River. Type in Leningrad.

12. **R. domesticus** Hartm. Scand. Fl. (1820) 148; Ldb. Fl. Ross. III, 2, 506; Kryl., Fl. Zap. Sib. IV, 825; Kom. and Alis., Opr. rast. Dal'nevost. kr. I, 459. — R. longifolius DC. in Lam. et DC. Fl. Fr., Suppl. (1815) 386. — Ic.: Rchb. Ic. Fl. Germ. XXIV, t. 161. — Exs.: Pl. Finl. Exs. No. 600.

Perennial; stem 40 — 100 cm long, erect, sulcate, branched; radical and lower cauline leaves petiolate, oblong-ovate or ovate-lanceolate, 2.5 — 4.5 times as long as broad, slightly undulate marginally, rounded or subcordate at base; upper leaves narrow-lanceolate, undulate-margined, cuneate at base; petioles flat above; flowers in approximate whorls; inflorescence dense, profusely branched, often from base, sparingly leafy; outer tepals reflexed toward the pedicel, much less than half the length of the valves; valves 4 — 6 mm long, rounded-reniform, cordate at base, entire or slightly undulate, not tubercled; achene brownish, sharp-angled, enlarged toward base, acuminate at apex, to 2.6 mm long and 1.6 mm broad. Rare. June — July. (Plate XXVI, Figure 8).

Cultivated fields, forest glades, roadsides, and wood margins. — Arctic: Arc. Eur., Nov. Z.; European part: Kar.-Lap., Lad.-Ilm., Dv.-Pech., U. V., V.-Dnp., V.-Don, V.-Kama, M. Dnp. (N. part), L. Don; Caucasus: Cisc., Dag.; W. Siberia: Alt., U. Tob.; E. Siberia: Yen., Ang.-Say. **Gen. distr.** - Centr. Eur., Scand., Atl. Eur., N. Am. Described from Scandinavia.

Note. A weed of ornamental and truck gardens, rarely of cultivated fields. Difficult to eradicate because of regeneration of the root system.

13. **R. pamiricus** Rech. f. in Fedde, Repert. sp. nov. XXXI (1933) 258. — R. domesticus Fedtsch. in Trav. Mus. Bot. Ac. Sc. Pétersb. I (1905) 161; Ej. Fl. Pamir. in A. H. P. XXI (1903) 414; Ej. Rast. Turk. (1915) 312. — ? R. rectinervis Rech. f. in Fedde, Repert. sp. nov. XXXIII (1934) 355.

Perennial; rhizome stout, vertical, yellow in section; stem erect, reddish, sulcate, branching in inflorescence, to 1 m long; leaves firm,
459 reddish, long-lanceolate, oval-lanceolate or linear-lanceolate, cordate at base, cuspidate at apex, in inflorescence very narrow and acute; all leaves glabrous beneath, the slender lateral veins spreading at nearly a right angle in the lower part of the blade; petioles shorter than the blade, dilated toward base; inflorescence ovaloid-paniculate, beginning almost at the base of the stem; outer tepals oval, narrow; valves rounded-reniform, 6 — 8 mm in diameter, cordate at base, tapering to a rounded tip, slightly toothed on the margin, flat, thin, reticulate, reddish-violet or orange, all without tubercles; pedicels slender, about twice the length of the fruit, jointed at base; achene light brown, 2 mm long, 1 mm broad, narrowed at both ends. July — August. (Plate XXVI, Figures 9, 10).

Wet mountain meadows and banks of mountain streams and lakes. — Centr. Asia: Pam.-Al., T. Sh., Dzu.-Tarb. **Gen. distr.**: Dzu.-Kash. Described from Shugnan, Kun-lal. Type in Copenhagen.

Note. R. rectinervis Rech. f. was described from a single specimen, containing one leaf and a portion of inflorescence with immature fruits, from the Chimkent area, i. e., from the most westerly point of the distribution area. It differs from typical R. pamiricus Rech. f. in its longer inflorescence branches. There are no grounds whatever for regarding it as a distinct species. It may be a hybrid specimen. In his description of R. pamiricus, Rechinger mentions one tubercled tepal, a feature not to be found in this species.

14. **R. kamtschadalus** Kom. in Fedde, Repert. sp. nov. VIII (1914) 166. — R. arcticus Hultén, Fl. of Kamtch. II (1928) 45 (pro parte). — Ic.: Kom., Fl. Kamch. II (1928) 59, t. X.

Perennial; roots firm, black, bitter-tasting; stem to 1 m long and to 1 cm in diameter, erect, sulcate, glabrous, hollow, firm; ocreae brownish-hyaline, large; all leaves petiolate, oblong, cuneate at base, gradually attenuate to a rounded apex, pinnately veined (in dry plants), 10—20 cm long, 3—4 cm broad, paler beneath, glabrous; inflorescence a large panicle consisting of subapproximate or sometimes distant whorls; pedicels nodding, about 3 times the length of the fruit, without distinct articulation; staminate flowers rather large, with dark violet or yellow oval tepals; valves to 1 cm long, broad-ovate, subtruncate or subcordate at base, reticulate, not tubercled; achene oblong, trigonous, dark, lustrous. July.

Bogs, old overgrown riverbeds and lakes, solitary or in groups. — Far East: Kamch. Endemic. Described from the Kamchatka River basin near the village of Kirganik. Type in Leningrad.

15. **R. sibiricus** Hultén, Fl. of Kamtch. II (1928) 48. — R. salicifolius
460 auct.; Ldb. Fl. Ross. III, 2, 504.

Perennial; rhizome stout, creeping, rufous, with several offshoots; stem erect or decumbent, bent at the nodes, branched from the leaf axils, finely sulcate, glabrous, reddish-lilac especially in lower part, 20—60 cm long; radical and cauline leaves alike, narrow-lanceolate, narrowed and pointed at both ends, 5—10 cm long, 0.5—1.5 cm broad, glabrous, rather firm, with a broad midrib and faint lateral veins, very slightly crisped, the blade 2—3 times as long as the reddish petiole; inflorescence leaves sublinear, short-petioled, in clusters together with 2—5 fairly large leaves in the axils of stem leaves from which often arise, especially on the upper part of the stem, short flowering branches; ocreae rather firm, brownish, somewhat inflated, to 1.5 cm long, with faint longitudinal nerves anastomosing in upper part; inflorescence pyramidal, with slender divaricate fairly long branches; flowers in many-flowered whorls; inner tepals purple or yellow, oval-triangular, to 1.5 mm long; outer tepals shorter and narrower; pedicels short, jointed near base; inflorescence branches elongating in fruit, the whorled fascicles distant, only at the ends of branches approximate; valves elongating to 2 mm, firm, not reticulate, entire, oval, the large oval orange-red tubercle occupying the greater part of the tepal; achene brownish, 1 mm long, sharp-pointed, lustrous. August.

Sands and clayey soil, shallows, and river floodplains. — Arctic: An., Chuk.; E. Siberia: Lena-Kol., Yen. (N. part); Far East: Kamch. **Gen. distr.**: Alaska. Described from Kamchatka. Type in Leningrad.

16. **R. Komarovii** Schischk. et Serg. in Animadvers. Syst. ex Herb. Univers. Tomsk. ed. Sect. Tomsk. Soc. Bot. Ross. No. 3 (1929).

Perennial; stem branched from base, 12—20 cm long; branches arcuately spreading; leaves oblong-elliptic, oblong, or lanceolate, short-acuminate or subobtuse, slightly undulate-margined, cuneately narrowed toward base, 1.5—2.5 cm long, 0.4—1 cm broad, the lower on petioles 5—9 mm long, the upper smaller on short pedicels; inflorescence whorls many-flowered, leafy to very top of the stem, starting from base, the lower distant, the upper approximate; pedicels jointed at base; outer tepals oblong-ovate, one-fourth to one-half the length of the inner, one of them shorter than the other two; valves oblongly triangular-ovate, acuminate, equal or one of them somewhat longer, to 2 mm long, entire, transversely rugose by prominent nerves, one with a large tubercle, on the others the tubercle obsolescent or none. July.

461 Roadsides. — W. Siberia: Irt. Endemic. Described from the village of Georgievskoe in the Semipalatinsk District. Type in Tomsk.

17. **R. crispus** L. Sp. pl. (1753) 335; Ldb. Fl. Ross. III, 2, 505; Kryl., Fl. Zap. Sib. IV, 827; Shmal'g., Fl. II, 398. — Lapathum crispum Scop. Fl. Carn. ed. 2, I (1772) 261. — R. patientia var. crispus O. Ktze. Rev. Gen. I (1891) 560. — Ic.: Rchb. Ic. Fl. Germ. XXIV, t. 163.

Perennial; root vertical, fusiform stout; stem erect, glabrous, coarsely but rather shallowly sulcate, 50 — 120 cm long; leaves lanceolate to oblong-lanceolate, long-acuminate, cuneately narrowed toward base, strongly undulate marginally, the blade 15 — 25 cm long and 1.5 — 4 cm broad, the petiole slightly shorter than the blade; upper leaves smaller, short-petioled; flowers hermaphrodite, in whorls of 20 — 30, forming a narrow-paniculate inflorescence to 60 cm long, with appressed branches; valves rounded-ovate, cordate at base, acuminate at apex, slightly concave, smooth or slightly denticulate on the margin, to 5 mm across, all (var. triangulatus Bosw. = β typicus Beck.) or only one of them (var. unicallosus Peterm.) with a rounded-oval tubercle only a fraction of the length of the valve. (Plate XXVI, Figure 12).

Var. japonica (Meisn.) Makino, reported for Kamchatka, differs in having less undulate leaves. Hultén (Fl. Kamtch.) claims that it is a distinct species, but we retain it as a variety, following in this respect the Japanese taxonomists.

Meadows, field borders, weedy places, slopes, and roadsides. — European part: all regions except Arctic; Caucasus: throughout; Far East: Kamch., Uss.; Centr. Asia: Ar.-Casp., Balkh., T. Sh., Amu D., Syr D., Pam.-Al. **Gen. distr.**: Scand., Centr. Eur., Atl. Eur., Bal.-As. Min., Mong., Jap.-Ch., N. Am. Described from Europe. Type in London.

Note. A common weed of field borders; difficult to eradicate because of the strong root system regenerating on cutting. Very often confounded with R. pseudonatronatus Borb. Hybridizes with R. maritimus (R. fallacinus Hausskn. Mitt. Geogr. Ges. Thür. III (1885) 74), R. confertus, R. conglomeratus and R. obtusifolius (= R. acutus L. Sp. pl. (1753) 395).

18. **R. Fauriei** Rech. f. in Fedde, Repert. sp. nov. XXXIII, 11 — 25 (1934) 358.

Perennial; stem to 1 m long, deeply furrowed, thickened at the nodes, glabrous, branched in upper part; branches slender, upright; leaves oblong-oval to lanceolate, rounded or cordate at base, acuminate at apex,
462 slightly dilated toward base, finely crisped on the margin, the petioles shorter than the blade; radical and lower cauline leaves larger, 3 — 4 times as long as broad; upper cauline and inflorescence leaves oblong-lanceolate, 6 times as long as broad, rounded or cuneate at base; all leaves glabrous, flat, with distinct lateral veins arising at an angle of 60°; petioles of radical leaves about as long as or longer than the blade, those of upper leaves shorter than the blade; inflorescence ovaloid-pyramidal, dense; slender, jointed in lower part; flowers in dense whorls; tepals obovate, 3 — 5 mm long, 3 — 5 mm broad; valves oval or cordate-triangular, acuminate, entire, light brown, thin, not tubercled but with midnerve thickened

in lower part; outer tepals linear in fruit, 1.5 mm long, reflexed; achene yellowish-brown, pointed at both ends, to 2.5 mm long. May—July.

Wet soil and riverbanks. — Far East: Sakh. Endemic. Described from Sakhalin, the village of Korsakovo. Type in Vienna; cotype in Leningrad.

19. **R. pseudonatronatus** Borb. in Kerner Fl. Austro-Hungar. exs. No. 3071; Schedae VIII (1899) 71; Grigor'ev in Fl. Yugo-Vost. IV, 116. — R. fennicus Murb. in Bot. Not. (1899) 17 et ib. (1913) 228. — R. crispus Ldb. Fl. Ross. III, 2, 505, p. pte; Kryl.; Fl. Atl. V, 1150, ex parte. — R. domesticus β pseudonatronatus Borb. Bekésmegye flórája XI (1881) 63. — Ic.: Rchb. Ic. Fl. Germ. XXIV, tab. 162.

Perennial; stem erect, glabrous, shallowly sulcate, simple, only in inflorescence with appressed branches; leaves lanceolate to narrow-lanceolate, acuminate, cuneately narrowed toward base, undulate marginally, 15—20 cm long, 1.5—3 cm broad, the upper ones smaller; inflorescence slenderly paniculate, rather dense, 20—50 cm long, 2—6 cm broad, the approximate whorls many-flowered; pedicels longer than perianth, jointed below the middle; valves twice as long as the outer tepals, rounded-ovate, cordate at base, rounded at apex, entire, 4—5 mm long, none tubercled. June—July. (Plate XXVI, Figure 18).

Meadows, wood margins, roadsides, and shores. — European part: Kar.-Lap., Dv.-Pech., U. V., V. Kama, U. Dnp., Transv., M. Dnp., V.-Don, L. Don; W. Siberia: Ob, U. Tob., Irt., Alt.; E. Siberia: Yen., Ang.-Say.; Centr. Asia: Ar.-Casp., Balkh. **Gen. distr.**: Scand., Centr. Eur. (Hungary). Described from Hungary.

20. **R. hydrolapathum** Huds. Fl. Angl. ed. 2 (1778) 154; Ldb. Fl. Ross. III, 507; Shmal'g., Fl. II, 397. — Rumex maximus Gmel. Fl. Bad. II (1806) 99. — R. acutus Wahlenb. Fl. Ups. (1820) 118. — Lapathum rivale Rénault, Fl. Dep. Orne (1804) 60. — L. giganteum Opiz, Sezn. (1832) 57. — Ic.: Rchb. Ic. Fl. Germ. XXIV, tab. 165.

463 Perennial; stem 1—2 m long, erect, finely sulcate, with upright branches arising in upper part; leaves large, lanceolate, narrowed at both ends, pointed at apex, rather firm, slightly undulate marginally, with a stout midrib and slender lateral veins, to 60 cm broad, the petiole much shorter than the blade; upper leaves shorter and narrower, denticulate, in inflorescence narrow-lanceolate or linear-lanceolate; inflorescence spreading, broadly paniculate, dense, sparsely leafy; outer tepals oblong-linear, to 2.5 mm long; valves ovate-triangular, subtruncate at base, rounded at apex, to 7 mm long, 4 mm broad, all with an oblong tubercle half the length of the tepal. July—August. (Plate XXVI, Figure 17).

Bogs, and banks of rivers, lakes, and brooks. — European part: all regions except the north. **Gen. distr.**: Centr. Eur., Atl. Eur., Scand., W. Med., Bal.-As. Min. Described from England. Type in London.

Hybridizing with R. aquaticus (= R. heterophyllus Schultz, Prodr. Fl. Starg. Suppl. I (1819) 21); with R. obtusifolius = R. lingulatus Jung. in Bot. Notis. (1885) 115, f. 6; with R. crispus — two forms: R. Schreiberi Husskn. and R. Murbeckii Beck. (see Rchb. Ic. Fl. Germ. XXIV, 32, tab. 174, f. 1, 4). — described from Sweden.

21. **R. confertus** Willd. Enum. Hort. Berol. (1809) 397; Ldb. Fl. Ross. III, 2, 509; Shmal'g., Fl. II, 398; Kryl., Fl. Zap. Sib. IV, 822. — R. *retroflexus* Lag. Elench. Pl. (1816) 13. — R. *pauciflorus* Campd. Monogr. Rum. (1809) 104. — R. *alpinus* α *subcalligerus* Boiss. Fl. Or. IV (1879) 1007. — Ic.: Rchb. Ic. Fl. Germ. XXIV, tab. 159. — Exs.: Fl. exs. Austro-Hung. No. 3074.

Perennial; stem erect, 60—120 cm long, sulcate, leafy; lower leaves oblongly triangular-ovate, obtuse, cordate at base, wavy-margined, hispid on the veins beneath, 15—25 cm long, 6—12 cm broad, the petiole as long as or longer than the blade, channeled above; inflorescence narrow-cylindric, dense, with approximate whorls, almost leafless; pedicels jointed at or just below the middle; valves rounded-cordate, obtuse or acutish (var. *cordifolius* (Horn.) Gürke), reticulate, crenate on the margin, mostly broader than long, 6—7 mm across, one of them with a more developed tubercle. May—June. (Plate XXVI, Figure 14).

Hybridizing with R. *crispus*, R. *obtusifolius*, and R. *conglomeratus*.

Meadows, slopes, groves, forest glades. — European part: all regions except the north; Caucasus: Transc.; W. Siberia: Ob, U. Tob., Irt., Alt.; E. Siberia: Ang.-Say., Dau.; Far East: Uss.; Centr. Asia: Ar.-Casp., Balkh., Syr D. **Gen. distr.**: Centr. Eur., Bal.-As. Min. Described from a
464 cultivated specimen in the Berlin Garden. Type in Berlin.

Note. R. *confertus* from Central Asia has more triangular leaves.

22. **R. Fischeri** Rchb. Ic. Pl. Crit. IV (1823) 58; Ldb. Fl. Ross. III, 505.

Perennial; stem erect, sulcate, leafy, bearing in upper part a slender inflorescence with short appressed branches; radical leaves broad-ovate, rounded at apex, broadly cordate at base, to 20 cm long, 14 cm broad, glabrous above, scabrous beneath, hirtellous on the veins beneath, the petiole longer than the blade; cauline leaves oblong-oval, obtuse, rounded or narrowed at base, 5—7 cm broad, also long-petioled; inflorescence leaves lanceolate; flowers in few-flowered approximate whorls; pedicels much longer than fruit, jointed slightly below the middle; valves rounded-cordate, slightly taper-tipped, prominently nerved, slightly toothed on the margin, 5 mm broad, 4 mm long, one of them with round tubercles, the others without tubercles. April—May.

Slopes. — Centr. Asia: T. Sh., Dzu.-Tarb. (Ayaguz). Endemic. Reichenbach described this species from material sent by Fischer and marked "from Siberia."

23. **R. patientia** L. Sp. pl. (1753) 333; Ldb. Fl. Ross. III, 2, 507. — R. *patienta* ssp. *patientia* Rech. f. in Fedde, Rep. sp. nov. XXXI (1933) 256. — *Lapathum hortense* Moench, Meth. (1794) 354. — Ic.: Rchb. Ic. Fl. Germ. t. 164; Hegi III. Fl. Mitteleur. III, 178.

Perennial; stem 80—120 cm long, erect, sulcate, stout, branched at the end; lower leaves 20—30 cm long, 7— 9 cm broad, ovate-lanceolate, slightly cordate at base, acuminate or obtuse, flat or slightly wavy at the margin, the long petioles channeled above; upper leaves short-petioled, smaller, lanceolate; pedicels slender, jointed in lower part, slightly thickened below the flower; whorls of inflorescence 10—16-flowered, approximate in almost leafy clusters disposed in a long compact panicle; valves entire

or minutely denticulate, light brown, rounded-cordate, 6—8 mm long, 5—7 mm broad, reticulate, rounded or slightly pointed at apex, one of them with a callosity 1.5 mm long; in var. *callosus* F. Schmidt. et Maxim. Prim. Fl. Am. (1859) 227. = R. *callosus* Rech. f. in Fedde, Rep. XXXI (1933) 257, all valves with callosities; seeds sharp-angled, ovaloid, pointed, light brown,
465 3 mm long, 1.5—2 mm broad. June—July. (Plate XXV, Figure 15).

Meadows, riverbanks, and wet soils. — European part: Crim., M. Dnp., L. Dnp., Bl., L. Don; Caucasus: Cisc., E., W. and S. Transc., Dag.; W. Siberia: Alt.; Far East: (var. *callosus* F. Schmidt) Uss., Uda, Sakh. **Gen. distr.**: Med., Centr. Eur., Bal.-As. Min., Arm.-Kurd. Described from Italy. Type in London.

Note. Cultivated and used like spinach under the name of English spinach. Crimean specimens of this species are distinguished by large fruits and, on this account, designated by Rechinger as subsp. *orientalis* (Bork.) Dom.

Hybridizing with R. *obtusifolius* (= R. *erubescens* Simonk. Term. Füz. I (1877) 239).

24. **R. Rechingerianus** A. Los. sp. nova in Addenda IV, p.547.— R. *crispus* Paulsen in Bot. Tidskr. XXIX (1909) 153, non L. — R. *orientalis* Paulsen l. c., non Bern. — R. *domesticus* Fedtsch. in Trav. Mus. Bot. Acad. Sc. Pétersb. I (1905) 161, pro parte.

Perennial; stem stout, 40 cm—2 m long, slightly sulcate, branched from the middle, pale yellow; inflorescence long, narrow-ovaloid to ovaloid, with rather short subdivaricate branches; leaves flat or slightly wavy marginally, glabrous, scabrous on the veins beneath; lower leaves oblong-oval, rounded or subtruncate or slightly cordate at base, attenuate toward apex, the petiole 1/2—3/4 the length of the blade; cauline leaves oblong-lanceolate to lanceolate, cuneately narrowed toward base, acute at apex, scaberulous; flowers in dense whorls of 12—20; valves reddish-brown, 4—6 mm in diameter, broadly rounded-cordate, slightly apiculate, with a smooth flat margin and a concave midnerve, slightly reticulate, rather firm, one or all three of them with rather large convex callosities 1.5—2 mm long and 1—1.5 mm broad; achene 2.5 mm long, pointed at base and apex. July—August. (Plate XXVI, Figure 13).

As a weed in fields, on grassy slopes, and in thickets. — Centr. Asia: Pam. Al., T. Sh., Amu D., Balkh. **Gen. distr.**: Kashg., Dzu. Described from Ol'gina Luga in the Alai Valley. Type in Leningrad.

25. **R. aschabadensis** A. Los. sp. nova in Addenda IV p. 547.

Perennial; stem long, stout, firm, finely sulcate, in lower part hollow, in upper part solid, yellowish-pink in section, unbranched, except for short branches in the inflorescence; cauline leaves glaucescent, rather firm, hirtellous beneath, lanceolate-oval to oblong-oval, acuminate, narrowed or subtruncate at base, rather coarsely wavy marginally, to 20 cm long, 3—5 cm broad, the short petioles finely sulcate; small distinct inflorescences in the axils of upper leaves forming together an ovaloid-
466 cylindric inflorescence, this leafy in upper part; flowers in approximate many-flowered whorls; pedicels slightly longer than perianth, jointed in lower part; valves rounded-reniform in fruit, cordate at base, 6 mm

across, thin, reticulate, with an uneven margin, one of them with a very small narrow oval upward pointed callosity, the others without callosities. June—July.

Slopes and ravines. — Centr. Asia: Mtn. Turkm. (vicinity of Ashkhabad). Endemic. Described from Saondakh Ravine. Type in Leningrad.

Note. Closely related to R. Paulsenianus Rech. f., from which it differs in the shape of inflorescence and of the tepals. The features distinguishing it from R. Rechingerianus A. Los. are the absence of callosity on all tepals and the firmer shaggy leaves.

26. **R. tianschanicus** A. Los. sp. nova in Addenda IV, p. 548.

Perennial; stem stout, firm, hollow, coarsely sulcate, branched; inflorescence broad-paniculate; cauline leaves broad-ovate, attenuate to a pointed apex, rounded-cordate at base, 17— 25 cm long, to 15 cm broad, slightly undulate marginally; blade thin, prominently veined, light green or glaucescent; petioles short, flat below the blade, keeled in lower part; flowers in loose few-flowered whorls; pedicels slender, infundibularly enlarged at summit, unequal, half as long again to 3 times as long as the fruit, jointed at base; tepals thin in fruit, cordate, acutish, deeply cordate at base, slightly compressed, uneven-margined, 6—8 mm broad, 6—7 mm long, one of them with a rather large oval callosity, the others without developed callosities; achene acuminate, light brown, 2 mm long. May—June.

Mountain streams. — Centr. Asia: T. Sh. (Aleksandrovskii [Kirghiz] Range). Described from the Alarcha River valley. Type in Leningrad.

27. **R. Paulsenianus** Reching. f. in Fedde Repert. sp. nov. XXIX (1931) 246. — R. orientalis Paulsen, Bot. Tidskr. XXIX (1909) 153, pro parte, non Bern. — R. domesticus auct. in sched.

Perennial; stem stout, to 2 m long, sulcate, branched from the middle; branches slender, nodding, forming a broad spreading inflorescence; leaves thin, flat or slightly undulate, glabrous, scaberulous on the veins beneath; lower leaves oblong-ovate, attenuate toward apex, 2—5 times as long as broad, cordate, the petiole about as long as the blade; cauline leaves lanceolate or oblong-oval, subtruncate or broadly cuneate at base, short-petioled; flowers in rather loose whorls of 10; pedicels slender, longer than perianth, jointed in upper part; outer tepals to 2 mm long; valves cordate, attenuate at apex, to 6 mm long and to 7 mm broad, pale reddish-brown, deeply cordate at base, acuminate, thin, reticulate, slightly
467 crenate, all without callosity, dilated in middle part. May— June. (Plate XXVI, Figure 11).

Grassy slopes. — Centr. Asia: Pam.-Al., T. Sh., Syr D., Mtn. Turkm., Amu D., Balkh. **Gen. distr.:** Iran., Ind.-Him. Described from Pamir, Ku-i-lal'. Type in Copenhagen.

28. **R. armenus** C. Koch in Linnaea XXII (1849) 209.

Perennial; stem erect, finely sulcate, ca. 80 cm long, glabrous; radical leaves long-petioled, oblong-oval, to 25 cm long and 7 cm broad, rounded-cordate at base, attenuate toward apex, acuminate; cauline leaves smaller, oblong-cordate or lanceolate, with shorter petioles; all leaves glabrous, slightly wavy-margined; inflorescence slenderly ovaloid, with short appressed branches, sparsely leafy; whorls loose, many-flowered; pedicels 1.5 times to twice as long as the fruit, jointed near base; fruits (not quite mature) 9 mm long, 6 mm broad, ovaloid; outer tepals rounded-oval; valves oval-ovate, yellow with reddish margin, reticulate, rounded slightly narrowed at base, subacuminate, uneven-margined, the midnerve very prominent in lower part; achene pale yellow, oblong. June—July.

Mountain meadows. — Caucasus: S. Transc. **Gen. distr.**: Arm.-Kurd. Described from Soviet Armenia (Darachichak). Type in Berlin.

Note. The description of fruit is given here for the first time and is based on a specimen collected by Litvinov in former Kars Region. A little-explored species.

29. **R. aquaticus** L. Sp. pl. (1753) 336; Ldb. Fl. Ross. III, 508; Shmal'g., Fl. II, 399; Kryl., Fl. Zap. Sib. IV, 823. — Ic.: Rchb. Ic. Fl. Germ. XXIV, tab. 160.

Perennial; stem erect, sulcate, appressed-hairy, with upward-pointing branches in upper part, 60—150 cm long; petioles channeled, lower leaves to 30 cm long and to 15 cm broad, ovate to oblong-ovate, acuminate, cordate at base; upper leaves smaller and narrower, oblong or broad-lanceolate, also cordate at base, the petioles shorter; whorls many-flowered, approximate, disposed at the end of the stem in a slender paniculate inflorescence with appressed branches and few leaves at base; outer tepals oblong, obtuse, about half as long and half as broad as the valves; valves oblong-ovate or broad-ovate, 5—6 mm long and about as broad, truncate at base, obtuse-tipped, entire or in lower part crenate, all without callosities. June—July. (Plate XXVI, Figure 16).

468 Riverbanks and lakeshores, floodplains, wet forest and boggy meadows, sparse woods, field boundaries, and as weed among crops. — European part: all regions except Arctic, Bl.? and Crim.; Caucasus: Cisc., Dag.; W. Siberia: U. Tob., Irt., Alt.; E. Siberia: Yen., Lena-Kol., Ang.-Say., Dau.; Far East: Ze.-Bu., Uss., Uda, Kamch.; Centr. Asia: Ar.-Casp., Balkh. **Gen. distr.**: Centr. Eur., Scand., Bal.-As. Min., Mong., Jap.-Ch. Described from Europe. Type in London.

Note. A recently established new species, R. protractus Rech. f. (in Fedde Repert. sp. nov. XXXIII (1934) 356), is distinguishable from R. aquaticus L. by its smaller size; leaves narrower and more elongated, less deeply cordate at base; looser, slender inflorescence; outer tepals smaller and more rounded in fruit; and smaller achenes. This species embraces all the plants from Siberia and the Far East that we have referred to R. aquaticus L. The relationship between these two species has not as yet been fully clarified and requires further study.

Hybridizes with R. obtusifolius (= R. finitimus Gürke in Richter-Gürke Pl. Eur. II (1897) 98), with R. conglomeratus (= R. ambigens Beck. in Rchb. Ic. XXIV (1904) 27 et tab. 168, f. 4, 5 and R. dumulosus Beck. ibidem f. 6), with R. crispus, with R. domesticus and with R. hydrolapathum (= R. maximus Schreb. in Schweigg. und Körte Fl. Erl., I (1811) 152 = R. heterophyllus Schultz, Prodr. Fl. Starg. Suppl. I (1819) 21).

30. **R. alpinus** L. Sp. pl. (1753) 334. — Acetosa alpina Moench, Meth. (1794) 387. — Ic.: Hegi III. Mitteleur. Fl. III (1912) 172, tab. 90; Rchb. Ic. Fl. Germ. XXIV, tab. 158.

Perennial; stem 30 cm — 2 m, erect, sulcate, dark lilac-brown, glabrous or diffusely pubescent, bearing in upper half an ovaloid inflorescence with short upward-pointing branches; radical and cauline leaves long-petioled, broad-ovate, cordate at base, obtuse, 8 — 12 cm long, 6 — 10 cm broad, glabrous above, puberulous on the veins beneath; upper stem leaves smaller; inflorescence leaves acute, oblong-lanceolate; flowers in dense whorls; pedicels slender, recurved, jointed in lower part; tepals greenish; anthers yellow; valves ovate-triangular, lilac, reticulate, subtruncate at base, acuminate, 4 — 6 mm long, 4 — 5 mm broad, all without callosity; achene lustrous, narrowed at both ends, light brown, ca. 2 mm long. June — August. (Plate XXVI, Figure 21).

Wooded slopes, coppices, banks of brooks. — Caucasus: Cisc., W., E. and S. Transc., Dag. **Gen. distr.**: Centr. Eur., Med., Bal.-As. Min. Described from Switzerland. Type in London.

31. **R. Schischkinii** A. Los. sp. nova in Addenda IV, p. 548.

469 Perennial, with a taproot; stem solitary, erect, ca. 30 cm long, dark lilac-red, leafy, slightly sulcate, glabrous; radical leaves lance-ovate, rounded-cordate at base, acute, 5 — 7 cm long, 1 — 3 cm broad, coriaceous, slightly undulate marginally, glabrous above, minutely papillose beneath; petiole sulcate, papillose, shorter than the blade; cauline leaves 1 or 2, smaller, with shorter petioles; inflorescence leaves narrow-oval, 15 — 17 cm long, with short and slender branches; fruiting pedicels slender, unequal, jointed at base; valves broad-ovate, 5 — 6 mm long, 4 — 5 mm broad, slightly erose-margined, subcordate at base, obtuse-tipped, reticulate, all without callosity; achene narrow-ellipsoid, pointed at both ends, light brown, 4 mm long, 2 mm broad. Fl. July; fr. August.

Wet places and woods. — W. Siberia: Alt. (Chuya Steppe, Kurai River). Endemic. Described from the Chuya Steppe, near Koshagach. Type in Leningrad.

Note. Closely resembling R. arcticus Trautv.; differing in the consistency of leaves and wingless petioles.

32. **R. arcticus** Trautv. in Midd. Reise 1 (1847) 29; Ldb. Fl. Ross. III, 2, 506; Hultén Fl. of Kamtch. II, 45 (pro parte). — Ic.: Kom. Fl. Kam. II, 58, tab. IX f. dextra.

Perennial, somewhat fleshy, with fibrous root; stem 15 — 50 cm long, erect, sulcate, dark brown, in lower part dark violet; leaves oblong, slightly crisped marginally, brownish-yellow above, pale green beneath; radical leaves obtuse, rounded or subcordate at base, long-petioled; cauline leaves acute, tapering into a short petiole; ocreae scarious, brownish, to 1.5 cm long, mostly lacerate; inflorescence a simple mostly leafless raceme (sometimes branched at base); flowers in rather loose many-flowered whorls, hermaphrodite, nodding; tepals entire, oblong, obtuse or acuminate, without callosities, in fruit 4 mm long. August.

Sandy seashores, in wet tundra. — Arctic: Nov. Z., Arc. Sib., Chuk., An; E. Siberia: Yen., Lena-Kol.; Far East: Kamch. **Gen. distr.**: N. Am. (Arc.). Described from the Taimyra River (75° N. lat.). Type in Leningrad.

33. **R. angustifolius** Campd. Monogr. Rum. (1819) 63, 73; Ldb. Fl. Ross. III, 2, 499; Grossg., Fl. Kavk. II, 40. — Ic.: Campd. l. c. t. 1, f. 2.

Perennial; stem erect, unbranched, 20—40 cm long, glabrous, ribbed, bearing in upper part a simple raceme consisting of distant 5—7-flowered
470 whorls; leaves rather firm, lustrous above; radical leaves lanceolate, acute, cuneately narrowed toward base, 4—8 cm long, 0.5—1.5 cm broad, the petiole as long as the blade; cauline leaves short-petioled, narrow-lanceolate; inflorescence leaves only at the base of lower whorls, linear-lanceolate; valves unequal, one larger cordate-triangular, to 1 cm broad and 8 mm long, broadly cordate at base, acuminate, coarsely reticulate, with thickened midnerve, without callosity, the other two oblong, shorter and much narrower; pedicels short, not jointed at the middle, recurved; achene to 5 mm long, 2.5 mm broad, oblong-ovate, narrowed toward base, strongly acuminate, dull light brown. May—June. (Plate XXVI, Figure 22).

Wet meadows and slopes. — Caucasus: E. and S. Transc. **Gen. distr.**: Arm.-Kurd., Iran., Asia Minor. Described from Cappadocia.

34. **R. conglomeratus** Murr. Prodr. Fl. Goett. (1770) 52; Shmal'g., Fl. II, 396. — R. nemolapathum Ehrh. Beitr. I (1787) 181. — R. ferrugineus Willd. in Spreng. Syst. II (1825) 158. — Lapathum glomeratum Gilib. Exerc. Phytol. II (1792) 444. — Ic.- Rchb. Ic. Fl. Germ. XXIV (1909) t. 166.

Perennial; stem 0.5—1 m long, erect, branched, reddish-green, coarsely sulcate; branches divaricate, forming a broadly ovaloid-pyramidal inflorescence; leaves short-petioled, oblong, rounded or slightly cordate at base, 6—10 cm long, 2—3.5 cm broad; upper leaves lanceolate, slightly repand, smaller and narrower; inflorescence leaves small, acute, 1—2 cm long; pedicels nodding, jointed in lower part; flowers in rather dense distant whorls borne in the axils of inflorescence leaves; valves 3 mm long, oval-rectangular, crenate, reticulate, each with a large oval callosity. June—July. (Plate XXVI, Figure 23).

Shores, wet places, bogs, roadsides, and weed-infested places. — European part: U. Dnp., V.-Don, Bl., M. Dnp., Crim.; Caucasus: Cisc., E. and W. Transc.; Centr. Asia: Pam.-Al., Syr D., Mtn. Turkm. **Gen. distr.**: Centr. Eur., Scand., Atl. Eur., Med., Bal.-As. Min., Arm.-Kurd. Described from Switzerland.

473 Hybridizing with R. hydrolapathum (= R. hybridus Hausskn. Mitt. Geogr. Ges. Thür. III (1885) 69), with R. aquaticus, R. crispus, R. obtusifolius, R. maritimus and R. pulcher.

Note. Specimens from Gissar are characterized by smaller fruits with entire valves.

35. **R. sanguineus** L. Sp. pl. (1753) 333. — R. nemolapathum Ehrh. Beitr. I (1787) 181, pro parte; Ldb. Fl. Ross. III, 2, 503. — Ic.: Rchb. Ic. Fl. Germ. XXIV (1909) t. 167.

Perennial, with a taproot; all roots reddish; stem 40—100 cm long, reddish-yellow, sulcate, slender, branched, each branch terminated by a rather long raceme consisting of loosely 5—8-flowered whorls; radical

(471)

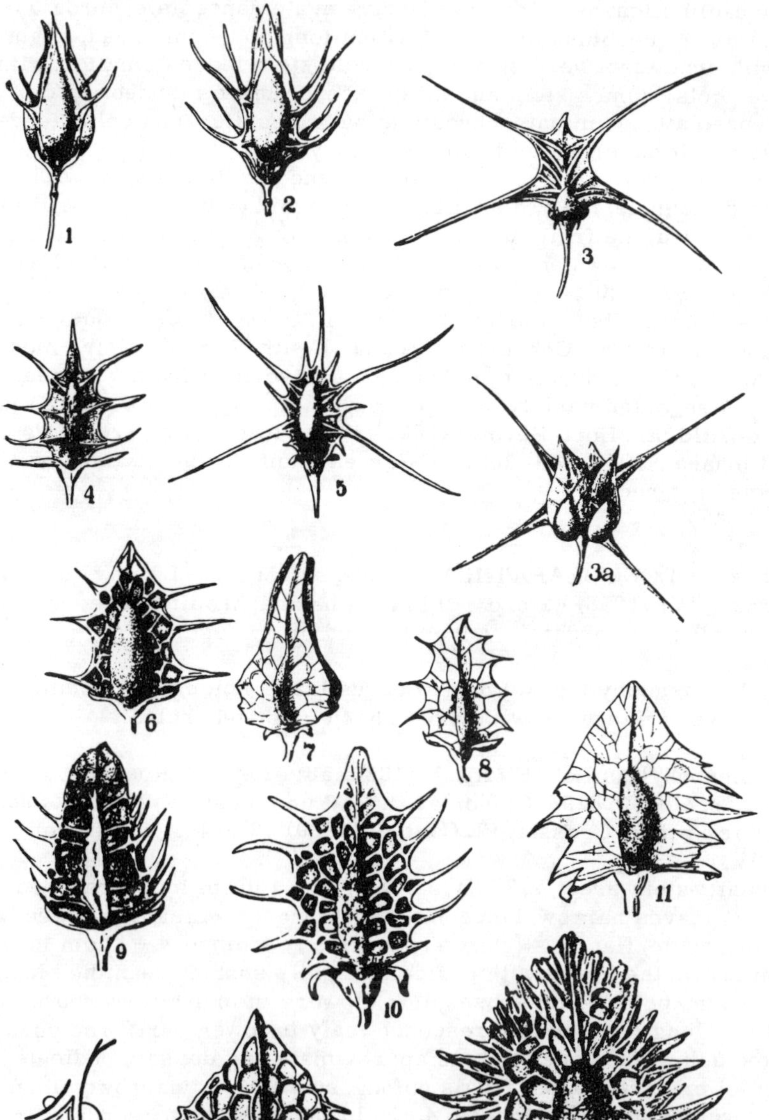

PLATE XXVII. 1. Rumex rossicus Murb.– 2. R. maritimus L.– 3. R. amurensis F. Schmidt.– 4. R. ucranicus Fisch.– 5. R. Marschallianus Rchb.– 6. R. Halaczi Reching.– 7. R. obtusifolius L. f. silvester.– 8. R. obtusifolius L. f. agrestis.– 9. R. pulcher L.– 10. R. reticulatus Bess.– 11. R. stenophyllus Ldb.– 12. R. syriacus Meisn.– 13. R. foveolatus A. Los.– 14. Schematic fruit section of this species.

and lower cauline leaves oblong to oblongly ovate-lanceolate, cordate or rounded at base, subobtuse at apex, to 10 cm long and 4 cm broad, slightly repand, with prominent reddish veins; middle stem leaves smaller, with shorter pedicels; upper stem and inflorescence leaves lanceolate or narrow-lanceolate, acuminate, cuneate at base, sessile; pedicels slender, jointed almost at base; outer tepals oblong; valves triangular-ovate, obtusely ligulate, reticulate, 1.5 —2 mm long, unequal, the largest with a globular red tubercle, the two smaller ones devoid of tubercles; fruiting pedicels recurved, the fruit suspended and pointing inward; achene blunt-angled, dark brownish-violet, trigonous-cordiform. April — May. (Plate XXVI, Figure 20).

Woods and wet soils. — European part: Bl., U. Dnp.; Caucasus: Cisc., Dag., E. and W. Transc. **Gen. distr.**: Scand., Centr. Eur., Atl. Eur., Med., Bal.-As. Min., Iran. Adventive in America. Described from Virginia.

Note. A reported variety: var. viridis Smith (= R. nemorosus Schrad. ex Willd. En. Hort. Berol. I (1809) 397. — R. daghestanicus C. Koch, Linnaea XXII (1849) 209). — A green plant, devoid of anthocyanin, the tubercles not red.

Section 4. ODONTOLAPATHUM A. Los. sect. nov. — Lapathum Meisn. in DC. Prodr. XVI (1856) 42 p. p. — Flowers hermaphrodite; base of leaves cuneate, rounded, or cordate; valves toothed or spinose.

Series 1. **Stenophyllae** A. Los. — Leaves large, oblong-lanceolate, cuneate, undulate; valves triangular, with short sharp teeth, tubercled.

36. **R. stenophyllus** Ldb. Fl. Alt. II (1830) 58; Ldb. Fl. Ross. III, 2, 505; Kryl., Fl. Zap. Sib. IV, 829; Grigor'ev in Fl. Yugo-Vost. IV, 118, No. 588. — R. odontocarpus Grossh., Fl. Cauc. II (1930) 72. — Ic.: Ldb. Ic. pl. Fl. Ross. IV, tab. 399.

474 Perennial; stem erect, slightly sulcate, 40 — 150 cm long, branched in upper part; leaves narrow-lanceolate or oblong, cuneately narrowed toward base, acute, flat or slightly repand, finely toothed, 4 — 17 cm long, 0.6 — 4 cm broad, the lower with pedicels slightly shorter than the blade, the upper short-petioled or subsessile; flowers in many-flowered whorls disposed in a long slender inflorescence leafy in lower part; racemes interrupted below, the upper whorls approximate and denser; pedicels jointed in lower part; outer tepals oblong, smaller and narrower than the inner; valves 3 — 4 mm long, to 4 mm broad, triangular-cordate, acutish, sharp-toothed or with teeth only in lower part, the teeth shorter than the breadth of the valve, all valves with an oblong tubercle; achene light brown, sharply 3-angled, pointed at both ends, to 3 mm long and 2 mm broad. June — July. (Plate XXVII, Figure 11).

Solonetz soils, boggy and wet places in the steppes. — European part: M. Dnp. Bl. L. Don, Transv., L. V., Crim.; Caucasus: Cisc.; W. Siberia: U. Tob.; Irt.; Alt.; E. Siberia: Ang.-Say.; Centr. Asia: Ar.-Casp., Balkh., Kara K., Syr D., T. Sh. Endemic. Described from the vicinity of Belenka, at the foot of Mount Ku. Type in Leningrad.

Note. A rare weed of irrigated crops in Soviet Central Asia.

37. **R. ussuriensis** A. Los. sp. nova in Addenda IV, p. 549. — R. stenophyllus Kom., Opr. rast. Dal'nevost. kr. I (1931) 459, non Ldb.

Perennial; stem erect, slightly flexuous, coarsely sulcate, reddish-brown, to 1 m long (and more ?); leaves narrow-lanceolate, acute, cuneately narrowed toward base, slightly crisp-margined, short-petioled, to 15 cm long and 4 cm broad; pedicels jointed in lower part, longer than the fruit; flowers in dense whorls; inflorescence leafy, dense, slenderly ovaloid, with upward-pointing branches; valves broadly ovoid-pyramidal, not pointed at apex, reticulate, 4 mm long, 5 mm broad, short-toothed, all valves tubercled, the tubercles infundibular, strongly narrowed toward summit; achene as in the preceding species. June; fr. July.

Wet meadows and shores. — Far East: Uss., Ze.-Bu. **Gen. distr.**: Korea. Described from the Murav'ev-Amurskii Peninsula. Type in Leningrad.

Note. Differing from R. stenophylus Ldb. in the shorter-toothed obtuse-tipped valves.

38. **R. Regelii** F. Schmidt, Mém. Acad. Sc., sér. VII, No. 2 (1868) 167. — R. stenophyllus var. sachalinensis Rgl. in Index Sem. horti Petr.
475 (1864) 21.

Perennial; stem long, stout, ribbed, branched in upper part; leaves oblong-lanceolate, narrowed at both ends, acuminate, cuneate or rounded at base, slightly crisped marginally, to 25 cm long, to 4 cm broad, rather firm, glabrous on both sides, short-petioled; upper leaves small, narrower; bracts linear-lanceolate, sessile; inflorescence ovaloid, with short upward-pointing branches, the dense many-flowered whorls distant; pedicels nodding, jointed in lower part, longer than the fruit; valves broadly reniform-cordate, coriaceous, obtusish, strongly reticulate, toothed on the margin, 6 mm long, 8 mm broad, all tubercled, the strongly convex tubercle to 4 mm long and 2 mm broad. Differs from R. ussuriensis A. Los. in the shape of the coriaceous valves, with small teeth and large tubercles. August — September.

Seashores and riverbanks. — Far East: Sakh. Endemic. Described from Sakhalin, from the village of Tunai. Type in Leningrad.

Series 2. **Dentatae** A. Los. — Leaves small, lanceolate, cuneate or rounded at base; tepals with long, stiff, spinescent teeth; inflorescence consisting of distant whorls.

39. **R. reticulatus** Bess. Ind. Sem. Hort. Kremen. (1820); Spreng. Neue Entdeck. III (1822) 18; Ldb. Fl. Ross. III, 2, 501. — R. strictus Link, Enum. hort. Berol. (1821) 350; Grigor'ev in Fl. Yugo-Vost. IV, 118, No. 589. — R. dentatus ssp. reticulatus Reching. f. in Beihefte Bot. Centr. XLIX 1 (1932) 18. — R. dentatus β pleiodon Boiss. Fl. Or. IV (1879) 1013. — Ic.: Fl. Yugo-Vost. IV, 119, Figure 249.

Annual; stem branched, leafy, sulcate, glabrous, with upright and spreading branches; leaves oblong, rounded or slightly cordate at base, obtusish, slightly wavy-margined, to 10 cm long, 2.5 cm broad, the petiole slightly shorter than blade; upper leaves smaller and narrower; inflorescence leaves of all whorls narrow-lanceolate; whorls many-flowered, distant; pedicels jointed in lower part; valves oblong-triangular,

2—4 mm long, reticulate, with 3—5 spreading slender teeth exceeding the breadth of the valve, all with a large tubercle; achene 1.5—2 mm long, light-brown, pointed at both ends. June—July. (Plate XXVII, Figure 10).

Wet places and shores; a weed. — European part: L. V. (delta); Caucasus: E., W., and S. Transc., Dag. **Gen. distr.**: Arm.-Kurd., Iran. Described from cultivated specimens.

Note. Frequent weed of cultivated fields.

40. **R. Halaczii** Rech. Verh. Zool. Bot. Ges. Wien XLIX (1899) 105. — R. dentatus ssp. Halaczii Rech. f. in Beih. Bot. Centralbl. XLIX, I (1932)
476 18. — R. obtusifolius ssp. subulatus Halaczi, Consp. Fl. Gr. III (1904) 63; Asch. u. Gr. Synops. VI (1908) 711. — R. pulcher Grossh. Fl. Kavk. (1930) 42 (pro parte).

Annual or biennial glaucescent plants; stem 15 —50 cm long, branched from base, many-angled, glabrous, leafy, with straight upright branches; leaves petiolate, the petioles of lower leaves about as long as blade; blades oblong-oval, obtuse or acuminate, cordate at base, glabrous, rather firm, slightly undulate marginally, to 7 cm long and to 2.5 cm broad; upper leaves smaller, narrow, cuneately attenuate; whorls of inflorescence dense, the upper approximate, the lower distant; pedicels jointed at base, recurved, about as long as the fruit; valves triangular, subtruncate at base, acute at apex, with 4 or 5 sharp spinescent teeth not exceeding the breadth of the valve, alveolate-reticulate, all with a large tubercle; achene 1 mm long, lustrous, trigonous, with broad-elliptic angles. June—July. (Plate XXVII, Figure 6).

Riverbanks, field borders, weedy places, irrigation canals, and cultivated fields. — Caucasus: E. Transc.; Centr. Asia: Mtn. Turkm., Kara K., T. Sh., Pam.-Al., Balkh., Syr D., Amu D. **Gen. distr.**: Arm.-Kurd., Iran., Bal.-As. Min. Described from Thessaly, from Trikkala. Type in Vienna.

Note. A common weed infesting many crops both under irrigation and in dry culture. Rechinger, the author of this species, considers it to be a hybrid between the species R. pulcher L. and R. limosus Thunb. This conclusion needs checking, since the distribution area of R. Halaczii is geographically very isolated and it does not overlap that of R. limosus Thunb.

41. **R. pulcher** L. Sp. pl. (1753) 336. — Lapathum sinuatum Lam. Fl. Fr. III (1778) 5. — L. pulchrum Moench, Meth. Suppl. (1802) 121. — R. denticulatus et R. reticulatus C. Koch., Linnaea XXII (1849) 207. — Ic.: Rchb. Ic. Fl. Germ. XXIV (1909) tab. 189; Hegi Fl. Mitt. Eur. III, t. 91, f. 3.

Perennial; stem 20—60 cm long, erect, sulcate, often sparingly pubescent, branched in upper part or rarely simple; branches spreading at nearly a right angle or slightly drooping; lower leaves oval-oblong, cordate at base, obtusish at apex, slightly wavy-margined, twice as long as broad, the petiole shorter than to as long as the blade; inflorescence a long leafy raceme with distant dense whorls; fruiting pedicels shorter than the fruit, jointed in lower part, firm and stout, nodding; outer tepals oblong and as broad as the inner ones; valves firm, oblongly ovate-triangular, 4.5 — 6 mm long, 2.5 — 4.5 mm broad, coarsely

477 reticulate, fleshy, tapering to an obtuse tip, truncate at base, with 4 — 5 strong subulate teeth not exceeding the breadth of the valve, all tubercled, the tubercles equal or unequal; achene 3—4 mm long, pointed-ovoid. May— June. (Plate XXVII, Figure 9).

Boggy wet soils, and weedy places. — European part: Crim.; Caucasus: E., W. and S. Transc., Dag., Cisc. **Gen. distr.:** Centr. Eur., Med., Bal.-As. Min. Described from Europe. Type in London.

Note. A very frequent weed of dry and irrigated crops.

Series 3. **Obtusifoliae** A. Los. — Leaves large, flat, cordate at base, obtuse; valves oval-triangular, with long sharp teeth, all tubercled.

42. **R. syriacus** Meisn. in DC. Prodr. XIV (1856) 53. — R. dictyocarpus Boiss. et Buhse, Nouv. Mém. Soc. Nat. Mosc. XII (1860) 192. — R. obtusifolius Boiss. Fl. Or. IV, 1011, non L. — R. denticulatus C. Koch, Linnaea XXII (1849) 208, non Campd. — R. turkestanicus O. Pauls. Bot. Tidscr. XXIV (1909) 154. — R. Drobovi E. Korov. in Opred. rast. okr. Tashkenta (1924) 84.

Perennial; stem erect, sulcate, pale green, to 1 m long, branched in upper part; lower cauline leaves oblong-ovate to oblong-lanceolate, obtusish to acute, more or less cordate at base, flat or slightly minutely undulate marginally, minutely papillose, 2—3 times as long as broad; upper leaves smaller and narrower, oblong-lanceolate, subtruncate or cuneate at base; all leaves short-petioled; inflorescence branches arising at an angle of 45°, nodding; inflorescence whorls approximate; outer tepals 1.5 mm long, their upper margin appressed to the middle ones; valves to 5—7 mm long, 5—6 mm broad, reticulate, rounded-cordate, acute, on each margin with 4—9 teeth 1.25—2.5 mm long, in var. subinteger Rech. f. with very small sharply triangular teeth, all tubercled, the tubercle 2 mm long; achene brownish, 3 mm long, sharply angled, lustrous. June—July. (Plate XXVII, Figure 12).

In the subalpine zone of mountains, on gravelly and damp slopes, and along riverbanks. — Caucasus: S. Transc.; Centr. Asia: Mtn. Turkm., Pam.-Al., Syr D., T. Sh. **Gen. distr.:** Arm.-Kurd., Iran. Described from Transc., Karadag near Gori. Type in Geneva.

43. **R. obtusifolius** L. Sp. pl. (1753) 335. — Lapathum sylvestre Lam. Fl. Fr. III (1778) 4. — L. obtusifolium Moench., Meth. (1794) 356. — Ic.: Rchb. Ic. Fl. Germ. XXIV (1909) t. 180; Hegi Fl. Mittle. Eur. III, 175, f. 524. — Exs.: HFR No. 1132.

Perennial; stem 0.6—1.2 m long, sulcate, branching from the middle;
478 branches arched-recurved, forming a broad-paniculate inflorescence; radical leaves broadly oblong-ovate, twice as long as broad, thin, delicate, dark green, flat, glabrous or slightly pubescent, to 25 cm long and 12 cm broad, rounded-obtuse or rarely subacuminate, cordate at base, longer than the petiole; cauline leaves smaller, oval-lanceolate, obtusely attenuate at apex, cordate at base, with shorter petiole; inflorescence loosely few-flowered, leafy or leafless; pedicels slender, 2 — 2.5 times the length of the fruit, jointed below the middle, enlarged in upper part; valves

oval-triangular or ovate-triangular, 2 — 3 mm long, 1.5 — 2 mm broad (var. agrestis Fries, Novit. fl. Suec. ed. 2 (1829) 99. — R. obtusifolius L. ssp. agrestis (Fries) Danser (ex Rech. Beih. Bot. Centralbl. XLIX, H. 1 (1932) 45), valves oval-triangular, obtuse, to 6 mm long, all or one of them tubercled, the teeth equaling or exceeding the breadth of the valve (var. sylvestris (Wallr.) Rech. Oester. Bot. Zeitschr. XLII (1892) 51. — R. sylvestris Wallr. Sched. crit. II, 1 (1822) 161), valves narrowly ovate-triangular, ligulately elongated at summit, 3 — 5 mm long, 2 — 2.5 mm broad, obtuse or acuminate, with a smooth margin or in upper part crenate, with or without tubercle, equal or unequal. July — August. (Plate XXVII, Figures 7, 8).

Woods, glades, gullies, coppices, weedy places, ornamental and truck gardens. — European part: almost ubiquitous except in the north, L. V., Transv., SE part of V.-Kama, Crim.; Caucasus: Cisc., E. and W. Transc. **Gen. distr.:** Scand., Centr. Eur., Atl. Eur., W. and E. Med., Bal.-As. Min., Arm.-Kurd., Iran. Described from Europe. Type in London.

Note. The two varieties of this species, var. agrestis and sylvestris, differ rather significantly but, since their distribution is very scattered and mixed up, it is impossible to separate as species. R. obtusifolius hybridizes with R. maritimus, crispus, conglomeratus, hydrolapathum, aquaticus, and confertus.

44. **R. foveolatus** A. Los. sp. nova in Addenda IV, p. 549.

Perennial, glaucescent, glabrous or puberulous; stem diffusely branched, coarsely sulcate, flexuous, bearing in upper part a broad leafy inflorescence; lower leaves oblong-ovate, flat, obtuse, cordate at base, the petiole shorter than the blade; bracts small, 1.5 — 4 cm long, lanceolate or oblong, narrowed at both ends, slightly crisp-margined, glabrous; flowers in distant whorls of 10 — 15, forming long branched racemes; valves very firm, coriaceous, broadly ovate-triangular, foveolate, unequal, all tubercled; largest valve
479 8 mm long and broad, its rather long sharp teeth mostly in lower part and equaling or exceeding the breadth of the valve; other valves smaller and narrower, with shorter teeth; tubercle of the large valve oblong-ovate, 3 — 4 mm long, 2 — 3 mm broad, strongly convex, that of the other valves smaller and flat; fruits borne on recurved pedicels, the large valve facing outward, the strong stout pedicels as long as or shorter than the fruit, jointed in the middle, thickened toward the end; achene brownish, lustrous, sharp-angled, pointed at apex and obtuse at base. Fr. June. (Plate XXVII, Figures 13, 14).

Slopes and ravines. — Centr. Asia: Mtn. Turkm. (Kopet Dagh). Endemic. Described from Ai-dere Ravine. Type in Leningrad.

Note. Hybridizes with the related species R. syriacus Meissn.; differing from it in the unequal large-tubercled valves.

Series 4. **Maritimae** A. Los. — Leaves small, lanceolate-cuneate; valves with soft bristlelike spines. Annual. Inflorescence consisting of approximate whorls.

45. **R. amurensis** F. Schmidt ex Maxim. Prim. fl. Amur. (1859) 228.

Annual; stems branched, reddish, obscurely sulcate, sometimes prostrate; leaves 1.5 — 3 cm long, oblong-lanceolate to linear-lanceolate, acutish, cuneate at base, entire or slightly undulate marginally, the petiole half the length of the blade; whorls 9- or 10-flowered; pedicels slender, as long as the fruit; outer tepals small, oval; valves triangular, reticulate, acute, 1.5 mm long, one of them with 2 symmetrically disposed spines on each margin, these exceeding the breadth of the valve and scarcely dilated toward base, all valves or one of them tubercled; tubercle rounded-oval, at the base of the valve and up to one-third of its length. August. (Plate XXVII, Figure 3).

Shallows and riverbanks. — Far East: Ze.-Bu., Uda. **Gen. distr.**: Jap.-Ch. (Manchuria). Described from the vicinity of Nikolaevsk (Cape Chikrakh), and from the Amur River between the village of Katar and Davund. Type in Leningrad.

Note. Closely related to R. Marschallianus, from which it differs in the smaller size of the perianth.

46. **R. Marschallianus** Rchb. Ic. Pl. Cr. IV, 56; VI, 13 (1828); Ldb. Fl. Ross. III, 2, 499; Kryl., Fl. Zap. Sib. IV, 833; Grigor'ev, Fl. Yugo-Vost. IV, 120. — Ic.: Rchb. Ic. Pl. crit. tab. 516 (1828).

Annual; stem branched, sometimes from the very base, 5—50 cm long, with divergent branches; leaves oblong-oval to lanceolate, narrowed toward summit, cuneate or rounded at base, undulate marginally, the lower 3—5 cm
480 long, 0.5—1.5 mm broad, with petiole shorter than blade, the upper shorter and narrower; flower-whorls distant, disposed in spikelike leafy racemes; pedicels jointed near base; valves ovate-triangular, 2—2.5 mm long, without pointed portion, 1.5 mm broad, with 2—6 bristlelike teeth exceeding the breadth of the valve (in var. brevidens Bong. et Mey = R. soongoricus Fisch. Ind. Sem. H. Petrop. (1842) 85 — teeth equaling the breadth of the valve); one valve with an oblong tubercle twice as long as broad, with longer teeth than the other valves. June—July. (Plate XXVII, Figure 5).

Steppes, solonetz soils, solonetz meadows, riverbanks, shores of saline lakes. — European part: Bl., L. Don, V.-Kama, L. V., Transv.; W. Siberia: U. Tob., Irt., Alt.; Centr. Asia: Ar.-Casp., Balkh., Kara K., Kyz. K., Syr D. Endemic. Described from Astrakhan.

47. **R. ucranicus** Fisch. Cat. Hort. Gorenk. ed. 2 (1812) 16, nomen nudum; Schult. Syst. veg. VII (1830) 1393; Grigor'ev in Fl. Yugo-Vost., IV, 118; Ldb. Fl. Ross. III, 2, 501. — ? R. rubellus Steud. Nomencl. ed. 1 (1821) 710. — R. dentatus Hornem. Hort. Hafn. Suppl. (1819) 53, non L. — Ic.: Rchb. Ic. Fl. Germ. XXIV (1909) tab. 184.

Annual; stem 5—25 cm long, branched from base, with long spreading branches; leaves 0.5—7 cm long, 3—12 mm broad, long-petioled, ovate-lanceolate, attenuate toward apex, subtruncate at base, often with small rounded auricles or broad-cuneate; upper leaves lanceolate to linear-lanceolate, short-petioled, cuneate at base; flower-whorls distant, disposed in leafy spikelike racemes; pedicels jointed near base; valves ovate-triangular, long-acuminate, 2—3 mm long (point excluded), 1—1.5 mm broad, on each side with three bristlelike teeth, these equaling or exceeding by half the breadth of the valve; all valves tubercled; tubercle oval-triangular, obtusish at summit. June—July. (Plate XXVII, Figure 4).

Inundated meadows, bottomlands, roadsides, and solonetz soils. — European part: V.-Kama, Transv., U.V., U. Dnp., V.-Don, M. Dnp., L. Don, Bl., L. V.; W. Siberia (rare): Ob, U. Tob., Irt., Alt.; E. Siberia: Yen., Lena-Kol.; Centr. Asia: Balkh., Ar.-Casp. **Gen. distr.**: Centr. Eur., Mong. Described from a plant grown in a Gorenki garden from seed collected in Ekaterinopol'e. Type in Leningrad.

48. **R. rossicus** Murb. Bot. Notis. (1913) 221. — R. Maritimus ssp. rossicus Kryl., Fl. Zap. Sib. IV (1930) 830.

Perennial; stem glabrous or sparsely pubescent, sulcate, sometimes
481 branched from the very base, then mostly in second third; branches spreading, forming a broadly ovaloid-cylindric inflorescence; leaves pilulose on the veins; lower oblong-lanceolate, subobtuse, rounded-narrowed at base, 6 — 20 cm long, 0.5 — 2 cm broad, long-petioled; upper narrow-lanceolate, acute, narrowed toward base, shorter than the upper ones, short-petioled or sessile; inflorescence leaves linear-lanceolate; flower-whorls rather dense, in lower part of branches distant, in upper part approximate in a spikelike inflorescence; pedicels jointed near base, as long as or shorter than the fruit, thickened toward summit; outer tepals narrow, scarious, small; valves oval-triangular, acute, with 1 or 2 bristlelike teeth about equaling the breadth of the valve, all tubercled; tubercle oblong-lanceolate, rounded-obtuse at base, pointed at summit, occupying about three-fourths of the valve; achene 1.1 — 1.3 mm long, fuscous-brown, trigonous-fusiform. July. (Plate XXVII, Figure 1).

Riverbanks, lakeshores, and bogs. — European part: Dv.-Pech., Lad.-Ilm. (E. part), U. V., V.-Kama, Transv.; W. Siberia: U. Tob., Alt.; E. Siberia: Yen., Ang.-Say., Lena-Kol. **Gen. distr.**: Scand. Described from Karelia, Gak-ruchei. Type in Helsinki.

49. **R. maritimus** L. Sp. pl. (1753) 335; Ldb. Fl. Ross. III, 2, 500. — R. aureus Mill. Gard. Dict. ed. 8 (1768) No. 7. — Lapathum minus Lam. Fl. Fr. III (1778) 4. — L. maritimum Moench, Meth. (1794) 355. — Ic.: Rchb. Ic. Fl. Germ. XXIV (1909) t. 186. — Exs.: Fl. exs. austr.-hung. No. 1013.

Perennial; stem 15 — 60 cm long, erect, sulcate, branched in second third, the branches obliquely ascending; leaves lanceolate, acute, cuneate at base, short-petioled, 6 — 20 cm long, 0.5 — 2 cm broad; upper leaves narrower, short, short-petioled or subsessile; whorls densely many-flowered, in lower part distant, in the upper part forming a dense leafy spikelike raceme; pedicels jointed near base, longer than or as long as the fruit; outer tepals about as long as the inner; valves ovate-triangular, with 2 — 5 bristlelike teeth exceeding the breadth of the valve, tubercled; tubercle large, oblong, to 0.5 mm broad. June. (Plate XXVII, Figure 2).

Inundated meadows, riverbanks, lakeshores, and bogs. — European part: all regions except Dv.-Pech. and Kar.-Lap.; Siberia: all regions; Far East: Ze.-Bu., Uss., Uda, Okh., Sakh.; Centr. Asia: Ar.-Casp., Kyz. K., Balkh. **Gen. distr.**: Centr. Eur., Atl. Eur., Med., Scand., Bal.-As. Min., Jap.-Ch., Mong. Described from Europe. Type in London.

Hybridizes with R. ucranicus, pulcher, obtusifolius, conglomeratus (= R. palustris* Sm. Brit. Fl. I (1800) 394 = R. limosum Thuill. Fl. Paris. (1799) 182) and R. crispus.

482 Being wind-pollinated, Rumex species hybridize readily. Hybrids, as recorded above, refer to many European species. No material is as yet available in regard to Asian species, but hybridization among these plants is more than probable. The Central Asian species apparently form hybrids with species of the horse-sorrel type, which are widely distributed in that area as weeds, above all with R. crispus. The hybrids are characterized by a lack of consistency and instability of their characters, often within the same plant. West European authors (e. g., Ascherson and Graebner) describe also triple hybrids, such as R. maritimus × conglomeratus × crispus or R. obtusifolius × conglomeratus × aquaticus. Search for hybrids and their precise dctermination are also very desirable in the USSR.

Genus 391. **RHEUM**** L.†

Gen. Pl. ed. I (1737) 120; ed. 5 (1754) 174.

Inflorescence leafy or leafless, simple or branched, displaying great variability in shape and type of branching, ranging from spikelike to paniculate or spherical; pedicels jointed; perianth segments 6, the outer 3 usually larger; stamens commonly 9 (fluctuating between 5 and 10); pistil with 3 stigmas borne on a short style; fruit a winged triquetrous achene, varying greatly in size, 10 — 30 mm long, the wings varying in size and color.

Perennial herbs, with long stout roots; leaves simple, often very large, forming a basal rosette; stems to 2 m tall, leafy, hollow or solid, glabrous or pubescent, smooth or sulcate, or plants scapiflorous. Russian: "reven'."

Rhubarbs have many useful properties and have therefore been in cultivation for food and for medicinal use since remote antiquity. The Siberian forest species gave rise to many cultivated varieties now known in Europe and in America as very valuable vegetables. Chinese rhubarbs were introduced into Europe as medicinal plants (R. offinale Baill., R. palmatum L., R. tanguticum Maxim.).

Rhubarb cultivation has begun to spread very widely in the USSR and, following the example of western countries, rhubarb is now a generally accepted food article. It is one of the earliest vegetables and is also used in confectionery. Rhubarbs, especially those of the Central Asian mountain regions (e. g., R. macrocarpum A. Los.), contain tannids
483 in their roots, and they are exploited in this connection by local inhabitants. At present the plants are studied for possible use as tanning agents.

1. Stem leafy . . . . 2.
+ Stem leafless . . . . 14.
2. Stem bearing numerous leaves; inflorescence a panicle of several parts, each subtended by an ocrea . . . . 3.

* This hybrid was published as species, R. paluster Sm. in Flora Yugo-Vost., IV, 119.

** From Greek "Rha," Volga, the Volga root.

† Treatment by A. S. Lozina-Lozinskaya.

\+ Stem bearing 1 or 2 leaves; inflorescence with a single ocrea . . . . . 7.
3. Fruit with broad red wings . . . . 4. **R. Wittrockii** Lundstr.
\+ Fruit with narrow brown wings . . . . 4.
4. Leaves glabrous or minutely papillose above . . . . 5.
\+ Leaves covered on both surfaces with setiform hairs . . . . 5. **R. orientale** A. Los.
5. Leaves rather densely papillose, the margin strongly undulate . . . . 1. **R. undulatum** L.
\+ Leaves sparsely papillose or glabrate . . . . 6.
6. Leaves orbicular, thin . . . . 2. **R. compactum** L.
\+ Leaves ovate-triangular, rather firm . . . . 3. **R. altaicum** A. Los.
7. Inflorescence subspherical; leaves thin . . . . 8.
\+ Inflorescence not spherical; leaves rather firm . . . . 10.
8. Fruit broad-winged, the wings broader than the achene . . . . 9.
\+ Fruit narrow-winged, the wings not broader than the achene . . . . 6. **R. tataricum** L.
9. Fruit uniformly broad, the achene flat . . . 7. **R. turkestanicum** Janisch.
\+ Fruit narrowed toward summit, the achene convex . . 8. **R. rupestre** Litw.
10. Leaves glabrous . . . . 11.
\+ Leaves verrucose or on the veins spiny . . . . 12.
11. Fruit narrowed toward summit . . . . 9. **R. Korshinskyi** Titov.
\+ Fruit cordate at both ends . . . . 10. **R. lucidum** A. Los.
12. Leaves not lobed . . . . 13.
\+ Leaves lobed . . . . 20. **R. plicatum** L. Los.
13. Inflorescence narrow, pyramidal; stem unbranched . . . . 18. **R. cordatum** A. Los.
484 + Inflorescence broadly paniculate . . . . 19. **R. hissaricum** A. Los.
14. Stem simple . . . . 15.
\+ Stem branched . . . . 17.
15. Leaves pubescent on both sides . . . . . . 12. **R. rhizostachyum** Schrenk.
\+ Leaves pubescent only on the upper side or glabrous . . . . 16.
16. Scapes surpassing the leaves . . . . 12. **R. reticulatum** f. **alaicum** A. Los.
\+ Scapes not surpassing the leaves . . . . 13. **R. reticulatum** A. Los.
17. Leaves and stem glabrous; leaves oval . . . . 11. **R. darwasicum** Titov.
\+ Leaves and stem verrucose or papillose . . . . 18.
18. Leaves not lobed . . . . 19.
\+ Leaves lobed . . . . 22.
19. Leaves glabrous above . . . . 15. **R. Fedschenkoi** Maxim.
\+ Leaves verrucose above . . . . 20.
20. Stem averaging 20 cm in height; perianth segments large, accrescent, appressed to fruit . . . . 14. **R. nanum** Sievers.
\+ Stem to 1 m tall; perianth segments small, wilting in fruit . . . . 21.
21. Inflorescence spreading, subspherical; stem dichotomously branched . . . . 16. **R. ribes** L.
\+ Inflorescence pyramidal; stem simple . . . 17. **R. Maximowiczi** A. Los.
22. Leaves 5-lobed, small . . . . 22. **R. lobatum** Litw.
\+ Leaves 3-lobed, large . . . . 21. **R. macrocarpum** A. Los.

Section 1. RHAPONTICA A. Los. — Stem leafy; leaves delicate, commonly glabrous; fruits numerous, small, narrow-winged.

1. **R. undulatum** L. Sp. pl. ed. II (1762) 531; Ldb. Fl. Ross. III, 496; Turcz. Fl. baic.-dah. II, 48. — R. rhabarbarum L. Sp. pl. (1753) 372, pro parte. — Ic.: Arzn. Gew. 12 (1830) 78.

Perennial; stem erect, hollow, leafy, to 2 m tall, 4 cm in diameter, glabrous, finely sulcate, with persistent foliaceous and scarious sheaths; leaves thin, elongate-triangular, strongly tapering toward apex, cordate at base, with 5 principal veins, minutely papillose on both sides, the blade 15 to 60 cm long; petiole as long as the blade, glabrous, sulcate; cauline leaves small, the upper ones sessile; inflorescence long, paniculate, with several peduncles arising from the axils of upper leaves, each with its own ocrea; flowers in groups of 5 or 6; perianth segments 6, oblong-ovate, narrowed toward base, yellow, 1-nerved, the outer 2 mm, the inner 1.5 mm
485 long; stamens with very short filaments; pedicels jointed in upper part; fruit broad-oval, ca. 8 mm long, 6—7 mm broad, the light brown medially nerved wings as broad as the dull brown ovate achene. May—June. (Plate XXVIII, Figures 1, 2).

Borders in steppes, sparse woods, and sandy soils. — E. Siberia: Dau. **Gen. distr.**: N. Mongolia. Described from plants raised in Europe from seed imported from Siberia. Type in London.

Note. R. undulatum L. is one of the species which have given rise to numerous widely distributed varieties cultivated for food.

2. **R. compactum** L. Sp. pl. ed. II (1762) 531; Ldb. Fl. Ross. III, 497; Turcz. Fl. baic.-dah. II, 49. — R. rhaponticum Herder in A. H. P. XI, 2 (1890) 185; Kryl., Fl. Zap. Sib. 839. — R. nutans Pall. Fl. Ross. I, 2 (1788), index. — Ic.: Arzn. Gew. 12 (1830) tab. 9.

Perennial; stem erect, hollow, leafy, to 2 cm high, 2—3 cm in diameter, slightly sulcate, glabrous; leaves thin, orbicular, flat, obtuse, cordate at base, 5-nerved, rough on the veins beneath and on the margin with minute papillae, the blade 20—40 cm long; cauline leaves smaller, orbicular; inflorescence a congested broad ovaloid panicle with subspherical parts, each peduncle with its own ocrea; flowers in groups of 5—8; pedicels long, jointed in lower part; perianth segments oval, white, subequal, ca. 2 mm long; stamens with rather long filaments; fruit ovaloid, to 12 mm long, to 1 mm broad; achene ovate, finely rugose, dark brown, lustrous; wings reddish-brown, cordate at both ends, as broad as the achene, the nerve at the middle. June. (Plate XXVIII, Figures 3, 4).

Banks of rivers and brooks, woods, valleys, and slopes in the tundra (in Siberia). — W. Siberia: Irt.; E. Siberia: Yen., Ang.-Say., Dau., Lena-Kol.; Far East: Ze.-Bu. **Gen. distr.**: N. Mong. Described from cultivated plants raised in Europe from seed imported from Siberia. Type in London.

Note. Used as vegetable. The juice contains 3.5% malic acid and potassium oxalate. The rootstock contains anthocyanin.

3. **R. altaicum** A. Los. in Acta Inst. Bot. Ac. Sc. URSS ser. I, 3 (1936). — R. rhaponticum Kryl., Fl. Zap. Sib. 839, p. p.; Herder in A. H. P. (1890) 185, p. p.

Perennial; stem erect, finely sulcate, hollow, pink in section, leafy, 15—50 cm high, 1—1.5 cm in diameter; leaves oval-triangular, slightly attenuate toward apex, cordate at base, undulate or plicate on the margin,

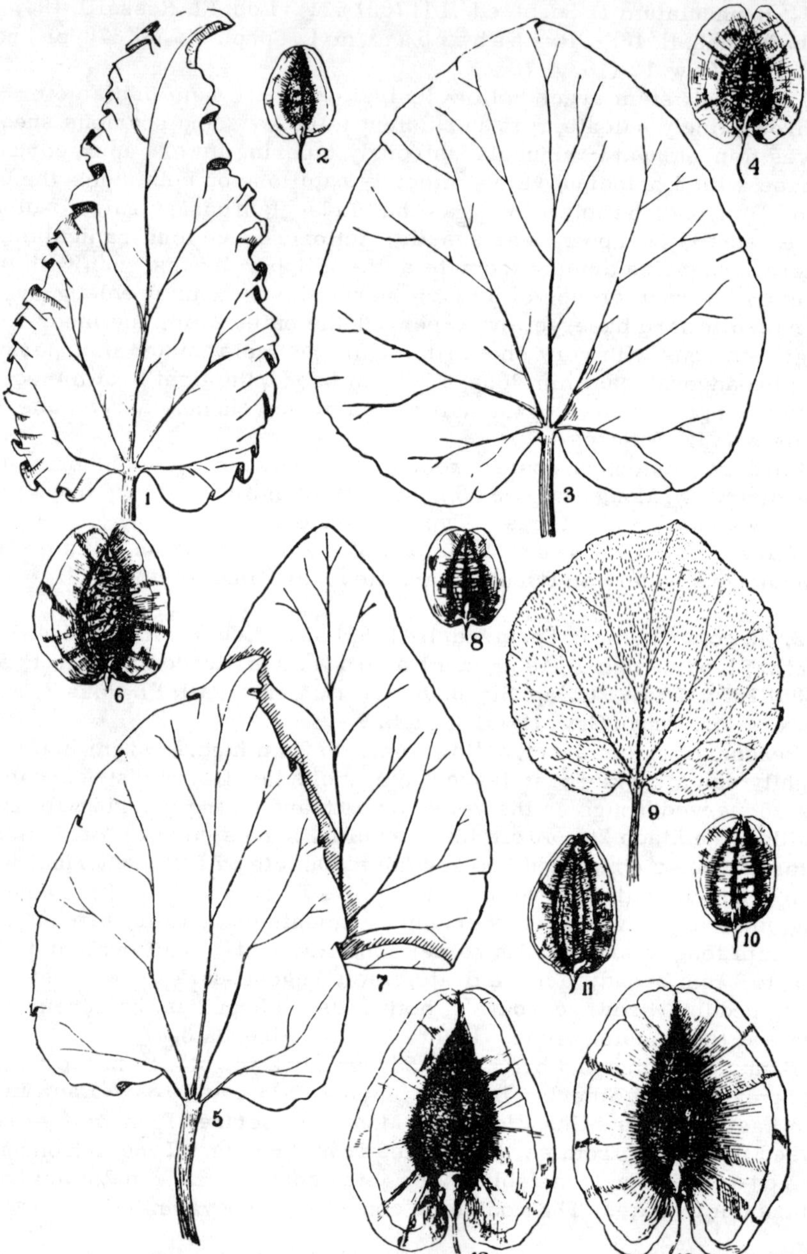

PLATE XXVIII. 1,2. Rheum undulatum L.— 3,4. R. compactum L.— 5,6. R. Wittrockii Lund.— 7,8. R. altaicum A. Los.— 9,10. R. orientale A. Los.— 11. R. tataricum L.f.— 12. R. rupeste Litw.— 13. R. turkestanicum Janisch.

486 • glabrous above, glabrous or slightly papillose beneath, 20—30 cm long; cauline leaves smaller; cauline and radical sheaths rufous-brown; inflorescence dense, slenderly ovaloid, erect, with upright peduncles; flowers small, yellowish, in groups of 4—7; pedicels equal, jointed in upper part; fruit 6 mm long, 5 mm broad, ovate; achene ovate, rugose, lustrous; wings narrow, light brown, rounded at both ends, the nerve marginal. June—July. (Plate XXVIII, Figures 7, 8).

Rocks, gravelly soil, stony steppes, and calcareous and stony slopes. — W. Siberia: Alt. **Gen. distr.**: N. and W. Mongolia. Described from Mongolian Altai. Type in Leningrad.

Note. A species related to R. compactum L.; distinguishable from it by lower growth, the firm glabrate triangular-ovate leaves, and the size of the fruit.

4. **R. Wittrockii** Lundstr. Acta horti Berg. III (1914) 23. — R. compactum Kar. et Kir., Enum. Pl. soong. (1841) No. 733. — R. rhaponticum Trautv., Enum. Pl. soong. (1860) No. 1060; Ldb. Fl. Ross. III, 2, 187; Fedchenko, Rast. Turk. (1915) 314. — Ic.: Lundstr. l. c. tab. 2

Perennial; stem erect, 50—100 cm high, sparsely leafy, hollow, glabrous, smooth; leaves oval-triangular or oblong-ovate, deeply cordate at base, slightly plicate and repand, with 5 straight principal veins, glabrous above, minutely papillose beneath and on the margin, the blade 40 cm long and 30 cm broad; cauline leaves smaller, often with red anthocyanin-colored papillae; petioles the length of blade, slender, slightly concave; ocreae reddish, heavily pubescent; inflorescence a loose spreading panicle of several often remote parts each with its own ocrea; flowers white or rose; perianth segments 6, to 2 mm long, the inner somewhat larger than the outer; stigmas white, exserted from perianth; pedicels short, jointed below the middle, strongly elongating in fruit; fruit round, 13 mm long, 15 mm broad; achene broad-ovate, finely foveolate, rugose, brown; wings as broad as the achene, red-hyaline, cordate at both ends, the nerve at the middle. May—June. (Plate XXVIII, Figures 5, 6).

Grassy and wooded slopes. — Centr. Asia: T. Sh. (Trans-Ili Ala Tau, Dzungarian Ala Tau, Aleksandrovskii Range, Fergana Range), Pam.-Al.
489 (Alai Range). **Gen. distr.**: Dzu.-Kash. (Tien Shan). Described from plants raised in Berlin from seed obtained from Soviet Central Asia.

Economic importance. The leaf stalks and young stems of this plant are used by the local population for food; they are eaten as vegetable in raw state. The plant is cultivated in botanical gardens of Europe for ornament on account of its large showy fruits.

5. **R. orientale** A. Los in Acta Inst. Bot. URSS ser. I, 3 (1936). — R. rhaponticum Herder, Pl. Raddeanae (1890) 185, p. p., non L.

Perennial; stem erect, hollow, reddish-brown, clothed with minute pointed hairs, 35—60 cm high, with one long-petioled leaf at the middle of the stem and several small floral leaves; radical leaves with stalks 1.5 times as long as the blade, orbicular, broadly cordate at base, 5-nerved, 10—20 cm long, glabrous above, the lower surface covered throughout with setiform hairs; petioles and principal veins spinous; inflorescence dense; pedicels jointed in upper part; flowers small, with oval segments 1–1.5 mm long; fruit ovate, small, the achene dark brown, the wings reddish; mature fruits unknown. June. (Plate XXVIII, Figures 9, 10).

Slopes and exposed stony places. — E. Siberia: Lena-Kol.; Far East: Okh. (Udskii Range). Endemic. Described from the Okhotsk Coast. Type in Leningrad.

Section 2. DESERTICOLA (Maxim.) A. Los. — Stem bearing 1 leaf; radical leaves large, delicate; inflorescence a spherical panicle.

6. **R. tataricum** L. f. Suppl. (1781) 22. — R. caspicum Pall. in Nova Acta Acad. Sc. Petrop. X (1797) 382. — R. soongoricum Schrenk in Bull. Phys. Mat. Acad. Pétersb. 2 (1844) 144. — Kazakh name: zhernak.

Perennial; rootstock vertical, strong, the old sheaths dark brown, the young darkish; stems 2 or 3, robust, hollow, sulcate, glabrous, branched from the middle at an angle of about 40°, the terminal branches drooping in maturity and thus inflorescence almost spherical; leaves to 35 cm long and 50 cm broad, rounded, cordate at base, with 3 prominent principal veins, the lower surface and the very short petioles minutely papillose, the upper surface glabrous; perianth segments 5, equal, 3 mm long, yellowish, with 3–5 brown nerves; fruit 10–12 mm long, 8–10 mm broad; achene obovate, dark brown or almost black, dull, minutely rugose; wings 1–1.5 mm broad, dark reddish-brown, cordate at base, attenuate at summit, the nerve running along the margin; perianth segments appressed to fruit.
490 April–May. (Plate XXVIII, Figure 11).

Steppes; on gravel, clay, grassy and dry slopes, and hillocks. — European part: Transv.; Centr. Asia: Ar.-Casp., Balkh. Endemic. Described from the Transvolga area.

7. **R. turkestanicum** Janisch. Schedae ad HFR VIII (1922) 92. — R. tataricum Boiss. Fl. Or. IV (1879) 1003, non L. — R. tataricum var. alatum Janisch., Izv. Nikol. Univ. II, 2 (1911). — R. latialatum Roshanez in O. et B. Fedtsch. Consp. Fl. Turk. VI (1916) 274. — Exs.: HFR No. 2570 (1916!).

Perennial; rootstock long, vertical, branched, covered with brown fibrous leaf sheaths; stem erect, 30–70 cm high, 1–3 cm thick, hollow in lower part, sulcate, glabrous, branching below the middle, forming a broad-pyramidal to almost spherical inflorescence; radical leaves flattened out upon the ground, short-petioled; rounded-reniform, to 1.5 m in diameter, obtuse, shallowly cordate at base, slightly undulate marginally, bulging at center, glabrous above, scabrous beneath particularly on the veins, with 3 palmately branched principal veins; cauline leaves 1 or 2, small, sessile; flowers white; perianth segments 6, oblong-oval, the outer 4 mm, the lower ones [?] 3 mm long; pedicels to 3 mm long, jointed at the middle; fruit to 25 mm long and 20 mm broad, ovate; achene narrow-ovate, dark brown, longitudinally striate; wings thin, scarious, very broad, greatly exceeding the width of the achene, deeply cordate at base, the nerve about 2 mm from the margin. April–May. (Plate XXVIII, Figure 13).

Sand dunes. — Centr. Asia: Kara K. Endemic. Described from Repetek. Type in Leningrad.

8. **R. rupestre** Litw. in sched. et in Acta Inst. bot. Ac. Sc. URSS ser. I, 3 (1936).

Perennial; stem erect, to 1 m tall, branched from the middle like the preceding species, hollow in lower part, sulcate, glabrous; leaves as in the preceding; inflorescence subspherical, the tips of branches drooping; inner perianth segments 4 mm long and 3 mm broad, the outer narrower and shorter; pedicels short, jointed at the middle; fruit broad-ovate, to 20 mm long and broad; achene broad-ovate, convex, dark brown, 7 mm broad; wings slightly wavy-margined, narrowed toward the subcordate summit, deeply cordate at base, rose-colored, the nerve 1.5—2 mm from the margin. April. (Plate XXVIII, Figure 12).

Gravelly mountain slopes. — Caucasus: S. Transc. (Nakhichevan); Centr. Asia: Mtn. Turkm. (Kopet Dagh). Endemic. Described from Kopet Dagh (Gaudan). Type in Leningrad.

491 Section 3. **GLABRIFOLIA** A. Los. — Stem bearing 1 leaf; plants glabrous throughout; leaves firm.

9. **R. Korshinskyi** Titov in sched. et in Acta Inst. Bot. Ac. Sc. URSS ser. I, 3 (1936).

Perennial; stem erect, glabrous, sulcate, rarely red, branched; inflorescence large, pyramidal, compound, the spikelike branches borne on slender peduncles; leaves broad-oval, to 30 cm long, cordate, form, glaucous beneath, pale green above, glabrous on both sides, the margin spinulose, the rufescent veins prominent on the lower side; petiole longer than the blade, glabrous, finely sulcate; cauline leaf subtending the inflorescence, smaller, short-petioled; flowers in groups of 3—5; pedicels slender, jointed near base; perianth segments thin, scarious, the outer obovate 2.5 mm long, the inner oblong 1.5 mm long; stamens 9, with exserted anthers; fruit ovate, narrowed toward summit, cordate at base; wings thin, scarious, narrower than the brown achene. June. (Plate XXIX, Figure 5).

Stony slopes. — Centr. Asia: Pam.-Al. (Darvaz). Endemic. Described from Darvaz (Kergovat). Type in Leningrad.

10. **R. lucidum** A. Los. in Bull. Jard. Bot. Princ. XXX, 3—4 (1931) 378.

Perennial; stem solitary, erect, ca. 80 cm tall, hollow, glabrous, branched in upper part, sulcate, reddish-green; inflorescence of several parts, each with own sheath; petioles long, glabrous, slightly sulcate, rufescent; leaf blades glaucous, coriaceous, with thick epidermis, suborbicular, slightly longer than broad, cordate at base, lustrous beneath, dull above, the margin smooth, the 5 rufous veins very prominent and slightly furrowed; remnants of cauline leaves persistent at the base of peduncles and below; flowers in groups of 4 or 5, forming a rather dense slender inflorescence; perianth segments oblong-obovate, 2—3 mm long, greenish-glaucous, with a firm epidermis; filaments of stamens and the appressed perianth segments persistent in fruit; fruits numerous; petioles short, slender, jointed at the middle; fruit 1 cm long and broad, short-ovate, light brown, strongly lustrous, foveolate, wings thin, about as broad as the achene, cordate, the nerve close to the margin. June—July. (Plate XXIX, Figure 7).

Stony slopes. — Centr. Asia: Pam. Al. (Panj River valley). Endemic. Described from the Panj River. Type in Tashkent.

492 11. R. **darvasicum** Titov in Acta Inst. Bot. Ac. Sc. URSS ser. I, 3 (1936).

Perennial; stem 40 cm tall, erect, light brown, finely sulcate, leafless, solid, with small racemiform branches below the middle; fruiting inflorescence a compact panicle; petiole about the length of the blade, glabrous, pinkish; leaf blade broad-ovate, glabrous on both sides, glaucous, inflexed at the tip, prominently nerved; foliar sheaths loose; fruit oval, 12 mm long, 9—10 mm broad; achene lustrous light brown, ovate; wings thin, 2.5 mm broad, pinkish-brown, the nerve at the very margin. July. (Plate XXIX, Figure 6).

Stony slopes. — Centr. Asia: Pam.-Al. (Darvaz). Endemic. Described from the Yazgulem River. Type in Leningrad.

Section 4. **SPICIFORMIA** A. Los. — Scapiflorous; inflorescence spikelike; leaves coriaceous.

12. **R. rhizostachyum** Schrenk, Bull. Acad. Sc. Pétersb. X (1842) 254 et in Enum. altera pl. nov. Soong. (1842) 17; Ldb. Fl. Ross. III, 2, 498. — R. a p l o s t a c h y u m Kar. et Kir. Enum. Pl. Soongor. (1842) 422.

Perennial; scapes surpassing the leaves; leaves broad-ovate, slightly asymmetric, averaging about 20 cm in length and breadth, cordate at base, slightly erose marginally, 5-nerved, covered on both sides, more heavily beneath, with stellate hairs; scapes to 1 cm in diameter, bearing a spikelike inflorescence, with lateral offshoots in lower part; fruits numerous, oval-ovate, 10 mm long, 6 mm broad, narrowed toward summit; achene ovate, 4—4.5 mm long; wings ca. 1 mm broad, yellowish-red, the nerve at the middle; pedicels shorter than fruit, jointed in lower part. Fr. July.

Mountain slopes. — Centr. Asia: Dzu.-Tarb. (Dzungarian Ala Tau). **Gen. distr.:** Kuldja. Described from the Ku-lya-su Pass, Dzungarian Ala Tau. Type in Leningrad.

13. **R. reticulatum** A. Los. in Acta Inst. Bot. Ac. Sc. URSS ser. I, 3 (1936). — R. s p i c i f o r m e auct. Fl. As. med.; Fedchenko, Rast. Turk. (1915) 314, non Royle.

Perennial, small plants, with firm coriaceous leaves; scapes numerous, to 15 cm tall, not surpassing the leaves; leaves ovate, acute, rounded or reniform at base, glabrous and smooth above, reticulate-veined beneath, reddish-lilac, with 5 principal veins, clothed with short often stellate papillose hairs; petioles lilac, or reddish, much shorter than blade; inflorescence simple, spikelike; flowers rather few; perianth segments 1.5 mm long, oval; stamens 7—9, surpassing the perianth; pedicels jointed at the middle; fruit broad-ovate, slightly narrowed toward
493 summit, 8—9 mm long, 6—8 mm broad, obtuse at apex, cordate at base; wings narrow; achene ovate. June. (Plate XXIX, Figures 1, 2).

Alpine zone; stony soil, pebbles, sand, and slopes. — Centr. Asia: T. Sh., Pam.-Al. Endemic. Described from the Pamir, Lake Kara-kul'. Type in Leningrad.

Note. A form with leaves glabrous on both sides, which occurs in the Pamir, is reported by Ruprecht in Sertum Tianschanicum as R. rhizostachyum β glabrum Rupr. (1869).

f. alaicum A. Los. f. nova. — Plants with large long-petioled leaves and solitary tall scapes; leaves ovate, cordate at base, acute, to 30 cm long and 22 cm broad, green, with 3—5 principal veins, sparsely and minutely papillose beneath, glabrous above; petioles glabrous, as long as or somewhat shorter than the blade; scapes equaling or slightly surpassing the leaves, to 1 cm in diameter; fruits numerous, densely covering the scape, orbicular, 10 mm in diameter, cordate at base, obtusish at apex; achene small, ovate, 2.5 mm broad, located in the lower part of the fruit, overgrown by the wings at the summit; wings orange-red, broader than the achene, the nerve at the middle. June. Grassy slopes. — Centr. Asia: Pam.-Al. (Alai Range and Valley). Described from the Alai Valley, from the banks of the Kizyl-su River.

Section 5. **ACAULIA** A. Los. — Scapiflorous plants; inflorescence branched; leaves coriaceous.

14. **R. nanum** Siewers in Pall. Neue Nord. Beitr. VII (1796) 264. — R. leucorrhizum Pall. in Nova Acta Acad. Petrop. X (1797) 381. — R. cruentum Mart. ex Pall. Neue Nord. Beitr. VII (1796) 294. — Ic.: Ldb. Ic. Fl. Ross. V, tab. 491.

Perennial; rootstock vertical, white in section, covered with dry dull dark brown sheaths; stem averaging 20 cm in height, coarsely sulcate, forking into 2 main branches forming a broad almost triangular inflorescence; leaves orbicular, often broader than long, with 3 principal veins, the upper surface verrucose, the lower with scattered short stellate hairs, the margin slightly undulate with a thin white warty rim, the short concave petiole sulcate; flowers subtended by scalelike bracts; pedicels short, stout, jointed at base; outer perianth segments yellow, brown toward base, broad-ovate, 4.5 mm long and 3 mm broad; inner segments yellow, red-tipped, long-lanceolate, 3 mm long, 1 mm broad; stamens with short filaments; fruit 10—12 cm [?] long and broad; wings rose-colored, cordate at base, emarginate at summit; the nerve 1—1.5 mm from the
494 margin, joined to the achene by 1 or 2 transversal nerves; achene oval, dull blackish-brown; inner perianth segments closely appressed to achene, the outer reflexed. May. (Plate XXIX, Figure 8).

Stony, sandy, or clayey soil; slopes, deserts, and steppes. — W. Siberia: Alt., Irt.; Centr. Asia: Ar.-Casp. (E. part), Balkh. **Gen. distr.**: N. Mong. Described from Altai (Dzhar-Gurban and Kurchum rivers). Type in Leningrad.

Section 6. **RIBESIFORMIA** A. Los. — Stem leafless; verrucose plants, with sour juice; fruit broad-winged.

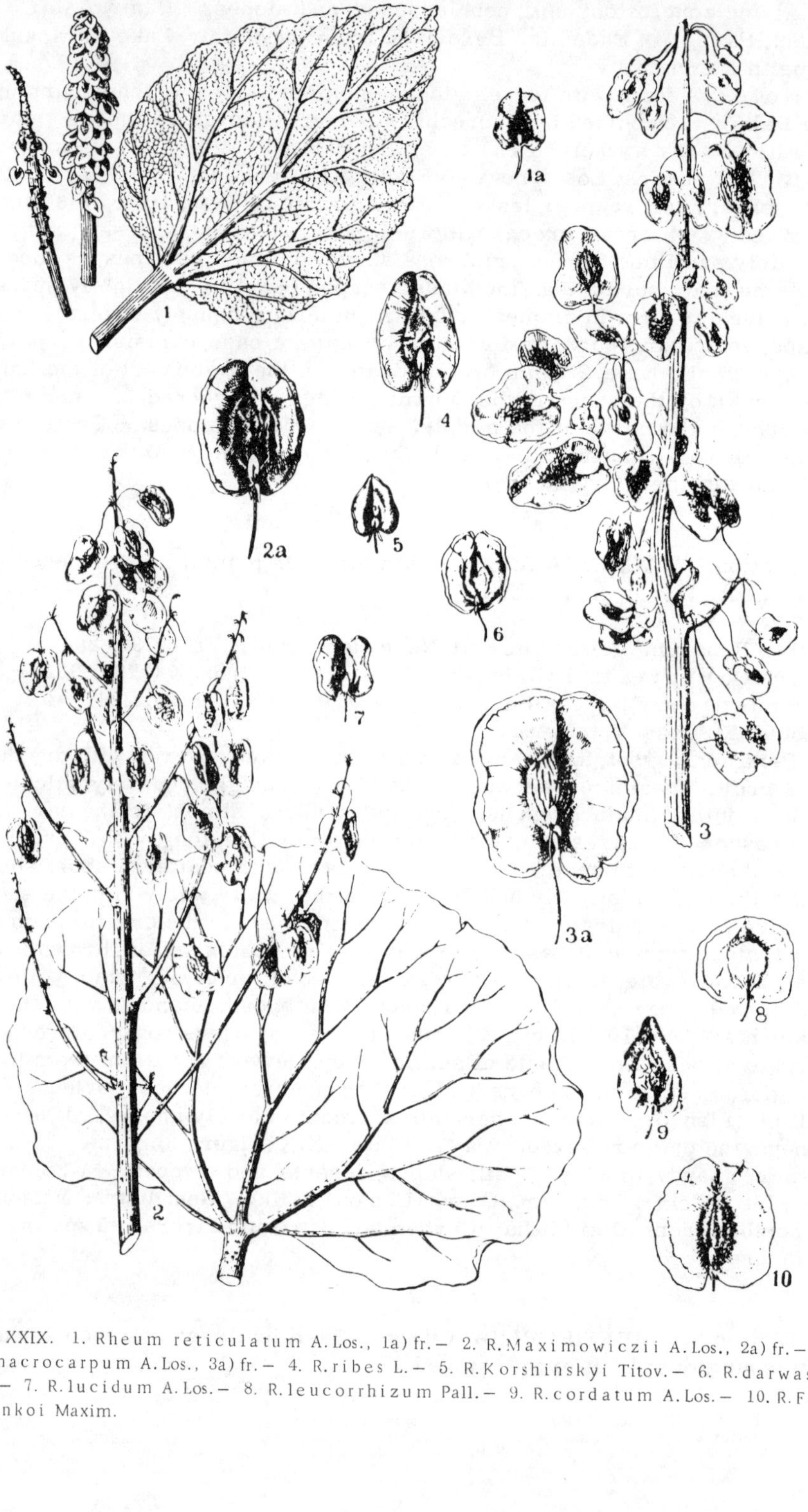

PLATE XXIX. 1. Rheum reticulatum A. Los., 1a) fr. — 2. R. Maximowiczii A. Los., 2a) fr. — 3. R. macrocarpum A. Los., 3a) fr. — 4. R. ribes L. — 5. R. Korshinskyi Titov. — 6. R. darwasicum Titov. — 7. R. lucidum A. Los. — 8. R. leucorrhizum Pall. — 9. R. cordatum A. Los. — 10. R. Fedtschenkoi Maxim.

15. R. **Fedschenkoi** Maxim. ex Regel in Izv. O-va Lyubit. Est. 34 (1882) 77.

Perennial; rootstock vertical, covered with old dark brown and new reddish-orange sheaths, stout, yellow in section; stems 1 or 2, scapose, to 30 cm tall, hollow, erect, strongly sulcate, minutely papillose, yellow, branched at an angle of 40 — 45°, forming an ovaloid inflorescence, the branches subtended by brown scarious collarlike bracts; leaves thick, coriaceous, oval or rounded-ovate, cordate at base, prominently 5-nerved, 15 — 20 cm long and broad, the margin slightly undulate, the upper surface glabrous; lower surface often anthocyanin-tinged, densely papillose, especially on the veins, the papillae interspersed with small epidermal excrescences; petiole shorter than or as long as the blade, finely sulcate, glabrous, reddish; flowers in groups of 6 — 8, subtended by small scarious brown bracts; pedicels fairly long, jointed in lower part; perianth segments 6, ovate, orange, the outer 2.5 mm long, the inner somewhat shorter, like the stem minutely papillose, in maturity almost as broad as long; fruit large, 16 — 17 mm long and broad; achene oblong-ovate, reddish-brown, foveolate-rugose; wings broader than the achene, orange-red, cordate at base and at apex, the nerve 2 mm from the margin. June. (Plate XXIX, Figure 10).

Mountain slopes and passes, on clayey and stony soil in the alpine zone. — Centr. Asia: Pam.-Al. Endemic. Described from the Zeravshan River valley. Type in Leningrad.

497 16. **R. ribes** L. Sp. pl. (1753) 372. — Ic.: Desfontaines in Ann. Mus. Hist. Paris, II (1803) tab. XLIX; Jaub. et Spach Ill. or. V (1853 — 57) tab. 470.

Perennial; stem to 1 m tall, almost dichotomously branched, rarely more or less erect, rough with warts, leafless, solid, flattened; leaves suborbicular, broader than long, with 3 principal veins, the lateral veins marginal up to the first ramification, the upper surface glabrous or verrucose, the lower more or less verrucose and in lower part spinulose on the veins, the petiole half as long as the blade; inflorescence broad, subspherical; flowers 5 — 12 together; pedicels 6 — 7 mm long, jointed near base; perianth segments 3 mm long, 1.5 mm broad, the inner narrower; stamens 9; fruit ovate, 17 mm long, 14 mm broad, narrowed toward summit, deeply cordate at base; wings reddish, rather firm, as broad as or narrower than the achene, the nerve marginal; achene oblong-ovate, dull brown. June. (Plate XXIX, Figure 4).

Clayey slopes and ravines in the subalpine zone. — Caucasus: S. Transc. (Nakhichevan). **Gen. distr.**: Iran., Arm.-Kurd. Described from Iran. Type in London.

Note. R. ribes L. was widely used in antiquity in Arabia and in Iran as a food and medicinal plant. The juice containing about half percent of potassium oxalate imparts a very pleasant taste. The species is little known in cultivation, even though it is particularly promising.

17. **R. Maximowiczi** A. Los. in Bull. Jard. bot. Princ. XXX, 3 — 4 (1931) 362. — R. megalacarpum Maxim. ex Fedchenko, Rast. Turk. (1915) 314, p. p., nomen nudum.

Perennial; stem 40 cm to 1 m tall, erect, reddish, branched, leafless, flattened, sulcate, variously rough with warts to smooth, the branches arising

at an angle of about 45° and forming a large pyramidal inflorescence; leaves suborbicular, broader than long, reniform at base, slightly undulate and spinous marginally, with 3 principal veins, the lateral veins marginal to first ramification; blade to 50 cm long and 60 cm broad, the upper surface glabrous, the lower rough-verrucose on the veins; petiole glabrous, less than half the length of the blade, slightly concave above, rough especially above and at base, rarely smooth; flowers in groups; petioles long, jointed in lower part; perianth segments subequal, 3 mm long, 1 mm broad, greenish; stamens with large long anthers; fruit large, 20 mm long, 15 mm broad,
498 broad-oval; achene ovate, rugose, fleshy, lilac-brown; wings as broad as the achene, cordate at both ends, at first bright red, becoming lilac-red, the nerve marginal; perianth segments appressed to the fruit. June. (Plate XXIX, Figure 2, a).

Grassy slopes. — Centr. Asia: Pam.-Al., Syr D., T. Sh. Endemic. Described from Chimgan. Type in Leningrad.

18. **R. cordatum** A. Los. in Bull. Jard. Bot. Princ. XXX, 3—4 (1931) 381. R. megalocarpum Maxim. in sched., p. p.

Perennial; rootstock horizontal, stout, with remnants of dark fibrous sheaths; stem solitary, solid, erect, 50—100 cm tall, sometimes forking at base into 2 or 3 stems, sulcate, glabrous, rufous, in upper part branched and forming a narrow paniculate inflorescence; leaves suborbicular, slightly produced at apex, undulate marginally, cordate or reniform at base, with 3 very prominent divergent veins, the lateral veins marginal to first ramification as in R. Maximowiczi A. Los.; blade to 30 cm long and 40 cm broad, glabrous, the sulcate veins smooth at base, in upper part sparsely spinous; petiole somewhat shorter than the blade, sulcate; radical leaves (1) 2; cauline leaves 1 or 2, smaller; flowers very small; perianth segments oblong-oval, light yellowish-green; stamens 10; pedicels lateral, slender, jointed below the middle, as long as the fruit; fruits in groups of 5—7, 13 mm long, 10—11 mm broad, cordate; achene broad-oval, strongly rugose, dark brown; wings reddish, drying light brown, to 5 mm broad, light-colored toward base, gradually attenuate toward the subacute summit, dilated toward the deeply cordate base, the nerve marginal. May. (Plate XXIX, Figure 9).

Gravelly and stony slopes. — Centr. Asia: T. Sh. (Chu-Ili Mountains and Karatau). Endemic. Described from Karatau. Type in Leningrad.

19. **R. hissaricum** A. Los. in Acta Inst. Bot. Ac. Sc. URSS ser. I, 3 (1936).

Perennial; stem erect, leafy, branched, coarsely sulcate, bearing an ovaloid-pyramidal inflorescence; leaves (cauline) firm, broadly triangular-elliptic; upper leaf 3-lobed, 20 cm long, 28 cm broad, rounded at apex and at base as in R. Maximowiczi A. Los., the slightly undulate margin spinulose, the upper surface glabrous, the lower very sparsely covered with minute spinules; petiole glabrous, flat, sulcate; ocreae reddish-lilac in lower part; inflorescence leaves very small; radical leaves unknown; inflorescence with several erect peduncles bearing short slender branches,
499 these densely covered with flowers; fruits (not quite mature) 13 mm long, 11—12 mm broad, broadly oval-triangular, acuminate at apex, cordate at base; achene broad-ovate, transversely striate, light brown; wings

purplish-lilac, dilated at base, narrower than the achene, with a dark marginal nerve and dark transversal subsidiaries. June—July.

Slopes in the subalpine zone. — Centr. Asia: Pam -Al. (Zeravshan-Gissar mountainous country). Endemic. Described from the Farab natural boundary area. Type in Leningrad.

20. **R. plicatum** A. Los. in Bull. Jard. Bot. Princ. XXX, 3—4 (1931) 360.

Perennial; stem 40—45 cm tall, erect, branched at the end, flattened, sulcate, rough with warts and minute simple or double papillae; radical leaves large, slightly 3—5-lobed, to 40 cm broad and 30 cm long, coarsely undulate and deeply plaited marginally, papillose and verrucose on both surfaces, reniform at base, with 3 principal veins; cauline leaves small, finely plaited, rounded-triangular; lateral veins marginal up to the first ramification; petioles only a fraction of the length of the blade, densely verrucose, flattened like the stem; inflorescence broad-paniculate, the composite parts each with its own sheath, easily disintegrating on the stem and in the herbarium; flowers small; perianth segments greenish, rose-margined, the inner 3 mm the outer 2 mm long, oblong-ovate; stamens with large anthers and short filaments; pedicels short, jointed in lower part; fruit ca. 15 mm long and broad, round, cordate at base; achene ovate, rugose, dingy brown, cordate at base; wings red, as broad as the achene. May.

Grassy slopes. — Centr. Asia: Pam.-Al. (Alai Range). Endemic. Described from the Kyrkdzhol natural boundary area. Type in Leningrad.

21. **R. macrocarpum** A. Los. in Bull. Jard. Bot. Princ. XXX, 3—4 (1931) 379.

Perennial; stem terete, branched, with a narrow lumen, deeply sulcate, covered with simple or stellate papillose hairs, bearing a broad spreading inflorescence; leaves large, to nearly 80 cm broad and 40 cm long, coarsely 5-lobed, cordate at base, with 3 branched principal veins; petiole shorter than blade, slightly sulcate, covered with warts of variable size, coarsely crenate, minutely papillose at base, the lateral veins marginal to first ramification; flowers large; perianth segments green, reddish-margined,
500 oval, unequal, the inner to 8 mm long and 4 mm broad, the outer 5 mm long and 2 mm broad; fruit orange-rose, very large, to 3 cm broad and long; achene minutely verrucose, slightly rugose; wings scarious, broader than the achene, deeply cordate at base, the marginal nerve joined to achene by several slender subsidiary nerves; fruiting pedicels slender, glabrous, jointed in lower part, to 3.5 cm long; perianth segments appressed to fruit. May. (Plate XXIX, Figure 3a).

Wheat fields and slopes. — Centr. Asia: T. Sh. Endemic. Described from Uzgen. Type in Leningrad.

Economic importance. The roots of R. macrocarpum contain tannins and are locally exploited for tanning.

22. **R. lobatum** Litw. ex A. Los. in Acta Inst. Bot. Ac. Sc. URSS ser. I, 3 (1936).

Perennial; leaves rather small, distinctly lobed with 5 sharply triangular lobes, shorter than broad, undulate-margined, rough beneath with a dense cover of small papillae, verrucose above; lateral nerves marginal to first ramification; petiole slender, much shorter than blade, strongly sulcate. Complete plants, with stem and fruits, unknown.

Mountain slopes. — Centr. Asia: T. Sh., Mogol-tau. Endemic. Herbarium record by Litvinov from Mogol-tau. Type in Leningrad.

Section PALMATA A. Los. — Leaves palmately lobed; fruits numerous, small.

***R. palmatum** L. Syst. ed. 10 (1759) 1010. — Ic.: Bot. Mag. tab. 214. — Russian: reven' dlanevidnyi, reven' lekarstvennyi [palmate or medicinal].

Perennial; root fusiform, bulky, up to 8—12 kg in weight; stem erect, to 2 m tall; leaves radical, to 75 cm in diameter, palmately divided into pinnatisect lobes, the petiole to 30 cm long, the entire leaf scabrous with short papillae; inflorescence to 50 cm long, branched, many-flowered; bracts short, coriaceous, half-clasping; pedicels capillary, 3—4 mm long, densely clothed with short papillae; perianth short, with 6 obtuse yellowish segments; stamens 9; ovary triquetrous, 1-ovuled, with 3 stigmas; fruit triquetrous, red, broadly rounded, 7—10 mm long. July.

Native in mountains of S. and W. China. Grown in the USSR for its roots. The commonly grown form is var. tanguticum Maxim. in Rgl. Gartenflora XXIII (1874) 305 et tab. 819, introduced by N. M. Przheval'skii in the years 1858—1873 from the Tatung Mountains, which lie west of Lake Kuku-nor [Ching Hai]. The Tangutian rhubarb differs from the type in less deeply lobed leaves and branches appressed to the stem. Substantial areas of cultivation are to be found in the European part: V.-Don, U. Dnp. and M. Dnp.

501 **Economic importance.** The brittle bright yellow roots yield rhubarb powder used as laxative. It contains a mixture of tannoglycosides and anthraglycosides, including derivatives of chrysophanic acid which impart to rhubarb its characteristic taste and odor.

***R. rhaponticum** L. Sp. pl. (1753) 371. — Ic.: Bot. Mag. tab. 215. — Russian: reven' chernomorskii, reven' cherenkovyi [Black Sea or pedunculate].

Perennial; stem 1—1.5 m tall; lower leaves rounded-ovate, slightly undulate-margined; upper leaves oblong-ovate, with shorter petiole; inflorescence leafy in lower part, many-branched; pedicels jointed below the middle; perianth greenish; fruit oval, emarginate at summit and at base.

Native in the Rhodope Mountains in Bulgaria (Rila Mountains). Widely distributed as a truck crop. Owing to its contents of malic, oxalic and citric acids, it somewhat resembles gooseberry in taste and provides material for jelly making and confectionery.

Genus 392. **ATRAPHAXIS*** L.**

L. Sp. pl. (1753) 333; Jaub. et Spach. Ill. pl. orient. II (1844—46) 11, emend. — Tragopyrum M.B. Fl. taur.-cauc. III (1819) 284. — Physopyrum M. Pop. in Ind. Sem. Horti Almaatensis No. 2 (1935) 23.

* The name, as used by Dioscorides, referred to garden orache; literally it signifies "inedible" (plant).

** Treatment by N. V. Pavlov (Moscow).

Flowers hermaphrodite; perianth simple, petaloid, persistent, somewhat accrescent in fruit, 4- or 5-partite; outer segments commonly 2, smaller, reflexed in fruit; inner segments 2 or 3, larger, enlarging into erect valves enclosing the achene; stamens 6 or 8, dilated at base and united into a glandular ring, the inner mostly somewhat longer, the outer with tuberclelike glands on either side at base; anthers small, ovaloid or almost globular, caducous; styles 2 or 3, short, distinct or united at base, thickened below the capitate stigmas; ovary unilocular, 1-ovuled, lenticularly compressed or trigonous; achene flattish lenticular round to oval or trigonous; sessile, wingless, falling together with the valves; seed albuminous, ovoid, lenticular or trigonous, the curved excentric embryo with linear cotyledons. Much branched shrubs or dwarf subshrubs; branches rigid, erect or spreading, often transformed into spines; leaves coriaceous or fleshy, rarely thin,
502 rather rigid, sessile or short-petioled, the petiole jointed to blade and attached to a callous cushion united with the base of the ocrea; ocreae membranous, sheathlike, more or less connate at base, rarely split to the base, usually produced at apex into 2 lanceolate to linear points, rarely irregularly lacerate; bracts abbreviated, cupuliform or campanulate, not leafy; flowers 1—3 from the ocreae, forming sessile fascicles or loose racemes; pedicels slender, capillary in upper part, jointed at the middle or near the base, nodding in fruit. Russian: "kurchavka" or "kolyuchaya grecha" [spiny buckwheat].

1. Flowers in axillary fascicles; rachis none; perianth of 4 segments; stamens 6; styles 2; achene compressed, lenticular (Subgenus Euatraphaxis Jaub. et Spach) . . . . 2.
\+ Flowers in capitate or elongated racemes with a distinct rachis; perianth segments 5; stamens 8; styles 3; achene trigonous (Subgenus Tragopyrum Jaub. et Spach) . . . . 6.
2. Leaves grayish, scabrous, covered on both sides with short thick bristles . . . . 3. **A. canescens** Bge.
\+ Leaves smooth, glabrous, green or glaucescent . . . . 3.
3. Shrubs 30—80 cm tall, with long slender branches; leaves alternate, solitary; internodes distinct . . . . 4.
\+ Dwarf undershrubs 15—30 or rarely 45 cm tall, with a very stout flexuous main stem and very short thick branches; leaves clustered; internodes short and indistinct . . . . 5.
4. The ends of all or most branchlets leafless and spinescent; perianth rather small in fruit, the inner segments 4—5 mm long and 5—6 mm broad . . . . 1. **A. spinosa** L.
\+ The ends of all or most branchlets leafy, unarmed; perianth relatively large in fruit, the inner segments 7—8 mm long and 8—9 mm broad . . . . 2. **A. replicata** Lam.
5. Leaves elliptic, ovate, or obovate, 5—10 mm long and 3—7 mm broad; the ends of some branchlets leafless and spinescent; pedicels jointed in upper part, the portion above the joint shorter than the reflexed outer perianth segments . . . . 4. **A. compacta** Ldb.
\+ Leaves orbicular or oval, 2—3 mm long and 1—2 mm broad; the ends of all branchlets unarmed, leafy; pedicels jointed at the middle, the portion above the joint twice as long as the reflexed outer perianth segments . . . . 5. **A. karataviensis** Lipsch. et N. Pavl.

503 6. Leaves distinctly reticulate-veined beneath, lanceolate, elliptic, ovate, or orbicular, always more than 3 mm broad . . . . 7.

\+ Leaves with a pronounced midrib and obscurely pinnate-veined beneath, linear or linear-lanceolate, not more than 3 mm broad . . . . 15.

7. Leaves 3—6 cm long and 1—2 cm broad, oblong-elliptic, thin, soft, subherbaceous; inflorescences terminal, ovaloid, not becoming woody; flowers very large; inner segments of fruiting perianth 7—8 mm long and 9—10 mm broad. . . . . 6. **A. Muschketovii** Krassn.

\+ Leaves to 3 cm long and 0.5—1 cm broad, leathery or fleshy, more or less rigid; inflorescences lateral or terminal and then the rachis becoming woody toward the end of flowering; flowers smaller; inner segments of fruiting perianth 5—6 mm long and 7—8 mm broad . . . . 8.

8. Twigs of the current year papillose-puberulent . . . . 9.

\+ Twigs of the current year glabrous . . . . 11.

9. Leaves elliptic or broad-elliptic, rounded or subtruncate at both ends, passing at base into a very short petiole, minutely point-tipped at apex, glabrous or slightly papillose-hairy on the margin, bright green; ocreae bipartite, with long slender points; inflorescences lateral; pedicels jointed below the middle, the portion above the joint twice as long as the reflexed outer perianth segments. (Altai) . . . . 7. **A. laetevirens** (Ldb.) Jaub. et Spach.

\+ Leaves oblong-elliptic, ovate, or suborbicular, cuneately narrowed toward base, slightly papillose-hairy at the base of the midrib and on the short pedicel; inflorescences terminal; pedicels jointed at or somewhat below the middle (Caucasus and Central Asia) . . . . 10.

10. Leaves oblong-elliptic, to 20 mm long and 5—8 mm broad, gradually cuneately narrowed at base to a very short petiole, slightly crisped-undulate at the margin, terminating in a short point; pedicels jointed at the middle, the portion above the joint slightly longer than the outer perianth segments . . . . 8. **A. Tournefortii** Jaub. et Spach.

\+ Leaves obovate or suborbicular, to 20 mm long and 9—12 mm broad, short-constricted at base into a distinct petiole (to 2 mm long), orbicular, slightly crisped-undulate on the margin, indistinctly crenate, terminating in a short point; pedicels jointed slightly below the middle, the portion above the joint 1.5 times as long as the reflexed outer
504 perianth segments . . . . 9. **A. caucasica** (Hoffm.) N. Pavl.

11. All racemes lateral on old branches; all or most woody branches leafless at the ends and spinescent . . . . 12.

\+ Racemes lateral and terminal or only terminal; all or most woody branches leafy soft and unarmed at the ends . . . . 13.

12. Leaves azure-glaucous, rounded or obtusely truncate at base, the blade 8—10 times the length of the petiole, faintly reticulate-veined beneath; racemes relatively loose, containing up to 20 flowers; pedicels jointed at or above the middle . . . . 10. **A. pungens** (M. B.) Jaub. et Spach.

\+ Leaves bright green, only beneath somewhat glaucescent, cuneately long-attenuate at base, the blade 3—4 times the length of the petiole, prominently reticulate-veined beneath; racemes 20—40-flowered; pedicels commonly jointed in the lower third . . . . 11. **A. pyrifolia** Bge.

13. Racemes lateral and terminal; leaves very thick, fleshy, orbicular, subreniform, or broadly elliptic, subsessile or truncately short-attenuate at base to a very short barely perceptible petiole, the margin callous-thickened, the apex obtusely rounded, retuse, or terminated by a very short obtuse point . . . . 12. **A. seravschanica** N. Pavl.

\+ All racemes terminal, long, linear, with a woody rachis; leaves lanceolate, rarely oblong-elliptic or ovate, gradually narrowed at base to a short petiole, the margin thin flat or revolute, the acuminate apex terminated by a short to fairly long cartilaginous point . . . . . . 14.

14. Branchlets of the current year rather short, projecting but little from the bush, their ends usually unbranched in the inflorescences; leaves of the current year's branchlets to 17 mm long, acuminate or terminated by a very short obtuse point; pedicels jointed at the middle . . . . 13. **A. frutescens** (L.) Ewersm.

\+ Branchlets of the current year much elongated, strongly projecting from the bush in all directions, their ends paniculately branched in inflorescences; leaves of the current year's branchlets 12—30 mm long, subobtuse or rounded at apex, terminated by a sharp spinule 0.5—2 mm long; pedicels jointed below the middle or in lower third . . .

505 . . . . 14. **A. virgata** (Rgl.) Krassn.

15. Herbaceous branchlets of the current year densely papillose-puberulous; leaves 15—40 mm long, the midrib papillose on both sides or merely beneath . . . . 16.

\+ Herbaceous branchlets of the current year glabrous, smooth; leaves 5—10 mm long, linear-lanceolate, glabrous or with papillose-ciliolate margin; a dwarf shrub, 5—15 cm in height, with a very stout flexuous main stem and short procumbent branches . . . . 15. **A. decipiens** Jaub. et Spach.

16. A tall shrub, 30—80 cm in height, with long slender branches; leaves densely papillose-pubescent on both sides; inflorescence long, linear, loose; flowers yellow or greenish-yellow; flowers commonly paired in the ocreae . . . . 16. **A. badghysi** M. Kult.

\+ A dwarf shrub, 5—25 cm in height, with short stoutish branches; leaves papillose only on the midrib beneath; inflorescence short, densely many-flowered; flowers whitish-pink, commonly in 3's . . . . 17. **A. angustifolia** Jaub. et Spach.

++ A dwarf shrub, 10—20 cm in height; leaves mostly cylindric, minutely verrucose-scabrous; inner segments of the fruiting perianth spherically enveloping the achene . . . . 18. **A. teretifolia** (M. Pop.) Kom.

Subgenus 1. **EUATRAPHAXIS** Jaub. et Spach, Ill. pl. orient II (1844—46) 12 (characters in the key).

1. **A. spinosa** L. Sp. pl. (1753) 475; Ldb. Fl. Ross. III, 514; Shmal'g., Fl. II, 388; Kryl., Fl. Zap. Sib. IV, 844. — A. karelini Bge (non. Jaub. et Spach) Mém. Acad. St.-Pétersb. sav. étrang. VII (1852) 482. — A. spinosa var. Linnaeana Meisn. in DC. Prodr. XIV (1856) 75. — A. replicata auct. (non Lam.). — Tragopyrum spinosum Presl, Bot. Bemerk. (1844) 109. — Polygonum crispulum var. Sims, Bot. Mag. (1808)

tab. 1065. — Ic.: Dill. Hort. Elth. (1774) tab. 32, fig. 47; L'Herit. Stirp. (1784—85) tab.14; Gaertn. De fruct. et sem. (1788—1807) tab. 119; Bot. Mag. (1808) tab. 1065 var.; Lam. Ill. (1823) tab. 265, fig. 2; Wats. Dendr. 2 (1825) tab. 119; Krassn. Script. Soc. Geogr. Ross. XIX (1888) tab. 3, fig. 3; C. K. Schn. Hdb. Laubholzk. I (1906) fig. 166.

Shrub 30—80 cm high; stem slender; branches long, spreading, slender, flexuous, woody, leafless at the ends, spine-pointed, covered with scabrous bark, the epidermis splitting longitudinally and peeling off; current year's
506 herbaceous branchlets of the second order soon becoming lignified, leafless at the ends, slender, straight or slightly flexuous, pointed, quite glabrous; ocreae short-cylindric, 1—2 mm long, brownish at base, membranaceous and obscurely nerved in upper part, much shorter than the internodes, on one side cleft into 2 short aristate teeth; leaves glaucous or glaucescent-green, 3—5 mm long and 3—4 mm broad, stiffish, thick, orbicular or rounded-elliptic, rarely broad-obovate, cuneately short-tapering at base to a very short petiole, obtuse or rounded at apex, slightly crisped-plicate at the margin, glabrous on both sides, smooth above, faintly reticulate-veined beneath; flowers in fascicles of 2—6, in the leaf axils at the ends of short branchlets of the current year, these borne laterally on old woody branches; pedicels long, jointed at or slightly below the middle; perianth segments bright pink with white margins or white; fruiting perianth rather small; inner segments rounded-cordate or reniform, 4—5 mm long and 5—6 mm broad, slightly longer and much broader than the achene; outer segments smaller, reflexed, oblong-ovate, slightly shorter than the above-joint portion of pedicel; achene compressed, ovate to broad-ovate, attenuate at apex, light brown or yellowish-green, smooth, lustrous. May—June.

Desert, clayey or gravelly steppes, sandhills and stony slopes. — European part: L. V.; Caucasus: W., E. and S. Transc.; W. Siberia: U. Tob., Irt.; Centr. Asia: Balkh., T. Sh., Dzu.-Tarb., Ar.-Casp., Syr D., Pam.-Al., Kyz. K., Kara K., Mtn. Turkm. **Gen. distr.**: Dzu.-Kash., Arm.-Kurd., Bal.-As. Min., E. Med. Described from Midye. Type in London.

Note. A rather stable species. The compoundness and polymorphism indicated in Russian floras has been due to arbitrary inclusion of completely unarmed forms which belong to the next species. The species displays slight variations in the shape and size of leaves and the size of perianth, but it has not been possible to establish any geographical correlations for these characters and the variations must therefore be considered as representing merely ecological forms.

Economic importance. In desert areas, thickets of this species are of some practical value as feed for camels.

2. **A. replicata** Lam. Encycl. I (1783) 329; Ej. Suppl. I (1810) 534. — A. crassifolia Agardh ex Roem. et Schult. Syst. veg. VII, 2 (1829—30) 1381?. — A. Laxmanni Agardh. l. c., p. 1382?. — A. Fischeri Jaub. et Spach, Ill. pl. or. II (1844—46) 12. — A. Karelini Jaub. et Spach, l. c. 12. — A. densiflora C. Koch in Linnaea XXII (1849) 212. — A. spinosa var. Fischeri Meisn. in DC. Prodr. XIV (1856) 75. — A. spinosa var. Karelini Meisn. l. c., p.75. — A. Calverti Boiss. Diagn. pl. or. ser. II,
507 4 (1859) 76. — A. spinosa var. mutica Rgl. in A. H. P. VI (1879) 395. — A. spinosa var. rotundifolia Boiss. Fl. Or. IV (1879) 1021. — A. spinosa var. glauca Boiss. l. c., p. 1021. — Ic.: Lam. Ill. (1823) tab. 265, fig. 1; Krassn. Script. Soc. Geogr. Ross. XIX (1888) tab. 2, fig. 3 et tab. 3, fig. 4.

Shrub 30 — 70 cm high, divaricately branched; stem slender; branches straight, long, slender, woody, leafy all the way up, unarmed, covered with yellowish-brown or reddish-brown bark, the whitish longitudinally fibrillose-splitting epidermis persistent; current year's herbaceous branchlets of second order soon becoming lignified, terminated by leaves or flowers, slender, straight, quite glabrous; ocreae short-cylindric, 2 — 3 mm long, brownish in lower part, white-hyaline and obscurely 2-nerved in upper part, much shorter than the internodes, cleft on one side into aristate-tipped teeth; leaves glaucescent or cinerescent, 3 — 8 mm long and 2 — 7 mm broad, thick, stiffish, obovate, orbicular, or subreniform, short-tapering toward base or gradually narrowed to a very short petiole, at apex short-attenuate to short-pointed or obtuse rounded or retuse, with revolute plicate or crisped-undulate margins, glabrous on both sides, smooth above, faintly reticulate-veined beneath; flowers in fascicles of 2 — 6 at the ends of short branchlets of the current year or in loose interrupted fascicled racemes at the ends of longer branchlets of the current year; pedicels long, jointed below the middle; perianth segments pink with white margins or white; fruiting perianth relatively large; inner segments rounded-cordate or subreniform, 7 — 8 mm long and 8 — 9 mm broad, slightly longer and much broader than the achene; outer segments smaller, reflexed, rounded-reniform, one-half to two-thirds the length of the above-joint portion of the pedicel; achene compressed, ovate to broad-ovate, attenuate at summit, light brown or brownish-green, smooth and glabrous, lustrous. May — June.

Clayey or stony desert steppes, sandhills, and stony mountain slopes. — European part: Bl., Crim., L. Don, L. V., Transv.; Caucasus: E. Transc., Dag.; W. Siberia: Irt.; Centr. Asia: Balkh., T. Sh., Ar.-Casp., Kyz. K., Kara K., Mtn. Turkm. **Gen. distr.**: Bal.-As. Min., Arm.-Kurd., Iran., E. Med. Described from Armenia. Type in Paris.

Note. This species has been reinstated following Fischer, under the name given by Lamarck, since both the description and the illustration
508 given by this author undoubtedly represent our plant with its few branches leafy throughout and its large perianth. On the basis of the disposition of inflorescences, Jaubert and Spach distinguish a species with unarmed branches, *A. Karelini* from the Caspian coast. However, Bunge implied under this name an exceedingly spiny form, *A. spinosa* L., with stout old branches. Similarly, Russian authors have erroneously defined, as var. *rotundifolia* Boiss., outstandingly spiny forms, even though the type of the variety originates from the Crimea where the named spineless species occurs exclusively. The situation is more complicated as regards Agardh's synonym, since the type of his *A. Laxmanni* originates from Siberia where *A. replicata* is unknown. It is possible that *A. Laxmanni* represents one of the forms of *A. spinosa*.

3. **A. canescens** Bge. Ind. sem. h. Dorpat. (1839) 3; Bong. et Mey. Suppl. II Fl. Alt., 62; Ldb. Fl. Ross. III, 513; Kryl., Fl. Zap. Sib. IV, 845. — *Tragopyrum canescens* Presl, Bot. Bemerk. (1844) 109. — Ic.: Bong. et Mey. l. c. tab. XIV.

A divaricately branched shrub, 30 — 60 cm high; stem stout, flexuous; branches stoutish, long, flexuous, woody, partly leafless at the ends, partly blunt unarmed and terminated by flowers, covered with grayish-brown bark, the epidermis splitting longitudinally and peeling off in fibers; current year's herbaceous branchlets of second order short, soon becoming lignified, terminated by flower-fascicles, bluntish or pointed and leafless at the ends, straight or slightly flexuous, quite glabrous; ocreae short-cylindric, 2—3 mm long, white-hyaline, obscurely nerved, much longer than the internodes, cleft on one side and split at summit into 2 slender aristate points; leaves gray or grayish-green, small, 3—4 mm long and 2—3 mm broad, stiffish, fleshy, ovate to suborbicular, cuneately short-tapering at base to a short petiole, acuminate or rarely obtusish and rounded at apex, with revolute entire or slightly crisped-undulate margins, scabrous on both sides, covered with short thick papillose bristles, the lower surface somewhat prominently reticulate-veined; flowers in fascicles of 2 or 3 in the leaf axils at the ends of short branchlets of the current year; pedicels long, jointed above the middle or in upper part; perianth segments bright pink, white-margined, turning brown, rather small; inner segments of fruiting perianth rounded-cordate, 4—5 mm long and 5—6 mm broad, slightly longer and broader than the achene; outer segments smaller, ovate, longer than the above-joint portion of the pedicel; achene compressed, broad-elliptic to almost round, obtusely rounded at summit, brownish-yellow, smooth and glabrous, lustrous. May—June.

509 Sandhills in valleys of desert and steppe rivers; stony and gravelly slopes. — W. Siberia: Irt.; Centr. Asia: Dzu.-Tarb. Endemic. Described from the shores of Lake Zaisan. Type in Leningrad.

Note. Most closely resembling A. spinosa L., but readily distinguishable by its vesture and never confounded with any other species.

4. **A. compacta** Ldb. Fl. Alt. II (1830) 55; Ej. Fl. Ross. III, 513; Kryl., Fl. Zap. Sib. IV, 844. — A. crassifolia Krassn. (non Agardh), p. p. Scripta Soc. Geogr. Ross. XIX (1888) 298. — A. spinosa auct., p. p. — A. spinosa var. compacta Trautv. in A. H. P. I (1871—72) 280. — Tragopyrum compactum Presl, Bot. Bemerk. (1844) 109. — Ic.: Ldb. Ic. pl. Fl. Ross. V, tab. 411. — Exs.: Kar. et Kir. (1841) No. 1919.

A dwarf undershrub, 10—35 cm high; stem short, stout, flexuous; branches spreading, stout, short, flexuous, woody, leafless at the ends, pointed and spinescent, covered with yellowish-gray bark, the epidermis splitting longitudinally into fibers and peeling off; current year's herbaceous branchlets of second order very short, stoutish, flexuous, soon becoming lignified, partly terminated by reduced leaves, partly pointed and leafless at the ends, quite glabrous; ocreae short-cylindric, 1—1.5—2 mm long, shorter than or as long as the internodes; in lower part brownish, in upper part white-hyaline, obscurely nerved, commonly cleft to base into 2 short curved teeth; leaves on old branches clustered, on herbaceous branchlets more or less remote, glaucescent, small, 5—8 mm long and 3—6 mm broad, oblong-obovate or broad-elliptic, obtusely cuneate-tapering at base to a very short petiole, obtusish or short-acuminate or slightly retuse at apex, muticous or obtusely point-tipped, with flat or slightly obtuse-crenate margins, glabrous on both sides, smooth above (only under strong magnification convexly puncticulate), prominently and stoutly reticulate-veined beneath; flowers in fascicles of 1—6, in the leaf axils at the ends of old woody branches or more rarely of annotinous branchlets; pedicels long, jointed above the middle in upper third or fourth; perianth segments bright pink with white margins or white, turning brown; fruiting perianth fairly large; inner segments rounded-reniform or cordate-reniform, 7—8 mm long and 8—9 mm broad, slightly longer and much broader than the achene; outer segments much smaller, ovate-oblong, reflexed, longer than or rarely as long as the above-joint portion of the pedicel; achene compressed, broad-elliptic or almost round, narrowed at summit, brownish-yellow, smooth and glabrous, lustrous. June—July.

Stony foothill slopes and boulder-trains, pebbly and sandy steppes, rarely
510 sandhills in the valleys of desert and steppe rivers. — W. Siberia: Alt.; Irt.; Centr Asia: T. Sh., Dzu.-Tarb., Ar.-Casp. **Gen. distr.**: Mong., Dzu.-Kash. Described from Lake Zaisan. Type in Leningrad.

Note. A species not very easily distinguishable from A. spinosa L. and coming into its own only in ecologically markedly contrasted stony mountain habitats.

5. **A. karataviensis** Lipsch. et N. Pavl. Animadv. syst. Herb. Tomsk. No. 5—6 (1933) 3. — Ic.: ibidem tab. sine numero.

A dwarf divaricately branched undershrub, 7—25 cm high; stem short, very stout, flexuous; branches very short, flexuous, woody, leafless but bluntish and unarmed, covered with dark gray bark, the epidermis splitting and peeling off in longitudinal fibers; current year's herbaceous branchlets of second order clustered at the ends of woody branches, very short, soon becoming lignified, terminated by leaves or flowers, not spinescent, quite glabrous; ocreae 0.5—1 mm long, scarious, squamiformly cleft, almost flat, longer than the internodes, broad-ovate, long-tapering at summit to an aristate point, with a faint nerve; leaves clustered at the ends of branchlets of the current year, green, minute, 2—3 mm long and 1—2 mm broad, oval to suborbicular, gradually cuneately narrowed at base to the petiole (this as long as the blade or nearly so), retuse or obtusish or minutely point-tipped at apex, fleshy, stiffish, flat, slightly thickened at the margins, glabrous on both sides, smooth above, faintly pinnate-nerved beneath; flowers in fascicles of 1—3 in the leaf axils at the tips of herbaceous branchlets of the current year; pedicels long, jointed at the middle; perianth segments bright or dark pink with narrow white margins, large; inner segments of fruiting perianth rounded-cordate or subreniform, 5—6 mm long and 7—8 mm broad, slightly longer and much broader than the achene; outer segments much smaller, ovate-oblong, one-half to two-thirds the length of the above-joint portion of the pedicel; achene compressed, broad-elliptic or almost round, narrowed at summit, brownish-yellow or yellowish-green, glabrous and smooth, lustrous. May—June.

Stony slopes and weathered peaks of desert mountains. — Centr. Asia: Syr D., Pam.-Al. Endemic. Described from the Syr Darya area of Karatau. Type in Moscow; cotype in Leningrad.

Note. An unusually distinctive species, first discovered by S. Korzhinskii
513 in Pamir-Alai. The plant is related to the Arabian A. sinaica Jaub. et Spach which has partially spinescent branchlets and short pedicels. It differs from A. compacta Ldb. in the markedly abbreviated and little-dissected ocreae.

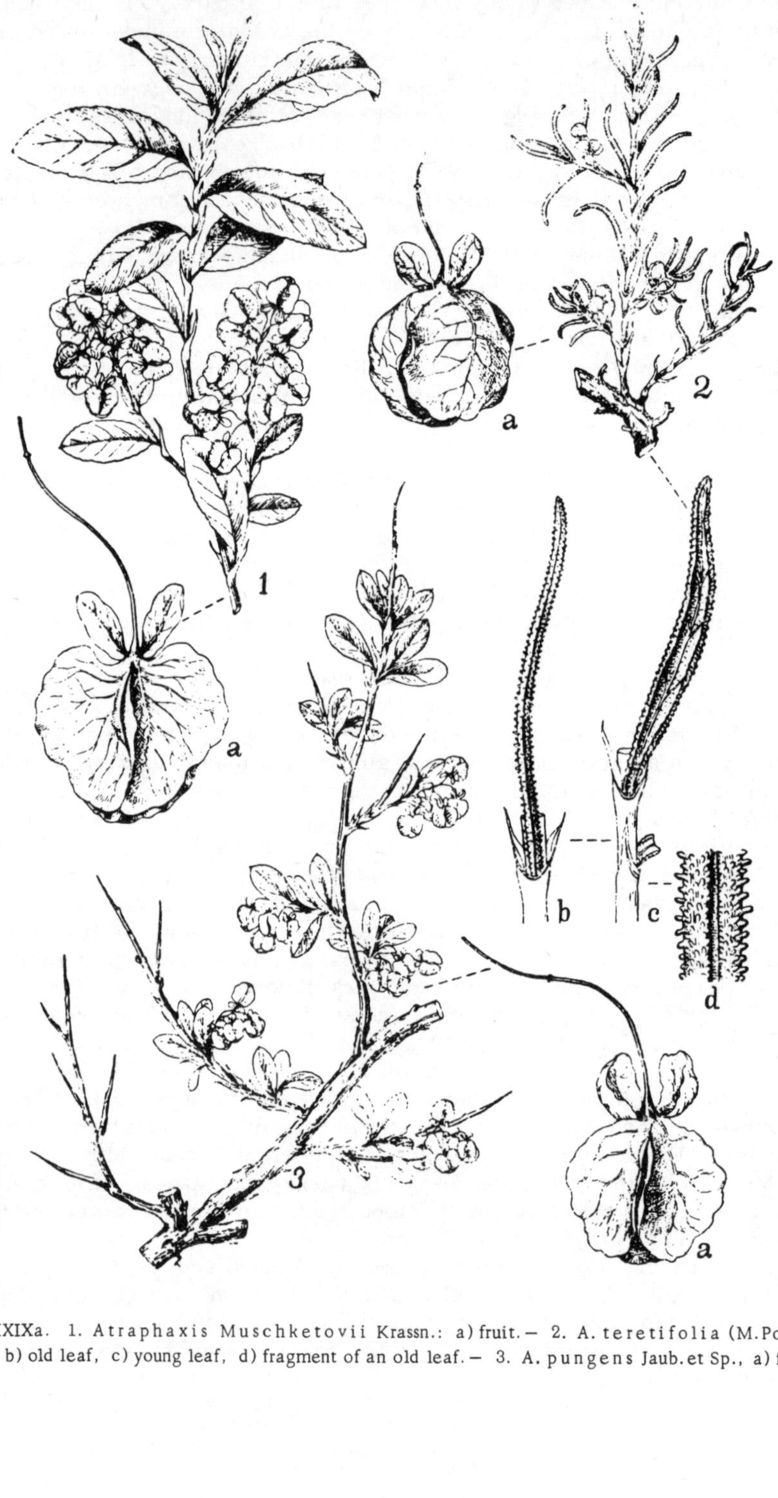

PLATE XXIXa. 1. Atraphaxis Muschketovii Krassn.: a) fruit. — 2. A. teretifolia (M.Pop.) Kom., a) fruit, b) old leaf, c) young leaf, d) fragment of an old leaf. — 3. A. pungens Jaub. et Sp., a) fruit.

Subgenus 2. **TRAGOPYRUM** Jaub. et Spach, l. c., p. 13 (characters in the key).

6. **A. Muschketovii** Krassn. Script. Bot. II (1886) 20. — A. latifolia Koehne, Mitt. Deutsch. Dendr. Gesellsch. III (1894) 13. — Tragopyrum lanceolatum var. latifolium Rgl. Gartenfl. XL (1891) 169. — Ic.: Curt. Bot. Mag. LI (1895) 609, tab. 7435; Krassn. Script. Soc. Geogr. Ross. XIX (1888) tab. 2, fig. 1. — Exs.: Koehne Herb. Dendr. No. 211.

Shrub, 50—100 cm high; stem stout; branches elongated, stout, straight or slightly flexuous, woody, obtuse, spineless, terminated by leaves or flowers, covered with deeply furrowed reddish-brown bark, the persistent yellowish epidermis splitting longitudinally into fibers; current year's herbaceous branchlets of second order long, flexuous, at length becoming lignified, terminated by leaves or flowers, unarmed, glabrous; ocreae long-cylindric, 5—6 mm long, in lower part brownish, in upper part thin scarious with 2 faint brownish nerves, much shorter than the internodes, clasping at base, cleft above into 2 long slender aristate-pointed teeth; leaves green, darker above, very large, 3—6 cm long and 0.8—2 cm broad, thin, soft, subherbaceous, oblong-elliptic, short-tapering at both ends, passing at base into a short petiole, obtusish at base or terminating in a very short obtusish point, flat, with entire or unequally crenulate, glabrous on both sides, reticulate-veined faintly above and prominently beneath; flowers at the ends of branchlets of the current year in ovaloid lateral (in relation to woody branches) leafless many-flowered racemes 3—6 cm long, the soft rachis tardily lignifying; pedicels arising from the ocreae, commonly in pairs, 5—8 mm long, jointed much below the middle or almost at base; perianth segments pale pink or white with pinkish margins, large; inner segments of fruiting perianth slightly unequal, rounded-cordate, the larger 7—8 mm long and 9—10 mm broad, greatly exceeding the achene in both length and breadth; outer segments much smaller, oblong-ovate, reflexed, one-fourth to one-third the length of the above-joint portion of the pedicel; achene trigonous, ovoid, acuminate, with bluntish ribs and broad-ovate faces, dark brown, glabrous and smooth, lustrous. May—June. (Plate XXIXa,
514 Figure 1, a).

Meadow and meadow-steppe foothill slopes, valleys of rivers and brooks. — Centr. Asia: T. Sh. Unknown elsewhere. Described from the vicinity of Alma-Ata (formerly Vernyi). Type in Leningrad.

Note. A very characteristic mesophyte species, apparently constituting sometimes an element of undergrowth in almost destroyed broad-leaved forests of Central Tien Shan. For this reason A. N. Krasnov considered this species to be a prototype of the entire genus.

Economic importance. The most beautiful species of the genus, long cultivated in W. Europe; of interest as an ornamental plant.

7. **A. laetevirens** (Ldb.) Jaub. et Spach, Ill. pl. or. II (1844—46) 14; Kryl., Fl. Zap. Sib. IV, 843. — Tragopyrum laetevirens Ldb. Fl. Alt. II (1830) 75; Ej. Fl. Ross. III, 516. — T. buxifolium Kar. et Kir. (non M. B.) Bull. Soc. Natur. Mosc. XIV (1841) 741. — Polygonella laetevirens Meisn. Comment. II (1836—43) 228. — Ic.: Ldb. Ic. pl. Fl. Ross. V, tab. 422. — Exs.: Kar. et Kir. (1840) No. 429; (1841) No. 1917.

Shrub 30—80 cm high; stem stout; branches spreading, slender, long, flexuous, woody, obtusish, spineless, terminated by leaves or flowers, covered with yellowish or dark gray bark, the persistent epidermis fibrillose-splitting; current year's herbaceous branchlets of second order long, flexuous, at length becoming lignified, terminated by leaves or flowers, spineless, rough with short papillose bristles; ocreae short-cylindric, 2—3 mm long, in lower part brownish, in upper part thin and membranaceous, much shorter than the internodes, with 2 faint nerves produced from the truncate slightly toothed summit of the ocrea into slender awns, these as long as or longer than the rest of the ocrea; leaves bright green, stiffish, coriaceous, small, 7—14 mm long and 5—8 mm broad, elliptic to broad-elliptic, rounded or bluntish at both ends, passing at base into very short petiole, terminating in a very short obtusish point, flat, with entire or slightly crisped and obscurely undulate-crenate margins glabrous and almost smooth above, prominently reticulate-veined beneath, the midrib and margins covered with short papillose hairs; flowers in few-flowered capitate leafless mostly lateral racemes 1.5—3 cm long, borne at the ends of branchlets of the current year; pedicel commonly in pairs, from the ocreae, long, jointed below the middle or in lower third; perianth segments pink with white margins or white, turning brown, fairly large; inner segments of fruiting perianth slightly unequal, rounded-cordate, the larger 5—6 mm long and 7—8 mm broad, greatly exceeding
515 the achene in both length and breadth; outer segments much smaller, broad-ovate, reflexed, half as long as the above-joint portion of the pedicel; achene trigonous-ovoid, acuminate, with bluntish ribs and ovate-rhombic faces, dark brown, smooth and glabrous, lustrous. May—June.

Gravelly and stony mountain slopes and stony steppes. — W. Siberia: Irt., Alt.; E. Siberia: Ang.-Say.; Centr. Asia: Dzu.-Tarb. Endemic. Described from the Semipalatinsk Altai. Type in Leningrad.

Note. A species very similar to A. caucasica and closely akin to it. However, the unique ecology and the complete area displacement in the east point to this species as being rather recently differentiated in analogy to A. compacta and A. spinosa.

8. **A. Tournefortii** Jaub. et Spach, Ill. pl. or. II (1844—46) 14. — Tragopyrum Tournefortii Endl. Gen. Suppl. IV, 2 (1847) 54. — Ic.: Jaub. et Spach, l. c. tab. 112.

Shrub, 30—80 cm high; stem stout; branches spreading, slender, flexuous, woody, obtusish, unarmed, terminated by leaves or flowers, covered with light-gray or yellowish-gray bark, the persistent epidermis splitting longitudinally into fibers; current year's herbaceous branchlets of second order long, flexuous, slender, at length becoming lignified, terminated by leaves or flowers, unarmed, densely covered with short spreading papillose bristles; ocreae short-cylindric, 3—4 mm long, in lower part brownish, in upper part scarious, much shorter than the internodes, with 2 faint brownish nerves, commonly clasping at base, cut at summit into 2 slender aristate points; leaves green, stiffish, coriaceous, fairly large, 12—20 mm long and 5—8 mm broad, oblong-elliptic, gradually narrowed at base to a very short petiole, terminating in a short obtusish point, flat, with slightly crisped-undulate entire or bluntly crenulate margins, glabrous and almost

smooth above, prominently reticulate-veined beneath, the midrib and the margins papillose-puberulous; flowers in few-flowered capitate mostly terminal leafless racemes 2—3 cm long, borne on branchlets of the current year; pedicels commonly in 3's or 2's from the ocreae, long, jointed at or slightly below the middle; perianth segments pink with white margins or white, turning brown, fairly large; inner segments of fruiting perianth slightly unequal, rounded-cordate, the larger 5—6 mm long and 7—8 mm broad, twice as long and broad as the achene; outer segments smaller, broad-ovate, reflexed, slightly shorter than the above-joint portion of the
516 pedicel; achene trigonous-ovoid, acuminate, with bluntish ribs and ovate-rhombic faces, dark brown, glabrous and smooth, lustrous. May — June.

Gravelly and stony mountain slopes. — Caucasus: W. and S. Transc., Dag. **Gen. distr.:** Arm.-Kurd. Described from Turkish Armenia. Type in Paris.

Note. A species most closely related to A. caucasica and apparently forming its barely distinguishable western race. However, even in the Caucasus, where the two species occur side by side, they differ in a number of minor characters.

9. **A. caucasica** (Hoffm.) N. Pavl. comb. nova. — Polygonum caucasicum Hoffm. Comm. soc. phys.-n. ed. mosq. I, 1 (1806) 10. — P. crispulum Sims. Bot. Mag. (1808) tab. 1065 excl. var. — P. frutescens Güldenst. It. I (1787) 224 non alior. — P. buxifolium M. B. Fl. taur.-cauc. I (1808) 300. — Polygonella buxifolia Meisn. Comment. II (1836—43) 228. — Tragopyrum buxifolium M. B. l. c. III (1819) 284; Ldb. Fl. Ross. III, 516. — Atraphaxis buxifolia Jaub. et Spach, Ill. pl. or. II (1844—46) 14. — A. laetevirens Krassn. Script. Soc. Geogr. Ross. XIX (1888) 298, p. p., quoad pl. ibericam. — Ic.: Bot. Mag. (1808) tab. 1065 excl. var. β; Krassn. l. c. tab. 2, fig. 2; C. K. Schn. Laubholzk. I (1906) fig. 167a—c. — Exs.: Herb. Fl. cauc. No. 64; Pl. Or. exs. No. 8.

Shrub, 30—80 cm high; stem stout; branches spreading, long, slender, flexuous, woody, obtusish, unarmed, terminated by leaves or flowers, covered with reddish-gray bark, the persistent epidermis splitting longitudinally into fibers; current year's herbaceous branchlets of second order long, flexuous, at length becoming lignified, terminated by leaves or flowers, obtusish, unarmed, covered with short thickish papillose bristles; ocreae cylindric, 2—3 mm long, much shorter than the internodes, in lower part brownish, in upper part membranaceous and almost pellucid, with 2 brownish nerves produced from the truncate slightly toothed upper margin into slender awns, these as long as or shorter than the rest of the ocrea; leaves bright green, rather large, 15—17 mm long and 9—12 mm broad, obovate, elliptic, or suborbicular, obtusely and cuneately short-tapering to a distinct petiole to 2 mm long, obtusish or short-acuminate at apex and produced into a very short obtusish point, with commonly slightly crisped-plicate bluntly and indistinctly crenate margins, glabrous and almost smooth above, somewhat prominently reticulate-veined beneath, papillose-puberulous on the midrib petiole and margins at base; flowers in rather loose and few-flowered capitate mostly terminal leafless racemes 1.5—3 cm long,
517 borne at the ends of branchlets of the current year; pedicels mostly in 2's or 3's from the ocreae, long, jointed at or slightly below the middle;

perianth segments pink with white margins or white, turning brown, rather large; inner perianth segments of fruiting perianth slightly unequal, rounded-cordate, the larger 5—6 mm long and 7—8 mm broad, greatly exceeding the achene in both length and breadth; outer segments smaller, broad-elliptic, reflexed, 1.5—2.5 times shorter than the above-joint portion of the pedicel; achene trigonous-ovoid, with bluntish ribs and ovate-rhombic faces, dark brown, glabrous and smooth, lustrous. May—June.

Dry, stony and steppe mountain slopes; dry, stony beds of rivers and brooks. — Caucasus: E. and S. Transc.; Centr. Asia: Balkh., T. Sh., Dzu.-Tarb. **Gen. distr.**: Arm.-Kurd. Described from Georgia. Type in Leningrad.

Note. Hoffman undoubtedly described the plant named earlier by Marshall-Bieberstein, and in Hortus Mosquensis (1808) he himself ceded priority to the latter, the reason being apparently that Marshall's epithet has Tournefort's pre-Linnaean priority "Polygonum orientale, Buxi folio rigido etc." Contemporary West European and Russian taxonomists oppose in this respect certain authorities tending to extend nomenclature "ad calendas graecas," and from this point of view the species should be renamed. It is the most widely distributed species of the group which also includes the two preceding species and overlaps their periphery both in the east and in the west.

10. **A. pungens** (M. B.) Jaub. et Spach, Ill. pl. or. II (1844—46) 14; Kryl., Fl. Zap. Sib. IV, 843. — Tragopyrum pungens M. B. Fl. taur.-cauc. III (1819) 284; Ldb. Fl. Ross. III, 515. — T. pungens var. latifolium Ldb. l. c. (1851) 516. — Polygonella pungens Meisn. Comment. II (1836—43) 228. — Ic.: Gmel. Fl. sib. III (1768) tab. 12, fig. 1; Ldb. Ic. pl. Fl. Ross. V (1834) tab. 426.

Shrub 20—50 cm high; stem stout; branches short, spreading, stoutish, flexuous, rigid, woody, leafless at the ends, pointed and spinescent, covered with brownish-gray rough bark, the epidermis splitting longitudinally into fibers, persistent; current year's herbaceous branchlets of second order short, stoutish, soon becoming lignified, flexuous, not terminated by leaves or flowers, pointed, whitish, quite glabrous; ocreae short-cylindric, 2—3 mm long, commonly shorter than the internodes, brownish except the membranaceous summit, cleft to base into 3 teeth, of these the lateral
518 long-acuminate and awn-tipped, the middle one short lanceolate; leaves azure-glaucous, rather large, 15—17 mm long and 5—10 mm broad, rather stiffly coriaceous, broad-elliptic or oblong-obovate, almost round or obtusely cuneate-tapering to a very short petiole (one-tenth to one-eighth the length of the blade), rounded-obtuse at apex or rarely terminating in a very short cartilaginous point, with flat subentire or slightly undulate margins, glabrous on both sides and on the petiole, almost smooth above, somewhat prominently reticulate-veined beneath; flowers only in lateral few-flowered (to 20) capitate racemes to 2—3 cm long borne laterally on woody branchlets; pedicels long, jointed at or above the middle; perianth segments pinkish-white or greenish-white, rather large; inner segments of fruiting perianth slightly unequal, rounded-cordate, the larger 5—6 mm long and 7—8 mm broad, greatly exceeding the achenes in both length and breadth; outer segments rather small, elliptic, reflexed, from slightly shorter than the above-joint portion of the pedicel to two-thirds the length

of that portion; achene trigonous-obovoid, acuminate, with bluntish ribs and ovate-rhombic faces, dark brown, glabrous and smooth, lustrous. May — June. (Plate XXIXa, Figure 2, a).

Stony slopes of desert mountains and hills; stony and gravelly saline desert steppes. — W. Siberia: Irt., Alt.; E. Siberia: Ang.-Say., Dau. **Gen. distr.**: Ind.-Him., Dzu.-Kash., Mong. Described from Transbaikalia. Type in Leningrad.

Note. A very characteristic species, confined to Siberia; erroneously reported by Regel for Central Asia instead of the easily distinguishable next species.

11. **A. pyrifolia** Bge. Mém. Acad. St. Pétersb. sav. étrang. VII (1852) 483. — A. pungens Rgl. A. H. P. VI (1879) 398 et auct. plur. fl. turkest., non Jaub. et Spach. — A. pungens var. elliptica Rgl. l. c. (1879) 398. — A. Billardieri Rgl. l. c., 398, non Jaub. et Spach. — A. buxifolia Krassn. Script. Soc. Geogr. Ross. XIX (1888) 297, non Jaub. et Spach. — Tragopyrum glaucum Capus, Ann. agron. VIII (1882) No. 3. — Ic.: Krassn. l. c. tab. 2, fig. 5. — Exs.: N. Pavl. et S. Lipsch. Pl. turkest. (1932) No. 112.

Shrub, 1 — 2 m high; stem stout, tortuous; branches rather short, spreading, rigid, flexuous, woody, leafless at the ends, pointed and spinescent, covered with brownish-gray rough bark, the epidermis splitting longitudinally
519 into fibers and persistent; current year's herbaceous branchlets of second order straight, stoutish, not terminated by leaves or flowers, soon becoming lignified, acuminate, whitish, quite glabrous; ocreae short-cylindric, 2 — 3 mm long, commonly shorter, except on short sterile twigs where longer than the internodes, in lower part brownish, in upper part membranaceous and subpellucid, obscurely 3-nerved, clasping at base, deeply cut at summit sometimes down to base into 3 teeth, of these the lateral long and slenderly aristate-tipped, the shorter middle one broad-lanceolate; leaves bright green above, glaucescent beneath, 15 — 17 mm long and 10 — 12 mm broad, rather stiffly coriaceous, orbicular or broad-obovate, cuneately long-tapering at base to a septiole one-fourth to one-third the length of the blade, short-acuminate at apex and produced into a short obtusish point, rarely obtusely rounded or retuse, with flat entire or slightly crenate margins, on both sides and on petiole quite glabrous, reticulate-veined faintly above and prominently beneath; flowers on short branchlets of the current year, in densely 20 — 40-flowered capitate racemes borne laterally on woody branches; pedicels long, jointed below the middle, usually in lower third; perianth bright pink or yellowish-pink, turning brown, fairly large; inner segments of fruiting perianth slightly unequal, rounded-cordate, the larger 5 — 6 mm long and 7 — 8 mm broad, greatly exceeding the achene in both length and breadth; outer segments smaller, elliptic, reflexed, one-half to two-thirds the length of the above-joint portion of the pedicel or rarely equaling that portion; achene trigonous, rounded-ovoid, acuminate, with bluntish ribs and ovate-rhombic faces, dark brown, glabrous and smooth, lustrous. May — June.

Stony and steppe mountain slopes. — Centr. Asia: Balkh., T. Sh., Syr D., Amu D., Pam.-Al. Endemic. Described from Zeravshan. Type in Leningrad.

Note. A very characteristic species, undoubtedly related to A. pungens and replacing it in the mountain areas of Soviet Central Asia.

12. **A. seravschanica** N. Pavl. Animadv. syst. Herb. Tomsk. No. 5—6 (1933) 3. — A. pyrifolia auct. fl. turkest., p. p., non Bge. — A. pungens auct. turkest., p. p., non Jaub. et Spach.

Shrub, 50—90 cm high; stem stout, flexuous; branches long, slender, spreading, woody, obtuse, unarmed, terminated by leaves or flowers, their bark gray rough, the persistent epidermis longitudinally splitting into fibers; current year's branchlets of second order long, straight, slender,
520 whitish, quite glabrous; ocreae short-cylindric, 2—3 mm long, shorter than the leaves and much shorter than the internodes, in lower part brownish, in upper part thin membranaceous and with 2 faint brownish nerves, on one side cleft and terminating in 2 elongated triangular-lanceolate points; flowering ocreae campanulate or scalelike, nerveless, erose-margined, much shorter than the pedicels, the uppermost in inflorescence leafless; leaves grayish-green or glaucous-green, thick, stiffly coriaceous, 5—7 mm long and 5 — 8 mm broad, rarely to 20 mm long and 12—20 mm broad, the upper ones smaller, rounded-subreniform or broad-elliptic, truncately short-tapering at base to a very short petiole, obtusely rounded or retuse at apex or terminating in a very short obtusish point, glabrous on both sides, with midrib and a prominent network of lateral veins, the flat entire margin slightly callous-rimmed; flowers on short branchlets of the current year, in terminal or laterally linear strict many-flowered racemes 4—6 cm long, borne laterally on woody branches; pedicels long, at length nodding, in 3's or at the base of inflorescence in pairs from the ocreae, jointed considerably below the middle slightly above the base; perianth pink or red with white margins, turning brown, fairly large; inner segments of fruiting perianth slightly unequal, cordate-orbicular, 6—7 mm long and 7—8 mm broad, obtuse, slightly crisp-margined, greatly exceeding the achene in both length and breadth; outer segments smaller, reniform or elliptic, reflexed, one-third to one-half the length of the above-joint portion of the pedicel; achene trigonous, rounded-ovoid, acuminate, with bluntish ribs and ovate-rhombic faces, dark brown, glabrous and smooth, lustrous. May— June.

Stony and steppe mountain slopes. — Centr. Asia: Syr D., Amu D., Pam.-Al., Mtn. Turkm. Endemic. Described from Zeravshan. Type in Leningrad.

Note. A distinctive species. Confounding it with A. pyrifolia unnecessarily complicates the approach to this beautiful and very consistent species. Our plant bears more resemblance to the next species which is widespread and well known; it differs from it in the characteristic leaf shape and some other characters indicated above.

13. **A. frutescens** (L.) Ewersm. Reise v. Orenb. nach Buchara (1823) 115; Kryl., Fl. Zap. Sib. IV, 841. — Polygonum frutescens L. Sp. pl. (1753) 359. — P. fruticosum Gmel. Fl. sib. III (1768) 60. — Tragopyrum lanceolatum M. B. Fl. taur.-cauc. III (1819) 285; Ldb. Fl. Ross. III, 515; Turcz. Fl. baic.-dah. II, 1, 57. — T. lanceolatum var. divaricatum Ldb. l. c., (1851) 515. — T. lanceolatum var. subspinosum Bong. et Mey. Suppl. II, Fl. Alt. (1841) 63. — T. glaucum Spreng, Syst. veg. II (1825) 251; Less. Linnaea IX (1834) 204. — T. pungens var. angustifolium Ldb. l. c. (1849) 516. — T. pungens var. inermis Ldb. l. c. 516. — Polygonella lanceolata Meisn. Comment. (1836—43) 228. — Atraphaxis

521 lanceolata Meisn. in DC. Prodr. XIV (1856) 78; Shmal'g., Fl. II, 388. — A. micrantha Jaub. et Spach, Ill. pl. or. II (1844—46) 15. — A. davurica Jaub. et Spach, l. c., 15. — Ic.: Gmel. Fl. Sib. III (1768) tab. 12, fig. 2; Krassn. Script. Soc. Geogr. Ross. XIX (1888) tab. 2, fig. 7, 8; C. K. Schn. Laubholzk. I (1906) fig. 167e, f. — Exs.: Kar. et Kir. (1840) No. 430; (1841) No. 1918.

A much branched shrub, 20 — 70 cm high; stem stout; branches slender, flexuous, spreading or ascending, woody, commonly terminated by leaves or flowers, obtuse and unarmed or leafless at the ends and spinescent (var. subspinosa Bong. et Mey.), covered with brownish-gray bark, the persistent epidermis splitting longitudinally into fibers; current year's herbaceous branchlets almost straight or slightly flexuous, slender, short, not projecting or but slightly projecting beyond the bushy woody branches, terminated by leaves or flowers, obtusish, at length becoming lignified, quite glabrous or very rarely papillose-puberulous; ocreae short-cylindric, 2—3 mm long, commonly shorter than the internodes, in lower part brownish, in upper part thin membranaceous and subpellucid, with 2 faint nerves produced from the truncate or slightly dentate-incised or lacerate upper margin into slender-pointed teeth; leaves commonly glaucous, more rarely glaucescent-green or grayish-green, fleshy, stiffish, 12—17 mm long and 2—8 mm broad, ranging from narrow-lanceolate or lance-linear to oblong-elliptic or oblong-obovate, gradually narrowed at base to a short petiole, acuminate at apex or terminating in a short cartilaginous whitish point, with flat or often slightly revolute entire or obscurely and obtusely crenulate margins, glabrous on both sides, smooth above, rather prominently reticulate-veined beneath except for very narrow leaves with sometimes faint nervature except for the prominent midrib; flowers in loose terminal racemes 1.5—6 cm long; pedicels in 2's or 3's from the ocreae, jointed almost at the middle; perianth bright pink with white margins or white or rarely greenish-white, rather small; inner segments of fruiting perianth slightly unequal, almost semiorbicular or slightly cordate, 4—5 mm long and 4—6 mm broad, as long as or slightly longer and somewhat broader than the achene; outer segments smaller, rounded, reflexed, about half as long as the above-joint portion of the pedicel; achene triquetrous with sharp almost winglike angles and ovate-triangular faces, acuminate, dark brown, glabrous and smooth, lustrous. May—June.

522 Desert, saline, clayey, or sandy steppes, stony and steppe trains and slopes of mountains and hills, pebbly and gravelly valleys of steppe and desert rivers, brooks, and gullies. — European part: Bl., Crim. (cultivated), V.—Don, L. Don, L. V., Transv.; Caucasus: Cisc.; W. Siberia: U. Tob., Irt., Alt.; E. Siberia: Ang.-Say., Dau.; Centr. Asia: Balkh., Ar.-Casp., Mtn. Turkm. Gen. distr.: Dzu.-Kash., Mong. Described from Dauria. Type in London.

Note. Most probably an aggregate species, containing some minor species, which cannot as yet be described because of lack of sufficient material. Jaubert and Spach considered it possible to separate Altai forms from those of Dauria which served as type for the Linnaean species; however, the characters indicated by these authorities are so insignificant that it is impossible to check them with the very scanty material. Ledebour apparently had doubts as to the occurrence of this species in Dauria and derived the spineless varieties from A. pungens. On the ground of Turchaninov's material and records we refer them to this species.

14. **A. virgata** (Rgl.) Krassn. Scripta Soc. Geogr. Ross. XIX (1888) 295. — A. stricta Krassn. l. c., p. 296. — A. lanceolata var. virgata Rgl. A. H. P. VI (1897) 397. — A. lanceolata var. stricta Rgl. l. c., p. 396. — Tragopyrum lanceolatum var. strictum Ldb. Fl. Ross. III (1846—1851) 515. — A. frutescens var. stricta Kryl., Fl. Zap. Sib. IV (1930) 842. — A. Tournefortii Rgl. l. c., p. 397, non Jaub. et Spach. — Ic.: Krassn. l. c. tab. 2, fig. 6. — Exs.: H. F. A. M. No. 102.

Shrub, 1—2 m high; stem stout; branches spreading, almost straight or slightly flexuous, long, slender, obtusish and unarmed, covered with dark brownish-gray bark, the persistent epidermis splitting longitudinally into fibers; current year's herbaceous branchlets of second order very long, straight or slightly flexuous, slender, strongly projecting in all directions, forking at the ends and terminated by leaves or a loosely paniculate cluster of floriferous branchlets, furrowed, whitish or grayish, quite glabrous; ocreae short-cylindric, 2—3 mm long, always shorter than the internodes, in lower part brownish, in upper part membranaceous and almost pellucid, with 2 distinct rather thick nerves, deeply cleft on one side into 2 triangular-lanceolate slenderly awned teeth; leaves grayish-green, thick fleshy, rather large especially on branchlets of the current year, 12—30 mm long and 5—12 mm broad, oblong-elliptic or oblong-obovate, narrowed at base to a short and broad petiole, obtusely acuminate or rounded at apex,
523 terminating in a whitish cartilaginous spinule 0.5—2 mm long, with flat or slightly revolute obscurely and obtusely crenate margins, glabrous on both sides, smooth above, faintly reticulate-veined beneath; flowers on branchlets of the current year in loose terminal racemes 5—15 cm long; pedicels commonly in pairs from the ocreae, jointed below the middle or in lower third; perianth bright pink with white margins or white, rather small; inner segments of fruiting perianth rounded-elliptic with crisped-undulate margins, 4—5 mm long and 5—6 mm broad, slightly surpassing or about as long as but much broader than the achene; outer segments much smaller, rounded, reflexed, two-fifths to one-half the length of the above-joint portion of the pedicel; achene with sharp almost winged angles and ovate-triangular faces, dark brown, glabrous and smooth, lustrous. May—June.

Clayey, gravelly, and sandy desert-steppes, steppes, stony and steppelike slopes of desert mountains and hills. — W. Siberia: Irt.; Centr. Asia: Balkh., T. Sh., Dzu.-Tarb., Syr D. **Gen. distr.**: Dzu-Kash., Mong. Described from the Mediterranean Region. Type in Leningrad.

Note. The present author combines A. virgata and A. stricta since, among the original specimens, it has been impossible to establish differences between them. As a whole, however, the type of A. virgata is sufficiently distinct from forms of A. frutescens and probably represents a barely distinguishable desert-dominated southern race.

15. **A. decipiens** Jaub. et Spach, Ill. pl. or. II (1844—46) 14. — A. lanceolata var. decipiens Trautv. in Bull. Soc. Natur. Mosc. III (1867) 78. — A. desertorum Krassn. Scripta Soc. Geogr. Ross. XIX (1888) 295, 296 in annot. — Ic.: Krassn. l. c. tab. 2, fig. 4.

A dwarf undershrub, 5—15 cm high; main stem very stout, short, flexuous; branches short, stoutish, flexuous, woody, leafy at the ends, unarmed, covered with brownish-gray bark, the persistent epidermis splitting longitudinally; current year's herbaceous branchlets of second order numerous and crowded, short, slender, straight or slightly flexuous, terminated by leaves or flowers, obtusish, at length becoming lignified, whitish, quite glabrous; ocreae short-cylindric, 1—2 mm long, in lower part slightly brownish, in upper part thin membranaceous, pellucid, on floriferous branchlets as long as the internodes, on leafy sterile branchlets always longer than the internodes, with 2 faint nerves produced from the truncate and slightly denticulate or deeply incised upper margin into slender-pointed awns; leafy green, 5—10 mm long and 0.5—2 mm broad, rather stiffly coriaceous, linear-lanceolate to linear, gradually narrowed
524 at base to a very short petiole, obtusely long-acuminate at apex, slightly revolute at the margins, glabrous on both sides except for the sparingly papillose-ciliate margin, smooth above, with a prominent midrib and faint pinnate veneation beneath; flowers on branchlets of the current year in few-flowered terminal racemes 10—12 mm long; pedicels commonly in pairs from the ocreae, long, jointed at or slightly above the middle; perianth pink with white margins, rather small; inner segments of fruiting perianth slightly unequal, rounded-elliptic, 4— 5 mm long and broad, as long as or slightly longer but much broader than the achene; outer segments smaller, rounded, reflexed, as long as or slightly longer than the above-joint portion of the pedicel; achene elongated, triquetrous, with sharp almost winglike angles and narrow sublanceolate faces, dark brown, glabrous and smooth, lustrous. May—June.

Stony peaks and slopes of steppe and desert mountains. — W. Siberia: Irt.; Centr. Asia: Balkh., Dzu.-Tarb., Ar.-Casp. Endemic. Type in Paris.

Note. As already noted by Trautvetter, our plants conform fully to the short description of Jaubert and Spach, but the origin of the plant described by these authorities is not quite clear, since their information is based on inadequately labelled purchased herbarium material. There is no reason to doubt the indication of these authorities that this plant originates from the USSR and the present author therefore retains the name given by them. The distribution area of the species corresponds to rather low and disjointed countryside known as the Kirghiz "Melkosopochnik." This species is related most closely to A. frutescens and represents a distinct diminutive race associated with mountain deserts.

16. **A. badghysi** M. Kult. Trans. Sc. Soc. Turkest. I (1923) 118. — A. Aucheri auct. nonn., non Jaub. et Spach. — Ic.: M. Kult. l. c. tab. sine numero.

Shrub, 30—80 cm high; stem stout; branches long, straight or flexuous, woody, terminated by leaves and flowers, unarmed, covered with brownish-gray bark, the persistent epidermis splitting longitudinally into fibers; current year's herbaceous branchlets of second order numerous, simple or sparingly branched, straight or flexuous, at length becoming woody, nearly all terminated by flowers, rarely by leaves, obtusish, whitish, obscurely furrowed, rough with a dense cover of short papillose hairs; ocreae short-cylindric, 2—3 mm long, in lower part brownish, in upper

part white-membranaceous, shining, scabrous, shorter than the leaves and the internodes, with 2 brownish nerves produced beyond the incised upper margin into 2 long slender lanceolate awns; leaves grayish-green,
525 2—4 mm long and 1.5—3 mm broad, stiffish, linear-lanceolate to sublinear, gradually narrowed toward base, sessile or nearly so, long-acuminate at apex and terminating in a short obtuse point, commonly revolute or rarely flat at the margins, densely papillose-puberulous on both sides, with a faint midrib above, with a prominent midrib and less prominent pinnate venation beneath; flowers on young branchlets in loose few-flowered terminal racemes 2—6 cm long; pedicels in pairs or rarely solitary from the ocreae, long, jointed at or slightly above the middle; perianth yellow or yellowish-green, fairly large; inner segments of fruiting perianth orbicular or rounded-ovate, slightly unequal, the larger 6—8 mm long and 5—6 mm broad, slightly longer and broader than the achene; outer segments smaller, orbicular or cordate-orbicular, to 5 mm long and 6 mm broad, at length reflexed, longer than the above-joint portion of the pedicel, slightly rough-papillose at base; achene trigonous, acuminate, with bluntish ribs and ovate-rhombic faces, dark brown, glabrous and smooth, lustrous. April—May.

Dry sandy and stony slopes of desert hills; clayey and sandy steppes. — Centr. Asia: Kara K., Mtn. Turkm. Endemic (probably occurs also in Iran). Described from the Transcaspian area. Type in Leningrad.

Note. A species readily distinguishable from all other *Atraphaxis* species by the coloring of its flowers. It is noteworthy that in the Kushka area this species occurs together with the shrubby *Polygonum arianum* Grig. which might easily be taken for *Atraphaxis* if it were not for its perianth which is not parted down to base.

17. **A. angustifolia** Jaub. et Spach, Ill. pl. or. II (1844—1846) 15. — *A. ordubadensis* Šir. ex Grossh. Fl. Cauc. II (1930) 45 in annot. — Ic.: Jaub. et Spach, l. c. tab. 115.

An undershrub 5—15 cm high; main stem short, very stout, flexuous; branches short, stoutish, spreading, flexuous, terminated by leaves or by flowers, obtusish and unarmed, covered with light gray bark, the epidermis splitting longitudinally into fibers and peeling off; current year's herbaceous branchlets of second order short, slender, slightly flexuous, approximate, simple, terminated by flowers or rarely by leaves, obtusish, at length becoming lignified, rough with dense short papillose hairs; ocreae short-cylindric, 2—3 mm long, commonly longer than the internodes, in lower part brownish, in upper part thin white-membranaceous, shining, with 2 brownish nerves produced beyond the slightly lacerate-toothed or
526 incised upper margin into long lanceolate awns; leaves bright green, 1.5—4 cm long and 1.5—3 mm broad, fleshy-coriaceous, linear-lanceolate to linear, the upper ones subulate, gradually narrowed toward both ends, subsessile, terminating in a very short point, papillose-ciliate on the revolute margins, glabrous above, the prominent midrib beneath covered with short papillose hairs; flowers on branchlets of the current year in many-flowered linear racemes; pedicels in 3's from the flowering ocreae, jointed at the middle; perianth pink with white margins, turning brown, rather small; inner segments of fruiting perianth cordate-orbicular, slightly

unequal, the larger 4—5 mm long and 5—6 mm broad, much longer and broader than the achene; outer segments smaller, orbicular or elliptic, reflexed, shorter than the above-joint portion of the pedicel; achene trigonous, acuminate, with bluntish ribs and ovate faces, dark brown, glabrous and smooth, lustrous. May—June.

Dry stony and gravelly mountain slopes. — Caucasus: S. Transc. **Gen. distr.**: Arm.-Kurd. Described from Turkish Armenia. Type in Paris.

Note. A species very rare in the USSR and apparently with an altogether very restricted distribution area; hence insufficiently explored. As pointed out by A. Grossgeim, the plant was identified in the herbarium of the Tiflis Botanical Garden by G. I. Shiryaev who distinguished it from A. angustifolia and labelled it with the name recorded here as a synonym. In studying a single specimen of this species, representing a poor duplicate of the Tiflis plant, in the Herbarium of the Academy of Sciences, the present author did not find sufficient deviations from the description given by Jaubert and Spach and retains the name given by these authorities.

18. **A. teretifolia** (M. Pop.) Kom. comb. nova. — Physopyrum teretifolium M. Pop. in Ind. Sem. horti Almaatensis No. 2 (1935) 24.

An undershrub, 10—20 cm high; stem flexuous; branches short, spreading, weakly spinescent, covered with fissured brownish-gray bark; branchlets of the current year 5—10 cm long, leafy, straight or slightly flexuous, the lower ones longer and sterile, all verrucose-scaberulous; ocreae somewhat longer than the internodes, the nerves produced into a pointed tooth; leaves grayish-green, 1.5—2.5 cm long and 1—1.5 mm broad, fleshy, cylindric due to convoluted margins, with a nerve beneath, when young sometimes to 3 mm broad, like the branchlets verrucose-scaberulous, often slightly falcate, acuminate; flowers on branchlets of the current year; pedicels in 2's or 3's from the ocreae; perianth brownish-
527 pink, in fruit to 5 mm in diameter; inner segments of fruiting perianth orbicular, concave and spherically enveloping the achene; outer segments much smaller, reflexed; achene trigonous, narrow, ca. 2 mm long, brownish-black, lustrous. Fl. May; fr. June. (Plate XXIXa, Figure 3, a—d).

Saline gravelly deserts. — Centr. Asia: Balkh. Endemic. Described from the northern shore or Lake Balkhash, near Bertys Bay. Type in Alma-Ata; cotype in Leningrad.

Genus 393. **CALLIGONUM*** L.**

L. Sp. pl. (1753) 530; Hort. Gliff. (1737) 212; Gen. pl. ed. I (1737) app. 345.

Flowers perfect; perianth simple, colored, 5-parted; segments subequal, not accrescent and not elongating in fruit, mostly reflexed, rarely spreading; stamens 12—18, the filaments connate at base and surrounded by dense tufts of short hairs; ovary superior, tetragonous, usually prominently ribbed at the angles; styles 4, short, with capitate stigmas; fruit a straight or coiled achene, with indurated almost woody pericarp, the ribs grown out into membranaceous or coriaceous wings or beset with numerous stiffish

* From Greek "callos," beautiful, and "gonos," knee, referring to the geniculately set shapely branchlets.

** Treatment by N. V. Pavlov.

filiform bristles, these sometimes overgrown at the ends by a thin bladder-like membrane or borne on the margins and on the surface of reduced wings; embryo straight, surrounded by endosperm. Much branched shrubs from 40 cm to 7 m high; branches flexuous, rarely straight or curved; herbaceous branchlets of the current year straight, jointed, almost leafless; leaves inconspicuous, very small, 5 — 7 mm long, linear or filiform, united with ocreae or distinct.

Note. The taxonomic nature of this genus is so far rather uncertain for the following reasons. While representing a number of very distinctive morphological systems, the taxonomic units are far from being geographically separated and often occur in exceedingly crowded stands consisting of a large number of forms. Further, as is evident from the following keys, the characters which taxonomists apply to determination of species are based entirely on reproductive organs and practically ignore the vegetative parts. Finally, right up to the present treatment, most of the already known species have not had their distribution areas determined.
528 Consequently, D. I. Litvinov already deemed it necessary to indicate that the morphological systems established by him were treated without the knowledge of their true nature and that the species were set up provisionally.

To explain the widespread polymorphism of the genus and the exceptional variability of the fruits, one is led to assume extensive hybridization of a number of principal forms, with subsequent splitting and regrouping of their characters in the progeny. There is, unfortunately, no experimental work whatever concerning hybridization and heredity in species of *Calligonum* and nothing is known about the nuclear complement or difference in the shape and number of chromosomes. The taxonomic definition of the entities separated merely by formal characters of external morphology must be considered as conditional to some extent. It deviates from the usual concept of species which calls for characterization in terms of geography and genotype in addition to morphological systematics.

The geographical distribution of *Calligonum* species follows some general lines which apply to whole sections rather than to individual species. The northern and northwestern part of the distribution area of the genus within the USSR is conspicuously dominated by winged species of the section *Pterococcus*; the color of the bark of the northern species is mostly dark, red or brown. Dominant in the southern and southeastern part of the distribution area are setaceous-fruited species of the section *Eucalligonum*, while *Pterococcus* species, as far as they occur in this part, have mostly white bark. Widely distributed along the line of contact between these sectional areas are species of the mixed section *Pterygobasis*, and this fact is strongly suggestive of hybrid derivation of species composing this section. Since, however, nobody has so far observed hybridization in our time, the presumably hybrid origin of *Pterygobasis* species must be so far removed in time that the existing forms may be safely considered as more or less fixed.

We now pass on to comparative evaluation of distinguishing characters previously used and those employed in the present work. A diagnostic character for the section *Pterococcus* long established in the literature is bark color, light or dark. This character, which at first bore a rather absolute character, is also used by us but it has lost by now much of its

value in view of the occurrence of parallel and almost convergent forms. Admittedly, even in this case, complete identity is not to be found and some slight differences can always be discerned. On the other hand, certain forms may be readily distinguished by the color of bark even though they are undoubtedly very closely related. Another somewhat inconclusive character is the degree of coiling of the fruit and of its wings. While varying to some extent within certain forms, this character is more or less constant for most species of the Pterococcus group. It should, however, be noted that the character in question is merely concerned with the degree of coiling of the achene and of the wings and not the direction of coiling which was also considered of taxonomic
529 significance by some authors in the course of early studies of the genus. The consistency of this character was already questioned by D. I. Litvinov. Our own observations on ample material of species considered by I. G. Borshchov as most stable, notably C. colubrinum Borszcz. and C. erinaceum Borszcz., have definitely established its inadequacy. And so the emphasis shifts essentially towards quite minor pliable characters among which we may mention the shape and size of fruit, the degree of rigidity of wings and bristles, flatness or flexion of wing margins and the degree of their dentation, the presence or absence of outgrowths on the wing surface, and, in species with setaceous fruits, the density and position of ramification of the bristles as well as characteristics of the branches.

A few words remain to be said in this connection about the technique of collecting species of Calligonum designed to facilitate accurate identification. The usual herbarium sheet with crumpled and flattened fruits is of little use for this purpose. It is desirable to collect ripe or almost ripe fruits, in supplementation to the herbarium sheet with the branches and taken, of course, from the same shrub, in rigid containers where they can be preserved without crumpling or flattening. When it comes to the worst, collection may be confined to such a container with fruits, but it is necessary under such circumstances to record on the spot the color of the bark, particularly for species of the section Pterococcus. It is necessary for identification, more than in any other genus, to take cognizance of the combination of characters as a whole rather than rely on any one random character. The excellent illustrations of fruits of all species of the genus and of nearly all varieties, drawn by M. A. Dobrov exclusively from authentic specimens, will undoubtedly prove of great help in this respect, although a certain amplitude of intraspecific variability must naturally be taken into consideration.

Economic importance. Species of Calligonum, growing mainly in the sandy deserts of Central Asia, often provide the only available source of fuel. They have also long been highly appreciated as planting material for fixation of shifting sand. According to data of research stations in sandy areas, they are raised very easily from seed. Young fruits of some species of the section Pterococcus are used for food, while ripe fruits have weak tanning properties. The hard wood of the taller species is sometimes used for production of miscellaneous articles.

Key to Sections

1. Fruit winged or setose . . . . 2.

\+ Fruit a globular or ovaloid bladderlike membranous sac formed by membranously dilated tips of the bristles united among each other and completely covering the achene and the bristles . . . . Section 4. **Calliphysa** Endl., p. 457.

530 2. Fruit bearing 4 double coriaceous or membranous wings with entire, incised-dentate or aculeate but not setiferous margins; wing surface glabrous or covered with lamelliform or setiform outgrowths . . . . Section 1. **Pterococcus** Endl.

\+ Fruit setose, the bristles borne on the achene or on reduced wings . . 3.

3. Fruit wings of two kinds, the ribs of the achene forming fairly prominent or rudimentary coriaceous secondary wings, the primary wings setiferous only on the margins or also on the surface, the bristles filiform or flattened at base . . . . . . Section 2. **Pterygobasis** Borszcz., p. 437.

\+ Fruit wingless, setose, the bristles quite free at base or more rarely some of them united at base but not forming wings . . . . Section 3. **Eucalligonum** Borszcz., p. 449.

Section 1. PTEROCOCCUS Endl. Gen. pl. (1836—40) 308. — Pterococcus (gen.) Pall. Iter II (1773) 332, app. 738 (excl. sp. nonn.). — Pallasia (gen.) L. fil. Suppl. (1781) 252.

1. Bark of mature woody branches dark: red, brown or blackish-gray . . . 2.

\+ Bark of mature branches light-colored: white, grayish, or yellowish . . . . 31.

2. Fruit cylindric, broader than long . . . . 3.

\+ Fruit ovaloid or globose, usually longer than or rarely as long as broad . . . . 4.

3. Fruit dark red; wings rather stiffly coriaceous, wide apart at the base and at the margin, their lobes united at the top of the fruit, at base often free or united, the margins of wings unequally and coarsely bluntly or rarely sharply double-toothed . . . 38. **C. coriaceum** N. Pavl.

\+ Fruit yellowish-brown; wings thin, membranous, distant at base but approximate at the margins, their lobes united at both ends or rarely at the top only, the margins of wings subentire or sparsely minutely and sharply denticulate . . . . 39. **C. cartilagineum** N. Pavl.

4. Fruit wings more or less strongly involute . . . . 5.

\+ Fruit wings flat, flexuous or crisped-plicate, not involute . . . . 8.

5. Fruit very strongly coiled for an almost full turn; wings rounded at both ends, their margins entire or almost entire with indistinct rounded teeth . . . . 1. **C. aphyllum** (Pall.) Gürke var. **Androssowii** Litw.

531 + Fruit not coiled, straight or nearly so, only in maturity slightly flexuous; wings notched, cordate, or pointed at base, their margins more or less sharp-toothed . . . . 6.

6. Fruit wings dark reddish, rather stiffly coriaceous, distant at the base and at the margin, their lobes united at both ends, the wing margin unequally and almost doubly sharp-toothed . . . 29. **C. involutum** N. Pavl.

\+ Fruit wings yellowish-brown, thin, membranous, distant at the base but approximate at the margin, their lobes free, united at base or at summit or rarely at both ends . . . . 7.

7\. Fruit wings cordately notched at base, their lobes free or united only at the summit, the wing margin entire or almost entire with sparse indistinct broad obtusish teeth . . . . 2. **C. alatum** Litw. var. **involutum** N. Pavl.

\+ Fruit wings pointed at base, their lobes united at both ends or merely at base, the margin with short often double sharp teeth . . . . 30. **C. Eugenii-Korovini** N. Pavl.

8\. Surface of wings covered at center or at the margins with spiniform or lamelliform outgrowths, these sometimes forked at the ends . . . . 9.

\+ Surface of wings glabrous, commonly without outgrowths (occasional fruits sometimes with very sparsely aculeolate wing surface) . . . . .16.

9\. Surface of wings almost glabrous, with spiniform forked or decussate outgrowths borne only on the margin and quite close to it . . . . 19. **C. tetrapterum** Jaub. et Spach.

\+ Surface of wings with lamelliform or spiniform outgrowths at center or nearer the margin, the margin of wings toothed or entire . . . . . .10.

10\. Fruit not coiled, the surface and margin of wings flat or nearly so, becoming slightly flexuous only in maturity . . . . .11.

\+ Fruit more or less coiled, in maturity almost for a full turn, the surface and margin of wings flexuous . . . . .14.

11\. Margin of wings entire or almost entire with sparse and indistinct obtusish or rarely acutish dentations . . . . .12.

\+ Margins of wings coarsely and sharply double-toothed or incised-toothed . . . . .13.

12\. Fruit dark red; wings very stiffly coriaceous, 2-lobed at base with
short acute lobes, the surface of wings strongly convex near the
532 margin and bearing here a single row of short stiff spiniform or
lamelliform outgrowths, or more rarely a row of very unequal
lamellae running only along the middle of the wing . . . .
. . . . 20. **C. Russanovii** N. Pavl.

\+ Fruit yellowish-brown; wings stiffly membranous to subcoriaceous, rounded at base, the flat surface with 1 or 2 rows of stiff and pointed lamelliform outgrowths at the middle . . . . 3. **C. rigidum** Litw. var. **aculeatum** Litw.

13\. Fruit cordate or truncate at base, slightly acuminate at summit; wings with coarsely deeply and unequally double-toothed margin, the surface covered with slender unequally forked outgrowths . . . . 23. **C. membranaceum** (Borszcz.) Litw.

\+ Fruit rounded at both ends; wings closely finely and almost equally single-toothed, the surface near the margin covered with thick stiffish lamelliform outgrowths . . . . 1. **C. aphyllum** (Pall.) Gürke var. **lamellatum** Litw.

14\. Fruit rounded at both ends; wings thin, membranous, their margin strongly flexuous, crisped-plicate, with short broad teeth or subentire . . . . 1. **C. aphyllum** (Pall.) Gürke var. **crispatum** Litw.

\+ Fruit rounded or cordate at base, acutish at summit; wings stiffly coriaceous, their margin stiffly sharp-toothed . . . . .15.

15. Wings along the margins with 1 row of lanceolate or linear lamelliform or subulate outgrowths, the margin sharply sinuate-toothed . . . . 21. **C. crispum** Bge.

\+ Wings along the margin with 1 or 2 rows of outgrowths, these flat at base and furcate-toothed at apex with 1—5 unequal subulate teeth, the margin deeply incised-toothed, the teeth flat at base and often furcate-toothed at the tips with 2—5 subulate teeth . . . . 22. **C. spinosissimum** N. Pavl.

16. Fruit uncoiled, the margins and surface of wings flat or nearly so, only in maturity slightly crinkled . . . . 17.

\+ Fruit more or less coiled, in maturity almost for a full turn, the margin and the surface of wings crinkled . . . . 25.

17. Fruit not more than 15 mm and mostly 10—12 mm long, the wing margin always toothed . . . . 18.

\+ Fruit more than 15 mm long and, if rarely shorter, then the wing margin entire . . . . 20.

533 18. Fruit almost round; wings square with rounded angles, their margin very unequally toothed, the teeth up to 2—3 mm long . . . . 4. **C. quadrapterum** Eug. Kor.

\+ Fruit cordate; wings triangular, broader at base and tapering to a point at the summit, their margin with short subequal teeth . . . . 19.

19. Fruit to 10 mm long; wings stiffish subcoriaceous, very distant at the base and at the finely and almost simply toothed margin; segments of fruiting perianth reflexed . . . . 5. **C. humile** Litw.

\+ Fruit 10—15 mm long; wings thin, soft, membranous, distant at the base but approximate at the rather coarsely unequally and doubly sharp-toothed margin; perianth segments spreading . . . . . 6. **C. tenue** N. Pavl.

20. Fruit wings notched or cordate at base; ribs of achene thin, winglike, very prominent . . . . 21.

\+ Fruit wings rounded at base; ribs of achene commonly thick, broad, not prominent . . . . 22.

21. Fruit wings closely and unequally incised-toothed to 1—3 mm, the teeth in their turn with sharp secondary dentations . . . . 23. **C. membranaceum** (Borszcz.) Litw. var. **nudum** Litw.

\+ Fruit wings almost entire, undulate or with indistinct rounded teeth or rarely with small close dentations . . . . 2. **C. alatum** Litw.

22. Margins of wings stiffish, entire or almost entire with sparse indistinct dentations . . . . 23.

\+ Margins of wings soft, closely and finely toothed . . . . 24.

23. Wings thick, stiffly subcoriaceous; teeth on the margin, if present, sharp . . . . 3. **C. rigidum** Litw.

\+ Wings thin, softly membranous; teeth on the margin, if present, indistinct and rounded . . . 1. **C. aphyllum** (Pall.) Gürke var. **typicum** Litw.

24. Ribs of achene bluntish; wing lobes distinct . . . . 1. **C. aphyllum** (Pall.) Gürke var. **commune** Litw.

\+ Ribs of achene thin, prominent; wing lobes slightly united at base and at summit . . . . 1. **C. aphyllum** (Pall.) Gürke var. **costatum** Litw.

534 25. Margin of wings unequally incised or dissected to at least one-third their width . . . . 26.

\+ Margin of wings entire or at most 1/5—1/10 their width, toothed . . . 27.

26\. Margin of wings incised to more than 1/3—1/2 their width or even to the very edges of the achene into acute lobes, these in turn minutely and sharply toothed; wings very thin, coriaceous . . . . 12. **C. dissectum** T. Popova.

\+ Margin of wings incised to one-third their width into unequal narrow pointed teeth, these in turn minutely and sharply crenulate; wings thin, membranous . . . . 11. **C. affine** T. Popova.

27\. Wing margin more or less crisped-plicate; wings thin, membranous . . . . 28.

\+ Wing margin sinuous but not crisped-plicate; wings rather stiffly coriaceous or subcoriaceous . . . . 29.

28\. Wing margin entire or indistinctly rounded-crenulate . . . . 16. **C. indulatum** Litw.

\+ Wing margin with close bluntish or sharp dentations . . . . 1. **C. aphyllum** (Pall.) Gürke var. **crispatum** Litw.

29\. Fruit dark reddish; wings thick, stiffly coriaceous, commonly folded longitudinally, the lobes often slightly united at both ends, the margins with broad bluntish teeth, the wing surface without outgrowths . . . . 13. **C. rubicundum** Bge.

\+ Fruit yellowish-brown; wings thinner, softly membranous, the margin with sharp lanceolate teeth, the wing surface sometimes with scattered lamelliform outgrowths . . . . 30.

30\. Wings commonly distant at the margin, this subequally and singly toothed, only at base the teeth sometimes longer and acute, almost subulate . . . . 14. **C. flavidum** Bge.

\+ Wings tightly enveloping the achene and approximate at the margin, this minutely sharply and doubly or almost doubly cartilaginous-toothed . . . . 15. **C. Borszczowii** Litw.

31 (1). Fruit cylindric, up to 1.5 times broader than long . . . . 32.

\+ Fruit oval or round, usually narrower than or rarely as broad as long . . . . 37.

32\. Wings much longer than the achene, commonly closely convergent at the summit and united with the base of the style . . . . 33.

\+ Wings about as long as the achene, wide apart at the summit and not united with the base of the style . . . . 35.

535 33. Fruit more or less coiled; wing margin strongly saccately involute, with sparse indistinct short and broad acutish teeth; segments of fruiting perianth reflexed . . . . 32. **C. physopterum** N. Pavl.

\+ Fruit uncoiled or almost uncoiled, but slightly crinkled; wing margin flat, not involute, distinctly and closely sharp-toothed; segments of fruiting perianth spreading . . . . 34.

34\. Wings distant at base, closely convergent at the summit, acuminate and concealing the style, the lobes slightly united at both ends, their margin with sharp indistinctly doubled teeth, the surface glabrous . . . . 36. **C. Lipskyi** Litw.

\+ Wings less distant at base, approximate at the margin, not convergent at the summit and not concealing the style, the lobes slightly united at both ends, their margin sharply and doubly toothed, the surface glabrous or with sparse lamelliform outgrowths along the margin . . . . 37. **C. patens** Litw.

35\. Wings united at both ends into round cups, the margin entire or (only under strong magnification) minutely and indistinctly toothed . . . 33. **C. batiola** Litw.

\+ Wings united only at the summit, rarely also slightly united at the base and then incompletely so, the margin toothed . . . 36.

36\. Wings very distant, discontiguous; achene produced from the depressed summit of the wings into a short ribbed cone; wing margin doubly sinuate-toothed to between one-fifth and one-fourth the width; wing surface glabrous or with sparse lamelliform outgrowths along the margin . . . 35. **C. Androssowii** Litw.

\+ Wings subdistant, the inner surfaces contiguous; achene not produced at the summit into a ribbed cone beyond the truncate upper margin of the wings; wing margin rather shallowly doubly serrate-dentate; wing surface glabrous . . . 34. **C. obtusum** Litw.

37\. Margin of wings very regularly patellately involute throughout, with very short sharp teeth, the lobes folded longitudinally, approximate at the margins, slightly crinkled, the surface glabrous or with sparse sharp outgrowths . . . 31. **C. lanciculatum** N. Pavl.

\+ Margins of wings flat, crinkled or crisped-plicate, not involute . . . 38.

536 38. Surface of wings with spiniform or lamelliform sometimes furcate-tipped outgrowths at the middle or near the margin . . . 39.

\+ Surface of wings glabrous, usually without outgrowths or some fruits occasionally sparsely lamellate-aculeolate . . . 45.

39\. Fruit very strongly coiled for nearly a full turn, the margins and surface of its wings strongly crinkled; wings stiffly subcoriaceous, dark reddish, with a single row of sparse acute subulate outgrowths . . . 18. **C. turbineum** N. Pavl. var. **subulatum** N. Pavl.

\+ Fruit not coiled or scarcely coiled, the margins and surface of its wings flat or but slightly crinkled; wings thinner, membranous, mostly yellowish or yellowish-brown . . . 40.

40\. Wings oblong-oval, at least twice as long as broad, pointed at base and at summit and closely convergent; surface of wings with a single row of long thin subulate or lamelliform furcate-tipped outgrowths, these longer than the distance from the wing margin . . . 24. **C. calcareum** N. Pavl.

\+ Wings triangular, oval or orbicular, not more than 1.5 times as long as broad, the outgrowth of the surface always shorter than the distance from the margin . . . 41.

41\. Wings triangular, attenuate to a point from a broad base, the margins deeply and unequally incised, doubly toothed, the terminal teeth often subulate . . . 42.

\+ Wings oval or round, the margins rather shallowly almost equally and indistinctly double-toothed . . . 43.

42\. Outgrowths of the wing surface forked . . . 25. **C. acanthopterum** Borszcz.

\+ Outgrowths of the wing surface simple, subulate or lamelliform . . . 25. **C. acanthopterum** Borszcz. var. **subnudum** N. Pavl.

43\. Fruit at least 15 mm long . . . 44.

\+ Fruit less than 15 mm long . . . 45.

44. Wings notched at base with obtuse lobes; margins of wings entire or rather closely and minutely sharp-toothed; outgrowths of the surface thin, simple, subulate or lamelliform; small prickles often present also on the outer surface of the wings . . . .
537 . . . . 9. **C. alatiforme** N. Pavl. var. **aculeolatum** N. Pavl.

\+ Wings often with lobes united at base or, if notched, the lobes acute; wing margin coarsely unequally and doubly sharp-toothed; outgrowths of the wing surface stiffish, lamelliform, flat, usually bidentate at the ends . . . . 26. **C. cristatum** N. Pavl.

45. Fruit oval, much longer than broad; wing margin minutely and doubly sharp-toothed; outgrowth of the wing surface 1 – 2 mm long, thin, bristle-shaped or subulate . . . . 27. **C. Dubjanskyi** Litw.

\+ Fruit round or nearly so, scarcely longer than broad; wing margins coarsely toothed, the teeth flattish at base, subulate and double at the tips, to 2 – 3 mm long; outgrowth of the surface to 3 – 5 mm long, filiform or lamelliform, subulate-tipped . . . . 28. **C. Bubyri** B. Fedtsch.

46. Fruit not coiled; margin and surface of wings flat or almost flat, only in maturity slightly crinkled . . . . 47.

\+ Fruit coiled, in maturity nearly for a full turn; margin and surface of wings strongly crinkled . . . . 50.

47. Fruit cordate-triangular in outline; wings attenuate to a point from a broad base, the margin sharp-toothed . . . . 48.

\+ Fruit oval or almost round in outline; wings obtusely rounded at both ends, the margins entire or subentire or rarely closely and minutely toothed . . . . 49.

48. Wings rather stiffly membranous to subcoriaceous, notched at base and at summit, the divergent lobes pointed; wing margin cartilaginous, subequally serrulate-denticulate . . . . 8. **C. aralense** Borszcz.

\+ Wings softly and thinly membranous, almost transparent, angular, the lobes at base and at summit acute; wing margin unequally and doubly sharp-toothed . . . . 7. **C. gracile** Litw.

49. Fruit to 22 mm long, broad-oval or almost round; wings thin, softly membranous; ribs of achene thin, prominent, almost winglike; the margin subentire or closely and minutely toothed . . . .
. . . . 9. **C. alatiforme** N. Pavl.

\+ Fruit to 15 mm long, oval or oval-triangular, slightly attenuate toward summit; ribs of achene broad, not prominent; wings rather stiffly
538 membranous, the margin entire or sparsely serrulate-denticulate . . . .
. . . . 10. **C. leucocladum** (Schrenk) Bge.

50. Wings rather stiff, subcoriaceous, united with the base of the style, the margin flat . . . . 51.

\+ Wings thin, membranous, not united with the base of the style, the margin crisped-plicate, undulate . . . . 52.

51. Fruit slightly coiled, oval or broad-oval in outline; wing margin entire or almost entire with indistinct obtusish or rarely acutish teeth . . . . . .
. . . . 3. **C. rigidum** Litw. var. **australe** N. Pavl.

\+ Fruit very strongly coiled, round or almost round in outline; wing margin incised to 2 – 3 mm deep into simple entire acute or flattish unequal biparted or sometimes double teeth . . . 18. **C. turbineum** N. Pavl.

52. Fruit 15—23 mm long; wing margin with short acutish teeth or rarely subentire . . . . .17. **C. plicatum** N. Pavl.
+ Fruit to 15 mm long; wing margin entire or indistinctly rounded-crenulate . . . . 16. **C. undulatum** Litw. var. **australe** N. Pavl.

1. **C. aphyllum** (Pall.) Gürke (s. ampl.) Pl. europ. II (1897) 111. — Pterococcus aphyllus Pall. Iter. II (1773) 332, app. 738. — Calligonum Pallasia L'Herit. Transact. of Linn. Soc. Lond. I (1791) 180. — C. polygonoides Pall. Iter III (1776) 530 (non L.). — Pallasia Pterococcus Pall. Fl. Ross. II (1788) 70, p. p. — P. caspica L. fil. Suppl. (1781) 252. — Ic.: Pall. Iter II (1773) 738, tab. 5; Pall. Fl. Ross. II (1788) tab. 77, 78 (excl. pl. songor.); Borszcz. Mém. Acad. de St. Pétersb. VII sér. III, I (1860) tab. I, fig. 1—1c; Litvinov, Tr. Botan. Muzeya Akad. Nauk XI (1913) tab. 7, fig. 1—9. — Exs.: HFR No. 1382, 2405, 2406, 2407, 2408; Fl. caus. exs. No. 306.

Shrub to 2 m high; bark of mature branches reddish-brown or purplish-brown, lenticellate; herbaceous branchlets slender, terete, bright green, in clusters of 2—6 from nodes of annotinous branches, jointed, the joints terminating in a sheathlike ocrea, this white-membranous, obtusish and finally bifid at summit; leaves not united with ocreae, semiterete, 2—3 mm long, slightly enlarged at apex, readily deciduous; flowers in pairs, rarely 1 or 3 from the axils; pedicels glabrous, 4—5 mm long, jointed below the middle; flowers honey-scented; perianth segments glabrous, unequal; the outer broad-ovate, 3 mm long, green on the back, broadly white-margined;
539 the inner narrower, short, white, on the back pinkish; fruit broad-oval or almost round, 15—20 mm long and 14—18 mm broad, exceedingly variable; achene elliptic-oval, almost uncoiled or more or less coiled anticlockwise, ribbed, the ribs bluntish or keen, slightly prominent; wings membranous, flat, rounded at both ends, united with the base of the style. In the typical form (var. typicum Litw.) the ribs of the achene are bluntish; wing margins entire, with indistinct rounded teeth, the surface glabrous and smooth. Other combinations of characters are: finely and sharp-toothed wings with glabrous surface, the achene ribs bluntish (var. commune Litw.) or thin and prominent (var. costatum Litw.); wings subentire or toothed with a single row of prickle-shaped or lamelliform flat-based outgrowths (var. lamellatum Litw.); wings with toothed and more or less crisped-plicate margins, with a row of lamelliform or prickle-shaped outgrowths, the achene more strongly coiled (var. crispatum Litw.); and finally, wings with entire or almost entire margins, strongly involute, the surface naked and smooth, the achene very strongly coiled (var. Androssowi Litw.). Fr. June—July. (Plate XXX, Figures 1—5).

Sandhills; sandy desert-steppes. — European part: L. V.; Caucasus: Cisc., Dag., E. Transc.; W. Siberia: Irt.; Centr. Asia: Balkh., Dzu.-Tarb., Ar.-Casp., Kyz. K., Kara K., Syr D., Amu D., Pam.-Al. **Gen. distr.**: Dzu.-Kash. Described from the Ryn Sands in former Astrakhan Province. Type in Moscow.

2. **C. alatum** Litw. Trav. Mus. Bot. Ac. Sc. XI (1913) 52. — Ic.: Litw. l. c. tab. 7, fig. 10—12.

Shrub to 2 m high; bark of mature branches dark, brownish; herbaceous branchlets slender, terete, grayish-green, in clusters of 2—6 from nodes

of woody branches, jointed; joints 2—4 cm long, terminating in a short obtusish transparent ocrea; leaves 2—4 mm long, filiform, obtusish at apex, united with ocrea, readily deciduous; flowers axillary, commonly in 3's, glabrous, unequal, 6—8 mm long, jointed below the middle; perianth segments glabrous, unequal; the outer broad-obovate, 4 mm long, rounded at apex, green on the back, widely white-margined; the inner shorter, narrow; all segments at first horizontally spreading, finally reflexed, anthers purple; fruit broad-oval, slightly blunted at both ends, (10) 13—15 mm long, (8) 11—13 mm broad; achene scarcely coiled, the sharp prominent ribs about the same width at base and at summit; wings cordate
540 at base, rounded at summit, not united with the base of the style and distant from it, approximate at the margin, flat, rather stiffly membranous, naked, with entire or indistinctly rounded-toothed, or rarely with close bluntish teeth (var. *dentatum* Litw.), flat or strongly involute (var. *involutum* N. Pavl.). Fr. May—June. (Plate XXX, Figures 6—8).

Sandhills and sandy desert-steppes. — Centr. Asia: Ar.-Casp., Kyz. K., Kara K., Balkh. Endemic. Described from the Aral Kara-Kum Desert. Type in Leningrad.

3. **C. rigidum** Litw. Trav. Mus. Bot. Ac. Sc. XI (1913) 53. — Ic.: Litw. l. c. tab. 7, fig. 16—18.

Shrub to 1.5 m high; bark of mature branches dark, brown, only at the southern extremity of the distribution area often light, whitish (var. *australe* N. Pavl.); flowers unknown; segments of fruiting perianth reflexed; fruit broad-oval to almost round, ca. 15 mm long and 12—14 mm broad; achene mostly uncoiled or (in southern specimens) more or less coiled, the ribs sharp; wings flat, stiff, prominently nerved especially at the middle, nearly always united with the base of the style, closed at base and at summit, contiguous at the margins; margin entire or with indistinct rounded or rarely sharp teeth, the surface naked or beset at the middle with acutish lamellae or prickles, these sometimes rather long (var. *aculeolatum* Litw.). Fr. June—July. (Plate XXX, Figures 10, 11).

Desert sandhills and sandy steppes. — Centr. Asia: Ar.-Casp., Kyz. K., Balkh., Dzu.-Tarb. Endemic. Described from the Aral Kara-Kum Desert. Type in Leningrad.

4. **C. quadraepterum** Eug. Kor. ex N. Pavl. in Fedde, Repert. sp. nov. XXXIII (1933) 147.

A spherical shrub to 0.7 m high, with prostrate branches; stem short, 5—6 cm in diameter, covered with blackish-gray bark; mature branches stiff, with short longitudinal fissures, tortuously discontinuous, nodose, the bark dingy brownish-yellow; flowers unknown; segments of fruiting
543 perianth reflexed; fruit almost round, 13—15 mm long and 10—12 mm broad; achene oblongly oval-cuneate, slightly coiled from right to left, anticlockwise, the sharp ribs scarcely developed at the middle and more prominent at base and at summit; wings stiffly membranous, rounded-quadrate, rosy honey-colored, slightly concave at base, deeply notched at summit, united with the base of the style; wing margin cartilaginous, pectinate-toothed, the teeth pricklelike and some 2—3 mm long. Fr. May—June. (Plate XXX, Figure 9).

(541)

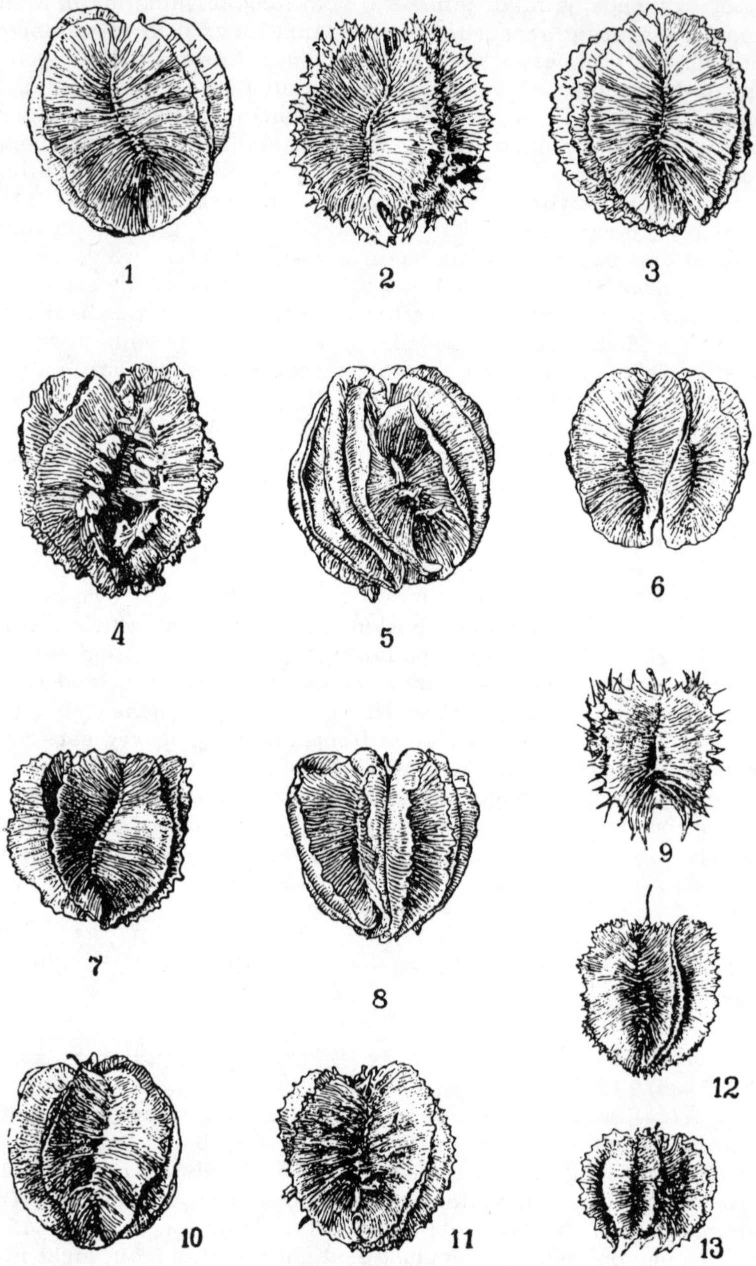

PLATE XXX. 1. Calligonum aphyllum (Pall.) Gürke var. typicum Litw. – 2. C. aphyllum var. commune Litw. – 3. C. aphyllum var. costatum Litw. – 4. C. aphyllum var. lamellatum Litw. – 5. C. aphyllum var. Androssowi Litw. – 6. C. alatum Litw. – 7. C. alatum var. dentatum Litw. – 8. C. alatum var. involutum N. Pavl. – 9. C. quadraepterum Eug. Kor. – 10. C. rigidum Litw. – 11. C. rigidum var. aculeatum Litw. – 12. C. tenue N. Pavl. – 13. C. humile Litw.

Desert sandhills. — Centr. Asia: Kyz. K., Kara K. Endemic. Described from Turkmenia. Type in Tashkent.

5. **C. humile** Litw. Trav. Mus. Bot. Ac. Sc. XI (1913) 54; Litw. l. c. tab. 7, fig. 28—30.

Shrub to 0.5 m high; bark of mature branches dark; flowers unknown; segments of fruiting perianth reflexed; fruit oval or subcordate, to 10 mm long and 9—11 mm broad; achene slightly coiled, with sharp ribs; wings soft, not united with the base of the style, very distant at the margin; wing margin sharp-toothed, the surface naked, the nerves evanescent near the margin. Fr. May— June. (Plate XXX, Figure 13).

Desert sandhills and sandy steppes. — Centr. Asia: Ar.-Casp., Balkh. Described from the Aral Kara-Kum Desert. Type in Leningrad.

6. **C. tenue** N. Pavl. in Fedde, Repert. sp. nov. XXXIII (1933) 147. — C. rubicundum var. humile Borszcz. Mém. Acad. de St. Pétersb. VII sér., III, 1 (1860) 30. — Ic.: Borszcz. l. c., tab. 1, f. 2—2d.

A much branched shrub to 0.7 m high; bark of mature branches dark, reddish-brown; flowers unknown; segments of fruiting perianth widely spreading; fruit oval, 10—15 mm long and 9—12 mm broad; achene almost plane or slightly coiled to the right [clockwise], the thin ribs rather sharp; wings membranous, soft, thin, delicate, reddish-brown, surpassing the achene and approximate at the margin, with prominent transverse nerves evanescent near the margin, broader and slightly cordate-notched at base with acutish lobes, attenuate at apex, short-acuminate, not united with the base of the style; wings naked, the margin with unequal often double sharp almost subulate teeth. Fr. May—June. (Plate XXX, Figure 12).

Centr. Asia: Ar.-Casp., Balkh. Endemic. Described from sands of the Ili River valley. Type in Moscow; cotype in Leningrad.

7. **C. gracile** Litw. Trav. Mus. Bot. Ac. Sc. XI (1913) 55. — Ic.: Litw.
544 l. c., tab. 8, fig. 1—3.

Shrub to 1.5 m high; bark of mature branches whitish; herbaceous branchlets slender, sulcate, with loosely disposed flowers at the ends, jointed; joints to 5 cm long, terminating in a brownish sheathing ocrea; leaves filiform, to 3—5 mm long, united with ocreae, readily deciduous; flowers in 2's or 3's in the ocreae; pedicels unequal, glabrous, to 5 mm long, jointed near base; perianth segments oval, subequal, to 3 mm long, green on the back, broadly white-margined; anthers purple; fruit cordate, 12—14 mm long and 13—15 mm broad; achene scarcely to slightly coiled, with sharp ribs; wings soft, thin, membranous, not united with the base of the style, flat, the lobes acute at summit and at base, the wing margin doubly sharp-toothed, the naked surface distinctly and prominently nerved, the nerves anastomosing at the margin; segments of fruiting perianth reflexed. Fr. May—June. (Plate XXXI, Figure 1).

Desert sandhills. — Centr. Asia: Ar.-Casp., Balkh., Dzu.-Tarb., Kyz. K., Syr D. Endemic. Described from the former Syr Darya Region. Type in Leningrad.

8. **C. aralense** Borszcz. Mém. Acad. St.-Pétersb. VII sér. III, 1 (1860) 31. — Ic.: Borszcz, l. c., tab. I, fig. 3—3b.

Shrub to 2 m high; bark of mature woody branches light dull yellow; herbaceous branchlets slender, striate, jointed; joints to 5 cm long, terminating in an ocrea, this 2—3 mm long, triangularly lacerate, membranous; leaves filiform, not united with ocreae, at the ends of terminal branchlets to 10—12 mm long, readily deciduous; flowers axillary, usually in 3's; pedicels unequal, 2—5 mm long; perianth segments unequal; the outer to 3 mm long, broad-oval, rounded at apex, green on the back, broadly white-margined; anthers rose; fruit almost round, 14—15 mm long and 12—13 mm broad; achene almost plane, oval-cuneate, spirally coiled anticlockwise, the ribs very sharp; wings stiff, membranous, crinkled, not united with the base of the style, bilobed at both ends, the acute lobes divergent; wing margin cartilaginous, finely serrate-dentate, the naked surface with thick prominent nerves. Fr. May—June. (Plate XXXI, Figure 2).

Desert sandhills and sandy steppes. — Centr. Asia: Ar.-Casp., Balkh., Kyz. K. Endemic. Described from the Aral Kara-Kum Desert. Type in Leningrad.

9. **C. alatiforme** N. Pavl. in Fedde, Repert. sp. nov. XXXIII (1933) 150.

Shrub to 2 m high; bark of mature branches whitish, the rounded-oval
545 lenticels rather sparse, the wood pale lemon-yellow; flowers unknown; segments of fruiting perianth reflexed; fruit broad-oval or almost round, to 22 mm long and 20—23 mm broad; achene scarcely coiled; ribs sharp, winglike, membranous, 1—2 mm broad, at the ends 2-winged, the same width at base and at summit; wings slightly notched at base, rounded at summit, not united with the base of the style, stramineous or reddish, approximate at the margins, softly membranous; wing margin subentire or rather closely sharp-toothed, the flat naked surface with a row of slender aculeiform or lamelliform outgrowths; aculeoles sometimes present along the margin and on the outer surface of the wing (var. aculeolatum N. Pavl.). Fr. June—July. (Plate XXXI, Figures 3, 4).

Desert sandhills and sandy steppes. — Centr. Asia: Ar.-Casp., Kyz. K., Balkh. Endemic. Described from the lower reaches of the Sary-su River. Type in Tashkent.

10. **C. leucocladum** (Schrenk) Bge. Mém. Acad. St. Pétersb. sav. étrang. VII (1851) 485. — Pterococcus leucladus Schrenk, Bull. phys. math. Acad. Pétersb. III (1845) 211. — C. anfractuosum Bge. l. c., p. 487 (in adnot. ad C. calliphysa Bge.). — P. aphyllus Kar. et Kir. Enum. pl. soong. (1842) No. 738 excl. synon. (non Pall.). — Ic.: Borszcz. Mém. Acad. St.-Pétersb. VII sér. III, I (1860) tab. I, fig. 3. — Exs.: HFR No. 2409.

Shrub to 2 m high; bark of mature branches white; branches strict, long, straight (var. strictus Bdb.) or prostrate, strongly flexuous, angularly refracted, nodose (var. flexuosus Ldb., C. anfractuosum Bge.); herbaceous branchlets slender, striate-sulcate, with loosely disposed flowers at the ends, jointed; joints to 3—4 cm long, terminating in a short sheathing brownish ocrea; leaves filiform, 2—5 mm long, not united with ocreae, readily deciduous; flowers axillary, commonly in pairs;

pedicels glabrous, 4 — 5 mm long, jointed below the middle; perianth segments oval, unequal; the outer to 4 mm long, green on the back, broadly white-margined; anthers purple; fruit suboval, 13 — 15 mm long and 11 — 13 mm broad; achene linear-elliptic, uncoiled or slightly coiled; wings soft, membranous, with slender transverse nerves, united with the base of the style, at summit and at base almost 2-lobed, the rounded lobes approximate; wing margins thin, entire or very rarely finely toothed (var. serratum Litw.), at length involute and splitting, the wing surface naked. Fr. May — June. (Plate XXXI, Figures 5, 6).

Desert sandhills and sandy steppes. — W. Siberia: Irt.; Centr. Asia: Ar.-Casp., Balkh., Syr D., Kyz. K., Kara K., Amu D., Dzu.-Tarb. Endemic. Described from the Lake Balkhash area. Type in Leningrad.

546 11. **C. affine** T. Popova in Animadv. system. ex Herb. Univers. Tomsk. (1918) No. 5; Kryl., Fl. Zap. Sib. IV, 848. — Ic.: T. Popova l. c. tab. 5 — 6.

Shrub to 2 m high; bark of mature branches reddish-brown with white lenticels; herbaceous branchlets green or glaucescent, terete, smooth, lustrous, jointed, in clusters from nodes of whitish annotinous branches covered with fibrillose remnants of epidermis; joints 1.5 — 4 cm long, the lower shorter, the upper elongated, all terminating in an ocrea this sheathing, with white membranous margin, at length turning brown and partly deciduous; leaves linear-filiform, to 10 mm long, united with ocreae, acuminate at apex, at first appressed to branchlets, finally recurved, turning brown and readily deciduous; pedicels axillary, in 3's or 4's, glabrous, 4 — 6 mm long, jointed about the middle; flowers unknown; segments of fruiting perianth persistent, reflexed; fruit 15 — 22 mm long and 15 — 20 mm broad, broad-oval to almost round; achene 9 — 10 mm long and 4 mm broad, scarcely or slightly coiled; wings thin, stiffish, at first pinkish or pale yellow, finally honey-brown, broad, half-open, united with the base of the achene and the base of the style; wing margins approximate, covering the achene, coarsely sinuate-dentate, the incisions sometimes to one-third the width of the wing, the teeth in turn minutely lanceolate-toothed; wing surface naked. Fr. June — July. (Plate XXXI, Figure 9).

Riverside sandhills. — Centr. Asia: Balkh. Endemic. Described from Lake Zaisan. Type in Tomsk.

12. **C. dissectum** T. Popova, Animadv. system. ex Herb. Univers. Tomsk. (1928) No. 5; Kryl., Fl. Zap. Sib. IV, 847. — Ic.: T. Popova l. c., tab. 1 — 4.

Shrub to 1.5 m high; bark of mature branches blackish-gray; younger branches and those of the current year covered with lustrous reddish-brown bark with whitish fibrillose remnants of epidermis; herbaceous branchlets slender, 0.3 — 0.8 mm thick, glaucescent, more or less branched, in clusters from branch nodes, jointed; leaves linear-filiform, 4 — 5 mm long, united with membranous ocreae, acuminate, at first appressed to branchlets, finally
549 recurved, turning brown, readily deciduous; pedicels axillary, in 2's or 3's, glabrous, 4 — 5 mm long, jointed about the middle; flowers unknown; segments of fruiting perianth persistent, reddish-brown, reflexed; fruit oval, 10 — 16 mm long and 10 — 13 mm broad; achene 9 — 12 mm long and ca. 5 mm broad, fusiform or cuneiform, scarcely or slightly coiled, abruptly constricted about the middle; wings very stiff, thickish, often unequal,

(547)

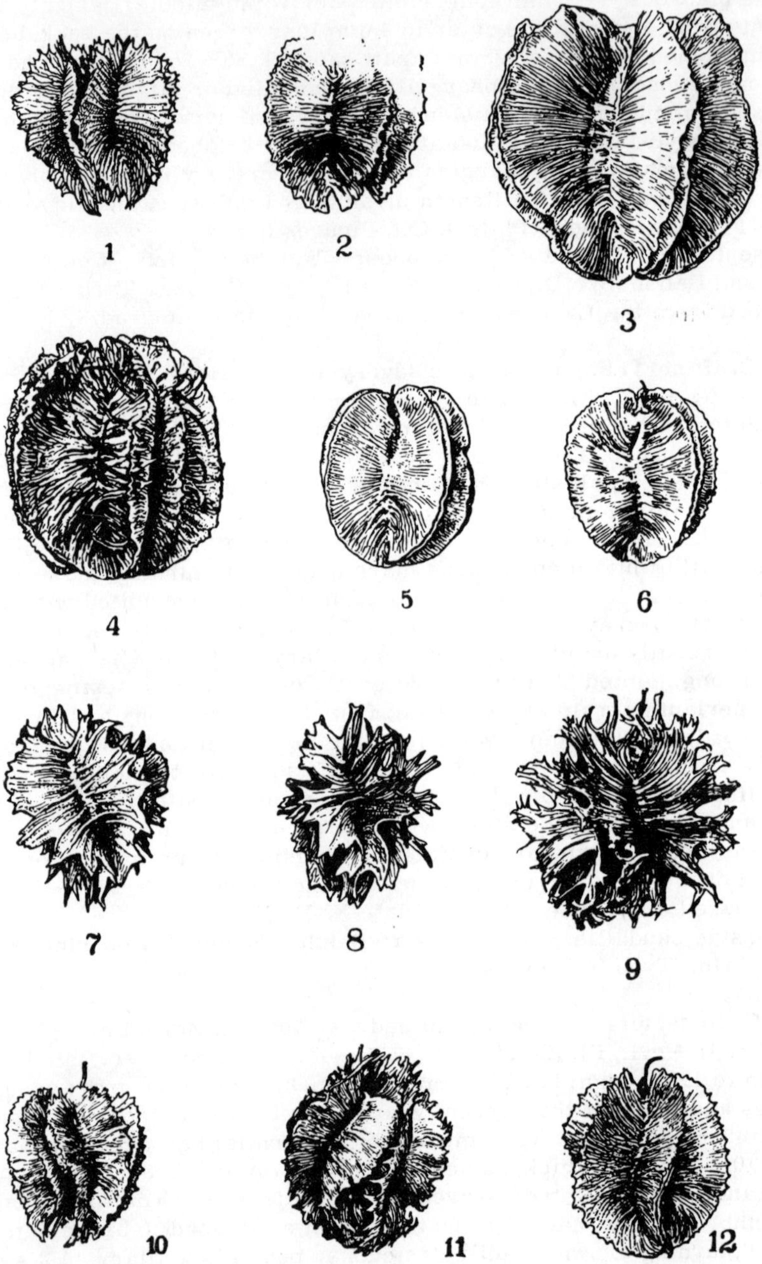

PLATE XXXI. 1. *Calligonum gracile* Litw. – 2. *C. aralense* Borszcz. – 3. *C. alatiforme* N. Pavl. – 4. *C. alatiforme* var. *aculeolatum* N. Pavl. – 5. *C. leucocladum* (Schrenk) Bge. – 6. *C. leucocladum* var. *serratum* Litw. – 7–8. *C. dissectum* T. Popova. – 9. *C. affine* T. Popova. – 10. *C. rubicundum* Bge. – 11. *C. flavidum* Bge. – 12. *C. Borszczowii* Litw.

almost free or sometimes partly united longitudinally, surpassing the base of the style, at first pale yellow, turning brown, attenuate toward summit, approximate or rarely distant at the margins; wing margin coarsely and unequally sinuate-dentate, the very deep incisions to between one-third and one-half the width of the wing, sometimes reaching the ribs of the achene and the wings dissected-lobate; teeth or lobes unequal, acute, ranging from flat and broad to lanceolate or even subulate, secondarily sharp-toothed; wing surface naked, without outgrowths, rather closely and very prominently straight-nerved. Fr. June—July. (Plate XXXI, Figures 7, 8).

Riverside sandhills. — Centr. Asia: Balkh. Endemic. Described from Lake Zaisan. Type in Tomsk.

13. **C. rubicundum** Bge. Delect. sem. horti Dorp. VIII (1839); Kryl., Fl. Zap. Sib. IV, 846. — Pterococcus soongoricus β rubicundus C. A. M. Bull. Acad. St. Pétersb. VIII (1841) 340; Ldb. Fl. Ross. III, 495. — Pallasia Pterococcus Pall. Fl. Ross. II (1788) 70, tab. 77, 78 pp. (excl. pl. wolgens.). — Calligonum songaricum Endl. Gen. plant. Suppl. IV, 2 (1847) 50. — Ic.: Pall. l. c. tab. 77, 78 (excl. pl. wolgensi).

Shrub to 1 m high; bark of mature branches dark, brownish-gray or reddish-gray; flowers unknown; segments of fruiting perianth reflexed; fruit ovate or rounded-ovate, 10—15 mm long and 9—13 mm broad; achene slightly coiled, with broad almost 2-winged ribs; wings mostly reddish; thick, stiffly coriaceous, somewhat folded longitudinally, broadly united at the middle with the achene ribs and nearly always united with the base of the style, the lobes often slightly united at both ends; wing margin unequally and rather shallowly obtusely and doubly sinuate-toothed; wing surface naked. Fr. June—July. (Plate XXXI, Figure 10).

Riverside and desert sandhills; sandy steppes. — W. Siberia: Irt.; Centr. Asia: Dzu.-Tarb. **Gen. distr.**: Mong., Dzu.-Kash. Described from Lake Zaisan. Type in Leningrad.

14. **C. flavidum** Bge. Delect. sem. horti Dorp. VIII (1839). — Pterococcus soongoricus α flavidus C. A. M. Bull. Acad. de St.-Pétersb. VIII (1841) 340.

Shrub to 1.5 m high; bark of mature branches dark, brownish-gray or
550 reddish-gray; flowers unknown; segments of fruiting perianth reflexed; fruit ovate or rounded-ovate, 13—20 mm long and 11—15 mm broad, commonly yellow or yellowish-red; wings membranous, rather thin, almost plane or slightly wavy, with very prominent anastomosing nerves; margins dissected into unequal lanceolate acutish teeth to 1.5—2 mm long, those near the base longer, acute, almost subulate; wing surface naked or rarely with few short outgrowths. Fr. June—July. (Plate XXXI, Figure 11).

Riverside and desert sandhills. — W. Siberia: Irt. Endemic. Described from Lake Zaisan. Type in Leningrad.

15. **C. Borszczowii** Litw. Trav. Mus. Bot. Ac. Sc. XI (1913) 54. — Ic.: Litw. l. c. tab. 7, fig. 22—24.

Shrub to 2 m high; bark of mature branches dark; flowers unknown; segments of fruiting perianth reflexed; fruit broad-oval, to 15 mm long and 12—13 mm broad; achene slightly coiled, with sharp ribs; wings rather stiffly membranous, flat, prominently nerved, subcordate at base, attenuate and convergent at summit, always united with the base of the style; wing margins cartilaginous, almost doubly toothed; wing surface naked or sometimes diffusely aculeate at the middle. Fr. June—July. (Plate XXXI, Figure 12).

Desert sandhills and sandy steppes. — Centr. Asia: Ar.-Casp., Balkh. Endemic. Described from the Aral Kara-Kum Desert. Type in Leningrad.

16. **C. undulatum** Litw. Trav. Mus. Bot. Ac. Sc. XI (1913) 53. — Ic.: Litw. l. c., tab. 7, fig. 13—15.

Shrub to 2 m high; bark of mature branches dark, brown, only at the southern extremity of the distribution area whitish (var. australe N. Pavl.); flowers unknown; segments of fruiting perianth reflexed; fruit round, to 13 mm long and 11—12 mm broad; achene slightly coiled, with sharp ribs; wings stiffly membranous, crisped-plicate or undulate along the margin, distant from and not united with the base of the style; margins entire or indistinctly rounded-toothed, approximate; wing surface naked. Fr. May—June. (Plate XXXII, Figure 1).

Desert sandhills. — Centr. Asia: Ar.-Casp., Kyz. K., Balkh., Dzu.-Tarb. Endemic. Described from the Aral Kara-Kum Desert. Type in Leningrad.

17. **C. plicatum** N. Pavl. in Fedde, Repert. sp. nov. XXXIII (1933) 151.

Shrub to 2 m high; bark of mature branches light, whitish; flowers unknown; segments of fruiting perianth reflexed; fruit almost round, 15—23 mm long and 13—20 mm broad; achene strongly coiled anticlockwise,
551 with sharp ribs; wings rather stiffly membranous, reddish, strongly coiled, approximate at the margin, closely convergent at summit and at base, not united with the base of the style; wing surface naked, with prominent transverse nerves, the margin plicate-crisped or undulate with short sharp teeth or rarely subentire. Fr. May—June. (Plate XXXII, Figure 2).

Desert sandhills and sandy steppes. — Centr. Asia: Kyz.K. Endemic. Described from sands of the Kyzyl-Kum Desert. Type in Moscow; cotype in Leningrad.

18. **C. turbineum** N. Pavl. in Fedde, Repert. sp. nov. XXXIII (1933) 152. — C. flavidum var. iliensis Trautv. in Bull. Soc. Nat. Mosc. XL, 3 (1867) 50, No. 1017.

Shrub to 2 m high; bark of mature branches light-colored, white or grayish; flowers unknown; segments of fruiting perianth reflexed; fruit almost round, 15—20 mm long and 14—19 mm broad; achene very strongly coiled clockwise or anticlockwise, the ribs obtusish, not prominent; wings stiffly membranous, reddish, surpassing the achene, approximate at the margin, strongly crinkled, narrowed at base and at summit, convergent and united with the base of the style; wings naked and smooth, prominently nerved, rarely with a single row of sparse sharp subulate outgrowths (var. subulatum N. Pavl.); wing margins incised to 2—3 mm into simple entire or unequal biparted sometimes double teeth. Fr. May—June. (Plate XXXII, Figures 3, 4).

Desert sandhills and riverside sands. — Centr. Asia: Balkh., Dzu.-Tarb. Endemic. Described from the sands beside the Ili River. Type in Moscow; cotype in Leningrad.

19. **C. tetrapterum** Jaub. et Spach, Illustr. pl. orient. V (1853—57) tab. 471. — Pterococcus tetrapterus Meisn. ex DC. Prodr. XIV (1856) 29. — Ic.: Jaub. et Spach., l. c. tab. 471.

Shrub to 2 m high; bark of mature branches dark gray; flowers unknown; segments of fruiting perianth reflexed; fruit ovate, 14—17 mm long and 12—15 mm broad; achene uncoiled or slightly coiled, with broad obtuse 2-winged ribs; wings stiffly coriaceous, flat, rounded at base, pointed at summit and nearly always united with the base of the style; wing margin doubly toothed, the stiff teeth subulate and sometimes forking into 2 or 3 teeth; wing surface naked, at the very margin with a single row of subulate outgrowths, these sometimes furcately toothed at the tips. Fr. June—July. (Plate XXXII, Figure 5).

552 Desert sandhills. — Centr. Asia: Ar.-Casp., Balkh. **Gen. distr.** - Iran., Bal.-As. Min. Described from Mesopotamia. Type in Paris.

20. **C. Russanovii** N. Pavl. in Fedde, Repert. sp. nov. XXXIII (1933) 146.

A much branched shrub to 1 m high; bark of mature branches dark, reddish-brown; herbaceous branchlets slender, grayish-green, furrowed when dry, more or less ramified, jointed; joints 2—3 cm long, terminating in a short and broad sheathing brownish ocrea; leaves linear-filiform, 2—4 mm long, acuminate, united with ocreae, deciduous; flowers unknown; segments of fruiting perianth reflexed; fruit oval in outline, 12—15 mm long and 11—12 mm broad; achene scarcely to slightly coiled anticlockwise, the ribs broad and blunt; wings very stiffly coriaceous, reddish, approximate at the margin, surpassing the achene, broader at base with 2 short acute lobes, narrowed toward summit, acuminate, united with the base of the style; wing surface with thick transverse nerves, strongly convex near the margin, a single row of sparse short stiff aculeiform or lamelliform outgrowths near the margin or occasionally at the middle; wing margin almost entire, undulate, or with sparse short broad obtuse unequal teeth. (Plate XXXII, Figure 6).

Riverside desert sands. — W. Siberia: Irt. Endemic. Described from Lake Zaisan. Type in Leningrad.

21. **C. crispum** Bge. Delect. sem. horti Dorp. VIII (1839); Kryl., Fl. Zap. Sib. IV, 846. — C. Pallasia Ldb. Fl. Alt. II (1830) 206, excl. synon. (non l'Herit.). — Pterococcus crispus C. A. M. Bull. Acad. St. Pétersb. VIII (1841) 340; Ldb. Fl. Ross. III, 495.

Shrub to 2 m high; bark of mature branches dark, brown or reddish; flowers unknown; segments of fruiting perianth reflexed; fruit ovate or rounded-ovate, 15—18 mm long and 10—13 mm broad, yellow to honey-red; achene coiled, ribbed, the thick ribs bluntish; wings stiff, thick, subcoriaceous, somewhat undulate, surpassing the achene at base and at summit, rounded at base, slightly attenuate at summit; margin sharply sinuate-toothed; wing surface near the margin with a row of prominent stiff lanceolate or linear or subulate outgrowths, these sometimes toothed, not exceeding the distance from the wing margin. Fr. June—July. (Plate XXXII, Figure 10).

Riverside desert sands. — W. Siberia: Irt.; Centr. Asia: Dzu.-Tarb. **Gen. distr.**: Mong. Described from Lake Zaisan. Type in Leningrad.

553 22. **C. spinosissimum** N. Pavl. in Fedde Repert. sp. nov. XXXIII (1933) 146. — C. crispum auct. (non Bge.) p. p.

Shrub to 1.5 m high; bark of mature branches dark, brown or reddish; flowers unknown; fruit ovate or rounded-ovate, 16—18 mm long and 12—14 mm broad, honey-red; achene slightly coiled, 4-ribbed, the ribs bluntish; wings stiff, thick, subcoriaceous, surpassing the achene at base and at summit, slightly cordate at base, convergent at summit and united with the base of the style; margin strongly and sharply incised-toothed, the teeth flat at base, subulate at the tips, unequal, ranging from simple to furcate-toothed with 2—5 secondary teeth; wing surface near the margin with 1 or 2 rows of outgrowths, these raised, stiff, flat at base, furcate-toothed at the tip, not exceeding the distance from the wing margin. Fr. June—July. (Plate XXXII, Figure 11).

Riverside sandhills. — W. Siberia: Irt. Endemic. Described from Lake Zaisan. Type in Moscow.

23. **C. membranaceum** (Borszcz.) Litw. Trav. Mus. Bot. Ac. Sc. XI (1913) 53. — C. flavidum var. membranaceum Borszcz. in Mém. Acad. St.-Pétersb. VII, sér. III, 1 (1860) 33. — Ic.: Borszcz. l. c. tab. I, fig. 4—4b; Litw. l. c. tab. 7, fig. 19—21.

Shrub to 2 m high; bark of mature branches dark, brown; flowers unknown; segments of fruiting perianth reflexed; fruit oval or almost round, to 20 mm long and 18—20 mm broad; achene slightly coiled or uncoiled, the sharp thin prominent ribs as broad at base as at summit; wings softly membranous, distinctly nerved, approximate at the margin, not united with and rather distant from the base of the style; lobes at base rounded or truncate, at summit acute; wing margin closely and unequally incised-toothed to 1—3 mm, the soft teeth with sharp and sometimes subulate secondary teeth; wing surface near the margin with a row of outgrowths, these 3—4 mm long, subulate or setiform or flat, those near the base or the summit of the wing dilated and sometimes forking, or else the wing surface naked (var. nudum Litw.). Fr. June—July. (Plate XXXII, Figures 7, 8).

Desert sandhills and sandy steppes. — Centr. Asia: Ar.-Casp., Balkh. Endemic. Described from the Aral Kara-Kum Desert. Type in Leningrad.

24. **C. calcareum** N. Pavl. in Fedde, Repert. sp. nov. XXXIII (1933) 149. — C. ferganicum Litw. in sched. Herb. Acad. Sc. URSS.

A much branched shrub to 0.5 m high; bark of mature woody branches
554 light, whitish; herbaceous branchlets in clusters of 3—6 from the nodes, slender, glaucous, furrowed when dry, jointed; joints 1—2 cm long, terminating in a short broad sheathing brownish membranous ocrea; leaves united with ocreae, linear, setiform, 1—3 mm long, acuminate, soon deciduous; pedicels 1—3 from the ocreae, jointed near the base; flowers

(555)

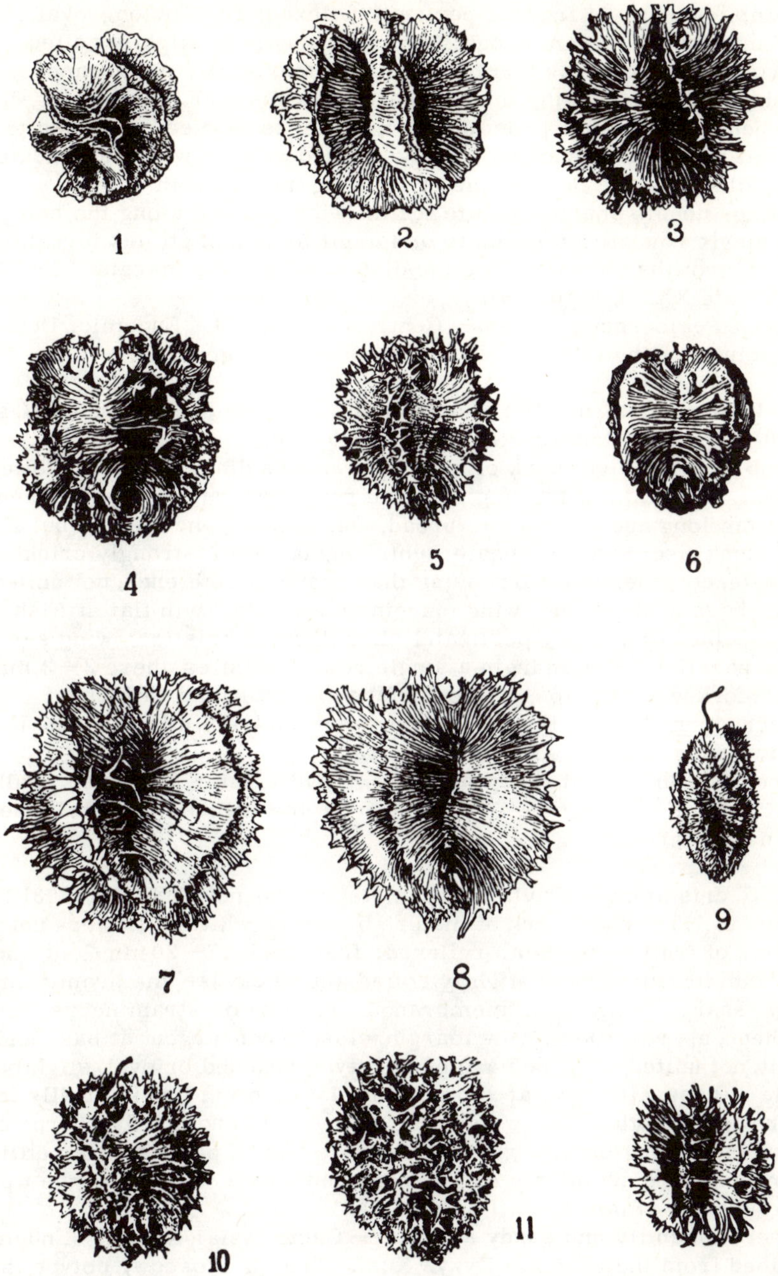

PLATE XXXII. 1. Calligonum undulatum Litw. – 2. C. plicatum N. Pavl. – 3. C. turbineum N. Pavl. – 4. C. turbineum var. subulatum N. Pavl. – 5. C. tetrapterum Jaub. et Spach. – 6. C. Russanovii N. Pavl. – 7. C. membranaceum (Borszcz.) Litw. – 8. C. membranaceum var. nudum Litw. – 9. C. calcareum N. Pavl. – 10. C. crispum Bge. – 11. C. spinosissimum N. Pavl. – 12. C. acanthopterum Borszcz.

unknown; segments of fruiting perianth reflexed; fruit oblong-oval, 10—15 mm long, 4—7 mm broad; achene scarcely to slightly coiled clockwise, the broad ribs bluntish; wings rather stiffly membranous, reddish-brown, surpassing the achene, approximate at the margins, closely convergent at both ends, prominently transverse-nerved, narrowed to a point at base and at summit, rarely notched at base with short acute lobes, united with the base of the style; wing margin with unequal sometimes double sharp subulate teeth; wing surface along the margin with a single row of thin subulate or lamelliform sometimes furcate-tipped outgrowths not exceeding the distance from the margin. Fr. May—June. (Plate XXXII, Figure 9).

Exposed calcareous slopes. — Centr. Asia: Syr D. Endemic. Described from vicinity of the city of Kokand. Type in Leningrad.

25. **C. acanthopterum** Borszcz. in Mém. Acad. de St.-Pétersb. VII, sér. III, 1 (1860) 34. — Ic.: Borszcz. l. c. tab. I, fig. 5—5e.

Shrub to 1.5 m high; bark of mature branches dingy grayish; flowers unknown; segments of fruiting perianth reflexed; fruit cordate-triangular, 13—20 mm long and 10—17 mm broad, achene oval, uncoiled or but slightly coiled clockwise; wings elongate; subtriangular, stiff, strongly crinkled, acuminate, considerably surpassing the achene at both ends, not united with the base of the style; wing margin on all sides with flat stiffish double-toothed or prickly pectinately lamelliform bristles; wing surface bearing parallel to the margin a single row of bristles, these 2—3 mm long, slender, forked at the tips, or the surface almost naked with short unbranched subulate or lamelliform outgrowths (var. subnudum N. Pavl.).
557 Fr. May—June. (Plate XXXII, Figure 12).

Desert sandhills and sandy steppes. — Centr. Asia: Ar.-Casp., Balkh., Kara K., Kyz. K. Endemic. Described from the Aral Kara-Kum Desert. Type in Leningrad.

26. **C. cristatum** N. Pavl. in Fedde, Repert. sp. nov. XXXIII (1933) 149.

Shrub to 1.5 m high; bark of mature branches whitish; flowers unknown; segments of fruiting perianth reflexed; fruit oval, 17—20 mm long and 15—18 mm broad; achene slightly coiled anticlockwise, the prominent wings thin and sharp; wings stiff, membranous, reddish or stramineous, surpassing the achene, approximate at the margin, closely convergent at base and at summit, not united with the base of the style, rounded or oval, slightly cordate at base, attenuate at summit, the flat or often longitudinally folded lobes united at both ends; wing margin unequally and doubly sharp-toothed; wing surface along the margin with a single row of pectinate-lamelliform outgrowths, these acuminate or enlarged and double-toothed at the tips. Fl. May—June. (Plate XXXIII, Figure 1).

Desert sandhills and sandy steppes. — Centr. Asia: Kyz. K. Endemic. Described from the sands of Kyzyl-Kum. Type in Moscow; cotype in Leningrad.

27. **C. Dubjanskyi** Litw. in Trav. Mus. Bot. Ac. Sc. XI (1913) 57. — Ic.: Litw. l. c. tab. 8, fig. 13—15.

Shrub to 2 m high; bark of mature branches whitish; flowers unknown; segments of fruiting perianth reflexed; fruit oval, to 15 mm long and 12 mm broad; achene strongly coiled with bluntish ribs; wings stiffish, flat, prominently transverse-nerved, surpassing the achene, united with the base of the style and closely approximate at the margins, rounded at base, acuminate at summit; wing margin doubly sharp-toothed; wing surface almost at the very margin with a row of aculeiform or subulate outgrowths. Fr. May — June. (Plate XXXIII, Figure 2).

Desert sandhills and sandy steppes. — Centr. Asia: Kyz. K. Endemic. Described from the Syr Darya River valley between Beiga-Kum and Chiili stations. Type in Leningrad.

28. **C. Bubyri** B. Fedtsch. ex N. Pavl. in Fedde, Repert. sp. nov. XXXIII (1933) 150.

A much branched shrub to 1 m high; bark of mature branches light, whitish-gray; herbaceous branchlets slender, terete, pale green, almost straight; flowers borne on slender pedicels; outer perianth segments oblong-elliptic, 3 mm long, broadly white-margined; fruit oval or almost round in outline, 15 mm long and 13 mm broad; achene slightly coiled
558 clockwise, 4-ribbed, the broad ribs 2-winged; wings 5 — 6 mm broad, pale, membranous, prominently transverse-nerved, truncate at both ends, obtusely rounded, adjacent ones with divergent margins; wing margin with teeth 0.5 — 2 mm long, enlarged at base, subulate-tipped; wing surface at the very margin with a row of flattish or filiform outgrowths 3 — 5 mm long. Fr. May — June. (Plate XXXIII, Figure 3).

Desert sandhills and sandy steppes. — Centr. Asia: Kara K. Endemic. Described from Turkmenia. Type in Leningrad.

29. **C. involutum** N. Pavl. in Fedde, Repert. sp. nov. XXXIII (1933) 145.

A much branched shrub to 1 m high; bark of mature branches dark, reddish-brown; herbaceous branchlets slender, grayish-green, furrowed when dry, slightly ramified, jointed; joints 2 — 5 cm long, terminating in a short broad sheathing brownish ocrea; leaves linear, filiform, 2 — 5 mm long, acuminate, united with ocrea, deciduous; flowers unknown; segments of fruiting perianth reflexed; perianth broad-oval in outline, 18 — 20 mm long and 14 — 18 mm broad; achene slightly coiled clockwise, the thin ribs prominent; wings stiffish, membranous, reddish, surpassing the achene, distant at base and at the margins, the lobes folded lengthwise, united at both ends, broader at base, narrowed and acuminate at summit, not united with the base of the style; wing surface naked, with slender transverse nerves; wing margin unequally and almost doubly sharp-toothed, usually strongly involute, the teeth short and broad. Fr. May — June. (Plate XXXIII, Figure 4).

Riverside desert sands. — Centr. Asia: Balkh. Endemic. Described from Lake Zaisan. Type in Moscow; cotype in Leningrad.

30. **C. Eugenii-Korovini** N. Pavl. in Fedde, Repert. sp. nov. XXXIII (1933) 145.

A much branched shrub to 1 m high; bark of mature branches dark, blackish-brown; flowers unknown; segments of fruiting perianth wide-spreading; fruit broad-oval in outline, 14 — 18 mm long and 13 — 16 mm broad; achene uncoiled or slightly coiled clockwise, the prominent ribs thin

and sharp; wings membranous, thin, yellowish-brown, surpassing the achene, very distant at base, approximate at the margin, rounded-oval, the lobes united at both ends or only at base or rarely only at summit, closely convergent at base, distant at summit and not united with the base of the style; wing surface naked or rarely with few sharp outgrowths,
559 prominently transverse-nerved; wing margin strongly involute, the short sharp teeth broad toward base, often double. Fr. May—June. (Plate XXXIII, Figure 5).

Riverside and desert sandhills. — Centr. Asia: Dzu.-Tarb., Balkh. Endemic. Described from Kunduzda on the Ili River. Type in Moscow; cotype in Leningrad.

31. **C. lanceolatum** N. Pavl. in Fedde, Repert. sp. nov. XXXIII (1933) 148.

Shrub to 1—2 m high; bark of mature branches whitish; flowers unknown; segments of fruiting perianth reflexed; fruit almost round in outline, 13—17 mm long and 12—16 mm broad; achene scarcely to slightly coiled clockwise, the prominent ribs sharp; wing membranous, thin, yellowish-brown, surpassing the achene, approximate at the margin, not united with the base of the style, the lobes slightly folded longitudinally; wing surface prominently transverse-nerved, naked or with few sharp outgrowths; wing margin patellately involute all around, the very short sharp teeth simple or sometimes double. Fr. May—June. (Plate XXXIII, Figure 6).

Desert sandhills. — Centr. Asia: Kyz. K. Endemic. Described from sands of Kyzyl-Kum. Type in Leningrad.

32. **C. physopterum** N. Pavl. in Fedde, Repert. sp. nov. XXXIII (1933) 148.

Shrub to 1.5 m high; bark of mature branches whitish; flowers unknown; segments of fruiting perianth reflexed; fruit cylindric in outline, 14—19 mm long and 18—22 mm broad; achene scarcely to slightly coiled anticlockwise, the prominent ribs thin and sharp; wings thin, membranous, light reddish or yellow, surpassing the achene, very distant at base, approximate at the margins, convergent at summit and united with the base of the style, the lobes united at both ends; wing surface prominently transverse-nerved, naked; wing margin rather strongly saccately involute, indistinctly and sparsely toothed, the short broad unequal sharp teeth sometimes double. Fr. May—June. (Plate XXXIII, Figure 7).

Desert sandhills. — Centr. Asia: Kyz. K. Endemic. Described from sands of Kyzyl-Kum. Type in Moscow; cotype in Leningrad.

33. **C. batiola** Litw. in Trav. Mus. Bot. Ac. Sc. XI (1913) 56; Eug. Kor. Sched. ad Herb. Fl. Asiae Med. V (1925) No. 104 (emend.). — Ic.: Litw. l. c. tab. 8, fig. 4—6. — Exs.: H. F. A. M. No. 104.

Shrub to 0.7 m high, much branched almost from base; mature woody
560 branches refracted-tortuous, knotted at the bends, the bark pale brown; herbaceous branchlets crowded in large numbers at the nodes, very slender, densely covered with flowers, nearly all deciduous; flowers paired in the ocreae; pedicels to 4 mm long, jointed near base; perianth segments unequal, the larger 3 mm long; fruit cylindric, 10 mm long and 15 mm broad; achene scarcely coiled, blood-red, narrow-ribbed, the ribs the same width throughout; wings stiffly membranous, cupuliformly

united at base and at summit, distant from each other at base, about equaling the achene, not united with the base of the style, honey-yellow, often blood-red within, pale elsewhere; wing margin entire, under strong magnification indistinctly and minutely toothed; wing surface naked; perianth segments reflexed in fruit. Fr. May—June. (Plate XXXIII, Figure 8).

Desert sandhills. — Centr. Asia: Ar.-Casp., Balkh., Kyz. K. Endemic. Described from sands around Tartugai in the Syr Darya River valley. Type in Leningrad.

34. **C. obtusum** Litw. in Trav. Mus. Bot. Ac. Sc. XI (1913) 56. — Ic.: Litw. l. c. tab. 8, fig. 7—9.

Shrub to 2 m high; bark of mature branches whitish; flowers unknown; segments of fruiting perianth reflexed; fruit cylindric, 15 mm long and to 20 mm broad; achene uncoiled narrow-ribbed, the ribs the same width throughout; wings rather stiffly membranous, lurid, equaling the achene, not united with the base of the achene, distant at the margin, the lobes often united at summit, free or almost free at base; wing margin almost doubly serrate-dentate; wing surface naked. Fr. May—June. (Plate XXXIII, Figure 9).

Desert sandhills. — Centr. Asia: Ar.-Casp., Balkh., Dzu.-Tarb., Kyz. K. Endemic. Described from sands around Tartugai in the Syr Darya River valley. Type in Leningrad.

35. **C. Androssowii** Litw. in Trav. Mus. Bot. Ac. Sc. XI (1913) 56. — Ic.: Litw. l. c. tab. 8, fig. 10—12.

Shrub to 1.5 m high; bark of mature branches whitish; flowers unknown; segments of fruiting perianth reflexed; fruit cylindric, 10—14 mm long and 18 mm broad, lurid; achene uncoiled, sharp-ribbed, produced beyond the depressed summit of wings into a ribbed almost narrow-winged cone; wings rather stiffly membranous, barely as long as the achene, not united with the base of the style, the lobes united at summit, truncate and almost free at base; wing margin closely and almost doubly sinuate-toothed; wing surface naked or sometimes along the margin with sparse lamelliform or occasionally forked outgrowths (var. lamellatum N. Pavl.).
561 Fr. May—June. (Plate XXXIII, Figure 10).

Sandhills. — Centr. Asia: Ar.-Casp., Kyz. K. Endemic. Described from sands around Tartugai. Type in Leningrad.

36. **C. Lipskyi** Litw. in Trav. Mus. Bot. Ac. Sc. XI (1913) 54. — Ic.- Litw. l. c. tab. 7, fig. 25—27.

Shrub to 1.5 m high; bark of mature branches whitish; fruit unknown; segments of fruiting perianth wide-spreading; fruit cordate in outline, 12—14 mm long and 13—15 mm broad; achene slightly coiled, sharp-ribbed; wings stiffly membranous, the lower lobes rounded and divergent, the upper acute contiguous and always united with the base of the style; wings with thin soft almost doubly toothed margin and naked surface. Fr. May—June. (Plate XXXIII, Figure 11).

Desert sandhills. — Centr. Asia: Ar.-Casp., Kyz. K. Endemic. Described from sands around Tartugai. Type in Leningrad.

(563)

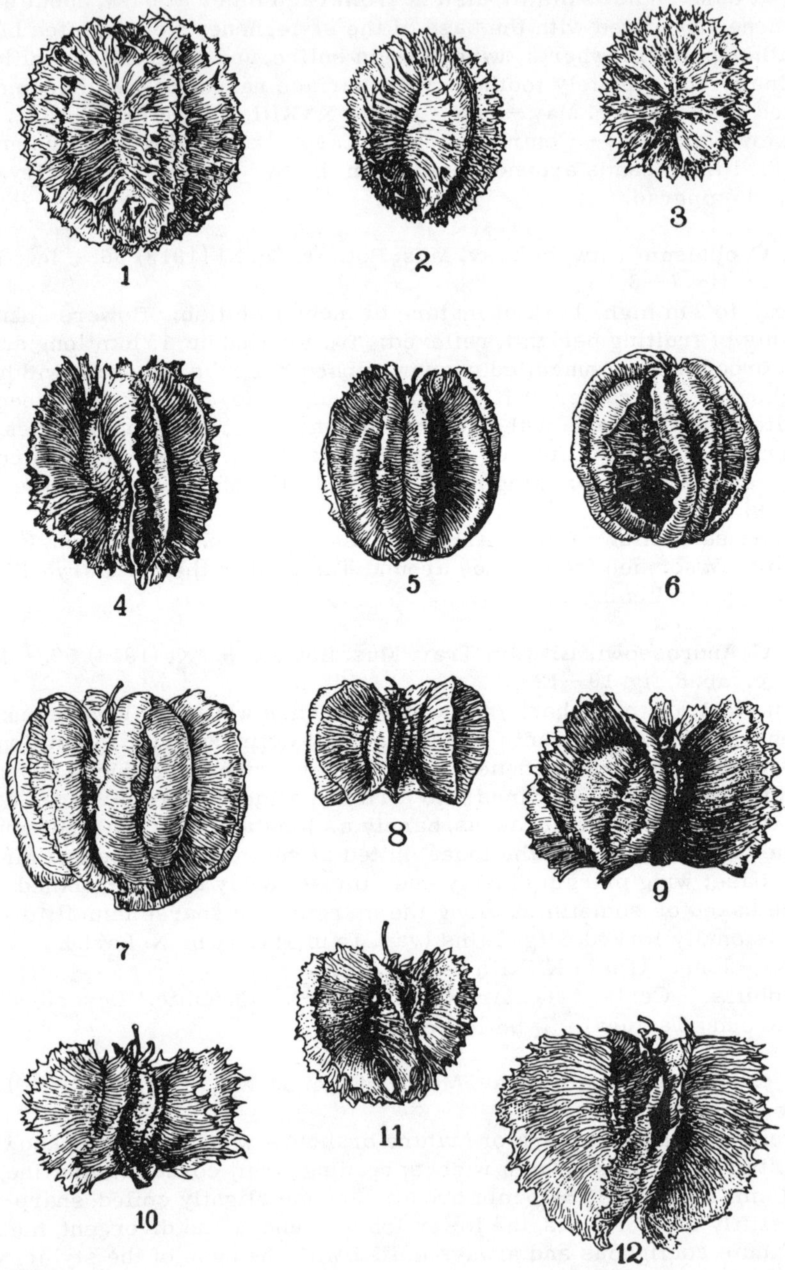

PLATE XXXIII. 1. Calligonum cristatum N.Pavl.– 2. C.Dubianskyi Litw.– 3. C.Bubyri B.Fedtsch.– 4. C.involutum N.Pavl.– 5. C.Eugenii-Korovini N.Pavl.– 6. C.lanceolatum N.Pavl.– 7. C.physopterum N.Pavl.– 8. C.batiola Litw.– 9. C.obtusum Litw.– 10.– C. Androssowii Litw.– 11. C.Lipskyi Litw.– 12. C.patens Litw.

37. **C. patens** Litw. in Trav. Mus. Bot. Ac. Sc. XI (1913) 55. — Ic.: Litw. l. c. tab. 7, fig. 31—33.

Shrub to 1.5 m high; bark of mature branches whitish; flowers unknown; segments of fruiting perianth spreading; fruit cordate in outline, to 18 mm long, 22 mm broad; achene slightly coiled, the sharp ribs more prominent at base, narrowed and barely distinct at summit; wings softly membranous, often crinkled, united with the base of the style, the lobes acute and in 1 or 2 wings sometimes united, rounded at base or at summit; wing margin doubly sharp-toothed; wing surface sometimes with sparse lamelliform outgrowth along the margin (var. lamellatum N. Pavl.). Fr. May—June. (Plate XXXIII, Figure 12).

Desert sandhills. — Centr. Asia: Ar.-Casp., Balkh., Kyz. K. Endemic. Described from sands around Tartugai. Type in Leningrad.

38. **C. coriaceum** N. Pavl. in Fedde, Repert. sp. nov. XXXIII (1933) 144.

A much branched shrub to 1 m high; bark of mature branches dark, reddish-brown; herbaceous branchlets slender, grayish-green, furrowed when dry, always ramified, jointed; joints 1—4 cm long, terminating in a short broad sheathing brownish ocrea; leaves linear, setiform, 1—3 mm long, acuminate, united with ocreae, deciduous; flowers unknown; segments of fruiting perianth spreading; fruit cylindric in outline, 10—15 mm long and 14—20 mm broad; achene uncoiled, the ribs broad and bluntish; wings rather stiffly coriaceous, reddish, about equaling the achene or surpassing
562 it at summit, not united with the base of the style, very distant at base and at the margin, the lobes united at the ends, united or often free at base, broader below, narrowed and acuminate at the ends; wing surface naked; wing margin unequally doubly and obtusely toothed or rarely some teeth acute. Fr. May—June. (Plate XXXIV, Figure 1).

Riverside desert sands. — Centr. Asia: Balkh. Endemic. Described from Lake Zaisan. Type in Moscow; cotype in Leningrad.

39. **C. cartilagineum** N. Pavl. in Fedde, Repert. sp. nov. XXXIII (1933) 144.

A much branched shrub to 1 m high; bark of mature branches dark, reddish-brown; flowers unknown; segments of fruiting perianth reflexed; fruit cylindric in outline, 13—15 mm long and 14— 16 mm broad; achene uncoiled or slightly coiled anticlockwise, the thin prominent nerves sharp; wings membranous, thin, yellowish-brown, surpassing the achene, very distant at base, approximate or often distant at the margin, rounded, the lobes united at both ends or only at summit, cup-shaped, not united with the base of the style; wing surface naked prominently transverse-nerved; wing margin subentire or with sparse very short acutish teeth. Fr. May—June. (Plate XXXIV, Figure 2).

Desert sandhills and riverside sands. — Centr. Asia: Dzu.-Tarb., Balkh. Endemic. Described from sands along the Ili River near the Kunduzda natural boundary area. Type in Moscow; cotype in Leningrad.

Section 2. **PTERYGOBASIS** Borszcz. in Mém. Acad. St.-Pétersb. VII sér. III, 1 (1860) 34; Litw. in Trav. Mus. Bot. Ac. Sc. XI (1913) 57 (excl. C. Dubianskyi). — Eucalligonum Bge. ex Boiss. Fl. Or. IV (1879) 1000, ex parte (quoad C. polygonoides); Litw. l. c. 58, ex parte (quoad C. molle) (characters in the key).

1. Fruit oval or ovate, small, to 15 (rarely 20—22) mm long including bristles; wings membranous; bristles on the margin and on the surface shorter than to barely exceeding the width of the wing . . . . . 2.

565 + Fruit round or rarely cordate, commonly large, together with bristles not less than 20 mm long or broad or, if smaller, then the rudimentary wings stiffly subcoriaceous; bristles on the margin or on the surface of wings greatly exceeding and up to several times the width of the wing or rarely equaling it . . . . 5.

2. Wings 1 — 2 mm broad; plants of the Caucasus . . . . 3.

\+ Wings 4—5 mm broad, conspicuous . . . . 4.

3. The flattish bases of bristles united into a barely perceptible wing or almost free; bark of mature branches whitish; herbaceous branchlets of the current year ramified . . . . 40. **C. polygonoides** L.

\+ Wings 1—2 mm broad, quite distinct; bark of mature branches grayish-black; branchlets of the current year simple, not ramified . . . . 41. **C. bakuense** Litw.

4. Fruit to 22 mm long and 18—20 mm broad, broad-oval; wing surface naked or almost naked with few bifurcate bristles near the margin; bark of mature branches light-colored: gray or somewhat reddish; plants of the Caucasus . . . . 42. **C. Petunnikowii** Litw.

\+ Fruit to 13 mm long and 6—10 mm broad, oblong-oval; wing surface along the margin densely beset with bristles, these 2- or 3-toothed at the tips; bark of mature branches dark, brownish-gray; plants of Soviet Central Asia . . . . 43. **C. santoanum** Eug. Kor.

5. Wings stiff, subcoriaceous, very narrow and rudimentary, rarely to 4 mm broad, rather resembling the united bases of bristles; bristles mostly confined to the wing margin, commonly stiffish, brittle, their short terminal branches rather thick and spiniform . . . . 6.

\+ Wings membranous, broader, quite distinct, entire or sometimes dissected; the bristles mostly distinct, commonly soft, elongated, borne on the margin and on the surface or only on the surface along the margin of the wing, their terminal branches mostly very slenderly long-capillary . . . . 10.

6. Bristles very dense, usually completely concealing the wings and the achene, their terminal branches elongated slender and more or less tangled . . . . 7.

\+ Bristles sparse, not concealing the wings or the achene, their terminal branches thickish short and not tangled . . . . 8.

566 7. Fruit spherical, 25—30 mm in diameter, reddish or rufous; wings to 4 mm broad, confluent at base; bristles few, borne also on the surface along the margin of the wing . . . . 44. **C. macrocarpum** Borszcz.

\+ Fruit spherical, to 15 mm in diameter, yellowish or yellowish-brown; wings to 2—2.5 mm broad, notched at base, free; bristles in 2 rows only at the margin . . . . 45. **C. densum** Borszcz.

8. Fruit to 15 mm in diameter; bristles thick, shorter than or equaling the width of the achene, their terminal branches short, subulate; achene mostly coiled clockwise . . . . 9.

\+ Fruit 16—18 mm in diameter; bristles more slender, long, equaling or exceeding the width of the achene, their terminal branches relatively long, spiniform; achene mostly coiled anticlockwise . . . . 46. **C. colubrinum** Borszcz.

9. Bristles rather dense, their terminal branches approximate; wings 1—2 mm broad; bristles branching nearly from base, the rather long terminal branches subulate . . . . 47. **C. erinaceum** Borszcz.

\+ Bristles very sparse, spreading, their terminal branches distant; wings 2—4 mm broad; bristles branching above the middle, the very short terminal branches stellate . . . . 48. **C. squarrosum** N. Pavl.

10. Wings asymmetric, deeply dissected nearly to the achene into lamelliform lobes, these gradually passing into teeth and forking marginal bristles; wing surface with few sparse forking bristles at the tips of the teeth, these and marginal bristles alike . . . . 49. **C. platyacanthum** Borszcz.

\+ Wings more or less symmetric, not dissected, the largest toothed and incised, the incisions far from reaching the achene; bristles usually rather strongly delimited from the wings, on the surface numerous and dense or none . . . . 11.

11. Fruit together with bristles cordate, rarely round and then the lobes of wings cupuliformly united at both ends; bristles equaling the width of the wings or nearly so . . . . 12.

\+ Fruit together with bristles round; bristles on the margin and on the surface exceeding sometimes manyfold the width of the wings . . . . . 13.

12. Fruit cordate in outline; wings flat or slightly folded longitudinally, their lobes not united, slightly incised, the bristles on the margins and on the surface of wings simple or almost so, only some of them with 1 or 2 prongs . . . . 50. **C. Muravljanskyi** N. Pavl.

567 + Fruit cordate or round in outline; wings strongly folded longitudinally and more or less cupuliformly united at both ends; bristles on the margin and on the surface of wings forking . . . . 51. **C. rotula** Borszcz.

13. Achene including wings 15—22 mm long and 12—17 mm broad, the wings 9—10 mm broad; bristles stiffish, brittle . . . . 14.

\+ Achene including wings 10—15 mm long and 7—10 mm broad, the inconspicuous wings 2—5 mm broad; bristles slender, soft . . . . 17.

14. Fruit including bristles 35—40 mm in diameter; bristles very dense, almost concealing the wings and the achene, densely paniculately branched nearly from base . . . . 52. **C. pulcherrimum** Eug. Kor.

\+ Fruit including bristles 20—30 mm in diameter; bristles sparser, not concealing the wings or the achene, forking or rarely simple . . . 15.

15. Fruit including bristles 25—30 mm in diameter; bristles borne only on the wing surface along the margin; wing margin with subulate or prickly teeth up to 2—4 mm long, those of adjacent wings intersecting . . . . 55. **C. kzyl-kumi** N. Pavl.

\+ Fruit including bristles 20—25 mm in diameter; bristles borne both on the surface and on the margins of the wings or rarely some of the marginal bristles reduced to subulate teeth . . . . 16.

16. Wings obtusish or truncate at base; all bristles more or less forking . . . . 53. **C. setosum** Litw.

\+ Wings cordate at base; all bristles simple or nearly so, only occasionally with 1 or 2 upright prongs . . . . 54. **C. cordatum** Eug. Kor.

17. Fruit including bristles to 25 mm in diameter; wings inconspicuous, 2—3 mm broad; bristles very soft, not brittle, borne on the surface along the margin; marginal teeth subulate, 3—4 mm long . . . . 56. **C. molle** Litw.

\+ Fruit including bristles 40 — 45 mm in diameter; wings stiffish, 3 — 5 mm broad; bristles very long, somewhat stiffish and brittle, in a single row on the margin, the wing surface naked or nearly so . . . . 57. **C. Paletzkianum** Litw.

568 40. **C. polygonoides** L. Sp. pl. (1753) 530.

Shrub to 1.5 m high; bark of mature branches whitish; flowers unknown; segments of fruiting perianth reflexed; fruit obovate, 12 — 16 mm long and 10 — 13 mm broad, rather densely setose; achene slightly coiled, the broad ribs narrowly 2-winged (or sometimes the wings rudimentary or obsolescent); wings 1 — 2 mm broad, always united with the base of the style, the margin bearing a single row of bristles; bristles dense, in 8 rows, flattish, membranous-margined, pale, terminating in 2 or 3 forking branches, these in turn with elongated brittle delicate branchlets. Fr. June — July. (Plate XXXIV, Figure 4).

Desert sandhills. — Caucasus: S. Transc. **Gen. distr.:** Iran., Arm.-Kurd., Bal.-As. Min. Described from Armenia (foot of Mount Ararat). Type in London.

41. **C. bakuense** Litw. Sched. ad Herb. Fl. Ross. VIII (1922) 7. — Exs.: HFR No. 2415.

A much branched shrub to 1.5 m high, the branches curved and flexuous; bark of mature woody branches grayish-black, sulcately fissured; herbaceous branchlets lateral, clustered at the nodes of annotinous branches, simple, sulcate, all floriferous, jointed; joints 2 — 3 cm long, terminating in a short pellucid cup-shaped ocrea; leaves filiform, 2 — 4 mm long, soon deciduous; flowers in fascicles of 2 — 5 from the ocreae; pedicels unequal, 2 — 4 mm long, glabrous; perianth segments 3 mm long, ovate-orbicular, reflexed in fruit; anthers pink; fruit oval to oblong-oval, 10 — 15 mm long and 9 — 13 mm broad, densely setose; achene slightly coiled, with bluntish ribs; wings (or bristles united at base) 1 — 2 mm broad, united with the base of the style, in 2 rows, beset on the margin with flat stiffish bristles, these twice ternately (or more simply) branched, the terminal prongs spreading subspinescent; surface of wings naked or with scattered short spreading bristles near the margin. Fr. May — June. (Plate XXXIV, Figure 5).

Stony desert slopes. — Caucasus: E. Transc. (Baku). Endemic. Described from the vicinity of Baku. Type in Leningrad.

42. **C. Petunnikowii** Litw. Sched. ad Herb. Fl. Ross. VIII (1922) 7. — Exs.: HFR No. 2413, 2414.

571 A divaricately branched shrub to 1.5 m high with curved branches; bark of mature woody branches light gray or somewhat reddish, striate; herbaceous branches of the current year lateral, clustered at the nodes, sulcate, brittle, simple or with short sterile branchlets, jointed; joints to 5 cm long, terminating in a cup-shaped pellucid membranous ocrea 2 — 3 mm long, the margin of the ocrea coarsely toothed or erose; leaves ovate, acute, 2 — 3 mm long, with colored midrib, often obsolescent and promptly deciduous; flowers in fascicles of 2 — 5 in the ocreae; pedicels

(569)

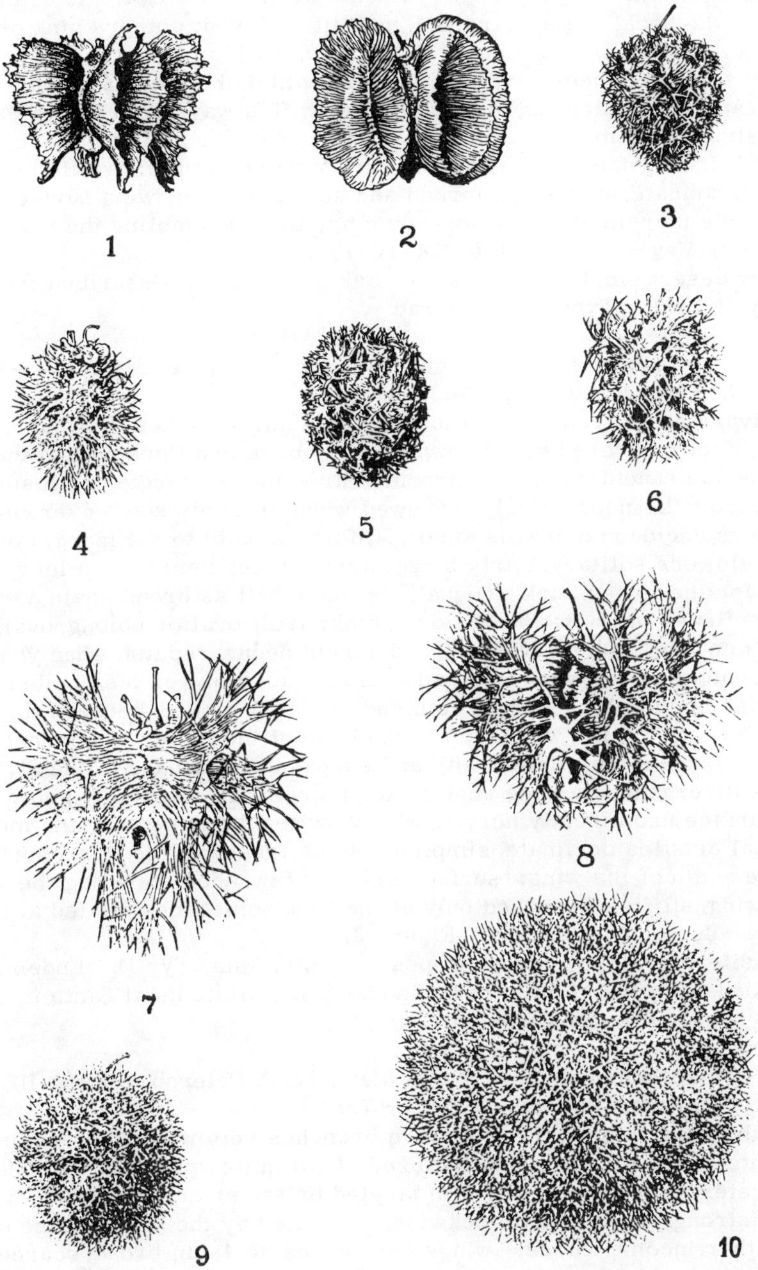

PLATE XXXIV. 1. Calligonum coriaceum N. Pavl. – 2. C. cartilagineum N. Pavl. – 3. C. santoanum Eug. Kor. – 4. C. polygonoides L. – 5. C. bakuense Litw. – 6. C. Petunnikowii Litw. – 7. C. Muravljanskyi N. Pavl. – 8. C. rotula Borszcz. – 9. C. densum Borszcz. – 10. C. macrocarpum Borszcz.

unequal, 1 — 5 mm long, glabrous; perianth segments obovate, to 5 mm long, green on the back, white-margined, in fruit reflexed; anthers pink or sanguine; fruit oval or broad-oval, to 22 mm long and 18 — 20 mm broad, densely setose; achene 4-ribbed, slightly coiled, the obtusish ribs 2-winged; wings rather conspicuous, to 4 mm broad, stiff, always united with the base of the style; wing margin with 2 rows of bristles, these flat at base, almost twice ternately branched or rarely more simple; bristle branches stiffly divaricate, at the tips forked and subspinescent; wing surface naked or near the margin with few spreading bristles resembling the marginal ones. Fr. May—June. (Plate XXXIV, Figure 6).

Stony desert slopes. — Caucasus: Baku. Endemic. Described from the vicinity of Baku. Type in Leningrad.

43. **C. santoanum** Eug. Kor. Sched. ad Herb. Fl. Asiae Mediae V — VII (1925) 30. — Exs.: H. F. A. M. No. 103.

A divaricately branched shrub to 0.5 m high; branches flexuous-refracted, arched, woody, covered with brownish-gray bark and thickened at the bends; herbaceous branchlets of the current year glaucous, terete, in clusters of 4 — 8 from the nodes, finely furrowed when dry, only some overwintering, the others deciduous; leaves small, filiform, adnate to the margin of ocrea; flowers solitary, fairly large; perianth segments 4 mm long, white except for the green back, unequal, the inner half as broad again as the outer, reflexed in fruit; anthers oval, pink; fruit oval or oblong-oval, 10 — 13 mm long and 6 — 10 mm broad, stramineous, reddish when young; achene oblong-oval, 9 mm long and 4 mm broad, more or less coiled clockwise or anticlockwise, at both ends pointed, ribbed, between the ribs deeply furrowed; ribs winged up to the base of the style; wings double, oblong to oval-oblong, united with and surpassing the base of the style, stiffish, divergent at a wide angle; wing margin almost doubly spiny-toothed;
572 wing surface prominently nerved, with very few bristles near the margin; marginal bristles declinate, simple or some forked at the tips, shorter than the width of the wing; surface bristles few, about equaling the width of the wing, stiffish, branched only at the tips, sometimes dilated at base. Fr. May—June. (Plate XXXIV, Figure 3).

Gypsiferous clayey desert slopes. — Centr. Asia: Syr D. Endemic. Described from the vicinity of oil industry installations at Santo in the Kokand area. Type in Tashkent.

44. **C. macrocarpum** Borszcz. in Mém. Acad. Pétersb. VII sér. III, 1 (1860) 35. — Ic.: Borszcz. l. c., tab. II, fig. 7 — 7c.

Shrub to 1 m high; bark of mature branches yellowish; flowers unknown; segments of fruiting perianth reflexed; fruit quite spherical, 25 — 30 mm in diameter, densely covered with tangled bristles; achene broad-oval, obtuse, strongly coiled anticlockwise, concealed by the bristles, the obtuse ribs rather inconspicuous; wings coriaceous, to 4 mm broad, scarcely surpassing the achene, pale, divergent from the center at an obtuse angle, almost contiguous at the margin; marginal bristles broad, flat, elongate-branched, the 2 — 5 straight divaricate branches branched in turn, the secondary branches dilated at base, subulate-capillary, brittle, delicate, strongly divaricate. Fr. May—June. (Plate XXXIV, Figure 10).

Desert sandhills. — Centr. Asia: Ar.-Casp., Balkh., Kyz. K., Kara K. Endemic. Described from the Aral Kara-Kum Desert. Type in Leningrad.

45. **C. densum** Borszcz. in Mém. Acad. St.-Pétersb. VII, sér. III, 1 (1860) 36. — Ic.: Borszcz. l. c., tab. II, fig. 8 — 8f. — Exs.: HFR No. 1432. —

Shrub to 2 — 3 m high; bark of mature branches blackish-gray; flowers unknown; segments of fruiting perianth reflexed; fruit quite spherical, 13 — 15 mm in diameter, very densely setose; achene obconical, coiled clockwise, the bluntish ribs very prominent and confluent toward summit; wings stiffly coriaceous, 2 — 2.5 mm broad, notched at the ends, free, their margins with 2 rows of bristles; bristles stiff, flattened, divaricately forked, the branches forked in turn into divaricate tufted branchlets, these terminating in forked stiff subfalcate capillary-spinescent setules. Fr. May — June. (Plate XXXIV, Figure 9).

Desert sandhills. — Centr. Asia: Ar.-Casp., Kyz. K., Balkh., Dzu.-Tarb., Kara K. Endemic. Described from the Aral Kara-Kum Desert. Type in Leningrad.

573 46. **C. colubrinum** Borszcz. Mém. Acad. de St.-Pétersb. VII sér., III, 1 (1860) 38. — Ic.: Borszcz. l. c. tab. II, fig. 10 — 10c.

Shrub to 2 m high; bark of mature branches light gray or yellowish; flowers unknown; segments of fruiting perianth reflexed; fruit subspherical, 15 — 17 mm long and 14 — 16 mm broad, rather densely setose; achene coiled anticlockwise or clockwise, obtusish, the ribs extending nearly to the summit; wings thinly coriaceous, narrow, almost united at base, free at summit, divergent at an acute angle; bristles in a single row, flat, stiff, branching from the middle upward into 2 or 3 divaricate branches, these in turn terminating in stiff divaricate almost subulate-spinescent slender branchlets. Fr. May — June. (Plate XXXV, Figure 2).

Desert sandhills and sandy steppes. — Centr. Asia: Ar.-Casp., Kyz. K., Balkh., Dzu.-Tarb. Described from the Aral Kara-Kum Desert. Type in Leningrad.

47. **C. erinaceum** Borszcz. Mém. Acad. St.-Pétersb. VII sér., III, 1 (1860) 37. — Ic.: Borszcz. l. c. tab. II, fig. 11 — 11c.

Shrub to 2 m high; bark of mature branches dark, grayish or brown; pedicels from ocreae mostly in 3's, 1 — 3 mm long, jointed below the middle; perianth segments broad-oval, rounded at apex, purple on the back, broadly white-margined; anthers purple; fruit subspherical, 12 — 14 mm in diameter, densely setose; achene oval-elliptic, coiled clockwise or anticlockwise, the obtuse ribs confluent below the summit, almost decurrent and thus the acutely pyramidal tip of the achene projecting from between the ribs; wings 1 — 2 mm broad, stiffly coriaceous, notched at base, truncate at summit; bristles flat, broad, stiff, divided nearly from base into 2 or 3 straight divaricate branches terminating in a tuft of short stellately spreading branchlets, these in turn terminating in furcate or stellately recurved rather sharp spinules. Fr. May — June. (Plate XXXVI, Figure 4).

Desert sandhills and sandy steppes. — Centr. Asia: Ar.-Casp. Endemic. Described from the Aral Kara-Kum Desert. Type in Leningrad.

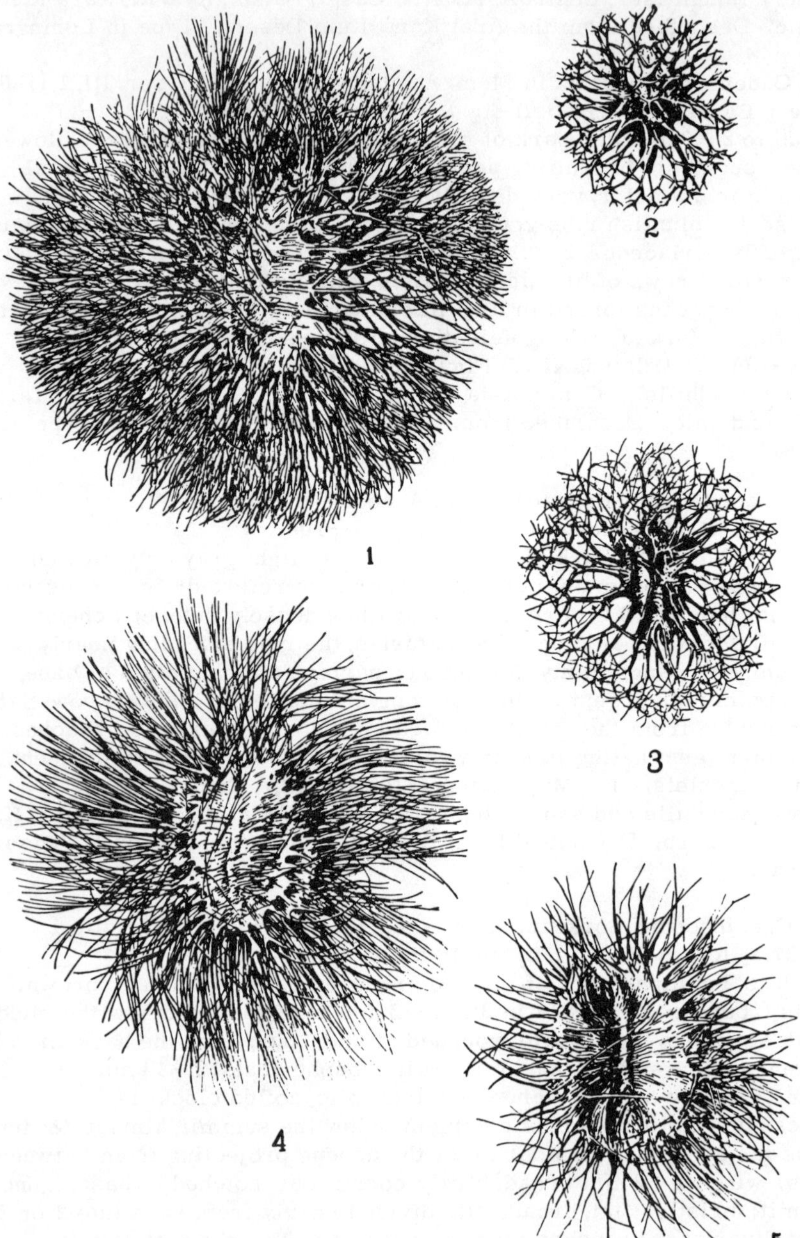

PLATE XXXV. 1. Calligonum pulcherrimum Eug. Kor. – 2. C. colubrinum Borszcz. – 3. C. platyacanthum Borszcz. – 4. C. setosum Litw. – 5. C. cordatum Eug. Kor.

48. **C. squarrosum** N. Pavl. in Fedde, Repert. sp. nov. XXXIII (1933) 152.

A divaricately much branched shrub to 1.5 m high; bark of mature branches light dull gray; flowers unknown; segments of fruiting perianth dark purple, white-margined, horizontally spreading; fruit broad-ovoid, 13—15 mm long and 11—14 mm broad, rather sparsely setose; achene
574 oval-elliptic, coiled clockwise, bluntly ribbed, produced above the base of uppermost bristles into a short alately ribbed cone; wings 2—4 mm broad, stiffly coriaceous, reddish, notched at the base of the achene, truncate at summit, not united with the base of the style; bristles flat, broad, stiff, irregularly branched above the middle into 2 or 3 spreading strongly flattened branches, these terminating in very short divaricate often curved lurid subulate sharply spinescent branchlets. Fr. May—June. (Plate XXXVI, Figure 5).

Desert sandhills and sandy steppes. — Centr. Asia: Kyz. K. Endemic. Described from the sands of Kyzyl-Kum. Type in Moscow; cotype in Leningrad.

49. **C. platyacanthum** Borszcz. in Mém. Acad. St.-Pétersb. VII sér. III, 1 (1860) 37. — Ic.: Borszcz. l. c., tab. II, fig. 9—9c. — Exs.: HFR No. 2410, 2410a.

Shrub to 2 m high; bark of mature branches gray; pedicels from the ocreae in 3's, unequal, 1—3 mm long, jointed below the middle; perianth segments subequal, broad-ovate, rounded at apex, 3 mm long, green or greenish-purple on the back, broadly white-margined; anthers pink; fruit 22—24 mm long and 20—22 mm broad, obovate in outline, sparsely and asymmetrically setose; achene oval, coiled clockwise, the inconspicuous rather sharp ribs narrowly 2-winged; wings coriaceous, distant, free, deeply and sharply incised-toothed nearly to the achene; bristles in a single row, curved, broad and flat at base, stiffish, elongated, often confluent at base, forking into 2 or 3 branches, these paniculately divaricately and almost 1-sidedly branched in turn, the secondary branches straight flexible and long. Fr. May— June. (Plate XXXV, Figure 3).

Desert sandhills. — Centr. Asia: Ar.-Casp., Kyz. K., Kara K., Amu D. Endemic. Described from the Aral Kara-Kum Desert. Type in Leningrad.

50. **C. Muravljanskyi** N. Pavl. in Fedde, Repert. sp. nov. XXXIII (1933) 153.

Shrub to 1 m high; bark of mature branches light dull gray; flowers unknown; segments of fruiting perianth spreading; fruit including bristles
577 cordate-oval in outline, 23—25 mm long and 22—24 mm broad, almost 4-angled, very loosely setose; achene scarcely to slightly coiled clockwise, fusiform, sharp-ribbed, for most of its length rather broadly 2-winged, produced at summit into a short alately ribbed cone; wings stiffly coriaceous, reddish, with thick transverse nerves, surpassing the achene, almost flat or slightly folded longitudinally, broader and cordately notched at base, attenuate to a point at summit; wing margin deeply and unequally incised nearly to half the width of the wing, the incisions irregularly divided and prolonged into simple or 2- or 3-pronged elongated subulate-tipped prickly bristles equaling or exceeding the width of the wing; wing

surface with a single row of bristles along the margin or at the base of the incisions; surface bristles spreading, subterete, simple or with 1 or 2 branches, spiniform, shorter than or as long as the margin bristles. Fr. April—May. (Plate XXXIV, Figure 7).

Desert sandhills. — Centr. Asia: Kyz. K. Endemic. Described from the sands of Kyzyl-Kum. Type in Moscow; cotype in Leningrad.

51. **C. rotula** Borszcz. in Mem. Acad. St.-Pétersb. VII sér. III, 1 (1860) 35. — Ic.: Borszcz. l. c. tab. 1, fig. 6 — 6g. — Exs.: HFR No. 390.

Shrub 1 — 1.5 m high; bark of mature branches dull gray; flowers unknown; segments of fruiting perianth reflexed; fruit cordate-cylindric in outline, 18 — 20 mm long and 25 — 27 mm broad, rather loosely setose; achene slightly coiled anticlockwise, subcylindric, tapering at both ends, 4-ribbed; wings sharp, broad-winged at the middle, abruptly evanescent at both ends; wings stiffly subcoriaceous, truncate at both ends, almost quadrangular, spreading from the ribs at a right angle, the adjacent parallel and distant; wing lobes united at summit and at base slightly or strongly and almost cupuliformly; marginal bristles flattish, broad, short, stiff, irregularly branched with short spreading spinescent endings; surface bristles elongated, subterete, shortly declinate-branched with very stiff spreading subulate-spinescent endings. Fr. May—June. (Plate XXXIV, Figure 8).

Desert sandhills. — Centr. Asia: Ar.-Casp., Balkh., Kyz. K., Kara K. Endemic. Described from the Aral Kara-Kum Desert. Type in Leningrad.

52. **C. pulcherrimum** Eug. Kor. ex N. Pavl. in Fedde, Repert. sp. nov.
578 XXXIII (1933) 153.

A spreading shrub to 2 m high; mature woody branches refracted-flexuous, divaricately branched, nodose, covered with dull whitish bark; branches of the current year creamy white, flexuous, slightly thickened at the nodes; branchlets of the current year mostly in clusters of 2 or 3 at the nodes, the roundish lenticels rather dense, the wood light sulfureous; flowers unknown; segments of fruiting perianth reflexed; fruit spherical in outline, large, together with bristles to 40 mm in diameter, very densely setose, the tangled dark honey-red bristles concealing the achene; achene oblong-oval, 12 — 13 mm long and 6 — 6.5 mm broad, cuneate, slightly coiled anticlockwise, ribbed, convex between the ribs; wings stiffly membranous, as broad as the achene, triangular-ovate, scarcely united with the base of the style, the margin covered with soft bristles twice the width of the wing and with stiff inward-pointing prickles; surface bristles densely disposed in a single row along the margin; wing lobes slightly notched at base, notched-toothed at summit; marginal bristles divaricate-branched above the base; surface bristles strongly paniculate-branched almost from base, in lower part dilated and flattened, in upper part capillary. Fr. May—June. (Plate XXXV, Figure 1).

Desert sandhills. — Centr. Asia: Kara K. Endemic. Described from the Kara-Kum Desert in Turkmenia. Type in Tashkent.

53. **C. setosum** Litw. in Trav. Mus. Bot. Ac. Sc. XI (1913) 57; Eug. Kor. Sched. ad Herb. Fl. Asiae Mediae X (1926) No. 228. — C. acanthopterum var. setosum Litw. Sched. ad. Herb. Fl. Ross. II (1900) 28. — Ic.: Litw. l. c. tab. 8, fig. 18. — Exs.: HFR No. 285; B. Fedtschenko Herb. Turkest. No. 33; H. F. A. M. No. 228.

A rather low shrub, barely reaching 1 m in height, spherically branching; bark of mature branches whitish; branches nodose, refracted-flexuous, with internodes to 3 cm long; branchlets of the current year numerous, some floriferous, finally all deciduous; leaves filiform, to 5 mm long, promptly deciduous; flowers whitish-green, remote, in 1 or 2 fascicles per branchlet; pedicels of upper flowers to 5 mm long, jointed below the middle; fascicles mostly particolored; perianth segments unequal; the inner broad, obovate, white with a green dorsal stripe, 3 mm long; the outer smaller, ovate; all reflexed in fruit; anthers rounded-ovate, yellow; fruit excluding bristles oval in outline, 15—20 mm long and 12—17 mm broad; achene more or less coiled, sharp-ribbed; wings not united with the base of the style, stiff, slightly surpassing the achene, prominently transverse-nerved, the margin studded on all sides with long stiff subulate
579 often twice forking bristles equaling or exceeding the width of the achene; surface bristles in 1 or 2 (rarely 3) rows, equaling or more often exceeding the marginal ones, always forking. Fr. May—June. (Plate XXXV, Figure 4).

Desert sandhills and sandy steppes. — Centr. Asia: Kara K., Kyz. K., Amu D. Endemic. Described from the Kara-Kum Desert in Turkmenia. Type in Leningrad.

54. **C. cordatum** Eug. Kor. ex N. Pavl. in Fedde, Repert. sp. nov. XXXIII (1933) 154. — C. Litwinowii Eug. Kor. olim (non Drob. 1916).

A spreading shrub to 2 m high; mature woody branches declinate, flexuous, nodose, covered with rather dark brownish-yellow slightly fissured bark; branches of the current year lighter-colored, clustered at the nodes, when fully grown to 40 cm long, refracted-flexuous near the ends, covered with creamy white bark, horizontally branching, rather densely lenticellate, the wood sulfureous; flowers unknown; segments of fruiting perianth reflexed; fruit excluding bristles cordate-oval, to 18 mm long and 16—17 mm broad; achene oval-oblong, 9 mm long and 4—4.5 mm broad, cuneate, slightly coiled anticlockwise, sharp-ribbed; ribs 4 or sometimes 3, broad; wings stiffly membranous, lustrous, honey-rose-colored, oval-triangular, subcordate at base, flat or slightly involute at the margins, with stiff anastomosing transverse nerves, united with the base of the style; wing surface convex; surface bristles in a single row near the margin; flexuous, simple or slightly forking at the tips; wing margin bluntly incised-toothed, the teeth terminating in short or fairly long stiffish prickles. Fr. May—June. (Plate XXXV, Figure 5).

Desert sandhills. — Centr. Asia: Kara K. Endemic. Described from the Kara-Kum Desert in Turkmenia. Type in Tashkent.

55. **C. kzyl-kumi** N. Pavl. in Fedde, Repert. sp. nov. XXXIII (1933) 154.

Shrub to 2 m high; bark of mature branches whitish; herbaceous branchlets slender, sulcate, solitary, declinate, jointed; joints 2—5 cm long, terminating in a narrow membranous cup-shaped ocrea; leaves filiform, 2—9 mm long, acuminate, united with ocreae, readily deciduous; wood white; flowers unknown; segments of fruiting perianth reflexed; fruit oval, slightly cordate at base, winged, including bristles 30—36 mm in diameter; achene oval-oblong, 9 mm long and 4—5 mm broad, cuneate,
580 slightly coiled anticlockwise; thinly and sharply ribbed; ribs 4, rarely 3,

(581)

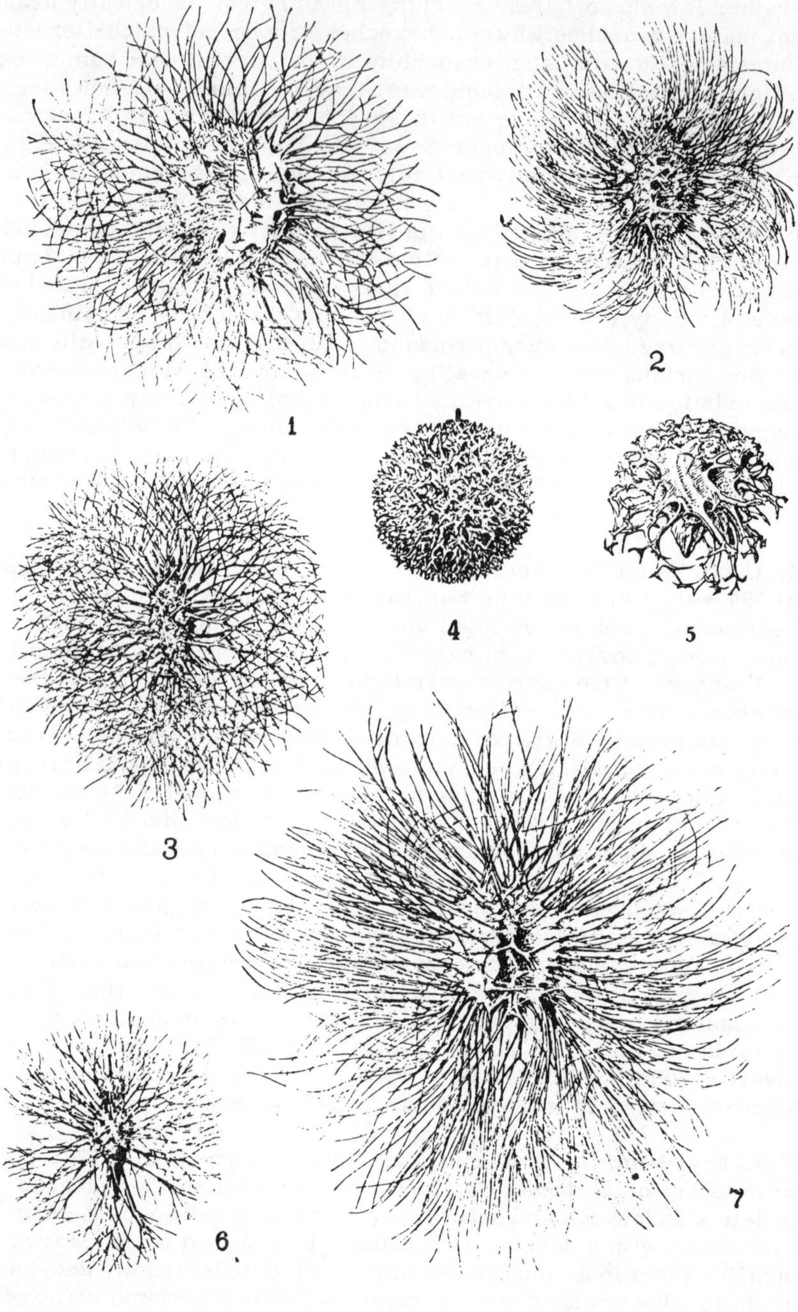

PLATE XXXVI. 1. Calligonum kzyl-kumi N.Pavl. – 2. C.molle Litw. – 3. C.eriopodum Bge. – 4. C.erinaceum Borszcz. – 5. C.squarrosum N.Pavl. – C.orthotrichum N.Pavl. – 7. C.Paletzkianum Litw.

rather broad; wings lustrous, at first sanguineous, finally yellowish, oval-triangular, 15 mm long and 9—10 mm broad, adnate to the base of the style; wing surface slightly crinkled, distinctly transversely reticulate, with 1 or rarely incomplete 2 rows of bristles near the margin, the bristles long, stiffish, dilated at base, forking from the middle, the tips of the branches stiffish and nonbrittle; wing margin doubly toothed, the teeth 2—4 mm long, subulate, prickly, sometimes furcate. Fr. May—June. (Plate XXXVI, Figure 1).

Desert sandhills and sandy steppes. — Centr. Asia: Kyz. K. Endemic. Described from the sands of Kyzyl-Kum. Type in Moscow.

56. **C. molle** Litw. in Trav. Mus. Bot. Ac. Sc. XI (1913) 58. — Ic.: Litw. l. c. tab. 8, fig. 20.

A shrub of medium size, to 1.5 m high; bark of mature branches whitish; flowers unknown; segments of fruiting perianth reflexed; fruit almost round, to 25 mm in diameter including bristles, rather densely setose; achene oval, 10 mm long and 8 mm broad, slightly coiled; ribs rather sharp, rather inconspicuous; wings 2—3 mm broad; surface bristles dense, in 4 rows, long, dilated and sometimes united at base, in upper part capillary, forked, with straight upright branches; marginal bristles short, subulate or acicular. Fr. May—June. (Plate XXXVI, Figure 2).

Desert sandhills. — Centr. Asia: Kara K., Amu D. Endemic. Described from the vicinity of Repetek station. Type in Leningrad.

57. **C. Paletzkianum** Litw. in Trav. Mus. Bot. Ac. Sc. XI (1913) 57. — Ic.: Litw. l. c. tab. 8, fig. 19.

Shrub to 3 m high; bark of mature branches yellowish; flowers unknown; segments of fruiting perianth reflexed; fruit excluding bristles 12—15 mm long and 10—13 mm broad; achene slightly coiled, the obtuse ribs inconspicuous; wings (or bristles alately united at base) stiff, prominently nerved, always united with the base of the style, 3—5 mm broad, naked except for
583 a single row of rather sparse bristles along the margin; bristles 15—30 mm long, flat at base, in upper part stiffly capillary, those at the summit of the fruit longer and denser, almost tufted. Fr. May—June. (Plate XXXVI, Figure 7).

Desert sandhills. — Centr. Asia: Kara K. Endemic. Described from the vicinity of Repetek station. Type in Leningrad.

Section 3. **EUCALLIGONUM** Borszcz. in Mém. Acad. St.-Pétersb. VII, sér. III, 1 (1860) 39. — C a l l i g o n u m Endl. Gen. pl. (1836—40) 308 (excl. C. p o l y g o n o i d i). (Characters in key to sections).

1. Fruit including bristles 20—30 mm in diameter . . . . 2.
+ Fruit including bristles to 15 mm in diameter . . . . 5.
2. Fruit quite spherical; bristles covering the fruit uniformly on all sides; terminal branches of bristles divaricate, tangled, spiniform; bristles fairly loose with achene showing through or dense and almost concealing the achene . . . . . 70. **C. caput Medusae** Schrenk, s. l.
+ Fruit more or less oval; bristles partly crowded about the somewhat elongated summit of the achene, united at base or tufted . . . . 3.

3. Achene densely covered with truncate crinkled chafflike scales; pedicels and the back of perianth segments densely papillose; fruit 25—30 mm in diameter, with flexible capillary bristles . . . . 58. **C. eriopodum** Bge.

\+ Achene not chaffy; pedicels and perianth segments glabrous . . . . . . 4.

4. Fruit oblong-oval; bristles at the base of the achene often absent, at the summit very long, rather stiff, brittle, yellowish, with strongly divaricate branches, the terminal branches somewhat tangled; stem arborescent . . . . 69. **C. arborescens** Litw.

\+ Fruit almost round; bristles at the summit of the achene barely longer than the rest, yellow or rufescent, very slender and soft, with almost upright branches, the terminal branches upright, not tangled; a much branched shrub . . . . 59. **C. orthotrichum** N. Pavl.

5. Bristles in 12—16 rows, soft, slender, flexuous, forked with upright branches . . . . 60. **C. turkestanicum** (Eug. Kor.) N. Pavl.

\+ Bristles in 4—8 (rarely 10) rows, mostly rather stiff, divaricately branched . . . . 6.

6. Fruit spherical, very small, to 10 mm in diameter . . . . 7.

\+ Fruit ovaloid, rarely spherical, larger, 12—15 mm long . . . . 8.

584 7. Achene very strongly coiled for almost one and a half turns; ribs very prominent, rounded, carinate; bristles in 8 rows, branched almost from base, dense and concealing the achene, rarely sparse . . . . 61. **C. microcarpum** Borszcz.

\+ Achene slightly coiled; ribs less prominent, sharp-angled; bristles branched from the middle or higher up, mostly in 4—6 rows, very loosely disposed with achene showing through . . . . 62. **C. pellucidum** N. Pavl.

8. Bristles loosely disposed, their terminal branches rather loose, not concealing the achene . . . . 9.

\+ Bristles loosely disposed, their terminal branches tangled and almost eompletely concealing the achene . . . . 12.

9. Bristles at the summit of the achene elongated, tufted, united at base into membranous longitudinal crests terminating the achene; bristles very diffuse . . . . 10.

\+ Bristles at the summit of the achene crowded, not elongated and not united at base, the achene projecting as a short ribbed cone . . . . 11.

10. Ribs of achene conspicuous; bristles soft, delicate, slender, capillary; stem arborescent . . . . 63. **C. elatum** Litw.

\+ Ribs of achene rounded, inconspicuous; bristles rather stiff, brittle, the terminal branches spiniform; a branching shrub . . . . 63. **C. triste** Litw.

11. Ribs of achene sharp; bristles soft, delicate, capillary, rather dense; terminal branches of bristles tangled . . . . . . 64. **C. griseum** Eug. Kor.

\+ Ribs of achene broad and bluntish; bristles stiffish, brittle, rather sparse; terminal branches of bristles subulate or spiniform, not tangled . . . . 68. **C. ferganense** N. Pavl.

12. Bristles shorter than the width of the achene, very thick, stiff and brittle, often united at base in groups of several, branched only at the ends; terminal branches of bristles forked, very short, thick, conoidal, obtuse-tipped; fruit ovaloid . . . . 66. **C. murex** Bge.

+ Bristles exceeding the width of the achene, slender, not brittle, free or rarely some approximate or united at base, branched from the middle or from above it; terminal branches of bristles rather elongated, slender, acute, subulate or spiniform . . . . 13.

585 13. Fruit ovaloid, longer than broad . . . . 65. **C. Litwinowii** Drob.

+ Fruit spherical . . . . 70. **C. caput Medusae** Schrenk.

58. **C. eriopodum** Bge. in Mém. Acad. St.-Pétersb. sav. étrang. VII (1851) 486. — Ic.: Borszcz. Mém. Acad. St.-Pétersb. VII sér. III, I (1860) tab. III, fig. 13—13d. — Exs.: HFR No. 391; H. F. A. M. No. 227.

An arborescent shrub to 5 m high; bark of mature branches whitish-gray, rarely reddish; branchlets of the current year and leaves often covered with short papillose hairs; flowers axillary, in pairs, rarely solitary; pedicels as long as or 1.5—2 times as long as the perianth; perianth segments 3 mm long, pink, broad-ovate, obtuse, broadly white-margined at apex, reflexed in fruit, like the pedicels papillose-puberulent; anthers pink; fruit subspherical, 25—30 mm in diameter including bristles, diffusely setose; achene subelliptic, similarly narrowed at both ends, slightly coiled clockwise, often profusely covered with truncate imbricated glumaceous scales, the ribs obsolescent, obtuse; bristles in 8 rows, quite free, flexible, flattish, 2—4-branched from base, the branches terminating in long capillary flexible spreading or recurved branchlets. Fr. May—June. (Plate XXXVI, Figure 3).

Desert sandhills and sandy steppes. — Centr. Asia: Ar.-Casp., Kara K., Amu D. Endemic. Described from the Kyzyl-Kum Desert. Type in Leningrad.

59. **C. orthotrichum** N. Pavl. in Fedde, Repert. sp. nov. XXXIII (1933) 155.

A much branched shrub to 0.5 m high; bark of mature branches light grayish; flowers unknown; segments of fruiting perianth reflexed; fruit including bristles oval in outline, 20—22 mm long and 18—20 mm broad, loosely setose; achene fusiform, 9—10 mm long and 3—4 mm broad, slightly coiled clockwise, rather inconspicuously and bluntly ribbed, produced above the base of apical bristles into a short cone; bristles in 8 rows, quite free, subterete or at base slightly dilated and flattened, slender, soft, rufescent, forked from the middle or from higher up into strict upright capillary branches; bristles at the base of achene shorter, those at the summit longer and gathered in a dense tuft. Fr. May—June. (Plate XXXVI, Figure 6).

Desert sandhills. — Centr. Asia: Amu D. Endemic. Described from "Bukhara." Type in Moscow; cotype in Leningrad.

586

60. **C. turkestanicum** (Eug. Kor.) N. Pavl. in Fedde, Repert. sp. nov. XXXIII (1933) 155. — C. c o m o s u m var. t u r k e s t a n i c u m Eug. Kor. Sched. ad Herb. Fl. Asiae Mediae X (1926). 3. — C. c o m o s u m auct. fl. turkest. (non L'Herit.). — Exs.: HFR No. 484; H. F. A. M. No. 226a, 226b.

Shrub to 1 m high; bark of mature branches whitish; herbaceous branchlets slender, densely covered with flowers at the ends; flowers axillary, mostly in pairs; pedicels jointed at base, glabrous, 2—3 mm long, as long as or longer than the perianth; perianth segments subequal, broad-oval, obtusish and membranous-margined at apex, 2 mm long, white or whitish-green or faintly violet; anthers pink; fruit oval or almost round in outline, 12—13 mm in diameter including bristles, rather loosely setose; achene slightly coiled anticlockwise, linear-elliptic, enlarged at the middle, cylindrically tapering toward both ends, the ribs bluntish rounded; bristles in 12—16 rows, never dilated at base, quite free, soft, divaricately branched from base, the branches many-branched in turn, the branchlets long and capillary; fruit yellow (var. flavum Eug. Kor.) or, especially when young, red (var. rubicundum Eug. Kor.). Fr. May—June. (Plate XXXVII, Figure 4).

Desert sandhills. — Centr. Asia: Kara K., Amu D. **Gen. distr.**: Iran. Described from the Kara-Kum Desert in Turkmenia. Type in Tashkent.

61. **C. microcarpum** Borszcz. in Mém. Acad. St.-Pétersb. VII sér. III, I (1860) 41. — C. minimum Lipsky, Zap. Kievsk. Obshch. Estestv. XI, 2 (1891) 59. — Ic.: Borszcz. l. c. tab. III, fig. 14—14c. — Exs.: HFR No. 485.

A low shrub, ca. 0.5 m in height, rarely up to 3 m; bark of mature branches grayish or reddish; herbaceous branchlets slender, densely covered with flowers; flowers axillary, mostly in groups of 5; pedicels jointed below the middle, glabrous, as long as or 1.5 times as long as the perianth; perianth segments subequal, broadly obovate, whitish-green, to 2 mm long; anthers pink; fruit ovaloid to subspherical, 7—9 mm in diameter, rather densely setose or rarely loosely setose with achene showing through the bristles (var. pellucidum N. Pavl.); achene spirally coiled clockwise or anticlockwise, linear-elliptic, enlarged at the middle, cylindrically tapering toward both ends, the rounded bluntish ribs keeled; bristles in 8 rows, quite free, stiffish, narrow, flattened, divaricately and almost 1-sidedly 4—6-branched, the branches repeatedly branched into
587 long capillary branchlets. Fr. May—June. (Plate XXXVII, Figure 1).

Desert sandhills. — Centr. Asia: Ar.-Casp., Kyz. K., Kara K., Amu D. Endemic. Described from the Aral Kara-Kum Desert. Type in Leningrad.

62. **C. pellucidum** N. Pavl. in Fedde, Repert. sp. nov. XXXIII (1933) 157.

Shrub to 2 m high; strongly and loosely branched; bark of mature branches dark gray, rough; herbaceous branchlets clustered, slender, grayish-green, simple, furrowed when dry, glabrous; leaves very small, 1—2 mm long, subulate, united with ocreae, deciduous; ocreae membranous, whitish, sheathing, slightly lacerate; pedicels commonly in pairs from the ocreae, glabrous, 1—2 mm long, jointed below the middle; flowers unknown; segments of fruiting perianth reflexed; fruit including bristles round in outline, 10 mm in diameter, rather loosely setose; achene fusiform, slightly coiled clockwise, prominently sharp-ribbed, produced into a short ribbed cone above the base of uppermost bristles; bristles rufescent, mostly in 4 (rarely 6—8) rows, rather stiff, brittle, dilated and flattish at base, free or some occasionally united, divaricately 2- or 3-branched from the middle or from higher up, the branches terminating in short slender branchlets. Fr. May — June. (Plate XXXVII, Figure 2).

Desert sandhills. — Centr. Asia: Amu D. Endemic. Described from the vicinity of Farab. Type in Leningrad.

63. **C. elatum** Litw. in Trav. Mus. Bot. Ac. Sc. XI (1913) 59; Litw. Sched. ad Herb. Fl. Ross. VIII (1922) 3. — Ic.: Litw. l. c. tab. 8, fig. 17. — Exs.: HFR No. 2417, 2417a.

Arborescent shrub to 7 m high; bark of woody mature branches light grayish; young herbaceous branchlets glabrous or scaberulous, strongly branched; ocreae sheathing, pellucid; leaves united with ocreae, small, to 4 mm long, setiform, deciduous; pedicels in fascicles of 2—4, glabrous, unequal, 1—3 mm long; perianth segments 3 mm long, ovate, rounded at apex, sanguineous, broadly white-margined, reflexed in fruit; anthers purple; fruit including bristles oval in outline, 10—12 mm long and 9—10 mm broad, diffusely setose; achene strongly spirally coiled, ovaloid, attenuate and tapering toward summit, the obtuse rounded ribs wingless; bristles in 8 rows, yellowish, capillary, soft, with upright branches, free or mostly at the summit of the achene the free bases approximate and forming barely perceptible crests, the terminal bristles somewhat
588 elongated. Fr. July—August. (Plate XXXVII, Figure 3).

Desert sandhills. — Centr. Asia: Ar.-Casp., Kara K., Amu D. Endemic. Described from the Syr Darya River valley, from Chiili station. Type in Leningrad.

64. **C. griseum** Eug. Kor. ex N. Pavl. in Fedde, Repert. sp. nov. XXXIII (1933) 156.

A low loosely branched shrub, 0.5 m in height; woody mature branches glabrous, often nodding or procumbent and then assurgent, strongly nodose, covered with blackish-gray bark; herbaceous branchlets in clusters of 10—15, divaricate, glaucescent, jointed, the joints terminating in sheathing membranous brownish erose-toothed ocrea; pedicels from ocreae mostly in groups of 2 or 3, nodding, glabrous, 2 mm long, jointed below the middle; bark rather densely covered with minute rounded lenticels, the wood yellowish; flowers unknown; segments of fruiting perianth reflexed, ovate; fruit including bristles obovate in outline, 13—15 mm long and 8—10 mm broad; achene oblong-fusiform, 8.5 mm long and 4.2 mm broad, rather gently coiled anticlockwise, ribbed nearly to the base of the style, produced at summit above the base of uppermost bristles into a short cone, the grooves between the ribs impressed; ribs at the base of the achene obsolescent, somewhat prominent above, bluntish and not furrowed, loosely setose; bristles in 2 rows, slender, brittle, slightly dilated and flattish at base, yellowish, at length turning faintly reddish, quite free, only those at the summit of the achene elongated crowded and some united at base, all bristles divaricately forked from the middle or from higher up. Fr. May—June. (Plate XXXVII, Figure 5).

Desert sandhills and stony slopes. — Centr. Asia: Kyz. K., Amu D. Endemic. Described from Bukhara. Type in Tashkent.

65. **C. Litwinowii** Drob. in Trav. Mus. Bot. Ac. Sc. XVI (1916) 140. — Exs.: HFR No. 2786.

Shrub 1—2 m high; bark of woody branches white or faintly grayish; herbaceous branchlets slender; leaves very small, to 1 mm long, short-pointed at apex, united with ocreae. Pedicels from ocreae solitary or in pairs, unequal, 1—3 mm long, jointed below the middle; perianth segments broad-ovate, 3 mm long, obtuse, green, broadly white-margined; anthers

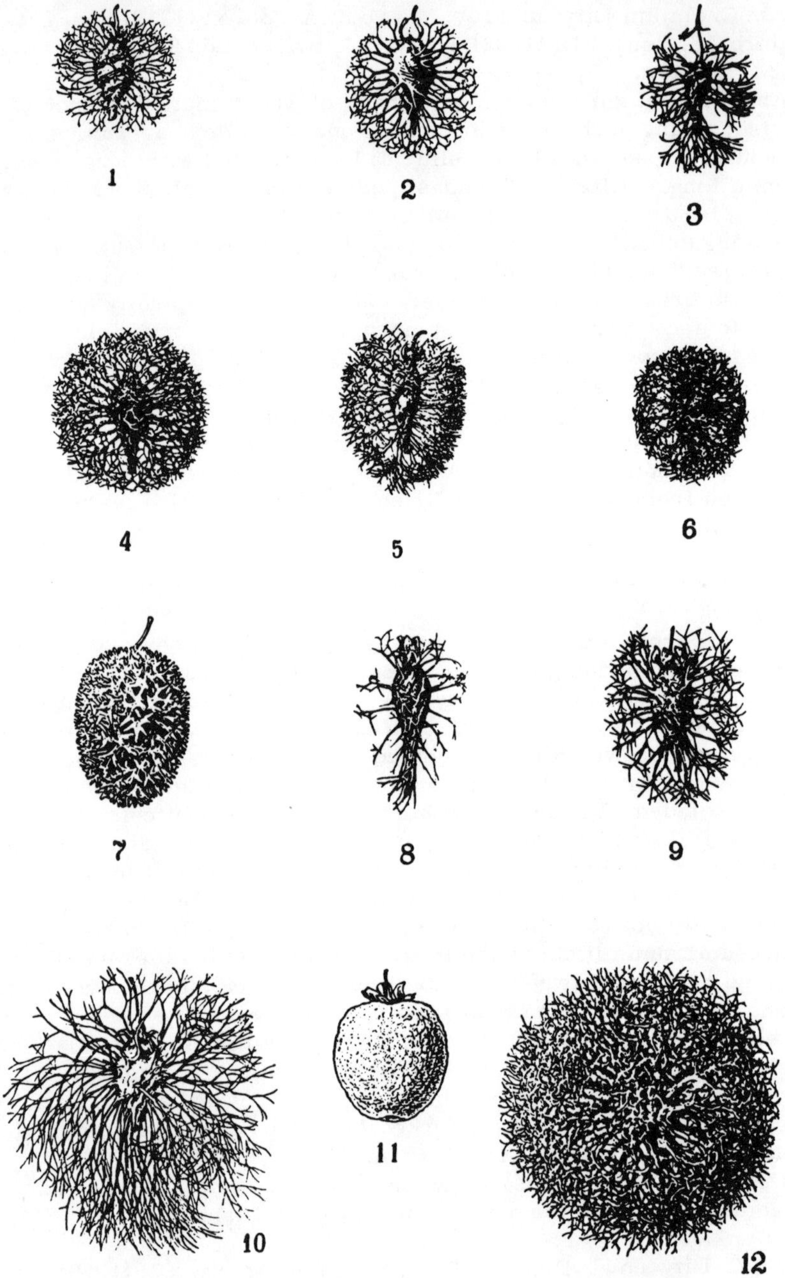

PLATE XXXVII. 1. Calligonum microcarpum Borszcz. – 2. C. pellucidum N. Pavl. – 3. C. elatum Litw. – 4. C. turkestanicum (Eug. Kor.) N. Pavl. – 5. C. griseum Eug. Kor. – 6. C. Litwinowii Drob. – 7. C. murex Bge. – 8. C. triste Litw. – 9. C. ferganense N. Pavl. – 10. C. arborescens Litw. – 11. C. junceum (Fisch. et Mey.) Litw. – 12. C. caput Medusae Schrenk.

591 pink; fruit oval in outline, 9—12 mm long and 7—9 mm broad, rather densely setose; achene elliptic, 7—8 mm long, 3—4 mm broad, slightly coiled anticlockwise, bluntly ribbed, produced above the base of uppermost bristles into a short ribbed cone; bristles in 8—10 rows, rather stiff, not brittle, flattish and dilated at base, free or occasionally some united at base but not forming wings, above the middle about the upper third divaricately branched, the slender branches reddish-yellow or brownish. Fr. May—June. (Plate XXXVII, Figure 6).

Desert sandhills. — Centr. Asia: Syr D. Endemic. Described from sands around Mel'nikovo station in the Kokand area. Type in Leningrad.

66. **C. murex** Bge. in Mém. Acad. St.-Pétersb. sav. étrang. VII (1851) 486. — C. *paniculatum* Borszcz. in Mém. Acad. St.-Pétersb. VII, sér. III, 1 (1860) 41. — Ic.: Borszcz. l. c. tab. III, fig. 15—15e. — Exs.: HFR No. 2411, 2412.

Shrub to 2 m high; bark of mature branches dark gray; flowers unknown; segments of fruiting perianth reflexed; fruit oval in outline, 12—14 mm long and 9—10 mm broad, rather densely setose; achene elliptic-oval, coiled anticlockwise, bluntly ribbed; bristles in 8 rows, stiff, stout, dilated toward apex and toward base, sometimes united in groups of several, branched at the ends; branches short, stout, membranous-margined and again divaricately 2- or 3-branched, the endings obtusish; fruit reddish (var. *typicum* Litw.) or yellowish (var. *pallidum* Litw.) Fr. May—June. Plate XXXVII, Figure 7).

Desert sandhills. — Centr. Asia: Ar.-Casp. Endemic. Described from the Aral Kara-Kum Desert. Type in Leningrad.

67. **C. triste** Litw. in Trav. Mus. Bot. Ac. Sc. XI (1913) 58; Litw. Sched. ad Herb. Fl. Ross. VIII (1922) 158. — Ic.: Litw. l. c. (1913) tab. 8, fig. 16. — Exs.: HFR No. 2416.

A divaricately branched shrub to 0.75 cm high; woody mature branches pale, with corky-fissured bark; young herbaceous branchlets clustered, simple or slightly branched, bright green, striate-furrowed when dry, the internodes long; ocreae hyaline, half-clasping; flowers in ocreae solitary or in pairs; pedicels 3—5 mm long, jointed below the middle; perianth segments obovate, ca. 4 mm long, broadly white-margined, brownish or reddish on the back, in fruit mostly spreading; anthers pink; fruit oval in outline including bristles, 12—15 mm long and 8—9 mm broad, loosely
592 setose; achene strongly coiled, oblong-oval, 3—4 mm broad, the rather inconspicuous ribs round-obtuse; bristles in 2 rows, yellow, quite free at base, capillary, at the summit of the achene elongated and tufted, elsewhere sparse, spreading, stiff, brittle, branched at a right angle from the middle or from higher up, with sharp subulate endings. Fr. June—July. (Plate XXXVII, Figure 8).

Desert slopes and outcrops of gypsiferous clay. — Centr. Asia: Ar.-Casp. Endemic. Described from the Aral Kara-Kum Desert. Type in Leningrad.

68. **C. ferganense** N. Pavl. in Fedde, Repert. sp. nov. XXXIII (1933) 157.

Shrub to 2 m high, strongly and loosely branched; bark of mature branches light reddish; herbaceous branchlets slender, glaucescent, subequal, strongly forked, furrowed when dry, glabrous; leaves very small, to 1 mm long, subulate, united with ocreae, deciduous; ocreae membranous, brownish, sheathing, the margins slightly lacerate; pedicels glabrous or scaberulous, usually in pairs from the ocreae, 1—3 mm long, jointed below the middle; perianth segments broad-ovate, rounded, 3 mm long, green, broadly white-margined, reflexed in fruit; anthers pink; fruit including bristles ovate in outline, 12—15 mm long and 10—12 mm broad, rather loosely setose; achene slightly coiled anticlockwise, with blunt ribs, produced above the base of uppermost bristles into a ribbed cone; bristles lemon-yellow or rarely rufous and then the achene coiled clockwise (var. rufescens N. Pavl.), in 8 rows, stiffish, brittle, dilated and flattish at base, free or occasionally some united, divaricately 2- or 3-branched from the middle or from higher up or rarely from below the middle, the declinate branches branched in turn into slender somewhat elongated branchlets. Fl. and fr. September—October. (Plate XXXVII, Figure 9).

Desert sandhills. — Centr. Asia: Syr D., Amu D. Endemic. Described from the Fergana Valley. Type in Tashkent.

69. **C. arborescens** Litw. Sched. ad Herb. Fl. Ross. II (1900) 28. — Exs.: HFR No. 284.

Arborescent shrub to 3 m high; bark of mature branches whitish, diffusely brownish-tinged; young herbaceous branchlets elongated, in paniclelike clusters from the nodes of and appressed to the woody branches, simple or slightly branched, jointed; joints sulcate, the lower shorter, the
593 others to 3 cm long; ocreae sheathing, scarious-margined, pellucid; leaves small, united with ocreae and terminating in a short brown persistent point; pedicels mostly in 3's from the ocreae, of these two with caducous flowers and one fruiting, glabrous, ca. 3 mm long, jointed below the middle; perianth segments broad-linear, rounded at apex, reflexed in fruit; anthers pink; fruit including bristles ovate in outline, 20—22 mm in diameter; achene strongly coiled anticlockwise, 10 mm long and 5 mm broad, 4-ribbed; ribs prominent, often evanescent toward base, extending at summit to the base of the style, separated by narrow grooves, these sometimes puckered along the bottom; bristles rather sparse, in 8 rows, 2 rows per rib, often absent near the base of the achene, denser and longer at the summit, all yellow, dilated and not united at base, divaricately or furcately branched from the middle or from higher up or occasionally only at the tip, the branchlets elongated and slender. Fr. May—June. (Plate XXXVII, Figure 10).

Desert sandhills. — Centr. Asia: Kara K., Amu D. Endemic. Described from the Kara-Kum Desert in Turkmenia. Type in Leningrad.

70. **C. caput Medusae** Schrenk, Enum. pl. nov. I (1841) 9. — Ic.: Borszcz. in Mém. Acad. St.-Pétersb. VII sér. III, 1 (1860) tab. III, fig. 12—12e; Engl. et Prantl, Natürl. Pflanzenfam. III Abt., Ia (1894) 23. — Exs.: HFR No. 437, 737.

Shrub to 2 m high; bark of mature branches light grayish or pinkish-yellow; leaves small, to 2 mm long, subulate, recurved, membranous-margined at base, united with ocreae; perianth segments ovate, purple, 3 mm long, broadly white-margined, reflexed in fruit; anthers purple; fruit obovoid or often subspherical, (10) 20—25 mm in diameter including bristles, these usually rather dense and concealing the achene; achene elliptic, coiled clockwise, terminating in a cylindric point, rather sharply ribbed; bristles in 8 rows, almost free or some united at base, flattish or subterete, divaricately 3-branched almost from base, these 2- or 3-branched and divided at the ends into spreading stiff spiniform somewhat elongated branchlets; bristles yellow or blood-red (var. rubicundum Herd.). Fr. May—June. (Plate XXXVII, Figure 12).

Desert sandhills and sandy steppes. — Centr. Asia: Ar.-Casp., Balkh., Syr D., Kyz. K., Kara K., Dzu.-Tarb., Amu D. **Gen. distr.**: Dzu.-Kash. Described from the shores of Lake Balkhash. Type in Leningrad.

594 Section 4. CALLIPHYSA Endl. Gen. pl. (1836—40) 308. — Calliphysa (gen.) Fisch. et Mey. Ind. sem. horti Petrop. II (1835) 24 (characters in the key to sections).

71. **C. junceum** (Fisch. et Mey.) Litw. Schedae ad Herb. Fl. Ross. VIII (1922) 9; Kryl., Fl. Zap. Sib. IV, 849. — Calliphysa juncea Fisch. et Mey. Ind. sem. horti. Petrop. II (1835) 24. — C. calliphysa Bge. Delect. semin. horti Dorpat. (1839) VIII; Ldb. Fl. Ross. III, 495. — C. horridum Borszcz. in Mém. Acad. St.-Pétersb. VII sér. III, I (1860) 43. — Ic.: Borszcz. l. c., tab. III, fig. 16—16e.— Exs.: HFR No. 2418.

A strongly branched shrub to 1 m high; bark of mature branches gray or brownish; herbaceous branchlets in clusters of 5 — 10 from nodes of annotinous branches, green, slender, branched, jointed; joints 0.5 —3 cm long, terminating in a short truncate white or brownish membranous ocrea; leaves not united with ocreae, linear to linear-oblong, obtusish, 3—8 mm long, readily deciduous; pedicels 1 —4 from the ocreae, slender, 3—4 mm long, thickened toward apex, jointed below the middle; perianth segments glabrous, ovate, unequal, the outer 3 mm long, green on the back, with a broad white scarious margin; fruit round in outline, rarely somewhat oblong, 8 —9 mm in diameter, membranous-saccate; achene uncoiled, elliptic, long-tapering at apex, 4-ribbed, the broad ribs obscurely keeled; bristles numerous, in 12 rows on the ribs, capillary, soft, slightly enlarged at the tips and supporting a thin utricular membrane investing the fruit on all sides. Fr. May—June. (Plate XXXVII, Figure 11).

Desert sandhills; clayey and sandy steppes. — W. Siberia: Irt.; Centr. Asia: Ar.-Casp., Balkh., T. Sh., Syr D., Dzu.-Tarb., Kyz. K., Kara K. **Gen. distr.**: Mong., Dzu.-Kash. Described from the shores of the Caspian Sea. Type in Leningrad.

Genus 394. **POLYGONUM*** L.**

L. Gen. pl. ed. I (1737) 116.

Perianth of 5 (4—6) segments, often corolloid, slightly accrescent in fruit; stamens 4—8, the outer alternating with perianth segments, the inner opposite the angles of the ovary, this often surrounded at base by a glandular

* From Greek "polygonon," a name occurring in the writings of Dioscorides; "polys," many, abundant, and "gonos," knee, referring to the prominent nodes of many species.

** Treatment by V. L. Komarov, except for the section Aconogonon treated by Y. S. Grigorev. [The reader's attention is drawn to the fact that the Russian terminology in descriptions of Polygonum species differs from the familiar usage. Thus, for instance, there is no term corresponding to ocreoles. To avoid excessive deviation from and possible misinterpretation of the original text, the generally adopted equivalents for such terms as bract, stipule, etc., have been adhered to.]

595 ring with incision alternating with stamens; styles 2 or 3, often united at base; fruit enclosed by the perianth, lenticular or 3-angled; embryo lateral, curved in the seed; cotyledons long, flat, straight. Annual or perennial herbs or rarely undershrubs; stem prostrate, erect, or scandent; leaves often sparsely rough-verrucose or smooth; flowers in few-flowered cymes disposed in racemose or paniculate inflorescences. About 200 cosmopolitan species most widely distributed in warm-temperate (subtropical) lands. Russian name: "gorets."

Key to Sections

1. Leaf blade jointed to petiole; ocrea typically 2-lobed, often with dissected lobes; filaments of stamens, at least the inner ones, dilated . . . . Section 1. **Avicularia** Meisn., p. 458.
\+ Leaf blade not jointed; ocrea not 2-lobed; filaments slender . . . . . . 2.
2. Ocreae cylindric, truncate at apex . . . . 3.
\+ Ocreae oblique, more or less open toward the leaf . . . . 4.
3. Inflorescence capitate; perianth segments 4 or 5; stamens 6—8; ocrea obliquely truncate, not ciliate . . . . Section 2. **Cephalophilon** Meisn., p. 491.
\+ Inflorescence spiciform; perianth segments 3—5; stamens 4—8; ocrea truncate, the margin ciliate, rarely without cilia . . . . Section 3. **Persicaria** Meisn., p. 492.
4. Inflorescence paniculate, spreading or contracted and then the branches appressed; stamens 8; ocrea not ciliate . . . . Section 4. **Aconogonon** Meisn., p. 506.
\+ Inflorescence not paniculate . . . . 5.
5. Rootstock stout, fleshy or woody; floriferous stems simple; leaves mainly radical; inflorescence simple, spikelike; style elongated; ocrea not ciliate . . . . Section 5. **Bistorta** Adans., p. 514.
\+ Rootstock none or slender and not fleshy; stems branched; leaves mainly or exclusively cauline; inflorescence branched . . . . 6.
6. Perianth segments mostly prominently keeled or winged . . . . 7.
\+ Perianth segments not keeled; stems densely echinate on the angles . . . . Section 6. **Echinocaulon** Meisn., p. 525.
7. Stigmas capitate; styles short and erect or none; stems clinging, trailing or scandent . . . . Section 7. **Tiniaria** Meisn., p. 530.
\+ Stigmas broad, toothed; styles divergent; stems sturdy, erect . . . . Section 8. **Pleuropterus** Benth. et Hook., p. 533.

596 Section 1. **AVICULARIA** Meisn. Polyg. Prodr. (1826) 43 et 85, tab. I, f. 36.— Polygonum Tourn. Instit. (1719) 510, tab. 290.— Annuals, perennials and undershrubs; stems terete, often sulcate, erect or prostrate; ocreae white, pellucid, or ferruginous; inflorescences axillary, very small, enclosed before anthesis in the ocreae; perianth segments colored, with a green keel; stamens 8 (rarely 5 or 6); styles 3, very short, sometimes obsolescent; achene included in the dry perianth, trigonous or compressed-trigonous; bracts resembling the ocreae.

1. Perennials with a more or less developed rhizome . . . . 2.
+ Annuals without rhizome . . . . 20.
2. Rhizome strongly developed, woody . . . . 3.
+ Rhizome inconspicuous . . . . 16.
3. Rhizome woody, multicipital; branches whitish, woody . . . . 4.
+ Rhizome single-headed; branches often prostrate, herbaceous . . . . . 7.
4. Undershrubs to 0.5 m high, with erect often branched stems . . . . 1. **P. arianum** Grig.
+ Smaller undershrubs with simple or rarely branched stems . . . . . . 5.
5. Leaves long, broadly lanceolate, the bracteal subulate . . . . 3. **P. luzuloides** Jaub. et Spach.
+ Leaves narrow, lance-oblong to lance-linear, the bracteal abbreviated . . . . 6.
6. The entire plant smooth; stems erect or ascending, often crowded; leaves linear-lanceolate; racemes strongly interrupted, slender . . . . 2. ***P. setosum** Jacq.
+ The entire plant rough; stems relatively short, divergent; leaves lanceolate to lance-oblong; racemes simple, loose . . . . 2. **P. thymifolium** Jaub. et Spach.
7. Rhizome fusiformly enlarged or prolonged into a caudex bearing rather numerous green or reddish stems, these not lignifying at base . . . . 8.
+ Rhizome branched, rather slender; stems flagelliform . . . . 9.
8. Branches short; plants pulvinate, grayish-green; leaves greatly surpassing the ocreae . . . . 5. **P. pulvinatum** Kom.
+ Plants relatively loose, with long or short branches . . . . 10.
9. Stems curved, much branched, apparently densely cespitose; the entire plant often reddish; ocreae 2-parted, terminating in 2 long points . . . . 6. **P. biaristatum** Aitsch. et Hemsley.
597 + Stems with elongated almost trailing branches, their ends densely leafy and beset with large silvery ocreae, hence the plant whitish; ends of branches very brittle and readily disintegrating at the nodes . . . . 4. **P. paronychioides** C. A. M.
10. Ocreae of upper leaves entire or only at apex more or less toothed . . . . 11.
+ Ocreae of upper leaves deeply bifid . . . . 16.
11. Ocreae oval-lanceolate, elongated, separating at the margins into conspicuous fibrous threads; leaves narrow, much longer than the internodes . . . . 7. **P. fibrilliferum** Kom.
+ Ocreae usually without threads or only in age with short threads; leaves shorter and more fleshy than in the preceding species . . . . . 12.
12. Leaves broadly lanceolate, acute, much longer than the internodes, these densely covered with flowers; ocreae broadly oval and point-tipped, scarcely covering the flowers . . . 8. **P. ammanioides** Jaub. et Spach.
+ Leaves distinctly petioled, oblong or oval, with midnerve and no lateral veins . . . . 13.
13. Ocreae narrow, cylindric or lanceolate, leaves yellowish-green, their margins ciliolate; flowers 1—3 in the axils, red . . . . 9. **P. myrtifolium** Kom.
+ Ocreae broad at base, acuminate, with divergent tips . . . . 14.

14. Stems very densely leafy; leaves small, less than 1 cm long, terminating in a spinule; perianth elongate-infundibular, the green segments with rosy or white margins . . . . 10. **P. pamiroalaicum** Kom.

\+ Stems less densely leafy; leaves 1.5 — 3 cm long or longer . . . . 15.

15. Pedicels shorter than perianth . . . . 11. **P. alpestre** C. A. M.

\+ Pedicels, or at least some of them, longer than perianth . . . . 12. **P. rupestre** Kar. et Kir.

16. Ocreae much shorter than the internodes, brownish at base; stems decumbent or erect; leaves oblong, acute; flowers 2 or 3 in the axils, disposed in almost leafless terminal racemes . . . . 13. **P. equisetiforme** Sibth. et Sm.

\+ Ocreae elongated, brown at base, pellucid in upper part; flowers 1 — 3 in the axils, the prostrate stems leafy throughout . . . . 17.

598

17. Ocreae 12-nerved, longer than the internodes; stems decumbent, thickened; achene as long as or but slightly exceeding the perianth . . . . 14. **P. maritimum** L.

\+ Ocreae 6-nerved, much shorter than the internodes; stems more slender, much branched; achene half as long as the perianth . . . . . . 18.

18. Leaves linear to narrowly lanceolate, acute; flowers up to 4 in the axils . . . . 16. **P. oxyspermum** Meyer et Bge.

\+ Leaves lanceolate to elongate-oboval; flowers 1 — 3 in the axils . . . . 19.

19. Leaves more lanceolate; flowers disposed closer to branch tips . . . . 17. **P. Roberti** Lois.

\+ Leaves more oboval; flowers often from the stem base . . . . 15. **P. Rayi** Babingt.

20. Perianth hyaline; ocreae silvery, almost covering the flowers; slender whitish plants . . . . 54.

\+ Perianth herbaceous, green, white, roseate or red; flowers exceeding the ocrea . . . . 21.

21. Leaves rounded at base; inflorescence often paniculate; pedicels about the length of the flower; achene flattened . . . . 31. **P. floribundum** Schlecht.

\+ Leaves always cuneate at base; achene distinctly 3-angled . . . . 22.

22. Bracteal leaves little smaller than the rest, thus the whole inflorescence leafy; flowers mostly disposed over the entire plant almost from base . . . . 23.

\+ Bracteal leaves much smaller than the rest, thus the inflorescence consisting of leafless raceme or racemes and conspicuously differentiated above the leafy branches . . . . 39.

23. Ocreae deeply bifid . . . . 24.

\+ Ocreae undivided, only at apex more or less lacerate or toothed . . . . 36.

24. Roots relatively stout; stems robust, branched, prostrate or erect, 20 — 60 cm long; leaves more fleshy than in species of the next group, ovate or elongate-lanceolate, obtusely rounded or short-acuminate at apex; achene lustrous . . . . 25.

\+ Roots slender; stems ascending or decumbent, rarely erect, less robust; leaves obovate, lanceolate, or sublinear . . . . 28.

25. Stems erect or in lower part ascending; leaves oblong, obtusish, obscurely veined; flowers 1 — 3 in the axil; achene shorter than perianth . . . . 20. **P. fusco-ochreatum** Kom.

\+ Stem prostrate as a rule . . . . 26.

599 26. Stems rather short, firm; leaves linear-oblong, obtusely rounded or elliptic; flowers 1 or 2 in the axil; achene barely exceeding the perianth . . . . 19. **P. buxifolium** Ldb.
+ Stems long, slender, branched; leaves acute . . . . 27.
27. Leaves oblong, ribbed by the prominent veins; achene shorter than the perianth . . . . 21. **P. lencoranicum** Kom.
+ Leaves oval or linear-lanceolate; flowers 3—6 in the axils; achene shorter than the perianth. . . . . 18. **P. litorale** Link.
28. Plants glaucous, whitened by tubercles; leaves fleshy, oblong-linear, obtuse, without prominent veins; pedicels short; fruit small, with oval faces, finely puncticulate, almost smooth . . . . . . 28. **P. acetosum** M. B.
+ Plants dark green; leaves linear-lanceolate to elliptic; achene broader . . . . 29.
29. Perianth segments as long as or less than twice the length of the basal portion. . . . . 30.
+ Perianth segments more than twice as long as the turbinately fused basal portion of the perianth . . . . 31.
30. Stems mostly prostrate; cauline leaves resembling the rameal; perianth segments less than half the length of the fused basal portion; achene black, ca. 2 mm long, its faces pyramidal from an ovate base . . . . 23. **P. aviculare** L.*
+ Differing from the preceding in the oblong leaves strongly auriculate at base and the green narrow flowers . . . . 29. **P. caspicum** Kom.
31. Achene smooth, lustrous, barely 2—3 mm long. . . . . 32.
+ Achene almost dull, puncticulate-roughened under the microscope . . 33.
32. Achene longer than perianth; flowers 4—10 in the axil . . . . 26. **P. humifusum** Pall.
+ Achene shorter than perianth . . . . 34.
33. Leaves narrow, linear to linear-lanceolate, acute . . . . 26. ***P. graminifolium** Wierzb.
+ Leaves elliptic or oval to oboval. . . . . 24. **P. calcatum** Lindm.
34. Cauline leaves broadly ovate-lanceolate to obovate-lanceolate; rameal leaves slightly more than half as long, lanceolate or oblong; leaves of branchlets much narrower, oblong-lanceolate to linear-lanceolate; perianth segments several times the length of the short turbinate tube . . . . 22. **P. heterophyllum** Lindm.
600 + All leaves more or less alike. . . . . 35.
35. Leaves broadly elliptic or oblong-obovate, rarely narrow to sublinear; achene slightly striate . . . . 25. **P. neglectum** Bess.
+ Leaves oblong, sessile; ocreae shorter than in related species; achene smooth, lustrous . . . . 27. **P. propinquum** Lbd.
++ Leaves tomentose-cobwebby; achene with granular faces and gibbous angles . . . . 27. **P. araraticum** Kom.
36. Leaves oval to lanceolate, the lateral veins prominent beneath . . . . 30. **P. venosum** Steward.
+ Leaves linear-spatulate or linear-subulate . . . . 37.
37. Leaves small, linear-spatulate; stems numerous, procumbent, 5—20 cm long; flowers 4—6 in the axil; pedicels slender, 3—4 times the length of the perianth; perianth segments shorter than the tube; seeds small, smooth, lustrous . . . . 32. **P. corrigioloides** Jaub. et Spach.

* Closely related is a form described from Kharkov as a distinct species, P. acetosellum Klok., No. 28, p. 477.

\+ Flowers 1 or 2 in the axil; pedicels very short; flowers in narrowly cylindric terminal racemes . . . . 38.

38\. Stems rough with minute prickles; leaves linear, sessile, scarcely as long as the ocreae . . . . 34. **P. acerosum** Ldb.

\+ Resembling the preceding, but leaves subulate-linear; flowers 1—2 in the axil; ocreae shorter; perianth and achene ca. 2 mm long . . . . 33. **P. polycnemoides** Jaub. et Spach.

39.(22). Flowers at pollination contracted, cup-shaped . . . . 40.

\+ Flowers at pollination wide open, funnel-shaped . . . . 52.

40\. Plants branching from the base; branches procumbent or ascending, at the ends erect; main stem not evident . . . . 41.

\+ Stems erect from base, more or less branched, the main stem clearly evident among the branches . . . . 43.

41\. Branches varying in length, decumbent, some very short, others long; flowers approximate; perianth segments one-third as long as the tube . . . . 37. **P. Aschersonianum** H. Gross.

\+ Branches slender, all about the same length, ascending . . . . 42.

42\. Pedicels shorter than the perianth; perianth segments shorter than the tube; achene somewhat lustrous . . . . . 36. **P. samarense** H. Gross.

\+ Pedicels longer than perianth; perianth segments longer than the tube; achene dull, punctate; branches of inflorescence very slender . . . . 35. **P. salsugineum** MB.

43\. Fruit 4—5 mm long . . . . 38. **P. Kitaibelianum** Sadl.

\+ Fruit not more than 2—3 mm long . . . . 44.

44\. Stems erect, rather sturdy, with elongated internodes; leaves early deciduous . . . . 45.

601 + Stems more slender, erect, declinate, recurved, or spreading; internodes not elongated; leaves not deciduous before the end of flowering . . . . 46.

45\. Leaves oblong or linear-oblong, acute; pedicels shorter than perianth; achene slightly exceeding the perianth . . . . 46. **P. junceum** Ldb.

\+ Leaves lanceolate to linear-lanceolate, acute; pedicels as long as or longer than perianth; achene enclosed in the perianth . . . . 47. **P. argyrocoleum** Steudel.

46\. Plant whitish or gray, branched from base; branches prostrate; flowers mostly solitary . . . . 41. **P. cretaceum** Kom.

\+ Plants green, with main stem and lateral branches . . . . 47.

47\. Branches slender, upcurved; stems rather densely leafy in lower part; racemes interrupted in lower part, more compact above; flowers 2—3 mm long; young unbranched specimens sometimes occur . . . . 48.

\+ Branches few, often as thick as the stem; leaves narrower; flowers larger . . . . 50.

48\. Branches declinate; leaves oblong to oblong-linear, acute, prominently nerved; flowers 1—3 in the axils; achene finely striate, as long as the perianth . . . . 39. **P. patulum** M. B.

\+ Branches erect; leaves sessile, oblong-linear; flowers in pairs; achene striate only in lower part . . . . 49.

49\. Ocreae tubular, appressed to stem, brown in lower part, paler above; leaves to 4 mm broad; flowers white or roseate . . . . 40. **P. gracilius** (Ldb.) Klokov.

\+ Ocreae ferruginous or brown in lower part; leaves 2 — 6 mm broad; flowers with roseate or red lobes . . . . . 42. **P. novo-ascanicum** Klokov.

50. Branches arched or falcate; flowers 1 — 5 in the axils; achene smooth, lustrous . . . . .43. **P. inflexum** Kom.

\+ Branches straight, erect; flowers 1 — 3 in the axils . . . . 51.

51. Leaves yellowish-green, the rameal much shorter than the cauline; perianth with nerves strongly accrescent in fruit . . . . 45. **P. tiflisiense** Kom.

\+ Leaves green, smaller and more equal than in the preceding, without distinct lateral veins beneath; nerves of perianth enlarging but slightly . . . . 44. **P. oxanum** Kom.

602 52.(39). Stems procumbent or ascending, branched; leaves mostly linear-lanceolate; flowers 3 — 5 in the axils; pedicels longer than perianth; perianth segments white to roseate; achene ca. 2 mm long, lustrous . . . . 49. **P. arenarium** Waldst. et Kit.

\+ Stems ascending or erect, branched; inflorescence paniculate; perianth segments yellowish-white, rarely roseate . . . . 53.

53. Perianth 1.5 — 2.5 mm long; achene lustrous, quite smooth . . . . 50. **P. pseudoarenarium** Klokov.

\+ Perianth 2.5 — 3.25 mm long; achene punctate-roughened, slightly lustrous . . . . 51. **P. Janatae** Klokov.

54. (20). Plants 3 — 7 cm high; stems smooth, brittle, more or less branched; ocreae rather large, silvery-white; leaves setiform; flowers solitary; perianth pellucid; achene oval, flat . . . . .52. **P. molliiforme** Boiss.

\+ Achene white-margined . . . . 53. **P. Bornmülleri** Litw.

Series 1. **Fruticulosae** Boiss. Fl. Or. IV (1879) 1026. — Rhizome woody; stems numerous; branches more or less woody.

1. **P. arianum** Grig. in Acta Inst. Bot. Ac. Sc. ser. I, 1 (1933) 104.

Shrub 20 — 40 cm high, strongly branched; old branches 2 — 2.5 mm in diameter, with gray longitudinally splitting bark, light brown, terete, sulcate, all minutely but rather densely puberulent; inflorescence branches slightly zigzag (sympodium); ocreae brownish in lower part, pellucid above, white, bifid to the base, lacerate at the tips with long narrow pointed divisions; petioles short; leaves broadly lanceolate, long-acuminate, 10 — 36 mm long, 2.5 — 5 mm broad, rough on the margins and on the midrib beneath with short sharp hairs; inflorescence a leafless panicle with broad ocreae; flowers broadly funnel-shaped, with pedicels 2 — 3 mm long, white or stramineous, ca. 5 mm in diameter, 2 mm long, elongating in fruit to 6 — 7 mm, the perianth segments 2 — 3 times as long as the fused base; achene trigonous, black, lustrous, the concave faces oval-acuminate. April — May. (Plate XXXVIII, Figure 1).

Sandy hilly steppes. — Centr. Asia: Mtn. Turkm. Endemic. Described from the vicinity of Kushka (Shur-sufi Ravine). Type in Leningrad.

N o t e. Students who do not seriously consider the structure of the achenes are liable to mistake this plant for A t r a p h a x i s.

603 2. **P. thymifolium** Jaub. et Spach, Illustr. pl. or. II (1844—46) 116. — P. Kotschyanum Boiss. in Ky. Schedulis (1845) ex Boiss. Fl. Or. IV, 1040.

A small undershrub to 40 cm high, rough with short hairs; rhizome strong, oblique, easily disintegrating; woody stems 12—18 cm long; young branchlets whitish, densely leafy; ocreae pellucid, oblong, 1- or 2-nerved, often lacerate with narrow pointed divisions; leaves light green, lanceolate or lance-oblong, revolute-margined, with 2 or 3 lateral veins on each side of the midrib, short-petioled, 7—15 mm long and ca. 3 mm broad, the bracteal much smaller; racemes interrupted; flowers 1—3 in the axil, funnel-shaped, dark at base, the stramineous perianth segments with a reddish keel, oval-angular and reticulate in fruit; achene lustrous, oval-trigonous, point-tipped, ca. 2.2 mm long. June—July. (Plate XXXVIII, Figure 2).

In stony soil on rather steep mountain slopes, among juniper thickets. — Centr. Asia: Pam.-Al., Syr D., T. Sh. **Gen. distr.**: Iran. Described from the Kuh Daena Mountains (alpine zone) and from mountains near Shiraz in Iran. Type in Paris.

2. *P. **setosum** Jacq. Observ. III (1770) 8; Boiss. Fl. Or. IV, 1038; Grossg., Fl. Kavk. II, 48. — Ic.: Jacq. ibid. tab. 57.

Shrub, smooth throughout; stems numerous, woody, erect or ascending, sulcate, 20—40 cm long; ocreae pellucid from base, tubular, many-nerved, dissected at apex; leaves narrow, linear-lanceolate, 20—40 cm long, 2—3 mm broad, with numerous lateral veins beneath, the bracteal abbreviated, the uppermost shorter than the flowers; flowers sessile, 1—3 in the axil, disposed in slender interrupted spikes; perianth infundibular; segments white, with a red median stripe enlarged toward the base; achene lustrous, with oval faces. July—August. (Plate XXXVIII, Figure 3).

Stony meadows and slopes in the alpine zone. — Recorded for the former Kars Region. Expected in the adjoining part of the Caucasus, i. e., S. Transc. (Aleksandropol' area). Described from Asia Minor (Smyrna area). Type in Paris.

3. **P. luzuloides** Jaub. et Spach, Ill. pl. or. II (1844) 127; Boiss. Fl. Or. IV, 1039; Grossg., Fl. Kavk. II, 48. — P. dracunculifolium Boiss. in Pl. Noë ex Boiss. Fl. Or., ibidem (1879). — P. pachyrrizum Trautv. in A. H. P. II, 2 (1873) 483. — Ic.: Jaub. et Spach, ibid. tab. 126.

Perennial, smooth throughout; rhizome woody; branches numerous, erect, virgate, indurated; ocreae colored, pellucid, tubular, many-nerved,
604 fringed at apex; leaves long, broadly lance-linear, prominently veined beneath, the bracteal much abbreviated, the uppermost obsolescent; flowers pale roseate, 2—4 in the axil, with indistinct tube and nodding free segments; achene oblong, smooth, 3-angled, as long as the perianth. July—August.

Dry mountain slopes. — Caucasus: S. Transc. (Karabakh, Armenia from Daralagez to Zangezu). **Gen. distr.**: As. Min., Iran. Described from the mountains of Asia Minor. Type in Paris.

4. **P. paronychioides** C. A. M. ex Hohenacker in Bull. Soc. Nat. Moscou IV (1838) 336; Meisn. in DC. Prodr. XIV, I, 89; Boiss. Fl. Or. IV, 1040; Grossg., Fl. Kavk. II, 47; Fedch., Consp. Fl. Turk. VI, 285. — P. paronychia C. A. M. Enum. pl. Cauc., (1831) 158 (non Cham. et Schlecht.);

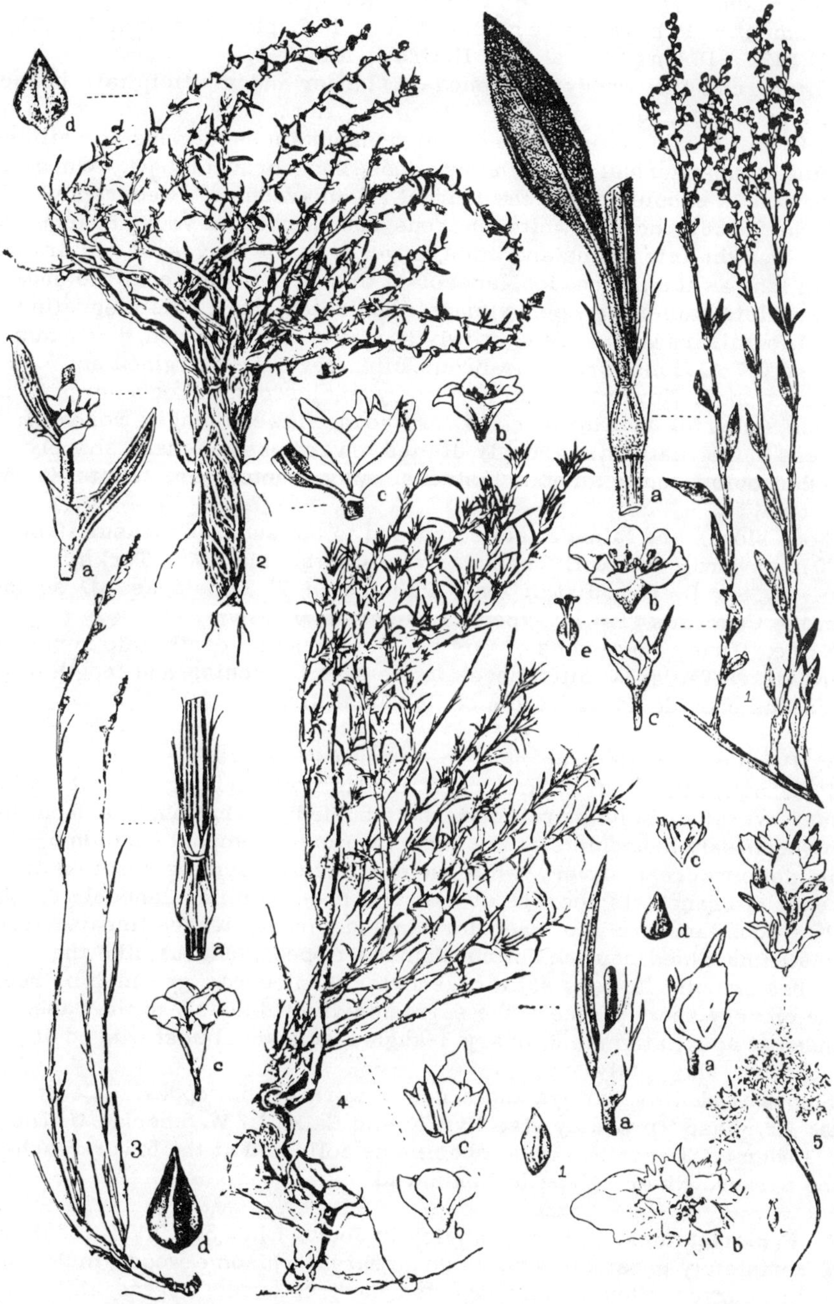

PLATE XXXVIII. 1. Polygonum arianum Grig. – 2. P. thymifolium Jaub. et Sp. – 3. P. setosum Jacq. – 4. P. paronychioides C.A.M. – 5. P. paronychioides var. compactum B. Fedtsch.
Annotation to all species: a) ocrea and leaf, b) flower, c) flower and bracteole, d) achene, e) ovary.

Ldb. Fl. Ross. IV, 534. — P. Meyeri Steudel, Nomencl. Bot. ed. 2, II (1841) 376 (nomen). — P. himalaiense H. Gross. in Engler. Bot. Jahrb. XLIX (1913) 343. — P. Englerianum H. Gross, ibid. 344.

Shrub; rhizome woody, thickened or slender and multicipital; branches mostly prostrate, very rough, 5 — 30 cm long, herbaceous or woody and then erect (f. suffruticosum Kom.); high-mountain specimens sometimes forming a subspherical top with very dense short branchlets (f. compactum B. Fedtsch. in supplement to the journal "Zemlevedenie" (Geography), 1900, No. 9, 186); ocreae white, lustrous, subpellucid, on young branchlets longer than the internodes and often longer than the leaves, on old branches half as long as the internodes, lanceolate to oval-lanceolate, at the ends of branchlets numerous, approximate, imbricated, finally disintegrating into filiform shreds; leaves narrowly linear, acute, 1-nerved, 8 — 15 mm long and 0.7 — 1.5 mm broad, scabrous, often revolute-margined and spinous-tipped; flowers axillary, solitary, enclosed in the ocreae; perianth segments shorter than the tube (in very young flowers longer, roseate); stamens 8, the filaments strongly dilated toward base; achene sharply 3-angled, oblong, quite smooth, lustrous. May—September. (Plate XXXVIII, Figures 4, 5).

Stony slopes and taluses; sites exposed to the sun. — Caucasus (from 1,200 m upward); S. and E. Transc., Tal.; Centr. Asia: Mtn. Turkm., Pam.-Al., Syr D. Gen. distr.: Iran, Afghanistan, W. Himalayas. Described from the Caucasus (Tal.). Type in Leningrad.

Note. Dried specimens of this species break up easily into separate internodes. The great differences in the size, branching, and length of the internodes provide an indication of its variability.

607 5. **P. pulvinatum** Kom. sp. nova in Addenda IV, p. 549.

Perennial, with a tuft of firm dark fibrous roots arising from the collar; stems several, closely approximate, half-buried underground, ca. 5 cm long, with short leafy branchlets at the ends, forming a compact cushion 5 — 10 cm in diameter; ocreae silvery-white, semipellucid, forming rosettes at the ends of branchlets; ocreae at the stem nodes similar, lanceolate, fully partite, soon lacerate into irregular tangled shreds; leaves linear, narrow, revolute-margined, acutish but not spinous-tipped, glabrous, like the branches smooth; flowers solitary, shorter than ocreae, greenish or red, borne on very short pedicels, the segments divided nearly to the base, rounded at apex; stigmas 3; ovary 3-angled, smooth, almost winged at the angles. May — July.

Slopes of shallow valleys and sands in wormwood steppes. — Centr. Asia: Ar.-Casp. (probably also Kyz. K. and Balkh.); W. Siberia: U. Tob., Irt. Endemic. Described from specimens collected at the lower reaches of the Sary-su River. Type in Leningrad.

6. **P. biaristatum** Aitch. et Hemsley in Journ. Linn. Soc. XVIII (1881) 90.

A completely prostrate smooth undershrub; rhizome woody, multicipital; branches curved, often short, congested, dark-colored; ocreae white, large, 2-parted, each with 2 ferruginous awns; leaves approximate, sessile or short-petioled, subcoriaceous, to 10 mm long and 1 — 2 mm broad, elliptic or obovate, sometimes minutely mucronulate, the slightly revolute margins beset with small 1-celled papillae; flowers sessile or borne on a very short pedicel, solitary or in pairs, axillary; perianth red, 5-parted, with equal somewhat fleshy segments; stamens 8, the filaments dilated toward base; styles cylindric; achene 3-angled. June—September. (Plate XXXIX, Figure 5).

Small alpine meadows, stony slopes and taluses. — Centr. Asia: Syr D., T. Sh., Pam.-Al. (Pamir). **Gen. distr.**: Afghanistan, Himalayas. Type in London; cotype in Leningrad (Aitchinson, Kurrum Valley, No. 816).

Note. Diameter of the shrub 6—36 cm. As compared with the preceding species, the plant is characterized by its dark color, with a reddish tinge.

Series 2. **Cognatae** Kom. — Rhizome woody; stems form a caudex, herbaceous, densely leafy; leaves firm, somewhat fleshy.

7. **P. fibrilliferum** Kom. sp. nova in Addenda IV, p. 550.

Perennial; rhizome not thickened with strongly lignified branches or thickened; branches buried in lower part, with developed ocreae but
608 leafless, densely leafy above; ocreae pellucid, ferruginous in lower part, hyaline above, oval or oval-lanceolate, 2-parted nearly to the base, acuminate, with slender white threads splitting off at the sides; leaves oblong-lanceolate, cuneate at base, 5—40 mm long, 2—3 mm broad, short-petioled, the lateral veins visible only on the largest leaves, the margins slightly revolute and beset with small papillae; flowers axillary, subsessile, 1—3 in a glomerule; perianth 3—4 mm long, green, the deeply incised segments white- or roseate-margined, or roseate throughout; achene ca. 2 mm long, lustrous, trigonous-ovoid, acute at apex. July—August. (Plate XXXIX, Figure 4).

Dry stony slopes in the juniper and maple zone, at 1,800—2,000 m, clayey-sandy fields; a noxious weed. — Centr. Asia: Syr D., Pam.-Al. Endemic. Described from the Sanzar River basin on the northern slopes of the Turkestan Range. Type in Leningrad.

8. **P. ammanioides** Jaub. et Spach, Illustr. pl. or. II (1844) tab. 119; Grossg., Fl. Kavk. II (1930) 48. — P. cognatum γ ammanioides Meisn. in DC. Prodr. XIV (1856) 96. — P. alpestre β ammanioides Boiss. Fl. Or. IV (1879) 1038.

Perennial; rhizome woody, branched; branches decumbent, elongated, with short internodes; ocreae oval-lanceolate, acuminate, pellucid-white, as long as ocreae; leaves short-petioled, oblong-lanceolate, ca. 12 mm long and 3 mm broad, the margin rough with papillae, 1-nerved, often recurved; flowers in dense axillary glomerules; pedicels as long as perianth; perianth segments relatively large; achene ca. 3 mm long, lustrous, trigonous, acute at apex. July—September. (Plate XXXIX, Figure 7).

Paths, pebbles, and stony slopes.— Caucasus: S. and E. Transc. **Gen. distr.**: Iran. Described from Iranian Azerbaijan. Type in Paris.

9. **P. myrtillifolium** Kom. sp. nova in Addenda IV, p. 550.

Perennial; rhizome straight, undivided, ca. 5 mm thick, gray, giving rise to a tuft of erect or curved branches; ocreae deeply bifid, hyaline, acuminate, entire; leaves petiolate, elliptic or oval, obtusely rounded or minutely cuspidulate, prominently veined beneath, ca. 10 mm long and 5—7 mm broad; flowers mostly solitary; pedicels exposed, equal, ca. 2 mm long; perianth minute, dark red, dissected nearly to the base; achene narrow, trigonous, lustrous, ca. 2 mm long. July—August. (Plate XXXIX, Figure 3).

Taluses. — Centr. Asia: Pam.-Al. Endemic. Described from Shugnan. Type in Leningrad.

609 10. **P. pamiroalaicum** Kom. sp. nova in Addenda IV, p. 551.

Perennial; rhizome multicipital; branches decumbent, green; ocreae broad, obtusely rounded or truncate, pellucid, sometimes toothed at summit; leaves longer than internodes, less than 1 cm long, 2—4 mm broad, short-petioled, flat-margined, 1-nerved; flowers in axillary glomerules of 2—5, red with green, oval, almost rounded at apex; pedicels half as long as perianth; perianth 4—5 mm long, the segments one-third the total length; achene lustrous; trigonous, narrow, ca. 3 mm long, acute.

Stony banks of small rivers, near glaciers, etc. in the subalpine and alpine zones. — Centr. Asia: Pam.-Al. Endemic. Described from the Zeravshan Valley and from the Shchurovskii glacier (3,000—4,000 m). Type in Leningrad.

11. **P. alpestre** C. A. M. Enum. pl. Cauc. (1831) 157; Grossg., Fl. Kavk. II, 48; Boiss. Fl. Or. IV, 1087. — P. pluriflorum C. Koch in Linnaea XXII (1849) 202. — P. cognatum α alpestre Meisn. in DC. Prodr. XIV (1857) 96. — Ic.: Jaub. et Spach, III. pl. or. II, tab. 118.

Perennial; rhizome woody, multicipital; stems prostrate, 10—13 cm long, with short internodes, hence ocreae as long as and leaves longer than the internodes; leaves petiolate, horizontally spreading, oblong, rarely oval or lanceolate, minutely mucronulate, on the average 2 cm long and 5—8 cm broad, the margins rough with 1-celled papillae; pedicels much shorter than perianth; perianth carneous or roseate, deeply lobed, 5—7 mm long, in fruit with prominent longitudinal and transverse nerves; achene lustrous, narrow, ovaloid-trigonous, ca. 3.5 mm long. June—September. (Plate XXXIX, Figure 2).

Stony and gravelly slopes and near paths in the alpine zone. — Caucasus: Greater Caucasus, Dag., E. and S. Transc. **Gen. distr.**: As. Min., W. Iran. Described from Georgia. Type in Leningrad.

12. **P. rupestre** Kar. et Kir. Enum. pl. Alt. in Bull. Soc. Nat. Mosc. XIV (1841) 740; Kryl., Fl. Zap. Sib. IV, 852. — P. alpestre β rupestre Fisch. et Mey. Ind. VIII sem. Horti Petrop. (1842) 19. — P. cognatum β rupestre Meisn. in DC. Prodr. XIV (1857) 96.

Perennial; root to 10 mm thick, multicipital; stems smooth, prostrate, 5—50 cm long; ocreae deeply bifid, hyaline, oblong-ovate from a broad base, acuminate, 5—8 mm long, mostly longer than internodes; leaves short-petioled, elliptic or ovate, thickish, rounded at apex, often minutely mucronulate, 8—20 mm long and 5—13 mm broad; flowers almost all along the stem in fascicles of 3—6; pedicels unequal, longer than in
610 other "cognata" species, sometimes hamately recurved; perianth 3—3.5 mm long, campanulate, mostly green, more rarely roseate, divided to one-third

or to the middle into 5 segments, the margins of segments roseate or white; stamens 8; achene trigonous-ovoid, almost black, lustrous, shorter than perianth. June — September. (Plate XXXIX, Figure 6).

Stony and clayey slopes, weathered gravel, pebbles, etc., in high mountains, at altitudes of 1,000 — 3,000 m. — W. Siberia: Alt. (Lake Marka-kul'); Centr. Asia: Dzu.-Tarb., T. Sh., Pam. Al. **Gen. distr.**: Tibet, W. Him., Dzu.-Kash., Described from Tarbagatai (Chegerak-Assu River). Type in Leningrad.

Series 3. **Perennes** Boiss. (partim). — Rhizome and lower part of stems woody but character of plants not suffruticose; ocreae distinctly 2-parted to the base.

13. **P. equisetiforme** Sibth. et Smith, Fl. Graeca I (1806) 269; Boiss. Fl. Or. IV, 1036; M. Popov in Tr. Penzen. Obshch. Lyub. Estv. III — IV (1918) 271; Grossg., Fl. Kavk. II, 48. — Ic.: Sibth. et Smith, l. c. tab. 364. — Exs.: H. F. A. M. No. 108.

Perennial; stems sturdy, ca. 3 mm thick and 2 m long, decumbent, ascending, or erect; ocreae many-nerved and colored in lower part, lacerate at the tips into slender to capillary division, much shorter than the internodes; leaves subsessile, oblong to oblong-lanceolate, acute, often crisp-margined, ca. 4 cm long and 0.5 — 1 cm broad, glaucous; flowers white or roseate, longer than ocreae, disposed in longer slender leafless terminal racemes, like the bracteal leaves small and inconspicuous; flowers infundibular, cleft to the middle; achene ca. 2.5 mm long, ovaloid-trigonous, lustrous. May — October. (Plate XL, Figures 1a — d).

Gardens, bluffs, ditches, weed-infested and slightly saline places, pastures, wormwood steppes, in irrigated areas. — Caucasus: S. and E. Transc.; Centr. Asia: Syr D., Amu D., Mtn. Turkm. (piedmonts). **Gen. distr.**: Bal.-As. Min., N. Afr., Iran. Described from Crete. Type in London.

14. **P. maritimum** L. Sp. pl. (1753) 519; Boiss. Fl. Or. IV, 1037; Ldb. Fl. Ross. III, 534; Shesterikov, Opred. p. okr. Odessy 159; Shmal'g., Fl. II, 392. — Ic.: Sibth. et Smith, Fl. Graeca, tab. 363.

Perennial; rhizome multicipital; stems ascending from base or half-buried in the sand, erect at the ends, with few erect branches; ocreae broad, many-nerved, silvery-white, commonly longer than internodes, at length torn longitudinally; leaves short-petioled, glaucous, elliptic, prominently veined, revolute-margined, 1 — 2 cm long and 0.5 — 1 cm broad; flowers short-pediceled, 1 — 3 in the leaf axil, disposed in long
611 rather loose leafy racemes; perianth ca. 5 mm long, urceolate, white or whitish-roseate, green at base, divided to the middle; achene 4 — 5 mm long, ovaloid-trigonous, sharp-angled, longer than perianth. July — October. (Plate XL, Figure 2a — c).

Sands on seashores; solitary or in groups. — European part: Bl. (Odessa), Crim. (Sudak); Caucasus: Cisc., W. Transc. **Gen. distr.**: Atl. Eur., Med., Bal.-As. Min., N. Afr., Atl. N. Am. Described from S. France (Marseilles). Type in London.

Economic importance. An infusion of this plant serves as a popular remedy for lithiasis. The plant consolidates coastal sands.

15. **P. Rayi** Babingt. in Trans. Linn. Soc. XVII (1836) 458; Asch. et Gr. Synopsis IV, 863; Hjelt, Consp. Fl. Fenn. II, pars. I, 220. — P. Rayi subsp. norvegicum Sam. in Acta H. Bergiani XI (1931) No. 3, 67. — Ic.: Babingt. Suppl. Engl., Bot. III (1843) tab. 2805; Sam. l. c., tab. 2, f. 1, 5, 6.

Perennial; stems decumbent, to 0.5 m long or longer, more herbaceous than in the preceding species, few-branched, the branches capable of elongation; internodes to 1 — 1.25 cm long; ocreae shorter, covering 0.3 to 0.5 of the internode length, colored in lower part, 6 — 8-nerved, pellucid above, entire or lacerate; leaves thickish, short-petioled, oblong-lanceolate or with transitions toward elongate-oboval (subsp. norvegicum Sam.), pinnately veined, 1 — 2 cm long; flowers often disposed from the base to the end of the stem, 1 — 3 in the axil, the pedicels about the length of the perianth; perianth segments broad-elliptic with a green keel and white margins, to two-thirds the flower length; achene greatly surpassing the perianth, to 5.3 mm long and 3.5 mm broad, ovaloid-triangular, acute, rounded at base, dark brown, lustrous. July — September. (Plate XL, Figure 5a — d).

Seashore sands and pebbles. — Arctic: Arc. Eur. (Rybachii Peninsula, in herbarium only on Murman Coast — Schrenk). **Gen. distr.**: seacoasts of Scandinavia (W.), Atl. Eur. Described from England. Type in London.

16. **P. oxyspermum** Meyer et Bunge in Ind. Sem. Horti Dorpat. (1824) 5; Ldb. Fl. Ross. III, 530. — Ic.: G. Samuelson in Acta H. Bergiani XI, No. 3 (1931), tab. 1, f. 2 et tab. 2, f. 3, 4.

Perennial; branched from base; branches prostrate or ascending, elongated, canaliculate, the internodes 1.5 — 4 cm long; leaves linear to narrowly lanceolate, acute; petioles short; ocreae scarcely covering 0.25 — 0.5 the length of internodes, lacerate at apex; flowers 4 in the axil, disposed in long leafy racemes; pedicels as long as or longer than perianth; perianth segments green with roseate margins; achene ovaloid-
612 oblong, smooth, 1.5 — 2 times as long as perianth, ca. 6 mm long and 3 mm broad, olivaceous to pale brown. The plant reddish throughout. July — September.

Gen. distr.: S. coasts of Norway and Sweden. In Finland often between Vyborg and the USSR border. Likely to be found also on the Soviet coast of the Gulf of Finland on seashore sands. Described from Abro Island in Estonia.

17. **P. Roberti** Lois. in Mém. Soc. Linn. Paris VI (1827) 409 (ex parte). — P. maritimum Lloyd Fl. l'Ouest Fr. éd. 4 (1886) 300. — P. Rayi C. Koch in Linnaea XXII (1849) 202, non Babingt.

Perennial; branches prostrate, 10 — 50 cm long, green or glaucescent; root rarely woody; ocreae markedly shorter than internodes, white, dry, little torn, faintly 6-nerved; leaves elliptic or lanceolate, prominently veined; flowers 1 — 4 in the axil, roseate or white; achene 3 — 4 mm long, trigonous, lustrous, about twice as long as perianth. May — October.

European part: Bl., Crim. (Kerch); Caucasus: Cisc., W. Transc. **Gen. distr.**: Mediterranean coasts. Described from France.

Note. According to Meisner (in DC. Prodr. XIV (1857) 89), the specimen of P. Roberti which he received from Loiseleur himself, was found to consist of branches of P. aviculare, P. litorale, and P. Rayi, and he therefore decided that the species P. Roberti ought to be cancelled. All contemporary authors are, however, in agreement that P. Rayi does not grow on Mediterranean coasts and is replaced there by a distinct race. We therefore assume that the component of P. Roberti which Meisner identified with P. Rayi represents the true P. Roberti Lois. The species P. Roberti is a Mediterranean complex of forms corresponding to the Atlantic complex known under the name P. Rayi Babingt.

Series 4. **Aviculariformes** Kom. — Annuals with flower in the axils of normal leaves, most closely related to the principal type P. aviculare L.

18. **P. litorale** Meisn. in DC. Prodr. XIV (1857) 98 (non Link); Grossg., Fl. Kavk. II (1930) 40.

Annual plant, branched from the base; root thicker and more indurated than in the other species of the group; stems decumbent, divergent, branched, ca. 1 mm in diameter, 20—60 cm long, the internodes 5—7 mm long; numerous short axillary densely leafy branchlets, absent in other maritime species; ocreae on main branches short, colored, on axillary and terminal branchlets silvery, broadly long-tapering at apex and often strongly lacerate; leaves thickish, oblong-lanceolate to ovate-spatulate, obtuse or mucronate, sometimes somewhat crisp-margined (f. crispum Kittel, Taschenb. 3 Aufl. (1853) 304), prominently veined beneath; flowers small, greenish, with almost white tips, subsessile, 3—5
613 in the axil; achene to 2 mm long, acute, lustrous, the surface puncticulate. July—October. (Plate XL, Figure 3).

Saline meadows near the sea and sands within reach of the surf, e. g., in depressions among dunes. — Caucasus: Cisc., W. Transc. (Khosta). **Gen. distr.**: Scand., Atl. Eur., Med. Reports for North America relate to a plant which is not identical with the Caucasian. Described from W. Europe. Type in Berlin.

19. **P. buxifolium** Nutt. ex Bong. Veget. d. ins. Sitcha (1831) 161. — P. littorale β buxifolium Meisn. in DC. Prodr. XIV (1857) 98; Maxim. Primit. Fl. Amur. 230. — P. aviculare ε buxifolium Ldb. Fl. Ross. III, 2 (1851) 532.

Annual; stems decumbent, shorter than in the preceding; leaves linear-oblong, obtuse at apex, strongly narrowed toward base, without distinct veins; flowers 1 or 2 in the axil; stamens 5, the filaments strongly dilated at base, surpassing the perianth; achene dull, with punctate surface. August—September. (Plate XL, Figure 6a—b).

Sandy and pebbly seacoasts. — Far East: Uda, Uss., Sakh. Described from Sikhta.

20. **P. fuscoochreatum** Kom. sp. nova in Addenda IV, p. 551.

Annual; stems rather thick, ca. 3 mm in diameter, ascending from base, curved at the nodes, ca. 50 cm long, with few erect branches; ocreae broad, brown up to the apex, many-nerved, at length strongly lacerate; leaves narrow, oblong, often obtusish at apex, elongate-cuneate toward base, often somewhat crisp-margined, ca. 15 mm long, 2 mm broad; flowers 1—3 in the axil; pedicels slender, ca. 2 mm long; perianth oblong, green, closed, cleft to two-thirds, the segments roseate-margined; stamens 5; achene shorter than perianth, ovaloid-trigonous, acute at apex, ca. 3 mm long, dull, light brown. August—October. (Plate XL, Figure 4a—c).

Sandbanks along seacoasts. — Far East: Uss. Endemic. Described from the shores of Ussuri (village of Rechitsa). Type in Leningrad.

21. **P. lencoranicum** Kom. sp. nova in Addenda IV, p. 552. — P. litorale Grossh. in Fl. Kavk. II (1930) 49.

Annual; stems branched from the root collar, trailing in all directions, sulcate, to 25 cm long; branches variegated as in P. maritimum but very much thinner; ocreae nerved, ferruginous-brown from base, white above, broadly acuminate, at length lacerate, cup-shaped, 4—8 mm long;
614 leaves oblong, short-petioled, acute, 6—12 mm long, 2—4 mm broad, dull, pinnately veined, slightly ribbed; flowers axillary, but slightly exserted from ocreae, mostly solitary, ca. 5 mm long; perianth tube broadly obconical, the oblong round-tipped segments green, white-margined; achene black, dull, broadly ovaloid-trigonous, acute, to 2 mm long. May—October.

Wet sands and pebbles in the estuaries of rivers which flow into the sea or near the sea. — Caucasus: Tal. Endemic. Described from the Buzei-dagun River mouth. Type in Tiflis (Botanical Garden).

22. **P. heterophyllum** Lindm. in Svensk Bot. Tidskr. Bd. 6, No. 3 (1912) 960. — P. aviculare L. Sp. pl. (1753) 362 (pro parte). — P. aviculare var. vegetum et var. erectum Meisn. in DC. Prodr. XIV (1857) 97. — P. aviculare Ldb. Fl. Ross. III, 531 (pro parte); Shmal'g., Fl. II, 392 (pro parte, var. exclusis). — P. Fauriei Lévl. et Vant. Bull. Soc. Bot. Fr. LI (1904) 423. — Ic.: Lindman l. c. tab. 23—25. — Exs.: HFR No. 1183.

Annual; stems erect, with erect or slightly divergent branches and branchlets, or more rarely plants prostrate, all 20—60 cm long; leaves of several kinds; cauline larger, broadly lanceolate or obovallanceolate or narrowly lanceolate or sublinear, 2—4 cm long; leaves of branches half as long, lanceolate or oblong; leaves of branchlets smaller still, narrower, linear-oblong or linear-lanceolate; all leaves acuminate, bright green; bracteal leaves smallest; flowers often racemiformly or spiciformly approximate; perianth parted from patellate base into oblong green segments with white or roseate or purple margins; achene included in the perianth, trigonous-ovaloid to subellipsoid, dull or slightly lustrous, castaneous, rarely blackening in maturity. April—November; farther north May—October. (Plate XLI, Figures 1, 7).

Ornamental and truck gardens, arable land, roadsides, city streets, etc. — European part; Caucasus; W. Siberia; E. Siberia; Far East; Centr. Asia. **Gen. distr.**: all cultivated areas. Described from Sweden.

Note. In Soviet Central Asia apparently hybridizing with P. petalum.

23. **P. aviculare** L. Sp. pl. (1753) 362, pro parte; Ldb. Fl. Ross. III, 531 (pro parte); Shmal'g., Fl. II, 392 (pro parte). — P. aequale Lindm. in Svensk. Bot. Tidskr. VI (1912) 692; Grossg., Fl. Kavk. II, 49. — Ic.: Lindm. l. c. tab. 23, 26. — Russian names: gorets ptichii [bird knotweed], sporysh, gusyatnitsa, burkun, etc.

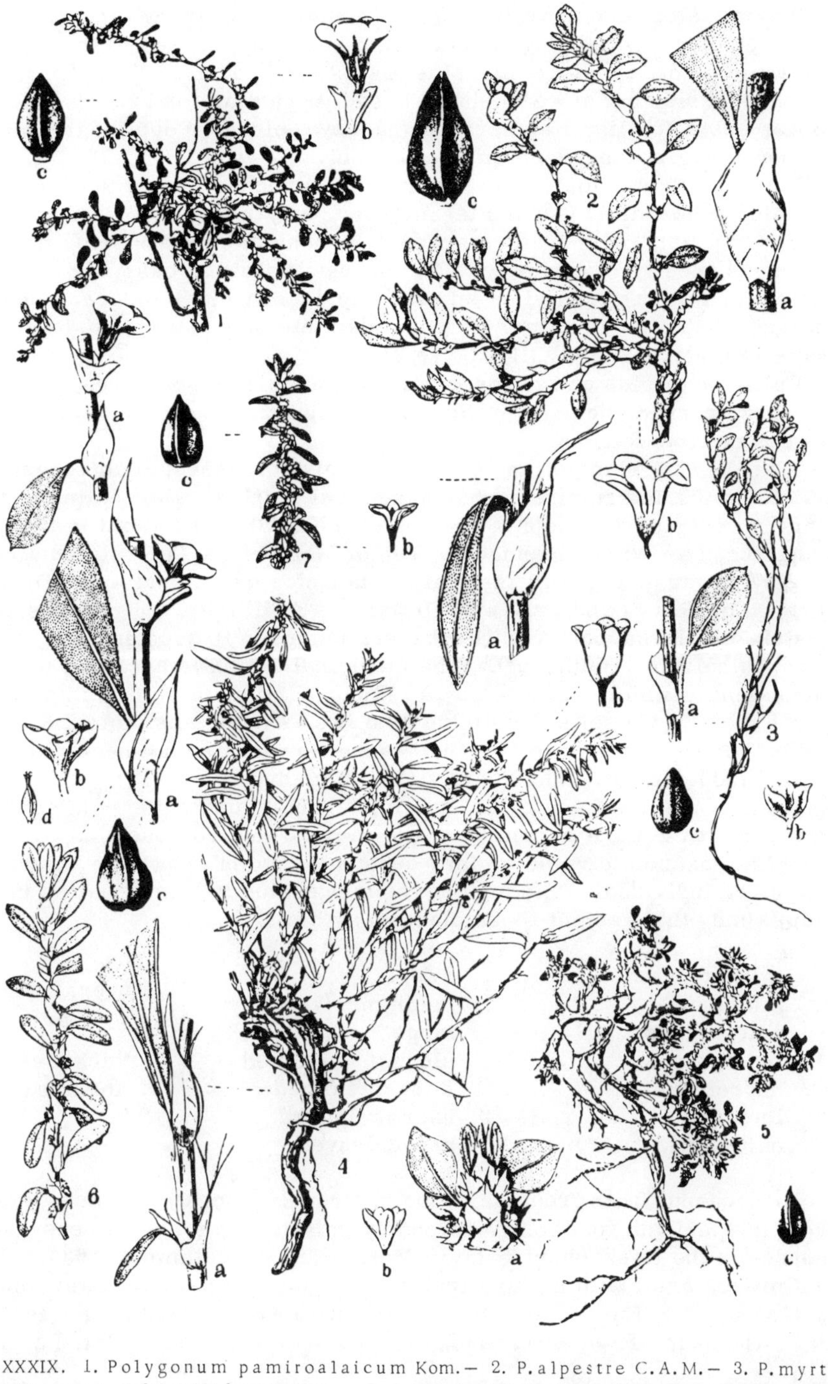

PLATE XXXIX. 1. Polygonum pamiroalaicum Kom. – 2. P. alpestre C.A.M. – 3. P. myrtillifolium Kom. – 4. P. fibrilliferum Kom. – 5. P. biaristatum Aitch. et Hemsley. – 6. P. rupestre Kar. et Kir. – 7. P. ammanioides Jaub. et Spach. – Annotation of details to all drawings: a) ocrea and leaf, b) flower, c) achene.

617 Annual; stem erect, ascending, or decumbent; branches mostly long, suberect, divergent, those on decumbent stems appressed to the ground; the whole plant 10—40 cm high, the lowest internodes 1—3 cm long; leaves variable in size but alike in shape, broadly spatulate or broadly elliptic to oboval-oblong or liguliform, rarely narrow-oblong or sublinear, obtuse to subacute, grayish-green or glaucescent; flowers 1—5, all distinctly axillary, not approximate near the ends of branches; perianth divided to the middle, obconical (turbinate) in lower part, the pale green segments narrowly white- or roseate-margined; achene often surpassing the perianth, trigonous, mostly 2 or rarely to 3 mm long, ovaloid at base, the straight angles forming a pyramid or an acute tip, mostly black or occasionally castaneous, the faces rather obscurely striate or punctate, slightly lustrous. May—October. (Plate XLI, Figure 2).

Cultivated fields and roads, riverside sands, flats, pastures, etc. — European part; Caucasus; W. Siberia; E. Siberia; Far East; Centr. Asia. Described from Sweden.

Note. This plant, often occurring in great masses, is an accepted article on the international pharmaceutical market. Harvesting of the plant is reported from some regions of the USSR. It is sold in German drug stores as Wiedermannscher Tee or Homeriana Tee; it contains 2—2.5% sugar, traces of essential oil, tannoids, resin, and wax. Dried herbage of P. aviculare, with flowers, is sold in the drug stores of Austria as Herba polygoni (Herba centumnodii, H. sanguinariae, H. sanguinalis, or Blutkraut). The medicinal action is ascribed to tannoids. The extent to which P. heterophyllum participates in this connection is not known. On soils containing zinc, the ash contains up to 2.9% $ZnCO_3$, and on normal soils up to 15% SiO_2. The flowers are not nectariferous. A good fodder plant. The roots yield a blue dye.

In analyzing the form cycle known previously under the common name P. aviculare L., it should be borne in mind that various characters are of no taxonomic value as they depend solely on growth peculiarities of a given individual. Lindman gives expression to this situation in maintaining that each of these plants may be:

microtypus	dwarf	overall plant size
miotypus	medium-sized	
macrotypus	large	
trachytypus	slender-stemmed	thickness of internodes
hadrotypus	thick-stemmed	
leptotypus	narrow-leaved	
eurytypus	broad-leaved	

oedocarpus, with greenish, smooth, strongly elongated achenes, exserted from the perianth, and hollow (probably galls); this form was observed in
618 masses in the Neskuchnyi Sad* in Moscow, in the autumn of 1934.

Only material with mature fruits lends itself to precise determination.

Note. In S. Europe and in Soviet Central Asia hybridizing with P. acetosum, P. gracilius, P. patulum, and other related species.

* [Part of the Park of Culture and Rest.]

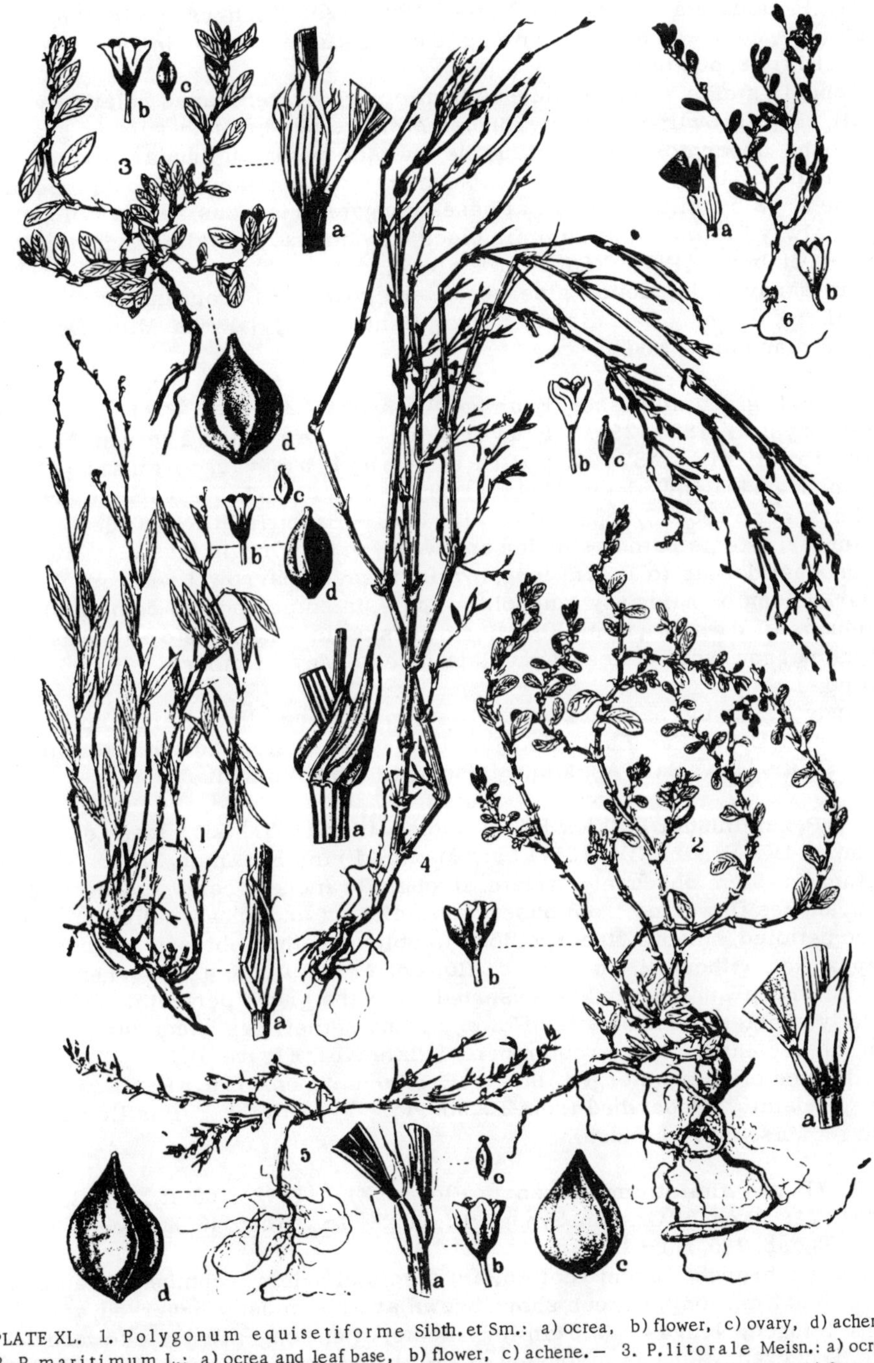

PLATE XL. 1. Polygonum equisetiforme Sibth. et Sm.: a) ocrea, b) flower, c) ovary, d) achene.— 2. P. maritimum L.: a) ocrea and leaf base, b) flower, c) achene.— 3. P. litorale Meisn.: a) ocrea and leaf base, b) flower, c) ovary, d) achene.— 4. P. fuscoochreatum Kom., a) ocrea, b) flower, c) ovary.— 5. P. Rayi Babingt.: a) ocrea and leaf base, b) flower, c) ovary, d) achene.— 6. P. buxifolium Nutt.: a) ocrea and leaf base, b) flower.

24. **P. calcatum** Lindm. Bot. Notis. (1904) 139; Aschers. et Graebn. Synops. IV, 861. — P. aviculare ε depressum Meisn. in DC. Prodr. XIV (1857) 98, pro parte.

Annual; stems weak, semidecumbent or prostrate; leaves relatively small, elliptic, oval, or obovate, prominently veined beneath, pale or grayish-green, the uppermost resembling the lowest; perianth tubular, divided to the middle or less, the light green segments white-margined; stamens 5; achene 2 — 2.5 mm long, broad at base, elongate-trigonous, rarely oblong-ovoid, always 3-angled at summit, black, smooth, rarely faintly puncticulate. May — October. (Plate XLI, Figure 6).

Dry sandy soils, slopes, pine woods, and roads. — European part: Kar.-Lap., U. Dnp. **Gen. distr.**: Scand., Centr. Eur., Bal.-As. Min. Described from Finland and Sweden.

25. **P. neglectum** Besser, Enum. pl. Volhyn. (1820) 45. — P. flagellare Spreng, Syst. II (1825) 255. — P. aviculare δ angustissimum Meisn. in DC. Prodr. XIV (1857) 98. — P. heterophyllum var. angustissimum Lindm. Svensk Bot. Tidskr. VI (1912) 691. — P. aviculare var. neglectum Aschers. et Graebn. Synopsis IV (1913) 858. — Exs.: HFR No. 1629.

Annual; stems rather slender and weak, with more or less elongated internodes; ocreae to 1.3 cm long, hyaline; leaves narrow, linear or linear-oblong or narrowly lanceolate, acute, in some species completely deciduous (f. denudatum Rouy); flowers as in P. heterophyllum; achene narrow, with almost lanceolate faces. May — October. (Plate XLI, Figure 4).

Sandy fields and pastures. — European part: Lad.-Ilm., Kar.-Lap.; E. Siberia: Dau.; Far East: Ze.-Bu., Uss. **Gen. distr.**: Scand., Atl. Eur., Med., Centr. Eur., Bal.-As. Min. Described from Volhynia.

26. **P. humifusum** Pall. ex Ldb. Fl. Ross. III, 2 (1851) 531; Merk ex
621 Meisn. in DC. Prodr. XIV (1857) 95; Maxim. Prim. Fl. Amur. 229.

Annual; stem obscurely striate, strongly branched from base, decumbent, the branches flowering from base; ocreae short and broad, pellucid; leaves short-petioled, smooth, linear-oblong to oblong, rather obtusely rounded at apex, flat, without distinct veins; flowers 4 — 10 in the axil; achene smooth, sharp-angled, slightly exserted from the green perianth; pedicels initially short, elongating after flowering and sometimes becoming longer than the perianth. July — September. (Plate XLI, Figure 5).

Silty and often submerged flats. — E. Siberia: Lena-Kol.; Far East: Uss., Uda. Endemic. Described from E. Siberia. Type in the Pallas Herbarium (British Museum in London).

26. ***P. graminifolium** Wierzb. in Flora XXI, I (1842) 280 et XXVIII, I (1845) 321; Asch. et Graebn. Synops. IV, 862. — Ic.: Rchb. Ic. Fl. Germ. XXIV, 63, tab. 205, f. 1—4.

Annual; branched from root collar; stems slender, decumbent, branching, curved, ca. 3 cm long; ocreae short, brown at base, mostly 6-nerved, at length fringed; leaves to 2.5 cm long, acute, often long-attenuate toward base, 1-nerved; flowers in groups of 1 — 3; pedicels including perianth

1.5 mm long, roseate, white, or red; achene surpassing the perianth, 2 mm long, acute, smooth, brown. August—October.

Gen. distr.: according to Ascherson "Siberia, Rumania, S. and SE Russia." Notwithstanding such an authoritative report, the plant has not so far been recorded for the USSR. Described from sands of the Danube Valley in Hungary.

Note. Related to P. salsugineum and possibly confounded with it.

27. **P. propinquum** Ldb. Fl. Ross. III, 2 (1851) 532.

Annual; stem branched, decumbent from base, sulcate, like the branches leafy to the end; leaves sessile, oblong, acute or obtuse, narrowed toward base; flowers 2 or 3 in the axils of fully developed leaves; pedicels shorter than flower; achene smooth, surpassing the perianth. (Plate XLI, Figure 3).

European part: L. V. Endemic. Described from the vicinity of Astrakhan.

Note. Differing from P. aviculare in having very short ocreae and smooth lustrous achenes surpassing the perianth. According to Ledebour, the plant had not been mentioned by any earlier authority. It is distinguishable from P. acetosum, also recorded for Astrakhan, by the thin prominently veined leaves.

27. ***P. araraticum** Kom. sp. nova in Addenda IV, p. 552.

Annual; root straight, sturdy, multicipitate; stems to 15, erect or slightly curved, in lower part without chlorophyll and leafless for about
622 4 cm, grayish-green above, to 20 cm long, unbranched or with small axillary branchlets; ocreae short, distinctly 2-parted, pellucid, brownish at base, strongly lacerate; leaves distinctly petiolate, elliptic, obtusely rounded at apex, on both sides tomentose-cobwebby, grayish, distinctly veined beneath; flowers solitary, axillary, more approximate toward the ends of branches, nearly always solitary; perianth 2—2.5 mm long, united to not more than one-third, the oblong green segments white-margined; achene to 2 mm long, dark brown, trigonous, sharp-beaked, the flat faces gibbous at the middle, granular all over the surface.

Differing from P. aviculare L. and other related species above all in the leaf vesture and the granular gibbous-faced achene.

Apparently growing on stony taluses. — Caucasus: S. Transc. Collected by A. Grossgeim, 14 March 1913, on Little Ararat (mountain) in Turkey; also occurring on the left bank of the Araks River on stony mountain slopes. Type in Leningrad.

28. ***P. acetosellum** Klokov in Journ. of Agricult. Botany Ucr. vol. I, 3 (1927) 171.

Annual; stems branched from the root collar, slender, sulcate, 15—30 (33) cm long; ocreae 2-parted, short, dull, lacerate; leaves oblong-elliptic, 4.5—6.5 (7) mm long, 1.25—1.5 mm broad, rounded at apex, cuneately narrowed at base to a petiole, somewhat fleshy, without distinct veins; flowers 2 or 3 together, disposed in dense leafy racemes; perianth at least 1.5 mm long, 2 mm in fruit, green, white-margined; achene completely included in the perianth, 1.5—1.75 mm long and to 0.75—1.25 mm broad, black, lustrous. July—October.

Railroad embankments. — European part: V.-Don. Described from the vicinity of Kharkov.

Note. Described as "sp. nova provisora" and designated as a problematic form, with a note stating that it differs from P. propinquum Ldb. in the small leaves without visible veins and from P. acetosum M. B. in the much smaller leaves and smooth achene. Probably little different from P. aviculare (P. aequale).

28. **P. acetosum** M. B. Fl. taur.-cauc. I (1808) 304; Ldb. Fl. Ross. III, 533; Grigor'ev in Fl. Yugo-Vost. IV (1930) 108, Figure 246.

Annual, glaucous, decumbent, branched from base; branches uniformly leafy, decumbent or ascending; ocreae white, almost nerveless, 2-parted, the divisions remaining long united; leaves somewhat fleshy, veinless, covered with small whitish tubercles, oblong-linear, obtusely rounded at apex or more rarely acute, gradually narrowed toward base, 2—3 mm long;
623 flowers 3—7 in the axil; pedicels 1—3 mm long; perianth 2—3 mm long, with obconical tube; segments closely convergent at apex, glaucous-green, very narrowly whitish-margined; achene 1.7—3 mm long, light-colored, with rounded angles, slightly lustrous, obscurely punctate. May—June.

Riverbanks and valley slopes, in gritty and loamy soils. — European part: L. V.; W. Sibera: U. Tob.; Centr. Asia: Ar.-Casp., Balkh., Syr D., T. Sh. **Gen. distr.**: reported for the Caucasus and for Afghanistan. Described from the vicinity of Astrakhan. Type in Leningrad.

29. **P. caspicum** Kom. sp. nova in Addenda IV, p. 553.

Perennial; branched from the root collar; some branches 3—7 cm long in fall, others erect unbranched and 40—60 cm long, finely sulcate, with exposed internodes 5—25 mm long; ocreae broad, pale brown, pectinately lacerate; leaves oblong, ca. 5 mm long, very prominently pinnate-veined beneath, slightly coriaceous; flowers solitary or in pairs, borne on very short pedicels, altogether ca. 3 mm long, equaling the ocrea and half as long as the leaf, green, narrow; perianth segments slightly longer than the tube, roseate-margined; achene trigonous, dark brown, with minutely tuberculate faces, ca. 2 mm long and 1—1.5 mm broad. July—October.

Sands and pebbles. — Caucasus: E. Transc. Endemic. Described from the vicinity of Baku (Balakhany, Zabrat). Type in Leningrad.

30. **P. venosum** Steward in Contrib. Gray Herb. LXXXVIII (1930) 26.

Perennial, smooth; stems sulcate, arcuately ascending; ocreae white-hyaline, lanceolate, shorter than internodes and leaves; leaves sessile, coriaceous, oval-lanceolate (the upper lanceolate, the lower oval) entire, becoming somewhat ferruginous in drying, 3—8 mm long, with slightly thickened revolute margins, prominently veined beneath, the lateral veins arising at such an acute angle as to appear almost parallel to it; flowers axillary, short-pediceled, crowded at the ends of lateral branches in spikelike inflorescences; bracts leaflike; perianth 2 mm long, apparently roseate, 5-parted to below the middle, the broadly rounded segments prominently 1-nerved; achene trigonous, about as long as the perianth, with finely granular faces.

Described from specimens collected by Gandoger in Dzungaria, without precise indication of location. Apparently belonging to the P. patulum group and distinguishable by the very compact inflorescence. More definite identification has not been possible. Type in the herbarium of the Missouri Botanical Garden, under No. 124, 702.

624 31. **P. floribundum** Schlecht. ex Meisn. in DC. Prodr. XIV (1857) 97; Ldb. Fl. Ross. III, 529; Grigor'ev in Fl. Yugo-Vost. IV (1930) 108. — Ic.: F. Yugo-Vost., Fig. 247.

Perennial*; stems branched from base; branches divergent, bearing leaves and flowers uniformly throughout their length; ocreae pellucid, colored in lower part, several-nerved, split to the middle into numerous slender divisions; leaves 7—22 mm long, oblong or ovate, rounded at apex, revolute at the margins, 1-nerved; flowers 3—10 in the axils; pedicels often longer than perianth, always very distinct; perianth ca. 2 mm long, infundibular, roseate, divided to two-thirds; achene not surpassing the perianth, slightly lustrous, ovaloid, almost flat. June— September.

Saline places in semideserts, in wormwood thickets. — European part: L. V.; W. Siberia: U. Tob.; Centr. Asia: Ar.-Casp. Endemic. Described from specimens collected by Pallas at Lake Inder. Type in Leningrad.

Series 5. **Acroanthae** Kom. — Annuals with flowers disposed in dense terminal raceme; stems prostrate or ascending, rather short.

32. **P. corrigioloides** Jaub. et Spach, Illustr. Pl. or. II (1844) 33; Boiss. Fl. Or. IV, 1033; B. Fedtsch. Consp. Fl. Turk. VI, 287. — Ic.: Jaub. et Spach. l. c. tab. 124. — Exs.: H. F. A. M. No. 107. — Kazakh names: Kizyl-tamyr, Kiik-chub (O. Fedch.).

Annual, smooth throughout, glaucous or reddish-violet, branching from the root collar; branches at first appressed to the ground, spreading in all directions, flexuous, at length ascending; internodes longer than leaves; ocreae short-oval, pellucid, nerveless, the margin toothed or lacerate; leaves linear-spatulate, narrowed toward base, 1-nerved, slightly fleshy, ca. 5 mm long and 1—1.5 mm broad; flowers disposed in long filiform racemes, 4—6 in the axils of bracts; pedicels jointed above the middle, 3 times the length of perianth; perianth whitish-roseate or bright rose, more than 2 mm in diameter, the segments much longer than the tube; achene small, lustrous, almost black, triquetrous, 1.5 mm long, shorter than the perianth. Dense bushy plants ca. 30 cm in diameter. April—May. (Plate XLII, Figure 3).

Saline soils in lowlands and on riverside terraces. — Centr. Asia: Syr. D., Amu D., Pam.-Al., Kyz. K., Kara K., Mtn. Turkm. **Gen. distr.**: Iran. Described from the vicinity of Baghdad. Type in Paris.

33. **P. polycnemoides** Jaub. et Spach, Illustr. Pl. or. II (1844) 30; Boiss. Fl. Or. IV, 1033; Meisn. in DC. Prodr. XIV, 92; B. Fedtsch. Consp. Fl. Turk. VI, 286. — Ic.: Jaub. et Spach, l. c. tab. 120 et 121.

* Species 29—31, known only from the herbarium, may prove to be annuals upon further study.

625 Annual; stem prostrate, branched from base; branches filiform, weak, slightly angled, smooth; internodes longer than leaves; ocreae shorter than leaves but longer than internodes, pellucid, nerveless, oval-lanceolate, entire or often fringed-incised; leaves somewhat fleshy, subulate-linear, almost smooth, 1-nerved, 5—10 (12) mm long, ca. 1 (1.5) mm broad; flowers 1 or 2 in the axil; pedicels very short; spikes long, slender, loose; perianth segments shorter than the tube, roseate, white-margined; achene lustrous, puncticulate, less than 2 mm long, rather gently trigonous; bracts spreading, conspicuous throughout the inflorescence. June—August. (Plate XLII, Figure 4).

Taluses, pebbles, and dry stony slopes or more rarely clayey slopes; in mountains mostly between 1,200 and 2,500 m. — Caucasus: S. and E. Transc.; Centr. Asia: Syr D., Pam. Al., Mtn. Turkm. (including the Balkhan mountain ranges). **Gen. distr.**: Asia Minor, Iran. Described from Baghdad. Type in Paris.

34. **P. acerosum** Ldb. ex Meisn. in DC. Prodr. XIV (1857) 92; B. Fedtsch. Consp. Fl. Turk. VI, 286. — Exs.: H. F. A. M. No. 105.

Annual; stem 20—25 cm long, branched from base; branches nodding or obliquely divergent; ocreae silvery-white, nerveless, acute or 2-parted at apex, half as long as and near the ends of branches exceeding the internodes; leaves erect, appressed, lanceolate, acute, owing to the revolute margins resembling conifer needles, to 1 cm long; flowers solitary, sessile, shorter than leaves and ocreae; perianth segments short, whitish or more or less brightly roseate, under the microscope prominently papillose on the keel; anthers dark purple, round; bracts pellucid, silvery, 3-toothed at apex; stamens 8; achene included in the perianth, less than 2 mm long, sharply trigonous, the lustrous surface with dotlike and streaklike depressions. May—July. (Plate XLII, Figure 5).

Wormwood steppes, pebbles, dry stony slopes, and rocks. In mountains up to 3,000 m. — Centr. Asia: Ar.-Casp., Balkh., Dzu.-Tarb., T. Sh., Syr D., Pam.-Al. Endemic. Described from the Sary-su River in Dzungaria — near the Irtysh (Schrenk). Type in Leningrad.

Series 6. **Salsugineae** Kom. — Annuals with flowers disposed in loose often interrupted racemes in upper part of stems and branches; leaves concentrated mainly in the middle part of the stem and none in inflorescence.

35. **P. salsugineum** M. B. Tableux d. prov. Caspien. (1798) 169 et Appendix No. 41; Ej. Fl. taur.-cauc. I (1808) 304; Ldb. Fl. Ross. III, 530; Grigor'ev in Fl. Yugo-Vost. IV (1930) 109; Grossg., Fl. Kavk. II, 50; Fedtsch. Consp. Fl. Turk. VI, 287. — P. tenuifolium Steph. ex Ldb. l. c. — P. Bellardi
626 β salsugineum Shmal'g., Fl. II (1897) 393. — Exs.: HFR No. 787.

Annual; stems very slender, erect from base, strongly branched, finely sulcate, forming low spreading bushes 12—20 cm high, sparingly leafy; ocreae narrowly cupuliform, toothed, ferruginous-brown to the apex; leaves narrowly linear, acuminate, often almost subulate, ca. 3 mm long; flowers axillary solitary, sessile, except the lowest, to six, one above another, forming a strongly interrupted terminal raceme; perianth tubular-infundibular, yellow or reddish; fruit sharply triquetrous, mat with punctate surface. July— September. (Plate XLII, Figure 2).

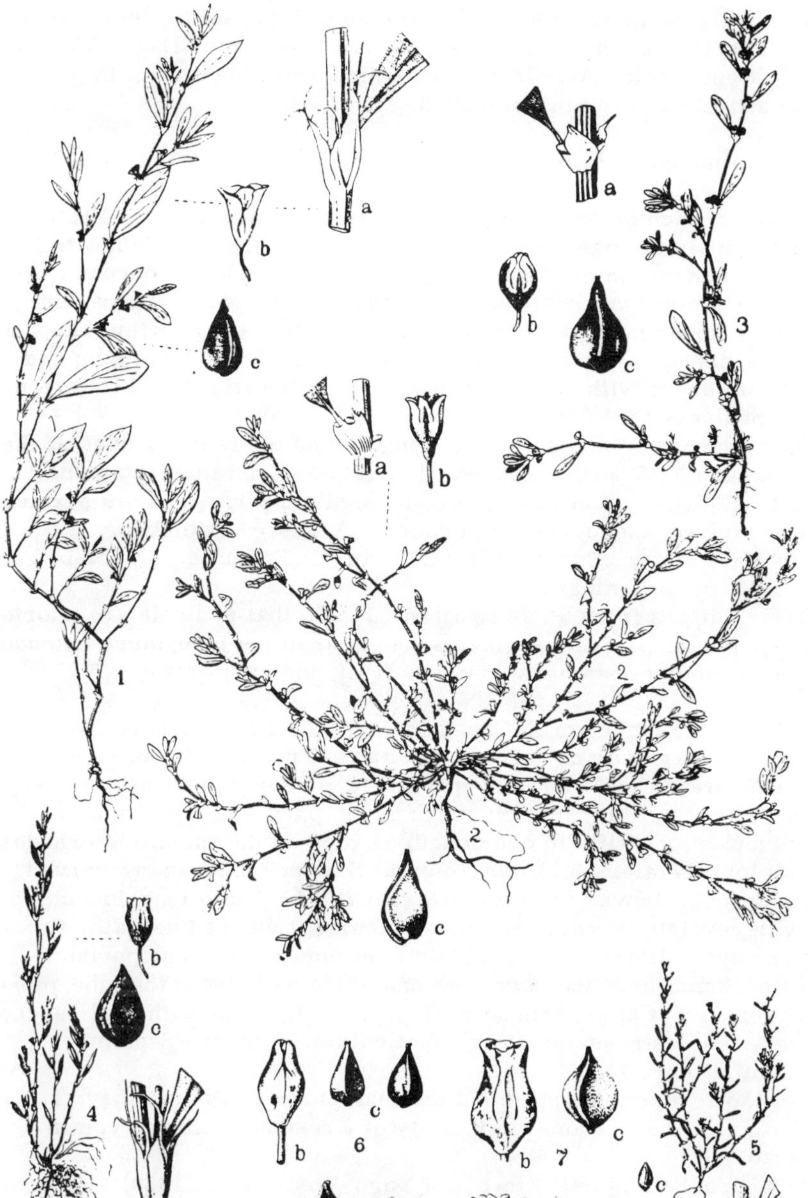

PLATE XLI. 1. Polygonum heterophyllum Lindm. – 2. P. aviculare L. – 3. P. propinquum Ldb. – 4. P. neglectum Ldb. – 5. P. humifusum Pall. – 6. P. calcatum Lindm. (from Bot. Notis. (1904) 140). – 7. P. heterophyllum Lindm. – 8. P. aequale × heterophyllum. – 9. P. aequale Lindm. – The last three drawings are reproduced from Sv. Bot. Tidskr. 6 (1912) tab. 23. – Designation of details to all drawings: a) ocrea and leaf base, b) flower, c) fruit.

Saline soils, bituminous shale, gypsiferous clays, saline depressions. — European part: L. Don (E.), Transv., L. V.; Caucasus: Dag.; W. Siberia: U. Tob.; Centr. Asia: Ar.-Casp. Endemic. Described from Dag. (M. B. No. 753) and L. V. Type in Leningrad.

36. **P. samarense** H. Gross. in Engler's Bot. Jahrb. 49 (1913) 340. — Exs.: HFR No. 787.

Annual; branching from a root neck, stems numerous, 30—40 cm long, ascending or almost erect, slender, slightly channeled, smooth, with filiform straight branches; internodes elongated, 2—2.8 cm long; ocreae narrow, tubular, 6—10 mm long, rusty-brown in their lower part, transparent-white above, their upper margin toothed or fringed-lacerate, shorter than the internodes, narrowly linear or almost subulate, 1—1.5 (2) cm long and 0.7—1.8 mm broad with one prominent vein on lower side. Flowers mainly solitary, pedicels ca. 1.5 mm long, perianth tubular, 1.3 mm long, its segments convergent and somewhat longer than the tube, rounded at the apex; stamens 3—7, with extended filament bases on inner stamens; fruit flat-triquetrous, castaneous or black, slightly lustrous, its surface dotted or striped, exceeding the perianth. August—September.

Saline meadows. — European part: Transv. Endemic. Described from Samara. Type in Leningrad.

Note. Differs from P. salsugineum in that pedicels are shorter than the perianth, perianth segments longer than the tube, much extended bases of the shorter stamens, lustrous fruit, and stouter stems.

629 37. **P. Aschersonianum** H. Gross in Engler's Bot. Jahrb. vol. 49 (1913) 341. — P. salsugineum var. prostrata Trautv. in herb.

Annual; stems 5—18 cm long, prostrate, strongly branched at base, with very short axillary branchlets in upper part, slightly channeled; ocreae tubular, one-fifth to one-fourth as long as the middle internodes, 3—4 mm long, semipellucid, dark only at the very base, many-nerved, toothed at apex; leaves sessile, linear, acute, 4—7 mm long, to 1 mm broad, often with revolute margins and then acicular, 1-nerved beneath; flowers nearly always solitary, their pedicels less than 1 mm long; perianth tubular, ca. 2 mm long, the segments one-third as long as the tube, more or less rounded at apex; achene castaneous, trigonous, with oval-lanceolate concave faces, lustrous, minutely puncticulate. August—September. (Plate XLII, Figure 1).

Saline soils among steppes. — European part: L. Don, Transv., L. V. Endemic. Described from the vicinity of Krasnoarmeisk. Type in Leningrad.

Note. Yu. S. Grigor'ev (in Flora Yugo-Vostoka, IV (1930) 110) refers P. Aschersonianum H. Gross. to the cycle of hybrids between P. salsugineum and P. aviculare, without however offering any support for his assumption.

Series 7. **Bellardiiformes** Kom. — Flowers disposed in slender interrupted terminal racemes.

38. **P. Kitaibelianum** Sadler, Fl. Pest. I (1825) 287; Klokov in Journal of Agricult. Botany, vol. I, part. 3 (1927) 176. — P. patulum A. Kitaibelianum Asch. et Graebn. Synops. IV (1913) 865. — P. Bellardi α typicum Beck. in Rchb. Ic. Fl. Germ. XXIV (1905) 67. — Ic.: Rchb. l. c. tab. 2091.

Annual; stem slender, erect, to 60 cm long, with elongated virgate branches; ocreae semipellucid, 6 — 8-nerved, developing with age; leaves oblong to lanceolate, flat, pinnate-veined; flowers 3 — 5 in lower and 1 — 3 in upper axils; racemes elongated, all internodes including the uppermost long, hence the whole inflorescence apparently interrupted; perianth roseate or red, 2 — 2.5 mm long; achene 3 — 5 mm long, completely covered by the accrescent perianth. July — October.

Pastureland and fallow grazing land. — European part: Bl., Crim.; Caucasus: Cisc. **Gen. distr.**: Centr. Eur. (SE and France), Bal.-As. Min. Described from Hungary.

Note. Distinguished among all related species by the largest fruits.

39. **P. patulum** M. B. Fl. taur.-cauc. I (1808) 304; Aschers. et Graebn. Synops. IV, 864; Grossg., Fl. Kavk. II, 50; Grigor'ev in Fl. Yugo-Vost. IV,
630 110; Kryl., Fl. Zap. Sib. 854. — P. Bellardi (non All. Fl. Ped. II (1787) 205) Ldb. Fl. Ross. III, 530; Shmal'g., Fl. II, 393, p.p. — P. aviculare var. Bellardi Duby Bot. Coll. (1828) 405. — P. Bellardi β patulum Meisn. in DC. Prodr. XIV (1857) 99. — P. spectabile Lehm. Ind. Sem. h. Hamb. 1822. — P. Kotovi Klokov in Journ. of Agricult. Bot. Ucr. vol., I, 3 (1927) 67.

Annual; stem erect, sturdy though more slender than in the preceding species, with elongated internodes and obliquely ascending branches (hence the name), 20 — 80 (120) cm long; ocreae short, semipellucid, 6- or 7-nerved, ferruginous at base, finally fringed-lacerate; achene commonly to 8 mm long (15 mm in P. Kotovi Klokov); leaves long-elliptic to lance-linear, acuminate, subsessile or short-petioled, prominently veined, 1 — 4 (6) cm long, 1.5 — 7 mm broad, the uppermost short and narrowly linear; flowers 1 — 4 together, pedicellate, the pedicels as long as perianth; perianth subtubular, closed, green, roseate-margined; racemes regularly interrupted, leafy only in lower part, erect; achene trigonous-ovoid, acuminate, lustrous, punctate, ca. 2 mm long. May — October. (Plate XLIV, Figure 3).

Feathergrass and wormwood-feathergrass steppes, solonetz and steppe meadows, sands, and alluvial riverside deposits, as weed of roadsides and cultivated fields, and in fallows. — European part: M. Dnp., Bl., Crim., V.-Don, Transv., L. Don, L. V.; Caucasus: Cisc., Dag., S. and E. Transc.; W. Siberia: U. Tob., Irt., Ob (S.), Alt.; Centr. Asia: Ar.-Casp., Balkh. T. Sh., Syr D., Amu D., Pam.-Al., Mtn. Turkm. **Gen. distr.**: Atl. Eur., N. Eur., Med., Bal.-As. Min., Arm.-Kurd., Iran., Dzu.-Kash. Described from the Crimea and Cisc. Type in Leningrad.

Note. Ascherson and Graebner point out that, according to Rouy's investigations, the description presented by Allioni (P. Bellardi) refers to P. rurivagum Jord. of the group of species related to P. aviculare and it should therefore be named P. patulum M. B. They split it up into A. Kitaibelianum Asch. et Gr., with inflorescence interrupted also near the end and large flowers and fruits 4 — 5 mm long, the fruit completely

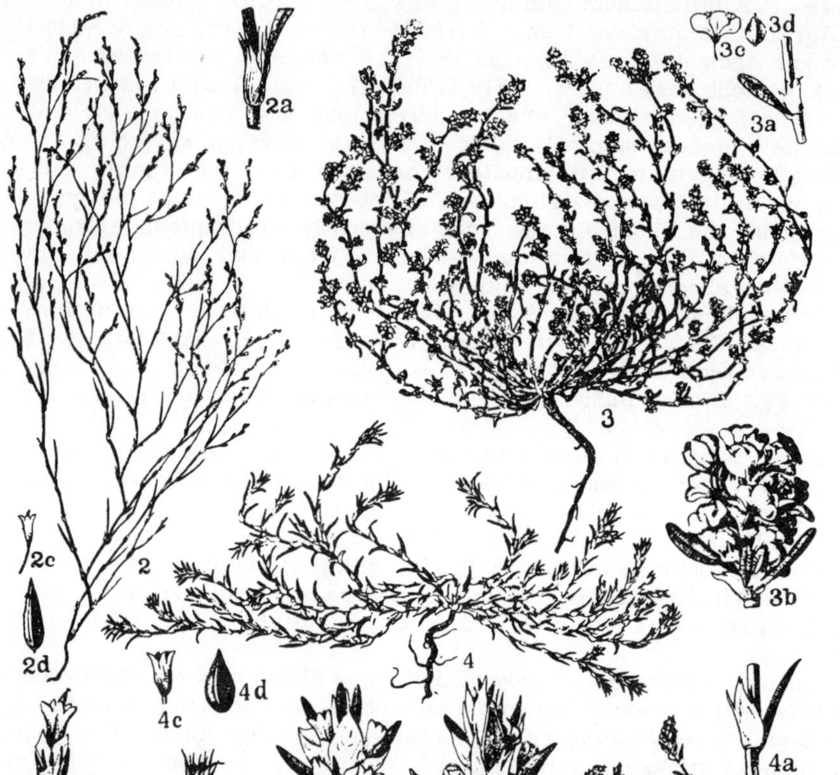

PLATE XLII. 1. Polygonum Aschersonianum H.Gross.– 2. P.salsugineum M.B.– 3. P.corrigioloides Jaub.et Sp.– 4. P.polycnemoides Jaub.et Sp.– 5. P.acerosum Ldb.– Designation of details to all drawings: a) ocrea and leaf, b) portion of inflorescence, c) flower, d) achene.

included in the perianth, and B. genuinum Rouy (= our P. patulum) with approximate terminal inflorescence nodes and fruits 2.5 — 3 mm long but exceeding the perianth. The large-fruited form is either completely absent in our material or, if it has been accidentally overlooked, then its occurrence must be very rare indeed.

Hybrids occurring: P. patulum × P. aviculare; P. patulum × gracilius; P. patulum × arenarium.

40. **P. gracilius** (Ldb.) Klokov in Journ. of Agric. Bot. I, part. 3 (1927) 169. — P. Bellardi β gracilius Ldb. Fl. Ross. III (1849) 530; Meisn. 631 in DC. Prodr. XIV, 99; Turcz. Fl. baic.-dah. II, 2, 70. — P. strictum Ldb. Fl. Alt. (1830) 86. — Ic.: Ldb. Ic. pl. Fl. Ross. tab. 444. — Exs.: HFR No. 1630.

Annual; stem slender, erect, 10 — 50 cm long; lateral branches always upright; ocreae appressed to stem, tubular, 4 — 10 mm long, brown in lower part, paler above, at length strongly lacerate, with several prominent nerves; leaves few, acute, elongate-lanceolate or lance-linear, 2 — 2.5 mm long and to 4 mm broad, the uppermost linear; flowers 2 or 3 in the axils, disposed in slender interrupted racemes; pedicels to 3 mm long, mostly shorter; perianth tubular-infundibular, green, the segments white- or roseate-margined; stamens 8; achene somewhat lustrous, winged-trigonous, with strongly concave faces and rounded ribs, the punctate surface appearing glandular under the microscope. July — September. (Plate XLIII, Figure 1).

Solonetz soils in semidesert areas, dry stony beds, riverside sands and pebbles, and solonetz pastures. — European part: M. Dnp., Bl., L. Don, V.-Don, Transv., L. V.; Centr. Asia: Ar.-Casp., Balkh., Dzu.-Tarb.; W. Siberia: U. Tob., Irt., Alt.; E. Siberia: Ang.-Say. (Minusinsk), Dau. (along the Selenga River). Gen. distr.: Dzu.-Kash., Mong. Described from the Ul'ba and Bukhtarma rivers in Altai. Type in Leningrad.

Note. The specific name R. strictum cannot be retained since it was given much earlier to another species, see Allioni Fl. Pedemont. (1785) No. 2051, tab. 68, f. 2, apparently P. minus Huds.

Hybrid produced: P. gracilius × P. aviculare.

41. **P. cretaceum** Kom. sp. nova in Addenda IV, p. 553.

Annual, branched from base, grayish throughout; stems erect, sulcate, smooth, with appressed upright branches; all ocreae separated into narrow mostly filiform divisions; leaves small, oblong, due to revolute margins subacicular, readily deciduous, 1-nerved; flowers mostly solitary, disposed in loose filiform interrupted racemes, the inflorescence leaves transformed into minute bracts or absent; perianth subtubular, the roseate segments half as long as the tube; achene very small, with lanceolate faces, narrow, acute, punctate-rugose, light brown, lustrous. August — October.

Chalk outcrops. — European part: V.-Don. Endemic. Described from the Derkul' River (Stepanov, 9 September, 1895, a chalky slope in Gorodishche). Type in Leningrad.

Note. According to Klokov, this is a P. gracilius × P. aviculare hybrid.

42. **P. novo-ascanicum** Klokov in Journ. of Agricult. Bot. of the Ucraine, v. I, 1 (1926) 168.

632 Annual; stem branching, 8 —80 cm long; branches upright, appressed to stem or slightly divergent or horizontally nodding; ocreae dull or somewhat lustrous, in lower part ferruginous or brown, narrow, finely dissected, to 8 mm long; leaves oblong, linear, or lance-oblong, to 35 or even up to 50 mm long and 2 —6 (to 7.5) mm broad, cuneately narrowed at base to a very short petiole, long-tapering toward apex; racemes long, slender, interrupted; flowers in 2's to 4's; pedicels slightly longer than perianth but shorter than ocreae; perianth green at base, the segments roseate or reddish (only exceptionally white); achene slightly longer than perianth, black, dull, punctate-ribbed, 1.75 —2 mm long, 1.25 mm broad. July—October.

Sub-Pontic steppes and stony slopes. — European part: M. Dnp., Bl., L. Don, L. V., Crim.; Caucasus: Cisc. Endemic. Described from the vicinity of Poltava and Dnepropetrovsk. Type in Kharkov (Klokov's material collected from Askaniya-Nova, 27 August, 1931, should be regarded as type).

Note. A plant closely related to P. gracilius (Ldb.) Klokov, but differing in the ferruginous coloring of ocreae, the more numerous narrower leaves, and the more slender inflorescence branches.

43. **P. inflexum** Kom. sp. nova in Addenda IV, p. 553.

Annual; stems 10 —30 cm long, erect, sparingly dichotomously branched; branches arcuately or falcately upcurved, their tips inflexed; ocreae on young branches silvery-white, tubular, 2-toothed at the margin, finally lacerate, green at base, becoming darker and but slightly brown; leaves lanceolate, oblong, or linear-lanceolate, somewhat glaucous, fleshy, with prominent midrib and faintly translucent 2 or 3 lateral veins, acute or subacute, 5 —25 mm long, 3 —7 mm broad, the petiole 2 —5 mm long; flowers 1 —5 in the axils of minute green narrow and often conduplicate green bracteal leaves, relatively approximate especially toward the ends of branches; pedicels ca. 1 mm long; perianth tubular, dissected nearly to base; segments green, broadly white- or roseate-margined, elongating in fruit to 2.5 mm, without distinct nerves, almost smooth; achene trigonous, lustrous, acute, to 2 mm long and 1.5 mm broad, under the microscope quite smooth. April—June. (Plate XLIII, Figure 3).

Sandy and pebbly banks of rivers and irrigation ditches. — Centr. Asia: Syr D., Pam.-Al., Mtn. Turkm. Endemic. Described from the Kara-sai River in the Margelan area. Type in Leningrad.

Note. Distinguishable from other "Bellardiiformes" by the unique disposition and shape of branches as well as the completely smooth fruit.

633 44. **P. oxanum** Kom. sp. nova in Addenda IV, p. 554.

Annual; stems simple or mostly branched from base, erect, 20 —50 cm long, finely sulcate, rather densely leafy; branches ascending or obliquely upright; ocreae faintly colored, silvery-white at base, with broad pointed teeth; pedicels to 5 mm long, slender; leaves grayish-green, somewhat fleshy, with almost imperceptible lateral veins, oblong, 1 —4 cm long, 3 —10 mm broad; inflorescence an erect interrupted raceme, sometimes caudiform or in large specimens almost paniculate; flowers 1 — 3 in the axils; pedicels 1 —2 mm long, jointed at the very end; perianth roseate or red, at first tubular, finally campanulately expanding, ca. 1.5 mm long, elongating in fruit to 3 mm, divided nearly to the base; achene completely included in the perianth, trigonous, with strongly convex faces, lustrous, 2.5 mm long and 1—1.5 mm broad. April—September. (Plate XLIII, Figure 2).

Foothills, loess hills, solonetz meadows, sands, weed-infested places, pastures, and alfalfa and wheat fields. — Centr. Asia: Syr D., Kyz. K., Amu D., Pam.-Al., Kara K., Mtn. Turkm. **Gen. distr.:** Iran. Described from the vicinity of Novaya Bukhara [Kagan]. Type in Leningrad.

Note. Related to P. patulum; distinguishable by the fleshy leaves without distinct veins, the marked delimitation between inflorescence and the leafy part of the stem, and the broader obtusish achene not distinctly punctate. Possibly of hybrid derivation P. patulum × P. aviculare?

Economic importance. A widespread weed of unirrigated crops in Uzbekistan, Tadzhikistan, and Turkmenistan.

45. **P. tiflisiense** Kom. sp. nova in Addenda IV, p. 552.

Annual; stem erect, 20—40 cm long, finely sulcate, with appressed upright branches; ocreae deeply bifid, white, at length turning brown; leaves yellowish-green, 1—6 cm long, 0.3—3.2 cm broad, greatly exceeding the internodes, broadly oval-lanceolate, acute or obtusish at apex, the distinct petiole 1—4 mm long; racemes strongly interrupted, with exposed internodes to 2 cm long; flowers 1—3 in the axils of linear or (the upper) almost subulate bracts; flowers tubular, temporarily expanding campanulately, 2—3 mm long, deeply divided; segments green with red tips or stramineous to almost white; fruiting perianth with enlarged longitudinal nerves, hence apparently rugose; achene broadly trigonous,
634 ca. 3 mm long and 2 mm broad, punctate-granular. May—July.

Caucasus: E. Transc. Endemic. Described from Zerran station in Akhaltsikhe District. Type in Leningrad.

Note. Differing from other species of the series Bellardiiformes in its heterophylly characterized by the large middle cauline leaves and the strongly developed nervature of the perianth.

46. **P. junceum** Ldb. Fl. Ross. III, 2 (1849) 529; Meisn. in DC. Prodr. XIV (1857) 100. — P. divaricatum Lessing in Linnaea IX (1834) 204, non Willd.

Annual; stems erect, slender, strongly paniculately branched, channeled, with very long internodes; ocreae strongly lacerate; leaves oblong to linear-oblong, acute, cuneately narrowed to petiole; flowers solitary or in 2's or 3's, approximate at the ends of branches and branchlets; pedicels shorter than perianth; achene slightly exserted from the perianth.

Steppe hills. — W. Siberia: U. Tob. Endemic. Described from Lessing's specimens collected in the vicinity of Iletskaya Zashchita [Sol'-Iletsk]. Type in Berlin.

Note. Approaching P. patulum but distinguishable by its strong ramification, the long internodes, the sparsely leafy stem shedding nearly all its leaves before flowering, and the completely leafless inflorescence branches. Also reminiscent of P. floribundum from which it differs in the slender branches, the narrower petiolate leaves, and looser racemes.

(641)

PLATE XLIII. 1. Polygonum gracilius (Ldb.) Klok. – 2. P. oxanum Kom. – 3. P. inflexum Kom.
Designation of details to all drawings: a) ocrea and leaf base, b) portion of inflorescence, c) flower, d) achene.

47. **P. argyrocoleum** Steud. in Kotschy Pl. Alepp. No. 440 ex Kunze in Linnaea XX (1847) 17 et ex Meisn. in DC. Prodr. XIV (1857) 99; Boiss Fl. Or. IV, 1035; Fedtsch. Consp. Fl. Turk. VI, 287; Grossg., Fl. Kavk. II, 50.

Annual; stems strongly branched, slenderly virgate, erect or ascending, to 1 m long and to 4 mm in diameter, with elongated internodes; branches obliquely ascending, sometimes subappressed to stem; ocreae short, cupuliform, dark brown at base, silvery above, 6—8-nerved, the truncate margin at length fimbriate; leaves few, lanceolate to linear-lanceolate, acute, flat, 1—3 (4) cm long, 4—8 mm broad, soon deciduous; inflorescence branches virgate, with long internodes; flowers 1—3 together; pedicels slender, the length of the perianth; perianth ca. 2 mm long, expanded,
637 campanulate, bright roseate or green, the rounded segments white-margined; achene brown, quite smooth, very lustrous, included in the perianth. July — September. (Plate XLIV, Figure 1).

Sandhills, solonetz soils, and weed-infested places; also reported among crops. — Caucasus: S. and E. Transc.; Centr. Asia: Ar.-Casp., Balkh., Mtn. Turkm., Kara K. **Gen. distr.**: Arm.-Kurd. Described from cultivated fields near Mosul and Baghdad. Type in Geneva.

Note. Indistinguishable from other species of the P. Bellardii All. group by the virgate branches leafless at the flowering stage, the racemes 10—20 cm long and interrupted right up to the end, and the very lustrous completely smooth achenes.

48. **P. deciduum** Boiss. et Noe in Schlecht. Bot. Zeit. (1853) 734 (nomen solum); Meisn. in DC. Prodr. XIV (1857) 99.

Annual; stem branched, sturdy from base, ca. 2 mm in diameter and 50 cm long, sulcate; ocreae infundibular, colored, many-nerved, strongly disintegrating with age; leaves soon deciduous, flat, elongate-lanceolate, acute, 1-nerved; flowers in long interrupted racemes, 2—4 in the axil; upper leaves transformed into bracts barely surpassing the flowers; fruiting perianth with strongly developed nerves; achene very slightly shorter than perianth, quite smooth, lustrous. May—October.

Stony slopes and terraces. — Centr. Asia: Syr D., Amu D. **Gen. distr.**: Iran. Described from Iranian Kurdistan.

Note. Distinguishable from the related P. argyrocoleum Steud. by its more slender branches, narrow flowers, and less developed ocreae.

49. **P. arenarium** Waldst. et Kit. Pl. rar. Hungar. (1802) 69; Meisn. in DC. Prodr. XIV, 100; Ldb. Fl. Ross. III, 531; Shmal'g., Fl. II, 393. — Ic.: Waldst. et Kit. l. c., tab. 67. — Exs.: HFR No. 1086.

Annual; stems branched from base, divergent, in lower part terete, in upper part angled, in inflorescence square in cross section, sulcate, slender; internodes much elongated; slender branchlets flexuous, often zigzag; ocreae infundibular, dark at base, pellucid-white above, lacerate, with 6 or fewer nerves; leaves numerous, linear-lanceolate, acute, 1-nerved, flat, 2—2.5 cm long and 5 mm broad; inflorescence leafless, divaricately many-branched, the individual glomerules distant; flowers 1—3 together; pedicels as long as perianth; perianth campanulate or infundibular, expanded, 2—3 mm long; the rounded segments white, yellowish-white, or roseate; anthers yellow; achene to 2 mm long, exposed, trigonous, lustrous, dark brown, smooth. June — October.

Sands, sandy deposits, low-lying places in steppes, solonetz and loess soils, often among crops, both inland and near the coast. — European part:
638 M. Dnp., Bl., L. Don; Caucasus: W. and E. Transc. **Gen. distr.**: Centr. Eur. (only S.), Bal.-As. Min. (W.). Described from Hungary. Type in Prague.

50. **P. pseudoarenarium** Klok. in Beiträge zum Naturschutz (Materiyali okhoroni prirodi na Ukraini), I (1928) 74. — P. arenarium (non Waldst. et Kit.) Shmal'g., Fl. II, 393 (ex max p.); Grigor'ev in Fl. Yugo-Vostoka, IV, 110; Kryl., Fl. Zap. Sib. 855; Ldb. Fl. Ross. III, 531 (ex maxima parte). — Exs.: HFR No. 1086 sub P. arenarium.

Annual; stem 30—90 cm long, branched from base, the branches declinate; leaves oblong-lanceolate or elliptic, to 5 cm long, 4—9 mm broad; inflorescence leafless, branched, mostly paniculate, the divergent branches straight; flowers 2 — 4 in the axils; perianth broadly infundibular, yellowish-white except for the greenish base, rarely roseate, 1.5 — 2.5 mm long; achene lustrous, with perfectly smooth faces, 1.5 — 3 mm long. July—September. (Plate XLIV, Figure 2).

Coastal sands, solonetz meadows and wet places in steppes. — European part: M. Dnp., Bl., Crim., V.-Don, L. Don, L. V., Transv.; Caucasus: Cisc., Dag.; W. Siberia: U. Tob., Ob (S.), Irt.; Centr. Asia: Ar.-Casp., Balkh., Syr D., Amu D., Mtn. Turkm., Pam.-Al. Endemic. Described from the shores of the Sivash lagoons, from spits and islands in the Dnieper River estuary. Type in Kharkov.

Note. Pedicels of some flowers reach up to 5 mm in length, i. e., twice the length of the perianth.

51. **P. Janatae** Klokov in Beiträge zum Naturschutz (Materiyali okhoroni prirody na Ukraini), I (1928) 73.

Annual; stem erect, 15—40 cm long, terete, sulcate; the branches arising well above ground level, inflorescence branches often 3-angled, the uppermost flat; ocreae on the stem cupuliform, fulvous-brown, the pointed tips of the teeth lighter-colored; ocreae of slender branchlets cyathiform or obconical, brownish with broad white tips, nerveless; leaves few, petiolate, oblong or linear-lanceolate, acute, slightly revolute-margined, pinnately veined, 5—25 mm long, 1—2 mm broad; flowers crowded at the ends of branches of a paniculate inflorescence, mostly longer than the internodes; pedicels 2—3 mm long, jointed at the end; perianth broadly infundibular or campanulate, yellowish-white (stramineous), rarely roseate, divided nearly to the base, 2.5 → 3.25 mm long; achene punctate-rugose, dull, with oval back and rounded ribs, acute. July—October.

Solonetz and coastal sands. — European part: Bl., Crim. (northward to Kerch). Endemic. Described from the Kherson area (Dolgii and Kruglyi Islands in the Black Sea). Type in Kharkov.

639 Note. Related to P. arenarium Waldst. et Kit., but differing in having larger and notably broader achenes, and flowers crowded toward the ends of panicle branches.

Series 8. **Molliiformes** Kom. — sect. Pseudomollia Boiss. Fl. Or. IV, 1043. — Perianth pellucid, membranaceous, deeply divided, the two outer divisions alately keeled; achene flat.

52. **P. molliiforme** Boiss. Diagn. Ser. I, 7 (1846) 84; Fl. Or. IV, 1043; Fedtsch. Consp. Fl. Turk. VI, 289.

Annual; stems short, simple or slightly branched; branches furcately divergent, short, filiform; ocreae pellucid, silvery, nerveless, somewhat inflated from base, lanceolate, acuminate, setaceously disintegrating; leaves linear, setiform, 3—5-nerved, surpassing ocreae and flowers; flowers solitary, subsessile, slightly exserted from ocreae; perianth divided nearly to base into 5 segments; stamens 5, of these only 2 or 3 antheriferous; styles divergent; achene oval, flattened, lustrous, sharp-beaked. June—September.

Dry stony slopes and pebbles. — Centr. Asia: Syr D., Pam.-Al. (including Pamir). **Gen. distr.**: Iran. Described from mountains of N. Iran. Type in Geneva.

53. **P. Bornmülleri** Litw. in Trav. Mus. Bot. Acad. Pétersb. VII (1910) 82.

Annual; stems not more than 3 — 7 cm long, brittle, smooth except for scabrous nodes, more or less branched in upper part or sometimes also from base; ocreae pellucid, lustrous, oblong-lanceolate, acute or rounded at apex, entire; leaves setiform, 1-nerved, ca. 1 cm long, 0.5 mm broad, acute, the uppermost surpassing ocreae and flowers; flowers solitary, subsessile, included in ocreae; perianth ca. 2 mm long, divided to two-thirds into 5 acute segments; stamens 5; styles erect; achene to 1.75 mm long, broadly lanceolate, black, lustrous, white-marginate or narrow-winged on the ribs. May—June.

Stony mountain slopes. — Centr. Asia: Pam.-Al. Endemic. Described from the Vancha Valley in Darvaz at altitudes of about 1,500 m. Type in Leningrad.

Note. A form growing in the Kugitang Range (S. Nevskii) has fringed ocreae.

Section 2. **CEPHALOPHILON** Meisn. in Wallich, Pl. As. rariores 3 (1832) 59. — Inflorescences capitate; perianth semipetaloid, 4- or 5-parted; stamens 8 or more rarely 6; perigynous glands none; styles 3, filiform, semiconnate, with capitate stigmas; ocrea inclosed in a slightly accrescent dry or somewhat fleshy perianth, trigonous or more rarely lenticular; pubescence absent; ocreae membranaceous, cylindric, with ciliate or naked
640 margin; leaves entire or sinuate-lobed, the petioles often with foliaceous bracts at base; flowerheads globose, solitary or paired or numerous, disposed in a paniculate inflorescence; bracts scarious, flat or subtubular.

54. **P. alatum** Hamilt. ex D. Don, Prodr. Fl. Nepal. (1825) 72; Kom. Fl. Mansh. in A. H. P. (1903) 128; Hook. Fl. Brit. Ind. V, 41. — P. punctatum Hamilt. in Don, Prodr. Fl. Nepal. (1825) 72. — P. nepalense Meisn. Monogr. Pol. prodr. (1826) 84 et in DC. Prodr. XIV, 128; B. Fedtsch. Consp. Fl. Turk. VI, 292. — Persicaria alata (Hamilt. ex D. Don) Nakai Fl. Quelpaertensis (1914) 40. — P. nepalensis Miyabe Journ. Fac. Agr. Hokk. XXVI, 4 (1934) 514. — The name referring to the shape of the petiole.

Annual; stems sometimes creeping at base, commonly erect or slightly ascending, simple or branched, 2—50 cm long; stems and branches slender, weak, somewhat watery, smooth or sparsely glandular-hairy; ocreae more or less tubular, the upper margin obliquely truncate; leaves broadly or narrowly oval or deltoid, 1—9 cm long and 0.5—3 cm broad, minutely puncticulate when dry owing to protrusion of numerous druses present in the mesophyll; the broadly winged petioles amplexicaul at base; pedicel varying in length, with a tuft of rather long glandular hairs terminated by blackish-purple heads; inflorescences in the axils of and often surpassing the sessile upper leaves, capitate, compact, ca. 1 cm in diameter; bracts oval-lanceolate, greenish, brown, or red, white-margined, surpassing the flowers; perianth green or white or rosy-purple, membranaceous; anthers blackish-purple; styles relatively long, bifid, with capitate stigmas; achene completely included in the perianth tube and crowned by its segments, rounded-oval, with a rounded keel and tuberculate surface, the tubercles disposed in tightly approximate parallel rows. July—October.

Banks of forest rivulets, riverside sands and pebbles, dampish stony slopes, and as weed of ornamental and truck gardens and of fields. — Centr. Asia: Pam.-Al. (altitudes of 1,500—3,500 m in mountains); Far East: Uss. (S.) near to the sea. **Gen. distr.:** Jap.-Ch., Ind.-Him., Iran. (Afghanistan), Abyssinia. Described from Nepal (Him.). Type in London.

55. P. runcinatum Hamilt. in D. Don, Prodr. Fl. Nepal. (1825) 73; Meisn. in DC. Prodr. XIV, 130. — Ic.: A. Makashvili, Sornye r. Gruzii (1934), tab. 26.

643 Perennial; rhizomes slender, horizontally creeping, branching; stems ascending, soft, sulcate, slightly branched, smooth, 30—40 cm long; leaves pinnately lobed, with rounded-triangular terminal lobe and 2—6 rounded lateral lobes, glabrous or short setiform hairs especially beneath, somewhat fleshy; petiole with 2 auricles at base; ocreae ciliate-margined; pedicels covered with bristles sometimes interspersed with glandular hairs; flowers in compact heads ca. 1 cm in diameter; bracts oblong, acute, scarious; perianth segments 5, bright roseate, white, or greenish; stamens 6, with dark blue anthers; stigmas 3; achene rounded-trigonous, with a granular surface. May— October.

Caucasus: tea plantations of W. Transc. (Salibauri). Plants introduced from China. **Gen. distr.:** Ind.-Him., Jap.-Ch. (S.). Described from Nepal (Him.). Type in London.

Note. USSR plants are much smaller than the tropical.

Section 3. PERSICARIA* Meisn. Polyg. Prodr. 66 (1826) excl. § 2, tab. I, fig. 18, 27—29 et tab. 3 fig. L. — As distinct genus Persicaria Tourn. Instit. 5—9, tab. 290; Mill. Gard. Dict. Abr. Ed. 4 (1754). — Annuals, rarely perennials; stems erect, uniformly leafy; ocreae membranaceous, rarely green, truncate at apex; leaves entire, often lanceolate, with or without glandular dots beneath; flowers disposed in simple spikelike racemes; perianth mostly colored, the 4 or 5 segments convergent at apex; stamens

* The name Persicaria alludes to the similarity between the leaves of this group and those of the peach.

4—8; styles 2 or 3, mostly connate; stigmas capitate; achene included in a slightly accrescent wingless perianth, lenticular or trigonous; endosperm horny.

1. Perennials with creeping branched rhizomes; petioles arising above the middle of ocrea . . . . 2.
\+ Annuals with an erect or rarely prostrate stem; petioles arising below the middle or almost at the base of ocrea . . . . 3.
2. Aquatic plant with smooth long-petioled floating leaves and more or less solitary emersed flowers . . . . 56. **P. amphibium** L. f. aquaticum Leyss.
\+ Terrestrial plant with an erect densely leafy stem covered with appressed setiform hairs; inflorescence terminal; leaves lanceolate,
644 acute . . . . 56. **P. amphibium** L. f. terrestre Leyss.
3. Individual inflorescences spikelike, more or less compact and stout, cylindric . . . . 4.
\+ Inflorescences slender, sparsely flowered, often interrupted . . . . . . 11.
4. Ocreae with green foliaceous or dry lobes . . . . 59. **P. orientale** L.
\+ Ocreae with entire upper margin, not lobed . . . . 5.
5. Ocreae narrow, firmly clasping the stem especially in its lower part . . . . 6.
\+ Ocreae broad, loosely adherent to stem . . . . 8.
6. Ocreae glabrous, with or without very short cilia on the margin; leaves acute . . . . 62. **P. linicola** Sutulov.
\+ Ocreae with long straight cilia on the margin . . . . 7.
7. Ocreae glabrous; leaves broad, obtusely rounded at summit . . . . 61. ***P. tinctorium** Lour.
\+ Ocreae covered with appressed hairs; leaves mostly lanceolate, acute . . . . 63. **P. persicaria** L.
8. Stem rather densely clothed with spreading setiform hairs . . . . 59. **P. viscosum** Hamilt.
\+ Stem devoid of setiform hairs . . . . 9.
9. Stem sparsely covered with sharp retrorse prickles . . . . 58. **P. Bungeanum** Turcz.
\+ Stem smooth or more rarely covered with soft hairs . . . . 10.
10. Peduncles densely covered below the inflorescence with glandular hairs; margin of fruiting perianth prominently nerved . . . . 61. **P. scabrum** Moench.*
\+ Peduncles eglandular or with very few glands . . . 62. **P. nodosum** Pers.
11. Achenes trigonous; styles 3 . . . . 12.
\+ Achenes flat or terete; styles 2 . . . . 14.
12. Upper internodes viscous in their upper part . . . . 73. **P. viscoferum** Makino.
\+ All internodes smooth and dry . . . . 13.
13. Leaves broadly lanceolate; achene 2—2.5 mm long . . . . 72. **P. posumbu** Hamilt.
\+ Leaves narrowly lanceolate; achene 1—1.5 mm long . . . . 70. **P. excurrens** Stes.
++ Leaves linear; achene 1 mm long . . . . 71. **P. intricatum** Kom.

* The dense gray or white cobwebby vesture occurs both in P. scabrum and P. nodosum, but it is more frequent in P. scabrum.

645 14. Perianth covered with distinct flat glands . . . . 69. **P. hydropiper** L.
\+ Perianth smooth, eglandular . . . . 15.
15. Ocreae glabrous or almost glabrous with sparse short cilia . . . . . . 16.
\+ Ocreae with appressed hairs and long cilia . . . . 18.
16. Leaves broadly oblong-lanceolate, more than 2 cm broad . . . . 65. **P. imeretinum** Kom.
\+ Leaves lance-linear or linear . . . . 17.
17. Individual flowers in inflorescence approximate, more or less contiguous; shore plants . . . . 67. **P. foliosum** Lindb.
\+ Individual flowers in inflorescence clearly discontiguous; plants partly submerged . . . . 67. **P. foliosum** var. **aquaticum** Kom.
18. Leaves oval to oblong-lanceolate, with distinct lateral veins; fruiting perianth 2.5 — 3.5 mm long . . . . 68. **P. mite** Schrank.
\+ Leaves linear or linear-lanceolate, with indistinct lateral veins; fruiting perianth not more than 2.5 mm long . . . . . . 66. **P. minus** Huds.

Series 1. **Amphibiae** Kom. — Perennial plants with creeping rhizome; leaf petioles inserted on the stem above the middle of the ocrea; inflorescence dense, spikelike.

56. **P. amphibium** L. Sp. pl. (1753) 361; Meisn. in DC. Prodr. XIV, 115; Shmal'g., Fl. II, 390; Grossg., Fl. Kavk. II, 51; Fedtsch. Consp. Fl. Turk. VI, 290; Kryl., Fl. Zap. Sib. 862; Turcz. Fl. baic.-dah. II, 67; Kom. in A. H. P. XXII, 123; Kom. Fl. Kamtsch. II, 66; Ldb. Fl. Ross. III, 520. — P. purpureum Gilib. Exerc. Phytol. II (1792) 433. — P. amurense Niewland in Amer. Midl. Nat. II; (1912) 183. — Persicaria amphibia S. F. Gray, Nat. Arr. Brit. Pl. II (1821) 268. — Ic.: Rchb. Ic. Fl. Germ. XXIV, 77; Kom. et Al. Key Plants Far East. Reg. tab. 142. — Exs.: HFR No. 937 et No. 1238. —Russian names: vodnaya [aquatic] grechikha, utevnik, shchuch'ya trava; in Soviet Central Asia: tamyr-dari.

Perennial with creeping branched rhizome rooting at the nodes; stem erect, simple, to 1 m long, covered with stiff appressed hairs (var. terrestre Leyss. Fl. Hal. (1761) 391), or submersed, smooth, much elongated, flexible and branched (var. natans Leyss., ibidem); leaves in the former oblong-lanceolate, subsessile, appressed-hairy; in the latter floating, lustrous, long-petioled, oblong, at base rounded or cordate (var. amurense Korsh. in A. H. P. XII (1892) 382); ocreae elongated, truncate at summit; inflorescences solitary, ovaloid or cylindric compact spikes 3 — 5 cm long, at the ends of stem or of its branches; perianth bright roseate or white
646 (var. amurense Korsh.); stamens 5, more rarely 4 or 8; some flowers with conspicuous long stamens and well developed anthers, others with nonfunctional short-filamented stamens; achene black, lustrous, biconvex, ca. 2.5 mm long. June — September. (Plate XLV, Figures 1, 2).

The aquatic form in slow-moving and standing waters, in backwaters of large and small rivers, oxbows, lakes, ponds, etc., often forming extensive thickets; the terrestrial form on sandy and clayey banks, in cultivated fields, riverside meadows, roadsides, etc. There is also an intermediate form, var. decumbens Klett. u. Richter Fl. Lipps. (1830) 351, growing on the bottom of drying temporary water bodies, with aerial shoots arising from rhizomes immersed in the silt that formerly gave rise to typical f. natans Leyss. — European part: Kar.-Lap., Dv.-Pech., Lad.-Ilm.,

U. Dnp., U. V., V.-Kama, M. Dnp., Bl., V.-Don, L. Don, L. V., Transv.; Caucasus: Cisc., W., S. and E. Transc.; W. Siberia: Ob, U. Tob., Irt., Alt.; E. Siberia: Yen., Lena-Kol., Ang.-Say., Dau.; Far East (var. amurense Korsh.): Ze.-Bu., Uda, Uss., Sakh. and (var. typicum) Kamch.; Centr. Asia: Ar.-Casp., Balkh., Dzu.-Tarb., Amu D., Syr D., Pam.-Al., T. Sh. **Gen. distr.**: Scand., Centr. and Atl. Eur., N. Med., Bal.-As. Min., Arm.-Kurd., Iran., Ind.-Him., Dzu.-Kash., Mong., Jap.-Ch., Ber., N. Am. Described from Europe. Type in London.

Economic importance. The fruits may be used as food for domestic fowl. An infusion of the roots is used in Soviet Central Asia as a substitute for the expensive imported roots of greenbrier (Smilax) owing to its diuretic properties. The rhizomes and stems yield 18% tannins.

Note. The northern limit runs from Nuot-ozero on the Tulom, to the Varzug River, to the Solovetskie Islands, and Polovinka on the Yenisei River (67°35'); Kamchatka River.

Series 2. **Pinguiores.***— Annual plants; inflorescences compact, cylindric; achenes rounded; petioles inserted near the base of ocrea.

57. ★**P. tinctorium** Ait. Hort. Kew. II (1789) 31; Lour. Fl. Cochinch. I (1790) 241; Meisn. in DC. Prodr. XIV, 102; Partanskii, Prakt. Botan. (1894) 280. — Persicaria tinctoria Spach, Syst. Veg. X (1841) 536.

Annual; stem erect, simple or few-branched, relatively stout, 30—80 cm long; leaves oval or ovate, obtusely rounded at apex, slightly pubescent on the margin, glabrous elsewhere with translucent glandular dots; ocreae narrow, glabrous, the margin ciliate; spikes compact, ovaloid or oblong, red, disposed in a raceme; bracts the length of pedicels; stamens 6—8; achene biconvex, lustrous, smooth, 2—3 mm long. July—September.

647 The plant was cultivated at one time in the Ukraine, in the Caucasus, and in Manchurian villages beside the Amur River, and has occasionally become naturalized in weed-infested places. — **Gen. distr.**: Indochina, China, and sporadically as a weed in Centr. Eur. and in Siberia. Described from Indochina.

Yields 4—5% indigo dye by hydrolysis and acidification of indican, $C_{14}N_{17}NO_6$, present in its tissues, this being achieved by soaking in shallow tanks.

58. **P. Bungeanum** Turcz. in Bull. Soc. Natur. Moscou XIII (1840) 77; Meisn. in DC. Prodr. XIV, 120; Kom. in A. H. P. XXII, 125. — Persicaria Bungeana Nakai ex Mori Enum. of pl. from Korea (1922) 131. — Exs.: HFR No. 2841.

Annual; stem 25—75 cm long, simple or more or less branched, ascending from base, flexuous, erect above, sparsely covered with recurved spiny prickles; branches clothed near the ends with red glandular hairs; leaves short-petioled, oblong-lanceolate, acute, 3—15 cm long, 0.8—2.5 cm broad; ocreae cup-shaped with truncate summit, not ciliate but with stiff hairs on the surface as also on the leaf midrib; racemes solitary or numerous, rather loose, often curved, with few rather small oblong round-

* The three species treated here do not form a natural genus, and occasionally they appear as three different genera, not represented within the USSR.

tipped often red bracts; perianth white or roseate, smooth, somewhat flattened down, ca. 3 mm long; achene subglobose, spinous-beaked, black, dull. August — October.

Sands among osier-beds, near ricefield canals, cultivated fields, gardens, and roadsides. — Far East: Uss. **Gen. distr.**: Jap.-Ch. (only N. China, Manchuria, and Korea). Described from the vicinity of Peking, from the village of Kantai. Type in Kharkov.

59. **P. viscosum** Hamilt. ex Don, Prodr. fl. Nepal. (1825) 71; Meisn. in DC. Prodr. XIV, 102; Kom. in A. H. P. XXII, 120. — Persicaria viscosa H. Gross ex Mori, Enum. of pl. from Korea (1922) 134.

Annual; stem in lower part ascending or creeping and rooting at nodes, erect above, simple or with several axillary branches, rather densely covered with spreading white setiform hairs; leaves oval, oval-lanceolate, or broadly lanceolate, acute, 4 — 10 cm long and 1.5 — 4 cm broad, cuneately narrowed to a short petiole, on the midrib with hairs to 0.5 cm long, elsewhere with sparse shorter hairs; ocreae narrow, densely clothed with setiform hairs, long-ciliate on the margin; peduncles rather heavily glandular; racemes 1 — 3 cm long, fairly compact; bracts subimbricate at
648 the onset of flowering, ciliate and hairy, about the length of the pedicels; perianth bright violet-rose, smooth, eglandular; stamens 8, short; achene ovaloid-trigonous, short-beaked, obscurely punctate, somewhat lustrous, dark brown. July — October.

Marshy riverbanks and lakeshores, low-lying places in meadows, valleys and low ridges, cultivated fields, and fallows (the depressed form). — Far East: Uss. **Gen. distr.**: Ind.-Him., Jap.-Ch. Described from Nepal.

Series 3. **Amblygonium** Meisn. (pro sectione). — Large annuals; perianth corolloid; stamens 7; achene with obtuse rounded ribs.

60. **P. orientale** L. Sp. pl. (1753) 362; Meisn. in DC. Prodr. XIV, 123; Asch. et Graebn. Synopsis IV, 839; Shmal'g., Fl. II, 391; Grossg., Fl. Kavk. II, 52; Fedtsch. Conspectus Fl. Turk. VI, 291; Kom. in A. H. P. XXII, 126. — Russian vernacular names: buryachki, turetskaya lipka; in Soviet Central Asia: guli-kamchi.

Annual; stem erect, branched, to 1 m or even up to 2.5 m long, rather densely covered with spreading or appressed hairs, rarely almost smooth; ocreae cup-shaped, more rarely cylindric, the lobed upper margin crowned by long bristles, the lobes green and foliaceous in Far Eastern and scarious in Central Asian plants; leaves short-petioled, oval, more rarely orbicular or broadly lanceolate, gradually attenuate to a point, 3 — 25 cm long and 1 — 15 cm broad, with hairs and with small round glands inserted in depressions and appearing as dark dots under a magnifying glass; spikes solitary or numerous at the ends of branches, oblong, to 8 cm long and 2 cm in diameter; bracts broad-oval, hairy, mucronate, ciliate-margined, each covering 1 — 5 relatively large bright red or roseate or white flowers; styles with capitate caplike stigmas; perianth broadly infundibular; achene lustrous, black, rounded from base, the flat faces deeply concave at the middle. June — September.

Grown in gardens for ornament; wild in weed-infested places and (in the Far East) on bottomland among osier beds. — European part: M. Dnp., V.-Don, Bl., Crim.; Caucasus: W. Transc.; Far East: Uss.; Centr. Asia: Mtn. Turkm., Amu D., Syr D., Pam.-Al. **Gen. distr.**: Ind.-Him., Malay Archipelago, Iran., Jap.-Ch., Dzu.-Kash. Described from the East. Type in London.

4. Series 4. **Persicariaeformes** Kom. — Annuals; leaves short-petioled, inserted on the stem nearly at the base of ocreae; achenes flat; styles 2.

61. **P. scabrum** Moench, Meth. (1794) 629; Fernald in Rhodora, 259. — P. tomentosum Schrank, Baier. Fl. (1789) 669; Aschers. et Graebn. Syn. IV, 812; Grossg., Fl. Kavk. II, 52; Kryl., Fl. Zap. Sib. 860. — P. lapathi-
649 folium L. Sp. pl. (1753) 360 (ex parte); Fedtsch. Consp. Fl. Turk. 6 (1916) 291 (ex parte); Turcz. Fl. baic.-dah. II, 67; Kom. in A. H. P. XXII, 124. — P. parviflorum Gromow, Enum. stirp. Charkov. in Trudy Obshch. Nauk pri Khar'k. Univ. I (1817) 145. — P. tenuiflorum Presl ex Nyman, Consp. Fl. Eur. (1881) 637. — Persicaria maculata S. F. Gray, Nat. Arr. Brit. Pl. II (1821) 270. — Exs.: HFR No. 1631, 2847.

Annual; stem erect, ascending, or decumbent (a form with prostrate branched stems — v. prostratum Asch. et Graebn. = P. decumbens Schur), often branched, 30—60 cm long, although dwarf or very large specimens up to 1 m tall also occur; leaves short-petioled, 4—10 cm long, oblong or oblong-elliptic to lanceolate, often obtusish, glandular-punctate beneath, often with a crescent-shaped black blotch above, green or with dense light gray or white cobwebby pubescence (P. incanum Schmidt, 1794), distinctly petiolate; ocreae glabrous or arachnoid, the margin ciliolate; racemes obtuse, to 4 cm long and 1.5 cm thick, the central peduncles long, the lateral ones short; perianth mostly greenish, ca. 3 mm long, the two outer segments strongly 3-nerved; inflorescence branches, pedicels and the outside of perianth covered with yellow glands; achene compressed laterally, with depressions on both sides, to 2.5—3 mm long, included in the perianth until fully mature. June—October.

Weed-infested places, cultivated fields, and fallows (usually associated with field crops), sands, and meadowland gullies. — European part: Kar.-Lap. (northward to about 67° N. lat.), Lad.-Ilm., Dv.-Pech. (Shenkursk), U. V., U. Dnp., M. Dnp., Bl., V.-Don, L. Don, L. V., Transv., V.-Kama; Caucasus: Cisc., Dag., W., S. and E. Transc.; W. Siberia: U. Tob., Ob (up to 65° N. lat.), Irt., Alt.; E. Siberia: Yen. (up to 70°20' N. lat.), Ang.-Say., Dau., Lena-Kol. (S.); Far East: Ze.-Bu., Uss., Uda, Sakh., An. (Penzhinsk, sporadically). **Gen. distr.**: Scand., Atl. Eur., Centr. Eur., Med., Bal.-As. Min., Iran., Ind.-Him., Mong., N. Am., etc. Described from Bavaria.

Economic importance. Together with P. persicaria and P. nodosum, used under the common name of peachwort [Russian "pochechuinaya trava" or "pochechui"], in the form of aqueous or alcoholic extracts together with buckthorn bark, for treatment of hemorrhoidal constipation. A widespread remedy.

Note. Produces the following hybrids: P. lapathifolium × nodosum, Rchb. Fl. Germ. Exc. (1832) 572; P. scabrum × hydropiper; P. scabrum × persicaria. The first of these is apparently widely distributed in the USSR.

62. **P. nodosum** Pers. Syn. I (1805) 440; Asch. et Graebn. Syn. IV, 814; Ldb. Fl. Ross., III, 521; Grossg., Fl. Kavk. II, 52; Kryl., Fl. Zap. Sib. 860; Kom. in A. H. P. XXII, 124. — P. glandulosum Kit. in Linnaea XXXII (1863) 363. — P. lapathifolium L. Sp. pl. (1753) 360, ex p.; Fedtsch.
650 Consp. Fl. Turk. 6 (1916) 291, p. p. — P. paniculatum Andrz., Ischisl. r. Podol'sk. gub. (1862) 136. — Persicaria lapathifolia S. F. Gray, Nat. Arr. Brit. pl. II (1821) 270. — Ic.: Rchb. Ic. Fl. Germ. XXIV, tab. 215, 1—5; Kom. and Alis., Opred. rast. Dal'nevost. kr. I, Plate 124. — Exs.: HFR No. 2848.

Annual; stem erect, ascending, or decumbent, 3—10 cm long, often thickened at the nodes, simple or branched, glabrous, red or speckled; leaves distinctly petiolate, oblong or lanceolate, nearly always acuminate, more rarely oval or linear-lanceolate, glandular-dotted beneath, often with a dark blotch above; ocreae mostly glabrous, their margin ciliolate; racemes elongated, often attenuate toward ends, nodding, the lateral shorter than the terminal; peduncles and pedicels eglandular; perianth white or roseate, ca. 2 mm long; achene lenticular, caving in on both sides, dark brown, ca. 2 mm long. June—October. (Plate XLV, Figure 4).

Wet pastures, canals and roadsides, gardens, etc.; also on the shores of lakes and oxbows with the base of the plant in water (outstandingly large specimens); also floating specimens in lakes (var. natans Schröter, 1902), with thick hollow nodes. — European part: Lad.-Ilm., U. V., V.-Kama, U. and M. Dnp., V.-Don, Transv., Bl., Crim., L. Don, L. V.; Caucasus: Cisc., Dag., W., S. and E. Transc.; W. Siberia: Ob, Irt.; E. Siberia: Ang.-Say., Dau.; Far East: Ze.-Bu., Uss., Sakh. **Gen. distr.**: Scand. (S.), Atl. and Centr. Eur., Med., Bal.-As. Min., Ind.-Him., Dzu.-Kash., Jap.-Ch. Described from Europe.

Note. Linnaeus conceived under the name P. lapathifolium three USSR species — P. tomentosum, P. nodosum, and P. linicola. Later, the most widespread Polygonum of northern and Central Europe, with glandular peduncles and perianth, received the name P. tomentosum Schrank (1787), but this appellation is unsatisfactory in view of the existence of P. tomentosum Willd. (1798) which grows in India and on the islands of the Malay Archipelago. Attempts that have been made to find a more appropriate name are therefore understandable. Japanese authorities picked on P. glandulosum R. Br. Prodr. Fl. Nov. Holland. I (1810) 419 and changed it to Persicaria glandulosa Nakai et Ohwi (Tokyo Bot. Mag. vol. XL (1926) 52) which, however, corresponds more particularly to P. nodosum Pers. (1805) that undoubtedly has priority.

Produces the following hybrids: 1) P. scabrum × P. nodosum, 2) P. nodosum × P. persicaria, 3) P. nodosum × P. hydropiper, 4) P. nodosum × P. mite, and 5) P. nodosum × P. minus.

63. **P. linicola** Sutulov in Izv. Sem. Kontr. ct. Mosk.-Sel'sko-khoz. Obshch. (1914) 1—12; Nenjukov Mitt. zur Fl. Estl. I in Sitzungsber. d. Naturforscher.
651 Gesellsch. d. Univ. Tartu XXXIV (1927) 157; Kryl., Fl. Zap. Sib. 859. — P. lapathifolium f. linicola Schwarz, Fl. v. Nürenb. Erlangen (1900) 687. — Persicaria linicola Nenjukov l. c. (1927) 156.

Annual; stem simple, glabrous, 40—60 cm long; leaves lanceolate, narrowed at both ends, 3—7 cm long, 5—13 mm broad, with petiole 3 to 15 mm long, finely grayish tomentose when young, finally quite smooth, dotted, the margin ciliolate; ocreae narrow, tightly clasping the stem, with sparsely ciliolate margin; racemes fairly thick, erect, 1—2 cm long; pedicels never jointed; flowers greenish-white or roseate, eglandular; stamens 5; achene brownish-black, smooth, lustrous, flat, caved in on the sides, rounded-ovate, beaked, 2—2.5 mm long. June—September. (Plate XLV, Figure 6).

Linseed fields. The achenes of the weed resemble linseed so closely in size and shape that they are not usually separated in winnowing and are therefore sown together with linseed. — European part: Kar.-Lap. (S.), Lad.-Ilm., U. V., V.-Kama, U. Dnp.; W. Siberia: Ob (Turinsk District). Not traced in other regions of the Soviet Union; further observations are needed.

Note. A. S. Sutulov subsequently separated from P. linicola the closely related species P. sublinicola which differs in having jointed fruit-stalks; the achene does not, however, break off at the joint as in P. scabrum L. but just below the joint. The seeds of both linicole species are larger and heavier than those of P. scabrum and P. nodosum.

The diagnosis of f. linicola Schwarz does not mention the presence or absence of a joint on the pedicel; the omission is more likely due to oversight than to actual absence of joints in the German plant.

64. **P. persicaria** L. Sp. pl. (1753) 361; Asch. et Graebn. Synopsis IV, 821; Shmal'g., Fl. II, 391; Grossg., Fl. Kavk. II, 52; Fedchenko, Conspectus Fl. Turk. 6, 291; Grigor'ev in Fedchenko, Flora Yugo-Vost., No. IV, 104; Kryl., Fl. Zap. Sib., 858; Turcz. Fl. baic.-dah. II, 68; Ldb. Fl. Ross. III, 522. — Persicaria mitis Gilib. Exerc. phytol. II (1792) 431. — Russian: gorets pochechuinyi, pochechuinaya trava, bloshnaya trava; Armenian: matiteg.

Annual; stem ascending in lower part and erect above, or erect from base, simple or branched, 20—80 cm long; leaves lanceolate to linear-lanceolate, long-acuminate, subsessile, cuneate at base, the lower ones with a more developed petiole, smooth or sparsely hairy, with a brown blotch above (Persicaria maculosa S. F. Gray, 1821) or without a blotch (P. persicaria var. immaculatum Peterm. 1838), 3—10 cm long, 0.5—2 cm broad; ocreae tightly clasping the stem, appressed-hairy, the upper margin long-ciliate; racemes terminal, compact and relatively stout, 5—8 mm across, 2—3 cm long; perianth roseate or white, like the pedicel glandular, 2.5—3.5 mm long; stamens 6; pistil with 2 or more
652 rarely 3 styles; achene broadly ovate, flat on both sides, rarely trigonous, black, lustrous. June—October. (Plate XLV, Figure 9).

Banks of rivers and canals, wet fields, gardens, etc. — European part: Lad.-Ilm. (N. limit 60°30'), U. V., V.-Kama, U. Dnp., M. Dnp., Bl., Crim., V.-Don, L. Don, Transv., L. V.; Caucasus: Cisc., W., S., and E. Transc.;

(653)

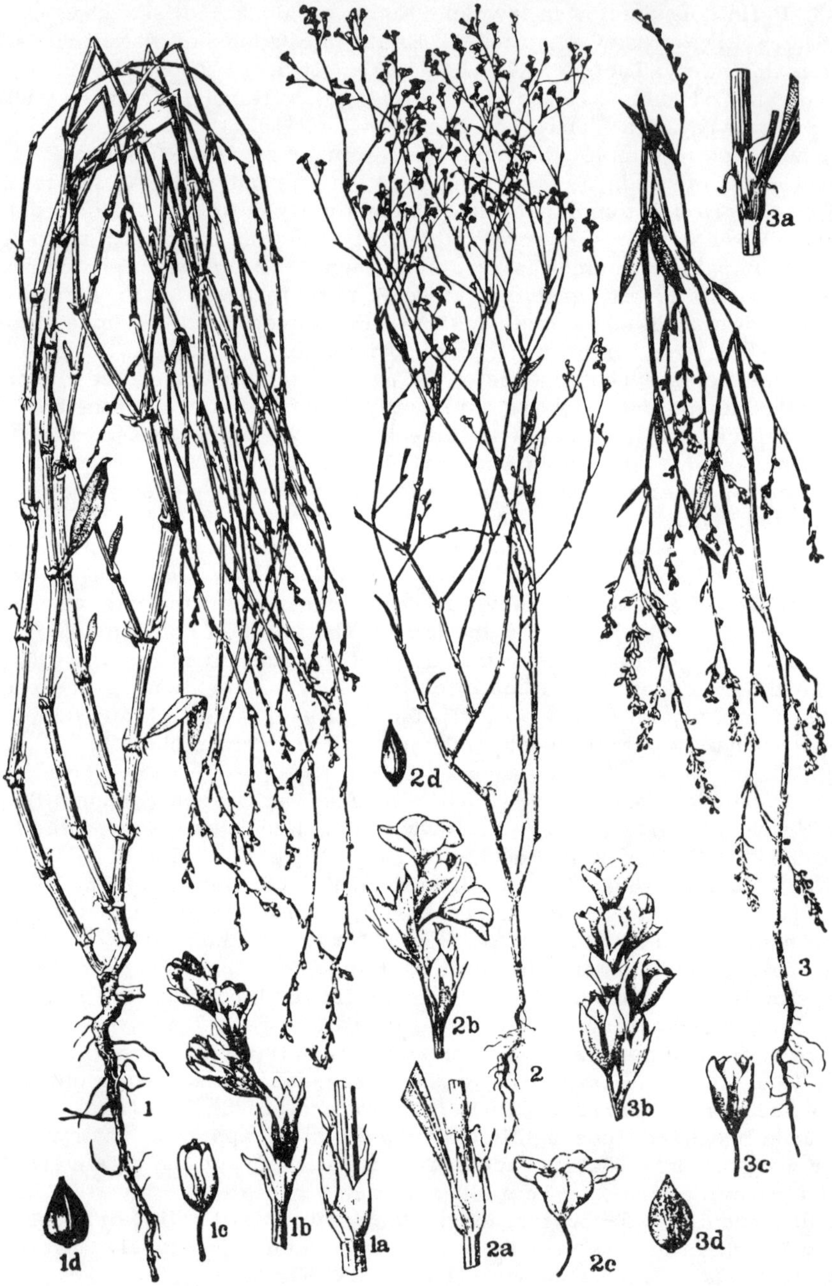

PLATE XLIV. 1. Polygonum argyrocoleum Steud.– 2. P. pseudoarenarium Klok.– 3. P. patulum M.B.– Annotations to all species: a) ocrea and leaf base, b) portion of inflorescence, c) flower, d) achene.

W. Siberia: U. Tob., Irt., Alt.; E. Siberia: Yen. (S.), Ang.-Say., Dau.; Far East: Ze.-Bu., Uss.; Centr. Asia: Balkh., T. Sh., Syr D., Pam.-Al. **Gen. distr.**: Scand. (S.), Atl. and Centr. Eur., Med., Bal.-As. Min., Ind.-Him., Jap.-Ch. Described from Europe as a weed. Type in London.

Economic importance. Of medicinal use and sold in drugstores under the name "Herba Persicariae" for its diuretic action. The plant is slightly nectariferous. The roots yield a yellow dye for yarn.

Series 5. **Hydropiperiformes** Kom. — Annual plants with slender and often interrupted inflorescences.

65. **P. imeretinum** Kom. sp. nova in Addenda IV, p. 554.

Annual; stems strongly branched, erect, 0.5—1.5 m long, smooth; ocreae smooth, the truncate summit not ciliate; ocreae of flowering branches sometimes bearing a few short cilia; leaves short-petioled, lance-oblong or oblong-lanceolate, acute, 10—15 cm long, 1.5—2.5 cm broad, green, the two surfaces almost concolor, without a blotch; racemes elongated, loose, slender, ca. 4 cm long, 0.8 cm broad; peduncles smooth, eglandular; flowers greenish, the perianth eglandular; achene flat, dull, ca. 2 mm long, the faces convex or obtusely keeled.

Brooks and canals, in groups and forming thickets. — Caucasus: W. Transc. Endemic. Described from Tskhaltuba in Lower Imeretia. Type in Leningrad.

66. **P. minus** Huds. Fl. Angl. I (1762) 148; Aschers. et Graebn. Synopsis IV, 830; Ldb. Fl. Ross. III, 523; Shmal'g., Fl. II, 392; Grossg., Fl. Kavk. II, — 52; Kryl., Fl. Zap. Sib. 857; Turcz. Fl. baic.-dah. II, 68; Kom. in A. H. P. XXII, 122. P. daphniphyllum Andrz., Ischisl. r. Podol'skoi gub. (1862) 137. — Persicaria pusilla S. F. Gray, Arr. Brit. pl. II (1821) 269. — Exs.: HFR No. 838.

655 Annual; stem slender, decumbent or ascending, more rarely erect, glabrous, 15—30 cm long; leaves linear-lanceolate with transitions to linear, rounded or but slightly narrowed at base, 5—7 cm long, 3—5 mm broad, quite glabrous or more or less hairy on the veins and on the margin, sometimes with a black blotch; ocrea long-ciliate on the margin; spikes erect or slightly nodding, loose but mostly uninterrupted, 1—4 cm long, eglandular; stamens mostly 5, rarely 8; achene biconvex, lustrous, rarely subtrigonous and then 3-stylar, 2—2.5 mm long. July—October. (Plate XLV, Figure 3).

Forest paths and roadways, villages, canals, banks of brooks, among hillocks, meadowland mochezhina,* sandy and pebbly lacustrine or fluvial flats. — European part: Kar.-Lap. (to the south of 62° N. lat.), Dv.-Pech., Lad.-Ilm., U. V., V.-Kama, U. Dnp., M. Dnp., V.-Don, Transv., Bl., Crim., L. Don, L. V., Urals; Caucasus: Cisc., W. and E. Transc., Tal.; W. Siberia: Ob, U. Tob., Irt., Alt.; E. Siberia: Yen., Ang.-Say., Dau.; Far East: Ze.-Bu., Uda, Uss.; Centr. Asia: Dzu.-Tarb. **Gen. distr.**: Scand., Centr. and Atl. Eur., Med., Bal.-As. Min., Ind.-Him., Jap.-Ch. Described from England.

Note. It is possible that the following reports should be referred here: A report for the Caucasus in Grossgeim Fl. Kavk. II, 52 sub P. mite Schrank, and a report in Meissn. in DC. Prodr. XIV, 110 sub P. serrulatum Lag.

* [Land permanently wet from outflow of underground water.]

67. **P. foliosum** H. Lindberg, Meddel. Soc. Faun. et Fl. Fenn. XXVII (1900) 3; H. Hjelt in Acta Soc. Faun. et Fl. Fenn. XXI (1902) No. 212; Kryl., Fl. Zap. Sib. IV, 858. — Ic.: H. Lindberg, l. c. fig. 1, 2.

Annual; stems more or less curved, branched from base or simple, 5—70 cm long, smooth; leaves with petioles 1.5—2.5 mm long, linear-lanceolate to linear, narrowed at both ends, obtusish, 2—4 cm long, 1.5—6 mm broad; ocreae narrowly campanulate, glabrous or with few appressed hairs, the margin ciliolate; racemes 2—4 cm long (up to 15 cm in var. aquaticum Kom. Bull. Jard. Bot. (1916) 166) and 2 mm thick, often beside the terminal also from all the leaf axils; perianth 2—2.5 mm long, eglandular; stamens 5; styles 2, short; achene ovate, compressed laterally, smooth, lustrous, ca. 2 mm long. July—September. (Plate XLV, Figure 7).

Flats and shores of rivers and lakes. — European part: Lad., Ilm., U. V. ?, V.-Kama ? ; W. Siberia: Ob; Far East: Ze.-Bu., Uss. Sometimes occurring also in intervening regions. **Gen. distr.**: Scand. (up to 65°38' N. lat.). Described from W. part of Finland. Type in Helsinki.

Note. Often growing together with P. minus Huds. from which it is not readily distinguishable and easily overlooked.

656

68. **P. mite** Schrank, Bayr. Fl. I (1789) 668; Ldb. Fl. Ross. III, 522; Aschers. et Graebn. Synopsis IV, 829; Shmal'g., Fl. II, 391; Kryl., Fl. Zap. Sib. 857. — Persicaria laxiflora Opiz, Sezn. (1852) 72. — Exs.: HFR No. 638.

Annual, related to P. hydropiper, but smaller, 15—30 cm high; roots and stems slender; leaves oblong-lanceolate, short-petioled, 5—10 cm long, 1—2 cm broad, the margin and often the veins beneath appressed-hairy; ocreae covered with short hairs, the margin long-ciliate; racemes long, slender, more or less nodding at summit; flowers mostly reddish; perianth 3—3.5 mm long, sometimes sparingly glandular; achene convex on one side, applanate on the other, punctate, smooth, almost black, slightly lustrous. July—October.

Wet places, mostly on clayey soil, canals, alder groves, damp thickets, shores of brooks and oxbows, and fallows. — European part: Lad.-Ilm., U. V., U. Dnp., M. Dnp., V.-Don, Transv., L. V., V.-Kama. **Gen. distr.**: Scand. (S.), Centr. Eur., Atl. Eur., W. Med. Described from Germany.

Note. A West European species which penetrates into the USSR but tends to fade out and becomes increasingly scarce in the direction of the Ural Mountains. Reports for the Caucasus (Grossgeim, Fl. Kavk. II, 52) are dubious, Talysh being most likely.

69. **P. hydropiper** L. Sp. pl. (1753) 361; Meisn. in DC. Prodr. XIV, 109; Shmal'g., Fl. II, 391; Ldb. Fl. Ross. III, 523; Grossg., Fl. Kavk. II, 51; Fedchenko Conspectus Fl. Turk. 6, 290; Kryl., Fl. Zap. Sib. 856; Turcz. Fl. baic.-dah. II, 69; Kom. in A. H. P. XXII, 122. — Persicaria hydropiper Opiz, Sezn. (1852) 72. — Exs.: HFR No. 2842, 2843. — Russian names: gorets perechnyi, vodyanoi perets [water pepper], lyagushech'ya trava.

Annual; stem more or less erect, moderately branched from the base, 30—60 cm long; leaves oblong-lanceolate, acute or obtusish (var. obtusifolium A. Br., 1824), narrowly cuneate at base, 3—6 cm long and 0.7—1.5 cm broad, often dotted with pellucid glands and often with dark spots below the middle, the margin slightly pubescent, the petioles very short but always quite distinct; ocreae cylindric, reddish, short-setose at summit, glabrous below; racemes loose, often interrupted, slender, nodding, 4—6 cm long; perianth 3—4 mm long; segments with golden-yellow glands, obtusish, green, the tips reddish or roseate or white; stamens 6, rarely 8; styles 2 or 3, erect; achene ovate-elliptic, plano-convex, 2.5—3.5 mm long, dotted-roughened, black. July—October. (Plate XLV, Figure 5).

657 Shores, canals, flats, swampy places and forests, low-lying places in meadows, etc. — European part: Kar.-Lap. (to the south of 65° N. lat.), Dv.-Pech., Lad.-Ilm., U. V., V.-Kama, U. Dnp., M. Dnp., V.-Don, Transv., Bl.; Caucasus: Cisc., Dag., W., S., and E. Transc., Tal.; W. Siberia: Ob, U. Tob., Irt., Alt.; E. Siberia: Yen. (to the south of Eniseisk), Ang.-Say., Dau.; Far East: Ze.-Bu., Uda, Uss.; Centr. Asia: Ar.-Casp., Dzu.-Tarb., Syr D., Pam.-Al., T. Sh. **Gen. distr.**: Scand., Centr. and Atl. Eur., Med., Bal.-As. Min., Iran., Jap.-Ch., N. Am. Described from Europe.

Economic importance. The leaves, when chewed, have a strong peppery taste and they can therefore be used for seasoning food. For the same reason the fresh herbage is avoided by livestock. Crushed herbage is applied as poultice and as analgesic for tumors. A yellow dye is obtained by boiling.

Note. Hybrids formed: 1) P. hydropiper × scabrum; 2) P. hydropiper × nodosum; 3) P. hydropiper × mite; 4) P. hydropiper × persicaria; 5) P. hydropiper × minus.

Series 5. **Caespitosae** Kom. — Annual plants with stems rooting in lower part; achenes trigonous.

70. **P. excurrens** Steward in Contr. Gray Herb. LXXXVIII (1930) 65. — Persicaria trigonocarpa Nakai in Tokyo Bot. Mag. XLV (1931) 115, 154.

Annual; stem erect, 0.5—1 m long, with excurrent main axis and at first pubescent later glabrescent branches, 4—10 cm long; leaves sessile, lanceolate, covered with spreading hairs; ocreae tubular, hispidulous, the margin ciliate; inflorescences slender, elongated, the flowers in groups of 2—4; bracts ciliate; pedicels as long as or slightly surpassing the bracts; perianth 1—2 mm long, 5-parted to the middle, the segments commonly 5-nerved; achene trigonous, smooth, lustrous, shorter than perianth.

Near water; banks of ponds and canals. — Far East: Uss. (Amur). **Gen. distr.**: Jap.-Ch. (E. China, Japan, Korea). Described from specimens originating from Hupeh Province. Type in Harvard University.

Note. Varying greatly in development of vesture and in the diameter of stems and inflorescences. Differing from the closely related species P. caespitosum Blume and its varieties in the restricted growth of the central axis.

71. **P. intricatum** Kom. sp. nova in Addenda IV, p. 555. — P. caespitosum Korsh. in A. H. P. XII (1892) 382 (non Blume, nec Meisn.).

Annual; stem erect, mostly simple, 20—50 cm long, smooth; leaves narrow, linear to lance-linear, acute, smooth, the slightly revolute margins

59)

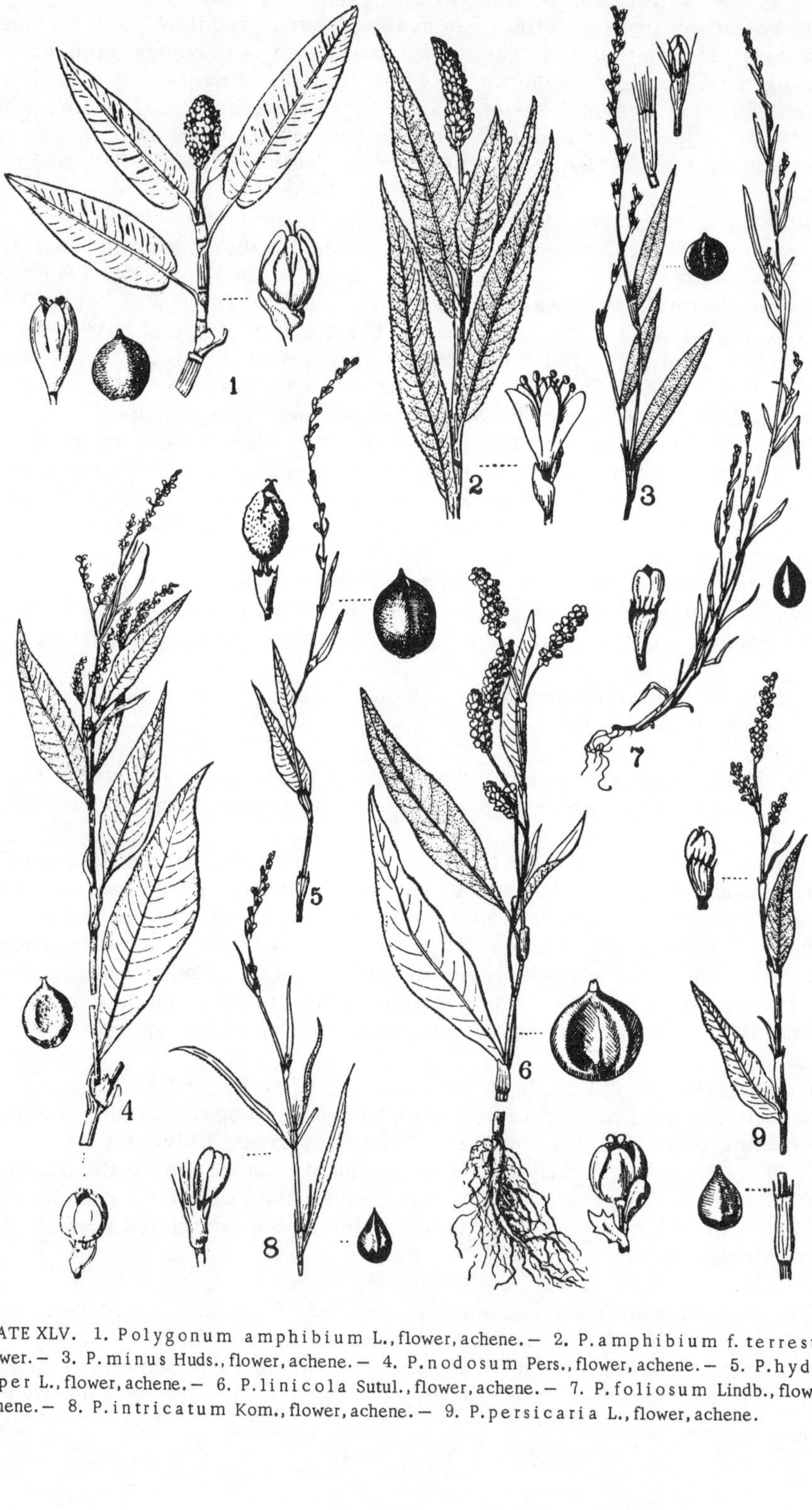

PLATE XLV. 1. Polygonum amphibium L., flower, achene.— 2. P. amphibium f. terrestre, flower.— 3. P. minus Huds., flower, achene.— 4. P. nodosum Pers., flower, achene.— 5. P. hydropiper L., flower, achene.— 6. P. linicola Sutul., flower, achene.— 7. P. foliosum Lindb., flower, achene.— 8. P. intricatum Kom., flower, achene.— 9. P. persicaria L., flower, achene.

sharply scabrous; petioles barely perceptible; ocreae cylindric, sparingly
658 ciliolate; racemes terminal and axillary, slender, sublinear, 1—2 cm long, 0.3 cm thick; bracts funnel-shaped, smooth, ciliate, almost contiguous, perianth greenish, smooth, closed; achene not more than 1 mm long, rarely longer, trigonous, lustrous, black, sharp-beaked. July—October.

Meadow bogs and reedgrass meadows. — Far East: Ze.-Bu., Uss. It possibly occurs also in Manchuria and in Korea, but has not so far been reported and it is thus endemic. Described from Amur Region (Il'inovka village). Type in Leningrad.

Note. This species differs from P. minus Huds. chiefly in the trigonous achene, and from P. posumbu Hamilt. in the narrow leaves and the very small achene.

72. **P. posumbu** Hamilt. in Don, Prodr. Fl. Nepal. (1825) 61; Meisn. in DC. Prodr. XIV, 105; Kom. in A. H. P. XXII, 121. — Persicaria posumbu Nakai, Veg. Isl. Quelpart (1914) 41, No. 555a. — P. caespitosum Blume, Bijdr. (1825) 532. — P. Yokusaianum Makino, Tokyo Bot. Mag. XXVIII (1924) 116.

Annual; stem decumbent or ascending at base, then erect, rooting from the lowest nodes, 45—70 cm long, smooth, sulcate; leaves broadly lanceolate, acute, 3—12 cm long, 1—2.5 cm broad, smooth, thick, narrowly cuneate at base, the petiole less than 1 cm long; ocreae of upper leaves subcylindric, narrow, smooth or with appressed bristles on the veins and sparse longer cilia on the margin; racemes solitary or numerous, disposed in a loose panicle; peduncles slender, smooth, eglandular; racemes slender, often slightly interrupted, sublinear; bracts narrow, contiguous, mostly subtending 1 or 3 flowers, long-ciliate; pedicels longer than bracts, glabrous; perianth green, white, or roseate, eglandular, rarely more than 2 mm long; stamens much shorter than the perianth; styles 3; achene commonly trigonous, very lustrous, black. August—October.

Inundated and mountain woods, forest paths, near rocks, on the site of fallen trees, near springs, and in thickets; generally a shade-loving plant. — Far East: Uss. **Gen. distr.**: Ind.-Him., Jap.-Ch., Malay Archipelago. Described from Java.

Note. Not tufted but growing in rather large groups and, on shaded bottomland, often apparently cespitose. The branches interfere with the development of the main axis; pedicels longer than in other related species.

661

73. **P. viscoferum** Makino in Tokyo Botan. Mag. XVII (1913) 115. — Persicaria viscofera Nakai ex Mori Enum. of pl. from Korea (1922) 134.

Annual; stem erect, 50—75 cm long, the upper 2—4 internodes viscous in upper one-third, apparently resinous (not glandular); leaves lanceolate to linear-lanceolate, acute, 5—7 cm long, 1—1.5 cm broad, with stiff appressed hairs on the midrib and on the margins; petioles barely 0.5 cm long; ocreae hairy, the upper margin ciliate; racemes 2—8, disposed paniculately slender, loose, sublinear; bracts ciliate, markedly discontinuous; perianth greenish, in fruit with prominent longitudinal nerves, rarely smooth; achene trigonous, lustrous, black, to 2 mm long. August—October.

Forest roadways, shores of brooks among shrubs, thickets, and seashores, more rarely near rocks in oak woods or in neglected fields. — Far East: Uss. (S.). **Gen. distr.**: Jap.-Ch. (Japan, Korea, Manchuria). Described from Japan.

Section 4. **ACONOGONON** Meisn. Monogr. Polyg. (1826) 43. — Aconogon Tourn. — Acontogonum Asch. et Gr. Synops. IV (1913) 840. — Aconopogon Mori Enum. pl. from Korea (1922) 129. — Pleuropteropyrum H. Gross in Bull. Geogr. Bot. XXIII (1913) 9. — Russian name of the section: "taran." — Perianth corolloid, 5-parted, eglandular; stamens 8; achene trigonous, with sharp angles; ocreae and bracts membranaceous, often pellucid, obliquely truncate, without cilia; flowers in a loose paniculate inflorescence, white or yellowish; style 3-fid, very short, with capitate stigmas.

1. Pedicels jointed at or below the middle; mature achene thin-walled, broadly 3-winged; leaves lanceolate or oblong-lanceolate, smooth . . . . 90. **P. tripterocarpum** A. Gray.
\+ Pedicels jointed at the perianth base, rarely at the middle; mature achene trigonous, thick-walled, wingless . . . . 2.
2. Leaves silvery-gray beneath, densely pubescent; perianth more or less pubescent on the outside . . . . 83. **P. sericeum** Pall.
\+ Leaves glabrous or more or less pubescent beneath but not silvery-gray; perianth glabrous . . . . 3.
3. Leaves gradually narrowed toward base, more or less hastate, with 2 short obtusish or rarely acute auricles; achene black . . . . 4.
\+ Leaves cuneate or rounded at base, always exauriculate; fruit buff or brown . . . . 5.
4. Stems 10—30 cm long . . . . 88. **P. sibiricum** Laxm.
\+ Stems 0.5—5 cm long . . . . 89. **P. pamiricum** Korsh.
662 5. Flowers red, the segments often white-or greenish-margined . . . . 84. **P. songoricum** Schrenk.
\+ Flowers white or yellowish (rarely greenish) . . . . 6.
6. Stems 1—2 m long, simple, rarely with 1 or 2 short lateral branches; leaves broad, ovate, glabrous or sparsely hairy; inflorescence branches upright in maturity; plants of inundated forests of the Far East . . . . 7.
\+ Stems more or less branched, rarely simple and then plants less than 1 m tall and leaves mostly heavily appressed-hairy (at least beneath) . . . . 8.
7. Leaves firm, rounded at base; panicle spreading, not longer than broad; pedicels jointed at the end . . . . 86. **P. limosum** Kom.
\+ Leaves thin, cuneately narrowed at base; panicle with few very slender branches; pedicels jointed at the middle . . . 87. **P. relictum** Kom.
8. Leaves glabrous, narrowly linear or oblong-linear (0.5—4 mm broad), margins often revolute; achene 2—2.5 (3) mm long . . . . 79. **P. angustifolium** Pall.
\+ Leaves more or less hairy or glabrous, broader (at least 7 mm to several cm), rarely sublinear; achene (2.5) 3—6 (7) mm long . . . . . . 9.

9. Stems simple, branched only in inflorescence (rarely with 1 or 2 vegetative branches); leaves thickish, elliptic to ovate-elliptic; inflorescence branches nodding in fruit . . . . 85. **P. hissaricum** M. Pop.
\+ Strongly branched plants to 1 m tall; leaves oblong to oblong-linear; achene large, 4 — 6 (7) mm long . . . . 77. **P. divaricatum** L.
++ Differing in these characters . . . . 10.
10. Plants 60 — 150 cm tall, divaricately branched from the base; leaves broad, ovate or ovate-lanceolate . . . . 11.
\+ Plants less than 1 m tall; leaves not ovate . . . . 13.
11. Inflorescence branches nodding in fruit (Soviet Central Asia) . . . . . . 12.
\+ Inflorescence branches upright in fruit; achene mostly exserted from the perianth (Far East) . . . . 78. **P. luxurians** Grig.
12. Perianth (2) 2.5 — 3.6 mm long; achene (2.8) 3 — 3.5 (4) mm long . . . . 75. **P. coriarium** Grig.
\+ Perianth 3.8 — 5.5 mm long; achene 3.5 — 5 mm long . . . . 76. **P. bucharicum** Grig.
13. Plants weakly branched, 30 — 40 cm long; leaves lanceolate to linear-lanceolate, relatively short; spikes axillary and not disposed in a panicle; flowers small . . . . 81. **P. ochreatum** L.
663 + Plants strongly branched; spikes gathered in a panicle . . . . 14.
14. Stems 10 — 30 (rarely 40 — 50) cm long, rather strongly branched, rarely almost simple; leaves elongate-lanceolate, oblong, or sublinear; achene 3 — 4 mm long, not exserted from the perianth . . . . 80. **P. Laxmanni** Lepech.
\+ Plants differing in some of these characters . . . . 15.
15. Stem 10 — 25 (30) cm long, more or less divaricately branched from the base; achene 2.5 — 3 mm long, not exserted from the perianth . . . . 82. **P. ajanense** (Nakai) Grig.
\+ Stem 20 — 90 cm long, with short branches; achene 3.5 — 4 (5) mm long, commonly more or less exserted . . . . 74. **P. alpinum** All.

74. **P. alpinum** All. Fl. Pedem. II (1785) 206. — P. undulatum Murr. in Comm. Götting. V (1774) 34; Kryl., Fl. Zap. Sib. IV, 869. — P. polymorphum Ldb. Fl. Ross. III, I (1847—49) 542, pro maxima parte; Turcz. Fl. baic.-dah. II, 1, 61. — Ic.: All. l. c. tab. 68; Murr. l. c. tab. 5. — Russian: kislets, bashkirskaya kapusta, taran; Ossetian: zhakaya-kolo.

Perennial; stem 20 — 90 cm long; branches short, glabrous or more or less hairy; leaves ovate-lanceolate to elongate-lanceolate, 5 — 13 cm long and 1 — 5 cm broad, acuminate, cuneately narrowed at base, more or less pubescent on both sides or merely beneath, more rarely glabrous; inflorescence a rather dense leafless panicle; perianth 3 — 3.5 mm long, in fruit 3.5 — 4 (5) mm; achene more or less exserted from or as long as the perianth. June — August.

Meadows and mixed-grass steppes; wood margins. — European part: V.-Kama (S. part), Transv., L. Don; Caucasus: all regions; W. Siberia: Irt., Alt., sporadically in U. Tob.; E. Siberia: all regions; Far East: Ze.-Bu.; Centr. Asia: Dzu.-Tarb., T. Sh. **Gen. distr.**: Mong., Jap.-Ch., W. Eur.

Economic importance. The sour leaves serve as a substitute for sorrel; an infusion of the plant is used against dysentery; the roots have a high tannin content and are used for tanning.

75. **P. coriarium** Grig. in Acta Inst. Bot. Acad. Sc. URSS, ser. I, 1 (1933) 101. — P. alpinum auct. non All. — P. polymorphum auct., non Ldb.

Perennial; stem to 1 — 1.5 m long, profusely divaricately branched; leaves ovate to ovate-lanceolate, ca. 6 — 10 cm long and 2.5 — 5 cm broad, broadly cuneate or rounded at base, rather densely appressed-hairy beneath or rarely on both surfaces, rarely glabrous; inflorescence a large, dense, many-branched panicle, the branches nodding in fruit; perianth white,
664 (2) 2.5 — 3.6 mm long, elongating in fruit to 4 mm; achene 3 — 4.5 (4) [?] mm long, exserted to not more than 0.5 mm or more rarely as long as the perianth. June — August.

The forest and subalpine zones of mountains. — Centr. Asia: Dzu.-Tarb., T. Sh., more rarely in N. part of Pam.-Al. **Gen. distr.**: probably occurring in Dzu.-Kash. Described from the Dzungarian Ala Tau, from the Kora River. Type in Leningrad.

Economic importance. The roots provide a highly concentrated tanning agent.

76. **P. bucharicum** Grig. in Acta Inst. Bot. Acad. Sc. URSS, ser. I, 1 (1933) 102. — P. polymorphum auct., non Ldb.

Perennial; stem to 1 — 1.5 m long, profusely divaricately branched; leaves ovate to ovate-lanceolate, ca. 6 — 10 cm long and 2.5 — 5 cm broad, acute, more rarely acuminate, broadly cuneate or rounded at base, rather densely appressed-pubescent on both sides or merely beneath; inflorescence a large dense many-branched panicle, its branches nodding in fruit; perianth white, 3.8 — 5.5 mm long, elongating in fruit to 6 — 7 mm; pedicels 2 — 5 mm long; achene 3.5 — 5 mm long, sometimes not exserted. June — July. (Plate XLVI, Figure 4).

Slopes in the subalpine zone and in the upper wood and scrub zone of mountains. — Centr. Asia: Pam.-Al., sporadically in W. part of T. Sh. **Gen. distr.**: Dzu.-Kash. (Kashgaria); probably occurring in Afghanistan. Described from the Yagnob River valley (Tadzhikistan). Type in Leningrad.

Economic importance. As the preceding species. The foliage is used (in Tadzhikistan) as a source of yellow dye for woolen goods.

77. **P. divaricatum** L. Sp. pl. (1753) 363; Ldb. Fl. Ross. III, 1, 526; Turcz. Fl. baic.-dah. II, 1, 63; Steward, Contrib. Gray Herb. Harv., Univ. LXXXVII (1930) III.

Perennial, divaricately branched plant forming a large subspherical bush to 1 m tall; leaves oblong-linear to oblong, gradually narrowed toward base, acute, glabrous, more rarely puberulous beneath or on both sides, ca. 5 — 12 cm long and 7 — 25 mm broad; inflorescence a large spreading many-flowered panicle; perianth ca. (2.5) 3 — 3.5 mm long, in fruit 4 — 5 mm; achene ca. 4 — 6 (7) mm long, rather markedly exserted. July — August.

Steppes and dry meadows. — E. Siberia: Dau.; Far East: Ze.-Bu., Uda, Uss. **Gen. distr.**: Mong., N. part of Jap.-Ch. Described from Siberia. Type in London.

78. **P. luxurians** Grig. sp. nova in Addenda IV, p. 555.

Perennial; stems 60 — 100 cm long, strongly divaricately branched from
665 the base; leaves broad-ovate, 5 — 11 cm long and 2 — 8 cm broad, acute, rounded or broadly cuneate at base, glabrous, more rarely puberulent

beneath or on both sides; petiole 3—12 mm long; inflorescence a broad open spreading panicle; perianth 2.5—3 mm long, more or less exserted. July—August. (Plate XLVI, Figure 6).

Broad-leaved woods and meadows. — Far East: Uss. Endemic. Described from the vicinity of Ternei Bay in the Maritime Territory. Type in Leningrad.

Note. This species resembles the related P. divaricatum L. in disposition of branches and in shape of inflorescence; it is distinguishable by the broad ovate leaves as well as the smaller flowers and fruits.

79. **P. angustifolium** Pall. Reise III (1776) 230. — P. polymorphum v. angustissimum Ldb. Fl. Ross. III, 1 (1847—49) 524, ex parte. — P. acidulum Willd. Enum. H. Berol. (1809) 429. — P. alpinum var. angustissimum Turcz. Fl. baic.-dah. II (1856) 453. — Persicaria montana Amman, Sc. rar. descr. (1739) 167.

Perennial; stems ca. 20—50 (60) cm long, rather strongly divaricately branched, rarely with but few short branches; leaves glabrous, narrowly linear to oblong-linear, ca. 2.5—6 (7) cm long and 0.5—3 (4) mm broad, commonly revolute-margined, more rarely flat; inflorescence a panicle, almost leafless or in lower part more or less leafy; perianth ca. 2—2.5 (3) mm long; pedicels ca. 1—2 mm long; achene 2—2.5 (3) mm long, sometimes included in the perianth. June—July. (Plate XLVI, Figure 1).

Steppes and dry gravelly slopes. — E. Siberia: Dau.; Far East: W. part of Ze.-Bu. **Gen. distr.**: Mong. Described from the Ingoda River. Type unknown.

80. **P. Laxmanni** Lepech. in Nova Acta Acad. Petrop. X (1797) 414, t. 13; Ldb. Fl. Ross. III, 1, 526, Turcz. Fl. baic.-dah. II, 1, 63; Kryl., Fl. Zap. Sib. IV, 870. — P. Gmelini Steud. Nomencl. bot. II (1843) 375. — Ic.: Gmelin Fl. Sib. III, 58, tab. XI, f. 2.

Perennial; stems 10—30 (rarely to 40—50) cm long, strongly divaricately branched from base, rarely few-branched or almost simple, glabrous or more or less hispid; leaves elongate-lanceolate, oblong, or sublinear, 3—7 cm long and 3—8 mm broad, sometimes with revolute margins, rather densely hairy beneath or on both sides, more rarely glabrous; inflorescence a loose spreading panicle, all its branches (except sometimes the uppermost 1 or 2) arising from leaf axils; perianth (2.5) 3—3.5 (4) mm long, in fruit to 4.5 mm; pedicels ca. 2—2.5 mm long; fruit 3—4 mm long, completely included in the perianth. July—August.

Sands, pebbles, and recent outcrops, chiefly in river valleys and along seashores. — Arctic: Arc. Eur. (?), Arc. Sib., An.; European part: Dv.-Pech.; W. Siberia: Ob; E. Siberia: Yen., Lena-Kol., Ang.-Say., Dau; Far East: Ze.-Bu., Okh., Uda (?). Endemic. Described from the Urakan River.

666 81. **P. ochreatum** L. Sp. pl. (1753) 361, non Houtt.; Gmel. Fl. Sib. III (1768) 51. — P. alpinum var. saligna Trautv. in A. H. P. V (1877) 101. — Ic.: Gmel. l. c. tab. VIII. — Exs.: HFR No. 2845.

Perennial; stem flexuous, 30—40 cm long, few-branched, glabrous, slightly sulcate; leaves lanceolate to linear-lanceolate, 2—5 cm long,

(667)

PLATE XLVI. 1. Polygonum angustifolium Pall. — 2. P. ajanense Grig., portion of inflorescence, a) flower. — 3. General aspect of same. — 4. P. bucharicum Grig., portion of inflorescence. — 6. P. luxurians Grig., portion of inflorescence. — 7. P. songoricum Schrenk, portion of plant. — 8. P. pamiricum Korsh., portion of plant. — 9. P. limosum Kom., portion of inflorescence. — 10. P. tripterocarpum A. Gray. — 5. Achene of same.

2—6 mm broad, entire, short-petioled, the margin and the veins beneath hairy; ocreae oblong, white or darkening, with few dark nerves; racemes terminal or axillary short; bracts resembling ocreae but much shorter; pedicels at anthesis slightly shorter than flower; perianth funnel-shaped, to 3 mm long, white. July—August.

Damp riverbanks. — E. Siberia: Yen. (Krasnoyarsk—Turukhansk area, Dudino village), Lena-Kol. Endemic. Type in London.

Note. Differing from related P. angustifolium Pall. and P. laxmanni Lepech. in the broader lanceolate leaves, the short scattered axillary flower clusters, the small flowers, and the habitat. A forgotten species, reestablished by D. I. Litvinov in HFR.

82. **P. ajanense** (Nakai) Grig. comb. nova. — Pleuropteropyrum ajanense Nakai Veg. Mt. Daisetsuran Mts. 46, 63. — Polygonum polymorphum var. ajanense Rgl. et Til. Fl. Ajan. (1858) 116. — P. angustifolium var. diffusum Skw. Contrib. Gray Herb. LXXXVIII (1930) 108. — P. polymorphum var. diffusum Ldb. Fl. Ross. IV (1849) 525.

Perennial; stems 10—25 (30) cm long, capillary, more or less divaricately branched from base; leaves lanceolate to ovate-lanceolate (rarely elongate-lanceolate), ca. 3—5 cm long and 7.15 mm broad, rather densely appressed-pubescent on both sides or merely beneath; inflorescence a panicle leafy in lower part, leafless above; perianth ca. 2.5—3 mm long, in fruit to 4 mm; achene 2.5—3 mm long, sometimes not exserted. July—August. (Plate XLVI, Figure 2).

Dry tundras, stony slopes of barren heights, and rarely sands. — E. Siberia: Dau., Lena-Kol.; Far East: Ze.-Bu., Uda, Okh., Sakh. Described from the vicinity of Ayan harbor.

Note. P. Komarovii Lévl. in Fedde Repert. sp. nov. VIII (1910) 171, described from Sakhalin (roadsides in the Korsakovo village area) as a species closely related to P. alpinum All., has achenes which would
669 refer it rather to P. nodosum Pers. than to the section Aconogum.

83. **P. sericeum** Pall. Reise III (1776) 286, 320; Ldb. Fl. Ross. III, 2, 527; Turcz. Fl. baic.-dah. II, 65. — Persicaria sericea H. Gross in B. G. B. XXIII (1913) 31.

Perennial; stems ca. 20—40 cm long, erect or ascending, commonly branched, rarely simple, the stems and branches rather densely hairy; leaves elliptic to oblong-elliptic, ca. 3—8 cm long and 2—4 cm broad, cuneate at base, acute or rounded at apex, rather densely appressed-pubescent above, silvery-gray beneath with dense pubescence; flowers in rather dense axillary racemes disposed in a loose leafy panicle; perianth ca. 3.5—4.6 mm long, densely pubescent outside. July—August.

Sands on riverbanks and lakeshores. — E. Siberia: Dau. Gen. distr.: Mong. Described from the Baikal area. Type unknown.

Note. P. sericeum apparently hybridizes with P. Laxmanni. The hybrids display characteristics intermediate between the parental species.

84. **P. songoricum** Schrenk in Fisch. et Mey. Enum. pl. a cl. Schrenk lect. I (1841) 8. — P. polymorphum f. songoricum Ldb. Fl. Ross. III, 2 (1851) 524. — P. angustifolium v. songaricum Stew. Contrib. Gray Herb. Harv., LXXXVII (1930) 108.

Perennial; stems ca. 20 — 70 cm long, simple except in inflorescence or rarely with few short vegetative branches, often with reduced shoots in the leaf axils; leaves thin, broad-ovate or ovate, ca. 6 — 10 cm long and 3.5 — 6 cm broad, gradually acuminate, rounded or cordate at base, sparsely hairy beneath, the petiole 2 — 4 cm long; inflorescence a rather loose slender panicle, its branches often spreading horizontally and in fruit nodding; perianth 2.5 — 3 mm long, in fruit to 4 mm, red, the segments often white or greenish at the tips; achene ca. 3.5 — 4 mm long (Plate XLVI, Figure 7).

Woods and subalpine meadows, rare in the alpine zone; a shade-loving plant. — Centr. Asia: Dzu.-Tarb., T. Sh., more rarely Pam.-Al. (Alai and Trans-Alai mountain ranges). **Gen. distr.**: Dzu.-Kash. (Kuldja). Described from the Dzhil-Karagai Mountains. Type in Leningrad.

85. **P. hissaricum** M. Pop. in Not. Syst. H. B. Reip. Ross. V (1924) 38. — P. rumicifolium auct., non Royle.

Perennial; stems ca. 20 — 40 (50) cm long, simple except in inflorescence, rarely with 1 or 2 vegetative branches, often with abbreviated shoots in the
670 leaf axils; leaves thickish, elliptic to ovate-elliptic, ca. 5 — 10 cm long and 3 — 5 (6) cm broad, acute or obtuse or rounded at apex, broadly cuneate or rounded at base, densely puberulous on both sides (rarely merely beneath); the petiole 1 — 3 cm long; inflorescence a rather dense panicle, leafy in lower part, leafless above, longer than broad, the branches nodding in fruit; perianth white, ca. 3 — 4 mm long, in fruit to 6 — 7 mm; achene ca. 3.5 — 4.5 mm long, sometimes not exserted. June — August.

Stony slopes in the alpine zone. — Centr. Asia: Pam.-Al., W. part of T. Sh. Endemic. Described from the sources of the Karatal River in the Gissar region. Type in Tashkent.

Economic importance. The young shoots of this species, like those of **P. bucharicum** Grig., are eaten raw (Tadzhikistan).

86. **P. limosum** Kom. Bull. Jard. Bot. Pétersb. XVI (1916) 165. — P. divaricatum var. limosum Kom. Fl. Mansh. II (1904) 140.

Perennial; stems to 180 cm long, simple except in inflorescence, rarely with 1 or 2 short vegetative branches; leaves thin, ovate to ovate-lanceolate, ca. 6 — 15 cm long and 3 — 9 cm broad, gradually acuminate, rounded at base, glabrous or with scattered hairs beneath (rarely on both sides), the petiole to 2 — 4 cm long; inflorescence a loosely spreading more or less leafy panicle, usually longer than broad; perianth ca. 1.5 — 2 mm long, in fruit to 3 mm; pedicels ca. 1 — 2 mm long; achene nodding, 3 — 4 mm long, exserted to about half its length. August. (Plate XLVI, Figure 9).

Bottomland in river valleys. — Far East: Uda, Uss. **Gen. distr.**: Jap.-Ch. Described from the basin of the Tsimukhe River (Shkotovo village in Ussuri Territory). Type in Leningrad.

87. **P. relictum** Kom. sp. nova. — P. polymorphum ε lapathifolium Maxim. Prim. Fl. Amur. (1859) 234 (non Ldb.).

Perennial from creeping rhizome, branched, the stems tufted, erect, 0.75 — 1.25 m long, to 5 mm thick, in upper part flexuous, sulcate,

inconspicuously covered with simple hairs, the internodes hollow; branches in upper part all axillary, short; leaves thin, soft, oval to oval-lanceolate, acute, cuneately narrowed toward base, glabrous above, ciliate on the margin and on the veins beneath, irregularly undulate-margined, on the average 10 — 12 cm long and 5 — 6 cm broad, the petiole 0.3 — 3 cm long; ocreae open, elongated, brown, ca. 2 cm long, long-ciliate at base, often acute at summit; panicles terminal, very poor, with few very slender filiform branches, glabrous; bracts ca. 1 mm long; pedicels 2 — 3 mm long, jointed at the middle; perianth greenish or whitish, funnel-shaped, decurrent on the pedicel, 1 — 2 mm long, the inner segments cucullate;
671 stamens 6; ovary sharply trigonous; styles 3; stigmas peltate; achene often abortive. July — September.

Forest brooks and highly humified soil in bottomland. — Far East: Uda. Endemic. Described from the vicinity of Nikolaevsk on the Amur River. Type in Leningrad.

88. **P. sibiricum** Laxm. in Nova Com. Acad. Petrop. XVIII (1773) 531, t. 7, f. 2; Ldb. Fl. Ross. III, 2, 527; Turcz. Fl. baic.-dah. II, 1, 64; Kryl., Fl. Zap. Sib. IV, 868. — P. hastatum Murr. in Nov. Comm. Götting. V (1774) 37, t. 6. — P. crassifolium Murr. in Linn. Syst. Veget. ed. XIV (1789) 378. — Exs.: HFR No. 2846a. and b.

Perennial; stems ca. 10 — 30 cm long, branched from base, the lower branches long and prostrate or ascending, or more rarely branches few or none; leaves thickish, elongate-elliptic to linear-oblong, ca. 3 — 11 cm long and 4 — 25 (30) mm broad, acute, with 2 obtusish or more rarely acute or sometimes obsolescent auricles at base, the petiole short; flowers in dense terminal and axillary racemes, several of these forming a panicle; perianth 2 — 2.5 mm long, in fruit to 3 mm; achene 2 — 3 mm long, black, smooth, lustrous, not exserted.

Salt marshes, more rarely sands or as a homestead weed. — W. Siberia: Irt., Alt.; E. Siberia: Ang.-Say., Dau., sporadically in Lena-Kol. and Yen.; Centr. Asia: N. part of Balkh., sporadically in T. Sh. Gen. distr.: Dzu.-Kash., Mong., Jap.-Ch. Described from Altai.

Note. Highly variable as regards extent of branching, leaf shape, and inflorescence characteristics.

89. **P. pamiricum** Korsh. in Mém. Acad. Petrop. sér. VIII, IV, No. 4 (1896) 98; O. Fedchenko, Flora Pamira (1903) 183.

Perennial; stems ca. 0.5 — 5 cm long, simple or with several short decumbent branches; leaves thickish, linear to oblong-linear, ca. 2 — 4 cm long and 2 — 3 (5) mm broad, often minutely auriculate at base; flowers in dense racemes 3 — 10 mm long, these solitary or 2 or 3 gathered in compact panicle; perianth ca. 1 — 2.5 mm long; achene ca. 2 — 2.5 mm long, black, smooth, lustrous, not exserted. July — August. (Plate XLVI, Figure 8).

High-mountain zone, riverbanks and lakeshores, solonetz soils, sands, and stony or pebbly plains. — Centr. Asia: Pam.-Al. (only in Pamir), T. Sh. (in the Karakol area). Described from Pamir. Type in Leningrad.

Note. Closely related to P. sibiricum and not readily distinguishable
672 from it. P. pamiricum should possibly be considered merely as an econological form of P. sibiricum and not as a distinct species. It resembles most closely the high-mountain Tibetan P. sibiricum var. Thomsoni Meisn. ex Stew.

90. **P. tripterocarpum** A. Gray ex Rothrock in Ann. Rep. Smiths. Inst. (1868) 453. — P. Pawlowskianum Glehn in A. H. P. IV (1876) 77. — Pleuropteropyrym Pawlowskianum (Glehn) H. Gross et Pleuropteropyrum tripterocarpum (A. Gray) H. Gross in Bull. Geogr. Bot. XXIII (1913) 9. — Polygonum frigidum Kudo, Contr. Fl. N. Sagh. (1923) 33.

Perennial; stems 15 — 40 (50) cm long, often reddish, branched only in inflorescence or rarely with 1 or 2 vegetative branches, reduced shoots arising from the leaf axils; cauline leaves elongate-lanceolate to lanceolate, acute, cuneate or almost rounded at base, ca. 4 — 7 cm long and 8 — 20 mm broad; leaves of reduced shoots narrower; inflorescence a slender panicle, usually longer than broad, leafy in lower part, leafless above; perianth ca. 2.5 — 3 mm long; pedicels ca. 1.5 — 2 mm long, jointed approximately at the middle; achene nodding, ca. 6 — 7 (8) mm long, included in the perianth only at base, with thin often pellucid walls, 3-winged, the wings to 1.5 mm broad. July — August. (Plate XLVI, Figures 10, 5).

Dry tundras and stony alpine meadows. — Arctic: Arc. Sib. (?), Chuk., An.; E. Siberia: Lena-Kol.; Far East: Kamch., Okh., Sakh., rarely in the N. part of Ze.-Bu., Uda (?). Endemic. Described from Arakamchechon Island in Chuk.

Section 5. BISTORTA Tourn. Instit. rei Herb. (1719) 511 (genus). — Adans. Fam. des plantes (1827) 277. — Perianth corolloid, deeply 5-parted, not accrescent in fruit; stamens 8; styles 3, long, with very small stigmas; achenes trigonous; perennials or undershrubs, always 1-spiked; spike uninterrupted, compact; bracts scarious, oval or lanceolate; ocreae thin, membranaceous, cylindric, never ciliate; leaves with narrow revolute margin, obscurely undulate or crisp-crenate.

1. Spike oblong-linear, slender, often with flowers replaced by bulbils from the base; leaf petioles always wingless . . . . 91. **P. viviparum** L.
\+ Spike broader, rather fleshy, without bulbils; leaf petioles more or less winged in upper part, thus the blade decurrent . . . . 2.
2. Radical leaves large, 4 — 5 or more cm broad, truncate or rounded or cordate at base; lower bracts broad . . . . 3.
\+ Radical leaves usually smaller, 2 — 3 cm broad or narrower, rounded
673 or cuneate at base, rarely subcordate; bracts broad or narrow . . . . 6.
3. Leaves coriaceous, prominently reticulate-veined; some leaves distinctly cordate . . . . 101. **P. pacificum** V. Petr.
\+ Leaves thin, and if firm and with prominent lateral veins then not reticulate-veined, oblong even if cordate at base . . . . 4.
4. Stems 5-nodal; lower bracts often broad, awned from apical notch between the prolonged margins . . . . 100. **P. carneum** C. Koch.
\+ Stems 6-nodal; lower bracts rather gradually tapering to a long point . . . . 5.
5. Uppermost 1 — 3 ocreae leafless or with leaf reduced to a small often filiform appendage; leaves always pubescent beneath. . . . . 102. **P. Regelianum** Kom.

\+ Uppermost ocreae as a rule with a small leaf; lower surface of leaves glaucous, glabrous with waxy bloom or else densely pubescent (var. *griseum* Beck.) . . . . 99. **P. bistorta** L.

6\. Stems 6—8-nodal . . . . 7.

\+ Stems 4-nodal, rather short . . . . 10.

++ Stems 5-nodal rarely with a sixth node; radical leaves lanceolate . . 11.

7\. Leaves densely pubescent beneath, tapering to a long point . . . . 105. **P. attenuatum** V. Petr.

\+ Leaves glabrous beneath . . . . 8.

8\. Cauline leaves subfiliform, narrow . . . . . 98. **P. alopecuroides** Turcz.

\+ Cauline leaves more or less amplexicaul, broadly auriculate at base . . . . 9.

9\. Stems slender, 1—2 mm in diameter; cauline leaves sagittate, to 10 cm long . . . . 97. **P. manshuriense** V. Petr.

\+ Stems stouter, 2—5 mm in diameter; cauline leaves strongly elongated, more than 15 cm long, their margins undulate . . . . 93. **P. subauriculatum** V. Petr.

10\. Leaves cuneately decurrent on the petiole; spikes cylindric; lower bracts short-awned . . . . 92. **P. ellipticum** Willd.

\+ Radical leaves rounded or cordate at base, with almost wingless petiole, more leathery; lower bracts long-awned . . . . 96. **P. abbreviatum** Kom.

11\. Lower bracts gradually tapering to a tail-like point; leaves softer, dark green . . . . 106. **P. ochotense** V. Petr.

674 + Lower bracteal scales broad, the midrib excurrent into a short point . . . . 12.

12\. Radical leaves crowded, the petioles not more than 2 cm long, the blades obliquely or even falcately recurved . . . . 104. **P. schugnanicum** Kom.

\+ Radical leaves long-petioled, upright . . . . 13.

13\. Anthers roseate, same color as perianth; bracts acuminate but not awned . . . . 95. **P. zeaense** Kom.

\+ Anthers red or violet, much more intensely colored than perianth; bracts awned . . . . 14.

14\. Radical leaves oblong, to 20 cm long; bracts always 1-awned . . . . 94. **P. intercedens** V. Petr.

\+ Radical leaves oval-oblong to oval-lanceolate, no more than 15 cm long; bracts sometimes 3-awned . . . 103. **P. nitens** (F. et M.) V. Petr.

Series 1. Apterae Kom. — Petioles wingless.

91\. **P. viviparum** L. Sp. pl. (1753) 360; Ldb. Fl. Ross. III, 519; Shmal'g., Fl. II, 389; Grossg., Fl. Kavk. II, 51; Kryl., Fl. Zap. Sib. 863; Turcz. Fl. baic.-dah. II, 2, 60; Kom. Fl. Kamch. II, 64. — *Bistorta vivipara* S. F. Gray, Nat. Arr. Brit. plants II (1821) 268. — *Colubrina vivipara* Montand. Fl. Jur. sept. (1856) 268. — *Polygonum bulbiferum* Royle ex Bah. in Trans. Linn. Soc. XVIII (1838) 94. — *Bistorta bulbifera* Greene, Leafl. I (1904) 21. — Ic.: Gmelin Fl. Sib. III, tab. VII, f. 2. — Exs.: HFR No. 337. — Vernacular names: makarsha, makarshino koren'e, serdechnyi koren'; Azerbaijani: mukez; Kamchatkan: iikum, sikui.

Perennial; rhizome indurated, tuberlike or sometimes contorted, blackish-brown, covered with remnants of old leaves; stems 5—30 cm long, rarely longer, erect; radical leaves long-petioled, oval or oblong

PLATE XLVII. Polygonum pacificum V.Petr., flower with bract.— 2. P. bistorta L., flower with bract, achene.— 3. P. viviparum L., flower with bract, achene.— 4. P. alopecuroides Turcz., flower with bract.— 5. P. abbreviatum Kom., bract, radical leaf.— 6. P. ochotense V.Petr., inflorescence, bract, radical leaf.— 7. P. ellipticum Willd., bract, radical leaf.

or lanceolate, 1 — 12 cm long, cordate or cuneate at base, pale glabrous pruinose beneath, coriaceous, prominently reticulate-veined, the thickened margins revolute, the petiole not winged; upper cauline leaves narrowly lanceolate, acute, sessile; ocreae tubular, long, the lower ones leafless; spike slenderly cylindric, erect, ca. 10 cm long and 15 mm thick, the lower flowers transformed into bulbils, the upper ones normally developed; bracts
677 ovate, strongly pointed; perianth white or roseate or rarely red, 5-parted to base, 3 — 3.5 mm long; anthers dark violet; achene trigonous (occasionally biconvex), dark brown, to 3 mm long. (May) June — September. (Plate XLVII, Figure 3).

Forest meadows, dry meadows and pastures on riverside terraces and water divides, subalpine and alpine meadows, moss tundras and stony tundras. — Arctic: Nov. (up to 74°30' N. lat.), Arc. Eur., Arc. Sib., Chuk., An.; European part: Kar.-Lap., Dv.-Pech., Lad.-Ilm., V.-Kama, Urals (S. to Mt. Iremel'); Caucasus: Cisc., Greater Caucasus (2,000 — 3,000 m), Dag.; W. Siberia: Ob, Alt., Irt.; E. Siberia: Yen. (up to 75° N. lat.), Ang.-Say., Lena-Kol., Dau.; Far East: Ze.-Bu., Uss., Uda, Sakh. (only barren heights), Okh., Kamch.; Centr. Asia: T. Sh., Pam.-Al. **Gen. distr.**: Arctic, Scand., Atl. Eur. (mountains), Centr. Eur. (mountains), Bal.-As. Min., Ind.-Him., Tibet, Dzu.-Kash., Mong., Jap.-Ch., Ber., N. Am. Described from mountain pastures of Europe. Type in London.

Economic importance. The rhizomes ground into a meal or boiled whole are used for food. An infusion of the rhizomes is sometimes used in the Caucasus as a substitute for tea.

Note. Groups of plants are encountered in which the whole spike is bulbiferous or, conversely, the spike entirely floriferous, without bulbils. Both pistillate and hermaphrodite flowers may be found; the plants may be monoecious or dioecious.

Series 2. **Bistortiformes** Kom. — Petioles of lowermost leaves winged in upper part.

92. **P. ellipticum** Willd. ex Spreng. Syst. Vegetab. ed. 16, vol. II (1825) 253; V. Petr. in Herb. 1917; Kom. Fl. pen. Kamtsch. II, 65. — P. bistorta Ldb. Fl. Ross. III, 518 (ex parte). — P. bistorta var. ellipticum Turcz. Fl. baic.-dah. II, 2 (1856) 60. — Bistorta elliptica Kom. comb. nova.

Perennial; rhizome stout, contorted, covered at the top with long fragments of old leaves; stem 5 — 40 cm long, erect, sulcate, glabrous; basal leaves cordate or rounded or cuneate at base, oblong or elliptic, 2 — 10 cm long, 1 — 4 cm broad, the winged petiole 2 — 8 cm long; upper leaves short-petioled or sessile, reduced, narrow; all leaves dark green and dull above, glaucescent rather densely pilose and rufous-veined beneath; ocreae dark rusty-brown, the lower ones leafless, sometimes overlapping; spike compact, 2 — 5 cm long, 1 — 1.6 cm broad; bracts brown throughout, without perceptible keel at summit, caudate; perianth segments ca. 5 mm long, violet-rose; anthers almost black; achene 3 — 5 mm long, trigonous or exceptionally tetragonous. July — September. (Plate XLVII, Figure 7).

Stony tundras, alpine meadows, and mountain tundras. — Arctic: Arc. Sib.,
678 Chuk., An.; E. Siberia: Lena-Kol. (E.); Far East: Kamch., Okh. **Gen. distr.**: NW Am. (Arc.). Described from the Lena River (Pallas Herbarium). Type in London or Berlin.

93. **P. subauriculatum** V. Petr. in Herbario IX 1918, nomen; Kom. in Not. syst. ex Herb. H. B. P. VI (1926) 4, diagnosis. — Bis:orta subauriculata Kom. ibid.

Perennial; rhizome relatively small, tuberlike, 5—20 mm long; stem 60—85 cm long, sulcate, smooth, with 6—8 nodes; leaves oval-oblong, regularly attenuate toward apex; radical leaves ca. 15 cm long, 2 cm broad, cuneately decurrent on the petiole, this ca. 20 cm long; cauline leaves 14—18 cm long, sessile, cordate at base, strongly auriculate; all leaves prominently reticulate-veined, quite glabrous; spike 4—8 cm long, scarcely more than 1 cm broad, compact; bracts oval to oblong-lanceolate, silvery, dark-keeled, awn-tipped; pedicels slender, terete; perianth segments bright pink, scarcely more than 2 mm long; anthers rose-violet; achene less than 3 mm long, lustrous, trigonous, beaked. June—August.

Wet meadows, shores, etc. — Far East: Ze.-Bu., Uss. Endemic. Described from the Bureya Mountains. Type in Leningrad.

Note. Related to P. Regelianum Kom., from which it differs in its narrow leaves, the radical being cuneate and not cordate at base. Another distinguishing feature is provided by the hyaline lower bracts.

94. **P. intercedens** V. Petr. (in schedis Herb., nomen nudum) sp. nova in Addenda IV, p. 555.

Perennial; rhizome strongly curved, black, reddish at the curvature, bearing 1—3 stems; radical leaves long-petioled, oblong, to 20 cm long, 1—2 cm broad, uniformly acuminate or obtusely rounded at apex (var. obtusifolium Kom.), firm, prominently reticulate-veined beneath, covered on the veins beneath with crisp hairs, only the ends of veins near the margin visible on the glabrous upper surface; stems 20—52 cm long, erect, 5-nodal, smooth; cauline leaves short-petioled; upper leaves sessile, oblong-lanceolate to lanceolate, rarely sublinear; ocreae slightly 2-lobed, cup-shaped, with a straight upper margin, those near the inflorescence destitute of green blades; inflorescence 3—4 cm long, at first compact and cylindric, finally loose by extension of pedicels; bracts brownish, broad, dark-keeled, awn-tipped; perianth open, campanulate; anthers dark violet; pedicels as long as or longer than perianth. July—August. (Plate XLV, Figure 8 [sic]).

Distinguishable from related species by the long radical leaves resembling those of some Rumex species. Its features place it halfway between P. abbreviatum Kom. and P. ochotense V. Petr.

679 Over moss undergrowth among thickets of Pinus pumila and Rhododendron chrysanthum or in damp places on slopes of barren heights. — Far East: Ze.-Bu. Described from mountains near Lake Toko on the upper Zeya River. Endemic. Type in Leningrad.

95. **P. zeaense** Kom. sp. nova in Addenda IV, p. 556.

Perennial; rhizome rather small, round or rounded-hamate, dark brown, pinkish at the curvature; stem erect, 28—68 cm long, 5—7-nodal, often hairy below the uppermost node; radical leaves long-petioled, the blade thin, as long as or longer than petiole, oblong-linear or oblong-lanceolate, regularly narrowed toward both ends, with only the midvein visible, the lower surface rather densely pubescent; ocreae of cauline leaves nearly all tubular; the upper ones without a green blade; middle and upper cauline leaves sagittate at base, the uppermost linear; inflorescence compact, cylindric; bracts broad-oval, point-tipped but not awned; perianth broadly campanulate, roseate; anthers roseate. June.

Wet valley meadows, shallow hummocky bogs, low ridges sparsely wooded with larch and birch, and sparse thickets of Alnus fruticosa. — Far East: Ze.-Bu. Endemic. Described from Ulanga on the Zeya River. Type in Leningrad.

A species occupying an intermediate position between P. subauriculatum V. Petr. and P. abbreviatum Kom.

96. **P. abbreviatum** Kom. sp. nova in Addenda IV, p. 556. — P. bistorta var. cordifolium Turcz. Fl. baic.-dah. II, 2 (1856) 60. — P. bistorta var. alpinum Turcz. in herbario. — Bistorta abbreviata Kom. comb. nova.

Perennial; rhizome short, hamate, densely covered with leaf remnants; stems 7—30 cm long, sulcate, smooth, 4-nodal; leaves oval-lanceolate, coriaceous, decurrent at center though usually rounded at base; the radical with petiole 4—8 cm long, the cauline narrow with very short petiole, all covered with short crisp hairs beneath, dark green and dull above; ocreae of the normal type; spike ovaloid or cylindric, 1—3 cm long; lower bracts forming at the base of the spike a kind of involucre, this often recurved and then particularly characteristic; individual bracts oblong, light brown, gradually tapering to a tail-like point, all surpassing the flowers; pedicels cylindric, perianth segments roseate, ca. 3 mm long, anthers blackish-violet; achene oblong, trigonous, brown, lustrous, beaked. June—August. (Plate XLII, Figure 5).

Alpine meadows and alpine tundras. — E. Siberia: Ang.-Say., Dau. **Gen. distr.:** N. Mongolia. Described from the Kentei Mountains in Mongolia.
680 Type in Leningrad.

Note. Peduncle just below the spike rather densely covered with hairs visible with the aid of a magnifier; this feature has not been observed in other bistorts.

97. **P. manshuriense** V. Petr. in herbario (1917) nomen; Kom. in Not. syst. ex Herb. H. B. P. VI (1926) 2 (diagnosis). — Bistorta manshuriensis Kom., ibidem.

Perennial; rhizome short, rather stout, almost tuberlike, black, reddish at the curvature; stem rather slender, 60—80 cm long, smooth, 8-nodal; leaves oblong or lanceolate, glabrous, glaucous beneath, gradually acuminate; radical leaves with petioles ca. 15 cm long, more than 15 cm long and 2—3 cm broad, undulate-margined, cuneate at base, decurrent on the petiole; middle and upper cauline leaves sessile, amplexicaul, strongly auriculate; ocreae long, tubular, the lower brownish membranaceous, the upper green with a narrow brown rim at the upper margin; spike 4—7—12 cm long and ca. 1 cm broad, cylindric; bracts broad, distinctly point-tipped or obtuse and then the upper margin toothed to almost fringed; perianth segments oblong, white or roseate, ca. 3 mm long; anthers the color of perianth; achene less than 3 mm long, lustrous, with narrow almost winglike ribs. May—July.

Wet valley meadows, wood margins, thickets, and mountain meadows up to 1,600 above sea level. — Far East: Ze.-Bu. (E. border), Uss. **Gen. distr.**: Jap.-Ch. (Manchuria). Described from the Ningan area in Manchuria. Type in Leningrad.

98. **P. alopecuroides** Turcz., Spis. rast. Baik. Fl. in Shchegl., Ukazat. otkrytii VIII, 1 (183 ?) 409; Besser Fl. d. Baic. in Beibl. zur Fl. 1 (1834) 23; Meisn. in DC. Prodr. XIV, 141. — P. bistorta var. longifolium Fisch. et Mey. in Ind. V. Sem. H. Petrop. (1838) 40. — P. bistorta var. graminifolium Turcz. Pl. exs. (1831) et in Fl. baic.-dah. II, 2 (1856) 59. — P. bistorta b. foliis angustioribus Ldb. Fl. Ross. III, 519. — Bistorta alopecuroides Kom. in Not. syst. ex Herb. H. B. P. VI (1926) 1, 3.

Perennial, smooth throughout; rhizome short and tuberlike or elongated, the old dying end upturned; stem terete, to 1 m long, with 8 nodes, the internodes subequal; radical leaves oblong, lanceolate, or lance-linear, with petiole ca. 20 cm long, gradually narrowed toward both ends, the slightly revolute margins obscurely toothed; lower cauline leaves rounded at base, the upper smaller, linear to subulate, all glaucescent beneath; ocreae short, the leaves arising at their upper margin; spike 5—6 cm long, dense; bracts oval or lanceolate, gradually tapering to a point, silvery-white, with darker keel; perianth funnel-shaped, pale roseate; segments
681 ca. 3 mm long, narrowed toward base, rounded toward apex; anthers pale; achene ca. 3 mm long, light brown, trigonous, acutish, slender-ribbed. June—July. (Plate XLVII, Figure 4).

Mixed-grass meadows, grassy mountain slopes, steppes, etc. — W. Siberia: Alt.; E. Siberia: Yen., Ang.-Say., Dau.; Far East: Ze.-Bu., Uss. Described from the Ingoda River. Type in Leningrad.

99. **P. bistorta** L. Sp. pl. (1753) 360; Ldb. Fl. Ross. III, 518 (exclusis varietatibus); Shmal'g., Fl. II, 389 (only from European part); Fedchenko, Conspectus Fl. Turk. 6, 292 (in part); Kryl., Fl. Zap. Sib. 864. — P. bistorta var. vulgare Turcz. Fl. baic.-dah. II, 2 (1856) 59. — Bistorta major S. F. Gray Nat. Arr. Brit. plants II (1821) 267. — B. officinalis Rafin. Fl. Tellur. III (1836) 12. — Colubrina intorta Montand. Fl. Jur. sept. (1856) 268. — Russian names: gorets zmennyi, gorets aptechnyi, gorlets, cherevnye koren'ya, zmennyi koren', rach'i sheiki.

Perennial; rhizome serpentine, sometimes with constrictions, covered in upper part with leaf remnants, 1—1.5 cm thick, black, reddish at flexure; stems 30—100 cm long, 5-nodose; radical leaves with long winged petioles, often large, oblong to oblong-lanceolate, cuneate or subcordate at base, glaucous and glabrous or with short crisp hairs beneath, 4—20 (30) cm long, 1—7.5 cm broad, the margins flat or slightly undulate; upper leaves subsessile, smaller and narrower; the uppermost sublinear; ocreae tubular, brown, glabrous, lance-tipped; spike compact, ovaloid or lanceolate, 1.5—6 cm long and 10—15 mm thick; bracts scarious, strongly pointed; perianth pale roseate, ca. 3.5 mm long, 5-parted nearly to base; stamens 8, violet, longer than perianth; pistils 3; stigmas capitate, small; achene trigonous, lustrous, ca. 4 mm long, acuminate, castaneous-brown. May—June. (Plate XLVII, Figure 2).

Forest meadows, inundated and watershed meadows, wood margins, and thickets; confined to acid soils with a high humus content. — Arctic: Arc. Eur.; European part: Kar.-Lap., Lad.-Ilm., Dv.-Pech., V.-Kama, U. V., V.-Dnp., M. Dnp., V.-Don, Transv., Urals up to 62° N. lat., Crim.; W. Siberia: U. Tob., Ob to 71°30' N. lat., Alt.; E. Siberia: Yen. to 74°15', Ang.-Say. **Gen. distr.**: Scand., Atl. and Centr. Eur. (mountain plant, even subalpine), Med. (only high mountains), Bal.-As. Min. Described from Switzerland. Type in London.

Economic importance. The rhizome contains tannoids (Polygonum rhizo-tannid), gallic acid, glucose, pararabin, up to 30% starch, oxymethyl-anthraquinone, and other tanning substances. Owing to its strongly astringent properties, it is used in the form of powders or infusions for treatment of diarrhea and dysentery. A very good honey plant. In ancient times the rhizomes was considered as an effective remedy for snake bites.

682

100. **P. carneum** C. Koch in Linnaea XXII (1849) 197; Grossg., Fl. Kavk. II, 51. — P. bistorta auct. Fl. Caucasi (non al.). — Bistorta carnea Kom. comb. nova. — Georgian: dvalura or dvakhuri; Armenian: mandik, agir; Azerbaijani: kizilband, agir.

Perennial; rhizome shorter than in the preceding, tending to tuberiform; stem 25—75 cm (to 1 m) long, sulcate, 4-nodose, 3-angled; leaves oblong to oblong-ovate, more rarely cordate-lanceolate, more or less acuminate, glaucescent and prominently veined beneath, 12—22 cm long, 2—7 cm broad, the basal long-petioled and cuneate or subcordate at base, the middle ones subsessile or sessile and then amplexicaul, the uppermost small linear acutish, all sparsely pilose beneath; ocreae scarious, brownish, often 2-parted at summit; spike cylindric, compact, 2—5 cm long, 1—2 cm thick; bracts brownish, broad, 2-lobed at summit, the lobes point-tipped (sometimes with a third point in between); perianth roseate, deeply 5-parted; stamens longer than perianth; anthers violet or dark red; achene trigonous, lustrous, elongate, to 5 mm long. June—August.

Subalpine and alpine meadows, the upper limits of the forest zone at altitudes of 1,500—3,500 m. — Caucasus: Cisc., Greater Caucasus, Dag., W., S. and E. Transc. **Gen. distr.**: Bal.-As. Min., Arm.-Kurd. Described from the Caucasus. Type in Berlin.

101. **P. pacificum** V. Petr. in herbario Petrop. (1917) nomen; Kom. in Not. syst. ex Herb. H. B. P. VI (1926) 2, diagnosis. — Bistorta pacifica Kom. comb. nova.

Perennial; rhizome stout, indurated, black, serpentine, prickly, 5—8 cm long and to 2 cm in diameter, with numerous adventitious roots; stems 1—3, erect, smooth, 6-nodose; radical leaves with petioles long-petioled, 4—12 cm long, 2—8 cm broad, firm, subcoriaceous, cordate-oval or elongate-oval, acute, prominently veined, glaucous beneath, the margin obscurely papillose, the petiole 8—30 cm long, winged in upper part; cauline leaves with shorter petiole, the upper ones sessile, the uppermost subfiliform; ocreae ferruginous, to 5 cm long, narrow, tubular, both the lower and upper ones leafless, the uppermost split with point-tipped divisions; spike 3—5 cm long and 1.2—1.5 cm broad, roseate; bracts ferruginous, rather broadly oval, caudate-tipped; pedicels as long as or longer than

perianth; perianth segments ca. 4 mm long, rounded at apex; anthers roseate, external; achene trigonous, with strongly convex faces and narrow
683 ribs, ca. 3 mm long, lustrous. June—September. (Plate XLVII, Figure 1).

Grassy sections among thickets, and wood margins near the sea. — Far East: Uss. **Gen. distr.**: Jap.-Ch. (only Korea). Described from specimens originating from Amerika Bay*. Type in Leningrad.

102. **P. Regelianum** Kom. nomen novum. — P. ussuriense V. Petr. in herb. (non P. ussur. (Rgl.) Nakai (1922); Kom. in Not. Syst. ex Herb. H. B. P. VI (1926) 2. — P. bistorta var. ussuriense E. Regel Tentamen Fl. Ussur. (1861) 126. — Bistorta ussuriensis Kom. ibidem. — Ic.: Kom. et Al. Key plants Far East. Reg. tab. 41.

Perennial; rhizome straight, 6 cm long, 1 cm thick; stem to 1 m long, 6-nodose; radical leaves long-petioled, oval to oval-oblong, acute, membranaceous, rounded or truncate or cuneate at base, the winged petiole 20—30 cm long; cauline leaves short-petioled or sessile, amplexicaul, with prominent rounded auricles, all with rather dense fugacious pubescence beneath, this sometimes disappearing except diffusely on the veins; ocreae tubular, the middle ones with a free ligulate flange, the upper green with free margin forming a rather narrow brownish collar; spike 1.5—10 cm long, 1—1.5 cm broad; bracts oblong, light brown, acute; perianth segments ca. 3 mm long, pale rose; anthers the color of perianth; achene ca. 4 mm long, winged-trigonous, lustrous. June—August.

Meadows in mountain valleys, forest meadows on slopes, oak groves, and margins of deciduous woods. — Far East: Uss. (S.), Sakh. (to the N. extremity). **Gen. distr.**: Jap.-Ch. (Korea). Described from the vicinity of Vladivostok. Type in Leningrad.

Note. Although this species had been mentioned by V. A. Petrov as far back as 1917, it was described and published by the present author only at the beginning of 1926. However, as early as 1922 Nakai presented the diagnosis of Persicaria ussuriensis (Rgl.) Nakai and the name Bistorta incana Nakai, the former representing an inacceptable combination of P. ussuriense (Rgl.) V. Petr.; the latter cannot for the present be identified with ours, since Nakai did not offer a description.

103. **P. nitens** (Fisch. et Mey.) V. Petr. in herb. (1918); Addenda IV, p. 556. — P. bistorta var. nitens Fisch. et Mey. Index semin. Horti Petrop. V (1838) 40. — Bistorta nitens Kom. comb. nova.

Perennial; rhizome hooked, dark brown, densely covered with roots and with leaf remnants; stems 20—35 cm long, 5- or 6-nodose, smooth; leaves ranging from oval through oval-oblong to oval-lanceolate, the radical decurrent on the long petiole (8—15 cm), the lower cauline short-petioled, the upper sessile and amplexicaul at base, rather densely clothed with appressed crisp hairs and hence canescent; ocreae ferruginous-brown, the two lowest long and leafless, the uppermost
684 reduced and 2-fid; spike 2—5 cm long, compact; lower bracts broad, 3-pointed, light brown; perianth segments light rose, ca. 3 mm long; anthers dark red; achene trigonous. June—July.

* [West of Cape Povorotnyi, Maritime Territory.]

(685)

PLATE XLVIII. 1. Polygonum Sieboldii Meisn., fragment of stem; achene. – 2. P. Maackianum Rgl., ocrea, flower. – 3. P. Thunbergi Sieb. et Zucc., flower, bract, stipule, ovary. – 4. P. paludosum Kom., ovary, achene, stipules. – 5. P. ussuriense (Rgl.) Nakai, fragment of stem, achene. – 6. P. strigos R. Br., flower, ovary. – 7. P. belophyllum Litw., achene.

Subalpine stony meadows, limestone screes, etc. W. Siberia: Alt.; E. Siberia: Yen., Ang.-Say., Dau.; Centr. Asia: T. Sh., Pam.-Al. **Gen. distr.**: Dzu.-Kash., N. Mong. Described from Altai. Type in Leningrad.

104. **P. schugnanicum** Kom. sp. nova in Addenda IV, p. 557.

Perennial; rhizome stout, sturdy, ca. 3 cm long and 1.8 cm thick, strongly bent at the middle, the two halves tightly appressed to each other; radical and lower cauline leaves 4—6, crowded, their petioles to 2 cm long, shorter than blade; leaves oblong, acutish, to 8 cm long, ca. 2.5 cm broad, glabrous or covered with very short hairs, often subfalcately curved, the margins revolute; stem but twice the length of radical leaves, slender, 4- or 5-nodose, 15—35 cm long; ocreae short, funnel-shaped, the uppermost fimbriate; cauline leaves sublinear, auriculate at base, the upper ones small and almost rod-shaped; spike 2—2.7 cm long and 1.4 cm thick compact; bracts brown, broad, awn-tipped; perianth roseate; anthers carneous. July—August.

The species is distinguishable from P. nitens (Fisch. et Mey.) V. Petr. by the short-petioled radical leaves, short stem, small perianth, and bracts longer than pedicels.

Small alpine meadows. — Centr. Asia: Pam.-Al. Endemic. Described from Shugnan (Ababad at the sources of the Aksu River — A. Regel') at an altitude of more than 3,000 m.

105. **P. attenuatum** V. Petr. in herbario (1922) sp. nova in Addenda IV, p. 557. — P. bistorta subsp. attenuatum V. Petr. in herbario (1918). — Bistorta attenuata Kom. comb. nova.

Perennial; rhizome hooked, short, black crowned by rather light-colored remnants of dead stems and leaves; stems ca. 50 cm long, less sulcate than in the other species, glabrous, 6-nodose; leaves oblong to lance-oblong, narrowed toward the ends, pale green above, canescent beneath with rather compact cover of short hairs, the lower 6—12 cm long and 1.5—3 cm broad, with rounded decurrent base, the middle cauline leaves sessile, cordate at base; ocreae tubular, puberulous in lower part; spike 4—7 cm long; bracts lanceolate terminating in a slender point, the
687 keel dark; pedicels terete; perianth segments 3 mm long, violet-rose; anthers slightly darker than perianth; achene 3 mm long, trigonous, relatively long-beaked, more strongly narrowed at base than in other bistorts. June—July.

Meadow slopes, dwarf birch meadows, small grassy plots in forests, and wood margins. — E. Siberia: Ang.-Say., Lena-Kol. (W. part), Dau. (W. part). **Gen. distr.**: N. Mong. Described from the upper reaches of the Lena River. As far north as the upper course of the Olenek River. Type in Leningrad.

106. **P. ochotense** V. Petr. in herb. (1918), nomen; in Addenda IV, p. 558. — P. bistorta Rgl. et Til. Fl. Ochot. (non al.). — Bistorta ochotensis Kom. comb. nova.

Perennial; rhizome slightly curved; stems erect, slender, sulcate, glabrous, 5-nodose, to 45 cm long; basal leaves with petioles 7—10 cm long, the upper subsessile, all lanceolate or oval-lanceolate or oblong-lanceolate, dull dark green above, covered beneath with short appressed hairs; ocreae as in related species; spike compact, 2—6 cm long; bracts awned, with dark keel and lighter margins; pedicels flat; perianth bright pink, campanulate, ca. 4 mm long; anthers violet; achene trigonous, lustrous, apiculate, ca. 4 mm long. July—August. (Plate XLVII, Figure 6).

Mountain meadows and damp valley woods. — Far East: Okh. (Ayan pluvial region). Endemic. Described from the vicinity of Ayan. Type in Leningrad.

Section 6. **ECHINOCAULON** Meisn. in Wallich Pl. As. rar. 3 (1832) 58. — Perianth corolloid, 5-parted, wingless in fruit, slightly accrescent; stamens 5 or 6, or 8; styles 2 or 3, connate to the middle, with large stigmas. Erect or creeping herbs, rather densely covered with sharp retrorse prickles; ocreae membranous or with a green herbaceous flange; leaves truncate, cordate, sagittate, or hastate at base; peduncles forking; inflorescences capitate, compact, more rarely loose.

1. Plants trailing or clinging . . . . 2.
+ Stems erect or semidecumbent . . . . 3.
2. Petiole attached to the deltoid blade away from the margin [leaves peltate]; ocrea foliaceous, amplexicaul; perianth bluish, succulent . . . . 107. **P. perfoliatum** L.
+ Petiole attached at the margin of the blade; ocrea funnel-shaped; perianth dry . . . . 108. **P. senticosum** Meisn.
3. Leaves sagittate, the auricles pointing straight down . . . . 4.
+ Leaves hastate, the auricles spreading . . . . 8.
4. Stem pubescent below the inflorescence . . . . 7.
688 + Stem glabrous below the inflorescence, quite smooth . . . . 5.
5. Leaves less than 1 cm long; inflorescence solitary, rarely paired . . . . 110. **P. paludosum** Kom.
+ Leaves larger, inflorescences at the end of the stem and of its branches . . . . 6.
6. Leaves 2—7 cm long; petioles with 2 rows of prickles . . . . 112. **P. belophyllum** Litw.
+ Leaves 7—10 cm long; petioles with 3 or 4 rows of prickles . . . . 109. **P. Sieboldii** Meisn.
7. Indument under the inflorescence distinctly glandular, the glands purple . . . . 111. **P. ussuriense** Rgl. Nakai.
+ Indument under the inflorescence of short hairs, velutinous . . . . 113. **P. strigosum** R. Br.
8. Stem under the inflorescence with glandular indument . . . . 9.
+ Stem under the inflorescence glabrous . . . . 10.
9. Indument under the inflorescence strongly glandular, ruby-red; basal leaf lobes broad and short . . . . 116. **P. dissitiflorum** Hemsley.
+ Indument of short hairs, velutinous, shorter than the prickles; basal lobes of leaves relatively long and narrow; plants grayish throughout with stellate hairs . . . . 115. **P. Maakianum** Rgl.
10. Leaves distinctly hastate, apparently 3-lobed . . . . 114. **P. Thunbergii** Sieb. et Zucc.
+ Leaves with rather indistinct lobes or even simply oval; a very low plant . . . . 114. **P. Thunbergii var. radicans** Franch. et Sav.

Series 1. **Chylocalycinae** Kom. — Annuals with creeping stems, foliaceous ocreae, triangular leaves, and globose achenes.

107. **P. perfoliatum** L. Sp. pl. ed. 2 (1762) 521; Grossg., Fl. Kavk. II, 47, Kom. in A. H. P. XXI, 135. — Chylocalyx perfoliatus Hasskarl ex F. Schmidt in Maxim. Prim. Fl. Amur. (1859) 236. — Persicaria perfoliata H. Gross in Loesln. in B. B. C. XXXVII, 2 (1920) 113.

Perennial; stem creeping, branched, covered with sharp retrorse prickles to 3—4 mm long; petioles 2—7 cm long, slender, covered with sharp prickles; leaf blades peltate, somewhat fleshy, glaucescent, 2—4 cm long, 3—5 cm broad, glabrous above, prickly on the veins beneath; ocreae green, leaflike, amplexicaul, almost horizontally spreading, 1—2 cm in diameter; flowers in short compact solitary terminal or axillary racemes; stipules small, acute, erect, longer than pedicels; perianth greenish, the rounded overlapping segments becoming blue and succulent, thus imparting to the fruit a berrylike appearance; achene subglobose, black, dull, 2—3 mm in diameter. July—October. (Plate XLIX, Figure 3).

689 Riverside osier beds, shallows, and sands. — Caucasus: W. Transc.; Far East: Ze.-Bu., Uss. **Gen. distr.**: Jap.-Ch., Malay Archipelago, India. Described from Cochin China and Japan. Type in Uppsala.

108. **P. senticosum** Meisn. ex Franch. et Sav. Enum. pl. Japon. I (1875) 401 (pro parte); Kom. in A. H. P. XXII, 136. — Chylocalyx senticosus Meisn. in Ann. Mus. Bot. Lugd.-Batav. II (1865—66) 65. — Persicaria senticosa Nakai Veget. Isl. Quelpart. (1914) 41. — Persicaria truellum Honda in Inui et Honda, List pl. Itsukushima, II. — Truellum japonicum Houtt. Naturl. Hist. Pl. VIII, 4 (1777) 427 et tab. 48.

Annual; stems weak, semidecumbent, retrorse-prickly, often turning red, up to ca. 1 m long; leaves with long prickly petiole, hastate-triangular, with a deep basal sinus, the petiole attached at the margin and not away from it as in the preceding species; ocreae funnel-shaped, very short, with toothed margins; racemes few-flowered; peduncles densely glandular-hairy; perianth bright pink, slightly accrescent in fruit, dry; achene globose, dull, black. July—October.

Among rocks, rock streams, sandy and pebbly shallows of river shores. — Far East: Uss. **Gen. distr.**: Jap.-Ch.

Series 2. **Sagittatae** Kom. — Annuals with weak ascending stem and sagittate leaves.

109. **P. Sieboldii** Meisn. in DC. Prodr. XIV (1857) 133. — P. sagittatum var. Sieboldii Maxim. ex Kom. in A. H. P. XXII (1903) 132. — Persicaria Sieboldii Ohki in Tokyo Bot. Mag. XL (1926) 54.

Annual; stem 4-angled, semidecumbent, with retrorse-prickly angles, to 1 m long, the internodes elongated; branches sparse, axillary; petioles short, with 3 rows of prickles; leaf blades sagittate-lanceolate, acute, 7—10 cm long, 2—3 cm broad, with a deep sinus at base and acute basal lobes; racemes compact, rather few-flowered, capitate, the short peduncles glabrous; bracts oval-oblong, acute; perianth white or roseate; stamens 8, shorter than perianth; styles 3; achene dark brown, strongly trigonous, point-tipped, lustrous, puncticulate, ca. 2.5 mm long. Plants glabrous throughout. July — October. (Plate XLVIII, Figure 1).

Banks of rivulets and springs, canals, swampy strips at the shores of lakes, oxbows, wet meadows among hillocks, riverside osier-beds, and shallows. — Far East: Uss. **Gen. distr.**: Jap.-Ch. Described from Japan.

Economic importance. Contains a small amount of anthraquinone derivatives (Wehmer).

110. **P. paludosum** Kom. sp. nova in Addenda IV, p. 558. — P. sagittatum var. paludosum Kom. in A. H. P. XXII (1903) 133.

Annual; stems slender, erect, 4-angled, minutely prickly on the angles, 7 — 20 cm long; leaves oval, obtusish, rather indistinctly sagittate at base,
690 0.4 — 2 cm long, 2 — 7 mm broad, the short petiole with a single row of prickles; ocreae narrow, brownish, ciliate at summit; flowers solitary or in groups of 2 or 3 in a capitate terminal inflorescence, occasionally with solitary flowers lower down; pedicels glabrous, smooth, slender; perianth roseate, ca. 3 mm long; achene black, 2.3 mm long, with rounded angles and convex faces, acute, somewhat lustrous. July — September. (Plate XLVIII, Figure 4).

Riverside and spring swamps; in great profusion. — Far East: Ze.-Bu., Uss. **Gen. distr.**: Jap.-Ch. (Korea, Manchuria). Described from the Amur. Type in Leningrad.

Note. Leaf shape clearly distinguishes this plant from other species of the section of Sagittate but, unfortunately the material collected for this species is limited and does not make it possible to determine the degree of its distinctiveness. The present author used to consider it as merely an ecological form of P. sagittatum depressed by excessive acidity of the soil on which it grows.

111. **P. ussuriense** (Rgl.) Nakai ex Mori, An Enumer. of plants from Korea (1922) 134, sub Persicaria ussuriensis (nomen). — P. sagittatum var. ussuriense Rgl. Tentamen Fl. Ussur. (1861) 126; Kom. in A. H. P. XXII, 133. — P. Korshinskyanum Nakai in Journ. Coll. Sc. Tokyo XXXI (1911) 169. — Ic.: Mats. Ic. Pl. Koisik. IV, V (1919) tab. 237.

Annual; stem 40 — 75 cm long, erect, simple, 4-angled, with short prickles or bristles on the angles; leaves oblong to sublinear, acute, 5 — 10 cm long, 1 — 1.4 cm broad, short-sagittate at base; petioles 0.5 — 1 cm long, smooth or with 2 or 4 rows of prickles scarcely extending to the midrib, the leaf margins densely beset with very short retrorse bristles; ocreae narrow, ciliate; rachis, pedicels and peduncle densely covered with ruby-colored glandular hairs; inflorescence short, capitate, 3 — 7-flowered; bracts oblong, rounded and densely ciliate at summit, glandular on the back; perianth roseate; achene trigonous, acute, ca. 2 mm long, lustrous. July — October. (Plate XLVIII, Figure 5).

Shores of lakelets and oxbows, often with the base in water, the lower nodes rooting in the silt; shallows and wet meadows among hillocks. — Far East: Ze.-Bu., Uss. **Gen. distr.**: Jap.-Ch. (Korea and Manchuria). Described from specimens collected at Lake Khanka (Maak). Type in Leningrad.

112. **P. belophyllum** Litw. in schedis ad HFR IX (1932) 25. — P. sagittatum var. sibiricum Meisn. in DC. XIV (1857) 132; Korsh. in A. H. P. XII, 383. — P. sagittatum Ldb. Fl. Ross. III, 529, non L.; Turcz. Fl. baic.-dah. II, 66. — Persicaria sagittata H. Gross, ex Mori An. Enum. of plants from Korea (1922) 133. — Ic.: Gmelin Fl. Sib. III, 65, tab. 13, f. 2. — Exs.: HFR IX (1932) 25 No. 2839, 2840.

691 Annual; stem slender but often erect, less than 1 m long, 4-angled, the angles prickly, the faces corrugated; petioles to 5 mm long, with a single row of prickles on the back extending to the midrib; leaves oblong-sagittate, acutish, 2—7 cm long and 1—2 cm broad, with a shallow sinus at base, the basal lobes apparently clasping the stem; ocreae short and narrow; inflorescence rather small, capitate solitary or 2—4 together; peduncles long, naked or with 1 or 2 prickles; bracts oval or lanceolate, some point-tipped; stamens 8; style 3-fid; achene trigonous with narrow ribs, point-tipped, lustrous, dark brown, 3—4 mm long. June—October. (Plate XLVIII, Figure 7).

Marshy meadows, lakeshores, shallows, and near water. — W. Siberia: Ob (Tomsk), Alt.; E. Siberia: Yen. (S.), Ang.-Say., Dau.; Far East: Ze.-Bu., Uss., Uda, Sakh. **Gen. distr.**: Jap.-Ch., Mong. Described from specimens collected by S. I. Korzhinskii in Ze.-Bu.

Note. P. sagittatum L. Sp. pl. (1753) 363 was first described from America. Plants collected in the Irkutsk area by Gmelin and Turchaninov were later referred to this species. They differ, however, from the American plants in complete absence of indument, while the closely related American species has bristly-ciliate leaf margins which are their consistent feature. Moreover, the bracts of the American plant terminate in a needle-shaped point. The suggestion to rename the American P. sagittatum as Persicaria americana Ohwi in Tokyo Bot. Mag. XL (1926) 55 is essentially misplaced since Linnaeus indicated that he had in mind only the American plant.

113. **P. strigosum** R. Br. Prodr. Nov. Holl. (1810) 420. — Tracaulon strigosum Greene Leafl. I (1904) 22. — Persicaria strigosa (R. Br.) Nakai in Sasaki List. Pl. Formosae (1928) 170.

Annual; stem weak, erect or flexuous, finely prickly, to 0.5 m long; leaves 0.5—4 cm long, 0.4—1.6 cm broad, oval, acute, rounded or slightly cordate at base, glabrous, sessile or short-petioled, the petiole with 4 rows of minute prickles; inflorescence paniculate; peduncles with short velutinous indument; flowers in small glomerules at the ends of few branches; bracts oblong, hairy; perianth segments bright pink, ca. 3 mm long; achene trigonous, smooth, lustrous. (Plate XLVIII, Figure 6).

Spring bogs, near water, wet pastures, and ricefields. — Far East: Uss. (Kondratenkova). **Gen. distr.**: Jap.-Ch., Malay Archipelago, Australia. Type in London.

Note. P. strigosum contains a number of geographic races. It is possible that the USSR plant may be closely related to P. nipponense Mak. in Tokyo Bot. Mag. XXIII (1909) 89, which has been referred here.

Series 3. **Hastatae** Kom. — Plants with erect stems and hastate leaves, often stoloniferous.

692 114. P. **Thunbergii** Sieb. et Zucc. in Abh. München Akad. III, Abt. 4 (1846) 84 No. 729; Kom. in A. H. P. XXII, 129. — P. stoloniferum F. Schmidt in Mém. Acad. Petrop. 7 sér. XII, No. 2, 168. — Persicaria Thunbergii Nakai Veg. Isl. Quelp. (1914) 41. — Polygonum hastato-trilobum Meisn. in Ann. Mus. Bot. Lugd. Batav. II (1865) 62. — P. stoloniferum F. Schmidt. Fl. Sachal. (1874) 168, No. 382.

Annual; stem mostly erect, more rarely ascending or decumbent (var. radicans Franch. et Sav.), 4-angled, retrorsely prickly on the angles, 0.4—1 m long, giving rise in lower part to numerous filiform runners (var. stoloniferum Maxim.); petioles 1—7 cm long; leaves hastate, subtruncate at base, constricted above the basal lobes, acute, 3—12 cm long, 2—10 cm broad, diffusely pilose, densely ciliolate on the margin, with a straight prickle on the petiole; ocreae membranous, cylindric, the green apical flange reflexed; flowers white or bright pink, in compact glomerules at the ends of branches; peduncles glandular-hairy; bracts green, oval or oblong, ciliate; perianth segments to 7 mm long; achene bluntly trigonous, smooth, dull, terminating in a very short point. July—October. (Plate XLVIII, Figure 3).

Springs, banks of rivulets and small rivers, bottomlands, wet places near roads, forest paths, etc., often in great profusion. — Caucasus: W. Trans.; Far East: Uss.; Uda, Sakh., Kamch. Gen. distr.: Jap.-Ch. Described from Japan. Type in Leyden?.

Note. Leaves dark green in plants growing in shade in the taiga or in bottomland, light green or even yellowish in open places. Cleistogamic flowers are formed at the ends of the runners; as a result of elongations of peduncles, these become buried in the ground and there produce fruits (geocarpy).

115. P. **Maackianum** Rgl. Tentamen Fl. Ussuriensis in Mém. Acad. Petrop. 7 sér. IV, No. 4 (1861) 127. — P. Thunbergii var. Maackianum (Rgl.) Maxim. ex Kom. in A. H. P. XXII (1903) 130. — Persicaria Maackiana Nakai ex Mori Enum. of plants from Korea (1922) 132. — Ic.: Rgl. l. c. tab. X; Kom. et Alis. Key of plants of the Far East. region 473, tab. 143 (f. dextra). — Exs.: HFR No. 2844.

Annual; stem decumbent, clinging, ribbed, rather strongly retrorse-prickly, like other parts grayish with small stellate hairs; petioles 2—4 cm long, with 2 rows of prickles; leaf blades hastate-sagittate at base, lance-oblong, acute, 4—7 cm long, the basal lobes horizontally divergent; upper leaves often sublinear, with little-developed base; ocreae green, open, the angle-toothed margin strigose-ciliate; flowers in groups of 2—5 at the ends of branches, in most vigorous specimens forming a loose panicle;
693 peduncles densely setose and glandular-hairy; bracts lanceolate, acute; perianth segments whitish, lanceolate, 3-nerved near the base, bright pink in fruit; achene quite smooth, lustrous, trigonous, with slightly convex faces, 3—4 mm long. July—October. (Plate XLVIII, Figure 2).

Oxbow lakes and other overgrown water bodies, the base of the plant in water or in silt, with bunches of long adventitious roots from the nodes;

occasionally in ricefields. — Far East: Uss. **Gen. distr.**: Jap.-Ch. Described from the Sungari River and Lake Khanka. Type in Leningrad.

116. **P. dissitiflorum** Hemsley in Forbes a. Hemsley, Index Fl. Chinensis in Journ. of Linn. Society XXVI (1889—1902) 338; Kom. in A. H. P. XXII, 134. — P. glanduliferum Nakai in Journ. of the Coll. Soc. Univ. of Tokyo XXIII (1908) 20, tab. 1, fig. 1 — Ic.: Nakai, l. c.

Annual; stem erect, forking, 50—70 cm long, terete, in lower part glabrous, in upper part minutely prickly, in the inflorescence densely covered with ruby-red glandular hairs; leaves including petiole 3—5 cm long, with slender setaceous prickles extending to the midrib, hastate or oblong-cordate, acute, divergent-lobed at base, 10—16 cm long and 3—8 cm broad, the surface minutely pilulose, the margin ciliate; ocreae light brown, long-ciliate at summit; inflorescence divaricate, loosely paniculate; bracts ciliate at summit; flowers solitary or the upper ones more approximate; perianth 5-parted, the lanceolate or rounded segments bright red with a white margin; stamens 8; styles 3; achene trigonous-globose, lustrous, dark red. August—October.

Riverside sands and wood margins. — Far East: Uss. **Gen. distr.**: Jap.-Ch. Described from the surroundings of Mukden. Type in London.

Section 7. **TINIARIA** Meisn. Monogr. (1826) 43 et 62. — Bilderdykia Dumort. Fl. Belg. (1827) 18. — Perianth semiherbaceous, 5-parted, accrescent in fruit, the 3 outer segments keeled or membranous-winged on the back; stamens 8, shorter than perianth; ovary eglandular at base; styles 3, very short; stigmas capitate or rarely fringed; achenes trigonous, included in the perianth. Mostly trailing herbs or undershrubs, without prickles; ocreae and bracts membranous, subcylindric obliquely truncate; leaves cordate, hastate, or sagittate; petioles often with a nectariferous gland at base.

1. Fruiting pedicel shorter than perianth; perianth segments wingless, bluntly keeled; achene dull . . . . 117. **P. convolvulus** L.
+ Fruiting pedicel as long as or longer than perianth; perianth segments winged on the back; achene lustrous . . . . 2.
694 2. Wing rounded at base . . . . 119. **P. dumentorum** L.
+ Wing cuneately decurrent at base . . . . 3.
3. Wing broad, its margin often toothed in upper part . . . . 120. **P. dentato-alatum** F. Schmidt.
+ Wing narrow, entire, green; achene dull . . . 118. **P. pauciflorum** Maxim.

117. **P. convolvulus** L. Sp. pl. (1753) 364; Ldb. Fl. Ross. III, 528; Shmal'g., Fl. II, 393; Turcz. Fl. baic.-dah. II, 65; Kryl., Fl. Zap. Sib. 866; Fedchenko Conspectus Fl. Turk. 6, 293; Kom. in A. H. P. XXII, 137. — Fagopyrum convolvulus H. Gross in Bull. Géogr. Bot. XXIII (1913) 21; Grossg., Fl. Kavk. II (1930) 53. — P. volubile Gilib. Exerc. phyt. II (1792) 435. — Bilderdykia convolvulus Dumort. Fl. Belg. Prodr. (1827) 28. — Helxine convolvulus Raf. Fl. Tellur. III (1836) 94. — Tiniaria convolvulus Montand., Webb. u. Moq. in Webb. u. Berth. Phyt. Canar. III (1836—1847) 221.

Annual; stem 0.1 to 1 m long, covered with short hairs, curved, trailing, sulcate, often more or less tinged with red; leaves petiolate, rotund or oblong-ovate, acute, cordate or hastate at base, with deltoid basal lobes, varying greatly in size; flowers 3—6 in axillary fascicles; pedicel shorter than perianth, jointed just below the perianth base; perianth 2—2.5 mm long; segments green, white-margined, the outer prominently blunt-keeled, slightly accrescent in fruit; achene black, dull, minutely tuberculate. July—October.

Fields and weedy places, pebbles and sands, thickets, etc. — European part: Kar.-Lap. (to 67° N. lat.), Lad.-Ilm., U. V., Dv.-Pech., V.-Kama, U. Dnp., M. Dnp., V.-Don, Transv., Bl., Crim., L. V., L. Don, Urals (to 61°30' N. lat.); Caucasus: Cisc., Dag., W., S., and E. Transc.; W. Siberia: U. Tob., Ob (to 60° N. lat.), Irt., Alt.; E. Siberia: Yen., Ang.-Say., Lena-Kol. (spreading with field crops to 64° N. lat.); Far East: Ze.-Bu., Uss., Uda, Okh., Kamch., Sakh.; Centr. Asia: Ar.-Casp., Balkh., T. Sh., Mtn. Turkm., Amu D, Syr D., Pam.-Al. **Gen. distr.**: Scand. (to 68° N. lat.), Atl. and Centr. Eur., Med., Bal.-As. Min., Arm.-Kurd., Iran., Dzu.-Kash., Mong., Ind.-Him., Jap.-Ch., Ber., N. Am., etc. Described from Europe. Type in London.

Economic importance. A weed of all kinds of field crops, appearing wherever its seeds are blown. Often included in the food for domestic fowl. As a weed of cereals, known under the names "yalovaya povelika" or "berezka." A nectariferous plant.

118. **P. pauciflorum** Maxim. Ind. seminum Horti Petrop. (1866) 3; Kom. in A. H. P. XXII, 136. — Fagopyrum pauciflorum H. Gross in Bull. de Géogr. Bot. XXIII (1913) 25. — Tiniaria pauciflora (Maxim.) Nakai
695 in Mori Enum. Pl. Kor. (1922) 136.

Annual; stems 20—80 cm long, trailing, terete, slightly sulcate; leaves hastate or subcordate at base, oblong-ovate, acute, smooth, 3—4 cm long, 1.5—2.5 cm broad; petioles 0.5—1.5 cm long; inflorescences axillary, leafy, of 3—5 remote small flowers 1—3 mm long; pedicels jointed at the middle; fruiting perianth to 5 mm long, green, the narrow wing somewhat decurrent on the petiole; achene dull black, to 4 mm long, puncticulate. August—October.

In the taiga at the foot of rocks or on taluses, in deciduous woods, on stony soil, and on riverside pebbles. — Far East: Uss. **Gen. distr.**: Jap.-Ch., Ind.-Him. Described from the Amur. Type in Leningrad.

Note. According to Steward Contr. Gray Herb. LXXXVIII (1930) 94, this species should be named P. pterocarpum Wall. Cat. (1828) No. 1690, nom. nudum.

119. **P. dumetorum** L. Sp. pl. (1762) 522; Shmal'g., Fl. II, 394; Ldb. Fl. Ross. III, 528; Fedchenko Conspectus Fl. Turk. 6, 293; Kryl., Fl. Zap. Sib. 867; Turcz. Fl. baic.-dah. II, 66; Kom. in A. H. P. XXII, 137. — Fagopyrum dumetorum (L.) Schreb. Spic. Fl. Lips. (1771) 42; Grossg., Fl. Kavk. II (1930) 53. — Tiniaria dumetora (L.) Nakai in Mori Enum. Plants from Korea (1922) 136. — Exs.: HFR No. 231.

Annual; stem 0.5—3 m long, trailing, smooth, angled, branched; leaves cordate, acute, glabrous, with rounded or deltoid basal lobes, the petiole 0.5—3 cm long, the blade 2—10 cm long and 1—7 cm broad; flowers in

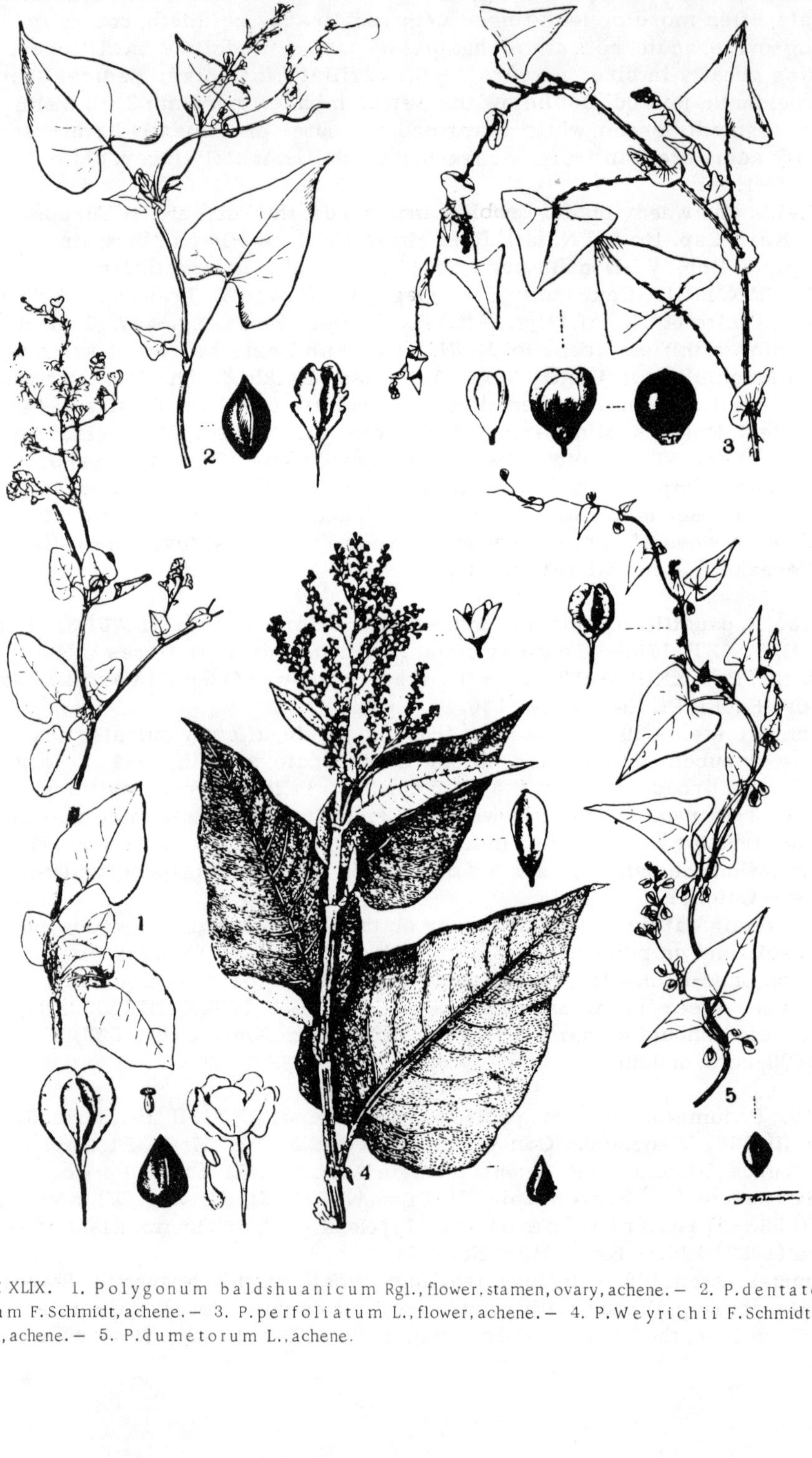

PLATE XLIX. 1. Polygonum baldshuanicum Rgl., flower, stamen, ovary, achene. – 2. P. dentato-alatum F. Schmidt, achene. – 3. P. perfoliatum L., flower, achene. – 4. P. Weyrichii F. Schmidt., flower, achene. – 5. P. dumetorum L., achene.

loose terminal racemes and also in axillary fascicles of 2—5; pedicels 2—4 mm long, the length of perianth, jointed somewhat below the middle; fruiting perianth to 7—9 mm long, the outer segments enlarging into coriaceous wings along the midrib and rounded at base, the fruit falling together with the upper part of the pedicel; achene to 3 mm long, black, lustrous, the faces concave, the ribs rounded. July—October. (Plate XLIX, Figure 5).

Riverside osier-beds, sands, damp thickets, rock streams, pebbles, broad-leaved woods, occasionally a weed of spring rye and spring wheat. — European part: Kar.-Lap. (S.), Lad.-Ilm., U. V., V.-Kama, U. and M. Dnp., V.-Don, Bl., L. Don, L. V.; Caucasus: Cisc., W. and E. Transc., Tal.; W. Siberia: Ob (as in Urals up to 58° N. lat.), U. Tob., Irt., Alt.; E. Siberia: Yen., Ang.-Say., Dau.; Far East: Ze.-Bu., Uss., Uda (S.); Centr. Asia: Balkh., Gen. distr.: Scand., Atl. and Centr. Eur., Med., Bal.-As. Min., Iran., Dzu.-Kash., Mong., Jap.-Ch. Described from N. Europe. Type in London.

696 120. **P. dentato-alatum** F. Schmidt in Maxim. Prim. Fl. Am. (1859) 232. — P. scandens Kom. in A. H. P. XXII, 138, non L. — Tiniaria scandens Nakai var. dentato-alata Nakai in Mori Enum. of Plants from Korea (1922) 137. — Fagopyrum scandens (L.) H. Gross var. dentato-alatum (Maxim.) H. Gross in Bull. Geogr. Bot. XXIII (1913) 23.

Annual; stem trailing, to 2 m or longer, smooth, branched, 4-angled; young branches 8-angled, minutely prickly on the angles; leaves cordate to cordate-oval, acute, 4—6 cm long, 3—5 cm broad, minutely papillose on the margins and on the veins; ocreae membranous, short, glabrous; racemes terminal and axillary, leafy, the flowers in fascicles of 3—5; pedicels jointed below the middle, 1—12 mm long; perianth purple, 5-parted, decurrent, the 2 outer segments flat, the 3 inner ones larger and winged on the back, the wing entire or toothed; fruiting perianth yellowish-green, the broad wings purple or rose; achene to 5 mm long, sharply trigonous, with strongly concave faces, somewhat lustrous, black, punctate. August—October. (Plate XLIX, Figure 2).

Riverside osier-beds, thickets, rocky bluffs, sandy and pebbly shallows, and sands; occasionally a weed of field crops. — Far East: Uss. Gen. distr.: Jap.-Ch. Described from the banks of the Amur below Khabarovsk. Type in Leningrad.

Note. Paralleling the North American P. scandens L., the latter being a perennial plant with larger leaves and longer petioles, its pedicels jointed near base, the perianth greenish-yellow, the achenes not more than 4.5 mm long.

Economic importance. Ornamental.

Section 8. PLEUROPTERUS (Turcz.) Benth. et Hook.; Turcz. in Bull. Soc. Nat. Mosc. I (1848) 587, as a separate genus. — Reynoutria Houttuyn Nat. Hist. VIII (1777) 640 et tab. LI, fig. 1. — Large plants; racemes numerous, disposed in terminal and axillary panicles; flowers small; perianth of 5 or 6 segments, the inner ones with a decurrent wing on the wings; stamens 7 or 8; styles 3.

1. Stem trailing, woody on lower part . . . . 2.
699 + Stem erect, firm, tall, not woody . . . . 3.
2. Leaf margins with 1 or 2 sinuses and minutely toothed; flowers 5—8 mm in diameter . . . . 121. **P. baldshuanicum** Rgl.
\+ Leaves entire; flowers small, 2—3 mm in diameter . . . . 121. ***P. multiflorum** Thunb.
3. Leaves densely tomentose beneath, almost white . . . . 122. **P. Weyrichii** F. Schmidt.
\+ Leaves glabrous or sparsely pubescent beneath, green; stems to 3 m long; leaves oblong-ovate, to 30 cm long . . . . 123. ***P. sachalinense** F. Schmidt.

Series 1. **Scandentes** Kom. — Stems climbing on trees or creeping upon rocks.

121. **P. baldshuanicum** Rgl. in A. H. P. VIII, 3 (1883) 684; Fedchenko Conspectus Fl. Turk. 6, 293. — Ic.: ibid., tab. 10.

Perennial; stem trailing, to many meters long, scandent on trees, the lower woody part with brownish-gray bark and numerous lenticels; young branches angled with prominent ribs, these bearing few small glands; petioles 2—4 cm long; leaf blades oblong-ovate, cordate at base, 2.5—9 cm long and 1.5—5.5 cm broad, the margin with 1 or 2 sinuses, obsoletely crenulate; ocreae pellucid, cylindric, appressed to stem, the margin somewhat cartilaginous; inflorescence paniculate, to 50 cm long; flowers 5—8mm in diameter, white, turning reddish; stamens 8, the filaments pubescent in lower part; perianth deeply cut, its base decurrent; pedicels 4—12 mm long, jointed at the middle or nearer the base; bracts oval, acute, ca. 2 mm long; fruiting perianth oboval or obcordate, ca. 12 mm long, to 10 mm broad, white or roseate, the wings ca. 3 mm broad; achene ca. 4 mm long, acute, with strongly concave faces, rather dull, dark brown. April—June. (Plate XLIX, Figure 1).

Thickets, wood margins, and rocks, at altitudes of 1,200—1,600 m. — Centr. Asia: Pam.-Al. (Bal'dzhuan, Gissar). Endemic. Described from the Vakhsh River, where it was collected at the foot of Mount Seistan. Type in Leningrad.

121. ***P. multiflorum** Thunb. Fl. Jap. I (1784) 169. — *Pleuropterus cordatus* Turcz. in Bull. Soc. Nat. Moscou XXI (1848) No. 1, 587, — *Pleuropterus multiflorus* Turcz. ex Nakai in Fedde Repert. sp. nov. XIII (1914) 267.

Perennial, smooth throughout; stem trailing, not woody; rhizome thickened; leaves cordate-oval; panicles large, many-flowered; flowers rather remote, small, 2—3 mm in diameter, white or roseate; filaments
700 glabrous; achene yellowish-brown, the broad wings entire. August—October.

A cultivated ornamental plant, suitable for the entire Black Sea coast. Native in Japan, in Korea up to 42°30' N. lat., closely related to *Pleuropterus ciliinervis* Nakai (*Polygonum multiflorum* Thunb. v. *ciliinervis* (Nakai) Steward), with annual stem, nonthickened roots, and leaves setose on veins beneath, producing enormous runners climbing up trees. Reported in cultivation for the same latitude or somewhat further north. Both these species are related to P. baldshuanicum Rgl. and are mentioned here as examples of the phytogeographical parallelism between our area and the subtropics.

Series 2. **Elatae** Kom. — Stems erect, sturdy, very tall.

122. **P. Weyrichii** F. Schmidt in Maxim. Prim. Fl. Amur. (1859) 234. — Pleuropteropyrum Weyrichii H. Gross in Bull. de Géogr. Bot. XXIII (1913) 9.

Perennial; stems to 1 m long, erect, weakly branched, deeply sulcate, with solid internodes, rooting and covered with dark hairs at the lower internodes, tomentose in upper part; leaves oval to broad-oval, acute, 12—20 (30) cm long, 10—15 cm broad, glabrous above, almost white with dense tomentum beneath; petioles ca. 2 cm long; ocreae elongated, membranous, strongly nerved, hairy, soon partly or fully disintegrating; panicles axillary and terminal, the lower short, the upper surpassing the leaves; bracts pellucid, 2-lobed; flowers in groups of 3—6; pedicels filiform, jointed at the middle; staminate flowers much larger than the hermaphrodite; stamens 8; ovary 3-angled, half as long as the perianth; styles 3; stigmas peltate; glands between the stamens relatively large; fruit cup-shaped by the 3 outer segments of the perianth, trigonous-obovaloid, lustrous, to 1 cm long; achene trigonous, with oblong-oval faces, blackish-brown, with dull uneven surface. July—October. (Plate XLIX, Figure 4).

Grassy mountain slopes. — Far East: Sakh., Uda (a dubious specimen from Cape Chikrak near the Amur River mouth). **Gen. distr.**: Jap.-Ch. (only N. Japan and S. Kurile Islands). Described from Sakhalin (Due). Type in Leningrad.

Economic importance. The hard roots interfere with the raising of virgin soil. Valuable as an ornamental plant.

Note. P. Weyrichii F. Schmidt var. alpinum Maxim. ex Franch. et Sav. Enum. pl. Jap. II (1875) 402. — Pleuropteropyrum alpinum (Maxim.) Koidz. in Bot. Mag. Tokyo XXX (1916) , with smooth leaves, was claimed for Kamchatka on the basis of Zeeman's specimens in Gray's Herbarium, but this is very unlikely.

701 ***P. cuspidatum** Sieb. et Zucc. Fl. Jap. Fam. nat. II (1844) 84. — Reynoutria japonica Houttuyn Nat. Hist. VIII (1777) 640. — R. elata Nakai in Cat. Sem. Hort. Tokyo (1914) 12. — Pleuropterus cuspidatum (Sieb. et Zucc.) H. Gross in Lees. in Beih. Centralbl. XXXVII (1919) 114.

Perennial; stem 2—3 m long, arising from a creeping branched rhizome; leaves large, broad-ovate, firm, 13 cm long, 10 cm broad; panicle in the axils of upper leaves; flowers white; fruit winged. August—September.

Very hardy even in Leningrad, this vigorous plant, native of central and southern Japan, beside being ornamental, provides good forage both fresh and in silage. It has been observed already in Vladivostok that the plant becomes readily naturalized. Recommended for regions where the soil does not freeze in winter.

123. *P. sachalinense F. Schmidt in Maxim. Primitiae Fl. Amur. (1859) 233. — Reynoutria sachalinensis Nakai, Veg. Dagelet Isl. (19) 18.

Perennial; rhizome producing long stolons; stems 2—3 m long, erect, stout, with hollow internodes, sulcate, green or brown, glabrous; petioles, 3—4 cm long; leaf blades broad-oval or oval-oblong, cordate or rounded (the uppermost cuneate) at base, terminating in a short point 10—30 cm long, 4—25 cm broad, dry, very thin, glabrous or with scattered short hairs above, sometimes finely tomentulose but mostly glabrous and sometimes pruinose beneath (quite glabrous in cultivation), the margins slightly undulate; ocreae oblong, smooth, membranous, prominently nerved, not ciliate, soon disintegrating; panicles axillary, short, solitary or fascicled; bracts small, oval, cuspidate, pilose, the branches densely dark-pubescent; pedicels slender, longer than perianth, jointed below the middle; perianth funnel-shaped, whitish, the 3 inner segments with winglike outgrowths strongly enlarging in fruit; stamens 8; styles 3; stigmas peltate; fruit trigonous, obovaloid; achene trigonous, lanceolate, dark brown, lustrous, point-tipped. August—October.

Bushy, spreading in long strips, on mountain slopes. — Far East: Sakh. (S.). Gen. distr.: Jap.-Ch. (only Hokkaido and N. Honshu). Described from the Noto-sama area on the W. coast of Sakhalin. Type in Leningrad.

Economic importance. Flourishing in cultivation even in the Leningrad area. The young shoots are readily eaten by horses and cows. Producing strikingly profuse herbage; suitable for ensilage.

702 Genus 395. **FAGOPYRUM*** Gaertn.**

Fruct. II (1791) 182, tab. 119. — Fagotriticum L. in Vet. Acad. Hand. (1744) 117, tab. 4.

Flowers bisexual; perianth 5-parted, not accrescent in fruit; stamens 8; styles and stigmas 3; ovary 3-angled, surrounded by a glandular ring; fruit a trigonous achene; seeds with axial embryo, the cotyledons transversely folded. Annual or rarely perennial herbs with glabrous branching stems; leaves alternate, sagittate-triangular. Russian: "grechikha."

1. Perennial plants with a woody root . . . . 3. **F. suffruticosum** F. Schmidt.
\+ Annual plants with slender roots . . . . 2.
2. Achene sharply trigonous, the faces smooth . . . 1. **F. sagittatum** Gilib.
\+ Achene sharply trigonous only in upper part, the faces obtuse and wrinkled at base . . . . 2. **F. tataricum** (L.) Gaertn.

1. **F. sagittatum** Gilib. Exerc. phyt. II (1792) 435; Kryl., Fl. Zap. Sib. IV, 871. — F. esculentum Moench, Meth. (1794) 290; Ldb. Fl. Ross. III, 515. — Polygonum Fagopyrum L. Sp. pl. (1753) 364. — Fagopyrum fagopyrum Karsten. Deutsch. Fl. (1883) 522. — Ic.: Rchb. Ic. Fl. Germ. XXIV, t. 226; Hegi Fl. III, t. 94.

Annual; stem 15—70 cm long, erect, branching, smooth, rarely minutely papillose in upper part, reddish, ribbed; leaves cordate-triangular or

* Name for buckwheat used by ancient writers, derived from Fagus, beech, because of the resemblance between beech nuts and the achenes of buckwheat.

** Treatment by A. S. Lozina-Lozinskaya.

cordate-sagittate, with rounded or acuminate basal lobes, glabrous, with minutely papillose veins, yellowish-green, somewhat fleshy, the lower long-petioled, the uppermost subsessile, the blade 2—6 cm long and 1.5—5 cm broad; flowers in racemes, the long axillary peduncles forming a corymbose inflorescence; perianth 5-parted, red or roseate or white, the ovate segments ca. 3 mm long and 2 mm broad; stamens 8, alternating with glands, these half as long as the perianth segments; styles 3, as long as the 3-sided ovary; achene sharply trigonous, brown, the smooth dull faces pointed at both ends. Fl. July; fr. August.

Throughout the central belt and the southern regions of Europe, cultivated as a field crop and growing as a weed in fields and along roads. Also cultivated and ruderal in the Temperate zones of the New World. Described from Asia.

703 Economic importance. Cleaned nutlets are used for food under the name "grechnevaya krupa" [Greek groats]. A honey plant. Widely cultivated; yields groats and buckwheat flour.

Note. The flowers are dimorphic, with long or short styles. Red-flowered varieties give higher yields than those with pink or white flowers.

2. **F. tataricum** (L.) Gaertn. Fruct., II (1791) 182, tab. 119; Ldb. Fl. Ross. III, 517; Kryl., Fl. Zap. Sib. IV, 873. — Polygonum tataricum L. Sp. pl. (1753) 364. — Ic.: Hegi, Fl. III, tab. 29; Rchb. Ic. Bot. XXIV, tab. 227.

Annual; stem 30—80 cm long, erect, branching, commonly green, glabrous, smooth; leaves sagittate-cordate, attenuate to a point, with acutish basal lobes, 3—8 cm long and broad, light green, the lower long-petioled, the upper short-petioled, glabrous; inflorescence as in the preceding; flowers smaller; perianth 5-parted, the greenish oblong-ovate obtuse segments 1.3—1.7 mm long and 0.5—1 mm broad; stamens 8, alternating with round glands; styles 3, about half as long as the 3-angled ovary; achene 3—4 times as long as perianth, blackish-brown, trigonous, the angles sharp in upper part, blunt wrinkled and apparently crenate below, the oblong-ovate faces narrowed upward, wrinkled and dull. Fl. June—July; fr. August.

In fields as a weed, among crops, and along roads. — European part: Lad.-Ilm., Dv.-Pech. (S.), U. Dnp., U. V., V.-Kama, M. Dnp., Transv., Bl., L. Don; Caucasus: Cisc.; Centr. Asia: Balkh., Pam.-Al.; W. Siberia: U. Tob., Irt., Alt.; E. Siberia: Ang.-Say., Dau.; Far East: Ze.-Bu. **Gen. distr.**: Centr. Eur. (Poland, Galicia), Ind.-Him., Jap.-Ch., Tib., introduced in America, reported in cultivation for the Himalayas. Described from "Tataria." Type in London.

Note. Commonly infests crops of buckwheat, together with which it is distributed in areas in the Ukraine, W. Siberia, and Kazakhstan. Most common in E. Siberia. In the European part of the USSR it has the character of an adventive plant.

One should mention the depressed small-leaved form from Shugnan and the bushy-branched from the Zeravshan Range. The lack of mature fruits on the available herbarium specimens precludes a complete comparison of this form with typical plants of the species.

3. **F. suffruticosum** F. Schmidt, Reis. in Amurl. u. Sach. Mém. Acad. Sc. Pétersb. VII, 12 (1869) 170; Kom., Fl. Manchzh. II, 1 (1903) 145. — F. tataricum var. suffruticosum (F. Schmidt) Miyabe in Journ. Fac. Agr. Hokk. Univ. XXVI, 4 (1934) 502.

Perennial with stout woody perennial root and numerous grayish ribbed
704 herbaceous shoots, to 30 cm tall; petioles long and slender; leaf blades ovate-triangular, broadly subcordate at base with broad obtuse lobes; ocreae rufous, scarious, broad; flowers few in remote whorls forming a racemose terminal inflorescence; pedicels short, filiform; segments of fruiting perianth narrowly lanceolate; fruit broad-ovoid, with 2 sharp upright toothlike outgrowths in upper part, the ribs in lower part sometimes minutely toothed-sinuate.

Far East: Sakh. Endemic.

Note. A species related to F. tataricum but readily distinguishable by such features as the perennial root, the shape of inflorescence, and the toothed fruit. Described from Sakhalin. Type in Leningrad.

Genus 396. **KOENIGIA*** L.**

L. Mantissa I (1767) 1241.

Flowers bisexual, with 3-parted perianth; stigmas 2; stamens 3, opposite the perianth segments; ovary flattened. A small annual plant.

1. **K. islandica** L. Mant. I (1767) 35; Ldb. Fl. Alt. I, 124; ej. Fl. Ross. III, 535; Turcz. Fl. baic.-dah. II, 71; Kryl., Fl. Zap. Sib. IV (1930) 819. — Macounastrum islandicum Small in Britt. and Br. Ill. Fl. N. Am. I (1896) 542. — Ic.: Fl. Dan. III, tab. 418 (1768); Engl. u. Pr. Pflanzenfam. III, 1a (1893) 9.

Annual, glabrous, branched, 1.5 — 7 cm tall, with spreading branches and short broad brown scarious sheaths; leaves petiolate, elliptic to rounded-elliptic, obtuse, the blade 1.5 — 6 mm long and 1 — 5 mm broad, the upper petioles short, the lower ones the length of the blade; flowers bisexual, in loose heads at the ends of stem and branches, bracteolate; perianth greenish-white, cut to three-fourths into 3 broad-elliptic to almost round segments 1 mm long, elongating in fruit to 1.5 mm; stamens 3, shorter than perianth segments; stigmas commonly 2, capitate, subsessile; fruit ovoid, narrowed at both ends, 1.25 mm long, more convex on one side than on the other, dull, minutely granular. July.

Alpine and polar-arctic zones, tundras, and bogs; banks of rivers and brooks. — Arctic: Arc. Eur., Nov. Z., Arc. Sib., Chuk.; European part: Kar.-Lap.; W. Siberia: Alt.; E. Siberia: Ang.-Say.; Far East: Kamch.; Centr. Asia: Dzu.-Tarb., T. Sh. Gen. distr.: Centr. Eur., Arc. Eur., Scand., Mong.-Ch., Him. Described from Iceland.

* Named for Johann Koenig, missionary-naturalist, pupil of Linnaeus.

** Treatment by A. S. Lozina-Lozinskaya.

DIAGNOSES PLANTARUM NOVARUM

IN TOMO V FLORAE URSS COMMEMORATARUM

(DIAGNOSES OF NEW SPECIES MENTIONED IN VOLUME V)

Junio 1936.

SALIX

707 **1. S. liliputa** Nas. sp. nova (sect. *Herbaceae* Borr.) — *S. herbacea* f. *pygmaea* Laksch. in sched.

Fruticulus parvulus, subherbaceus, ramulis tenuissimis, brevibus subterraneis apice parum foliaceis; folia brevipetiolata infrequentia 2-3-5 in numero, minuta, rotundata, ovalia, obovato-cordata ca. 0.3—0.9 cm lg., 0.3—0.7 cm lata, apice obtusata, venis primariis utrinque 3; stipulae adsunt. Amenta ♀ coaetanea, 2—5-flora, subsessilia, in pedunculo tomentoso, bas, bracteis 1—2 stipata; squamae apice obtusae vel truncatae, sordide-purpureaei margine parum albo-pilosae; germina ovato-conica ad apicem attenuata glaberrima, pedicellis usque ad 0.5 mm lg. nectarium aequantibus; nectaria 2, internum elongatum, 2—3-lobatum, externum oblongo-lineatum rubicundum. stylus brevis; stigmata bifida, arcuato-patentia vel patula; capsula ca. 4 mm longa.

H a b. In alpibus Transbaicaliae septentrionalibus.

T y p u s: Burjato-Mongolia. In alpibus; mons Kiren ad ost. flum. Angara Superior. 26 VII 1912 n° 295a, leg. Sukaczev et Poplawska; in Herb. Inst. Bot. Ac. Sc. URSS conservatur.

S. herbaceae et *S. Turczaninovii* proxima, sed differt: foliorum forma, foliis minoribus, amentis paucifloris, omnibus in partibus fere 2—4-plo minor.

2. **S. Raddeana** Laksch. sp. nova; in sched. 1913 (sect. *Capreae* Bluff.). — *S. villosa* Siuz. in sched. (nom. tant.).

Frutex videtur altus aut arbuscula; rami glabri, obscure-fusci, hornotini dense-ravide-pilosi; gemmae magnae, appressae, acutae costatae, obscuro-fuscae, pilosae; stipulae magnae ca 6—7 mm longae, acutae, glabrae vel pubescentes; petioli 5—10 mm lg. pilosi; folia crebra, coriacea, ca 3.7—7 cm lata, obovato-rotundata vel obovato-elliptica, apice breviter acuminata, basi rotundata vel acuta, integerrima, revoluta vel emarginata et grosse glanduloso-dentato-emarginata, supra obscure viridia subtus molliter velutino-tomentosa, novella albo-argenteo-tomentosa; nervi primarii circiter 8, utrinque prominentes, cellulas latas ad margines, quam *S. caprea* L., non formantes. Amenta praecocia, sessilia, basi bracteis squamosis sericeis praedita; squamae elliptico-ovatae, apice acuminatae vel obtusae, seminigrae vel antice brunne-

* [This appendix has been reproduced photographically from the Russian original.]

scentes, inferne splendidae, ca. 2.5 mm lg., 0.6—0.8 mm lat., pilis griseis longis tectae; germen anguste-conicum, attenuatum, 3.5—4 mm lg. cinereo-pilosum n pedicello 2—2.5 mm lg., nectarium 3—4-plo superante; stylus glaber, brevis, ca. 1 mm lg.; stigmata fusca,recta vel patentia; capsula ca. 6.5 mm lg. (Descr. M. Nasarov).

Hab. in sylvis et fruticetis Sibiriae transbaicalensis, amurensis et ussuriensis.

Typus: Nicolsk-Ussurijsk. 21—27 IV 1905. Leg. P. Siusev; in Herb. Inst. Bot. Ac. Sc. URSS conservatur.

„*S. capreae* proxima, differt squamis brunnescentibus, stylo longiore, stigmatibus paulo recurvatis, germinis pedicello breviore nectarium triplo superante“ (P. Lakschewitz).

3. **S. Olenini** Nas. sp. nova (sect. *Capreae* Bluff.).

Frutex altus vel arbor mediocris; ramuli novelli dense cinereo-tomentosi, veteri glabri, rubicundi vel castanei; gemmae ovato-semiconicae, acutae, costatae, ca. 5 mm lg., 3 mm lat., incarno tomentosae; petiolus 8—10 mm lg., piloso-villosus; stipulae nullae; folia 5—8 cm lg., 2.5—4.5 cm lata, crebra, rigida, ovata, utrinque attenuata vel obovata, ad basin versus attenuata, margine grosse et remote glanduloso-serrata, supra saturate viridia, glabra, subtus glauca vel glaucescentia, pilis adpressis laxis nitidis tecta; costis fere tomentosis; nerviis primariis utrinque 10—12, utrinque prominentibus, a costa sub angulo 40—50° divergentibus, in apice folii convergentibus et cellulas latas ad margines formantibus. Flores ignoti.

Hab. in fruticetis ripariis Sibiriae orientalis.

Typus: Jacutia, prope Shigansk, in valle fl. Lena. 4 VIII 1901 n° 1058. Leg. P. Olenin; in Herb. Inst. Bot. Ac. Sc. URSS conservatur.

S. capreae similis, sed differt: ramulis novellis, gemmis, petiolisque spisso-tomentosis, foliis subtus glaucis, disperse pilosiusculis, margine grosse serrato-glandulosis.

4. **S. Litwinowii** Görz. sp. nova (sect. *Hastatae* E. Fries).

Frutex ramulis novellis pubescentibus, anniculis glabris, brunneis; gemmae vix pilosae, semiovatae, brunneae; stipulae nullae; petiolus brevis (2—3 mm) parce pilosus dein glabrescens; folia adulta elliptico-lanceolata (ca 1 : 3) marginibus subparallelis, basi rotundata, apice breviter contracta, margine trite crenato-dentata (dentibus glandulis subcarentibus), supra glabra, costa subinde pilosa, excepta atroviridia, subtus pallidiora, pilis longis acroscopicis leviter adpressis aequaliter ad autumnum obsita, nervis primariis ± 10, aeque ac reticulum maculis amplis constitutum utrinque distincte prominulis. Flores desunt (R. Görz.).

Typus: Turcomania. Prope Tschuli ad scaturigines IX 1897 n° 2411. Leg. D. Litwinow; in Herb. Inst. Bot. Ac .Sc. URSS conservatur.

709 5. **S. brachypoda** (Trautv. et Mey.) Kom. sp. nova; in A. H. P. XXXIX (1923) 49 (nomen tantum). (Sect. *Incubaceae* Dum.). — *S. repens* L. var. *brachypoda* Trautv. ex Mey. in Middend. Sibir. Reise II, 2 (1856) 79; Schedae ad HFR VIII n° 2466. — *S. repens* L. var. *flavicans* Anderss. in DC. Prodr. XVI, 2 (1868) 238.

Frutex humilis, 0.75—1 m altus, ramis tenuibus, rectis, virgatis, fulvis, vel rufis, glabris, ramulis flavescentibus vel chryseo-tomentosis, infima glabris; gemmae semiovatae, erectae, rostratae, initio pubescentes, vel sericeae, demum glabrae, fulvae vel rufescentes; stipulae in vegetis validae, lanceolatae, 3—7 mm lg.; folia trivialiter nigricantia, in petiolis brevissimis, anguste-elliptica, oblongo lanceolata vel lineari-lanceolata, 2.5—5.7 cm lg., 0.4—1.5 cm lata, margine subrevoluta, intergerrima, supra viridia, pubescentia, demum glabrescentia, subtus flavicantia vel argenteo-sericea vel glaucescentia, rarius folia utraque argenteo-sericea; venis lateralibus 5—10, a costa sub angulo aucto abscendentibus. Amenta praecocia, lateralia, sessilia, basi bracteis caducis parvis praedita, oblonga vel cylindrica, densiflora, ♂ 2—3 cm lg., 0.6—0.9 cm lata, ♀ 1—1.5 cm lg. et circiter 0.5 cm lata; bracteae lingulatae vel obovatae, seminigrae, dense flavescenti-villosae, rachidibus sericeis; nectarium internum, oblongum vel lineare ca. 0.6 mm lg., pedicello paulum brevius; germen ovato-conicum, 2—2.5 mm lg., flavicanti-tomentosum; stylus brevis; stigmata 2—4-fida, erecta, patula; capsula ca. 4—5 mm longa.

Hab. In paludibus turfosis, in vallibus paludosis, in betuleto-fruticetis Sibiriae orientalis.

Typus: Sibiria orientalis, Ochotsk., leg. Middendorf; in Herb. Inst. Bot. Ac. Sc. URSS conservatur.

A *S. rosmarinifolia* L. differt ramulis, foliis amentisque flavicanti-villosis germinibus brevius pedicellatis et area geographica: in Sibiria occidentalis et in Europa absens.

6. **S. turanica** Nas. sp. nova (sect. *Viminales* Bluff.). — *S. viminalis* L. *β. splendens* (Turcz.) Anderss. 1° *songarica* Anderss. in DC. Prodr. XVI, 2 (1868) 265.

Frutex altus ramis testaceis vel vitellinis, tomento albido deciduo tectis; stipulae saepe adsunt; folia petiolo brevi, albo-sericeo, rigida, oblonga vel ovato-oblonga, 11—14 cm longa, 2—4 cm lata, ad basin latiora, leniter cuneata, ad apicem breviter acuminata, margine subundulata, fere plana, integra vel subemarginata, subtus sordide-viridia vel incano-virescentia, pilis brevibus dispersis adpressis canescentia, interdum subglabra, subtus intense albo-tomentosa vel argentoso-cana, splendida, nervis tomentosis prominentibus lateralibus utrinque 16—18 (—20), a costa sub angulo 50—70° divergentibus. Amenta omnino sessilia, praecocia. Ceterum, quam *S. viminalis* L. coll.

Hab. ad ripas.

Typus: Asia media, Heptopotamia, fl. Tschu leg. Schrenk; in Inst. Bot. Ac. Sc. URSS conservatur.

7. **S. strobilacea** (E. Wolf) Nas. sp. nova (sect. *Wiminales* Bluff.). — *S. viminalis* L. var. *strobilacea* E. Wolf in Изв. Лесн. Инст. (С.-Петерб.) XIII (1905) 3—4, tab. I.

Frutex altus, ramulis novellis albido-velutinis, annotinis griseo-velutinus; stipulae falcatae, acuminatae, glandulosae, petiolo brevior vel aequantes; folia anguste lanceolata ad basin latiora ad apicem arcuato-acuminata, margine intégerrima, subrevoluta vel subemarginata, 6—8—14 cm lg. et usque, 2 cm lata, petiolo brevi, ca 1 cm lg., supra sordide-viridia, pilosiuscula, ad margines glandulis dispersis tecta, subtus albido-sericea, pube dense splendente tecta; costa valida, fuscescens vel straminea, nervis primariis utrinque 13—15—22, elevatis, a costa sub angulo 50—60° divergentibus. Amenta coaetanea, pedunculo brevi, basi bracteis praedita; squamae germinum superantes, lingulatae, acutae vel obtusatae, fuscentes apicem atrae, parce et breve pilosiusculae; ovarium sessile vel in pedunculo brevissimo, ovato-conicum, splendido-pilosum; stylus plus minus bifidus, saepe ad usque basin; stigmata integra vel bifida, lobis angustis; nectarium oblongum, plerumque promissus.

Hab. ad ripas.

Typus: Sibiria occidentalis, Semipalatinsk, Karkaralinsk; in Herb. Ac. Silvicult. Leningr. conservatur.

8. **S. sajanensis** Nas. sp. nova (sect. *Viminales* Bluff.). — *S. viminalis* L. *a genuina* Turcz. in schedis et in Fl. baic.-dah. II, 2 (1854) 379.

Frutex vel arbuscula 2—5 m alt.; truncus ad basin cortice rorido, glabro, ramis rubicundo fuscis vel obscuris, glabris; ramuli lividinosi, tortuli vel flexuosi, frequenter-intorti deorsum vel adunci, novelli pubescentes et dense foliati, veteres glabri, cicatriculis numerosis et pulvinulis grossis praediti; gemmae ad 7 mm lg., 4 mm latae, obsure-castaneae, obtusae, initio sericeae, demum glabrae; petiolus 3—5 mm lg. sicut rami novelli dense pilosus vel sericeus; stipulae parvae, setiformes, glandulosae, sericeae, caduces; folia oblanceolata, oblongo vel lineari-lanceolata, 4—9 (—15) cm lg., 0.5—1.5 (—3) cm lata, plerumque versus basin et apicem aequaliter attenuata, leviter curvata, margine revoluta, plerumque integerrima vel dentibus longe remotis glanduliferis irregulariter dentata, supra obscure-viridia, nervis distincte prominentibus, disperse breviter pubescentia, opaca, subtus pallide-viridia, fere glabra vel leniter pubescentia vel albescenti-velutino-tomentosa; folia novella utrinque lutescenti-sericea; nervi laterales utrinque 12—15, subtus distincte prominentes, tenues, sub angulo 45—50° divergentes. Amenta ♀ in pedicellis brevissimis, basi squamis foliaceis praeditis, compacta, cylindrica, paullo curvata, 2.5—3 cm lg. circa 0.5 cm lata, matura (capsulis dehiscentibus) 4—6 cm lg., 0.5—0.8 cm lata; squamae ovatae, apice angustatae, obscure brunneae vel seminigrae,

711 utrinque pilis densis rectis longis argenteis tectae; nectarium ca 1 mm, oblongum: germina sessilia vel pedicello brevissimo, ovoideo-conica, virescentia, dense tomentosa; stylus longus, stigmatibus longis linearibus, interdum bifidis longitudine stylum superantibus vel eum aequantibus; specimina ♂ adhuc non observata.

Hab. in regione subalpina Altaica et Sajanensi, arboribus singulis crescit in altitud. 1700—2100 m. s. m. in sabulosis saxosis.

Typus: Sibiria orientalis. Montes Sajanenses, in alpibus Tunkinensibus. 1929. Leg. M. Nasarov n° 12450; 12456; 12703; in Herb. Inst. Bot. Ac. Sc. URSS conservatur.

9. **S. jacutica** Nas. sp. nova (sect. *Viminales* Bluff.).

Arbuscula videtur vel frutex altus, ramulis novellis pubescentibus veteris glabris, obscure-brunneis, gemmis ca. 5—6 mm lg., ovato-conicis, obtusiusculis, superioribus pilosiusculis, inferioribus; glabris; stipulae parvae, 2—5 mm lg., lanceolatae vel falcatae, leniter serratae, caduces; petiolu ; 6—10 min lg., pilosus ad basin dilatatus; folia novella sericea, intro obversa, adulta ad 8 cm lg., 1.8 cm lata, lanceolata, oblongo vel lineari-lanceolata, apice acuminata, basi rotundata vel acuminata, margine minute et remote dentata vel leniter undulato-emarginata, saepe fere integerrima, revoluta, supra viridia vel cineraceo-viridia, nervis primariis prominentibus, glabra vel disperse pubescentia secundum nervos pilosiuscula, subtus sericeo-splendida, pube densa brevi tecta; costa straminea, ad basin impresse dilatata; nervi primarii utrinque 12—14, a costa sub angulo 40—80° divergentes. Flores ignoti.

Hab. in vallibus ad ripas.

Typus: Jacutia, distr. Jacutsk et Verchojansk, prope Bestjach et Shigansk. 1901—1902, leg. P. Olenin; in Herb. Inst. Bot. Ac. Sc. URSS conservatur.

10. **S. Michelsonii** Görz sp. nova; in sched., nom. tant. (sect. *Helix* Dumort.).

Frutex altus, ramis tenuibus, novellis brunneis, veteribus rubicundo-luteis, ramulis brunneis, tenuissimis, glabris, nitidis; gemmae parvae, acutae, adpressae, flavicanti-brunneae, pubescentes; stipulae minutae, anguste-lanceolatae, serrulatae, bis-quaterve brevior, quam petiolus; folia in petiolo glabro, brevi, circa 3—5 mm lg., basi dilatato, interdum contorto, rigida, utrinque glaucescenti-ceraria lineari-lanceolata, marginibus parallelis vel ad medium et apicem latissimis, basi attenuatis, integerrima vel serrulata, novelli utrinque leniter et breve pubescentes, adulta glabriuscula vel pube dispersa brevi tecta; costa tenuis, straminea; nervi laterales 13—15, a costa sub angulo 15—25° divergentes; nervi secundarii prominentes, cellulas minutas formantes. Amenta ♀ matura (capsulis dehiscentibus) ad 4 cm lg., 0.8 cm lata in pedunculo brevi, basi squamis foliaceis praedita, densiflora, cylindrica; rachide subpiloso; squamae caducae pallidae, parce-pilosae, ca. 1 mm lg., 0.7 mm latae;

712 germina ovoideo-conica, subglabra vel pilosiuscula, in pedi- cello breve (ca. 0.6 mm); stylus brevis, brunneus, 0.5—0.6 mm lg.; stigmata bifida, brunnea, ca. 0.5 mm; capsula usque ad 6 mm longa, glabra (Descripsit M. Nasarov).

Hab. ad ripas.

Typus: Asia media, Dsharkent, limen Dshijdelik 2 VIII 1910 n° 2494. Leg. A. Michelson; in Inst. Bot. Ac. Sc. URSS conservatur.

11. **S. issykiensis** Görz sp. nova; in sched., nom. tant. (sect. *Helix* Dum.).

Frutex ramulis tenuibus, virgatis, glabris, castaneis, vel rubicundo-brunneis; gemmae ovatae, obtusae, parce adpressae, glabrae, sordido-brunneae; petiolus tenuis, 4—5 mm lg.; stipulae nullae; folia ad 5 cm lg., 1 cm lata, lanceolata, supra medium aut ad medium latissima, basi attenuata, apice in acumen subulatum contracta, margine minute et acute serrata, utrinque viridia, glabra; costa flavicanti-brunnea, nervis primariis utrinque 10—12 costatis, a costa sub angulo 25—35° divergentibus; nervi secundarii subtus costate-elevati, cellulas minutas formantes. Amenta matura (capsulis dehiscentibus) pedunculata; rachide pubescente; squamae ovatae vel lingulatae, acutae, brunneae, membranaceae cum 3—5 nervis brachiatis, glabrae vel basi pilosiusculae; capsula glabra ca 5 mm longa, brunnea; stylus tenuis ca. 1 mm; pedicellus glaber ca. 1 mm; stigmata 0.5 mm, bifida, oblonga divergentia. (Descripsit M. Nasarov).

Hab. ad ripas.

Typus: Asia media, lacus Issyk-kul, 13 IX 1915. Leg. Titov; in Inst. Bot. Ac. Sc. URSS conservatur.

12. **S. Lipskyi** Nas. sp. nova (sect. *Helix* Dum.). — *S. macrolepis* var. *Lipskyi* Görz in sched.

Arbor (?) ramis tenuibus, strictis vel arcuatis, rubicundo-fuscis, epruinosis, novellis et veteris glabris; gemmae minutae 3 × 2 mm, perstrictae, adpressae, fuscae, sordidae, glabrae; stipulae nullae; folia novella subtus sericea, margine sursum revoluta, adulta 5.5 × 1 cm, superiora ad usque 8 × 1.5—2 cm, oblanceolata vel oblongo-oblanceolata, basi attenuata, apice rotundata in acumen subulatum contracta, utrinque glauca, subtus pallidiora, fragilia, integra vel minute et acute glanduloso-serrulata; nervis primariis 18—19, supra et subtus indistinctis, a costa sub angulo acuto divergentibus. Amenta ♂ ignota; fructifera praelonga, tenui-cylindrica, 10—12 cm longa, 0.7 cm lata, valde divergentia, non patula, curvata in pedunculis ad 1—2 cm lg. foliis ceteris similibus praeditis; rachis pilosa; squamae membranaceae, caducae vel consistentes, 5-nerviae pallide fuscescentes, margine glabrae dorso et intus basi disperse ciliatae; capsula glabra, viridia vel fuscescenti-viridia ca. 3 mm lg., panduraeformis vel cochleariformis, ex basi ovato-inflata cito contracta; pedicellus 0.6—0.8 mm lg., basi pilosiusculus, nectarium internum

713 ter-quaterve superans; stylus unus, brevis, non caducus stigmata brevia, bifida, lobis incrassatis.

H a b. in angustiis montosis (ad ripas fluviorum?).

T y p u s: Asia media, Tjan-Schan, prope urb. Alma-ata in angustiis „Kordonnoje“. 6 VI 1909 n° 880. Leg. V. Lipsky; in Inst. Bot. Ac. Sc. URSS conservatur.

13. **S. ferganensis** Nas. sp. nova (sect. *Helix* Dum.).

Frutex altus vel arbuscula, cortice trunci glabro, olivaceo-fusco, ramis pallide-fuscis, nitidis, glaberrimis; gemmae crassae, magnae, adpressae, glabrae; stipulae semicordatae vel semilunares, magnae, denticulatae; petioli 5—7—10 mm lg., glabri, fusci, basi dilatati, canaliculati; folia oblonga vel oblongo-lanceolata, apicem arcuato-attenuata in acumen brevem, ad basin cito attenuata, ca. 11.5 cm lg., 3.2 cm lata margine remote glanduloso-serrata, leniter erosa, undulata, utrinque pure viridia, subtus paulo obscuriora, leniter glaucescentia, fere omnino glabra, sub lente subtus ciliis albis dispersis tecta; stomata utrinque numerosa; costa valida, bene elevata, basi dilatata, pallido-fusca, nervis primariis 11—13, a costa sub angulo 40— 50° divergentibus, anastomosis utrinque elevata. Flores ignoti.

H a b. in montis ad ripas fluviorum.

T y p u s: Asia media, montes Ferganenses in ripa fl. Kug-art 4 VIII 1895 n° 5554. Leg. S. Korshinsky; in Inst. Bot. Ac. URSS conservatur.

14. **S. euapiculata** Nas. sp. nova (sect. *Albae* Borr.). — *S. aus- tralior* Anderss. var. *apiculata* Laksch. in Sched. ad HFR VIII (1913—1922)19 n°2454.

Arbor excelsa, ramis lutescenti-fuscis, glabris, subnitidis, ramulis novellis gemmisve puberulis, demum glabris; folia in petiolis brevibus astipulata, lanceolata, novelli albo-argentea vel sericea, adulta utrinque viridia, omnino glaberrima, margine grosse dentata, ceterum *S. albae* L. similia; folia in turionibus infima obovata, basi attenuata, antice apiculo brevi instructa, margine integerrima, utrinque glabra vel subtus sericea. Amenta coaetanea, ♀ in pedunculis floralia 3—4, folia oblanceolata vel obovata, basi attenuata, margine integerrimi, subtus sericea vel utrinque glabra; squamae late-ovatae vel obovatae, lutescentes, antice emarginatae, dorso nudae; germina subsessilia, glabra, sub maturitatem brevissime pedicellata; ceterum quam *S. alba* L.

H a b. ad ripas Asiae mediae et Transcaucasiae.

T y p u s: Turcomania. Prope st. Farab viae ferrae 2 III 1900 et 2 V 1901. Leg. N. Androssov; in Inst. Bot. Ac. Sc. URSS conservatur.

QUERCUS

15. **Q. imeretina** Stev. sp. nova, in schedis.

Arbor ad 20 m alt.; folia subsessilia petiolis brevissimis, basi cordata auriculis magnis petiolos occultantibus, 5—14 cm longa et 2.5—5 cm lata, saepe curvata, infra glabra pallide-virentia, supra obscure-viridia, irregu-

714 lariter lobata; lobi utrinque 4—6 valde inaequales in parte inferiore laminae parvi dein longi angusti et ± curvati obtusi; nervi laterales curvati et flexuosi, intercalares semper bene evoluti; ad 10 cm longi foliis subaequales vel longiores; glandes 1—2 et 1—2 inevolutae in parte terminali tenui pedunculi insidentes; cupula humilis solum partem glandis amplectans; cupulae squamae subplanae griseo-pubescentes lanceolatae appendice fusco subglabro; glandes anguste-cylindricae ad 3 cm longae. (Descripsit V. Maleev).

A cl. Ch. Steven in schedis notata sed hucusque sine descriptione manebat.

Hab. Transcaucasia occidentalis. In promontoriis Abchasiae et Iberiae occidentalis a fl. Kodor usque ad fl. Kvirila.

Typus: Imeretia, Steven; in Herb. Inst. Bot. Tifl. conservatur.

Differt a *Q. roburi* L., cui proxima foliis subsessilibus basi profunde cordatis, auriculis majoribus, lobis foliorum valde inaequa ibus et curvatis, pedunculis longioribus et tenuioribus. A *Q. pedunculiflora* C. Koch foliis subtus pallide-virentibus et glabris nec glaucescentibus et puberulis.

16. **Q. Woronowii** Maleev sp. nova. — *Q. dschorochensis* Maleev in Journ. Bot. URSS 1935 n° 2 164, 174, non. C. Koch.

Frutex aut arbor humilis (?); ramuli glaberrimi, annotini rubiginoso nitentes; gemmae subrotundae parvae; folia petiolis 1—2 cm longis firma subcoriacea glaberrima supra laete viridia infra rufescentia; folia basi ramulorum late lanceolata, ovata vel elliptica integra vel 1—2 dentibus grossis, 2—5 cm longa et 1—3 cm lata; folia sequentia late obovata 5—8 cm longa et 3—4 cm lata utrinque lobis 3—5 latis et obtusis interdum vix conspicuis; nervi laterales utrinque 4—8 plus minus curvati inter se non paralleli in lobos bene evolutos et vix conspicuos et in latera loborum euntes; nervi tertiarii valde curvati densissime et irregulariter reticulati; glandes juveniles terni-quaterni pedunculati pedunculis 0.5—1 cm longis; cupulae squamae in cupulis juvenilibus subtriangulares, subplanae, fuscae subnitentes vix pubescentes appendice brevi obtuso; glandes maturae ignotae.

A cl. G. Woronowio in schedis ad specimen unicum ex Adsharia notata est sine descriptione manebat.

Hab. Transcaucasia occidentalis — Adsharia inferior; distr. Artwin.

Typus: Adsharia, in faucibus fl. Tschwana ad pagum Chemlisi 4 VIII 1910, leg. Woronow et Popow; in Inst. Bot. Ac. Sc. URSS conservatur.

A *Q. araxina* (Trautv.) Grossh. cui proxima, differt ramulis et foliis glabris nen noc foliis subtus rufescentibus non pallide viridibus. A *Q. dschorochensi* C. Koch foliorum forma, innervatione etc. longius distat.

URTICA

17. **U. cyanescens** Kom. sp. nova.

Rhizoma horizontaliter repens multicaule; caules c. 4 mm in diametro, usque 80 cm alti; stipulae extraaxillares liberae binae fere lineares acutae

715 membranaceae, ca. 12 mm lg., paulum ultra 1 mm latae; folia ovato-elliptica, basi rotundata, late-cuneata vel sinuata, 8—14 cm lg., 6—8 cm lata, ad venas subtus puberula, marginibus grosse argute-serratis, apice longiter acuminata in sicco cyanescentia, cystolithis numerosis bacilliformibus, dentibus usque 2 cm altis. Dioica, racemi masculi axillares bini prominintes, satis compacti, axi puberula; bracteae parvae, oblongae vel lineares acutae, partim caduces, pedicelli brevissimi, perianthium 1.2 mm in diametro, quadrifidum, antherae amplae luteae, pistilli rudimentum globosum. Flores femineae fructusque usque adhuc ignoti.

Crescit in solo lapidoso humo adnato in silvis mixtis montanis, neque secus rivulos in locis umbrosis. In regione Ussuriensi.

Typus: montes inter fl. Suputinka et Maiche, non procul a pago Chotunitschi. Leg. V. L. Komarov 1935; in Inst. Bot. Ac. Sc. URSS conservatur.

A speciebus *U. dioicae* affinibus differt imprimis, cystolithis bacilliformibus (non globosis), ab affini autem *U. laetevirenti* Maxim. caule firmo, foliis multo longioribus saturate viridibus cyanescentibus, dentibus altioribus, amentis prominentibus compactis.

RUMEX

18. **R. Rechingerianus** A. Los. sp. nova.

Caulis rubustus, 40 cm — 2 m altus, paulo sulcatus, ramosus, pallideluteus; inflorescentia longa anguste-ovata, ramis brevibus, squarrosis; folia plana vel margine paulo undulata, supra glabra, subtus ad nervos scabra; folia inferiora ovalia, basi truncata rotundata vel cordata, apice attenuata, petiolis subduplo brevioribus; folia caulina oblongo-lanceolata, vel lanceolata, basi cuneato-attenuata, apice acuta, breve petiolata. Flores in axillis (12—20); perigonii phylla rufo-brunnea, 4—6 mm in diametro, late-rotundato-cordata, apice paulo prominentia, margine integerrima, plana paulo reticulata solida, una vel omnia callosa; callis 1.5—2 mm lg., 1—1.5 mm lt., convexis; nuculae 2.5 mm lg. apice et basi acutiusculae. VII—VIII.

Hab. inter segetes, in declivibus herbosis, in montibus Pamiro-Alaicusi Thian-Schan in Kasachstania or., Kaschgaria et Songoria.

Typus. Yugum alaicum, Olgin-lug, Sufi-Kurgan 1895, 16 VII Korshinsky; in Inst. Bot. Ac. Sc. URSS conservatur.

19. **R. aschabadensis** A. Los. sp. nova.

Caulis magnus, crassus, minute sulcatus in parte inferiore cavus in sectione luteo-roseus, simplex; inflorescentia ovato-cylindrica, ramosa, basi foliata; folia caulina cinerea, solida, infra minute papillosa, oblongo-ovato-lanceolata, apice acutiuscula, basi truncata vel attenuata, margine undulata; lamina ad 20 cm lg., 3—5 cm lt., petiolis brevibus sulcatis; pedunculi fructibus paulo longiores; verticilli pluriflori, approximati; perigonii phylla orbiculato-reniformia, basi cordata, 6 mm in diametro, membranacea, leviter emarginata, unum callo parvo anguste-ovato-acuminato instructum, 2 cetera callo destituta,

H a b. in mont. Kopeth-Dagh.

T y p u s: Turcomania in angustiis versus mont. Saondak, 1897 15 VI Litwinow; in Inst. Bot. Ac. Sc. URSS conservatur.

Ad *R. Paulsenianum* Rech. f. appropinquat, sed inflorescentiae habitu et perigonii phyllis forma differt; a *R. Rechingeriano* A. Los. phyllis perigonii duobus callis destitutis.

20. **R. thjanschanicus** A. Los. sp. nova.

Caulis elatus, solidus, cavus, grosse-sulcatus, ramosus, inflorescentiam late-racemosam gerens; folia caulina late ovoidea, apice attenuata, basi rotundato-cordata, 17—25 cm lg., ad 15 cm lat., margine paulo undulata, laminis tenuibus, pallide viridibus vel cinereis, nervis prominentibus, petiolis brevibus, supra planis, infra carinatis. Flores in verticillis paucifloris, laxis; pedicelli tenues inaequales, basi articulati, supra infundibuliforme-dilatati; perigonii phylla membranacea, cordata apice acutiuscula, minute reticulata, margine emarginata, 6—8 mm lt., 6—7 mm lg., inaequalia, unum macrocallosum, phylla cetera callis non satis evolutis; nuculae acutae, pallide brunneae, 2 mm lg.

H a b. Asia media, Tjan-Schan occ. in vallibus fluviorum.

T y p u s: Jugum Alexandrovsky, in valle fl. Alarczi 29 VI 1914 leg. Titov; in Inst. Bot. Ac. Sc. URSS conservatur.

Differt a *R. patientia* L. fructu minore foliorum forma et magnitudine totius plantae.

21. **R. Schischkini** A. Los. sp. nova.

Caulis solitarius, erectus, ad 30 cm altus, paulo sulcatus, foliosus, glaber; folia radicalia lanceolato-ovoidea, basi rotundato-cordata, apice acuta, 5—7 cm lg., 1—3 cm lat., coriacea, margine minute undulato-crispa, supra glabra, subtus minute papillosa; petiolis sulcatis papillosis, lamina brevioribus; folia caulina 1—2, breviter petiolata, minora; folia floralia 1—2 ad basin inflorescentiae disposita, lanceolata, sessilia. Inflorescentia anguste-ovata 15—17 cm longa, breve ramosa pedunculis tenuibus, basi articulatis inaequalibus; perigonii phylla late-ovoidea, 5—6 mm lg., 4—5 mm lata margine erosula basi cordata, apice obtuse-acutata reticulata, omnia ecallosa; nuculae anguste-ellipticae, basi et apice acutae, pallidae brunneae, 4 mm lg., 2 mm lt.

H a b. solo humoso, in sylvis Sibiria occ., Altai in steppa Tshuense.

T y p u s: in stepposis ad Tschuja, fl. Kurai 19 VIII 1931 leg. Schischkin et Czilikina; in Inst. Bot. Ac. Sc. URSS conservatur.

Proximus *R. arctico* Trautv. differt autem foliis coriaceis, caule angustiore non alato, foliorum forma.

717 22. **R. ussuriensis** A. Los. sp. nova.

Caulis erectus, paulo tortuosus, sulcatus rubro-brunneus ad 1 m altus; folia anguste-lanceolata, apice acuminata, basi cuneata, margine crispata, breve pedunculata ad 15 cm lg., 4 cm lt. Flores pedunculis basi articulatis, fructu longioribus, in verticillis; inflorescentia foliata dense anguste-ovata congesta ramis erectis; perigonii phylla interiora late ovoidea, deltoidea, apice obtusata, reticulata, 4 mm lg., 5 mm lt., margine minute denticulata, omnia callum fusiforme, apice angustum, gerentia; nuces eis *R. stenophylli* simillimi.

Hab. In pratis humidis, ad ripas regionis Ussuriensis.

Typus: Distr. Vladivostok prope Okeanskaja, in argillosis 1919 20 VII, leg. Transchel; in Inst. Bot. Ac. Sc. URSS conservatur.

23. **R. foveolatus** A. Los. sp. nova.

Planta cinerea, glabra, vel minute pubescens; caulis diffuse ramosus, grosse-sulcatus, tortuosus; folia inferiora oblongo-ovata, plana, obtusa, basi cordata, petiolis laminis brevioribus; folia superiora (bracteae) parva, lanceolata vel oblonga, basi et apice attenuata, margine crispa, omnia glabra et solida. Flores 10—15 in verticillis remotis, racemos longos ramosos formantes; perigonii phylla inferiora inaequalia solida et coriacea, late-ovato-triangulari, foveolata, omnia callosa; phyllum majus 8 mm lat. et long. in parte inferiore margine dentibus acutis ejus latitudine aequilongis donatum, phylla cetera minora et angustiora, dentibus brevioribus et callis minoribus, aucta; pedunculi post anthesin reflexi, rigidi, crassi, sursum clavato incrassati in parte media articulati, fructui subaequilongi; nuculae acute-trigonae lucidae, brunneae.

Hab. Crescit in declivibus lapidosis mont. Kopet-Dagh.

Typus: Kopet-dagh, in angustiis Ai-dere 27 VI 1925 leg. E. Bobrov; in Inst. Bot. Ac. Sc. URSS conservatur.

A. *R. dictyocarpo* Boiss. et Buhse differt autem perigonii phyllis foveolatis et inaequalibus.

POLYGONUM

24. **P. pulvinatum** Kom. sp. nova (sect. *Avicularia* Meisn., series *Fruticulosae* Boiss.).

Radices fibrosae numerosissimiae ex uno collo orti, ca. 30 cm lg.; caules indurati numerosi ca. 5 mm crassi breves 5 cm lg. partibus inferioribus subterranei, apicem versus breviter ramulosi, caespitem vel pulvinum 6—12 cm in diametro formantes: ochreae pellucidae albae ca 5 mm lg. lanceolato-acuminatae integrae serius autem fimbriatim marginibus secedentes; folia angusta linearia, marginibus revolutis apice acutata, sed apiculo destituta glaberrima univenia. Flores solitarii, quam ochrea breviores viridis vel rubri, pedunculis brevissimis, perigonii laciniae usque ad basin liberae apice rotundatae; stigmata tres, achenia trigona, laevia, angulis attenuatis, faciebus laevibus.

Crescit in declivibus stepposis convallium distr. Turgaj Kasachstaniae.

Typus ad decursum inferiorem fl. Sary-sou in loco Ir legit 24 V 1914 H. M. Krascheninnikov, n° 5133; in Herb. Inst. Bot. Ac. Sc. URSS conservatur.

Affinis *P. paronychioidi* C. A. M., differt autem pulvino denso, coma viridi, perigonii phyllis ad basin usque liberis.

25. **P. fibrilliferum** Kom. sp. nova (sect. *Avicularia*, series *Cognatae* Kom.).

Perenne, rhizomate lignoso incrasssato vel non incrassato ramoso, ramis induratis plus minusve elongatis rufidulis; caulibus e collo pluribus, parum ramosis 10—45 cm lg. prostratis vel adscendentibus tenuiter sulcatis cano-viridibus; ochreae basi ferrugineae, dein argentatae ovatae vel ovato-lanceolatae fere ad basin liberae, acuminatae, a lateribus fibrillas tenues albas secernendae; folia elongato-lanceolata basi cuneata 5—40 cm lg., 3—2 mm lata, breviter petiolata, venae laterales in foliis solum amplioribus insignes, foliorum margines leniter revoluti papillati. Flores axillares fere sessiles, 1—3 in glomerulo, per'anthium fere 3—4 mm lg., viride, apicem versus roseum vel album, laciniis profunde separatis; stigmata tres; achenium ca. 2 mm lg. nitidum trigono-ovoideum apice acutum angulis rotundatis, faciebus concavis, 1.5—1.8 mm lg.

Crescit in declivibus lapidosis, praecipue 1.800—2.000 m. supra mare vel in agris subargillosis. In omni fere Tadshikistania occurit.

Typus: Legit A. Michelson 14 VI 1914 in jugo Turkestanico ad riv. Terekli-saj, fl. Sanzar infl.; in Herb. Inst. Bot. Ac. Sc. URSS conservatur.

Affinis *P. ammanioidi* Jaub. et Spach, differt autem imprimis ochreis fibrillosis, pedunculis brevioribus, fructibus fere duplo minoribus, statu ampliore.

26. **P. myrtillifolium** Kom. (sect. *Avicularia*, series *Cognatae* Kom.).

Perenne, rhizoma erectum integrum, c. 5 mm crassum multiceps; caules plures partim subterranei aphylli, dein erecti vel adscendentes 10—15 cm elati parce ramosi; ochreae profunde bifidae albae pellucidae acuminatae integrae; folia petiolata, elliptica vel ovata, apice obtusa, rotundata, vel apiculata, subtus venis lateralibus impressis, ca. 15 mm lg., 5—7 mm lata. Flores precipue solitarii, pedunculis flore aequilongis, apertis; perianthium ca. 2 mm lg. rubrum, usque ad basin incisum campanulatum; achenium trigonum nitidium angustum ca. 2 mm lg.

Fruticuli *Vaccinium myrtillus* L. in mentem vocant.

Crescit in lapidosis mobilibus declivium in vicinitate glacierum aeternarum.

719 Typus: in valle Langlif in decursu superiore fl. Seravschan influentis legit V. L. Komarov; in Herb. Inst. Bot. Ac. Sc. URSS conservatur.

A proximis *P. rupestre* Kar. et Kir. et *P. pamiroalaico* Kom. differt habitu fruticuloso, caulibus erectis, foliis univeniis, floribus praecipue solitariis, perianthii parvi laciniis fere liberis.

27. **P. pamiroalaicum** Kom. sp. nova (sect. *Avicularia*, series *Cognatae* Kom.).

Perenne; radix unica multiceps erecta; caules plures basi dense congesta dein prostrati vel etiam recurvi densissime foliati, 5—15 cm lg.; ochreae pellucidae albae a basi late-ovatae apicem versus acuminate vel rarius oblique truncatae; folia internodiis longiora, sed parva ca. 1 cm lg., 2—4 mm lata, breviter petiolata marginibus planis univenia apice rotundata. Flores axillares 2—3 in glomerulo, partim rubra partim viridia, perianthium parum incisum apicibus rotundatis; pedunculi perianthio breviores; achenium ca. 3 mm lg., nitidum anguste trigonum, acutum.

Crescit in pascuis alpinis solo glareoso, saepe ad glacies aeternas; caespites format. In tota Tadshikistania.

Typus: 21 VIII 1892 in valle Kumar ad fl. Seravschan alt. 3300 m supra mare et in locis numerosis aliis, leg. V. L. Komarov; in Herb. Inst. Bot. Ac. Sc. URSS conservatur.

A proxima *P. alpestre* C. A. M. differt ramis densius foliatis floribundis, foliis multo minoribus marginibus planis; a *P. rupestre* Kar. et Kir. pedunculis perianthio duplo brevioribus.

28. **P. fusco-ochreatum** Kom. sp. nova (sect. *Avicularia*, series *Aviculariformes* Kom.).

Annuum, caulis fere a basi ramosus nodis paulo incrassatis tenuiter striatus, usque 4 mm crassus 50 cm lg.; ochreae omnes fusco-ferrugineae fimbriatis secernendae, venis prominulis; folia oblonga ca. 15 mm lg., 2 mm lata basi elongato-cuneata, apice sensim acutata marginibus planis leniter undulatis. Flores axillares, 1—3 in glomerulo, pedunculis ca. 2 mm lg., gracilibus, perianthium oblongum viride basi obconicum, dein ad 2/3 quinquefidum, laciniis apice rotundatis roseis; stamina 5, stigma trilobum; achenium perigonio brevius ovato-trigonum acuminatum, ca. 3 mm lg., angulis rotundatis, faciebus concavis opacis minutissime granulatis.

Hab. In arenosis glareosisque, neque in rupibus litoralibus circa oppidum Vladivostok.

Typus: litora sinus Ussuriensis inter pagos Schkotovo et Rietschiza in valle litorali 17 VIII 1913 legit V. L. Komarov; in Herb. Inst. Bot. Ac. Sc. URSS conservatur.

A proximo *P. litorali* Meisn. differt caulibus erectis, minus ramulosis, foliis multo angustioribus, univeniis, pedunculis longioribus, acheniis sesquilongioribus.

720 29. **P. lencoranicum** Kom. sp. nova (sect. *Avicularia*, series *Aviculariformes* Kom.).

Annuum, caules a basi ramosi prostrati sulcati, ca. 25 cm lg., ramulis variegatis modo *P. maritimi*, sed tenuioribus, ochreae ferrugineae venis prominulis apicem versus albae late acuminatae, serius fimbriatae, 4—8 mm lg., folia oblonga brevi-petiolata acuta, 6—12 mm lg., 2—4 mm lata, opaca, venis lateralibus prominulis. Flores saepius solitarii axillares ca. 5 mm lati, perianthium basi late obconicum, laciniis oblongis rotundatis, viridibus, margine albis; achenium nigrum opacum acutum, faciebus late ovalibus, ca. 2 mm lg., trigonum.

Crescit in arenosis humidiusculis maris Caspii ad ostia rivulorum.

Typus: Legit clar. A. Grossheim in distr. Talysch ad estium fl. Busejdagun; in Herb. Inst. Bot. Ac. Sc. URSS conservatur.

A *P. maritimo* differt caulibus multo gracilioribus ochreis fuscis brevioribus, foliis parvis, nuculis duplo minoribus. A *P. litorali* ramis brevibus eramulosis, floribus solitariis, achaeniis opacis.

30. **P. araraticum** Kom. sp. nova. (sect. *Avicularia*, series *Aviculariformes*).

Annuum, radix erecta multiceps, caules plures, basi aphylli subterranei, lutescentes, altius cano-virides, usque ad 18 cm lg., ramis destituti, tenuiter sulcati; ochreae bifidae, breves, pellucidae, basi fuscescentes apicibus valde laceratis; folia petiolata, elliptica, apice obtuso rotundatae, pube brevi supra et subtus obsita, canescentes, interdum araneosa. Flores axillares fere ses siles saepissime solitarii, 2—2.5 mm lg., sepala $^1/_3$ libera, viridia, marginibus albis, oblonga ;achenium 2 mm lg., trigonum acutum apiculatum, faciebus gibbosis granulatis.

Typus: In monte Ararat minor Turciae, in lapidosis mobilibus 14 VIII 1913 clar. A. Grossheim legit; in Herb. Horti Tiflisiensis conservatur.

In Armenia desideratur.

A congeneribus differt foliis breviter arachnoideo-tomentosis, fructuum faciebus gibbosis granulatisque.

31. **P. tiflisiense** Kom. sp. nova. (sect. *Avicularia*, series *Bellardiiformes* Kom.).

Annuum, caules erecti ramis adscendentibus 20—40 cm lg. striolatis; ochreae bifidae albae serius fuscescentes, laciniae longiter acuminatae univeniae; folia valde inaequalia lutescenti-viridia, media quam in congeneribus ampliora 1—6 cm lg., 0.3—3.2 cm lata, internodiis longiora elliptica vel late lanceolata breviter acutata,venis lateralibus insignis, petiolis evolutis, 1—4 mm lg. Racemi terminales interrupti internodiis apertis usque ad 2 cm lg., flores in axilla bracteae 1—3, tubulosi serius campanulati, 2—3 mm lg., tepalis liberis viridibus, albis vel roseis, serius accrescentibus rete venarum valde impresso; fructus trigoni 3 mm lg., faciebus 2 mm latis granulatis subnitidis, quam in congeneribus latiores.

721 T y p u s: In Transcaucasia, distr. Achalzych, ad stationem Sarran 30 VI 1913 legit clar. D. Litwinow; in Inst. Bot. Ac. Sc. URSS conservatur.

A congeneribus differt: heterophyllum perianthio fructificante venis prominulis instructo, colore lutescenti.

32. **P. caspicum** Kom. sp. nova (sect. *Avicularia*, series *Aviculariformes* Kom.).

Perenne?, radix erecta valida lignosa, caules plures prostrati firmi ca. 50 cm lg., internodiis elongatis, autumnales breves longioribus immixti; ochreae latae pallide coloratae cristatae vel fimbriatae; folia oblonga ca. 5 mm lg., subcoriacea, subtus venis lateralibus rugulosis, marginibus interdum revolutis, supra (in sicco) longitudinaliter rimosis. Flores axillares solitarii vel bini, brevissime pedicellati, ca. 3 mm lg. ochreae aequilongi, quam folia duplo breviores, virides angusti, perianthium $^1/_2$ incisum, laciniis roseo-marginati obtusis; achenium trigonum ca. 2 mm lg., 1—1.5 mm latum, faciebus minutissime tuberculatis opacis.

H a b. in arenosis et in lapidosis circa oppidum Baku.

T y p u s: legit clar. A. Petunnikov circa p. Zabrat, nec non circa Balachany; in Herb. Inst. Bot. Ac. Sc. URSS conservatur.

A congeneribus (*P. aviculare* et caet.) differt venis lateralibus foliorum subtus rugulosis, a *P. venoso* Stev. ochreis coloratis, neque floribus subsolitariis per totam plantam dispersis.

33. **P. cretaceum** Kom. sp. nova (sectio *Avicularia*, series *Bellardiiformes* Kom.).

Annum, omne canum, caules erecti usque 50 cm lg., ramis adpressis virgatis sulcati glabri; ochreae omnes albae fimbriatim dissectae, juniores nonnulli lanceolati; folia angusta oblonga marginibus revolutis, fere acerosa, univenia, cito caduca, acuta. Flores subsolitarii versus apices ramorum racemos longos praeruptos tenues formantes, perianthium parum incisum roseum; achenium angustum acutum trigonum, faciebus lanceolatis punctulato-rimosis, subnitidis, 2.3 mm lg.

Crescit in declivibus cretaceis Ucrainae orientalis.

T y p u s: legit 9 IX 1895 clar. Stepanov in valle fl. Derkul; in Herb. Inst. Bot. Ac. RSc. USS conservatur.

A proximo *P. graciliori* (Ldb.) Klokov differt epidermide cana, acheniis minutissime punctulatis non granulatis.

34. **P. inflexum** Kom. sp. nova. (sect. *Avicularia* series *Bellardiifor mes* Kom.).

Annuum, caulis erectus angulatus striatulus, ramis inflexis, interdun falcato-recurvis; ochreae argenteo-albae, tubulosae bifidae, serius seceden tes fuscescentes; folia lanceolata, oblonga vel lineari-lanceolata, glauca car nosula, venis subtus prominulis, lateralibus 4—6, acuta vel acutiuscula

5—25 mm lg., 3—7 mm lata, petiolis 2—2.5 mm lg. Flores axillares, 1—5 in glomerulo bracteis parvis angustis viridibus, saepe longitudinaliter plicatis, racemi terminales praerupti efoliati; pedicelli ca. 1 mm lg.; perianthium usque ad basin fissum, laciniis viridibus late albe vel roseo-marginatis, fructificatione tempore accrescentibus c. 2.5 mm lg., venis obsoletis; achenium trigonum nitidum acutum usque 2 mm lg., 1.5 mm latum, laevissimum.

Crescit in argillosis secus canales irrigatorios, neque ad ripas fluviorum in Asia Media (Usbekistania, Tadshikistania, Turkmenistania).

T y p u s: legit A. Budogosskij ad st. Kagan in vicinitate opp. Buchara, in agro Medicaginis sativae, n° 825 3 VIII 1913; in Herb. Inst. Bot. Ac. Sc. URSS conservatur.

A *P. arenario* neque *P. graciliore* differt ramis inflexis, a primo caule erecto angulato ab altero fructibus laevissimis.

35. **P. oxanum** Kom. sp. nova (sect. *Avicularia,* series *Bellardiiformes* Kom.).

Annuum, caules simplices vel a basi ramis lateralibus instructi erecti vel adscendentes striati, 20—50 cm alti dense foliati; ochreae pellucidae albae bifidae lobis longiter acuminatis, vetustiores parum secedentes; folia petiolis gracilibus ca. 5 mm lg., subcanescenti-viridia, crassiuscula venis lateralibus fere obsoletis oblonga apice rotundata vel breviter acutata, 1—4 cm lg., 3—10 mm lata; racemi terminales erecti interrupti vel sepius continui, solitarii vel fere paniculati. Bracteae parvae virides, pedicelli 1—2 mm lg., versus finem articulati, flores 1—3 in eadem axilla, ab initio tubulosi, dein campanulatim aperti ca. 1.5 mm lg., fructiferi accrescentes 3 mm lg., perianthii lacinia ad basin usque libera; fructus trigonus faciebus concavis nitidis 2.5 mm lg., 1—1.5 mm latus minutissime punctulatus.

H a b. in pratis ripariis, in declivibus argillosis, in pascuis et arvis, nec non ad margines segetum in Usbekistania et Tadshikistania, praecipue ad Oxus fl. saepe occurit.

T y p u s: legit clar. acad. V. Lipsky circa oppidum Buchara 23 VI 1911, n° 4323; in Herb. Inst. Bot. Ac. Sc. URSS conservatur.

A *P. gracilius* (Ldb.) Klokov differt foliis plus erassiusculis obtusioribus latioribusque racemis multo densioribus, acheniis obtusioribus non granulatis.

36. **P. imeretinum** Kom. sp. nova (sect. *Persicaria* Meisn., series *Hydropiperiformes* Kom.).

Annuum, caules erecti ramosi alti, 0.5—1.5 mm lg., glaberrimi; ochreae glabrae laeves margine minime ciliatae (in ramulis floridis ciliae nonnullae adsunt), folia breviter petiolata lanceolato-oblonga vel oblongo-lanceolata, acuta; 10—15 cm lg., 1.5—2.5 cm lata, viridia sine maculis, fere concoloria, marginibus petiolisque setulosis. Racemi elongati laxiusculi tenues, ca. 4 cm lg., 0.8 cm lati, pedunculi laeves glandulis destituti; flores viriduli, perianthium fructiferum glabrum, glandulis nullis, venulis tenuibus parum visis; nuculae stylis 2 coronatae, faciebus convexis opacis 2 mm lg.

Typus: circa curortum Zchaltubo Imeretiae inferioris in Transcaucasia occidentali, 28 X 1935, ad rivulos abundantem legit V. L. Komarov; in Herb. Inst. Bot. Ac. Sc. URSS conservatur.

A *P. hydropiper* glandulis in tepalis absentibus, a *P. nodoso* racemis gracilioribus nucularum faciebus convexis, plano-carinatis differt.

37. **P. intricatum** Kom. sp. nova (sect. *Persicaria*, series *Caespitosae* Kom.).

Caules erecti ramosi, 20—50 cm alti glaberrimi, ochreae tubulosae glabrae margine cilii paucis brevibus ornatae; folia angusta linearia vel lanceolato-linearia acuminata marginibus leniter revolutis. Racemi terminales neque axillares angusti, fere lineares 1—2 cm lg., 0.5 cm lati, bracteae infundibuliformes laeves, margine ciliatae, tepala viridia vel rosea; achenia usque 1 mm lg. trigona nitida nigra, apice apiculata.

Hab. in pratis paludosis et in calamagrostidetis humidis vallis fl. Amur. decursu medio, et in regione Ussuriensi.

Typus: in paludosis prope Iljinovka inter Zejam et Burejam, 26 VII 1891, legit clar. S. Korshinsky; in Herb. Inst. Bot. Ac. Sc. URSS conservatur.

A *P. posumbu* differt acheniis minutulis foliisque linearibus, a *P. minus* acheniis trigonis, a *P. decurrenti* ochreis foliisque glabris, racemis densioribus.

38. **P. luxurians** Grig. sp. nova (sect. *Aconogonon* Meisn.).

Caulis ca. 60—100 cm altus a basi valde divaricato-ramosus; folia late ovata vel ovato-lanceolata, ca. 5—11 cm longa et 2—8 cm lata, acuta, basi rotundata vel lato-cuneata, glabra, rarius subtus vel utrinque plus minusve pubescentia; petioli 3—12 mm longi. Inflorescentia late divaricato-paniculata; perianthium 2.5—3 mm in fructificatione 3—4 mm longum; pedicelli ca. 2 mm longi; achenium 3.5—4 mm longum e perianthio plus minusve exsertum.

Typus: Oriens Extremus. Regio Ussuriensis. Sinus Ternei 14 VIII 1914. Leg. N. P. Krylov; in Herb. Inst. Bot. Ac. Sc. URSS conservatur.

A *P. divaricato* qui affinis est foliis late-ovatis vel lato-laceolatis, statura humiliore et ceteris notis bene differt.

39. **P. intercedens** V. Petr. sp. nova (sect. *Bistorta*). — *Bistorta intercedens* V. Petr. in schedis Herbarii.

Perenne. Rhizoma nigrum valde inflexum; folia infima longi-petiolata oblonga, ca. 20 cm lg., 1—2 cm lata apice acutiuscula vel rotundata (var. *obtusifolium* Kom. nova), subcoriacea, subtus reticulato venosa, venis crispato-pilosis, supra glabra, venarum inconspicuarum apicibus solum impressis; caules 20—52 cm lg., recti quinquènodosi; folia caulina brevi-petiolata, superiora sessilia, oblongo-lanceolata vel lanceolata, rarius fere linearia; ochreae calyciformes apice bilobae, margine truncatae. Racemi 3—4 cm lg., initio florendi campactae, serius laxae; bracteae fuscescentes latae vena fusca, acumine aristato; perianthium campanulatum apertum, antherae atroviolaceae, pedicelli bracteis aequilongi vel longiores. (Descripsit V. L. Komarov).

Typus: in cacuminibus montium altiorum, inter Rhododendreta et Pineta pumilae circa lacum Toko ad fl. Zeja legit clar. O. I. Kuzeneva; in Herb. Inst. Bot. Ac Sc. URSS conservatur.

A congeneribus differt foliis infimis oblongatis modo *Rumicum* quasi medium tenet inter *P. abbreviatum* Kom. et *P. ochotense* V. Petr.

40. **P. zeaense** Kom. sp. nova (sect. *Bistorta*). — *Bistorta zeaenze* Kom.

Perenne. Rhizoma quam in congeneribus minus rotundatum vel rotundato-hamatum, fuscum; caulis erectus 23—68 cm lg., quinque-septem nodosus, sub nodo supremo saepe pilosus; folia infima longe-petiolata, oblongolinearia vel oblongo-lanceolata, venis lateralibus obsoletis, subtus dense puberula; folia caulina sagittata, suprema anguste linearia; ochreae tubulosae. Racemus compactus, bracteae late-ovatae acuminatae, arista destitutae; perianthium late-campanulatum roseum, antherae roseae.

Hab. In pratis himidis, ad margines paludum, in Betuletis herbosis etc.

Typus: ad fl. Ulanga, fl. Zeja influentem legit clar. O. I. Kuzeneva; in Herb. Inst. Bot. Ac. Sc. URSS conservatur.

A congeneribus differt caule usque septemnodoso, bracteis latis exaristatis, antheris pallidis roseis. Medium tenet inter *P. subauriculatum* V. Petrov et *P. abbreviatum* Kom.

41. **P. nitens** (Fisch. et Mey.) V. Petr. sp. nova. — *P. bistorta* var. *nitens* Fisch. et Mey. Index seminum Horti Petrop. V (1838) 40. — *Bistorta nitens* Kom. in Herbario.

Perenne. Rhizoma valde hamatum, dense radicibus et residuis foliorum obsitum; caulis 20—35 cm lg., 5—6-nodosus glaber; folia ovata, ovato-oblonga vel ovato-lanceolata, infima in petiolos decurrentia, longipetiolata, petiolis 8—15 cm lg., caulina breviter petiolata, superiora sessilia semiamplexicaulia, subtus p. m. dense pilis adpressis crispis tecta, incanescentia; ochreae ferrugineo-brunneae, inferiores duo longae aphyllae, supremae breves apice bifidae. Racemus 2—5 cm lg., compactus; bracteae infimae latae triaristatae dilute brunneae; perianthium laete roseum ca. 3 mm lg.; antherae saturate rubrae; achenium trigonum. (Descripsit V. L. Komarov).

Hab. in pratis lapidosis subalpinis Sibiriae occidentalis usque ad Dahuriam, montes Asiae Mediae, Mongolia septentrionalis.

Typus: imprimis e montibus Altai indicatum; in Herb. Inst. Bot. Ac. Sc. URSS conservatur.

A congeneribus imprimis bracteis latis triaristatis differt.

42. **P. abbreviatum** Kom. sp. nova. — *P. bistorta* var. *cordifolium* Turcz. Fl. baic.-dah. II, 2 (1856) 60. — *P. bistorta* var. *alpinum* Turcz. in Herbario. — *Bistorta abbreviata* Kom. im Herbario.

725 Perenne. Rhizoma breve hamatum dense residuis foliorum emortuorum obsitum; folia ovato-lanceolata subcoriacea in petiolum decurrentia, basi rotundata vel cordata, infima petiolata, petiolis 4—8 cm lg., caulina angusta, brevissime petiolata; omnes subtus pube brevi crispato obsita, supra saturate virides, opaca; ochreae ut in congeneribus. Racemus ovatus vel cylindricus, 1—3 cm lg., bracteae inferiores quasi involucrum formantes, saepe revolutum, oblongae laete brunneae, sensim in acumen caudiforme attenuatae, floribus longiores; pedicelli teretes; perianthium roseum, ca 3 mm lg. antherae atroviolaceae; achenia oblonga trigona nitida apice apiculata. (Vide tabulam nostram XLII, f. 5).

Crescit in pratis nec non in lapidosis alpinis. Montes Sajanenses, Transbaicalia nec non Mongolia septentrionalis.

Typus: in alpe Urgudei, legit clar. N. Turczaninow; in Herb. Inst. Bot. Ac. Sc. URSS conservatur.

A congeneribus omnibus differt pedunculo sub racemo dense breviter puberulo (piloso).

43. **P. schugnanicum** Kom. sp. nova (sect. *Bistorta*). — *Bistorta schugnanica* Kom. in Herbario.

Perenne. Rhizoma crassum validum ca. 3 cm lg., 1.8 cm latum medio valde inflexum; folia infima 4—6 congesta, modo rosulae, petioli lamina breviores usque 2 cm lg., folia oblonga acutata usque 8 cm lg., 2.5 cm lata glaberrima vel brevissime puberula, saepe obliqua, fere leniter falcata, marginibus revolutis; caules usque duplo vel rarius quintuplo quam folia longiores, graciles, 4—5 nodosi, 15—35 cm lg.; ochreae breves infundibuliformes supremae fimbriatae; folia caulina fere linearia, basi auriculata, suprema minutula fere bacilliformia. Racemus brevis 2—2.7 cm lg., 1.4 cm latus compactus; bracteae fuscae, latae, aristatae; periantium roseum; antherae carneae; achenia trigona.

Crescit in pratis alpinis montium Pamir occidentalis.

Typus: legit clar. A. Regel circa pagum Ababad ad fl. Aksu in prov. Schugnan, 3000—4000 m; in Herb. Inst. Bot. Ac. Sc. URSS conservatur.

A proxima *P. nitente* (Fisch. et Mey.) V. Petrov foliis infimis brevepetiolatis, floribus parvis, bracteis quam pedicelli longioribus.

44. **P. attenuatum** V. Petr. sp. nova. — *P. bistorta* var. *attenuatum* V. Petr. in Herbario. — *Bistorta attenuata* Kom. in Herbario.

Perenne. Rhizoma hamatum breve nigrum residuis foliorum caulinumque vetustiorum coronatum; folia et caulis ca. 50 cm lg., minus, quam apud congeneribus sulcatus, 6-nodosus glaber; folia oblonga vel lanceolato-oblonga apice attenuata, supra laete viridia, subtus cinerea pube densa brevi obsita, infima 6—12 cm lg., 1.5—3 cm lata, basi rotundato in petiolum decurrenti, media sessilia basi cordata; ochreae tubulosae basin versus puberulae. Racemus 4—7 cm lg., bracteae lanceolatae, anguste acuminatae, carina fusca;

 pedicelli teretes, perianthium ca. 3 mm lg., violaceo-roseum, antherae fere concolores; achenium 3 mm lg., trigonum longiter acuminatum. (Descripsit V. L. Komarov).

Crescit in pratis, in fruticetis pratosis, in pratulis silvaticis et ad margines silvarum. Ad Lenam fluvium, in Transbaicalia et Mongolia septentrionali.

Typus: Lena superior; in Herb. Inst. Bot. Ac. Sc. URSS conservatur.

A congeneribus differt caulibus nodosis, foliis subtus dense puberulis apice longiter acuminatis; acheniis longiter acuminatis.

45. **P. ochotense** V. Petr. — *Bistorta ochotensis* Kom. in Herbario.

Perenne. Rhizoma leniter inflexum, caules erecti graciles sulcati 4—5-nodosi usque 45 cm alti, glabri; folia infima petiolis longis 7—10 cm lg. adnata, superiora fere sessilia, omnia lanceolata, ovato-lanceolata vel oblongo-lanceo-ata supra saturate-viridia opaca, subtus cana pube brevi adpresse obsita; ochreae congeneribus similes. Racemus compactus 2—6 cm lg., bracteae longiter acuminatae aristatae, carina fusca, marginibus albidae, pedicelli applanati, perianthium roseum campanulatum ca 4 mm lg., antherae violaceae; achenium trigonum nitidum apiculatum ca 4 mm lg. (Descripsit V. L. Komarov).

In pratis montanis et in silvis humidis. In regione maris Ochotensis.

Typus: circa opp. Ajan nec non ad ripas freti Tatarici; in Herb. Inst. Bot. Ac. Sc. URSS conservatur.

A congeneribus differt caulibus 4—5-nodosis, foliis saturate viridibus bracteis sensim in acumina caudiformia productis.

46. **P. paludosum** Kom. sp. nova. — *P. sagittatum* var. *paludosum* Kom. in A. H. P. XXII (1903) 133. — *Persicaria sagittata* Kom. in Herbario.

Annuum; caules graciles erecti simplices tetragoni, 7—20 cm alti, secus angulos minutissime aculeolati; folia ovata 0.4—3 cm lg., 2—7 mm lata, basi leniter subsagittata, petiolis brevibus aculeolis uniserialibus; ochreae angustae fuscescentes margine superiore ciliatae. Flores solitarii vel 2—5 in capitulo terminali unico collecti, interdum cum flore solitario unico sub capitulo adnato; pedicelli glaberrimi graciles, perianthium roseum ca. 3 mm lg.; achenium trigonum nigrum 2—3 mm lg. angulis rotundatis, faciebus convexis, apice acutum nitidulum.

Crescit in paludibus ad ripas aquarum vel ad fontes. In provinciis Amurensi nec non Ussuriensi, neque in Manshuria.

A proximo *P. belophyllo* Litw. differt foliis basi fere exauriculatis floribus interdum solitariis vel paucis pedunculis glabris laevissimis.

INDEX ALPHABETICUS*

ORDINUM FAMILIARUM, SPECIERUM ATQUE SYNONYMORUM PLANTARUM IN TOMO V FLORAE URSS COMMEMORATARUM.

* [This index has been reproduced photographically from the Russian original.]
** [Russian page numbers appear in the left-hand margin of the text.]

5401

VEGETATION REGIONS OF THE USSR

Abbreviated name	Full name
I. Arctic	
1. Arc. Eur.	Arctic (European part)
2. Nov. Z.	Novaya Zemlya
3. Arc. Sib.	Arctic (Siberia)
4. Chuk.	Chukchi
5. An.	Anadyr
II. European part	
6. Kar.-Lap.	Karelia-Lapland
7. Dv.-Pech.	Dvina-Pechora
8. Balt.	Baltic States
9. Lad.-Ilm.	Ladoga-Il'men
10. U. V.	Upper Volga
11. V.-Kama	Volga-Kama
12. U. Dnp.	Upper Dnieper
13. M. D.	Middle Dnieper
14. V.-Don	Volga-Don
15. Transv.	Transvolga area
16. U. Dns.	Upper Dniester
17. Bes.	Bessarabia
18. Bl.	Black Sea area
19. Crim.	Crimea
20. L. Don	Lower Don
21. L. V.	Lower Volga
III. Caucasus	
22. Cisc.	Ciscaucasia
23. Dag.	Dagestan
24. W. Transc.	Western Transcaucasia
25. E. Transc.	Eastern Transcaucasia
26. S. Transc.	Southern Transcaucasia
27. Tal.	Talysh
IV. West Siberia	
28. Ob	Ob region (from the eastern slopes of the Urals to the Yenisei River)
29. U. Tob.	Upper Tobol
30. Irt.	Irtysh
31. Alt.	Altai

V. East Siberia

32. Yenis.	Yenisei
33. Lena-Kol.	Lena-Kolyma
34. Ang.-Say.	Angara River-Sayans
35. Dau.	Dauria

VI. Far East

36. Kamch.	Kamchatka
37. Okh.	Okhotsk
38. Ze.-Bu.	Zeya-Bureya
39. Uda	Uda River area
40. Uss.	Ussuri
41. Sakh.	Sakhalin

VII. Soviet Central Asia

42. Ar.-Casp.	Aral-Caspian
43. Balkh.	Lake Balkhash area
44. Dzu-Tarb.	Dzungaria-Tarbagatai
45. Kyz. K.	Kyzyl-Kum
46. Kara K.	Kara-Kum
47. Mtn. Turkm.	Mountainous part of Turkmenistan
48. Amu D.	Amu Darya
49. Syr D.	Syr Darya
50. Pam.-Al.	Pamir-Alai
51. T. Sh.	Tien Shan

Accepted Regions for Indication of General Distribution of Species Appearing in "Flora of the U.S.S.R."

I. Arc.	Arctic (Spitsbergen, Greenland and farther)
II. Scand.	Scandinavia (Norway, Denmark, Sweden, Finland)
III. Centr. Eur.	Central Europe (Germany, Poland, Czechoslovakia, Hungary, Austria, Switzerland)
IV. Atl. Eur.	Atlantic Europe (Netherlands, Belgium, England, France, Portugal)
V. Med.	Mediterranean (including North Africa) (V'Western, V"Eastern)
VI. Bal.-As. Min.	Balkan Peninsula and Asia Minor
VII. Arm.-Kurd.	Lesser Armenia and Kurdistan
VIII. Iran.	Iran and Afghanistan
IX. Ind.-Him.	India and Himalayas
X. Dzu.-Kash.	[Dzungaria-Kashgar area] Eastern or Chinese Turkestan (Sinkiang)
XI. Mong.	Mongolia

XII.	Jap.-Ch.	Japan and China
XIII.	Ber.	North American coast of the Bering Sea
XIV.	N. Am.	North America (U.S.A. and Canada)
XV.	Tib.	Tibet

Other Geographical Abbreviations

Afr.	Africa
Aust.	Australia
Centr.	Central
E.	East(ern)
Gr.	Great, Greater
I.	Island
Is.	Islands
Mt.	Mount
Mts.	Mountains
N.	North(ern)
R.	River
S.	South(ern)
W.	West(ern)

TRANSLATOR'S NOTE

1. The Russian term "Srednyaya Aziya" is, in English, Central Asia (or Soviet Central Asia). Therefore the term Middle Asia has been used for Russian "Tsentral'naya Aziya," which is non-Soviet inner Asia, comprising western China (Sinkiang and Tibet) and Mongolia.

2. According to Russian usage, the European part of the USSR is "eastern Europe." Therefore "western Europe" includes the whole of Europe outside the USSR.

EXPLANATORY LIST OF ABBREVIATIONS OF RUSSIAN INSTITUTIONS AND PERIODICALS APPEARING IN THIS TEXT

Abbreviation	Full name (transliterated)	Translation
Bot. Geogr. Aralo-Kasp. Kr.	Botanicheskaya Geografiya Aralo-Kaspiiskogo Kraya	Geobotany of the Aral-Caspian Region
Bot.-geogr. issled. v Turkest.	Botaniko-geograficheskie issledovaniya v Turkestane	Botanical and Geographical Investigations in Turkestan
Bot. Mat. Gerb. Bot. inst. AN SSSR	Botanicheskie Materialy Gerbariya Botanicheskogo instituta AN SSSR	Botanical Materials of the Herbarium of the Botanical Institute of the Academy of Sciences of the USSR
Bot. Mat. Gerb. Gl. Bot. Sada	Botanicheskie Materialy Gerbariya Glavnogo Botanicheskogo Sada	Botanical Materials of the Herbarium of the Main Botanical Gardens
Bot. zap. SPb. univ.	Botanicheskie zapiski Sankt-Peterburgskogo universiteta	Botanical Notes of St. Petersburg University
Bot. zhurn. SSSR	Botanicheskii zhurnal SSSR	Botanical Journal of the USSR
Byull. Glavn. Bot. Sada	Byulleten' Glavnogo Botanicheskogo Sada	Bulletin of the Main Botanical Gardens
Byull. Obshch. lyubit. estestvozn., antrop. i etnogr.	Byulleten' Obshchestva lyubitelei estestvoznaniya, antropologii i etnografii	Bulletin of the Naturalists', Anthropologists' and Ethnographers' Society
Dendr.	Dendrarii	Arboretum
Der. i kust.	Derev'ya i kustarniki	Trees and Shrubs
Der. i kust. Kavk.	Derev'ya i kustarniki Kavkaza	Trees and Shrubs of the Caucasus
Dikie polezn. i tekhnich. rasteniya SSSR	Dikie poleznye i tekhnicheskie rasteniya SSSR	Useful Wild Plants and Industrial Crops of the USSR
Dikorastushchie r. Kavkaza, ikh rasprostranenie, svoistva i primenenie	Dikorastushchie rasteniya Kavkaza, ikh rasprostranenie, svoistva i primenenie	Wild Plants of the Caucasus, Their Distribution, Properties and Uses

Dokl. AN Azerb. SSR	Doklady Akademii Nauk Azerbaidzhanskoi SSR	Reports of the Academy of Sciences of the Azerbaijan SSR
Ezhegodnik Ekzotlesa	Ezhegodnik Ekzoticheskogo Lesa	Yearbook on Exotic Forests
Fl.	Flora	Flora
Fl. Abkh.	Flora Abkhazii	Abkhazian Flora
Fl. Almat. zapovedn.	Flora Alma-Atinskogo zapovednika	Flora of the Alma-Ata Reserve
Fl. Alt.	Flora Altaya	Altai Flora
Fl. Alt. i Tomsk. gub.	Flora Altaiskoi i Tomskoi gubernii	Flora of Altai and Tomsk Provinces
Fl. Az. Ross.	Flora Aziatskoi Rossii	Flora of Asiatic Russia
Fl. Evrop. Rossii	Flora Evropeiskoi Rossii	Flora of European Russia
Fl. Gruzii	Flora Gruzii	Georgian Flora
Fl. Kamch.	Flora Kamchatki	Kamchatkan Flora
Fl. Kavk.	Flora Kavkaza	Caucasian Flora
Fl. Man'chzh.	Flora Man'chzhurii	Manchurian Flora
Fl. Pamira	Flora Pamira	Pamir Flora
Fl. Sev. Kraya	Flora Severnogo Kraya	Flora of the Northern Territory
Fl. Sib.	Flora Sibiri	Siberian Flora
Fl. Sib. i Dal'n. Vost.	Flora Sibiri i Dal'nego Vostoka	Flora of Siberia and the Far East
Fl. Sr. Ross.	Flora Srednei Rossii	Flora of Central Russia
Fl. Sr. i Yuzh. Ross.	Flora Srednei i Yuzhnei Rossii	Flora of Central and Southern Russia
Fl. Talysh.	Flora Talysha	Talysh Flora
Fl. Yugo-Vost.	Flora Yugo-Vostoka	Flora of the Southeast
Gerb. donsk. fl.	Gerbarii donskoi flory	Herbarium of Don Flora
Gerb. Orlovsk. gub.	Gerbarii Orlovskoi gubernii	Herbarium of Orel Province
Gerb. Ukr. fl.	Gerbarii Ukrainskoi flory	Herbarium of Ukrainian Flora
GRF	Gerbarii Russkoi Flory	Herbarium of Russian Flora (HRF)
Ill. Fl. Mosk. gub.	Illyustrirovannaya Flora Moskovskoi gubernii	Illustrated Flora of Moscow Province
Ischisl. r. Podol'sk gub.	Ischislenie rek Podol'skoi gubernii	Estimation of Rivers of Podolsk Province
Izv. AN SSSR	Izvestiya AN SSSR	Bulletin of the Academy of Sciences of the USSR
Izv. Bot. Sada	Izvestiya Botanicheskogo Sada	Bulletin of the Botanical Gardens
Izv. Bot. Sada. Petra Vel.	Izvestiya Botanicheskogo Sada Petra Velikogo	Bulletin of Peter the Great Botanical Gardens
Izv. Gl. Bot. Sada	Izvestiya Glavnogo Botanicheskogo Sada	Bulletin of the Main Botanical Gardens
Izv. Kavk. Muzeya	Izvestiya Kavkazskogo Muzeya	Bulletin of the Caucasian Museum

Izv. Kazakhst. fil. AN SSSR	Izvestiya Kazakhstanskogo filiala Akademii Nauk SSSR	Bulletin of the Kazakhstan Branch of the Academy of Sciences of the USSR
Izv. Kiev. Univers.	Izvestiya Kievskogo Universiteta	Bulletin of the Kiev University
Izv. Kievsk. Bot. Sada	Izvestiya Kievskogo Botanicheskogo Sada	Bulletin of the Kiev Botanical Gardens
Izv. Lesn. Ins.	Izvestiya Lesnogo Instituta	Bulletin of the Forest Institute
Izv. Nikol. Univ.	Izvestiya Nikolaevskogo Universiteta	Bulletin of the Nikolaev University
Izv. Obshch. lyubit. estest. vozn., antrop. i etnogr.	Izvestiya Obshchestva lyubitelei estestvo-znaniya, antropologii i etnografii	Bulletin of the Naturalists', Anthropologists' and Ethnographers' Society
Izv. O-va. Lyubit. Est.	Izvestiya Obshchestva Lyubitelei Estestvoznaniya	Bulletin of the Naturalists' Society
Izv. Sem. Kontr. St. Mosk.-Sel'skokhoz. Obshch.	Izvestiya Semennoi Kon-trol'noi Stantsii Moskov-skogo Sel'skokhozyaist-vennogo Obshchestva	Bulletin of the Seed Stock Control Station of the Moscow Agricultural Society
Izv. SPb. Lesn. Inst.	Izvestiya Sankt-Peterburg-skogo Lesnogo Instituta	Bulletin of the St. Peters-burg Forest Institute
Izv. Zap. Sib. Otd. Russ. Geogr. Obshch.	Izvestiya Zapadno-Sibir-skogo Otdeleniya Russ-kogo Geograficheskogo Obshchestva	Bulletin of the West Siberian Branch of the Russian Geographic Society
Konsp. rast. okr. Khar'kova	Konspekt rastenii okruga Khar'kova	Compendium of Plants of Kharkov
Korm. rast estestv. senoko-sov i pastb. SSSR	Kormovye rasteniya estest-vennykh senokosov i pastbishch SSSR	Fodder Plants of Natural Hay-meadows and Pastures of the USSR
Krit. Obz. Mosk. Fl.	Kriticheskii Obzor Moskov-skoi Flory	Critical Survey of the Moscow Flora
Lesn. Rast. v Turk.	Lesnaya Rastitel'nost' v Turkestane	Forest Vegetation in Turkestan
Lesn. zhurn.	Lesnoi zhurnal	Forestry Journal
Mat. (dlya) Fl. Kavk.	Materialy dlya Flory Kavkaza	Materials on Caucasian Flora
Mat. (dlya) Fl. Kryma	Materialy dlya Flory Kryma	Materials on Crimean Flora
Mat. Kryms. Zapovedn.	Materialy Krymskogo Zapovednika	Materials of the Crimean Reserve
Mat. Okhor. Prir. na Ukr.	Materiyaly Okhorony Prirody na Ukraini	Materials on Nature Protection in the Ukraine
Mosk. Fl.	Moskovskaya Flora	Moscow Flora
Nov. obozr.	Novoe obozrenie	New Review
Obzor. Rast. Kievsk. Uch. Okr.	Obzor Rastitel'nosti Kievskogo Uchebnogo Okruga	Survey of Vegation in the Kiev Educational District

Och. rast. i fl. Karpat	Ocherki rastitel'nosti i flory Karpat	Survey of Carpathian Vegetation and Flora
Ocherk. Tifl. fl.	Ocherki Tiflisskoi flory	Survey of Tiflis [Tbilisi] Flora
Ocherki po Fitozots. i Fitogeogr.	Ocherki po Fitozotsenozu i Fitogeografii	Outline on Phytocenosis and Phytogeography
Opis. Amur. obl.	Opisanie Amurskoi oblasti	Description of the Amur Region
Opis. Rast. Khers. Gub. Lesa	Opisanie Rastitel'nosti Khersonskogo Gubern-skogo Lesa	Description of the Forest Vegetation in the Kherson Province
Opred. der. i kust.	Opredelitel' derev'ev i kustarnikov	Key to Trees and Shrubs
Opred. rast. Dal'nevost. kr.	Opredelitel' rastenii Dal'nevostochnogo kraya	Key to Plants of the Far Eastern Territory
Opred. rast. Kavk.	Opredelitel' rastenii Kavkaza	Key to Caucasian Plants
Opred. Rast. Okr. Tashkenta	Opredelitel' Rastitel'nosti Okruga Tashkenta	Key to Vegetation of the Tashkent District
Opred. vyssh. rast.	Opredelitel' vysshikh rastenii	Key to Higher Plants
Opred. (vyssh.) rasten. Evrop. chasti SSSR	Opredelitel' (vysshikh) rastenii Evropeiskoi chasti SSSR	Key to Higher Plants of the European USSR
Perech. rast. Turk.	Perechen' rastenii Turkmenii	List of Turkmenian Plants
Prakt. Botan.	Prakticheskaya Botanika	Practical Botany
Putesh.	Puteshestviya	Travels
Putesh. po Dol. r. Ussuri	Puteshestvie po Dolinam reki Ussuri	Journey along the Valleys of the Ussuri River
Put. v Turkest.	Puteshestvie v Turkestan	Journey into Turkestan
Rast. i fl. Karp.	Rasteniya i flora Karpat	Plants and Flora of the Carpathians
Rast. letn. pastb. Gandzh.	Rasteniya letnikh past-bishch Gandzhi	Vegetation of Gandzha [now Kirovabad] Summer Pastures
Rast. res. Turkm.	Rastitel'nye resursy Turkmenii	Plant Resources of Turkmenia
Rast. resursy Kavkaza	Rastitel'nye resursy Kavkaza	Plant Resources of the Caucasus
Rast. Sib.	Rastitel'nost' Sibiri	Vegetation of Siberia
Rast. Sr. Az.	Rastitel'nost' Srednei Azii	Vegetation of Soviet Central Asia
Rastit. Kavk.	Rastitel'nost' Kavkaza	Vegetation of the Caucasus
Rastit. pokrov. vost. Pamira	Rastitel'nyi pokrov vostochnogo Pamira	Plant Cover of the Eastern Pamirs
Rastit. syr'e Kazakhst.	Rastitel'noe syr'e Kazakhstana	Plant Resources of Kazakhstan
Rastit. Turkm.	Rasteniya Turkmenii	Vegetation of Turkmenia

Rezul't. dvukh puteshestv. na Kavk.	Rezul'taty dvukh puteshestvii na Kavkaz	Results of Two Travels to the Caucasus
Russk. Fl.	Russkaya Flora	Russian Flora
Russk. lek. rast.	Russkie lekarstvennye rasteniya	Russian Medicinal Plants
Sbor, sushka i raz. lek. rast.	Sbor, sushka i razvitie lekarstvennykh rastenii	Gathering, Drying and Development of Medicinal Plants
Sorn. rast. SSSR	Sornye rasteniya SSSR	Weed Plants of the USSR
Sornye r. Gruzii	Sornye rasteniya Gruzii	Weed Plants of Georgia
Sov. Bot.	Sovetskaya Botanika	Soviet Botany
Spis. rast.	Spisok rastenii	List of Plants
Tr. Bot. inst. AN SSSR	Trudy Botanicheskogo instituta AN SSSR	Transactions of the Botanical Institute of the Academy of Sciences of the USSR
Tr. Bot. Muz. Ak. Nauk	Trudy Botanicheskogo Muzeya Akademii Nauk	Transactions of the Botanical Museum of the Academy of Sciences
Tr. Bot. Sada	Trudy Botanicheskogo Sada	Transactions of the Botanical Gardens
Tr. Bot. Sada Yur'evsk. Univ.	Trudy Botanicheskogo Sada Yur'evskogo Universiteta	Transactions of the Botanical Gardens of Yur'ev [now Tartu] University
Tr. Dal'nevost. bazy AN SSSR	Trudy Dal'nevostochnoi bazy AN SSSR	Transactions of the Far Eastern Base of the Academy of Sciences of the USSR
Tr. i Issled. po Lesn. Opyt. delu	Trudy i Issledovaniya po Lesnomu Opytnomu delu	Transactions and Research of Experimental Forestry
Tr. Inst. nov. lub. syr'ya	Trudy Instituta novogo lubyanogo syr'ya	Transactions of the Institute of New Fiber Raw Materials
Tr. Obshch. Estestv. Nauk Khar'k. Un.	Trudy Obshchestva Estestvennykh Nauk Khar'kovskogo Universiteta	Transactions of the Natural Science Society of the Kharkov University
Tr. Obshch. isp. prir. Khar'k. univ.	Trudy Obshchestva ispytatelei prirody Khar'kovskogo universiteta	Transactions of the Naturalists' Society of Kharkov University
Tr. Penzen. Obshch. Lyub. Estv.	Trudy Penzenskogo Obshchestva Lyubitelei Estestvoznaniya	Transactions of the Penza Society of Naturalists
Tr. Peterb. obshch. estestvoisp.	Trudy Peterburgskogo obshchestva estestvoispytatelei	Transactions of the St. Petersburg Naturalists' Society
Tr. pochv.-bot. eksp. Peresl. upr.	Trudy pochvenno-botanicheskoi ekspeditsii Pereslavskogo upravleniya	Transactions of the Soil-Botanical Expedition of Pereslavl Administration

Tr. po geobot. obsled. pastb. Azerb.	Trudy po geobotanicheskim obsledovaniyam pastbishch Azerbaidzhana	Transactions of Geobotanical Investigations of Azerbaijan Pastures
Tr. prikl. bot. (gen. i sel.)	Trudy po prikladnoi botanike, genetike i selektsii	Transactions of Applied Botany, Genetics and Selection
Tr. Ross. Obshch. sadov.	Trudy Rossiiskogo Obshchestva sadovodov	Transactions of the Russian Horticulturists' Society
Tr. SAGU	Trudy Sredneaziatskogo Gosudarstvennogo Universiteta	Transactions of the Soviet Central Asian State University
Tr. Sarat. obshch. estestvoisp.	Trudy Saratovskogo obshchestva estestvoispytatelei	Transactions of the Saratov Naturalists' Society
Tr. Sil'skogospod. komit. bot.	Trudy sil'skohospodar'skoho komiteta botaniky	Transactions of the Botanical Agricultural Committee
Tr. SPb. obshch. estestv.	Trudy Sankt-Peterburgskogo obshchestva estestvoispytatelei	Transactions of the St. Petersburg Naturalists' Society
Tr. Tadzh. bazy AN SSSR	Trudy Tadzhikskoi bazy AN SSSR	Transactions of the Tadzhikistan Base of the Academy of Sciences of the USSR
Tr. Tbil. bot. inst.	Trudy Tbiliskogo botanicheskogo instituta	Transactions of Tbilisi Botanical Institute
Tr. Tbil. (or Tifl.) bot. sada	Trudy Tbilisskogo (Tifliskogo) botanicheskogo sada	Transactions of theTbilisi (Tiflis) Botanical Garden
Tr. Turkmensk. bot. sada	Trudy Turkmenskogo botanicheskogo sada	Transactions of the Turkmenian Botanical Garden
Tr. Turk. nauch. obshch.	Trudy Turkmenskogo nauchnogo obshchestva	Transactions of the Turkmenian Scientific Society
Uchenye Zap. Gos. Sar. Univ.	Uchenye Zapiski Saratovskogo Gosudarstvennogo Universiteta	Scientific Reports of the Saratov State University
Vest. Akad. Nauk (or AN) Kazakhsk. SSR	Vestnik Akademii Nauk Kazakhskoi SSR	Bulletin of the Academy of Sciences of the Kazakh SSR
Vestn. estestv. nauk	Vestnik estestvennykh nauk	Bulletin of Natural Sciences
Vestn. Ross. Obshch. sadov.	Vestnik Rossiiskogo Obshchestva sadovodov	Bulletin of the Russian Horticulturists' Society
Vest. Tifl. bot. sada	Vestnik Tiflisskogo botanicheskogo sada	Bulletin of the Tiflis Botanical Garden
Vizn. (or Vznachn.) rosl. URSR	Viznachnyk roslyn USSR	Key to Plants of the Ukrainian SSR

V obl. polupustyni	V oblasti polupustyni	(In the) Semidesert Region
Yadov. rast. lugov i pastb.	Yadovitye rasteniya lugov i pastbishch	Poisonous Plants of Meadows and Pastures
Yubil. Sborn. P. Borodina	Yubileinyi Sbornik posvyashchennyi P. Borodinu	Jubilee Collection Dedicated to P. Borodin
Zam. po sist. i geogr. rast. Tbil. bot. inst.	Zametki po sistematike i geografii rastenii Tbilisskogo botanicheskogo instituta	Notes on Taxonomy and Geography of Plants of the Tbilisi Botanical Institute
Zap. Kievsk. Obshch. Estestv.	Zapiski Kievskogo Obshchestva Estestvoispytatelei	Reports of the Kiev Society of Naturalists
Zhurn. Bot. obshch.	Zhurnal Botanicheskogo obshchestva	Journal of the Botanical Society
Zhurn. opytn. agron. Yugo-Vost.	Zhurnal opytnoi agronomii Yugo-Vostoka	Journal of Experimental Agronomy of the Southeast
Zhurn. Russk. Bot. Obshch.	Zhurnal Russkogo Botanicheskogo Obshchestva	Journal of the Russian Botanical Society